SCHAUM'S OUTLINE OF

THEORY AND PROBLEMS

OF

ZOOLOGY

•

BY

NANCY M. JESSOP, Ph.D.
Associate Professor of Life Sciences
Palomar College
San Marcos, California

•

SCHAUM'S OUTLINE SERIES
McGRAW-HILL BOOK COMPANY
New York St. Louis San Francisco Auckland Bogotá Hamburg London
Madrid Mexico Milan Montreal New Delhi Panama Paris
São Paulo Singapore Sydney Tokyo Toronto

To my daughters, Christina and Laurel

NANCY M. JESSOP, currently Associate Professor of Life Sciences at Palomar College, received her Ph.D. from the University of California at Berkeley. For more than 27 years she has developed and taught introductory and advanced courses in college zoology and biology. She is the author of *Biosphere: A Study of Life* and research papers in the fields of experimental zoology, mammalian genetics, and animal behavior. Dr. Jessop has also served as consultant to the Commission on Undergraduate Education in the Biological Sciences and the National Science Foundation, and as secretary of the Animal Behavior Society.

Schaum's Outline of Theory and Problems of
ZOOLOGY

1 2 3 4 5 6 7 8 9 10 11 12 13 14 15 16 17 18 19 20 SHP SHP 8 9 2 1 0 9 8

ISBN 0-07-032551-0

Sponsoring Editor, Elizabeth Zayatz
Production Supervisor, Leroy A. Young
Editing Supervisor, Marthe Grice

Library of Congress Cataloging-in-Publication Data

Jessop, Nancy M. (Nancy Meyer)
   Schaum's outline of theory and problems of zoology.

   (Schaum's outline series)
   1. Zoology—Outlines, syllabi, etc. 2. Zoology—
Examinations, questions, etc. I. Title.
QL52.J47 1987   591   87-22650
ISBN 0-07-032551-0

Preface

Zoology, the scientific study of animal life, is a fundamental discipline within the biological sciences. It deals with the animal world from unicellular protozoans to complex invertebrates and vertebrates: a prismatic array of adaptations and patterns of bodily organization. In addition, zoology explores basic processes of nutrition and integration, and the genetic mechanisms of continuity and change by which animals both propagate their kind and undergo adaptive modifications through time.

This Outline of zoology is intended to serve either as a concise, economical text which can be used on its own for college and university courses in general zoology, or as a supplement and study guide to accompany more comprehensive textbooks in general biology, general zoology, invertebrate zoology, and vertebrate zoology.

In keeping with the scientific spirit of inquiry, this book takes the form of a series of questions and answers. As an aid to self-study, many of these text questions can be looked upon as sample examination questions, and their answers as representative short-essay examination responses. The answers given are based upon scientific methods of description or experimentation, but they should not be considered definitive, since all "answers" formulated by scientists must be held subject to revision in the light of new discoveries. Furthermore, when there is a practical need to formulate concise answers to complex questions, those answers cannot help being somewhat oversimplified and incomplete.

Writing this Outline has proven especially challenging, since stringent requirements of brevity and conciseness have necessitated unflagging exercise of personal judgment as to which vital areas of zoological lore must be woefully abridged or omitted. In making these difficult decisions, I have been guided by the conviction that a short text devoted to the study of animal life should place its greatest emphasis on *animals*. Accordingly, 16 chapters have been devoted to reasonably thorough considerations of the particular characteristics and biotic contributions of the major animal groups or phyla; each of these chapters includes the organismal biology and ecological considerations germane to the animal group in question. The remaining nine chapters of the Outline (1, 2, 3, 4, 5, 8, 23, 24, and 25) present concise introductions to scientific methodology, structural organization of cells and organisms, basic chemistry, paleontological data, genetic and evolutionary mechanisms, embryonic development, nutritional processes in cells and organisms, integrative mechanisms, and ecological principles.

Despite the impressive volume of zoological data amassed to date, we should keep in mind that zoology, like other sciences, is a discipline in motion, constantly subject to revision and reinterpretation, and that zoologists, like other people, often find themselves in lively disagreement. As a case in point, the taxonomic grouping of animals on the basis of relatedness has long been an area of spirited contention, so that both students and authors come to feel frustrated by the plethora of contradictory schemes of classification found throughout the zoological literature. In due time, this confusion will subside as biochemical analysis of each species' genetic material provides quantitative means of exact definition. Until then, both students and practitioners of zoology do well to cultivate a flexible outlook by which differing systematic constructs are seen as mere milestones along the tortuous route toward fuller understanding of the history and interrelationships of the animal world as we see it today. The taxonomic usages favored herein are not necessarily more correct than those favored by other authors; all may be subject to eventual modification on the basis of DNA analyses. The important thing is to perceive and understand animals themselves, and not to dwell unduly upon the labels which humans bestow upon them, useful as these may be.

A lifelong fascination with animals has enriched my own cognitive life, and I hope to have successfully conveyed this interest to my readers, whose lives can also be enriched by a knowledge and appreciation of our fellow creatures. Yet today this appreciation must be laced with anxiety for the fate of an appalling number of species threatened with imminent extinction by the expansion of human populations to an unprecedented 5 billion hungry bipeds. In many ways, humans need and benefit by the great diversity of animal life: now, animals need us too, for without the efforts of concerned individuals and groups throughout the world, future human generations will irrevocably be robbed of the wildlife which we ourselves have enjoyed. If this book not only informs its readers but promotes concern for the preservation of animal life, it will have achieved one of my most cherished aims.

I wish to express especial thanks to Elizabeth Zayatz, editor of the Schaum's Outline Series, for her encouragement and expert assistance in revising the manuscript, and to the McGraw-Hill Book Company for permission to reproduce a number of line drawings from that classic text, *General Zoology,* Sixth Edition, by T.I. Storer, R.C. Stebbins, R.L. Usinger, and J.W. Nybakken. I thank colleagues whom I have consulted regarding zoological particulars, but wish to state that I alone am responsible for errors of subject matter, which I hope will prove few.

Finally, I acknowledge my indebtedness to those who have been my students during my 27 years of teaching zoology at the college level. I sincerely hope that their minds are fuller and vision keener for their classroom experiences with me, since they have provided my inspiration for the writing of this book.

NANCY M. JESSOP

Contents

CHAPTER   PAGE

1.   INTRODUCTION TO ZOOLOGY   1

2.   ANIMAL ARCHITECTURE   9

3.   CHEMICAL BASIS OF LIFE 21

4.   THE HISTORY OF LIFE 37

5.   HEREDITY AND EVOLUTION 45

6.   CELLS AS ORGANISMS: MEET THE PROTISTS 62

7.   THE LOWER METAZOANS: PARAZOA AND MESOZOA 82

8.   EUMETAZOANS: DEVELOPMENT AND DIVERSIFICATION 93

9.   THE RADIATA: CNIDARIANS AND CTENOPHORES 107

10. THE ACOELOMATE BILATERIA 125

11. THE PSEUDOCOELOMATES 140

12. LARVAL LINKS, BRANCH POINTS, AND MINOR MYSTERIES 150

13. MOLLUSKS 162

14. SEGMENTED WORMS AND WALKING WORMS 178

15. JOINT-LEGGED PROTOSTOMES: THE ARTHROPODS 193

16. CRUSTACEANS AND UNIRAMIAN MANDIBULATES 210

17. ECHINODERMS AND LESSER DEUTEROSTOMES 236

18. PHYLUM CHORDATA 252

19. ANAMNIOTES I: PRIMITIVE AND CARTILAGINOUS FISHES 280

20. ANAMNIOTES II: BONY FISHES AND AMPHIBIANS 294

CONTENTS

21. AMNIOTES I: REPTILES AND BIRDS 312

22. AMNIOTES II: MAMMALS 340

23. ANIMAL NUTRITION 373

24. INTEGRATION AND RESPONSE 397

25. ANIMAL ECOLOGY 419

     INDEX 439

Chapter 1

Introduction to Zoology

Zoology is fun. It can enrich your life by helping you understand the fascinating diversity of creatures that share this planet with us. It can also save your life.

1.1. What is zoology?

Zoology is the scientific study of animal life.

1.2. What is "scientific study"?

The basic commitment of science is to collect objective *data* (facts that are observable and measurable) and then reach conclusions and formulate generalizations by analyzing such data. A generalization that represents a cohesive statement of principle is known in scientific parlance as a *theory*. (Since in nonscientific usage "theory" may connote fanciful imaginings, this unfortunate semantic difference can lead to considerable misunderstanding between scientists and other people.) No matter how firm the data base upon which a scientific theory rests, the theory must always remain subject to revision in the light of additional data.

> **Example 1:** The "theory of evolution" is actually a well-established biological principle based on an extensive array of data that indicate that living things change through successive generations: the *fact* of such change is well established, but new data continue to elucidate its *mechanisms*.

1.3. How do scientists go about collecting data?

Scientific methods include (*a*) collecting data by observation and (*b*) collecting data by controlled experimentation.

(*a*) When *collecting data by observation*, without experimental intervention, scientists must ensure that the data are as free as possible of subjective bias (a real pitfall since scientists are only people), recorded and analyzed instrumentally when possible, and extensive enough so that such factors as range of variability can be defined, preferably statistically.

> **Example 2:** The song of a chipping sparrow is easy to recognize by ear, but it can be recorded more objectively and accurately by a sound spectrograph, which provides a visual record of frequency, duration, and pattern to augment acoustic recordings. Enough recordings must be made to determine how much or how little variability occurs in (1) the songs of each individual bird, (2) the songs of different individuals living in close proximity, (3) the songs of chipping sparrows from different geographic regions.

(*b*) When *collecting data by controlled experimentation*, scientists begin by asking questions, which they then try to answer. A testable question is called a *hypothesis*. Hypotheses are often tested by means of a *controlled experiment*, in which one or more *experimental* groups are compared with one or more *control* (often tantamount to "normal") groups, under conditions that are held standard except for one factor, the *variable*. The number of organisms used is important: an experiment based on only a few test organisms is apt to be nonpredictive and unreliable.

> **Example 3:** *Hypothesis:* Does calcium deficiency have detrimental effects on animal growth, health, and longevity?
>
> *Experiment:* The subject organisms (preferably siblings of inbred strains to accentuate genetic likeness), selected for similarity in age and original weight, are maintained under identical conditions *except* that the controls are given a diet with normal (but precisely measured) amounts of calcium, whereas the experimental groups are fed diets in which the calcium content is reduced by some known amount.
>
> *Conclusion:* Measurable and reproducible differences appearing between the control and experimental groups indicate effects of calcium deficiency upon the latter.

1

A verified conclusion becomes a *datum* (fact) that, summated with enough other compatible data, may contribute to the formulation of a theory (cohesive statement of principle). *Theory:* Calcium is widely required in the animal world for skeletal growth and the normal functioning of nerves and muscles.

Example 4:   Venomous American coral snakes are conspicuously banded in red, yellow, and black; they appear to be avoided by most predators. Why?

Two hypotheses suggest themselves: (*a*) the avoidance may be a learned response based on the predators' having recovered from a nonlethal bite; (*b*) the avoidance may be instinctive and shown by animals that never have been bitten. For testing these hypotheses, the major aspects of "control" are that predators of known life history must be used as subjects, and these must be offered sample objects of several patterns.

Experiment: Motmots (a tropical American bird species known to eat snakes in the wild) were raised in captivity so that they would have no experience with any snakes. Offered a selection of differently painted rods (as an artificial version of snakes), the naive motmots readily pecked at rods that were longitudinally striped in yellow and red or ringed in green and blue, but consistently avoided rods ringed end to end in red and yellow. (Note that the striped model served as a *control* to see if the birds would respond to red and yellow, regardless of pattern, as "warning colors," whereas the green-and-blue model served as control to determine whether the birds would avoid a ringed model even if it were not red and yellow.)

Conclusion: Naive motmots innately avoid rod-shaped models painted with alternating red and yellow rings.

This leads to the *tentative* conclusion that inexperienced motmots would also avoid attacking live coral snakes; if you don't mind risking your tame motmots, the next step would be to go collect some coral snakes.

Example 5:   Sometimes the control and experimental aspects of an experiment must be carried out sequentially, rather than being conducted simultaneously on two different groups. An example of this is Niko Tinbergen's classic field experiments on the homing behavior of digger wasps.

The female wasp digs burrows in which she lays her eggs, after which this diligent mama works ceaselessly to bring fresh insects to feed her young. Upon leaving a burrow, the wasp makes a localized reconnoitering flight, which looks as if she were fixing in her tiny brain landmarks that would enable her to find the burrow again. Such a conclusion cannot be assumed: it must be tested.

The investigator first recorded the wasp's behavior free of interference (the control); then, after the wasp had left the scene, he modified the setting to see if this impaired her ability to find the burrow when she returned with food (the experiment). For instance, while the mother wasp was within the burrow, the investigator would simply place a ring of pinecones around the burrow, and after she had flown away to gather more prey, he would shift the circle of cones a short distance to one side. On her return the mother would home on the ring of pinecones, losing track of the burrow because the landmarks had been moved. Since this behavior was verified in a number of these wasps, it could be concluded that, yes, such wasps really do acquire a "photographic memory" of landmarks during the brief circling flight they engage in after exiting the burrow.

Did these examples raise new questions in your mind? If so, you are coming to understand how science progresses; every question answered seems to lead to more, so that scientific investigation may have all the excitement of detective work.

1.4.   What constitutes "life"?

Life is defined by its properties; all animals *and* plants share certain fundamental characteristics of life, which are listed in Table 1.1.

Example 6:   Bodily changes between rest and exercise exemplify *adaptability* of the individual: the heart beats more rapidly and forcibly, the lungs take in deeper and more frequent breaths, the circulatory system shunts blood away from digestive organs to the muscles and skin, extra red blood corpuscles enter the bloodstream from the spleen, and so forth. These changes are temporary, but with continued training, growth of muscle fibers, a slower, stronger heartbeat, and other more lasting benefits accrue. Such adaptive changes are *not* genetically transmissible to the offspring.

Table 1.1 Characteristics of Living Organisms

Reproduction	The process of producing new generations of genetically similar organisms; may be *asexual* (single parent) or *sexual* (recombination of genes from two interacting parents).
Adaptation	The process of changing to promote survival; includes (*a*) *adaptability* of individual organism in direct response to some specific challenge (Example 6) and (*b*) *mutability* (alteration) of genes and chromosomes, which occurs at random and not in response to specific need, producing a range of variability in offspring (Examples 7 and 8).
Irritability	Ability to sense and respond adaptively to external and internal stimuli.
Endogenous (self-generated) motility	Not all organisms are capable of locomotion (mobility) during all stages of their lives, but no matter how stationary an organism itself may be, *self-generated* movement of body parts does occur (e.g., plants' foliage turns toward light and roots turn toward gravity).
Nutrition	The process of obtaining and using substances (nutrients) for growth, maintenance, and reproduction.
Ingestion	Taking in nutrients.
Digestion	Mechanical and chemical breakdown of nutrients.
Absorption	Uptake into tissues of products of digestion.
Internal transport	Distribution of foods, wastes, etc., through the body, often by means of a circulatory system.
Metabolism	Chemical conversion of nutrients within the cells; includes (*a*) *catabolism*, which is the breakdown of large molecules into smaller ones, with liberation of energy, and (*b*) *anabolism*, which is the synthesis of larger molecules from smaller ones, with the use of some of this energy.
Respiration	Exchange of gases involving, for animals, net uptake of oxygen (O_2, used in making water as a catabolic end product) and net removal of carbon dioxide (CO_2, another product of catabolism).
Excretion*	Removal of metabolic wastes, such as ammonia and urea, or any substance present to excess, as needed to maintain the body's state of dynamic chemical equilibrium (*homeostasis*).
Nucleic acids	DNA (deoxyribonucleic acid) and RNA (ribonucleic acid), which are macromolecules synthesized by all living organisms and on which all biological reproduction depends.
Proteins	Macromolecules of nearly infinite variety that perform many functions: structural components (collagen in connective tissue), hormones, oxygen-binding molecules (hemoglobin), enzymes that catalyze (facilitate) chemical reactions (e.g., pepsin, a protein-digesting enzyme).

*Excretion should not be confused with *egestion*, which is removal from the body of undigested residues that have never actually been absorbed into the cells.

Example 7: *Mutations* are changes that occur in the hereditary material, especially when genes and chromosomes are reproducing themselves. They occur spontaneously and can also be induced by *mutagenic agents* such as certain chemicals and forms of radiation. These changes *are* transmissible to offspring, provided they take place in reproductive cells. If this is the case, then all body cells of the new individual will carry the mutation; many mutations are *recessive* and neither help nor harm that individual, although the next generation may be affected (e.g., albinism).

Example 8: Environmental circumstances may define the survival value of a mutant trait. The very restricted habitat of the white gypsum sand dunes of White Sands National Monument, New Mexico, is successfully inhabited by whitish varieties of both mammals and reptiles, whose darker relatives inhabit the surrounding desert but lack protective coloration for the dunes. In this case a minor genetic change affecting coloration has opened up an entire new habitat for exploitation by the mutant organisms.

1.5. What is an animal?

The word itself, from the Latin *anima* ("breath," "soul"), implies a being that is animated (lively), mobile, and sentient, a bit much to ask of a sponge! Yet sponges are animals. So, more conservatively, let us say that an *animal* is an *organism* (living thing) that as a rule ingests organic materials (proteins, fats, carbohydrates, etc.) and digests them internally.

1.6. How does animal life differ from plant life?

The major differences between most plants and animals are outlined briefly in Table 1.2.

Although the basic unit of structure and function of both plants and animals is the *eukaryotic cell* (i.e., nucleated cell; see Chapter 2) and plant and animal cells are so much alike as to strongly suggest a common ancestor, there are two salient points of difference: (*a*) animal cells lack *chloroplasts*, which are plants' photosynthetic factories, and (*b*) animal cells are not enclosed in the nonliving *cell walls*, composed mainly of *cellulose*, which furnish structural rigidity to plants.

Table 1.2   Some Major Differences between Plants and Animals

	Plants	Animals
Mode of nourishment	Autotrophic (carry out photosynthesis; contain chloroplasts)	Heterotrophic (do not photosynthesize; lack chloroplasts)
Extent of growth	Indeterminate	Determinate
Cell wall	Cellulose (long glucose chain); rigid, inert	Absent
Nervous system	Absent	Present in most
Mobility	Mostly immobile	Mostly mobile
Primary food reserve	Starch (unbranched glucose chain); unsaturated oils	Glycogen (multiply branched glucose chain); saturated fats
Waste products	O_2 from photosynthesis, CO_2 from metabolism; kidneys not needed since nitrogenous wastes not generated	CO_2 and nitrogenous wastes; kidneys needed in most animals

Owing to the presence of chloroplasts, most plants carry on *autotrophic* (self-nourished) *nutrition* through the process of photosynthesis; by converting radiant energy into chemical bonding energy, autotrophs build simple sugars out of carbon dioxide and water, giving off oxygen as a waste product. Although this process will not be dwelt upon in a book about animals, we should pay it due homage, because in the absence of autotrophic plants, animal life could not exist, for want of both food to eat and atmospheric oxygen to breathe. Animals exhibit *heterotrophic nutrition:* they cannot construct organic molecules out of inorganic substrates. This lamentable deficiency has had prodigious consequences, for it is fundamental to the evolutionary divergence of animals and plants. A lineage of organisms that has had to search out, apprehend, and often subdue dinners that may have their own means of resisting such fate is bound to come up with descendants remarkably different from those of autotrophic ancestry that need simply bask in the sun and absorb minerals, carbon dioxide, and water.

As a rule, plants grow throughout life (*indeterminate growth*), whereas most animals attain a certain adult size and cease to grow (*determinate growth*), compact body form being more compatible with the mobility prompted by the quest for food; all but the simplest plants are immobile.

Most animals have nervous systems and sense organs; nothing of this nature is demonstrable in plants, but some plant cells, especially at growing tips, are sensitive to certain stimuli.

Plants store much of their food reserves as *starch*, a long unbranched chain of glucose (simple sugar) molecules; the main carbohydrate food reserve of animals is *glycogen*, a highly branched chain of glucose, which yields free sugar molecules much more readily than starch does.

Animals' metabolic rates are so high that proteins are formed and broken down rapidly enough that toxic nitrogenous wastes (ammonia, urea, etc.) are produced and must be excreted as components of *urine;* plants do not produce urine and require no kidneys.

1.7.   Why study animals?

Animals are involved in many aspects of human existence, both contemporary and historical, and concern with animals is by no means limited to the scientific world. Directly and indirectly, more than we may realize, we rely on animals to maintain the complex, interdependent web of life that is our planetary ecosystem. The unprecedented expansion of human populations and the resultant pollution, transformation, and degradation of habitats are threatening the survival of nonhuman species as never before in historical times, making it imperative for our own long-term survival that we investigate and understand the interdependencies of species, even when these do not appear to be of direct or immediate benefit to humanity.

1.8.   How do animals contribute positively to human existence?

The many ways in which humans rely on animals is summarized in Table 1.3.

Table 1.3   Human Reliance on Animals

Benefits to Humans	Examples
Food sources	Livestock, game, fish, shellfish; animal products such as honey, eggs, and dairy products; exotic fare such as insects, grubs, and highly relished palolo worms.
Nonedible economic products	Leather, down, fur, silk, wool, ivory, biogenic limestone, chalk, and marble
Transport and labor	Horses, donkeys, llamas, camels, dogs, oxen, Asiatic buffalo, and elephants
Biomedical uses	
Products, extracts	Venoms, insulin, pig heart valves (chemically treated to destroy living pig cells and used to replace diseased human heart valves when judged superior to artificial devices), antibodies for protective inoculations (e.g., against tetanus)
Research	Subjects used to create animal models of human diseases and their treatment
Ecological value	Essential links in food chains (Example 9); essential for the pollination of most flowering plants (Example 10); agents of *biocontrol* (the maintenance of natural population balances that hold potentially detrimental plant and animal species in check) (Examples 11 and 12)
Psychological benefits	
Esthetic value	Subjects and inspiration for works of art, from cave paintings to present-day creations; some cultures revere totem animals and cultivate in themselves the positive attributes they perceive in animals.
Affectional bonds	Pets and residents in wildlife parks, fulfilling various noneconomic human needs; used by some psychotherapists in their work with patients (Example 13)

Example 9:   Microscopic plants known as diatoms are the major photosynthesizers of the open sea, but they are so minute that only extremely tiny animals can feed upon them. These wee creatures, such as copepods, are themselves mostly too small to be collected for human consumption, but they in turn are food for infant fish and other small zooplankton. Thus, through several levels of consumption, the energy and mass trapped by diatoms eventually benefits us in the form of millions of tons of tuna, cod, and other marketable fish. Should anthropogenic pollutants (i.e., those of human origin) decimate key species of zooplankton small enough to feed directly on diatoms, the higher levels of oceanic food chains would collapse even if diatoms themselves remained abundant.

Example 10:   Animals are essential for the *pollination* of most flowering plants. Unwise use of pesticides can result in widespread crop failures, leaving us little to eat but grains (which are wind-pollinated). But far beyond this, if pollinator species (most of which are insects) die off globally, visualize a world depleted of most types of higher vegetation, and therefore a world in which we ourselves might be unable to survive.

Example 11:   To avoid possible chemical insecticide injury to its valuable animal collection, the San Diego Zoo is said to rely entirely on biocontrol to protect its equally valuable plant collection, for instance, by release of laboratory-raised insectivorous (insect-eating) insects such as ladybird beetles.

Example 12:   Hippopotamuses in Africa and manatees in Florida are important agents in clearing waterways of obstructing vegetation such as the water hyacinth.

Example 13:   The tremendous popularity of pets and wildlife parks and the annual roster of injuries sustained when people get chummy with park bears give us a clue that animals not only are interesting to many persons but also provide some positive psychological reinforcement that may be of incalculable significance to the mental health of humanity as a whole. In fact, it has been shown that suicide rates are lower among pet owners. Empathy for nonhuman species may wear thin in regions where human nutrition is so marginal that nearly anything living is eyed with a view to edibility, but even in extreme circumstances of deprivation, people of some cultures and religions consider animals sacred and endowed with an inherent right to share the earth with humans.

Should human population levels soon stabilize and living standards rise globally, more than ever we may experience a psychological need to relate to nonhuman species. Even today the burgeoning of animal-oriented tourism attests to the importance to many people of affectional bonds with animals, even undomesticable sorts. Such psychological considerations rather than purely scientific ones often undergird the sometimes heroic measures taken to salvage species on the brink of extinction, even while human pressures on the environment are increasingly at the root of their predicament.

1.9.   Do animals pose any dangers to human life?

Although few animals consider humans a regular dietary item, many people still suffer death or injury from animal attacks, some predatory (e.g., by tigers, crocodiles), but most defensive (e.g., bites of venomous snakes and spiders, scorpion and insect stings). A number of animal species are dangerous human parasites that either inflict illness and death in their own right (e.g., the malaria parasite) or endanger their hosts by transmitting other disease agents (e.g., anopheline mosquitoes that carry malaria). Relatively few animal species are economically destructive, but these few (mainly insects) cause millions of dollars in damage to food crops, trees and lumber, and livestock.

1.10.   How can we *best* control "pest" species and guard ourselves against potentially detrimental animals?

Solutions have often been sought through ill-advised means that create more problems in the end, such as promiscuous use of chemical insecticides to which most insect pests by now have developed immunity, while the nonbiodegradable chemicals travel through food chains to harm many other life-forms. The better option seems to be to study each potentially harmful species and find out how to control it or protect ourselves through our knowledge of that species' behavior and needs.

Example 14:   In the late nineteenth century, a parasitic insect, cottony cushion scale, was accidentally introduced into California's San Joaquin Valley, where it happily fell upon the citrus crops, with severe economic consequences. Introduction of ladybird beetles reduced the scale population to a level well below economic significance and held it there for decades. After World War II, local farmers embraced the "wonder" pesticide DDT, one immediate effect of which was the first resurgence of cottony cushion scale to economically significant levels.

By now this is an old, familiar story: broad-spectrum pesticides may unleash the very pests they seek to control by killing off the agents of biocontrol, including insectivorous insects, birds, amphibians, reptiles, and mammals. In this case, the DDT also killed the ladybird beetles that had been keeping the cottony cushion scale under control; the pest species apparently recuperated more successfully from the insecticide than the predator could, resulting in a resurgence of the pest to significant levels.

Example 15:   Female insects of certain species are monogamous and mate only once in their lives. If they have the bad fortune to accept semen from an infertile male, they lay only unfertilized eggs. Discovery of this aspect of animal behavior is contributing to successful control of such pest species as the screwworm fly, whose larvae form hideous lesions in the hides of livestock. Screwworm flies are propagated in the laboratory, so that large numbers of males sterilized by X-radiation can be set

Table 1.4   Major Subdivisions of Zoology

Subdivision	Area of Study
TAXONOMIC	
Invertebrate zoology	Animals without backbones
Protozoology	Animals that are basically unicellular
Helminthology	Worms (mainly parasitic)
Malacology	Mollusks (snails, clams, oysters, squid, octopus)
Entomology	Insects
Vertebrate zoology	Animals with backbones
Ichthyology	Fishes
Herpetology	Amphibians and reptiles
Ornithology	Birds
Mammalogy	Mammals
NONTAXONOMIC	
Morphological zoology	Animal structure
Gross anatomy	Nonmicroscopic anatomy
Paleontology	Fossils
Histology	Organs and tissues at the microscopic level
Cytology	Cell structure, often at electron microscopy level
Physiological zoology	Animal function
Organismal physiology	Body functions of the entire organism; may be studied under standard laboratory conditions or in the field
Organ and cell physiology	Vital activities of organs and cells; may be monitored *in vivo* (within the living body) or *in vitro* (maintained outside the body, such as cells propagated in nutrient media as "tissue cultures")
Zoological genetics	Hereditary traits and their transmission; is the basis of selective breeding; contributes significantly to the knowledge of human genetics
Evolutionary zoology and systematics	Evolutionary relationships among animals; attempts to develop improved models of animal classification
Zoological ecology	The relationship between animals and their *biotic* (living) and *abiotic* (physicochemical) environments; is essential to successful management of animals in the wild; includes *population dynamics*, the analysis of population structure and trends (e.g., to regulate fisheries and hunting)
Animal behavior	Important to the management of wild, captive, and domestic animals and to the training of animals
Ethology	Species-typical (often instinctive) behaviors exhibited in the wild (or at least under quasi-natural conditions)
Animal psychology	Animal learning, particularly under controlled laboratory conditions
Animal pathology and epidemiology	The causes and effects of disease processes in animals; includes the practice of veterinary medicine in zoos and even in the wild

free wherever a screwworm outbreak is reported, swamping the wild population so that the females have a greater probability of breeding with sterile than with fertile males.

Example 16:   The life cycle of the human liver fluke, mainly endemic in the orient, includes a period of mandatory encystment (dormancy) in the flesh of certain freshwater fishes. People get infected with this parasite only by eating raw or pickled fish. Armed with this knowledge, self-protection is simple: cook the fish. *Knowledge*, therefore, is the key to understanding, appreciating, conserving, and controlling animal life, and much knowledge comes from scientific inquiry.

1.11.   What are some of the things zoologists do?

The major subdivisions of zoology are listed in Table 1.4. Some zoologists devote their time to studying animals belonging to one particular taxonomic category (i.e., one of a hierarchy of levels along which animals are classified); others study one or more aspects of animal structure, function, or behavior, often using a comparative approach.

Chapter 2

Animal Architecture

Animal life presents a dazzling array of creatures exhibiting many body forms and ranging in size from 30-meter (m) blue whales to unicellular protozoans only a few micrometers (μm) in length (Fig. 2.1). Confronted with such diversity, we need first to recognize that biotic organization is fundamentally the same in all. The differences that distinguish various kinds of animals largely disappear at cellular levels, and pervasive similarities abound. At the molecular level, we find that for all their remarkable complexity, living things are composed of the same basic elements that make up the nonliving world, transiently organized into molecules with unique properties, by virtue of the energy of light. Stolid and massive as an elephant may be, in the last analysis it is more space than mass, a latticework of atoms briefly bound together by fettered sunlight.

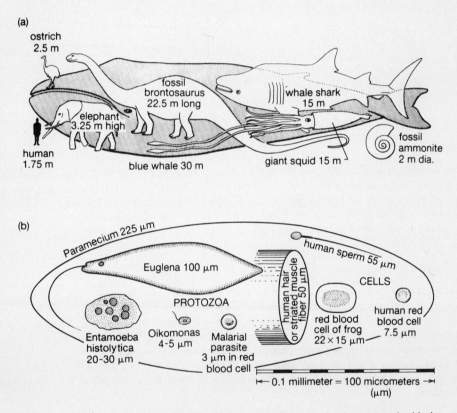

Fig. 2.1 Size in the animal world. (*a*) Some of the largest animals compared with the human body. (*b*) One-celled animals and animal cells, shown within the outline of a *Paramecium*. (*From Storer et al.*)

2.1. What basic factors determine body form?

Animal bodies are made up of from one to trillions of cells, all of which are fundamentally similar, but which must interact developmentally to produce a single, coordinated living body, the *organism*, exhibiting harmonious integration of form and function, together with adaptation to some particular mode of life. The basic features of organismal form include symmetry, axial polarities, and proportionalities.

2.2. What is meant by "symmetry"?

The animal body is usually *symmetrical,* which means that it can be cut into mirror-imaged halves along one or more planes of section (Fig. 2.2). Three basic types of animal symmetry, related to three different modes of life are spherical, radial, and bilateral. *Spherically symmetrical* animals are globular, with *appendages* (extensions or limbs) radiating in all directions (e.g., the unicellular *Actinosphaerium* and *Heleosphaera* (see Fig. 6.5a); they are usually buoyant drifters. *Radially symmetrical* animals display the symmetry of a wheel or umbrella with body parts radiating from a single central axis (e.g., jellyfish and sea anemones); these unambitious creatures may remain stationary or slowly creep or drift, contacting the environment equally on all sides.

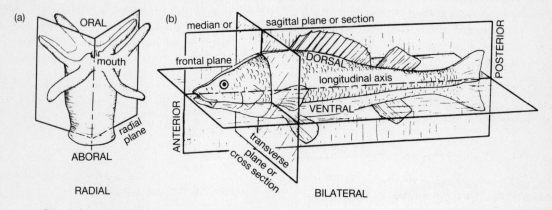

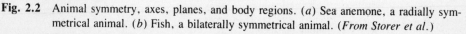

Fig. 2.2 Animal symmetry, axes, planes, and body regions. (*a*) Sea anemone, a radially symmetrical animal. (*b*) Fish, a bilaterally symmetrical animal. (*From Storer et al.*)

Bilaterally symmetrical animals are designed to live active lives, on the prowl for food and other amenities; their bodies can be divided along only one plane of section to yield mirror-image left and right halves, and the anterior end is specialized as a head loaded with sensory equipment. The development of bilaterally symmetrical body forms constituted a mighty evolutionary innovation for the animal kingdom, for it is this type of symmetry that characterizes most major animal groups, both today and in the fossil record.

2.3. What is meant by body axes?

Any symmetrical body form possesses at least one major *body axis,* an imaginary straight line that passes through the center of the body. The spherical *Actinosphaerium* enjoys a nearly infinite number of potential body axes, passing through its central core, provided that it does not care which part of its body is upward. But if it does care, then it shows *polarity* (as earth does) and it has only one axis, which passes right through its "up" and "down" poles.

Radially symmetrical animals show polarity, and have a single body axis known as the *oral-aboral axis:* the centrally located mouth locates the oral end of this axis, while the end opposite is aboral. In a starfish the mouth is located orally, and the anus aborally; a jellyfish also has a mouth and oral surface, but it lacks an anus and so has no alimentary opening on its aboral surface.

Bilaterally symmetrical animals have two major body axes: a *longitudinal,* or *anteroposterior,* axis that passes from head to tail and a *dorsoventral* axis that passes from top to bottom. The animal's dorsal surface, or "back," is usually uppermost, and its *ventral* surface, or "belly," is usually oriented downward, but the same terms apply to inverted animals such as tree sloths, which spend their lives suspended dorsum down, venter up.

2.4. What are the major anatomical planes of section?

An anatomical *plane of section* divides a body along a plane that intersects or parallels some body axis (see Fig. 2.2): (*a*) a *transverse* section bisects the body crosswise at right angles to the anteroposterior axis; (*b*) a *sagittal* section divides the body into left and right halves along the anteroposterior axis; (*c*) a *frontal* section is at right angles to the dorsoventral axis and divides the body into dorsal and ventral portions.

2.5.	What is an axis of polarity?

When we speak of an *axis of polarity*, or say that a body is *polarized* along a major body axis, we mean that the two ends of the axis are dissimilar. This is true of the oral-aboral axis of radially symmetrical animals and of both the longitudinal and the dorsoventral axes of bilaterally symmetrical organisms.

> **Example 1:**	The polarity of the longitudinal axis has resulted in *cephalization*, the development of a *head* at the end that leads in locomotion, thus meeting the environment first. This leading end is termed anterior, and the opposite end *posterior*. Since a well-developed head is a complicated structure housing a brain and major sense organs, during embryonic life an *axial gradient* must operate, with maturational rates being most rapid at the anterior end, while the hind end develops more slowly. At one point, a human embryo is more than half head, and the rest of the body only gradually catches up.

2.6.	What additional terms of orientation pertain to animals' bodies?

(*a*) *Medial* means toward the body midline, *lateral* toward the sides; (*b*) *deep* means within the body's interior, *superficial* at or near its surface; (*c*) *cephalad* or *craniad* means toward the anterior end, *caudad* toward the tail or posterior end; (*d*) *proximal* refers to the part of an appendage that is closer to the trunk, and *distal* to a part farther away.

2.7.	What factors affect body size and shape?

Gravity, surface-to-volume ratios, relationship of body parts, and adaptive specializations are the major factors affecting the form and proportions of animals' bodies.

2.8.	How does gravity affect body proportions?

The force of gravity limits the size of an animal to the mass it can support, but with correct proportionality of body parts the mass supported can be considerable.

> **Example 2:**	The Asian elephant plods on sturdy tree trunk legs, each ending in a short, compact foot with five blunt toes, the whole appendage depicting adaptation for bearing great mass. The foot of a sheep is long and slender, providing mechanical advantage in running; it has a reduced number of bones and ends in only two functional toes, the nails of which form the divided hoof on which the sheep moves daintily: the foot structure as evolved lacks architectural soundness for supporting a great body mass.
>
> Although the possibilities of genetic engineering foster speculations of producing gigantic livestock, e.g., sheep the size of woolly mammoths, the problems involved in such ambitious husbandry include that of adequate support, unless you plan to half-suspend your megasheep from helium balloons so they can drift over the meadows to graze. The skeletons of mammoths and sheep are indeed basically made up of the same bones, but if a sheep's skeleton were simply magnified to elephantine dimensions, without any change in proportions, it would collapse under the mass of flesh. Even mild changes in body size may engender problems that threaten survival: excessive secretion of pituitary growth hormone in childhood may produce human "pituitary giants" so tall that they are likely to die young from circulatory problems of venous return from the feet.

2.9.	How does the surface-to-volume ratio affect body form?

This ratio sets both upper and lower limits on animals' size, but it can be modified by changes in body shape. The effects of growth upon the ratio of surface to volume is most easily calculated for a solid, spherical body. This hypothetical living sphere depends for survival upon exchanges of nutrients and wastes with the surrounding environment. As a sphere increases in mass, its *volume* increases according to the *cube* of its radius ($4\pi r^3/3$), while its *surface area* increases only according to the *square* of the radius ($4\pi r^2$).

> **Example 3:**	If a spherical body grows until its radius has *doubled*, its volume will have increased *eightfold*. A rather average animal body cell has a diameter of about 20 μm [a micrometer is 1/1000

of a millimeter (mm)]; a modest amoeba with a diameter of 100 μm would have a volume 125 times greater than the 20 μm "average" cell. Needless to say, a body having the form of a smooth-surfaced sphere would be severely limited in growth, because traffic of wastes, gases, and nutrients through the surface would quickly become inadequate to sustain the volume of internal matter.

Example 4:   Few cells or organisms *are* spheres with smooth surfaces: amoebas freely change shape and increase the surface-to-volume ratio by means of their armlike locomotory pseudopods, while the surfaces of many animal body cells are expanded by ultramicroscopic excrescences known as "microvilli" and "blebs." Certain macroscopic animals, such as the "sea gooseberry," a comb jelly, are essentially globose, but their living tissues are mostly restricted to a single layer of cells that covers the outside of the body and lines the internal digestive cavity, while their bulk consists of a nonliving "jelly" that requires little sustenance.

Surface-to-volume ratios also set *lower* limits on size: the smaller a bird or mammal, the more body heat is lost through the body surface and the higher the metabolic rate must be to maintain normal body temperature.

Example 5:   A 2.2-gram (g) shrew (the tiniest adult mammal) must consume nearly its own weight in prey each day and must hunt both night and day since it can die of hypothermia brought on by only 2–3 hours of food deprivation. Hummingbirds (the tiniest being the bee hummingbird of Cuba, with a body no larger than a bumblebee) cannot feed at night and would starve by dawn except that they can survive hypothermia and so simply become torpid and "cold-blooded" at night. The body temperature of a hummingbird that lives in the high Andes drops from 38°C (100°F) by day to 14°C (57°F) by night. Thermoregulation may also affect the relative size of particular body parts: the impressive ears of desert foxes and hares serve to dissipate excess body heat through radiation, while the ears of Arctic foxes and hares are quite reduced, minimizing heat loss.

2.10.   How does the relationship of body parts affect body form?

An unkind wit once observed that a camel looks as if it were put together by a committee. Actually, a camel's body parts relate proportionally to each other quite well in terms of functional adaptation, as is true of animals in general. In fact, evolutionary modification of one body part often requires compensatory changes in other parts.

Example 6:   The reptilian pterosaurs (Fig. 21.9) were the first vertebrates to sail the skies, and although birdlike in some respects, they failed to accomplish the weight reduction that birds achieved by developing flimsy, toothless jawbones covered by lightweight, horny beaks. Early pterosaurs counterbalanced the weight of their long, toothed, fish-catching jaws by retaining long tails tipped by a spatulate rudder. Later species dispensed with the heavy tail by virtue of a bony crest that extended backward from the skull and offset the weight of the jaws so that the head could be balanced, teeter-totter fashion, on the fulcrum of the neck vertebrae.

2.11.   How does adaptive specialization affect body form?

The animal world is replete with examples of this phenomenon.

Example 7:   The cetaceans (whales and dolphins) and sirenians (manatees and dugongs) are aquatic mammals that evolved from *separate* terrestrial lineages but have converged considerably in body form. The entire body is smoothly streamlined and furless, the forelimbs have become flippers and the hind legs have vanished, while the tail has become thickly muscular and ends in a horizontal blade (fluke) that is moved up and down for propulsion so that these mammals in effect "gallop" with their tails.

2.12.   What are the levels of organization of an animal body?

Bilaterally symmetrical animals exhibit five levels of structural and functional body organization: organism, organ system, organ, tissue, and cell.

2.13. What are organs and organ systems?

Organs are bodily structures of characteristic size and shape that are made up of integrated masses of cells called *tissues* and serve to carry out specific vital functions. Examples are the stomach, liver, brain, and heart.

Organ systems are coordinated groups of organs that work together in accomplishing major bodily functions. The 10 organ systems, not all of which occur in lower animals, are:

1. *Integumentary system:* the skin and such cutaneous derivatives as scales, fur, feathers, horns, armor, glands, serving many protective functions

2. *Skeletal system:* either an internal *endoskeleton* or an external *exoskeleton*, made up of calcareous and/or organic materials

3. *Muscular system:* sets of muscles involved in changes in body form, posture, and locomotion (as opposed to contractility of internal organs)

4. *Digestive system:* the alimentary canal and accessory structures such as teeth, tongue, salivary and digestive glands, which process and absorb ingested nutrients

5. *Circulatory system:* heart, blood vessels, and corpuscles (also, in vertebrates, lymph vessels and nodes, spleen, and blood-forming organs such as red bone marrow), engaged in internal transport of solutes

6. *Respiratory system:* internal or external gills, lungs, etc., which facilitate gaseous exchange

7. *Excretory system:* kidneys and their ducts, which function in removal of metabolic wastes, osmoregulation, and homeostasis (maintenance of the dynamic equilibrium of body chemistry)

8. *Nervous system:* brain, ganglia, nerves, sense organs, which detect and analyze stimuli and bring about appropriate responses by stimulating the body's effectors (mainly muscles and glands)

9. *Endocrine system:* hormone-producing glands that act harmoniously in the regulation of growth, metabolism, and reproductive processes

10. *Reproductive system:* gonads (testes and ovaries), which produce sperm and eggs, genital ducts, and such accessory organs as glands and copulatory apparatus

2.14. What are tissues?

Tissues (Fig. 2.3) are masses or layers of specialized cells of similar type and common embryonic origin that operate in a coordinated fashion (usually through the agency of nerve fibers). Although the precise appearance of the specialized cells that form tissues differs to some extent among various animal groups, the major types of animal tissues include:

(*a*) *Epithelial tissue:* sheets that cover body surfaces and line body cavities, the individual cells being *squamous* (flat), *cuboidal*, or *columnar* in shape; most *glands* are also epithelial.

(*b*) *Muscle tissue:* elongated contractile cells known as *myocytes*, or *muscle fibers*, which may be uni- or multinucleate. Vertebrates and certain higher invertebrates possess two types of muscle fibers: *striated* (banded) fibers that contract rapidly for bodily movement, and unstriated *visceral* (or "smooth") fibers that contract more slowly and rhythmically and in vertebrates are restricted to the walls of organs, ducts, and blood vessels. Only vertebrates possess a third type of muscle tissue: *cardiac* fibers in the heart, which exhibit the unique property of spontaneous pulsation.

(*c*) *Nervous tissue:* mainly nerve cells (*neurons*) and their elongated processes (*nerve fibers*), which are specialized for irritability, conduction, and transmission. Neurons communicate with other neurons and with effector cells by secreting chemicals known as *neurotransmitters*, or, in some cases, *neurohormones* that are disseminated by way of the circulatory system. Most sensory cells (*receptors*) are actually sensory neurons, but in vertebrates certain receptors (such as those involved in olfaction, taste, balance, and hearing) are modified epithelial cells.

(*d*) *Connective and supportive tissues:* a heterogeneous assemblage of tissues that consist mainly of a nonliving mineral and/or organic matrix secreted by a variety of cells that usually have at least a potential for amoeboid movement. In the vertebrates these tissues include *connective tissue fibers*, deposited by cells known as *fibroblasts: cartilage*, laid down by *chondroblasts;* and *bone*, secreted by *osteoblasts*.

(*e*) *Vascular tissues:* blood corpuscles and wandering cells found in the body fluids and tissue spaces. In a number of invertebrates, amoeboid *excretophores* engulf and remove waste materials. "White corpuscles," or *leukocytes*, are defensive in function, either by phagocytosis or by secretion of protective chemicals such as antibodies. Corpuscles in the bloodstream or body spaces, which are colored by the presence of *respiratory pigments* (e.g., hemoglobin in *erythrocytes*) store and transport oxygen.

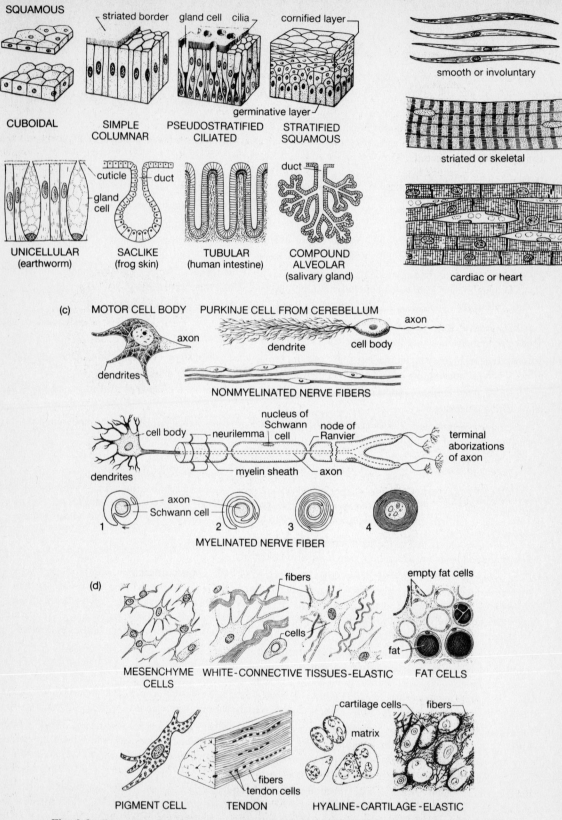

Fig. 2.3 Representative tissues. (*a*) Epithelial tissue, *above*, forming membranes; *below*, glandular. (*b*) Vertebrate muscle tissues. (*c*) Nervous tissue. (*d*) Supportive and connective tissues. (*From Storer et al.*)

14

2.15. How are tissues arranged to form organs?

> **Example 8:**   The mammalian intestine is an organ, and a section of its wall (Fig. 2.4) shows how different tissues are structurally and functionally integrated in an organ. The intestine is coated externally with a layer of squamous epithelium, the *serosa*, which secretes a watery (serous) lubricative fluid that reduces friction during digestive movements of the gut. Inward from the serosa lie first a longitudinal and then a circular layer of smooth muscle fibers, their contractions coordinated by a network (plexus) of nerve cells. The next layer, the *submucosa*, and the innermost layer, the *mucosa*, both contain connective fibers, blood and lymph vessels, nervous tissue, and myocytes. The mucosa is named for its mucus-secreting layer of *columnar epithelium* that lines the intestinal cavity and forms *intestinal glands*.

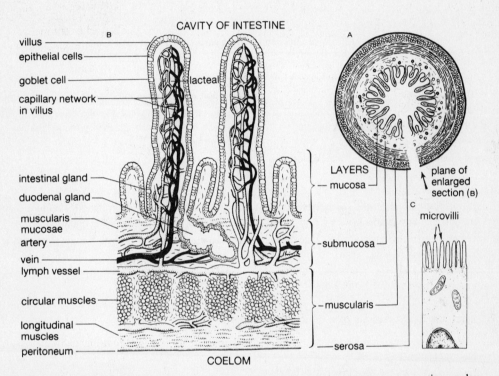

Fig. 2.4   Arrangement of tissues in an organ, the small intestine: A, transverse section; B, longitudinal section of wall; C, sketch from electron micrograph of surface of intestinal epithelial cell showing microvilli. (*From Storer et al.*)

> **Example 9:**   The human *biceps brachii*, another organ, has a definite shape, with a "belly" enclosed within a connective tissue sheath (*epimysium*), and fibrous cords (*tendons*) that anchor the muscle to the skeleton (see Fig. 23.9). This muscle is named "biceps" because it divides proximally into two "heads" that are anchored to the shoulder girdle by two separate *tendons of origin*. It is attached distally to the lower arm bone by way of a single *tendon of insertion*, so that when it contracts, the arm is bent at the elbow. Additional connective tissue subdivides the biceps internally into small bundles of fibers (*fasciculi*). Capillaries run lengthwise between the fasciculi. A major nerve enters the biceps and subdivides to terminate on every muscle fiber individually. All muscle fibers served by branches of a single nerve cell contract simultaneously and so are termed a *motor unit*. The number of motor units contracting at a given moment determines the strength of contraction of the muscle as a whole. Among its fibers and within its tendons, the muscle also contains small sense organs that monitor pressure and tension.

2.16. How are animal cells organized?

Although the specialized cells that make up body tissues differ greatly in appearance and activity, these cells arise from embryonic cells that look much alike, and even in maturity they retain the essential features that characterize all eukaryotic cells. The bodies of all animals (and plants as well) are made up of *eukaryotic* ("true nucleus") cells that are far more similar than they are different, suggesting that the eukaryotic cell may have evolved only once in earth's history. However, more primitive, *prokaryotic* ("before the nucleus") cells, which exist today as bacteria and blue-green algae, antedate unicellular eukaryotes in the fossil record by a full billion years. In fact, during the evolution of eukaryotic cells certain vital cellular components may have originated from prokaryotes living symbiotically within the larger cells.

2.17. How do prokaryotic and eukaryotic cells differ?

The major differences are summarized in Table 2.1.

Table 2.1 Differences between Prokaryotic and Eukaryotic Cells

Prokaryotic Cells	Eukaryotic Cells
1. Lack nuclear membrane	1. Nuclear membrane present
2. Usually singular, ring-shaped chromosome consists only of DNA, without associated proteins, and lacks centromere	2. Multiple, not ring-shaped, chromosomes consist of DNA together with attached proteins and have centromeres
3. Membrane-bound organelles (e.g., mitochondria) absent	3. Membrane-bound organelles present in cytoplasm
4. Diameter seldom exceeds 2 μm	4. Diameter typically 20 μm and more
5. Lack capacity to differentiate into specialized tissues in multicellular organisms	5. Great capacity to differentiate in structure and function within multicellular bodies
6. Occur only as bacteria and cyanophytes (blue-green algae)	6. Make up bodies of protists, fungi, plants, and animals

2.18. What are the major regions of a typical animal cell?

An animal cell has three major regions—cell surface, cytoplasm, and nucleus—each having its own components that carry out specific functions (Fig. 2.5).

2.19. What are the components of the cell surface?

(*a*) *Supportive materials* (e.g., the pellicle of many protozoans) may lie outside or just beneath the plasma membrane, maintaining a constant cell shape.

(*b*) The *plasma membrane* is the cell's *differentially permeable* boundary and regulates the traffic of materials into and out of the cell (Fig. 2.6). It consists of a double layer of two-pronged *phospholipid* molecules with fat-soluble "legs" pointing toward each other and water-soluble "heads" oriented outward toward the two surfaces of the bilayer. Fat-soluble molecules simply ooze between the phospholipid molecules, but other substances can cross the membrane only through pores or by way of special protein carrier-molecules embedded in the lipid bilayer. Only gases and water pass freely through the pores; most other substances must be transported by protein carriers since they either are too large to pass through the pores or, if electrically charged, are repelled by particles of like charge lining the pores.

(*c*) *Microvilli* are fingerlike projections of the plasma membrane that increase the surface area for absorption. They completely cover the exposed surfaces of certain epithelial cells, such as those of the intestinal lining.

(*d*) *Desmosomes* are oval disks with anchoring fibrils that lie just within the plasma membranes of epithelial cells subjected to being stretched. Each desmosome is matched with one on the adjacent cell, and the pair form a "rivet" or "spot-weld" that keeps the cells from being torn apart.

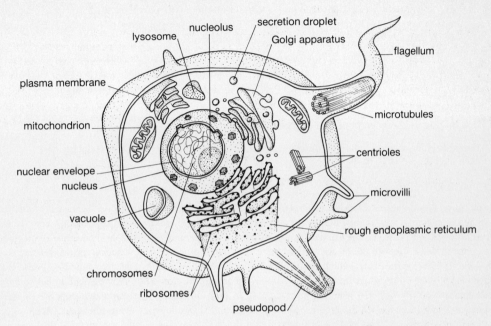

Fig. 2.5 Structure of a representative animal cell.

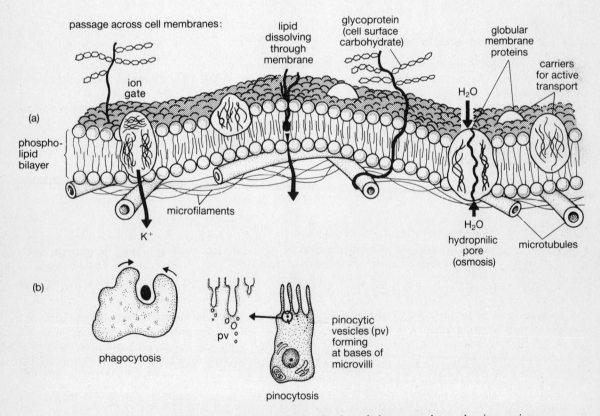

Fig. 2.6 How materials enter and leave cells. (*a*) Section of plasma membrane showing carrier molecules and an ion gate performing active transport, and water and lipids passing by diffusion through pores or between the phospholipid molecules. (*b*) Cells taking in materials by phagocytosis (*left*) and pinocytosis (*right*).

(*e*) *Gap-junction organelles* are hollow "pipes" formed by a ring of six dumbbell-shaped protein subunits that penetrate the plasma membrane of certain tissues and allow free flow of materials from cell to cell. The gap junction is formed by two such "pipes" in alignment between the conjoined cells.

(*f*) *Cilia and flagella* are motile fibrils (see Fig. 6.4) that protrude from the surface of certain types of cells, being covered by an extension of the plasma membrane. They arise from basal bodies (*kinetosomes*) that have the remarkable property of reproducing themselves, each daughter kinetosome then growing a new fibril.

2.20.   How do substances penetrate the cell surface?

Materials enter or leave cells by diffusion, mediated-transport, and exocytosis or endocytosis.

2.21.   What is diffusion?

Diffusion of substances across the plasma membrane is said to be *passive transport* because it costs the cell no energy since the substances simply follow their own *concentration gradients*, moving from regions of greater to regions of lesser concentration. *Osmosis* is the diffusion of water through a differentially permeable membrane, from the *hypotonic* (or *hyposmotic*) side (i.e., more dilute side, where there is a higher proportion of water molecules relative to solutes) to the *hypertonic* (*hyperosmotic*) side, which has a higher concentration of solutes and correspondingly less water. When the proportions of water and solutes on each side become equalized, the two solutions (now said to be *isotonic* or *isosmotic*) are in *osmotic equilibrium* and water diffuses equally in both directions. Diffusion rate depends on the concentration gradient: it is most rapid when the gradient is most steep, and slows down as the gradient declines.

Facilitated diffusion takes place when certain molecules within the cell bind the material diffusing inward, thereby maintaining a steep concentration gradient. For example, movement of oxygen across the membrane lining our lungs is facilitated by a protein known as a cytochrome, which binds oxygen as soon as it enters the lung cells.

2.22.   What is carrier-mediated transport?

Carrier-mediated transport is known as *active transport* because the cell expends energy in promoting interactions between carrier molecules and the materials that they transfer across the membrane. These carriers are specific with respect to the kind of substance they will transport, and they also usually operate unidirectionally, moving the substance either inward or outward, but not in both directions. By means of such carriers, cell membranes can even move substances *against* their concentration gradients.

2.23.   What are endocytosis and exocytosis?

Endocytosis and *exocytosis* are processes by which cells engulf and discharge large molecules, droplets, or actual particles by the formation of inpocketings and outpocketings of the plasma membrane, which may be pinched off internally or externally.

> **Example 10:**   *Endocytosis* of solid particles (*phagocytosis*), is accomplished by cytoplasmic projections (pseudopodia) that extend around the particle and fuse together, trapping the materials within a vacuole formed by a pinched-off section of plasma membrane. Water and large molecules in solution can be imbibed in a similar process (*pinocytosis*): ultramicroscopic *pinocytic vesicles* invaginate from the plasma membrane and pinch off into the cytoplasm.

Exocytosis may be an eliminative process by which cells (such as protozoans) discharge indigestible residues, or a secretory process by which glandular cells export such products as hormones, digestive enzymes, mucus, milk, and sebum.

> **Example 11:**   In *merocrine* glands (e.g., most human sweat glands) the secretory material gathers at the exit surface in ultramicroscopic vesicles and is liberated by fusion of the vesicles with the plasma membrane; alternatively, the secretory vesicles collect in tiny "blebs" that balloon outward and then pinch off, carrying away a bit of the plasma membrane without grossly rupturing it. In *apocrine* secretion the cell's exit surface or apex pops off like a volcano erupting (e.g., milk-secreting mammary cells), and the ruptured membrane must quickly regenerate. In *holocrine* secretion (e.g., sebaceous glands) the whole cell bursts and perishes in liberating its product.

2.24.   What are the components of cytoplasm?

Cytoplasm includes all material between the plasma membrane and the nucleus, including many bodies of definite form, referred to as *organelles*.

(*a*) *Hyaloplasm* is the clear, aqueous medium that bathes all cytoplasmic bodies and serves as a reservoir of solutes and water.

(*b*) *Microfilaments* are threadlike aggregations of protein molecules that serve to maintain cell shape (as part of an internal *cytoskeleton*), bring about changes in cell shape (as during embryonic development), and allow cells to contract. Most instances of contractility involve the interaction of microfilaments of actin and myosin, which slide past each other by forming a series of temporary, oscillatory cross bridges.

(*c*) *Microtubules* are hollow tubules, much stouter than microfilaments, made of a unique protein, tubulin. They too can maintain cell shape as part of a cytoskeleton, and also serve as spindle fibers that separate the chromosomes during cell division. Microtubules are also responsible for the motility of cilia and flagella, which contain a ring of nine longitudinal tubule doublets plus a central pair of larger microtubules.

(*d*) *Centrioles* occur as a single pair of tin can–shaped organelles in the cells of animals, fungi, and certain lower plants. During cell division the pair separate, move to opposite ends of the cell, and produce the spindle fibers that separate the chromosomes. Following cell division the single centriole in each daughter cell reproduces itself, with the new centriole growing at right angles out of the side of the old by a mechanism that remains obscure. Flagellated sperm cells develop from cells that originally have no flagella. In this case, one of the centrioles plays the role of a flagellary kinetosome and produces the sperm's "tail." Not surprisingly, both centrioles and kinetosomes share a nearly identical internal structure consisting of nine triplets of microtubules, suggesting a common evolutionary origin for the two organelles.

(*e*) *Ribosomes* occur in large numbers in all cells, both prokaryotic and eukaryotic. These minute bodies (about 0.02 μm) made of RNA (ribonucleic acid) and protein self-assemble from two subunits. The assembled ribosome engages a threadlike molecule of the messenger RNA that genes produce. Ribosomes contain enzymes that synthesize proteins, and bear attachment points for the molecules (transfer RNA) that transport the small amino acid molecules that will be incorporated into proteins. Each ribosome moves gradually along the length of the messenger RNA molecule, incorporating one amino acid at a time, in correct sequence to produce some specific protein.

(*f*) The *endoplasmic reticulum* (ER) is a system of membrane sheets enclosing fluid-filled channels, some of which extend all the way from the cell surface to the nucleus. These membranes are similar to the plasma membrane but are very labile, often disappearing and reforming. Rough ER is thickly studded with ribosomes and is most extensive in cells that generate proteinaceous secretions. Smooth ER lacks ribosomes (which in this event anchor to microfilaments in the hyaloplasm) but is thought to contain enzymes needed in synthesizing lipids and steroids.

(*g*) *Mitochondria* are bacteria-sized organelles that serve as energy-transfer "factories." Their internal membranes bear virtual assembly lines of enzymes and other substances that demolish sugars and other organic molecules in orderly fashion, transferring energy to a compound called ATP (adenosine triphosphate) which in turn provides energy to a wide variety of intracellular reactions. The possible origin of mitochondria from prokaryotic symbionts is supported by the fact that each mitochondrion contains its own prokaryote-like, ring-shaped chromosome, which allows it to reproduce, synthesize many of its own proteins, and even make its own ribosomes, which are distinctly different from the extramitochondrial ribosomes.

(*h*) The *Golgi complex* is a stack of flattened hollow saccules, rather like a stack of pancakes, located close to the nucleus and best-developed in glandular cells. This membrane complex is involved in "packaging" secretory materials, concentrating them in tiny, membrane-bound vesicles that bud off the rims of the saccules. Since the saccule membranes must constantly regenerate the membrane material lost in vesicle formation, the Golgi is considered to be an important site for the synthesis of cell membrane components. Golgi membranes also contain enzymes for synthesizing certain polysaccharides (complex sugars) that may be bonded to the proteinaceous secretions to form *glycoproteins* such as mucus.

(*i*) *Vacuoles* are membranous sacs that enclose a variety of substances, often for only temporary storage. Freshwater protozoans collect in *contractile vacuoles* the excess water that seeps into their bodies through osmosis; the vacuoles then contract to expel this water.

(*j*) *Lysosomes* are vacuoles that safely sequester potentially destructive hydrolytic enzymes within the cell, ordinarily liberating them in minute amounts to catalyze the breakdown of molecules larger than those dealt with by the mitochondria. Animals that engage in intracellular digestion (e.g., protozoans, sponges, and coelenterates) rely on lysosomes to release digestive enzymes into the vacuoles that contain food obtained by phagocytosis. Lysosomal enzymes also serve to digest aging organelles and (as a normal aspect of embryonic development) sometimes liberate their enzymes en masse, causing "cell suicide" (autodigestion). In keeping with the potency of their contents, lysosomes have remarkably tough membranes which resist disruption by ultrablenders which can make hash of other cell components.

(*k*) *Peroxisomes* are vacuoles distinguished by a unique solid core of crystalline tubules. They store the enzyme *catalase*, which is liberated in controlled quantities to convert the toxic hydrogen peroxide (H_2O_2), formed during certain metabolic processes, into harmless water and oxygen.

(*l*) The term *inclusion* is rather variably used but here will designate special materials that occur in certain cells and are not sequestered in vacuoles, such as pigment granules in cells called *chromatophores*, and granules of the waterproofing protein *keratin* in the outer skin cells of land vertebrates.

2.25.   What are the components of a nucleus?

The *nucleus*, so called because it usually lies toward the central core of a cell, encloses *chromosomes* and one or more *nucleoli* within a surrounding membrane, the *nuclear envelope*. Certain especially large cells (e.g., striated muscle fibers and some protozoans) have more than one nucleus.

(*a*) The *nuclear envelope* consists of two concentric phospholipid membranes with a space between. *Octagonal pores* penetrate through both layers, but each is plugged by a dense *annulus*, which appears to open only occasionally, letting large molecules such as messenger RNA and proteins leave the nucleus. During cell division the nuclear envelope usually disappears.

(*b*) A *nucleolus* is a dense mass of granules and fibrils that remains attached to the *nucleolar-organizing region* of one particular chromosome that produces ribosomal RNA and ribosomal proteins. The nucleolus thus seems to be a reservoir of constituents from which new ribosomes are assembled. It disappears during cell division, allowing dispersal of new ribosomes to both daughter cells.

(*c*) *Chromosomes* are darkly staining bodies that control cell growth, metabolism, and reproduction. Chromosomes can reproduce themselves and also serve as patterns for the assembly of RNA molecules. Each chromosome consists of a long thread of DNA wrapped tendril-fashion around a beadlike series of proteins called *histones* and bearing *nonhistone proteins* as well. The histone proteins appear to play roles in regulating the activity of the functional segments of DNA known as *genes*. The nonhistone proteins are enzymes that clip and mend the DNA strands, as needed, and aid the process of DNA replication. At some consistent point along its length each chromosome bears a proteinaceous *centromere*, to which spindle fibers attach during cell division. After the chromosome reproduces itself and at the proper stage of cell division, the centromere splits, allowing the daughter chromosomes to move apart.

Chapter 3

Chemical Basis of Life

The cell is the basic functional, structural, and reproductive unit of the living world, but cells in turn are made up of molecules, and these consist of fundamental particles known as atoms.

3.1. How are animal bodies organized at the molecular level?

Since living things are constructed entirely of materials ultimately obtained from the physicochemical environment, it follows that there can be no *uniquely biological* kinds of atoms. But living systems can use atoms of universal distribution to synthesize molecules that *are* biologically unique.

3.2. What is a molecule?

A *molecule* consists of two or more atoms joined by chemical forces called *bonds*. The atoms composing a molecule may be identical (e.g., H_2) or different (e.g., H_2O). A molecule composed of different atoms is called a *compound*.

3.3. What is an atom?

An *atom* is a fundamental unit of mass that consists of a positively charged *nucleus* about which move negatively charged *orbital electrons*. The atomic nucleus contains one or more positively charged subatomic particles known as *protons* and an approximately equal number of uncharged particles called *neutrons*. The number of protons represents the *atomic number* (a.n.); there are approximately 90 naturally occurring *elements*, each designated by a specific atomic number and made up of one type of atom. The smallest atom and lightest element is hydrogen (H), with an atomic number of 1; ordinary H has a nucleus consisting of a single proton, around which moves a single orbital electron (Fig. 3.1).

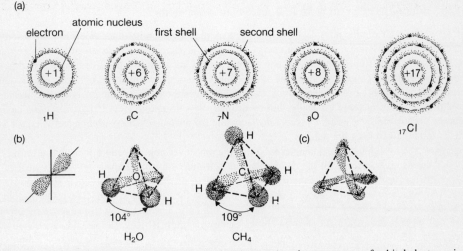

Fig. 3.1 Atomic structure. (*a*) Atomic number (protons) and arrangement of orbital electrons in the four elements that make up 99 percent of living matter, plus a heavier atom, chlorine. Each shell represents a quantum level reflecting the energy of electrons. (*b*) Bonding orbitals (*left*) and covalent bonding in water (*middle*) and methane (*right*). (*c*) Bonding orbitals of a carbon atom.

21

3.4.   Do organisms contain all 90 naturally occurring chemical elements?

Interestingly, most elements occur in living bodies only as scanty traces, if at all. Living matter consists 99.99 percent of only 12 elements; in fact, only 4 kinds of comparatively lightweight atoms (Fig. 3.1a)—oxygen (O, a.n. 8), carbon (C, a.n. 6), hydrogen (H, a.n. 1), and nitrogen (N, a.n. 7)—make up 99 percent of a typical living thing. Within living systems, atoms seldom exist in their elemental state, but as components of molecules or as electrically charged particles known as *ions*.

3.5.   What are isotopes?

The atomic number is a count of the number of protons in the elemental atom. Since in the elemental state an atom is electrically neutral, the number of electrons (each carrying a negative charge) must be equal to the number of protons (each carrying a positive charge). However, the number of neutrons may vary among atoms that have the same atomic number, in which case the different forms of the same element are known as *isotopes*.

> **Example 1:**   Two isotopes of ordinary hydrogen (which has one proton and no neutrons) are *deuterium* and *tritium*. Deuterium has one proton and one neutron in its nucleus; tritium has one proton and two neutrons. Isotopes share the same atomic number, but differ in *atomic mass*, which is the sum of the atom's protons *and* neutrons; thus, all H isotopes have the atomic number 1, but atomic masses of 1, 2, and 3, respectively.

Isotopes with extra neutrons are often unstable and undergo *radioactive decay* at typical and predictable rates, giving off subatomic nuclear particles until they achieve stability.

> **Example 2:**   Tritium is radioactive, with a *half-life* of 12.5 years, which means that 50 percent of it will decay within 12.5 years after it is produced, 50 percent of what is left will be gone by the end of another 12.5 years (leaving 25 percent of the original amount), and so forth. Tritium is very useful in biological research as a "radioisotope tag" that allows hydrogen-containing compounds to be traced through metabolic pathways.

During their decay, radioactive elements give off gamma rays, beta particles (*nuclear* electrons), and massive alpha particles consisting of two protons and two neutrons. When a radioisotope loses protons, its decay product will have a different atomic number and thus will be a different element.

> **Example 3:**   Medieval alchemists got headaches trying to transmute the element lead into gold; modern atomic physicists have not done a great deal better: radioactive gold produced in fission reactions decays into lead, but not vice versa.

3.6.   What determines whether an atom will react with other atoms?

The orbital electrons determine how, and whether, atoms will associate with other atoms to form *molecules*. Orbital electrons have negligible mass but considerable energy and move at incredible speeds about the atomic nucleus. Their kinetic energy prevents them from being drawn into the nucleus by its attractive positive charge, yet this same charge ordinarily prevents them from entirely escaping the atom. However, not all electrons possess the same amount of energy; those with less energy remain closer to the atomic nucleus, whereas those with more energy range farther away. The energy of electrons does not increase gradually, but by distinct steps or *quanta* (Fig. 3.1a). Electrons having the same amount of energy occupy the same *quantum level* (*shell*) around the nucleus. Those with a higher energy content occupy a quantum level farther away from the nucleus.

The innermost shell, or lowest quantum level, for any atom never contains more than two electrons. Each shell external to this contains *eight*, except for the outermost shell, or highest quantum level. It is the number of electrons in the outermost shell that determines the combining power (*valence*) of the atom. If the outer shell also contains eight electrons, the atom will be unable to bond with any other atom and the element is *inert*. The value of gold as currency and jewelry is enhanced by the fact that this inert element will not tarnish or rust away. However, most elements are not inert, because they have fewer than eight electrons in the outermost shell. If, for example, the outer shell contains only seven electrons, an atom needs one more electron to saturate this shell.

3.7.   How do atoms fill their outer shells?

An atom may "kidnap" an electron from some other atom, or the atom may share an electron pair with some other atom. The first alternative leads to *ionization*, which makes possible *electrostatic* (*ionic*) bonding between ions of opposite charge. The second alternative produces a *covalent bond*, the most stable type of interatomic association.

3.8. Do electrons ever leave their quantum level in ways other than being kidnapped?

Under certain conditions an electron can move from one quantum level to a higher one by absorbing a "packet" of energy (quantum) from some environmental source, usually the sun.

> **Example 4:** The essential feature of *photosynthesis* is that one particular electron in the chlorophyll molecule (perhaps an electron in chlorophyll's single atom of the metallic element magnesium) absorbs one quantum (*photon*) of light energy and becomes so "excited" that it entirely escapes from its atom, to be grabbed by nearby "electron-carrier" molecules, which use the excess energy of the excited electron to build ATP. This event (*photophosphorylation*) provides the energy needed for fixing carbon dioxide and water in the synthesis of sugars. Ultimately, the chemical bonding energy of all animals, from aardvark to zebra, is derived from photic energy trapped by electrons in chlorophyllous plants.

When an excited electron loses its excess energy, it drops back to its former quantum level.

> **Example 5:** When chlorophyll solution extracted from leaves is placed in a test tube, its electrons still absorb light photons and become excited, but they cannot hold on to this energy and have no electron carriers available to transfer it. As a result, the excess energy is given up all at once, and the test tube glows from the burst of visible light as the electrons return to their original energy state.

3.9. How do covalent bonds form?

Covalent bonding results from the tendency of electrons in the outermost shell to pair and spin around each other like frenetic dancers, each electron pair waltzing merrily within a volume of space known as an *orbital* (Fig. 3.1*b*). An inert element would have four orbitals in the outer shell, each containing one electron pair and thus would have room for no more. But in a reactive element, one or more orbitals will contain only a single electron and thus will constitute *bonding orbitals*, each capable of holding one more electron. When bonding orbitals of two adjacent atoms overlap, one electron from each atom becomes a member of a *shared* electron pair, and the two spin about each other within their common orbital.

A covalent bond does not automatically come about as a result of propinquity alone: *energy* is needed, and a *catalyst* is required to both make and break this bond. Various metals (e.g., platinum) and other inorganic materials operate as catalysts in the nonliving world. Biological catalysts are proteins known as *enzymes*, which reduce the amount of energy needed by holding the reacting atoms in positions that facilitate the reaction.

3.10. Are all compounds covalently bonded?

No. The different atoms of compounds may be bonded either covalently or electrostatically. Sugar is a *covalent compound*, formed by the covalent bonding of carbon, hydrogen, and oxygen atoms; in water, sugar dissolves but its atoms do not dissociate. In contrast, table salt, sodium chloride (NaCl), is an *ionic compound* formed by electrical attraction between a positively charged sodium ion (Na^+) and a negatively charged chloride atom (Cl^-). In water, NaCl both dissolves, and dissociates into its constituent ions. These ions do not behave at all like the elemental atoms from which they are derived. Elemental sodium (Na) is a gray metal, while elemental chlorine (Cl) is a toxic gas. Na^+ and Cl^- make your soup tastier. Na becomes positively charged by having one of its electrons abducted (leaving it with one more proton than electron); conversely, chlorine becomes Cl^- by acting as electron thief and coming to have one more electron than proton. In contrast, the salt sodium bicarbonate (baking soda), also an ionic compound, dissolves to yield only Na^+ and the polyatomic ion HCO_3^-. The bicarbonate ion, HCO_3^-, does not break down any further, because its atoms are bonded covalently, and a little thing like being dissolved in water does not bother a covalent bond.

3.11. What is the difference between organic and inorganic compounds?

Sodium chloride and sodium bicarbonate are both *inorganic* compounds, as are water and carbon dioxide. Inorganic compounds occur generally in the physicochemical environment, as well as within living things. Organic compounds always contain the element *carbon* and are usually products of *biotic* (living) processes. (That they may form *abiotically* under certain conditions has been demonstrated in laboratory experiments and is also shown by their occurrence in prebiotic sedimentary rocks on earth, as well as in meteorites from space.)

3.12.   Why is carbon prevalent in all organic molecules?

The carbon atom is particularly well-suited for engaging in the assembly of large, complex molecules. Its second (and outermost) shell contains only 4 electrons, which space out, one in each of four bonding orbitals that are arranged three-dimensionally, rather like the legs of a tripod with a fourth leg extending perpendicularly (Fig. 3.1c). As a result, each C atom can form covalent bonds with as many as four other atoms, producing straight or branching chains, rings, and interlocking rings.

3.13.   Why are organic compounds often shown in structural form?

The information furnished by an *empirical formula* such as Ca_2CO_3 is usually adequate to describe an inorganic compound, although it simply tells us that calcium carbonate consists of two atoms of calcium, one atom of carbon, and three atoms of oxygen. However, the arrangement of atoms even in rather simple organic compounds can vary enough to produce *isomers* having the same empirical formula but displaying very different properties. Therefore, organic compounds are better represented by *structural formulas* that show pictorially how the atoms are arranged in the molecule.

> **Example 6:**   The empirical formula C_2H_6O describes both ethanol and dimethyl ether. The latter is a gas that quickly puts you to sleep. Ethanol is a liquid that does nicely in margaritas and gin fizzes; it can put you to sleep too, but it takes a bit longer. Knowing the difference between the two could be important, and their structural formulas accomplish this (Fig. 3.2a). Ethanol and dimethyl ether are *structural isomers:* their atoms are quite differently arranged.
>
> Another type of isomerism results from the fact that organic molecules, being three-dimensional, may exist as mirror-image versions called *stereoisomers*. Stereoisomers can be discerned because the *dextro* isomer rotates the plane of polarized light to the right, while the *levo* isomer rotates it to the left. The simple sugar glucose ($C_6H_{12}O_6$) has right-handed (dextrose) and left-handed (levulose) versions. Amino acids, the building blocks of proteins, form equal amounts of their dextro and levo isomers when synthesized in a laboratory (Fig. 3.2b). Such a solution (called *racemic*) of synthetic amino acid does *not* rotate the plane of polarized light, because the effects of the two isomers cancel each other. However, in nature, only dextrorotatory amino acids occur. If an experimental animal were fed only synthetic levorotatory amino acids, the animal would soon die of malnutrition, being totally unable to use the *l* isomers for energy or protein building.

Structural formulas are also useful in distinguishing between single bonds and double bonds. A *single bond*, shown as $-$, represents one shared electron pair linking the two atoms. A *double bond*, $=$, indicates that the two atoms are joined by two shared electron pairs. Rather than being twice as strong as a single bond, double bonds are actually somewhat unstable. Molecules of animal fat contain few double bonds and so are known as *saturated fats;* because of this they are comparatively difficult to break down and hence are considered less healthful to eat than vegetable oils, which are described as *polyunsaturated* because they contain many double bonds at which their molecules can be easily degraded.

3.14.   What is molecular polarity?

Molecular polarity results from an asymmetrical distribution of electrons within the molecule, so that the molecule is "lopsided" and is at least a bit electropositive at one end and somewhat electronegative at the other. The behavior of such *polar molecules* can have tremendous effects upon the living world.

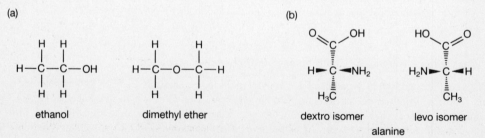

Fig. 3.2   Isomerism (*a*) Structural isomerism. (*b*) Stereoisomerism.

Example 7: Water (H_2O) is a polar molecule, a situation that yields two of its most important properties, namely, that water *dissolves* many things and that water molecules cling together by *hydrogen bonds*, forming drops when liquid and crystals when frozen. A water molecule is polarized, because its atoms are *not* arranged linearly, as H-O-H, but angularly (Fig. 3.3*a*). This angle results from the asymmetrical orientation of the two bonding orbitals of the oxygen atom, as shown. In addition, the nucleus of the O atom, with its eight protons, not only attracts that atom's own eight electrons, but exerts so much attractive force on the single electron of each of the two hydrogen atoms that the shared electron pair in each orbital is drawn closer to the O nucleus and farther away from the H nuclei (which are merely single protons). As a result, the angular water molecule is electronegative at its O end and electropositive at its H end. This makes water a good solvent, because H_2O molecules are attracted to any other polar molecule (or even to a nonpolar molecule that bears a few polar side groups) and stick to it to form a partial or complete *hydration shell*. Figure 3.3*b* shows how water molecules form a hydration shell around sodium and chloride ions, helping sodium chloride dissolve; note that in the hydration shell surrounding the chloride ion, the H end of the water molecule faces inward, attracted by the negatively charged Cl^-; in contrast, in the hydration shell around the positively charged sodium ion (Na^+), the O end of the water molecule faces inward. The coating of water molecules prevents the oppositely charged ions from coming together to form electrostatic (ionic) bonds. However, as the water evaporates, eventually the Na^+ and Cl^- *will* establish such bonds and solid salt crystals will appear.

Example 8: Sugar molecules do not ionize and are nonpolar, but they do bear *hydroxyl* ($-OH$) side groups that are weakly polar. This is enough to make sugar soluble in water, for the water molecules form weak hydration shells, just around the $-OH$ groups, which suffice to keep the sugar molecules from clumping together.

3.15. What are H bonds?

Hydrogen bonds, which form and break without the aid of catalysts, become established between polar molecules or the polarized side groups of nonpolar molecules. They involve the asymmetrical attraction of the electron of a hydrogen atom, away from its own nucleus, toward the electropositive end of a polar molecule or group. It is these weak bonds

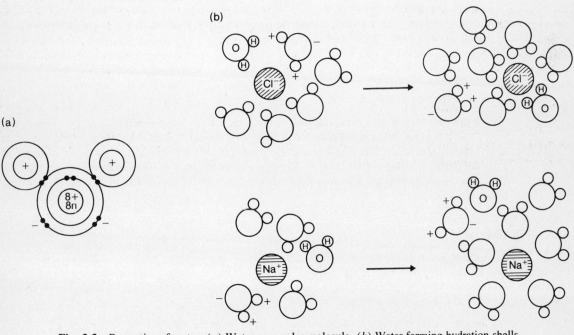

Fig. 3.3 Properties of water. (*a*) Water as a polar molecule. (*b*) Water forming hydration shells around chloride ions (*above*) and sodium ions (*below*).

that stabilize DNA and are essential to the capacity of DNA to reproduce itself and to form RNA. Life as we know it could not exist without H bonds!

3.16. How does H bonding affect the behavior of water?

Water molecules, being polar, readily form H bonds between the O end of one molecule and the H end of another, so that water is actually a "liquid crystal" and is accordingly highly *cohesive*. This cohesiveness affects the behavior of water in living systems in ways essential to life.

> **Example 9:** If water molecules were not cohesive, most mature trees would be too tall to raise water from the soil to their crowns, for this action depends on the upward pull exerted upon water columns in microscopic tubules as water evaporates from the leaf pores. Despite the downward pull of gravity, the columns of water molecules, held together by H bonds, are drawn upward to the tops of even giant trees. If water were not cohesive, osmotic pressure at the roots could push it upward to a height of no more than 9.7 m; thus forests would consist of rather stunted trees.

3.17. What is ionization?

Ionization takes place when the nucleus of one atom exerts enough attractive force to kidnap one or more electrons from some other atom(s). This "crime" causes both "thief" and "victim" to become electrically charged. The electron loser is left with a net positive charge (because its protons now exceed its electrons), while the electron grabber becomes negatively charged by adding the stolen electrons to its own, so that it now has more electrons than protons. Both electrically charged entities are now called *ions:* positively charged *cations* and negatively charged *anions*. When not dissolved in water (that is, when not clothed in hydration shells of water molecules), ions of opposite charge attract each other electrically, forming *ionic compounds* that sometimes build visible crystalline structures when the water evaporates. Compounds that ionize fall into three types: acids, bases, and salts.

3.18. What are acids?

Acids in solution give off the *hydrogen ion, H^+*, which is actually a proton unaccompanied by an electron.

> **Example 10:** Hydrochloric acid, HCl, a gas, is an *inorganic acid* that yields H^+ and Cl^- in solution; acetic acid (which makes vinegar sour) is an *organic acid* (Fig. 3.4) with a *carboxyl group* ($-COOH$) that in solution yields a proton, leaving the remainder of the molecule as a negatively charged anion: $H_3C-COO^- + H^+$. Acids have a pH less than 7.

3.19. What are bases?

Bases give off the *hydroxyl ion, OH^-*, in solution.

> **Example 11:** Sodium hydroxide, NaOH (a caustic alkali when concentrated), ionizes to Na^+ and OH^-. Ammonia (NH_3) also behaves as a base, even though it does not contain $-OH$; instead, it interacts with water, capturing a proton and leaving behind the remainder of the water molecule as OH^-: $NH_3 + H_2O \rightarrow NH_4^+ + OH^-$. Some organic molecules also ionize as bases, for example, the *nitrogenous bases* (e.g., adenine and thymine) that are parts of nucleic acid molecules. Bases have a pH greater than 7.

3.20. What are salts?

Salts, which yield neither H^+ nor OH^-, do not affect the acidity or alkalinity (i.e., pH) of a solution and are formed by reacting acids with bases. When an acid and a base are mixed together in solution, they will be *neutralized* as H^+ liberated by the acid unites with OH^- given off by the base, forming water, as in the following reaction between hydrochloric acid and sodium hydroxide to yield the salt NaCl and water:

$$(H^+ + Cl^-) + (Na^+ + OH^-) \rightarrow (Na^+ + Cl^-) + (H^+ + OH^-)$$

You would get painfully burned if you took a mouthful of either hydrochloric acid or sodium hydroxide solution, but a swig of NaCl in H_2O is simply saltwater, untasty but not corrosive.

3.21. What is pH?

pH means "potential hydrogen," where pH 7 indicates neutrality.

>　　　　**Example 12:**　Pure water if fully ionized to H^+ and OH^- would potentially yield a molar concentration
>　　　　of 10^{-7} H^+ and 10^{-7} OH^- (i.e., pH 7). A mildly acidic solution if fully ionized would yield 10^{-6} H^+
>　　　　and 10^{-8} OH^- and would be designated "pH 6," and so forth.

3.22. How are ions important to life?

Ions play many essential roles in living systems, as do organic acids and bases. For example, throughout the animal world the action of nerves and muscles depends on the movement across cell membranes of such ions as Na^+, Cl^-, K^+ (potassium ion), and Ca^{2+} (calcium ion).

With few exceptions, living systems do not tolerate strongly acidic or alkaline conditions, and their vital processes must take place within a range from pH 6 to pH 8.

>　　　　**Example 13:**　The pH of human blood must remain between 7.35 and 7.45, a slightly basic range
>　　　　zealously guarded by many plasma ions (such as bicarbonate ion, HCO_3^-) that can neutralize excess
>　　　　H^+ and OH^-; if human blood plasma merely becomes neutral, pH 7, this seemingly harmless deviation
>　　　　actually would represent a life-threatening *acidosis*!

3.23. How are inorganic compounds vital to animal life?

(*a*)　*Water* makes up 80–90 percent of cells and is essential as a solvent and medium in which chemical reactions can occur.

(*b*)　*Salts* furnish many essential ions, some already mentioned. In addition, solid, crystalline calcium carbonate forms many invertebrate skeletons, including mollusk shells, and calcium phosphate furnishes the mineral content of vertebrate skeletons and teeth. The nitrogen atoms needed for building proteins and nucleic acids mostly come to animals by their ingestion of plant tissues, but they enter plants from the soil mainly in the form of the *nitrate* ion (NO_3^-) and *ammonium* ion (NH_4^+).

3.24. What major types of organic compounds occur in animal bodies?

These include intermediate metabolites, carbohydrates, lipids, proteins, and nucleic acids.

3.25. What are intermediate metabolites?

Intermediate metabolites include a variety of small organic molecules that mostly are formed as "stepping-stones" in a metabolic pathway leading from original substrate to final end product. These include organic acids, organic bases, alcohols, aldehydes, ketones, and esters (Fig. 3.4). For example, one of the most important and universal energy-liberating metabolic pathways in both plants and animals, taking place within their mitochondria, is known as the "tricarboxylic acid (citric acid) cycle," because it includes as intermediates an entire series of organic acids.

3.26. What are carbohydrates?

Carbohydrates contain carbon, hydrogen, and oxygen atoms in the approximate C:H:O ratio of 1:2:1 (Fig. 3.5). The smaller, water-soluble varieties (*sugar*) mainly provide energy, but some sugar molecules can also be *polymerized* (chained together) to form giant *polysaccharide* molecules (e.g., glycogen) that serve in food storage or as structural components (e.g., *chitin*, which forms insect skeletons, is a polymer of nitrogen-containing "amino sugars," and the intercellular "cement" that helps hold tissues together is mostly made up of other nitrogenous polysaccharides such as *hyaluronic acid*).

3.27. What are the major types of carbohydrates?

(*a*)　*Monosaccharides*, "simple sugars," include six-carbon *hexoses* ($C_6H_{12}O_6$) such as *glucose* ("blood sugar") and five-carbon *pentoses* ($C_5H_{10}O_5$) such as *ribose*.

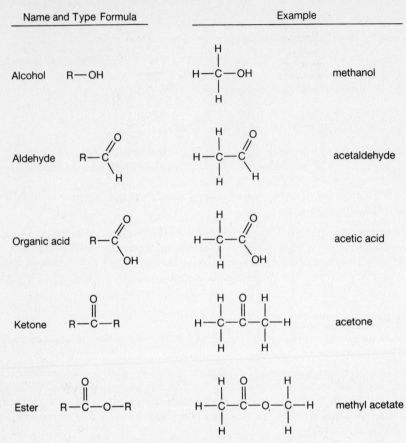

Fig. 3.4 Major types of intermediate metabolites.

(b) *Disaccharides*, "double sugars," consist of two simple sugar molecules covalently bonded together with the elimination of a molecule of water:

$$C_6H_{12}O_6 + C_6H_{12}O_6 \rightarrow C_{12}H_{22}O_{11} + H_2O$$

Disaccharides such as *sucrose* (the plant product that fills our sugar bowls) and *lactose* (milk sugar, a specialty of mammals) both have the empirical formula $C_{12}H_{22}O_{11}$ but are isomers with different arrangements of atoms and therefore have different structural formulas.

(c) *Polysaccharides* are long branching or unbranching chains of simple sugar molecules bonded with elimination of one water molecule for each bond made. Thus *glycogen*, *starch*, and *cellulose*—all made up of many glucose molecules—can be represented by the "type formula" $[C_6H_{10}O_5]_n$, where *n* is the number of glucose units present. Glycogen is a highly branched molecule used by most animals for food storage with quick energy release, while cellulose is a plant product that can be neither digested nor synthesized by most animals. Starch, another plant product, is an unbranched glucose chain that is readily digested by animals. Starch and cellulose are made up of slightly different glucose isomers and differ in how their glucose units are linked (Fig. 3.5d): the α-linkage (└O┐) of starch is readily broken by digestive enzymes, but the β-linkage (╵—O┐) of cellulose is not.

3.28. What are lipids?

Lipids include a variety of fat-soluble compounds (Fig. 3.6) made up of nonpolar molecules that clump weakly without electrostatic attraction. This weak attraction is called *hydrophobic* ("water-hating") bonding (van der Waals forces)

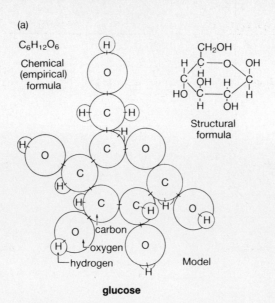

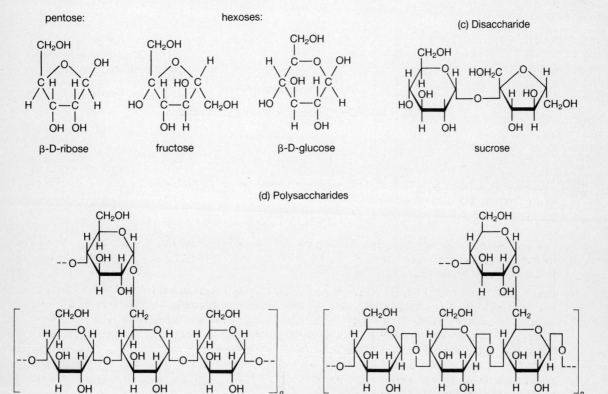

Fig. 3.5 Structural formulas of representative carbohydrates. (*a*) Three ways of representing an organic molecule. (*b*) Monosaccharides: a pentose (5 carbons), empirical formula $C_5H_{10}O_5$, and two hexoses (6 carbons each), $C_6H_{12}O_6$. (*c*) A disaccharide. (*d*) Polysaccharides: small portions of glycogen and cellulose, both polymers of glucose. [(*a*) *from Storer et al.*]

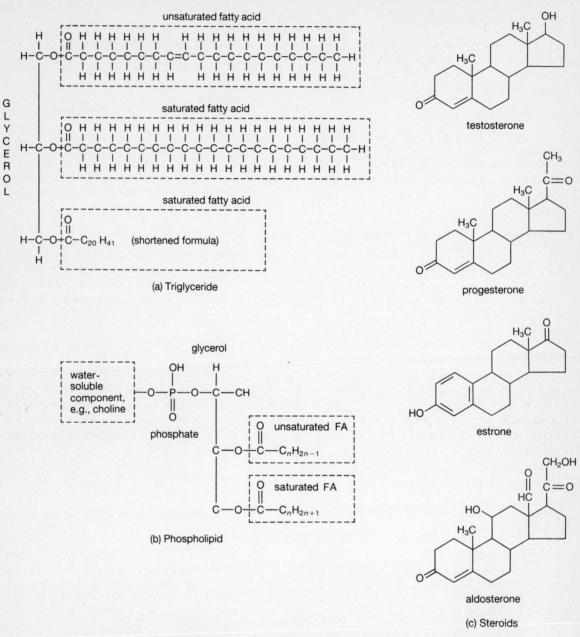

Fig. 3.6 Lipids. (*a*) Triglyceride, useful for food storage and insulation. (*b*) Phospholipid, a component of cell membranes. (FA = fatty acid.) (*c*) Steroids; (*above*) three vertebrate sex hormones; (*below*) a hormone from adrenal cortex.

and is responsible for the ability of oil to consolidate into droplets, lipids to soak readily through cell membranes (i.e., slip between the phospholipid components of the membrane), and nonpolar solvents such as gasoline and acetone to dissolve grease.

3.29. What types of lipids exist?

The three major types of lipids are triglycerides, phospholipids, and steroids.

3.30.   What are triglycerides?

Triglycerides consist of three hydrocarbon chains called fatty acids bonded through their $-COOH$ groups by way of a small alcohol known as *glycerol*. Fatty acids themselves are acidic because their terminal $-COOH$ gives off H^+, but when bonded to glycerol, they become parts of "neutral fat" molecules, which provide insulation and energy reserves.

3.31.   What are phospholipids?

Phospholipids consist of *two* fatty acids linked by a glycerol molecule to some phosphate-containing, *water-soluble* component. The latter is *polar* and *hydrophilic* ("water-loving"), while the two fatty acid "legs" are hydrophobic and cling to other lipid molecules. This behavior is fundamental to the formation of all cell membranes (Fig. 2.6), which are mainly phospholipid bilayers with inner and outer surfaces formed by the hydrophilic heads and with the hydrophobic legs making up the center of the sandwich. Phospholipids form such bilayers in vitro as well as in vivo, a behavior suggesting that these bilayers may have been fundamental to the origin of life.

3.32.   What are steroids?

Steroids are composed of four interlocking rings of carbon atoms. (In structural formulas, the C atoms in the rings are seldom shown, but are understood to occur at each angle of the ring.) Vertebrate sex hormones (e.g., estrone, progesterone, and testosterone) and adrenocortical hormones (e.g., aldosterone), as well as insect molting hormone (ecdysone), are steroids.

3.33.   What are proteins?

Proteins are polymers (chains) of some 20 different kinds of amino acids (Fig. 3.7) that theoretically can occur in any sequence or proportion (although only certain arrangements yield biologically active molecules). Although the structural formulas of representative amino acids can vary greatly in the side chain (R in Fig. 3.7) the type formula is the same for all. Amino acids are chained together by *carboxyamino* (*peptide*) linkages between the $-COOH$ (carboxyl) group of one amino acid and the $-NH_2$ (amino) group of the next, with elimination of H_2O. In vivo, the synthesis is less direct than this (see Question 3.43).

3.34.   What are nucleic acids?

Nucleic acids are linear polymers of *nucleotides*. There are two types of nucleic acids: DNA and RNA.

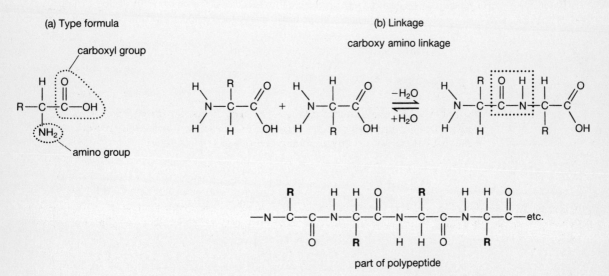

Fig. 3.7   Amino acids, the building blocks of proteins. (*a*) Type formula of an amino acid. (*b*) Carboxyamino linkage of two amino acids.

3.35.   What is DNA?

Deoxyribonucleic acid is a self-replicating molecule that is the basis of heredity. The most gigantic of all biological molecules, DNA is a polymer of four different *nucleotides,* of which the sugar *deoxyribose* is a component. The nucleotides are arranged in two chains that twist around each other to form a double helix (Fig. 3.9a). The chains—which are complementary, and not identical—are connected by hydrogen bonds. The two chains separate (a) when DNA replicates itself or (b) when specific portions of DNA (genes) synthesize RNA.

3.36.   What is RNA?

Ribonucleic acid is a polymer of nucleotides that contain the sugar *ribose* (which has one more oxygen atom than deoxyribose). RNA is involved in protein synthesis and in the structure of ribosomes. As is true for DNA, the nucleotides of RNA can theoretically be arranged in any order, although only certain sequences are biologically functional.

3.37.   What are nucleotides?

A *nucleotide* is a molecule made up of a pentose sugar (ribose or deoxyribose), from one to three phosphates, and a nitrogenous base (purine or pyrimidine; Fig. 3.9d). In addition to making up RNA molecules, free ribose nucleotides play roles in energy transfer (e.g., ATP). The nitrogenous bases are the variable portion of each nucleotide and come in two sizes: the larger *purines* and the smaller *pyrimidines*. The purines *adenine* and *guanine* and the pyrimidines *thymine* and *cytosine* are the bases occurring in *deoxy*ribose nucleotides. In contrast, the four most common bases in ribose nucleotides are adenine, guanine, cytosine, and *uracil*. The side groups (boldface in Fig. 3.9e) of the bases determine H bonding. Since it is the $=O$ (oxygen) and $-NH_2$ (amino group) that form H bonds upon interaction and since spatial constraints of the DNA molecule permit purines to bond only with the smaller pyrimidines, guanine will bond selectively with cytosine (G-C), and adenine with thymine (A-T) in DNA synthesis or with uracil (A-U) in RNA synthesis.

3.38.   How do nucleotides function in energy transfer and storage?

Free ribose nucleotides are vital in energy storage and transfer, especially the ubiquitous *adenosine triphosphate (ATP)* (Fig. 3.8a). Each phosphate group, $-PO_4$ (often represented as $-\textcircled{P}$), is bonded in tandem to the next, like the tail of a kite. This bonding affects electrons within the $-\textcircled{P}$ group in such a way that *high-energy phosphate bonds* (pictured as $\sim$) are produced. An unusually high amount of energy is required to attach inorganic phosphate (P_i) to an organic molecule, and part of this invested energy can later be transferred to some other molecule when ATP or another high-energy phosphate compound passes it one or two $\sim\textcircled{P}$ groups. The molecule thus *phosphorylated* is thereby made more reactive, as when glucose receives $\sim\textcircled{P}$ from ATP, becoming glucose phosphate. This generosity transforms ATP to *ADP* (adenosine diphosphate), which then can take up another P_i to become ATP once more. By analogy, energy stored in such compounds as fats and glycogen is like money in the bank; energy locked in the $\sim\textcircled{P}$ of such ribose nucleotides

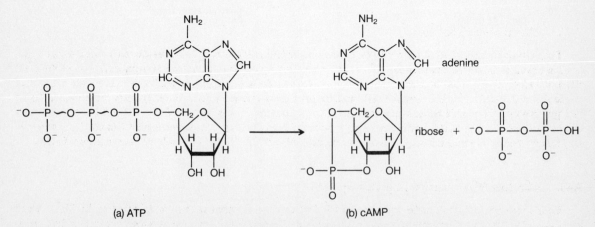

(a) ATP   (b) cAMP

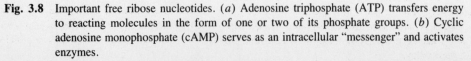

Fig. 3.8   Important free ribose nucleotides. (*a*) Adenosine triphosphate (ATP) transfers energy to reacting molecules in the form of one or two of its phosphate groups. (*b*) Cyclic adenosine monophosphate (cAMP) serves as an intracellular "messenger" and activates enzymes.

as ATP, GTP, UTP, and CTP is like money in the pocket, ready to be spent, the four different nucleotides serving as coins of different value.

Another essential free nucleotide is *cyclic AMP* (Fig. 3.8b) formed from ATP by action of the enzyme adenyl cyclase; this ubiquitous molecule functions widely as an intracellular "messenger" and is basic to the activity of most hormones.

3.39.   How do nucleotides function in DNA synthesis?

During DNA replication (Fig. 3.9b), the two complementary strands separate with the aid of an *unwinding protein*, and each of these old strands then serves as a pattern (*template*) for building a new, complementary strand. This is accomplished by formation of hydrogen bonds between nucleotides on the old strand and free deoxyribose nucleotides. The new nucleotides are then covalently bonded together in linear sequence by *DNA polymerase,* an enzyme, into chains of 100 nucleotides or more (*Okazaki fragments*) that are then tied together endwise by another enzyme, *DNA ligase*.

3.40.   How does DNA produce RNA?

When a given gene becomes active in RNA synthesis, an enzyme "clips" one strand of the DNA in just that particular location, allowing the two strands to unwind. Then, only one of the two strands, not both, serves as the template along which free ribose nucleotides are assembled by hydrogen bonding (Fig. 3.9c). The correctly positioned array of ribose nucleotides is then covalently bonded through the action of the enzyme *RNA polymerase*. The completed portion of the RNA molecule is progressively set free by dissolution of the hydrogen bonds holding it to the gene on which it was assembled. This permits additional free nucleotides to form H bonds with the gene segment to which the RNA strand is no longer attached, allowing *simultaneous transcription* of a number of RNA molecules along the same gene. When transcription is terminated, the completed RNA molecule is liberated and can pass out of the nucleus through the pores of the nuclear envelope.

3.41.   What determines whether DNA will reproduce itself or synthesize RNA?

Nobody knows for sure why DNA sometimes makes RNA and at other times makes new DNA, but the presence of the required enzymes and appropriate free nucleotides certainly affects the decision. Each cell type exhibits a characteristic *cell cycle* (Fig. 3.10), in which the period between divisions (*interphase*) is divided into an initial growth period (G_1) involving RNA and protein synthesis, a delimited (S) period in which DNA is synthesized and daughter chromosomes are produced, and a final period (G_2) during which the proteins making up spindle fibers are synthesized in preparation for cell division. In mammals, the cell cycle of actively multiplying cells takes about 18–24 hours from one division to the next, of which the S period of DNA synthesis lasts some 6 hours.

During periods when the cell is mainly concerned with metabolism and growth, but not with reproduction, certain genes along the chromosomes become active in the synthesis of various types of RNA.

3.42.   What types of RNA exist?

Four main categories of RNA are produced: (1) *messenger RNA(mRNA)*, which dictates the sequential order of amino acids in a protein (since cells make thousands of different proteins, there must be at least as many different mRNAs); (2) *ribosomal RNA* (*rRNA*), used in making new ribosomes; (3) *transfer RNA* (*tRNA*), which bonds to a specific amino acid at one end and to mRNA at the opposite end, thereby positioning that amino acid correctly upon the ribosome for its incorporation into the protein being synthesized (there are at least 20 different tRNAs, because 20 different amino acids occur in proteins and each must be carried by a unique kind of tRNA); (4) *heterogeneous nuclear RNA* (*hnRNA*), particularly long strands that never leave the nucleus but may fragment into mRNA, tRNA, or rRNA, or have unknown functions.

3.43.   How does RNA function in protein synthesis?

Each amino acid that will make up the protein is first covalently bonded to its own specific transfer RNA, which then forms H bonds with messenger RNA on the surface of a ribosome. This allows the ribosomal enzymes to link amino acids consecutively into a protein in a specific order (Fig. 3.11) while simultaneously freeing the tRNA molecules for reuse. One strand of mRNA can accommodate a whole series of ribosomes strung out along its length like beads, each spinning out a chain of amino acids. A molecule of mRNA that is 900 nucleotides long would encode a protein containing nearly 300 amino acids.

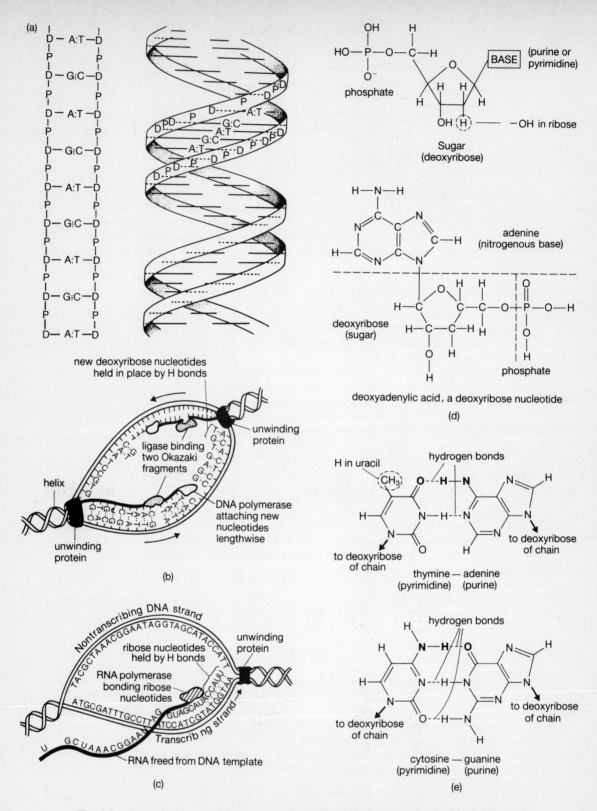

Fig. 3.9 DNA structure and functions. (*a*) Diagram of DNA: two strands of alternating deoxyribose (D) and phosphate (P) attached by H bonds (:) between complementary nitrogenous bases (A = adenine, T = thymine, G = guanine, C = cytosine). (*b*) DNA replicating. (*c*) DNA transcribing RNA (U = uracil). (*d*) Nucleotide structures. (*e*) H bonding between complementary nitrogenous bases. [(*a*) *from Storer et al.*]

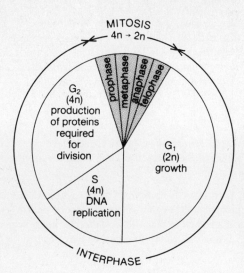

Fig. 3.10 Cell cycle, showing relative time spent in each phase. 2n = diploid state with two sets of homologous chromosomes and each chromosome consisting of one DNA molecule; 4n = each chromosome now consisting of two DNA molecules held together at the centromere. (*From Storer et al.*)

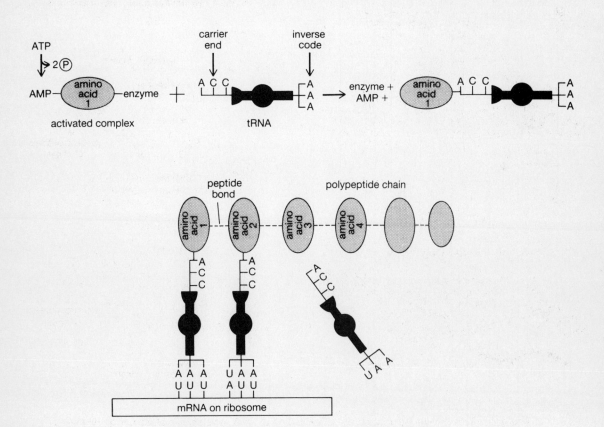

Fig. 3.11 Protein synthesis. *Above*, an amino acid is covalently bonded to a specific tRNA. *Below*, each amino acid–tRNA complex forms H bonds between an mRNA codon (base triplet) and a complementary tRNA anticodon, and ribosomal enzymes bond amino acids linearly into a polypeptide chain. (*From Storer et al.*)

Proteins as molecules are second in size only to nucleic acids, and their variety is staggering. Graft rejection and the success of bloodhound noses both result from the fact that every human being possesses certain proteins unique to that individual, in addition to which we also possess numerous other proteins limited to *Homo sapiens* but shared throughout our species, as well as others widely distributed throughout the animal world. Most serve as enzymes, some are hormones, and others are structurally essential as components of muscle and connective tissues.

3.44. What determines the sequential order of amino acids in protein synthesis?

The "genetic code" is essentially universal from bacteria to mammals. In brief, each three consecutive nucleotides in a strand of DNA or RNA represent a *codon,* or "letter" of the code, also known as a "base triplet." Each codon determines the location of one specific amino acid in a protein.

> **Example 14:** The base triplet ACC in DNA is transcribed as UGG in mRNA, and the sequence UGG can form H bonds with the particular tRNA that carries the amino acid tryptophan. This H bonding between an mRNA codon and a complementary tRNA *anticodon* holds the amino acid in the right position on the surface of the ribosome to allow ribosomal enzymes to bond it chemically into the lengthening protein chain.

At least one codon exists for each of the 20 amino acids that occur in proteins. If more than one codon specifies a given amino acid, ordinarily the first two bases in the triplet must remain the same, but the third often can vary, as if it were little more than a spacer. Thus, the four DNA codons CAA, CAG, CAT, and CAC all code for the amino acid valine. The DNA codon TAC (AUG in RNA) not only codes for the amino acid methionine, but is the starting point for every gene transcription; any one of three DNA codons (ATT, ATC, and ACT) terminates transcription and marks the end of the RNA strand.

Chapter 4

The History of Life

If you were to count from 1 to 3,000,000,000 at a rate of one number per second, not stopping even to eat or sleep, for how many years would you have to keep counting to reach 3 billion? Chances are, you wouldn't live long enough! It would take you more than *90 years* to count 3 billion *seconds*. Now, think of 3 billion *years!* Life on earth has been playing the gene-shuffling game for even longer than that and has had plenty of time to work itself up into a dazzling display of biotic diversity and versatility. Under favorable conditions bacteria reproduce three times an hour—how many generations of bacteria may have come and gone in 3 billion years?

4.1. What does the geologic record tell us of the history of life?

Rocks have much to tell us of their own age and origin, and sedimentary rocks in particular can also help us understand the history of life, for they often contain *fossils*. In the most fine-grained fossils all organic materials have been gradually leached away and in the process of *petrifaction* have first been permeated and then replaced, atom by atom, by silicate minerals. In such *replacement fossils* even microscopic cell structure may be faithfully copied. The most ancient *rocks* found to date occur in western Australia and are about 4.2 billion years old. The oldest well-preserved *fossils* known to date are unicellular prokaryotes, some even caught in the act of dividing, buried in silt that became the 3.4 billion-year-old sedimentary strata of the Fig Tree Group formation in South Africa.

4.2. How do we know the age of rocks and fossils?

If you have ever experienced the wonder of hiking down into that mile-deep gash in the earth's crust that is the Grand Canyon of the Colorado River, you have seen how most of the sedimentary layers lie flat atop each other without the buckling and distortion characteristic of mountainous regions. There is logic in assuming that the oldest layers lie toward the bottom of the canyon and are overlain by progressively younger strata. But relative position merely gives a clue to *relative* age. The fossils found in the Grand Canyon formations support the concept of relative age, for as our trail descends, land animals and plants disappear from the deeper strata, and still further down, vertebrates as a whole vanish and only invertebrates and algae remain. However, techniques do exist for establishing *absolute age* without relying on position; these involve measurement of the ratios of radioactive isotopes to their decay products in unweathered crystals.

> **Example 1:** The 4.2 billion-year-old Australian rocks mentioned above were dated by ion-microprobe analysis of the relative quantity of uranium 238 (^{238}U) and its decay product, lead 206 (^{206}Pb), in embedded zircon crystals. Lead 206 is an isotope of lead that is formed only by the decay of radioactive uranium and therefore could not have entered the crystal from other sources. The significance of the proportions of ^{238}U and ^{206}Pb in these crystals depends on the fact that radioactive materials decay at a very steady rate. Each radioactive substance has its own unique *half-life*, the time it takes for half the amount present to transform into its decay product (see Question 3.5). The half-life of ^{238}U is 4.5 billion years, so that it is especially useful for dating very ancient rocks.

> **Example 2:** Isotopes of somewhat shorter half-life, such as radioactive potassium (^{40}K), which decays to argon 40 and calcium 40 with a half-life of 1.3 billion years, are particularly useful in dating most fossils. Whenever possible, rock samples containing fossils are dated by using *three* different radioisotopes; this has been done with thousands of rock samples with such close concordance of results that the probable error is less than 5 percent.

4.3. Do radioactive materials decay at a steady rate?

Yes. In this nuclear age, we have available many radioactive isotopes whose half-lives *are* short enough to be tracked to complete decay.

> **Example 3:** Medically useful radioactive iodine (^{131}I) has a half-life of only 8.04 days, so its decay can readily be traced through successive half-lives until it is fully transformed. All the radioactive

products of atomic fission with half-lives short enough to be directly monitored have shown an invariable regularity of decay. As for corroborative evidence that radioisotopes of longer half-life behave with the same regularity, of all known radioisotopes with half-lives short enough to have allowed them to vanish completely in 4.5 billion years, only two occur in nature, these two being replenished by known processes.

4.4.   What are the oldest known fossils?

The oldest known sedimentary strata containing evidences of unicellular life are the 3.4 billion-year-old Fig Tree formation, bearing well-preserved prokaryotes, and the 3.5 billion-year-old strata of the "North Pole" region of Australia's western desert, containing what appear to be prokaryotic cells and filaments and even stromatolites (see Question 4.9). A variety of organic molecules occur in strata that predate even these.

4.5.   How could organic molecules form and accumulate in a world where no life yet existed?

They were probably formed abiotically, at a time when oxygen was lacking in the atmosphere. Should organic compounds form today by abiotic processes, they would mostly be degraded by interaction with atmospheric oxygen; their fossil presence attests to the likelihood that earth's primeval atmosphere lacked oxygen and may have been rich in gases found today in the atmospheres of giant planets such as Jupiter, to wit, hydrogen, methane, ammonia, carbon dioxide, nitrogen gas, and water vapor.

> **Example 4:**   Experimental enclosed primitive-earth microenvironments furnished with various mixtures of the gases noted above, and even hydrogen cyanide (HCN), and provided with energy sources ranging from mild heat to artificial lightning have, in fact, yielded an impressive array of organic compounds including amino acids, fatty acids, monosaccharides, purines, and pyrimidines. Such molecular building blocks tend to interact and bond when they are adsorbed onto the solid surface of crystalline clays. These clays exhibit a nonliving analog of growth and reproduction and thus could have passed along the fruits of such interactions long before self-replicating nucleic acid molecules came into being. Primitive-earth simulations are limited because, of necessity, they are very small, not global, and cannot be maintained for the half-billion years or so it may have taken for just one nucleotide chain to form that was long enough to replicate itself (approximately 50 nucleotides, judging from experiments with fragmented virus chromosomes).

Evidence that organic molecules and even cells arise quite early in the life of a planet that provides favorable conditions suggests that life may be a somewhat common phenomenon throughout the vastness of the universe, rather than being confined to earth alone.

4.6.   Does life violate the second law of thermodynamics?

Most of us shrink from lawbreaking, but when it comes to *natural laws* we don't even have the option. The *second law of thermodynamics* states that the free energy of a system (i.e., energy available to do work) tends to decrease inexorably, while the entropy (disorganization) of the system increases. In other words, everything goes downhill. (If you find this prospect depressing, the oscillating universe model suggests that whenever the universe becomes utterly disorganized, its mass is reconcentrated gravitationally to such an extent that another "big bang" ensues, providing fresh energy for another cycle of new heavens and earths. Whether or not this model proves true, for some billions of years to come the existing universe has abundant energy to power living systems.)

> **Example 5:**   The operation of the second law of thermodynamics is exemplified by the fact that only about 10 percent of the food an animal consumes is used for growth (*biomass* production), part of the rest is used to perform vital activities, and the balance is given off as heat, which dissipates and contributes to the entropy of the system.
>
>   In fact, living systems maintain themselves (temporarily) and advance in complexity, not in violation of the second law of thermodynamics, but in conformity with it, because the total *system* in which this law operates is not just the living entity, but also its surroundings, both planetary and universal. So long as energy is available to be shifted from its environment into a living or preanimate entity, that biotic part of the system *can* increase in organizational complexity while still contributing to the growing entropy of the system.

Example 6: It *has* been demonstrated that formation of organic molecules from inorganic substrates can take place by converting radiant or electrical energy in the inanimate environment into chemical bonding energy. The weight of scientific evidence indicates that, given planetary conditions compatible with life, organic molecules and even self-reproducing cellular systems could have originated without violation of known physicochemical laws. (Note that such a conclusion says nothing about first causes and final ends in a theological or philosophical perspective: science deals only with the understanding of processes and mechanisms and with phenomena that are objectively observable and measurable.)

4.7.   What were the first forms of life?

The heat of thermal springs may well have provided the first bonding energy for generating organic molecules and the first molecular or cellular unit that could be considered living in the sense of reproducing itself and taking up additional materials as nutrients from the environment. Since these nutrients originally would have been the accumulation of abiotically produced organic molecules in the habitat, such primordial living systems would have been *heterotrophic*, doomed to disappear when they had consumed the reservoir of organic compounds and each other.

4.8.   How, then, did organisms survive into the present?

One enduring and reliable flow of free energy was available for early organisms to tap: the sun. In conformity with the second law of thermodynamics, life persisted and flourished because some primordial prokaryotic cell became a little solar trap. Initially, solar energy may have been usable only for photophosphorylation, providing a source of chemical bonding energy by coupling ADP with inorganic phosphate to make ATP (see Question 3.38); the reactions by which ATP transfers energy to sugar-building pathways may have evolved even earlier in a reverse context: the building of ATP by degradation of preexisting organic compounds (see Question 23.38).

The simplest modern organisms that carry on photosynthesis are indeed prokaryotes, mostly those known as *cyano-phytes* (*blue-green algae*). When eukaryotic cells evolved, some are thought to have acquired as internal symbionts cyanophytes that survived ingestion to become the *chloroplasts* of these new autotrophic organisms.

4.9.   When did photosynthesizing organisms first appear?

One clue to this comes from the presence in the 3.5 billion-year-old North Pole Australian sedimentaries of structures that seem to be the same as modern *stromatolites*. Stromatolites are massive, often mushroom-shaped, calcareous structures built up of a series of thin laminations, bearing on their exposed surface a film of calcium-fixing cyanophytes that continue to deposit new layers of lime as the stromatolite grows. Stromatolites occur throughout the geologic record and into the present, but about 1 billion years ago blue-greens also existed in the form of huge floating mats at sea, so extensive that these self-reliant organisms, which can also fix nitrogen from the air, were the dominant life forms of that time and significantly influenced the future evolution of life.

4.10.   How did cyanophytes affect life on earth?

The entire period from 3.5 to 1 billion years ago, representing some two-thirds of the history of life, may be referred to as the "age of blue-green algae" because during their long reign the blue-greens not only flourished but changed forever the composition of earth's atmosphere, for they gave off great quantities of oxygen gas (O_2) as a waste product of photosynthesis. This doomed to extinction most of the *obligate anaerobes* of the time, to which O_2 is deadly (only a few kinds, such as tetanus bacteria, survive today), and made possible the success of a new breed of organisms (*aerobes*) that can use O_2 in breaking down molecules for energy. Blue-greens themselves can survive in the presence of O_2, but they do better under somewhat hypoxic (oxygen-deficient) conditions, so by liberating O_2 as a photosynthetic waste, they really undermined their own dominance.

4.11.   When did the first eukaryotic cells appear?

Cells with nuclei first appear in sedimentary strata about 1.5 billion years old. The 1 billion-year-old Bitter Springs formation of Australia contains beautifully preserved green algae (*chlorophytes*) showing nuclei and even nuclear division.

4.12.   Do all eukaryotes stem from one common ancestor?

Eukaryotes resemble one another at the chromosomal and cellular levels much more closely than they resemble prokaryotes. Furthermore, those eukaryotes tested to date exhibit such concordance of nucleotide sequences as to suggest that all may well have come from one original eukaryotic cell. Descendants of this common ancestor that became autotrophic probably did so by engulfing cyanophytes, which survived as endosymbionts and eventually evolved into *chloroplasts,* the photosynthetic organelles of plants. Those early eukaryotes that either remained heterotrophic, or later reverted to heterotrophism by loss of chloroplasts, qualified as the first animals (*protozoans*).

4.13.   What kingdoms make up the living world?

(*a*)   Prokaryotes traditionally have all been placed in the *kingdom Monera,* but recent genetic studies show that a group of anaerobic microbes, including methane-producing and purple sulfur bacteria, are as far removed from all other prokaryotes as they are from eukaryotes and therefore probably represent a separate kingdom, the *Archaebacteria.*

(*b*)   Eukaryotic organisms that are essentially unicellular (or at most form simple filaments or colonial clusters) are often grouped into a rather heterogeneous *kingdom Protista* including protozoans, certain simple funguslike forms, and a number of mostly autotrophic groups such as diatoms, dinoflagellates, and euglenoids. Enough diversity exists among modern protists to support the idea that already, at the unicellular level, separate ancestral lines may have existed that independently gave rise to multicellular plants (*kingdom Metaphyta* or *Plantae*), multicellular animals, (*kingdom Metazoa* or *Animalia*), and fungi (*kingdom Fungi*). Figure 4.1 contrasts different models of the origin and affinities of the five or six major lineages recognized today.

4.14.   Which model may be the more correct?

If genetic data now being accumulated continue to affirm that eukaryotes arose from a common ancestor, we should probably favor the three-kingdom model, in which prokaryotes make up kingdoms Archaebacteria and Monera, while eukaryotes are placed in a *kingdom Eukaryota.* On the other hand, if eukaryotic cells evolved more than once, from different prokaryote ancestors, combining the four kingdoms of modern eukaryotes into one would imply a common ancestry that did not exist. These relationships may seem too obscured in antiquity to permit taxonomic schemes to be other than speculative, but modern techniques of nucleic acid analysis may eventually resolve these uncertainties. Maybe you are thinking, who cares? Scientists do: they love to unravel puzzles, and if in doing so they make discoveries that directly benefit humanity, such felicitous applications are but a by-product of the curiosity that is the basic driving force of scientific investigation.

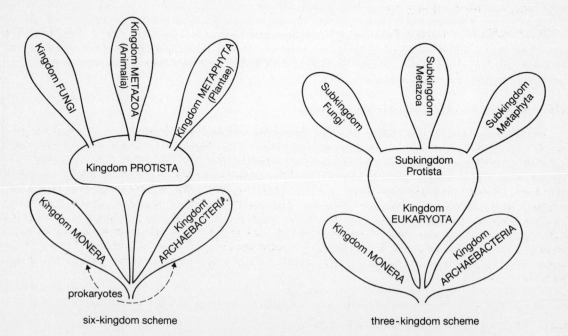

Fig. 4.1   Two models of the kingdoms of life: *left,* six-kingdom model showing two kingdoms of prokaryotes and four of eukaryotes; *right,* three-kingdom model showing all eukaryotes as members of a single kingdom, Eukaryota.

4.15.   When did multicellular organisms appear and diversify?

To date, no evidence of multicellular life (other than algae) has been found in rocks older than about 700 million years. The best-preserved fossil assemblage of this age comes from the Ediacara Hills of Australia and includes a variety of soft-bodied metazoans: jellyfish, corals, segmented worms (some with head shields that may have been ancestral to the early arthropods known as "trilobites"), together with a number of puzzling forms of unknown affinities. The paucity of fossiliferous strata older than this, which might contain clues to the origin of these already well diversified metazoan groups, seems to have been caused by a series of Precambrian glaciations that deeply eroded most continental surfaces.

4.16.   What was the Cambrian "explosion"?

Even relatively complex invertebrates such as segmented worms (annelids) were well established by 700 million years ago, but it was not until some 100 million years later, at the beginning of the Cambrian Period, that a remarkable proliferation and diversification of invertebrate life took place within what appears to have been only a few million years. As a result, all of today's major animal phyla, and several long extinct ones, are present in rocks of that age. (A *phylum* is the highest taxonomic subdivision of a kingdom and constitutes a major group of organisms of common descent that share a fundamental pattern of body organization.) Absent from Cambrian strata are land plants, terrestrial invertebrates such as insects, and vertebrates, which did not become established until about 100 million years after the start of the Cambrian.

4.17.   What caused this proliferation and diversification of life?

The causes of the Cambrian explosion remain obscure. For one thing, the buildup of atmospheric O_2 may have reached the concentration necessary for this gas to diffuse downward throughout water masses so that bottom-dwelling creatures could begin to flourish. Then too, perhaps the extensive erosion caused by the Precambrian glaciations raised marine concentrations of dissolved minerals, especially calcium, to some critical threshold necessary for the optimal functioning of nerves and muscles and the deposition of shells and skeletons.

4.18.   How has life fared over the past 600 million years?

The major biological events of the fruitful past 600 million years are summarized in the Geologic Time Table (Fig. 4.2). The division of this record into eras and periods is not arbitrary, but based on discontinuities caused by geologic events that had great effects on life, causing major extinctions followed by the appearance of many new species.

4.19.   What are the high points of the Paleozoic Era?

The Paleozoic Era spans 370 million years, from the beginning of the Cambrian Period (600 million years ago) to the end of the Permian Period (230 million years ago). During the 100 million years of the Cambrian, a number of major animal groups (some of quite unknown affinities) totally disappeared, while surviving groups underwent further diversification, as if benefiting from more viable body plans. Over the ensuing 75 million years of the Ordovician Period, invertebrates and multicellular plants colonized the land, and vertebrate fishes appeared. During the next 20 million years (Silurian Period), the first jawed fishes appeared and so flourished that the 50 million years of the ensuing Devonian Period are known as the "age of fishes." Amphibian fossils first appear in rocks of later Devonian age, as the earliest land vertebrates. These animals characterized the 65 million-year Carboniferous Period (sometimes called the "age of amphibians") but declined during the 50 million years of cooling and drying climates that marked the Permian Period. Reptiles, which diverged from early amphibian stock during the Carboniferous, were not so disadvantaged by these changes and began to proliferate and spread.

4.20.   What happened at the end of the Paleozoic?

Nearly half the known families of animal life became extinct as the Paleozoic Era came to a close, about 230 million years ago. (Taxonomically, a *family* is a fairly comprehensive group of related forms, such as the cat family, family Felidae, which includes lions, tigers, lynx, jaguar, bobcat, puma, and other felines.)

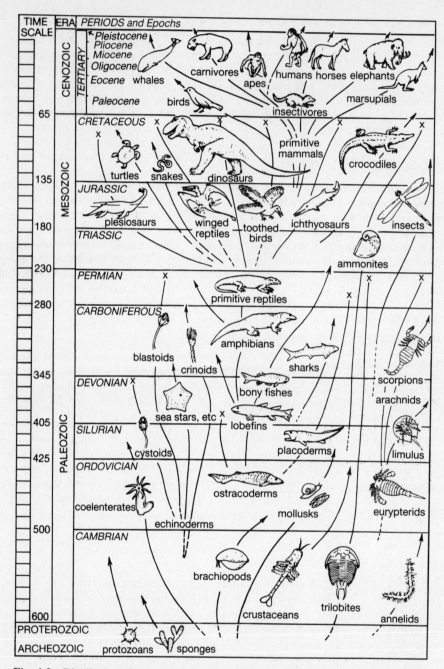

Fig. 4.2 Distribution of major animal groups in the geologic record. Solid curving lines commence at time when each group first appeared, with broken lines indicating presumed earlier origins. Lines terminating with an X indicate when certain groups became extinct; those ending in an arrow indicate that the group contains modern descendants. The time scale is in millions of years. (*Modified from Storer et al.*)

4.21.   Why did so many families die out in such a relatively short time?

Continents drift as seafloors expand by extrusion of molten rock along rifts called *seafloor spreading zones*. Toward the end of the Paleozoic, these drifting continents all collided to form a supercontinent we call *Pangaea*. (If this seems fanciful, right now North America is drifting westward at a measurable 2 cm a year: stick around long enough, and you'll witness a grand reunion with China!) The continental collisions that ended the Paleozoic obliterated intervening marine habitats, allowed the terrestrial biota of previously isolated land masses to come into competition, and triggered a period of mountain building that affected climate and drained shallow continental seas, while the Appalachians were thrust upward as grandly as today's Andes. The combined effects may have accounted for much extinction, but there may be still more to the story (see Questions 4.23–4.25).

4.22.   What were the major events of the Mesozoic Era?

Lasting for some 165 million years (from 230 to 65 million years ago), the Mesozoic is known as the "age of reptiles," for these became the dominant vertebrates and diversified into many forms, including the largest creatures that have ever walked the earth, marine species as massive as whales, and the most spectacular animals ever to soar the skies. The geographic spread of reptiles was facilitated by the fact that the land they so successfully invaded was the "one world" of Pangaea.

During the Triassic Period, dinosaurs and mammal-like reptiles (therapsids) spread throughout the conifer-dominated forests of Pangaea. The 45 million-year Jurassic Period saw the advent of birds and mammals and the start of the breakup of Pangaea, first into northerly and southerly land masses (Laurasia and Gondwanaland), carrying terrestrial organisms apart, to new climates and destinies.

Throughout the rest of the Mesozoic, the Cretaceous Period, the drifting continents fragmented further, while reptiles continued to dominate the earth and birds and mammals did little more than mark time in very subdued fashion in the face of formidably established reptilian competition. Then something catastrophic took place: almost suddenly some 25 percent of all existing animal families disappeared, not only the dinosaurs and other still-successful reptilian groups, but a wide variety of marine invertebrates from giant ammonites (a type of mollusk) down to microscopic zooplankton. So far as we can tell, no Mesozoic land animal with a body mass over 25 kg (60 lb) survived into the Cenozoic.

4.23.   What caused this mass extinction?

In various parts of the world a thin layer of clay has been found, which was deposited 65 million years ago and separates a rich Cretaceous fossil record from a very sparse fossil record marking the onset of the Cenozoic Era. This clay is rich in iridium, an element common in meteorites but almost lacking in the earth's crust. This suggests that a large meteorite (perhaps 6 km in diameter) impacted the earth, ejecting into the stratosphere such an enormous amount of particulate matter that before this iridium-laden dust settled, months of darkness ensued, with plummeting temperatures and suppression of photosynthesis. Most plants could recover from such a catastrophe by virtue of resistant seeds, spores, and rootstock, but many animals would die of starvation or cold during the long darkness. Furthermore, the elemental carbon (soot) found in this clay layer may have come from fire storms caused as continentwide forests were ignited by the passage through the atmosphere of such a massive extraterrestrial object. Smoke from such fire storms would have intensified and prolonged the crisis of darkness and cold.

4.24.   Have extraterrestrial bodies caused other mass extinctions?

This startling possibility has prompted intensive reanalysis of fossil data relating to mass extinction events of lesser magnitude. Surprisingly, it appears that such events have occurred with a periodicity of about 26 million years for as far back as such events can be traced.

4.25.   Why are the intervals for mass extinctions so regular?

An astronomical search is now under way to discover the cause of such periodic mass extinctions: if our sun, like many stars, is one of a binary pair, it may have a small companion star (already dubbed "Nemesis") with an orbit so eccentric that it passes through the solar system only once in 26 million years, towing a mass of comets collected from the dense Oort comet cloud that lies beyond Pluto. Alternatively, the unidentified celestial object may be a planet with a less far-flung orbit, which intersects Earth's orbit only every 26 million years with its gravitational train of comets. The severity of each mass extinction event would depend on the number and mass of cometary fragments impacting the Earth in the wake of the unknown planet or the companion star Nemesis. If the latter exists, it may not be readily found, for it should now be at about its farthest point from the sun, not due to return for another 13 million years.

4.26.   How do mass extinctions affect the history of life?

The most profound effect of mass extinctions is that the survivors enjoy virtual carte blanche, for they proliferate in a depopulated world providing opportunity for a great variety of genetic variants. Under these circumstances, evolution of a new biota can take place quite rapidly until the environment is again "saturated" with enough different life forms to maintain stable ecosystems over long periods of time. However, if a mass extinction is excessively severe, little may remain from which new forms can evolve.

> **Example 7:** We ourselves may be on the way to bringing about a mass extinction of quite unprecedented severity, at least among terrestrial species, by our continued multiplication and our conversion of habitats to our own uses. So far as we can tell, never before has one species expanded at the expense of so many. Already there are more than *5 billion* human beings on earth (1 billion more than in 1976), perhaps 250 billion kg of living flesh, an impressive environmental burden! If earth's biota should become simplified to little more than human beings and their domestic animals and plants, there would be scant genetic stock remaining from which the earth could ever be repopulated with its wealth of plant and animal diversity.

4.27.   What have been the major events of the Cenozoic Era to date?

The past 65 million years have witnessed the explosive proliferation of birds and mammals, freed (perhaps by a comet) from the yoke of reptilian dominance. On land, flowering plants have prospered over the dwindling ferns and conifers. Hominid (humanlike) fossils (mostly found in Africa) have been dated at an age of about 3 million years (*Australopithecus afarensis*), 1.7 million years (*Homo habilis*), and 1.5 million years for *Homo erectus,* with skulls transitional to *Homo sapiens* dating from 250,000 to 350,000 years ago. Although we have named ourselves "wise man," to most of the living world we are catastrophe personified, for countless animal and plant species have diminished into endangerment or extinction as *Homo sapiens* has proliferated. Awareness and concern can still turn the tide, if we really are wise enough to conserve our biological heritage and guard ourselves, too, from extinction.

Chapter 5

Heredity and Evolution

The history of life could never have taken place without the unbroken transmission, from cell to cell and generation to generation, of genetic material: chromosomes and the genes they contain. Eukaryotic *chromosomes* are essentially long double strands of DNA coiled at intervals around protein molecules, and *genes* are nucleotide sequences along the chromosomes that encode for specific molecules of RNA. Most genes encode messenger RNAs, each of which is a template molecule for building a particular protein with precision (see Chap. 3).

5.1. How is genetic material organized at the chromosome level?

The DNA of each chromosome consists of a linear sequence of genes, each having its own normal position (*locus*); between genes are lengths of DNA that do not transcribe any RNA at all, these "silent" regions (*heterochromatin*) being considered spacers. At one consistent location the chromosome bears a *centromere* to which spindle fibers attach during cell division.

5.2. How many chromosomes are there in an animal cell?

The number differs according to species. Within a given individual or species, the number of chromosomes is ordinarily the same in all body cells (except gametes) and represents a paired, or *diploid*, state. Individual chromosomes differ enough in length and centromere position that each can be distinguished microscopically. The typical "picture" of a species' chromosomes makes up that species' *karyotype*.

5.3. Does chromosome number reflect genetic complexity?

Chromosome number is no reliable indicator of the "superiority" of one animal over another. Human beings, for instance, have a normal karyotype of 46 chromosomes, while dogs have 78, and goldfish 94. On the other hand, the total number of *base pairs* (A-T, C-G) *is* a good indicator of relative genetic complexity: the fruit fly, *Drosophila melanogaster* (a favorite genetic research subject because of its small size, short life cycle, and gigantic chromosomes), has 80 million base pairs per cell, whereas a mouse cell contains 5 billion base pairs, giving it considerably greater protein-encoding capacity. Genetic complexity can also be defined by identifying the number of *different* mRNAs produced by cells in tissue culture: *Drosophila* cells synthesize some 5000 different mRNAs, while human cells produce 35,000 kinds.

5.4. What are the advantages of diploidy?

Diploid organisms have two sets of chromosomes per cell, one set from each parent. Both chromosomes of a homologous pair contain the same genes, so that each gene is present in two copies, known as *alleles*. Having two copies of a gene can double the amount of its product, since each allele can generate mRNA, and twice the mRNA can mean twice the protein production. Furthermore, if alleles come to differ by mutation, the unmutated allele may still function well enough alone to make up for the altered or reduced activity of the mutant one.

5.5. What is meant by dominant, recessive, and codominant alleles?

When one allele's expression completely obscures the effect of its homolog, the former is called the *dominant* allele and the latter the *recessive*. In cases where alleles differ and *both* exert an effect on the physical expression of the trait (*phenotype*), they are said to be *codominant*. An individual carrying two identical alleles is said to have a *homozygous* genotype for the trait in question; if the two alleles are not identical, the genotype is *heterozygous* and the phenotype (trait) will be determined either by the dominant allele only or by two codominant alleles producing an intermediate effect.

> **Example 1:** Persons who can roll their tongues into a trough carry a dominant allele (which we will call *R*) and can have the genotypes *RR* or *Rr*; those deprived individuals who cannot tongue-roll have the genotype *rr*.

Dominance is sometimes affected by sex hormones.

Example 2: The allele causing pattern baldness in humans is dominant in men and recessive in women, so that *Bb* produces a bald man but a normal woman. Baldness is therefore a *sex-influenced trait*.

Example 3: *Sex-limited* traits normally are expressed in only one sex, regardless of genotype. In poultry, hens and roosters have different plumage. "Hen-feathering" is due to a dominant allele, "cock-feathering" to a recessive. Males with the genotypes *HH* and *Hh* are hen-feathered, since normal male plumage develops only in roosters having the genotype *hh*. On the other hand, hens *never* develop male plumage, not even when they have the genotype *hh*—unless you remove their sex organs. Castrated birds of *both* sexes develop cock-feathering even when they carry the allele *H*. Apparently *H* suppresses male plumage, but can only exert its effect if *either* male or female hormones are present.

In cases of codominance, the phenotype of heterozygotes differs from that of either homozygote.

Example 4: Blue Andalusian chickens (Fig. 5.1) result from the mating of black fowl with ones that are splashed-white (white with little black spatters). Blue Andalusians do not "breed true": the offspring of blue fowl will be black, blue, and splashed-white in the ratio 1:2:1. If *B* stands for black and *b* for splashed-white, the genotype of black fowl would be *BB*; of splashed-white, *bb*; and of blue fowl, *Bb*. Why do you think the cross *Bb* × *Bb* would yield (statistically) 1/4 black, 2/4 blue, and 1/4 splashed-white?

5.6. How do chromosomes determine sex?

Diploidy provides animals with a double set of genetic instructions, except in one respect. Members of most homologous chromosome pairs (*autosomes*) look just alike, but in one pair, known as *heterosomes* (or as *sex chromosomes* because they determine the individual's sex), the two chromosomes may differ greatly in size and genetic content, or one may simply be lacking in individuals of one sex or the other. There are several major patterns of sex determination:

Fig. 5.1 Inheritance in Andalusian fowl of a monogenic trait showing codominance of alleles. P = parental generation; F_1 = first filial generation, the "blue" offspring of a cross involving parents true-breeding (homozygous) for black and splashed-white plumage, respectively; F_2 = second filial generation, the offspring of a cross between two hybrid (heterozygous) blue-plumaged fowl of the F_1 generation. Note characteristic 1:2:1 ($\frac{1}{4}:\frac{2}{4}:\frac{1}{4}$) phenotypic ratio in the F_2. (*From Storer et al.; modified from Hesse-Doflein.*)

(*a*) *XX-XO sex determination:* Females have two sex chromosomes (XX), and males have only one (XO). In other words, males have one less chromosome per cell than females do; this occurs in bugs (hemipterans).

(*b*) *XX-XY sex determination:* One sex chromosome (X) is full-sized and the other (Y) is a dwarf containing few genes. XX-XY sex determination has three versions: (1) the male has two full-sized chromosomes (XX), and the female is XY (e.g., birds, moths, and butterflies); (2) females are XX and males XY (e.g., *Drosophila*, in which experimentally produced XO individuals that have one X but lack a Y are still male, but sterile); (3) XX individuals are female and XY individuals are male, with the Y chromosome bearing a male-determining gene (e.g., mammals, in which XO individuals are female but usually sterile).

5.7.   What is sex-linked inheritance?

Genes that occur in the same chromosome are said to be *linked*. A gene in a sex chromosome is *sex-linked*. Except for the male-determining *Y-linked factor*, sex-linked genes in mammals are nearly all *X-linked*. This means that only XX individuals (females) have two alleles for each sex-linked trait; XY individuals (males) have only one, so the effects of that solitary allele will be expressed in males whether it is dominant or recessive.

> **Example 5:** In humans a recessive allele causing red-green color blindness occurs in the X chromosome. Identifying this gene as *c* and its normal allele as *C,* you can see that a woman would not be color-blind unless she received the allele *c* from *both* parents and was *cc.* By contrast, a man with only one *c* will be color-blind, because his Y chromosome carries no allele at all for this factor; his genotype would be stated *cY.* Now think: women, being XX, can produce only X-containing eggs, but men produce 50 percent X-bearing sperm and 50 percent Y-bearing sperm, so males determine the sex of their offspring. A man therefore is male because he received a Y chromosome from his father and an X chromosome from his mother. If a man is color-blind, from which parent did he inherit the allele for color-blindness? (You are right! His mother, of course.)

5.8.   How is gene dosage regulated in the case of sex chromosomes?

If an organism is diploid for all its autosomal genes, but can be *either* diploid (XX) or monoploid (XY) for sex-linked genes (*monoploid* meaning having only one set of genes or one chromosome of a pair), problems of *gene dosage* may exist: in XX individuals *two* functional alleles may be one too many for normalcy. The critical nature of correct gene dosage is seen in the gross abnormalities of Down's syndrome in humans, where just one short, extra chromosome is present. In other words, two copies of normal autosomal genes are just fine, but three can be disastrous. Now, when one sex is XX and the other XY or XO, the XX individual must cope with a dosage of X-linked genes that is twice that needed for normalcy by the opposite sex. In mammals this problem is solved by the extensive inactivation of one X chromosome or the other during embryonic development of the female. The mostly inactivated X chromosome becomes a dark, compact *Barr body.* Because it is a matter of chance as to *which* X chromosome is inactivated in any given embryonic cell, a female mammal is a *genetic mosaic* for any sex-linked trait for which she is heterozygous.

> **Example 6:**   You can tell the sex of a calico cat at a glance: it is female. Calico coloration (yellow and black patches) results from a sex-linked gene pair (which we will call *B* and *b*), in which *B* produces yellow fur, and *b* black fur. Yellow cats can be female (*BB*) or male (*BY*); black cats can be female (*bb*) or male (*bY*); but only *heterozygous* cats (*Bb*) are calico, and these of course are female. In 50 percent of the calico cat's skin cells the X chromosome carrying the allele *B* has become a Barr body, while in the other 50 percent the *b*-carrying X chromosome has been inactivated.

5.9.   How are chromosomes (and genes) transmitted?

Transmission of chromosomes and the genes they bear takes place by means of two types of cell division, *mitosis* and *meiosis,* which occur universally among eukaryotes. The self-replication of DNA that is prerequisite for either of these processes takes place during a cell's period of growth and metabolic activity (Fig. 3.10) and *not* while the cell is actually dividing. The daughter chromosomes (called *chromatids*) produced during DNA synthesis (Chap. 3) are held together by the undivided centromere, so that from the time of DNA synthesis until cell division each chromosome consists of *two* DNA double helices.

5.10. What is mitosis?

Mitosis is a process of asexual reproduction that produces genetically identical daughter cells. (Sometimes mitosis takes place without cytoplasmic division, creating two or more genetically identical nuclei within a single cell.) Mitosis has four phases (Fig. 5.2):

(*a*) *Prophase:* The nuclear envelope ordinarily disappears; each chromosome shortens and thickens by supercoiling until its two conjoined DNA molecules (the chromatids) are separately visible, held together only at the centromere. The centrioles divide and migrate toward opposite ends of the cell, developing between them a group of microtubules called *spindle fibers*. In animal cells, the centrioles also produce radiating fibers known as the *aster*, which may facilitate cytoplasmic division.

(*b*) *Metaphase:* The chromosomes migrate to the midline of the spindle, where spindle fibers extending from each pole of the spindle attach to the centromeres.

(*c*) *Anaphase:* The centromeres divide, and the attached spindle fibers progressively shorten, pulling the chromatids apart.

(*d*) *Telophase:* The migration of the two sets of daughter chromosomes to opposite poles of the spindle is completed, and spindle and asters disappear; the chromosomes uncoil as the nuclear envelope reforms around them. Ordinarily cytoplasmic division also occurs during telophase, so that the end product of mitosis is two daughter cells with the same number and kind of chromosomes as the original cell.

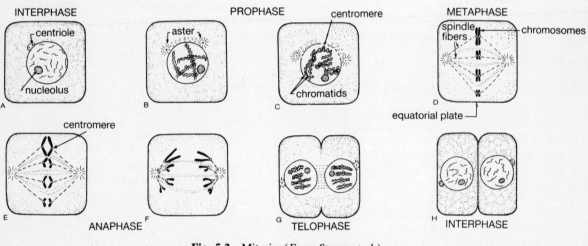

Fig. 5.2 Mitosis. (*From Storer et al.*)

5.11. What is meiosis?

Meiosis is a peculiar type of cell division that must take place at some point in the life cycle of any sexually reproducing eukaryotic species. In animals, it typically occurs during the maturation of sex cells (*gametes*). Sexual reproduction in animals involves the union of two gametes, each bearing one set of chromosomes; the resultant cell (usually a fertilized egg, or *zygote*) accordingly is diploid: it contains *two* sets of chromosomes. If no means existed for cutting the gametic chromosome complement in half, in each generation the chromosome number would double, which would spell a quick demise for sex! Reducing gametes to the *haploid* state (i.e., to half the normal chromosome number) must be accomplished with precision so that every gamete will still possess one chromosome of each homologous pair. In all eukaryotes, *meiosis* brings this about (Fig. 5.3). In humans, the normal chromosome number of 46 (23 homologous pairs) per cell is reduced meiotically to just 23 in the sperm and eggs; fertilization then brings together the two homologous sets, restoring the diploid state in the zygote.

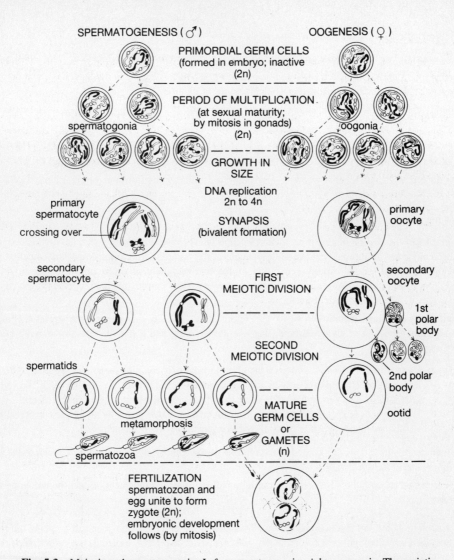

Fig. 5.3 Meiosis and gametogenesis. *Left,* spermatogenesis; *right,* oogenesis. The meiotic nuclear events are identical in both processes, but cytoplasmic division is equal in the former, yielding four sperm cells per spermatogonium, whereas in oogenesis the egg cell retains all the cytoplasm and the excess chromosomes are discarded in minute polar bodies. For simplicity, only three pairs of chromosomes are shown, one member of each homologous pair being depicted as black, the other as white. Note that after the first meiotic division only one member of each pair of homologous chromosomes remains in each daughter cell, but at that point each chromosome still consists of two daughter chromosomes (DNA molecules) held together at the centromere. These daughter chromosomes are separated during the second meiotic division. (*From Storer et al.*)

5.12. How is meiosis like mitosis?

Both types of cell division involve the four phases of prophase, metaphase, anaphase, and telophase, as described above.

5.13.   How does meiosis differ from mitosis?

(*a*)   Meiosis consists of *two* successive cell divisions, but the chromosomes reproduce (i.e., the DNA replicates) *only* before the *first* division (meiosis I).

(*b*)   Homologous chromosomes come together (synapse) only in meiosis.

(*c*)   Homologous chromosomes may exchange genetic information (i.e., by the crossing over of blocks of genes in meiosis I).

(*d*)   The centromeres do not divide during meiosis I.

(*e*)   At the end of meiosis I each chromosome still consists of two DNA molecules (chromatids).

(*f*)   The two meiotic divisions together produce *four haploid* daughter cells from one original diploid cell.

5.14.   What happens in meiosis I?

(*a*)   During prophase, the two members of each chromosome pair behave remarkably: they come together and "zip up" so that each allele lies in intimate contact with its allele in the homologous chromosome; this is called *synapsis* and is absolutely essential for the orderly separation of the two members of each chromosome pair in the ensuing anaphase. Since each chromosome already consists of two daughter chromatids held together at the centromere, the result is a *tetrad*, a bundle of four strands.

(*b*)   These intimately entwined strands can break and rejoin in such a way that part of one strand exchanges position with an equivalent portion of another strand; this is *crossing over*, and it has the effect of moving genes into new linkage groups.

> **Example 7:**   If one member of the chromosome pair contains the allelic sequence *ABCDEFGH* and its homolog contains the alleles *abcdefgh*, after crossing over the alleles might be redistributed as *abcDEFGH* and *ABCdefgh*. (Crossing over is a normal meiotic event and is useful in mapping the loci of genes in chromosomes: the closer together any two genes lie along the chromosome, the lower will be the incidence of a crossing over occurring between them.)

(*c*)   During metaphase, the tetrads are located at the midline of the spindle, and spindle fibers attach to the centromeres, *but the centromeres do not divide*.

(*d*)   As a result, when the spindle fibers shorten during anaphase, the two members of each homologous chromosome pair are separated.

Meiosis I is often called the "reductional division" because at its end each daughter cell contains only one member of each chromosome pair, although each chromosome still consists of *two* DNA molecules, or chromatids, held together by the undivided centromere.

5.15.   What happens in meiosis II?

Depending on species, meiosis II may begin at once or be delayed, but in either case DNA does *not* replicate between meiosis I and II.

(*a*)   When meiosis II commences, the chromosomes—which remain in their shortened condition between the two meiotic divisions—promptly move to the midline of the new spindle.

(*b*)   The centromeres finally divide, and one of the two chromatids of each chromosome passes to each daughter cell, so that meiosis II is often called the "equational division."

(*c*)   The result is four haploid cells, with each chromosome now consisting of only one DNA molecule.

5.16.   Where and when does meiosis take place?

In multicellular animals, meiosis takes place only in the gonads (testes and ovaries), during *spermatogenesis* and *oogenesis* (formation of sperm and egg, respectively); these processes differ mainly in how the cytoplasm is apportioned among the daughter cells. Spermatogenesis results in four haploid sperm cells from each original diploid cell, whereas in oogenesis the cytoplasm is retained by one daughter cell, the egg, or *ovum*, while the excess chromosomes are extruded into minute *polar bodies* (see Fig. 5.3).

5.17. What are the Mendelian laws of genetic recombination?

By hand-pollinating garden peas, the Austrian monk Gregor Mendel discovered that the hereditary factors we now call genes do not lose their identity when transmitted and recombined, but constitute discrete "particles" that may (if recessive) disappear from view in one generation only to reappear unchanged in a later one. Furthermore, his hereditary "particles" recombined according to certain reproducible phenotypic ratios in each generation of progeny. Although his findings antedated our modern knowledge of genes and chromosomes, his correct interpretations established two basic principles of inheritance that are still known as "Mendelian laws": the law of segregation and the law of independent assortment.

5.18. What is the law of segregation?

Stated in modern terms, only one member of each allelic pair of genes will end up in a gamete (because owing to meiosis, only one chromosome of each homologous pair will occur in each sex cell). When fertilization brings two sets of chromosomes back together, their genes may be *recombined* into new allelic combinations.

> **Example 8:** Two homozygous guinea pigs, one black and the other white, produce only black offspring. Although the "white" allele appears to be lost, it is not, for when these black hybrid offspring are interbred, only 75 percent of *their* offspring turn out to be black, while 25 percent are white. A "Punnett square" analysis of these matings (Fig. 5.4) shows that the 3:1 ratio obtained in the second progeny generation actually consists of 25 percent *BB* (homozygous black), 50 percent *Bb* (heterozygous black), and 25 percent *bb* (white). Because the white allele is fully recessive, no difference can be seen between individuals carrying one or two copies of the *dominant* allele for black. [This famous 3:1 ratio is best seen when the number of progeny is large; if a single pair of heterozygous (hybrid) guinea pigs had only four offspring, the probability that these four, considered alone, would show the 3:1 ratio is only 25 percent.]

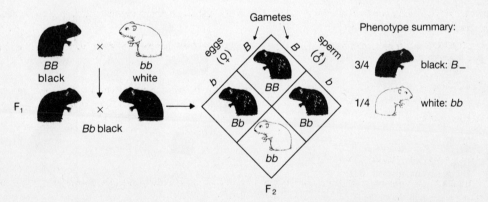

Fig. 5.4 Inheritance in guinea pigs of a monogenic trait (coat color) with dominance. A cross between a homozygous black and a homozygous white guinea pig yields only fully black progeny. When these F_1 black individuals are mated, 75 percent of their offspring are black and 25 percent white, a 3:1 Mendelian ratio for a trait determined by a difference in one pair of genes in which one allele is dominant and the other recessive. Note that a gamete can contain only one gene of an allelic pair.

5.19. What is the law of independent assortment?

Expressed in modern terms, *nonallelic* genes that are *not linked* (i.e., do not occur in the same chromosome) are segregated (during meiosis) and recombined (at fertilization) independently of each other. This is because it is a matter of chance as to *which* chromosome of each pair ends up in a particular sex cell, and which sperm fertilizes which egg.

Example 9:   Figure 5.5 shows a Punnett square summary of the four kinds of gametes, and the sixteen ways in which they can join in fertilization, in the second progeny generation of a cross involving two different nonlinked gene pairs in guinea pigs. The grandparents, as shown, were, respectively, homozygous black with rough coats and white with smooth coats. Since the *dihybrid* progeny were all black and rough, white and smooth are due to recessive alleles. When this heterozygous generation is inbred, their offspring show four phenotypic combinations: 9/16 black, rough; 3/16 black, smooth; 3/16 white, rough; and 1/16 white, smooth. Recombination has produced two new phenotypes: which are they? Take a good look at this 9:3:3:1 ratio and you will see that it is actually two separate 3:1 ratios multiplied together: $(3:1) \times (3:1) = 9:3:3:1$; each gene pair has obeyed the law of segregation, but quite independently of the other. Obviously, gene segregation and recombination increase phenotypic variability in a population, merely by reshuffling material that happens to be in the gene pool.

Example 10:   In poultry two pairs of independently assorting genes affect comb shape (Fig. 5.6). In this case, the heterozygous (dihybrid) offspring of a rose-combed fowl and a pea-combed fowl exhibit a new comb shape, "walnut," the phenotypic result of interaction between two different dominant genes. The offspring of two walnut-combed fowl ($RrPp \times RrPp$) exhibit a 9:3:3:1 ratio of 9/16 walnut-combed ($R\_P\_$), 3/16 rose-combed (R_pp), 3/16 pea-combed ($rrP\_$), and (surprise!) 1/16 single-combed ($rrpp$). Single-comb is the ordinary comb seen in most chickens, but it is recessive to *both* the mutant genes that produce rose and pea combs.

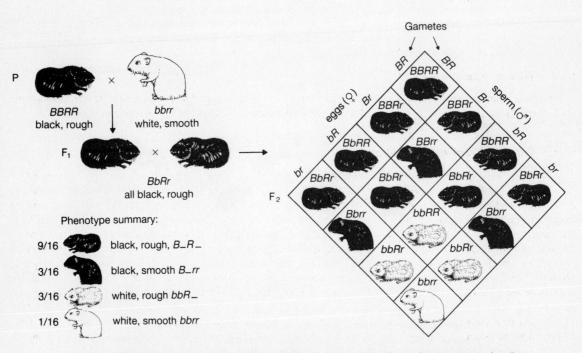

Fig. 5.5   Inheritance in guinea pigs of two monogenic traits (coat color and coat texture) showing dominance and independent assortment. A cross between a homozygous black, rough-coated guinea pig and one that is homozygous white, smooth-coated produces only black, rough-coated progeny. Thus, the allele for black (B) dominates that for white (b), and the allele for rough coat (R) dominates the one for smooth coat (r). When the hybrid F_1 individuals are crossed, each can produce equal proportions of four types of gametes (BR, Br, bR, br) random recombination of these yielding *four* types of progeny in the F_2: black, rough (9/16); black, smooth (3/16); white, rough (3/16); white, smooth (1/16). The 9:3:3:1 ratio, which is the product of two independent 3:1 ratios, tells us that the traits of coat color and texture are affected by two pairs of genes that do not occur in the same chromosome and therefore assort independently.

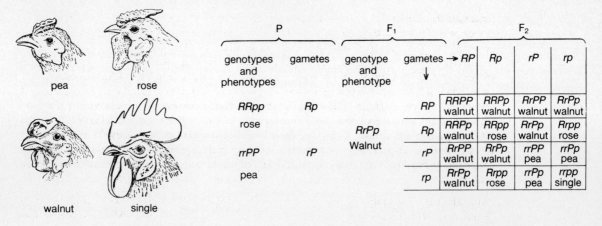

P			F$_1$		F$_2$			
genotypes and phenotypes	gametes		genotype and phenotype	gametes →	RP	Rp	rP	rp
RRpp rose	Rp			RP	RRPP walnut	RRPp walnut	RrPP walnut	RrPp walnut
			RrPp Walnut	Rp	RRPp walnut	RRpp rose	RrPp walnut	Rrpp rose
rrPP pea	rP			rP	RrPP walnut	RrPp walnut	rrPP pea	rrPp pea
				rp	RrPp walnut	Rrpp rose	rrPp pea	rrpp single

Fig. 5.6 Phenotypic interaction of two pairs of independently assorting genes in poultry that affect comb shape. When true-breeding strains of pea-combed and rose-combed fowl are mated, their offspring are all walnut-combed. Crossing walnut-combed poultry yields an F$_2$ of which 9/16 have walnut combs, 3/16 pea combs, 3/16 rose combs, and 1/16 single comb (a new phenotype). The 9:3:3:1 ratio in the F$_2$ tells us that comb shape is being determined by two nonlinked pairs of genes with dominance; *R__ pp* produces rose comb, *rrP__* pea comb, *R__ P__* walnut comb, and *rrpp* single comb. (*From Storer et al.*)

5.20.   What is organic evolution?

Evolution can be defined simply as changes in a population's gene pool through successive generations, a "gene pool" being the aggregate of all genes available for transmission and recombination in a breeding population.

5.21.   What characterizes the evolutionary process?

(*a*) Because accidental changes take place in genes and chromosomes, especially during DNA replication, evolution is both universal and inescapable. (*b*) Evolutionary change does not proceed at a constant rate but is strongly accelerated by environmental instability. (*c*) Evolution is an opportunistic, not a goal-oriented, phenomenon: the possibility of adaptive change in a population faced with a crisis is limited by the variability existing in the population's gene pool at that time, and there is no evidence that genes can mutate in response to need. (*d*) Although evolutionary change is ordinarily *adaptive,* at least at the time, in the long run it may lead to extinction, especially when it has tailored a population too precisely to a restricted habitat or overly specialized mode of life. (For example, yuccas appear to be pollinated by only one insect species, the yucca moth, which in turn lays her eggs only within yucca blossoms; although this symbiotic partnership seems felicitous for now, should the moth be exterminated, the plants may be doomed to extinction, and vice versa.)

Evolution is *not* synonymous with *speciation,* the origin of new species, but over a long enough period of time, evolutionary changes in a population's gene pool can lead to the accumulation of enough phenotypic change that systematists (taxonomists) would conclude that a new species had come into being.

5.22.   What constitutes a species?

A *species* is composed of organisms of one particular kind. Individuals belonging to the same species will demonstrate: (*a*) a characteristic constellation of traits (physical, behavioral, and physiological); (*b*) a specific karyotype; (*c*) a higher degree of protein and/or nucleic acid concordance than is obtainable with interspecific testing; (*d*) in the case of animals in the wild, *assortative mating* (i.e., mating only with one's own kind); (*e*) ability to produce offspring of unimpaired fertility. If hybrid young are produced, their reproductive success is usually impaired both because of problems encountered when nonhomologous chromosome sets try to pair up during meiosis and because of atypical breeding behaviors that prevent them from attracting mates of either parental species.

5.23.   What are punctuated equilibria?

Transition from one species to another may be greatly accelerated at times of environmental flux, when a population's previous state of adaptation is overthrown. After a new state of optimum or at least adequate adaptation has been attained, the new species may endure for some 5–10 million years with little further change. Thus, lengthy plateaus of phenotypic stability are interconnected by relatively short periods of rapid change, a phenomenon referred to as *punctuated equilibria*.

> **Example 11:**   An unusually continuous record of speciation showing punctuated equilibria is found in the 400-m-thick fossil beds of the Lake Turkana basin in Africa, which contain the shells of freshwater snails. Geologic and climatic instability caused marked fluctuations in water level that repeatedly isolated snail populations from one another, trapping them in residual portions of the lake. The morphology of the fossilized shells shows a history of rapid change during such periods of low water, with new species arising over periods ranging from no more than 5000 to 50,000 years. Transitional forms are found in only the few meters of sediment deposited during each of these periods of change; after each such episode, the new species, once established, endured for very much longer periods of time.

5.24.   What are the major modes of speciation?

Reproductive isolation is an important factor in the origin of new species, because changes in the gene pools of isolated populations cannot be shared. Impediments to gene flow include *geographic isolation* of subpopulations and *temporal isolation*, the passage of time that prevents descendant generations from breeding with their ancestors.

5.25.   What type of speciation occurs as a result of geographical isolation?

Schistic, or *cladogenetic, speciation* (Fig. 5.7a) is a result of *divergent evolution* between subpopulations that have become geographically isolated. It typically produces two or more contemporaneous *sibling species*. So long as incipient species remain *allopatric* (having nonoverlapping ranges), interbreeding is precluded. But should they spread to become partially *sympatric* (having overlapping ranges), the acid test of speciation is whether, in the sympatric zone, the two populations interbreed or practice *assortative mating*, and, if they do interbreed, whether or not their hybrid young are capable of reproducing successfully. If they fail on either of these counts, they are indeed separate species (Fig. 5.7b).

> **Example 12:**   Island clusters far from a mainland furnish numerous examples of sibling species that have evolved on separate islands and do not interbreed wherever they have become sympatric. Flocks of birds, storm-blown from their usual flyway, may make landfall dispersed among such islands and become successfully adapted to their new habitats with such attendant diversification that they come to represent an *adaptive radiation* of sibling species fitted for different diets and modes of life. The honeycreepers of the Hawaiian islands are products of this process of adaptation, so diversified in beak morphology and feeding behavior that they do not compete for food even when sympatric. Another example of the adaptive radiation of isolated castaways is the 13 species of unusual ground finches found only in the Galápagos Islands, often called Darwin's finches since in pondering their existence the great naturalist began to develop his theory that natural selection operates as a force in the origin of new species.

5.26.   What type of speciation occurs as a result of temporal isolation?

Phyletic speciation takes place when sufficient phenotypic change occurs through successive generations along a population's lifeline, so that if all generations were represented in the fossil record, we would be hard put to decide just when "species A" had changed enough to be called "species B." Geographic isolation need not be prerequisite to this mode of speciation since time is the barrier to gene flow (Fig. 5.7c).

> **Example 13:**   Shelled marine protozoans called foraminiferans have through many ages left droves of their limy shells on seafloors too deep to be exposed to erosion. When these submarine sediments are taken by coring, a long cylinder is withdrawn within which are found continuous sequences of fossil species and transitional forms. In the case of planktonic marine organisms of widespread distribution, speciation has often been probably less a product of geographic isolation than just the passage of innumerable generations.

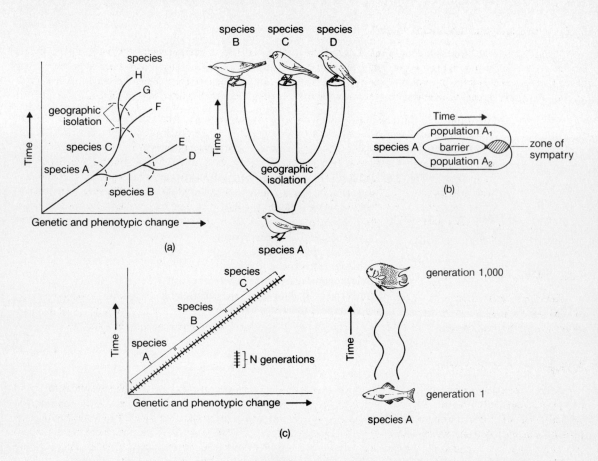

Fig. 5.7 Origin of new species. (*a*) Schistic or cladogenetic speciation follows the geographic isolation of subpopulations. (*b*) The criteria of speciation when two formerly separated subpopulations again become sympatric are whether or not they interbreed and, if they do so, whether the hybrid offspring are of unimpaired fertility. (*c*) Phyletic speciation occurs as a result of progressive shifts in a population's gene pool through successive generations, with time the barrier to reciprocal genetic exchanges.

Example 14:   The evolution of the horse family in North America, over a 60-million-year period, from four-toed, dog-sized *Eohippus* (*Hyracotherium*) to modern *Equus*, striding regally on the undivided hoof formed by the nail of its one remaining toe, took place at a sporadic rate of change and produced many branches that died out. But the historical mainline (in the sense of survival to the present) shows progressive adaptation toward fleet-footedness in open terrain: increased size, shortening of leg and lengthening of foot with improved mechanical advantage, reduction and loss of toes, and enlargement of the remaining functional digit with modification of its nail into a hoof (Fig. 5.8).

5.27.   Can evolution result in convergence as well as divergence?

Once speciation has occurred, the separate species cannot fuse to become one. In the sympatric portion of their range, they usually are forced to diverge even more, a phenomenon known as *character displacement,* which further reduces ecological competition and decreases the likelihood of infertile hybrid matings (by further divergence in calls, mating behavior, etc.). However, when different and even quite unrelated species adapt to similar modes of life, they may come to resemble each other closely in one or more respects, even to the point of uncanny similarity in overall body form and behavior. This evolutionary *convergence* simply results from the fact that successful pursuit of some particular mode of life may depend on certain specializations in form and function, which may originate and evolve quite independently

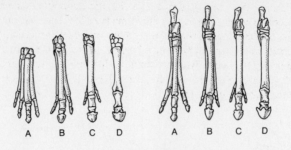

Fig. 5.8 Evolution of front feet (*left*) and hind feet (*right*) in the horse family as found in fossil beds of western North America.
A, *Eohippus* (Eocene Epoch);
B, *Miohippus* (Oligocene Epoch);
C, *Merychippus* (Miocene Epoch);
D, *Equus* (Pleistocene to present).
Note progressive reduction in number of toes, enlargement of third digit, and lengthening of entire foot (drawings not to scale).

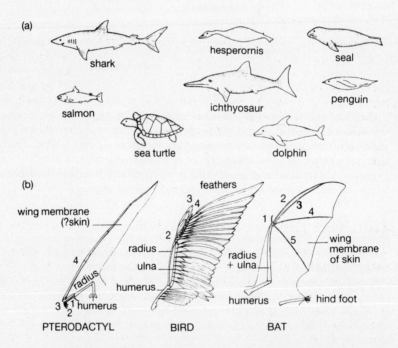

Fig. 5.9 Convergent evolution. (*a*) Convergence of air-breathing aquatic vertebrates [sea turtle and ichthyosaur (reptiles), hesperornis and penguin (birds), seal and dolphin (mammals)] toward the streamlined body form characteristic of fishes (shark and salmon). (*b*) Convergence for flight in pterodactyl (a reptile), bird, and bat (a mammal). Note that the bones of the forelimb are homologous but their modification toward support of a wing is very different: the skin web of a pterodactyl wing was supported by an enormously elongated fourth finger, that of a bat wing is supported by four long digits, while a bird wing consists of feathers and the hand bones are much modified, with digits 1 and 5 absent and 2 and 4 reduced. (*From Storer et al.*)

in unrelated groups (Fig. 5.9a). Species that play similar roles in different biotic communities are *ecological equivalents*. They may or may not be as convergent in form as they are in mode of life, but some are remarkably similar, such as the Australian marsupial mole and the "true" (placental) mole, and the marsupial Tasmanian "wolf" and the "true" (placental) wolf. Such convergence extends only to phenotypes involved in the adaptive process, while other attributes (internal anatomy, embryonic development, etc.) betray their true affinities. In addition, structures that perform equivalent functions in different organisms showing convergence may differ greatly in detail, as we can see in the wings of different flying vertebrates: birds, bats, and the reptilian pterodactyls (Fig. 5.9b).

5.28.   What of macroevolution?

Major evolutionary changes prerequisite to the origin of taxonomic groups above the species level can be explained in terms of the same processes and mechanisms as those that produce new species. However, whereas the diversification of the Galápagos finches into 13 *species* may have occurred in a few hundred thousand years, diversification of the class Aves into some 24 *orders* of birds has taken more than 100 million years. The diversification of vertebrates from the first small jawless armored fishes into the range of species seen today required over 400 million years, with certain transitional fossil forms enduring to mark the origin of amphibians from lobe-finned fishes, reptiles from early amphibians, and birds and mammals from different groups of reptiles.

The punctuated equilibrium model seems to hold true for macroevolutionary changes, which appear to have been most accelerated following mass-extinction events that reduced competition and opened habitats for exploitation by new life forms.

5.29.   Has animal evolution produced a "ladder" or a "tree"?

The old idea of a "ladder of life" ascending from simpler to more complex forms of life (with humans, of course, on the top rung) has some legitimacy, but it proves misleading, because many life forms of simple structure and relatively limited genome (i.e., an organism's total genetic material) have survived quite happily into the present, while many complex life forms, such as dinosaurs, have bitten the dust. It is more fruitful to think of the evolution of animal life as having produced a highly branching, flat-topped tree, of which the topmost twigs represent species existing today while twigs and branches ending lower down represent extinct offshoots. However, such a tree of life would look rather odd, for while a typical tree has a massive trunk, the evolutionary tree would have a very spindly trunk, because animal life has diversified and gained mass from the slender base provided by one or a few protistan ancestors. *Evolutionary systematics* seeks to construct such *phylogenetic trees* by comparing fossil and modern species and by utilizing not only anatomical but also behavioral, physiological, and biochemical comparisons to ascertain the degree of relatedness of living species.

5.30.   How does the Linnaean hierarchical system of classification relate to evolutionary systematics?

The eighteenth-century Swedish botanist Carolus Linnaeus, whose investigations were carried out a century before those of Charles Darwin and Gregor Mendel, believed that species never changed, but paradoxically he still worked out a hierarchical scheme of classification based on the idea that species show different degrees of mutual relatedness. Linnaeus originated the practice of *binomial nomenclature,* by which each newly described species is given a latinized dual name: generic and specific, so that, for example, the domestic dog is called *Canis familiaris.* This provides an immediate handle for grouping by relatedness, for closely similar species can be put together to make up a *genus.* Thus genus *Canis* includes also the wolf (*C. lupus*), coyote (*C. latrans*), golden jackal (*C. aureus*), and others. Genera that appear related are then grouped into a *family,* so that family Canidae includes such genera as *Canis, Urocyon,* and *Vulpes* (foxes). Related families constitute an *order;* e.g., families Canidae, Felidae (cats), Ursidae (bears), Mustelidae (weasels), Viverridae (civets), and others are grouped into order Carnivora, flesh-eating mammals with fanglike canine teeth. All orders of mammals are grouped into the *class* Mammalia: hairy, milk-producing vertebrates. The vertebrate classes, including four classes of living fishes plus the tetrapod ("four-footed") classes Amphibia, Reptilia, Aves, and Mammalia, are grouped into the *subphylum* Vertebrata, animals with backbones, which are allied with other animals (such as tunicates) that have an embryonic supportive rod called a *notochord,* to make up the *phylum* Chordata. Although *kingdoms* represent the most inclusive of taxonomic categories, into which phyla are clustered, taxonomists still disagree on how many kingdoms there should be, and just how these relate to each other (see Fig. 4.1). In fact, in this entire hierarchical scheme of classification, however useful, at the moment species are the only entities capable of objective delineation: all higher categories remain more *conceptual* than actual, at least until gene analyses come to provide definitive answers to such questions as exactly how much genetic concordance is needed for species to be properly assigned to the same genus, to what extent must genera differ genetically to deserve placement in different families, and so forth.

5.31. What are the mechanisms of evolution?

Evolutionary change relies on *changes in genes and chromosomes, selection pressure*, and *genetic drift*. The alterations in genes and chromosomes that provide the raw material on which selection and drift operate include *gene mutations*, *quantitative augmentation of the genome* through addition of genes and chromosomes, and *rearrangements of gene sequences* in the chromosomes.

5.32. How do gene mutations occur?

Gene mutations may involve *base substitution* or *frame shifting*. ("Base" as used here refers to an entire nucleotide and not just the nitrogenous base.) In brief, every sequence of three nucleotides in a gene (base triplet) dictates the presence and position of one specific amino acid in a protein; if one nucleotide in the triplet is changed, a different amino acid may be placed in the protein. This may seem minor, but it can greatly affect the activity of the protein: in humans, the substitution of one amino acid in a peptide chain contributing to the hemoglobin molecule changes the behavior of those molecules when oxygen is scarce, so that they line up and twist the whole red blood cell out of shape, causing sickle-cell anemia as the deformed cells are destroyed.

Frame shifts occur when a nucleotide is dropped from the sequence, or accidentally repeated. From that point on, all triplets will be misread; the resultant protein will be wildly transformed from the original gene product and, if biologically useful, will probably function very differently. If we had to read a sentence in groups of three letters, a frame shift would change a meaningful sentence, THE CAT ATE THE FAT RAT, to nonsense: THE CAA TET HEF ATR AT___ (deletion), or, THE ECA TAT ETH EFA TRA T (repetition). No wonder more gene mutations are harmful than helpful.

5.33. What is the importance of duplicate genes?

In the evolution of higher plants, duplication of entire sets of chromosomes (*polyploidy*) has played an important role in enriching the genome, but this seems unimportant in animal evolution. On the other hand, *duplicated genes* occur commonly, scattered at different loci among the chromosomes so that they are not only nonallelic, they may not even be linked. Duplicate genes are important in two ways: (*a*) if they remain unmutated, they act additively, resulting in *quantitative inheritance;* (*b*) should they mutate, the original function is not sacrificed, for it is still performed by another copy of the gene, and so the mutant duplicate gene can be carried along harmlessly through the generations, free to mutate further until it may attain a new biological function.

> **Example 15:** In the human X chromosome two genes lie right next to each other, a position that suggests accidental reiteration during chromosome replication; both are important for full color vision. One encodes a visual pigment that chiefly absorbs green light; the other, one that absorbs red light. One of these genes is most likely a copy of the other, which having mutated now provides us with the valuable bonus of an additional light-absorbing pigment that greatly improves color discrimination.

5.34. How does quantitative inheritance differ from single-gene inheritance?

Quantitative, or *polygenic*, *inheritance* produces traits that typically show a bell-shaped curve of *continuous variation* in a mixed population, rather than the picture of *discontinuous* variation presented by *monogenic* traits. The earlier examples of color blindness, tongue-rolling, and hair color and coat texture in guinea pigs illustrate monogenic traits, produced by a change in a single pair of allelic genes. The alternative traits are distinctly different (e.g., black versus white), and even when the alleles are codominant (e.g., black, blue, and splashed-white Andalusian fowl), the three phenotypic categories are distinct and discontinuous, without intergradations. In contrast, where *duplicate genes* are involved, each effective allele adds a bit to the phenotype. Many, perhaps most, phenotypes (including body size, height, etc.) are shaped by polygenic inheritance.

> **Example 16:** Four pairs or so of duplicate genes produce a range of human skin pigmentation from northern European to African. In the former, the eight genes (four different allelic pairs with codominance) are termed "ineffective"—they do not add to the pigmentation of the skin, which thus remains

pale and more likely to burn than tan. In the latter, the skin produces so much melanin that it is "black" (which is adaptive in guarding deeper tissues from damaging ultraviolet radiation), because the eight gene copies are all "effective" in producing melanin. If persons representing the genotypic extremes *aabbccdd* and *AABBCCDD* were to mate, their children would all have the genotype *AaBbCcDd* and be intermediate in skin color. You can see that matings of *AaBbCcDd* individuals could produce offspring with 0–8 effective genes, so that the phenotypic spectrum would grade smoothly from black to dark brown to medium brown to light brown to fair-skinned. In a large enough population practicing nonassortative mating with respect to skin color, the majority would fall into the intermediate part of the range, thus producing a bell distribution curve.

5.35.   What is the evolutionary importance of chromosomal rearrangements?

Chromosomal rearrangements can alter the sequence of genes within a given chromosome (*inversion*, e.g., from *ABCDEFGH* to *ABCHGFED*) or can occur when a broken-off chromosome fragment adheres to the end of another chromosome (*translocation*). These events can result in an embarrassing tangle when homologous chromosomes try to pair up during synapsis in meiosis I, so that few viable gametes are formed. However, should certain rearrangements spread throughout a population and become homozygous, they no longer cause meiotic difficulties. But if members of this population then mate with those of another in which those particular chromosomal rearrangements have not occurred, the hybrid offspring will probably be sterile because of meiotic interference. Thus chromosomal rearrangements are an important factor in speciation, since they can isolate populations reproductively even if intermatings do occur. In addition, translocation provides a source of duplicated genes as discussed above, for the translocated genes will replicate in their new position while their former alleles carry on at the original locus.

5.36.   Do allelic frequencies change or remain constant?

The familiar 3:1 or 1:2:1 Mendelian ratios are realized only when true-breeding (homozygous) strains are crossed, for only then are alternative alleles present in a 50:50 ratio. In actuality, *allelic frequency is seldom equal*. The percentage frequency of each member of an allelic pair in a population can be determined by the frequency of the recessive homozygotes, which can be told by direct observation. If a gene exists in only two alternative allelic states [e.g., albinism versus nonalbino (normal) pigmentation], the percentage frequency of the dominant allele can be represented as p and the percentage frequency of the recessive allele as q. These percentage frequencies can be calculated from the value of q^2 in the equation $p^2 + 2pq + q^2 = 1$, because q^2 is the percentage of recessive homozygotes, which can usually be determined by simple observation. Having established the percentage frequency of the recessive allele (as the square root of q^2), we can next determine the percentages of heterozygous individuals ($2pq$) and those homozygous for the dominant allele (p^2).

> **Example 17:**   The frequency of albino humans (*aa*) in America is 1 in 20,000, giving a value of 1/20,000 for q^2; the square root of 1/20,000 is about 1/141, and so the total frequency of the recessive allele *a* is 1 in 141 while the allele *A* makes up 140 in 141. The frequency of normal persons carrying a hidden gene for albinism is therefore $2pq = 2 \times 140/141 \times 1/141 = 1/70$. Surprisingly, although only one person in 20,000 is albino (*aa*), one person in 70 is a heterozygous carrier (*Aa*) of the allele for this trait.

As first applied to the genetics of populations, this equation seemed to contradict evidence of evolutionary change, for it showed mathematically that allelic frequencies remain stable as populations grow, even though recessive traits show up less frequently in larger populations. This *Hardy-Weinberg equilibrium principle* helps account for the long periods of phenotypic stability that characterize the history of well-adapted species, but it requires that other things be stable too. In actuality, allelic frequencies almost never remain stable, because of mutation pressure, selection pressure, and genetic drift.

5.37.   How does mutation pressure shift allelic frequencies?

Mutation pressure is the difference between the rates at which the allele *A* mutates to *a*, and *a* mutates to *A*. Since these values are seldom equal, one allele will gradually increase in frequency. Many mutant alleles are *neutral*, neither beneficial nor harmful, so these tend to build up in the gene pool, contributing to individuality and variety. However, if the allelic state favored by mutation pressure happens to be harmful, the effects of mutation pressure may be counteracted by selection pressure.

5.38.   How does selection pressure affect allelic frequencies?

Selection pressure represents the comparative reproductive success of the phenotypes produced by alternative alleles. *Natural selection* is a process whereby the frequency of detrimental alleles declines, while that of beneficial alleles increases, because of *differences in reproductive success*. Agents of natural selection (*selective factors*) include predators, parasites, food resources, climate, and the like. Although natural selection is often equated with "survival of the fittest," this is not strictly true: if a female cockroach produces a mutant pheromone that does not attract males, she can be perfectly healthy but will remain a virgin and fail to pass on her genes to offspring. In fact, *sexual selection*, or success in attracting mates, is a quite important factor in the reproductive success of many animals. Selection can operate upon the bell distribution curve of a polygenic trait in a *stabilizing, directional,* or *disruptive* mode (Fig. 5.10).

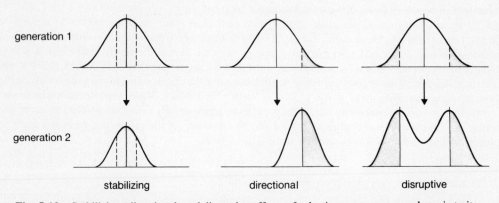

generation 1

generation 2

stabilizing   directional   disruptive

Fig. 5.10   Stabilizing, directional, and disruptive effects of selection pressure on a polygenic trait. Best reproductive success is denoted by the shaded areas. Note that under the influence of stabilizing selection, the mean for the trait remains the same through descendant generations, while directional and disruptive selection lead to progressive shifts in phenotype toward one or both extremes of the range of variation.

5.39.   What is stabilizing selection?

Stabilizing selection favors individuals that are "average" for the trait (e.g., average size) and preserves the phenotypic stability of well-adapted species.

5.40.   What is directional selection?

Directional selection occurs when individuals at one phenotypic extreme have greater reproductive success than those in the middle of the curve or at the opposite extreme, with the result that through successive generations the phenotypic mean continues to shift in the same direction until the limits of the gene pool have been reached or adaptive conflicts result. One example of this may be enlargement of the tail fan in peacocks: peacocks with short tails are apt to be unsuccessful in attracting mates, but peacocks with tails too cumbersome are more vulnerable to predators; the existing, still huge, tail fan may represent a compromise between directional sexual selection and predator pressure.

5.41.   What is disruptive selection?

Disruptive selection favors both extremes of the range of phenotypic expression, as may be the case when subpopulations invade differing habitats. The large ears of desert jackrabbits are effective in dissipating excess body heat; the reduced ears of snowshoe hares and pikas are less subject to frostbite.

Example 18: Disruptive selection, seemingly headed toward speciation, is seen in a species of swallowtail butterfly that escapes predation by mimicking one species of distasteful butterfly found in part of its geographic range and mimicking a quite different noxious model in another part of its range. Continued selection toward more faithful mimicry of the two different models should drive the two mimic populations and the original nonmimic population phenotypically ever further apart, until they no longer even recognize each other as potential mates.

5.42. How does genetic drift affect allelic frequencies?

Genetic drift is the elimination of the less common allele from a breeding population by chance alone, without regard for its effect on reproductive success. In nature, many breeding populations tend to be small, which accentuates the incidence of chance allelic loss. Since the frequency of *any* new mutation is bound to be low at the start, even alleles of potentially great benefit may get "nipped in the bud" by genetic drift.

Chapter 6

Cells as Organisms: Meet the Protists

Taxonomists who yearn to pigeonhole all organisms neatly are apt to confront the protists with contention and insomnia, for many of these organisms exhibit a slippery intermediacy that thwarts classification schemes, which have inadequate provision for such wayward creatures. Among living protists we find organisms that seem to reflect the common ancestry of higher animals, plants, and fungi, and others whose affinities remain quite baffling. There is even lack of agreement as to which eukaryotic groups belong in kingdom Protista, but the viewpoint adopted here is that the kingdom should be reserved for *unicellular* eukaryotes, together with a few kinds that form colonies or filamentous chains of cells.

6.1. What are the challenges of protistan systematics?

When dealing systematically with the protists, taxonomists find that certain groups present no identity crisis; however, for other groups there seems to be no right way of doing things! One way of classifying protists emphasizes their modes of nutrition, recognizing (1) plantlike protists that are autotrophic and possibly evolutionarily allied with multicellular phyla of algae and higher plants, (2) animal-like protists that ingest organic food for internal digestion and may have included the ancestors of multicellular animals, and (3) funguslike protists that absorb externally digested organic nutrients and may have been ancestral to multicellular fungi. Yet this approach proves inadequate: some autotrophic eukaryotes photosynthesize by day but feed like animals in the dark, and certain "twin" autotrophic and heterotrophic species look essentially identical, except that one has chloroplasts. Furthermore, the funguslike slime molds feed phagocytically like amoebas, even though they produce aerial spore-bearing bodies as fungi do.

Contradictory schemes of classification are imposed on the above three groups, resulting in confusing double-pigeonholing, as when the green flagellate *Euglena* is simultaneously classified by botanists as a plant in phylum (or *division*, the botanical equivalent of a phylum) *Euglenophyta* and by zoologists as a member of phylum *Protozoa* (or Sarcomastigophora, or Mastigophora, depending on how inclusive that phylum is seen to be). In favor of the zoological approach is the fact that biochemically *Euglena* has proved to be intimately related to a parasitic flagellate, *Crithidia* (allied to the disease-causing trypanosomes). Yet even when put together in the same phylum, *Euglena* and *Crithidia* still have been relegated to separate classes or subclasses, as if they were only distantly related.

Another approach to protistan classification emphasizes mode of locomotion, grouping together (1) protists that bear flagella, (2) those that have pseudopodia, and (3) those that are ciliated. Ciliates do seem to constitute a legitimate grouping, but schemes that classify all flagellates together take in dinoflagellates, which seem very far removed from *all* other protists, and also colonial and unicellular green algae that have cellulose cell walls and are traditionally classified in phylum *Chlorophyta* along with multicellular green algae. Worse still, some flagellates can reabsorb their flagella and go amoeboid, while some amoeboid forms can do the opposite. Grouping amoeboid organisms together invites similar problems, such as what to do with the slime molds.

The marvelous thing about the protists is that so many of them *do* defy pigeonholing. Even today, when protists have been in existence for a billion years, many have not settled down to a life of specialization, and some may not even have evolved very far from forms ancestral to the separate lineages of plants, animals, and fungi. They remain simple, yet are often surprisingly complex. Above all, protists in general have remained highly successful and prolific for a billion years and today constitute a gene pool out of which, given another billion years, the earth could be repopulated by who knows what manner of descendants, should the life-forms we consider "higher" than protists fail to make the grade in the long run.

6.2. What are protozoans?

Strictly defined, a protozoan ("first animal") would be a unicellular, heterotrophic organism displaying holozoic or saprobiotic nutrition. *Holozoic* nutrition includes ingestion of materials requiring internal digestion: solid particles engulfed by *phagocytosis* and complex organic molecules ingested in fluid-filled vesicles invaginated from the plasma membrane in the process of *pinocytosis*. *Saprobiotic* nutrition (referred to as *saprozoic* if carried out by an organism considered to be an animal and *saprophytic* if the organism is a plant or fungus) involves the taking in, by absorption or pinocytosis, of small organic molecules needing no further digestion. (Saprobiotic protozoans do not engage in extracellular digestion, but ingest small organic molecules such as sugars and amino acids produced by decomposers, namely bacteria and fungi.) The above definition would exclude from zoological consideration the plantlike Euglenophyta, Pyrrophyta, and simple green algae (Chlorophyta), which include mostly autotrophic species. However, protozoologists do treat a number of autotrophs as protozoans, so we will consider them here, as well as the slime molds, which some classify as protozoans and others as fungi.

6.3.   What are the general characteristics of protozoans?

(a)   *Size:* Protozoans range from being microscopic to being many centimeters in size.

> **Example 1:**   *Babesia,* an intracellular parasite within red blood corpuscles, is 2–3 μm in size, as is *Plasmodium* (malaria parasite) in its *merozoite* stage. Other protozoans are much larger. Some parasitic gregarines attain a length of 10 mm, and the fossil shells of nummulitic foraminiferans (seen in limestone blocks in the Egyptian pyramids) are dime-sized. The plasmodial slime molds may form a thin, creeping film of multinucleate cytoplasm extending across a diameter of many centimeters.

(b)   *Level of organization:* Most protozoans are unicellular, possessing characteristic eukaryotic organelles, as well as additional organelles (e.g., contractile vacuoles, among others) not present in metazoans. While some change shape freely, most maintain a definite shape owing to a rigid or elastic *pellicle* that underlies the plasma membrane. Many are multinucleate; some are colonial.

(c)   *Nutrition:* Some are strict autotrophs, possessing green chloroplasts or *chromoplasts* that contain yellow and red pigments as well. Certain autotrophic species are *facultative heterotrophs,* ingesting organic nutrients in the dark, and therefore are true "plant-animals." The rest are *obligate heterotrophs* with holozoic or saprobiotic nutrition. Some are *facultative anaerobes,* surviving in oxygen-poor environments, such as the intestinal tracts of hosts or oxygen-depleted water.

> **Example 2:**   Holozoic species may capture bacteria, diatoms, other protozoans, or even small metazoans. Herbivorous species even ingest entire filaments of blue-green algae, often distorting themselves comically in the process. Food is engulfed by cytoplasmic projections known as *pseudopodia* ("false feet") or through a *cytostome* ("cell mouth") located at one particular site at which the pellicle is interrupted and the plasma membrane can invaginate to pinch off *food vacuoles*. Digestive enzymes are secreted from lysosomes into the food vacuoles. As the food is digested and absorbed, the food vacuole shrinks and finally contacts the plasma membrane [at a special pore, the *cytopyge* ("cell anus"), when a pellicle is present], allowing indigestible residues to be egested by exocytosis. The main nitrogenous waste is ammonia, which readily diffuses through the cell surface. Other gases too are exchanged by diffusion across the plasma membrane.

(d)   *Osmoregulation:* Maintenance of water balance is accomplished by one or more *contractile vacuoles*.

> **Example 3:**   Contractile vacuoles are especially large and active in freshwater protozoans, in which they operate like bilge pumps to rid the body of water that constantly leaks inward from their hypotonic milieu. The spherical vacuoles swell with water, then contract periodically, expelling their contents to the exterior. They soon reappear as a coalescence of small vesicles formed at the conjunction of radiating tubules through which water is collected from the cytoplasm. The contractile vacuoles of marine protozoans pulsate much less frequently than those of freshwater species.

(e)   *Locomotion:* Pseudopodial (amoeboid), flagellar, ciliary, and wormlike movements occur in various species. Some life cycles include both amoeboid and flagellated forms, or a single organism may possess both pseudopodia and flagella, or be flagellated but also shorten and lengthen like a worm. Some are sessile (attached) and do not locomote at all when mature; others float without swimming actively.

(f)   *Reproduction:* All reproduce asexually, by mitosis in which the nuclear envelope typically remains intact and the spindle forms intranuclearly. Division may produce two daughter cells of equal size (*transverse* or *longitudinal binary fission*; Fig. 6.2c), two daughter cells of unequal size (*budding*), or a cluster of daughter cells when repeated nuclear divisions are followed by partitioning of cytoplasm (*multiple fission*). Many also reproduce sexually, by *conjugation* (exchange of haploid nuclei between temporarily conjoined individuals) or by *syngamy* (fusion of gametes). Sex cells may be morphologically alike (*isogametes*) or different (*anisogametes*). In the latter case, the larger gamete is considered to be the female egg and the smaller gamete, the male sperm cell.

(g)   *Ecology:* Some protozoans are free-living, while others are symbiotic in or on the bodies of host organisms.

6.4.   What are free-living protozoans?

Free-living organisms are those that are not symbionts. Although some can live actively in moist soil, the free-living protozoans are mostly aquatic, in freshwater and marine habitats, where they may be *planktonic* (floating) or *benthonic* (either sessile or creeping on the bottom). Autotrophic species are light-sensitive and exhibit *positive phototaxis,* moving into zones of favorable illumination. All protozoans exhibit positive and negative *chemotaxis,* moving toward food and away from injurious chemicals.

6.5.   What are symbionts?

Symbiotic associations involve organisms of two different species that live usually in *intimate physical contact* in a relationship that is *essential* to the survival of one or both of the species concerned. The species for which the relationship is always obligate is known as the *symbiont,* and the other as the *host*. Three types of *symbiosis* are commonly recognized—mutualism, commensalism, parasitism—according to the effects of the symbiont upon the host's bodily economy, although sometimes these categories intergrade.

6.6.   What is mutualism?

Mutualistic symbiosis is beneficial to both host and symbiont, and may be essential for both.

> **Example 4:**   Flagellated protozoans of certain genera (e.g., *Trichonympha*) occur in the intestines of termites and cannot live independently. Termites masticate wood, but cannot digest cellulose. The symbiotic flagellates ingest the microscopic fragments of wood and digest them, liberating sugars that sustain host as well as symbionts. Termites infect themselves with symbionts by eating each other's droppings.

> **Example 5:**   *Paramecium bursaria*, a ciliate protozoan, is host to unicellular green algae *(zoo-chlorellae)* that fill its cytoplasm and produce sugars that nourish both host and symbionts.

6.7.   What is commensalism?

Commensalism is a symbiotic relationship essential for the symbiont *(commensal),* but not measurably beneficial or harmful to the host.

> **Example 6:**   The marine ciliate *Foettingeria* uses two hosts commensally during its life cycle. As an adult, it inhabits the digestive cavity of sea anemones. The young encyst on small crustaceans, and when these are eaten by anemones, the protozoans excyst and settle down.

6.8.   What is parasitism?

Parasitism is a symbiotic relationship that is obligate for the *parasite* and more or less harmful to the host. Naturally it behooves a parasite to harm its host as little as possible, since an enfeebled host might die or fall victim to a predator; hence, selection favors parasites that are less *pathogenic* (disease-producing), and a parasitic relationship may well evolve into one that is commensal or even mutualistic. Human diseases caused by parasitic protozoans include malaria, African sleeping sickness, and amoebic dysentery.

6.9.   What is the evolutionary status of protozoans?

The occurrence of protozoan *tests* (shells) in Precambrian deposits indicates that certain lineages have been in existence for more than 600 million years, but with the exception of radiolarian and foraminiferan tests that show characteristic species assemblages in each period of the geological record from Precambrian times to the present, these soft-bodied, mostly microscopic, organisms have left no fossil trace. Nevertheless, modern protozoans cannot, as a group, be considered either simple or primitive. "Simple" and "primitive" are not synonymous. "Primitive" means closer to the supposed ancestral pattern, which would appear to be especially true of the phytoflagellate euglenids; "advanced" protozoans, such as many ciliates, possess newer patterns superimposed on the ancestral ones. "Simple" is the antonym of "complex"; by and large, protozoan cells are more complex than cells in metazoan bodies, even though protozoans are probably more primitive than metazoans.

6.10. How animal-like are the slime molds?

Slime molds ingest solid food phagocytically and exist as independent amoeboid cells in moist soil for most or some part of their life cycle. Protozoologists who classify slime molds as protozoans tuck both *cellular* and *plasmodial* types into a single class (Eumycetozoea = "true fungus-animal") within the phylum (or subphylum) Sarcodina. This is probably incorrect, for although it may be successfully argued that they are more animal-like than funguslike, the two types of slime molds are certainly not much like *each other* and deserve considerable taxonomic separation.

6.11. What are the characteristics of cellular slime molds?

Cellular slime molds (phylum/division Acrasiomycota) spend nearly their entire lives as independent, free-living soil amoebas, ingesting bacteria and other organic particles phagocytically. They divide by fission and multiply so long as the environment is favorable. But when moisture or food runs short, some amoebas emit a communicative chemical *(pheromone)* that entices others to drop everything and gather together (Fig. 6.1a). Different species exude their own special pheromones, so that when two species of these so-called *social amoebas* live sympatrically, they will not form mixed-species aggregations. One such pheromone is the nucleotide cyclic AMP. Once established, the aggregation arranges itself into what is essentially a multicellular, slug-shaped organism, the *pseudoplasmodium,* which creeps, sluglike, to the surface of the soil, where some of the former amoebas cooperatively form the base, stalk, and bulbous apex of an aerial *sporangium,* while other cells crawl up through the stalk to encyst as *spores* that can be blown away to more favorable areas. There, they excyst and resume existence as independent amoebas. Sexual reproduction appears to be rare, involving fusion of two haploid amoebas to form a diploid *macrocyst.* Through nearly their entire life cycle, the acrasiomycotes look and act like sarcodine protozoans, but their aggregative behavior is unique.

6.12. What are the characteristics of plasmodial slime molds?

Plasmodial slime molds (phylum/division Myxomycota) seem not at all closely related to the cellular slime molds. During part of their life cycle, they exist as haploid soil amoebas, which can proliferate by fission for many generations,

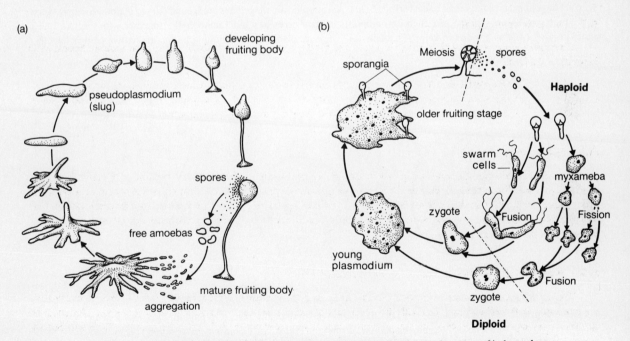

Fig. 6.1 Slime molds. (*a*) Life cycle of a cellular slime mold, showing coalescence of independent soil amoebas into a pseudoplasmodium that sends up spore-bearing bodies. (*b*) Life cycle of a plasmodial slime mold, showing development of a multinucleate syncytial plasmodium that bears spores that develop into either haploid amoebas or haploid flagellated swarm cells; fusion in pairs of either type of haploid cell produces a diploid zygote that develops into a new plasmodium.

feeding phagocytically. Some of these will fuse in pairs, forming diploid amoebas; alternatively, such diploid amoeboid zygotes can form from the fusion of flagellated isogametes that emerge from the spores of some species. Each diploid amoeba creeps about engulfing food, continues to grow and branch, becoming multinucleate by synchronous nuclear divisions, and may attain a diameter of several to many centimeters. When food or moisture fail, the multinucleate syncytium or *plasmodium* stops moving and sends up a number of erect stalks topped by *sporangia*. Meiosis occurs within the sporangia, and haploid spores are shed. The spores, which resemble hard-walled cysts, open under favorable conditions to liberate either amoeboid cells or flagellated gametes (Fig. 6.1*b*). Throughout its entire life cycle, a myxomycote feeds like an animal and, in its uninuclear state, strongly resembles other soil amoebas that lack the talent for syncytial (multinucleate) growth.

6.13.   How animal-like are the "plantlike protists"?

The plantlike dinoflagellates, chrysomonads, cryptomonads, volvocids, and euglenids all contain both autotrophic and heterotrophic species, although the majority are photosynthetic. Some autotrophic species also ingest prey by phagocytosis, imbibe organic molecules in solution or do both. Heterotrophs include both holozoic and saprobiotic feeders. Each of the above groups is classified alternatively as a separate phylum or division, or as an order within the class (or subclass) Phytomastigophorea of the phylum (or subphylum) Mastigophora. (When Mastigophora is considered a subphylum, it falls within phylum Protozoa. However, the close biochemical concordance of certain phytoflagellates and zooflagellates suggests that this taxonomy is more convenient than accurate.)

6.14.   What are the dinoflagellates?

Dinoflagellates (phylum/division Pyrrophyta or order Dinoflagellida) are unlike all other eukaryotes, including protists, in some fundamental respects and certainly deserve a phylum of their own, if not a kingdom. Some are holozoic predators, and many possess thread-discharging capsules (*trichocysts*) like those of ciliate protozoans. The spiny, armored autotroph *Ceratium*, weary of sunbathing, can slyly extrude pseudopodia, capture prey, and ingest it through a mouth region between its plates (Fig. 6.2*a*). The aberrant, colorless *Noctiluca* ("night light," for its bioluminescence) is exclusively predatory, with one flagellum and a fingerlike tentacle (Fig. 6.2*a*). Autotrophic dinoflagellates are second only to diatoms in importance as producers of biomass in the sea; their chromoplasts contain carotenes and xanthophylls, together with the two types of chlorophyll (*a* and *c*) found in kelps (Phaeophyta). They are planktonic, swimming by the movement of two flagella,

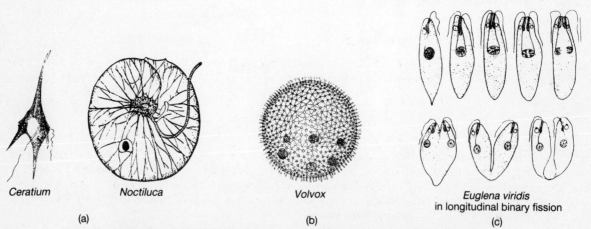

Ceratium   Noctiluca   Volvox   *Euglena viridis*
                                                                              in longitudinal binary fission
              (a)   (b)   (c)

Fig. 6.2   Phytoflagellates. (*a*) Dinoflagellates (taxonomically placed in phylum Pyrrophyta or in subclass Phytomastigophorea of phylum Protozoa): *Ceratium,* an armored species that can conduct photosynthesis but also ingests prey; *Noctiluca,* which is strictly predatory, (*b*) *Volvox,* a pinhead-sized, autotrophic colony, classified with green algae in phylum Chlorophyta or in subclass Phytomastigophorea of phylum Protozoa. (*c*) *Euglena viridis,* an autotrophic phytoflagellate placed in phylum Euglenophyta or in subclass Phytomastigophorea of phylum Protozoa; shown in longitudinal binary fission. (*From Storer et al.*)

one trailing and the other vibrating in an equatorial groove. The body is naked or encased in two thick cellulose plates. The pyrrophytes also include amoeboid species lacking cellulose walls and having contractile vacuoles.

The name "pyrrophyte" ("fire plant") refers to the bioluminescence of many dinoflagellates, but some also turn the sea red when they are numerous because of their accessory pigments. "Dinoflagellate" ("terrible whip-bearer") refers to the fact that some species are toxic, and when they multiply explosively (as a result of nutrient upwelling) to as many as 20–40 million per ml (as once measured at Monterey Bay), they may kill fish by the millions. The neurotoxin of *Gessnerium catenellum* is so powerful that a single gram could kill 5 million mice! Dinoflagellate toxins do not kill the shellfish that feed upon them, but render those shellfish (especially mussels) poisonous to humans.

Many dinoflagellates exist as mutualistic symbionts *(zooxanthellae)* within certain protozoans, corals, sea anemones, and clams; they help reef-building corals fix calcium out of seawater. Some heterotrophic species are parasitic in marine invertebrates. Curiously, a few dinoflagellate species contain stinging capsules very much like the *nematocysts* of coelenterates; this suggests the possibility that the nematocyst-forming cells *(cnidoblasts)* of coelenterates (corals, anemones, etc.) may have evolved from endosymbiotic dinoflagellates.

6.15. **How do dinoflagellates differ from other eukaryotes?**

The nucleus of a dinoflagellate is different from that of all other eukaryotes. The nuclear membrane is a *single* lipid bilayer, not a double envelope. The chromosomes, like those of prokaryotes, lack centromeres and have no or few histone proteins. They are attached to the nuclear membrane, much as bacterial chromosomes attach to the outer membrane of the bacterial cell. However, they remain permanently in the supercoiled state typical of eukaryotic chromosomes in mitosis. Cell division is bacterialike: no spindle forms, and the chromosomes separate as the nuclear membrane simply elongates and pinches in two. Meiosis, too, is unusual, consisting of only *one* division. The origins and affinities of dinoflagellates remain obscure; although nutritionally they may resemble plants, animals, or "plant-animals," their peculiarities preclude close relationship with any other eukaryotes and their taxonomic inclusion with protozoans seems unwarranted.

6.16. **What characterizes the plantlike chrysomonads?**

Chrysomonads (phylum/division Chrysophyta; or order Chrysomonadida in phylum Protozoa or phylum Mastigophora or Sarcomastigophora) are classified by botanists with the diatoms, but zoologists lay no claim to diatoms—they only want the chrysomonads! The latter are freshwater or marine, have one to three flagella and yellow or brown chromoplasts that contain the forms of chlorophyll found in kelps (*a* and *c*), and form *siliceous cysts*. Some species lack chromoplasts and feed holozoically. Chrysomonads can also shed their flagella and become amoeboid, ingesting solid food with their pseudopods like proper little amoebas.

6.17. **What characterizes the cryptomonads?**

Cryptomonads (phylum/division Cryptophyta; or order Cryptomonadida) have two flagella and an anterior depression (reservoir); most have two chromoplasts, but some are colorless, such as *Chilomonas*, a common form feeding saprobiotically in polluted waters.

6.18. **What are the characteristics of volvocids?**

Volvocids are classified with protozoans as order Volvocales, or with multicellular green algae (phylum/division Chlorophyta). Surely volvocids may be considered protists, for they are merely unicellular or colonial, but it is true that green algae present a spectrum from one-celled through colonial or filamentous to large branching or sheetlike organisms, and so all green algae are sometimes classified in kingdom Protista, or all in Metaphyta (Plantae). Volvocids may resemble the protistan ancestors of the more advanced green algae.

Most volvocids live in fresh water. They have two to four flagella, a single cup-shaped chloroplast, and a red, light-sensitive "stigma." Their sturdy cellulose cell wall is certainly a characteristic shared with metaphytes, yet some volvocids are colorless, permanent heterotrophs, while the autotrophic *Haematococcus* becomes saprobiotic in the dark.

Volvox globator (Fig. 6.2*b*) forms spherical colonies to 1 mm in diameter, containing up to 50,000 biflagellate cells united by cytoplasmic strands. Although each cell carries on photosynthesis independently, all flagella beat in coordination so that the colony rolls along like a little ball. Only a few cells engage in reproduction. During asexual reproduction several cells migrate into the hollow center of the colony, where each proliferates mitotically into a daughter colony. At first the cells of the daughter colony are oriented with flagella pointing inward; then they rotate so that the flagella end up on the outside. Eventually the young colonies escape by rupture of the parent colony. In sexual reproduction a few cells differentiate into anisogametes: eggs containing stored food and microgametes that divide into a bundle of flagellated sperm cells, which leave the parent colony and seek egg cells in other colonies. The zygote is adapted for overwintering: dormant within a hard protective shell, it survives freezes that kill adult colonies and in spring multiples within its shell to form a small colony that is then set free.

6.19. What are the characteristics of euglenids?

Euglenids, classified as protozoans in order Euglenida or as plants in phylum/division Euglenophyta, are probably the closest of all living protists to the common ancestor of plants and animals. The anterior end of the elongate body (Fig. 6.2c) is invaginated as a *reservoir,* from the bottom of which arise one long and one very short flagellum. Some euglenids have a rigid pellicle and can swim only by propellerlike rotations of the longer anterior flagellum, but those that can change body shape also creep by wormlike bodily contractions and extensions. Euglenids possess chloroplasts with chlorophyll types *a* and *b,* which are characteristic of green algae and land plants. Their red *stigma* is not truly an eyespot, but a shield over a photosensitive area that permits orientation toward light.

Table 6.1   Classification of Animal-like Protists*

Monophyletic Classification	Polyphyletic Classification
PHYLUM PROTOZOA	PHYLUM SARCOMASTIGOPHORA†
SUBPHYLUM SARCOMASTIGOPHORA (with pseudopodia and/or flagella)	
CLASS MASTIGOPHORA (flagellates)	SUBPHYLUM MASTIGOPHORA
Subclass Phytomastigophorea (mainly autotrophic groups also classified as separate phyla/divisions)	CLASS PHYTOMASTIGOPHOREA
Subclass Zoomastigophorea (heterotrophic flagellates not placed in any group of phytoflagellates; include types intermediate between flagellates and sarcodines, and between flagellates and sponges)	CLASS ZOOMASTIGOPHOREA
CLASS SARCODINA (having pseudopodia)	SUBPHYLUM SARCODINA
Subclass Rhizopodea (pseudopods lack central filament)	SUPERCLASS RHIZOPODA
Subclass Actinopodea (pseudopods have central filament)	SUPERCLASS ACTINOPODA
CLASS OPALINATA (possible intermediates between flagellates and ciliates)	PHYLUM OPALINIDA OR SUBPHYLUM OPALINATA
SUBPHYLUM SPOROZOA (parasitic; spores lack polar filaments)	PHYLUM APICOMPLEXA (Sporozoa)
SUBPHYLUM CNIDOSPORA (parasitic; spores have polar filaments)	PHYLUM MYXOZOA (Cnidospora)
	PHYLUM MICROSPORA (Cnidospora)
SUBPHYLUM CILIOPHORA (ciliated at some time of life; complex infraciliature; two kinds of nuclei)	PHYLUM CILIOPHORA
CLASS CILIATA (lumpers consider all ciliophores to be homogeneous enough to be grouped in one class)	

*Some minor phyla or subphyla (mostly parasitic) have been omitted.
†Some "splitters" further subdivide the Sarcomastigophora into PHYLUM MASTIGOPHORA and PHYLUM SARCODINA.

Within this mainly freshwater group are a number of colorless, heterotrophic forms. Some autotrophs feed saprobiotically in the dark. Furthermore, if the common, autotrophic *Euglena* is abused with heat or chemicals or kept in the dark for a long enough time, its chloroplasts disintegrate permanently, so that all its descendants are obligate heterotrophs. Some such event may have originated the genus *Astacia*, an apparent twin of *Euglena*, except for lacking chloroplasts. Still other euglenids ingest solid food, even other euglenids. The colorless *Peranema* can dilate its cytostome widely enough to ingest prey nearly as large as itself; its anterior *rod organ* (made of two parallel rods) is protruded to anchor the prey for swallowing, or if the latter is too large to be ingested whole, the rod organ pierces it and its contents are sucked out. With such existing heterotrophic species as *Astacia* and *Peranema*, and the previously mentioned biochemical evidence of the close kinship of *Euglena* with the trypanosomid zooflagellate *Crithidia*, it seems likely that animal-like protists classified as protozoa *in sensu stricto* originated from a euglenid phytoflagellate that lost its chloroplasts and developed an appetite, setting out upon that unrelenting hungry search that even today characterizes the animal world.

6.20. Are "true" protozoans monophyletic or polyphyletic?

Taxonomists seem to fall psychologically into two persuasions: (1) "lumpers," who stress similarities and conceive of fewer but larger natural groupings of organisms, and (2) "splitters," who stress differences and recognize more numerous and more homogeneous groupings. Lumpers subscribe to the *monophyletic concept*, which holds that all animal-like protists (i.e., excluding those discussed above) may have arisen from one common ancestor, probably a heterotrophic euglenid, which not only diversified into the various groups of protozoans, but also provided the ancestry of multicellular animals. Since the major groups of modern protozoans seem to be interlinked by a number of intermediate forms, there seems no compelling reason to subdivide phylum Protozoa into a number of separate phyla. Splitters subscribe to the *polyphyletic concept*, which holds that the major groups of protozoans came from *different* moneran or protistan ancestry and so belong in separate phyla, or that even if they were of common ancestry, they have diverged so greatly that they cannot legitimately be placed in a single phylum. The two means of classification are shown in **Table 6.1.**

6.21. What are some important zooflagellates?

These include the rhizomastigidans, choanoflagellates, kinetoplastids, diplomonads, trichomonads, and hypermastigidans.

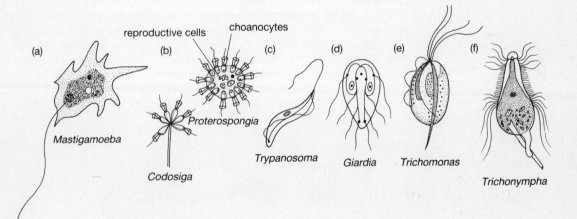

Fig. 6.3 Zooflagellates. (*a*) *Mastigamoeba*, a rhizomastigidan with both pseudopodia and flagellum. (*b*) Choanoflagellates: the colonial *Codosiga* and *Proterospongia*, with collar cells (choanocytes) as found in sponges. (*c*) *Trypanosoma*, a kinetoplastid, with one flagellum free and the other forming an undulating membrane. (*d*) *Giardia*, a diplomonad with two nuclei and bilaterally arranged flagella. (*e*) *Trichomonas*, a parasitic trichomonad; note rigid longitudinal rod, the axostyle. (*f*) *Trichonympha*, a hypermastigidan with numerous flagella, a mutualistic symbiont in the guts of wood-eating insects.

6.22. What are the rhizomastigidans?

Rhizomastigidans are permanently amoeboid, having one to many flagella, which they may reabsorb so that they look like sarcodines. Most occur in fresh water, others are symbionts (e.g., *Histomonas meleagris* causes blackhead in poultry). *Mastigamoeba* (Fig. 6.3*a*) is a representative genus. *Multicilia* (Fig. 6.4*d*) bears numerous short flagella, each with its own separate kinetosome; it may represent a possible transition form between flagellates and opalinids, ciliates, or both.

6.23. What are the characteristics of choanoflagellates?

Choanoflagellates have a single flagellum surrounded by a collar of long microvilli (Fig. 6.3*b*). They may be solitary or colonial. In the stalked colony *Codosiga*, all *zooids* (individuals in a colony) are alike, but in *Proterospongia*, the collar cells occur on the surface of a gelatinous matrix and amoeboid cells are found within the matrix, an arrangement reminiscent of sponges, the only other group of animals having collar cells. Sponges may well have descended from choanoflagellate ancestry.

6.24. What are the characteristics of kinetoplastids?

Kinetoplastids have a *single, large mitochondrion* with a DNA-containing *kinetoplast* within it. Most are parasitic members of family Trypanosomatidae, including *Trypanosoma, Leishmania,* and *Crithidia*. The latter is a gut parasite of insects. *Leishmania* attacks the lining of human blood vessels, destroying blood cells in a disease called kala azar; in tropical America, *L. brasiliensis* causes a disfiguring skin disease. Leishmanias are spread by dogs, bloodsucking flies, and interpersonal contact.

Trypanosoma species (Fig. 6.3*c*) have a single anterior flagellum, attached along the length of the body by an undulating membrane. They are transmitted by bloodsucking invertebrates: those parasitic in fish, amphibians, and reptiles are transmitted by leeches; those parasitic in mammals, by insects.

> **Example 7:** *Trypanosoma lewisi,* spread among rats by rat fleas, inhabits the bloodstream asymptomatically. In Africa, several species occur in native mammals without being pathogenic, but when transmitted to livestock or humans by the bite of tsetse flies *(Glossina)*, they cause fatal diseases: ngana in livestock and sleeping sickness in humans. *T. gambiense* produces Gambian sleeping sickness with prolonged coma and slow death; *T. rhodesiense* causes Rhodesian sleeping sickness that brings death in a few weeks. In tropical America and occasionally in the southwestern United States, *T. cruzi* is transmitted to humans from wood rats, armadillos, and other mammals by the bite of the "kissing bug" that favors the lips of sleepers; the resulting Chagas's disease is apt to be chronic and presents such a bewildering array of symptoms that it may elude diagnosis. *T. cruzi* often invades the heart muscle and causes death by congestive heart failure. The recurrent symptoms that afflicted Charles

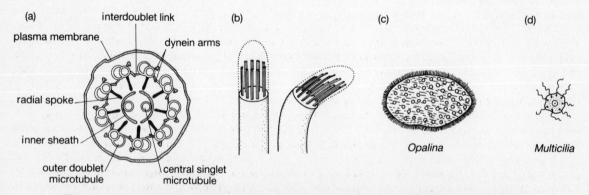

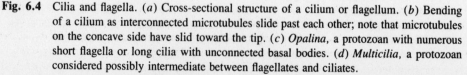

Fig. 6.4 Cilia and flagella. (*a*) Cross-sectional structure of a cilium or flagellum. (*b*) Bending of a cilium as interconnected microtubules slide past each other; note that microtubules on the concave side have slid toward the tip. (*c*) *Opalina*, a protozoan with numerous short flagella or long cilia with unconnected basal bodies. (*d*) *Multicilia*, a protozoan considered possibly intermediate between flagellates and ciliates.

Darwin during much of his life seem consistent with Chagas's disease, which he may have contracted while ship's naturalist on *HMS Beagle* circumnavigating South America. His log tells how he incurred a painful insect bite while engaged in inland exploration, which could have been the moment of transmission and the beginning of lifelong debility.

6.25.   What are the characteristics of diplomonads?

Diplomonads are bilaterally symmetrical, with one nucleus on each side, each associated with four flagella (Fig. 6.3*d*). Most are parasites, including *Giardia*, which can cause dysentery (giardiasis) when ingested in fecally polluted water.

6.26.   What are the characteristics of trichomonads?

Trichomonads include *Trichomonas vaginalis,* a parasite of the human vagina and male reproductive tract, which has four anterior flagella and a fifth bordering an undulating membrane; a rigid rod *(axostyle)* extends lengthwise (Fig. 6.3*e*). *T. foetus* causes abortion in cattle.

6.27.   What are the characteristics of hypermastigidans?

Hypermastigidans such as *Trichonympha* bear hundreds of flagella that spring from kinetosomes fused into rings or spiral or longitudinal rows (Fig. 6.3*f*). Mutualistic symbionts in the guts of termites and cockroaches (see Example 4), *Trichonympha* digests cellulose. The relationship is essential for insects that feed exclusively on chewed wood, for they themselves possess no enzymes capable of breaking down cellulose to sugars. Although we may not appreciate this partnership when our houses are at stake, in nature it is very important in recycling fallen trees.

6.28.   What is the nature of flagella and cilia?

If it is true that a flagellated protist was the common ancestor of all protozoans, flagella must have appeared a good deal earlier in the history of life than cilia. In fact, flagellated sperm occur in most animals and plants, while cilia occur only in animals, including ciliophorans and most metazoans. Both types of motile fibrils are *identical* internally (Fig. 6.4*a*), suggesting that cilia arose from shortened flagella. Both have a central pair of microtubules connected by spokes to a ring of nine peripheral microtubule doublets. Microtubules always consist of chains of identical globular protein units *(tubulin)* twisted into a hollow helix. A pair of arms form cross bridges from each doublet to the next; these arms consist of an enzyme with ATPase activity and may cause the doublets to slide past each other by a ratchetlike action (Fig. 6.4*b*). A ratchet action involving oscillating cross bridges also occurs in muscle tissue but involves quite different proteins, so that the two mechanisms are merely convergent. Nonmotile cilia lacking the central pair of tubules occur widely in metazoan sense organs, such as the rods and cones of the vertebrate retina, and the "hair cells" of the inner ear.

The basal bodies *(kinetosomes)* of flagella and cilia are also identical, short cylinders containing a ring of nine *microtubule triplets,* which also happens to be the structure of centrioles. In fact, in sperm cells and some flagellates, the basal body *is* a centriole. Flagellar kinetosomes are usually independent of one another, but may sometimes fuse into a ribbonlike structure, although the flagella they bear always remain separate. In ciliophorans, the cilia themselves may fuse as "membranelles" or bristles, and even when unfused, they arise from kinetosomes that are interconnected by a network of fibrils *(kinetodesma)* that form a complex *infraciliature.*

A flagellum may be as much as 150 μm long and undulates in one or two planes, propelling water parallel to its own main axis. If the wave of undulation begins at the base and propagates toward the tip, the flagellum exerts a pushing force; if undulation proceeds from tip to base, the flagellum pulls the cell along like an airplane propeller.

Cilia appear to differ from flagella only in being shorter (5–20 μm long) and usually more numerous. Their stroke pattern differs, but this may be mainly an artifact of length. They work like oars, with a rigid power stroke, followed by a recovery stroke in which the cilium rotates counterclockwise, flat to the cell surface. The two strokes are reversed when ciliates back up, but the beat cannot be reversed in the ciliated epithelia of metazoans. The effect of cilia is to propel water parallel to the surface to which they are attached. When many cilia are present, they must beat in coordination, but the mechanism of this remains obscure. Metachronous waves of ciliary contraction that follow each other down the length of a paramecium may be due to an action potential traveling along the plasma membrane; however, coordinated swimming can also be achieved in "ghost" paramecia killed by detergent, then treated with magnesium ions and ATP.

Whatever the mechanism, it is very effective, for some larger ciliates can pelt along at more than 2 mm/second (s), outdistancing flagellates like greyhounds versus dachshunds.

6.29.   Where do opalinids fit in?

Opalinids (Fig. 6.4c) are gut commensals or parasites, usually of frogs and toads, and are shed as fecal cysts that may be ingested by tadpoles. They are zoologically interesting because they seem intermediate between flagellates such as *Multicilia* (6.4d) and ciliates such as *Paramecium* (Fig. 6.7a). Opalinids are covered with motile fibrils that have been described as either cilia or shortened flagella. The "cilia" arise singly from kinetosomes arranged in longitudinal rows; unlike those in ciliates, these kinetosomes are not interconnected by kinetodesma, and an infraciliature of microfilaments is lacking. Opalinids possess from two to many nuclei that are *all alike,* whereas ciliates have two types of nuclei: a small micronucleus and a large macronucleus. Opalinids divide like flagellates, longitudinally, whereas ciliates divide transversely. Ciliates have a mouth (cytostome) opening through the pellicle, through which most ingest solid food; opalinids lack a cytostome and form pinocytic vesicles at the surface. Sexual reproduction in opalinids is by fusion of flagellated gametes that emerge from the infective cysts; such gametes are unknown among ciliates, which exchange haploid nuclei bidirectionally in an act of conjugation.

6.30.   What are sarcodines?

All sarcodines produce flowing cytoplasmic extensions, called *pseudopodia,* which are used for prey capture and for locomotion (in benthonic forms). Cytoplasmic streaming occurs in all sarcodines and is involved in the extension and shrinkage of pseudopods, but contact with a substratum is required for amoeboid movement. Sarcodines are classified according to the nature of their pseudopods, which may be soft *lobopods* or *reticulopods,* or rigid *axopods.* Some also bear flagella or produce flagellated gametes or do both. Sarcodines with flagella as well as pseudopods occur in both the Amoebida and the Heliozoa (helioflagellates). *Mastigamoeba* and other amoeboid zooflagellates, considered above as Rhizomastigida, can equally well be classified as flagellated amoebas and grouped with the sarcodines, for they are true intermediate forms.

Sarcodines include marine, freshwater, and soil forms; planktonic and benthonic types; and some parasitic species (Fig. 6.5). They are either spherically or radially symmetrical, or asymmetrical with no constant body form. Reproduction is usually asexual, by binary fission, or by multiple fission in multinucleate types, but sexual reproduction also occurs in some groups, by fusion of haploid nuclei (heliozoans) or biflagellated anisogametes (foraminiferans).

In overall structure, sarcodines appear to be the simplest protozoans (but not the most primitive, which distinction probably falls to the euglenids), with few specialized organelles other than those characteristic of eukaryotic cells in general, plus a contractile vacuole for osmoregulation. However, this apparent simplicity may be illusory, for the majority secrete skeletal structures (called *tests*), some of which are marvels of beauty and complexity, such as the siliceous (glass) tests of radiolarians. We have little idea how such symmetrical test architecture, which is also species-specific, can be encoded by the genes of such apparently simple, representative cells, but this developmental riddle is ubiquitous in biology: we have scant knowledge of how genes control the development of form (morphogenesis) in *any* organism. Furthermore, amoeboid movement itself is no mean feat and is still only partly understood today despite long acquaintance with *Amoeba proteus* as a microscopic subject.

6.31.   What is amoeboid movement?

Amoeboid movement involves formation of cytoplasmic extensions *(pseudopods)* into which the internal cytoplasm flows, lengthening the pseudopods so that the cell creeps in the direction of pseudopodial growth (Fig. 6.5c). The flow of sol-like (fluid) internal cytoplasm *(endoplasm)* can be reversed, so that a pseudopodium can also shrink and disappear or the cell can change direction. The first sign of a new pseudopod is a localized thickening of the *ectoplasm* (external, gel-like cytoplasm) as a *hyaline cap.* The flow of endoplasm is directed toward this cap, just beneath which the endoplasmic stream sprays outward like a fountain. At the outer edges of this *fountain zone,* the flowing cytoplasm changes colloidal state from sol to gel (rather like Jell-O℠ setting) and joins the ectoplasm. Thus the pseudopod lengthens, supported by an outer tube of gelated cytoplasm. The ectoplasm itself actually does not move, but grows at the leading tips and is destroyed at the posterior end, where it reverts to a sol state and moves inward as endoplasm once more. Small pseudopodia project downward beneath the amoeba, holding it free of the substratum so that the animal creeps on the tips of its pseudopods, rather than flowing blobbishly.

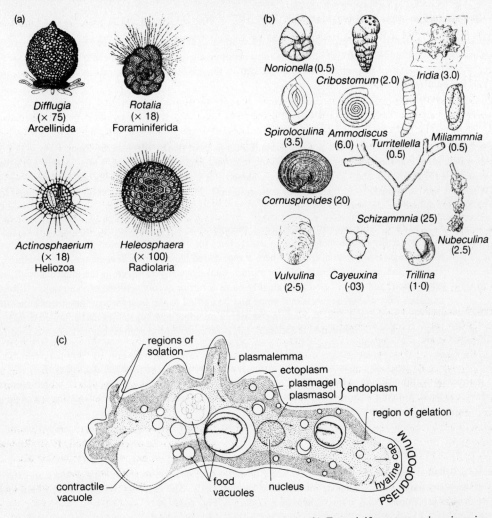

Fig. 6.5 Sarcodines. (*a*) Representatives of four orders. (*b*) Foraminiferan tests, size given in millimeters. (*c*) *Amoeba proteus*, structure and locomotion; internal cytoplasm moves to the surface at the leading tip of a pseudopod, undergoes gelation, and moves backward to posterior regions of solation, where it again liquefies and passes into the interior. [*From Storer et al.; (a) after Wolcott*, Animal Biology, *(b) after Galloway*, A Manual of Foraminifera, *(c) after Mast.*]

6.32.   But what makes the endoplasm flow directionally?

Evidently, the endoplasm is squeezed forward by the formation of a meshwork of protein microfilaments, only 3–6 nm in diameter [1 nanometer (nm) = 1/1000 micrometer (μm) or 1/1,000,000 millimeter (mm)], made up of a chain of globular molecules of the protein *actin*. If an amoeba is treated with a reagent (cytochalasin B) that blocks the activity of actin microfilaments, cytoplasmic streaming stops. These actin microfilaments are more abundant at the rear than at the leading edge, suggesting that they push the endoplasm forward from behind. Interestingly, the contraction of vertebrate skeletal muscle fibers results from an interaction between actin filaments and those of another protein, *myosin*. Myosin filaments have been identified in *Amoeba proteus* and may occur generally in amoeboid cells throughout the living world. Myosin molecules have projecting "heads" that can split ATP and form cross bridges with the actin filaments, so that in muscle fibers the two types of microfilaments slide past each other, ratchet-fashion. In amoeboid cells this sliding-filament interaction may operate to bring about directional cytoplasmic streaming. There may in fact be only two basic mechanisms of movement in the animal world: ratchet sliding of microtubules, as seen in cilia and flagella, and ratchet sliding of actin and myosin microfilaments during amoeboid movement and muscle contraction.

6.33.   What sarcodines have soft pseudopods?

Rhizopodeans ("root-footed") have pseudopods that lack supportive axial rods and may be bluntly rounded *(lobopods)*, pointy and sometimes branched *(filopods)*, or branching and anastomosing (uniting branches) into a network *(reticulopods)*. When slime molds are classified as animals, they are placed in this taxonomic group, which also includes amoebids, testacids, and foraminiferans.

6.34.   What are the characteristics of amoebids?

Amoebids are usually uninucleate, lack a skeleton, and change form freely, creeping by lobopods or filopods. Most live in fresh water, including the giant amoeba, *Pelomyxa*, that may attain a diameter of 5 mm. A few are marine, and many are symbionts. The tiny amoeba *Vampyrella* is aptly named, for it punctures algae cells and sucks out the contents. *Endamoeba* species are commensal in the intestine of certain insects. The genus *Entamoeba* includes the most important sarcodine parasite of humans: *E. histolytica*.

> **Example 8:**   *Entamoeba histolytica* is the causative agent of *amoebic dysentery*. It is ingested as cysts in fecally contaminated food or water, lives for a while in the colon phagocytizing bacteria, and then invades the intestinal lining, where it ingests red blood corpuscles, creates slow-healing ulcerous lesions, and provokes severe and sometimes fatal dysentery. Lodged in the intestinal wall, it resists eradication, and the infected individual remains a carrier even when free of symptoms. However, the amoebas may also penetrate small veins and be carried to the liver, where they form large abscesses that may prove fatal. Self-protection consists of avoiding raw vegetables and unboiled drinking water in areas where *E. histolytica* is endemic, but even cooked foods can be contaminated if food handlers are not periodically screened for cysts in their stools.

Other *Entamoeba* species live nonpathogenically in the human mouth *(E. gingivalis)* and colon *(E. coli)*, and still others live harmlessly within nonhuman hosts. The free-living amoeba *Naegleria* has a life cycle in which amoeboid and flagellated forms alternate; when it accidentally enters the human body, *Naegleria* can cause a fatal disease of the nervous system.

6.35.   What characterizes the testacids?

Testacids are shelled amoebas that live in fresh water or damp soil, or upon mosses. They produce an external shell (test) with a single opening for extension or retraction of the lobopods. The test may be siliceous or chitinoid, or an organic cement to which minute sand grains adhere (see *Difflugia*, Fig. 6.5a).

6.36.   What are the characteristics of foraminiferans?

Foraminiferans ("porebearers") are very abundant in the sea (*Rotalia*, Fig. 6.5a). Although most are benthonic, *Globigerina* and related forms are so numerous that their calcareous tests (of calcium carbonate) accumulate on seafloors as *Globigerina ooze*, which carpets some 30 percent of all ocean basins. "Forams" capture prey by a network of soft, interconnecting cytoplasmic filaments *(reticulopods)*, which in large forams is sticky enough to ensnare even small crustaceans, which are quickly paralyzed and attacked by enzymes even before being engulfed into food vacuoles.

Foram shells are characterized by numerous pores *(foramina)* through which the cytoplasm extends and from which the group takes its name. The shells assume many distinct forms, some snail-like, some *unilocular*, with a single chamber, and others *multilocular*, with additional chambers added to the first *(proloculum)* as the animal grows (Fig. 6.5b). The fossilized shells of giant forams over 1 cm in diameter formed nummulitic (from the genus *Nummulites*) limestone, quarried for Egyptian pyramids, and even today some tropical forams have shells at least 0.5 cm across. Forams are economically important: (*a*) Their abundance from the Cambrian Period on have made them important contributors to marine limestones, chalk, and marble; (*b*) since characteristic assemblages of fossil forams mark each geologic period, they are among the most useful indicator species to determine the relative age of sedimentary strata and are relied on for much stratigraphic correlation in petroleum exploration.

Foraminiferan life cycles typically alternate sexual and asexual generations. The shells produced by individuals of each generation may be identical or different. In the latter case, the sexual individual *(gamont)* secretes a shell with a large proloculum; the asexual one *(schizont)*, a shell with a small proloculum. The mature gamont liberates a large number of biflagellated anisogametes. These fuse in pairs, forming diploid zygotes that mature into schizonts. The schizont becomes multinucleate and undergoes multiple fission to produce small, uninucleate amoebas that become gamonts.

6.37.   What sarcodines have rigid pseudopods?

Actinopodeans ("ray-footed") possess stiff, radiating *axopods,* each supported by an axial rod. According to species, this rod consists of anywhere from three microtubules to a spiraling array of hundreds; these microtubules can lengthen or shorten speedily, allowing rapid axopodial extension and retraction. Large radiolarians with axopods fully spread can span several millimeters, and colonial forms may reach 20 cm in length. Most actinopodeans are planktonic, with their rays providing buoyancy by way of expanded surface area, but when they are on the bottom, the tips of their axopods adhere to the substratum and they roll merrily along by shortening the axopods in front and lengthening those behind. Mainly, the axopods serve to capture food, as small organisms stick to their surface and are either surrounded there by a food vacuole or carried to the central body by shortening of the axopod. Radiolarians and helizoans are actinopodeans.

6.38.   What are the characteristics of radiolarians?

Radiolarians (e.g., *Heleosphaera,* Fig. 6.5a) produce exquisite tests, usually of *silica* (but in one group, strontium sulfate), that are so resistant to dissolution that they accumulate as *radiolarian ooze* on deeper seafloors (where carbonic acid dissolves away the calcareous foram tests); today covering some 5 percent of the total ocean floor, these tests are also found in fossil deposits as old as the Precambrian. The cytoskeleton of radiolarians may also include siliceous spicules and always features a porous, spherical, *chitinoid capsule* that separates the inner cytoplasm from a frothy, vacuole-filled outer region that usually contains numerous symbiotic dinoflagellates. Radiolarians reproduce asexually by binary fission; in some species the test too divides in half, while in others one daughter cell gets the entire test and the other grows a new one. Although radiolarians also produce flagellated cells, the sexual role of these is uncertain.

6.39.   What are the characteristics of heliozoans?

Heliozoans ("sun animals") are usually freshwater plankton, but some are benthonic and may even be attached by stalks. They are spherical, with axopods radiating in all directions; if they form a skeleton at all, it is usually a siliceous mosaic sphere, or a set of radiating needles, or a beautiful spherical lattice of pectin (see *Actinosphaerium,* Fig. 6.5a). They are often multinucleate and reproduce by multiple fission. Sexual reproduction is known in some genera, such as *Actinophrys,* which withdraws its axopods and encysts, then divides within the cyst into two daughter cells that undergo meiosis, extruding all but two haploid nuclei, which then fuse to form a zygote nucleus.

6.40.   What are sporozoans?

Sporozoans are parasitic protozoans that produce an infective stage known as a *spore.* During most of their life cycle they lack special locomotory structures, but may move by bending or gliding. Once thought to constitute a single phylum, or even a single class in phylum Protozoa, these parasites are now classified into several quite different groups, of which only two, the *apicomplexans* and the *cnidosporans,* will be considered here.

6.41.   What are the apicomplexans?

Phylum (or subphylum) *Apicomplexa* (= Sporozoa) includes some exceedingly dangerous, mostly intracellular, parasites that afflict human and other vertebrate hosts, as well as certain invertebrates, mainly insects, which may serve as carriers as well as hosts. Malaria still kills or debilitates millions of people a year, and congenital toxoplasmosis is considered responsible for 2 percent of all mental retardation in the United States. Coccidiosis is often fatal to young poultry, and piroplasmids transmitted by ticks cause red water fever in cattle.

The apicomplexans—which include gregarines, eimeriids, and hemosporidians—derive their name from an *apical complex* of organelles at the anterior end. The complex consists of rings, bands, and microtubules that may serve to penetrate host cells.

The life cycle (here generalized) starts when the infective stage *(sporozoite),* free or within the spore (an oocyst formed around the zygote), enters the host. This sporozoite transforms into a feeding stage *(trophozoite)* that eventually undergoes multiple fission *(schizogony)* to produce a number of *merozoites* that are sometimes nearly as small as bacteria. Each merozoite grows into a new trophozoite, so that this asexual cycle can go on indefinitely. Persons who survive the initial acute phase of malaria may become asymptomatic except for occasional relapses, but their blood remains infectious and they must never serve as blood donors. Eventually, a few merozoites develop instead into *gamonts,* which according to species will mature into either isogametes or anisogametes (eggs and sperm). (In some sporozoans, the gamonts will not mature unless they are taken into the body of another host, such as a mosquito.) The gametes fuse, and the zygote encloses itself in an *oocyst*, undergoing meiosis and multiple fission *(sporogony)* to produce a cluster of sporozoites.

6.42.   What are the characteristics of gregarines?

Gregarines are parasites of invertebrates such as annelids and insects. Those that are intracellular parasites are only a few micrometers long, but those that are extracellular parasites in the host's gut or body cavities include enormous wormlike forms up to 16-mm long. The host becomes infected by ingesting a spore containing eight sporozoites produced by meiotic and mitotic divisions of the zygote. Each sporozoite bears an anterior attachment structure (hook, sucker, knob) with which it attaches to the host's intestinal wall. It may remain in the gut or migrate elsewhere. It grows into a trophozoite, which may or may not reproduce by schizogony. Eventually, the trophozoite transforms into a gamont, and the gamonts fuse in pairs. The zygote secretes a hard-walled oocyst (spore), within which schizogony will occur. These cysts or spores are expelled with feces.

6.43.   What characterizes the eimeriids?

Eimeriids include the agents of coccidiosis (mainly *Eimeria*) and toxoplasmosis *(Toxoplasma)*. The life cycle involves a single host. The sporozoites are transmitted within the oocyst. Those responsible for *coccidiosis* invade the intestinal lining of calves and poultry, causing dysentery. *Toxoplasma* species infect a number of birds and mammals, invading various tissues. Humans may become infected from cat droppings or undercooked beef. A woman who contracts *toxoplasmosis* in early pregnancy may transmit it to the fetus via the placenta, sometimes causing mental retardation. An estimated third of all human beings have sustained *Toxoplasma* infections, mostly congenitally.

6.44.   What characterizes hemosporidians?

Hemosporidians have a two-stage life cycle, with asexual sporogony taking place in a vertebrate host and fertilization and schizogony in an invertebrate host, a bloodsucking insect. The infective sporozoites break out of the oocyst before being transmitted to the vertebrate host by the bite of the insect host. The malarial parasite, *Plasmodium*, is chief culprit within this group (Fig. 6.6). *Plasmodium* species afflicting birds are carried by mosquitos of the genus *Culex*, whereas mammalian malaria is transmitted by *Anopheles* mosquitos.

> **Example 9:**   Humans are host to four species of *Plasmodium*, of which *P. falciparum*, the agent of malignant tertian malaria, is most serious. The life cycle closely follows the generalized scheme given above. Infective sporozoites migrate from the mosquito's salivary glands, down her salivary tube into the human bloodstream. Initially, they invade cells in organs such as the liver, multiply, and reenter the circulation as tiny merozoites, each of which penetrates a red blood cell (erythrocyte). Within the erythrocyte, the parasite grows as an amoeboid trophozoite, which shortly fills the interior of the corpuscle and undergoes *schizogony*. The infected red cells burst synchronously, each releasing 16 merozoites, and the victim suffers a chill-fever episode. These episodes occur every 48 hours (except for *P. malariae*, which has a 72-hour cycle). Eventually some merozoites grow into gamonts of two sizes (macrogamonts and microgamonts) that cannot mature further unless ingested by the anopheline mosquito. Within her stomach, the macrogamonts become eggs and the microgamonts produce sperm, fertilization occurs, and the motile zygote *(ookinete)* penetrates the gut wall to grow and form an oocyst in the body cavity. *Sporogony* takes place, the oocyst ruptures, and the freed sporozoites migrate to the host's salivary glands to await transmission.

Because mosquitos have developed resistance to insecticides and *Plasmodium* to antimalarial drugs, malaria remains a persistent threat, especially in tropical regions. Interestingly, some human populations native to regions where *P. falciparum* is endemic carry a gene which, although it produces the sometimes lethal sickle-cell anemia in the homozygous state, improves the malaria resistance of heterozygotes, who at most suffer from the less serious "sickle-cell trait." The hemoglobin of homozygous individuals is 100 percent of the mutant type that under hypoxic (oxygen-poor) conditions distorts the erythrocytes so that they block capillaries and are attacked by defensive white corpuscles. In heterozygotes only 50 percent of the hemoglobin in each red cell is of the sickling type, and the erythrocytes therefore sickle only under extreme conditions. *Plasmodium* does not flourish in corpuscles containing sickle hemoglobin, so this otherwise detrimental allele is selected for by the malaria resistance of heterozygous individuals, especially in areas where malignant tertian malaria poses a threat to human survival.

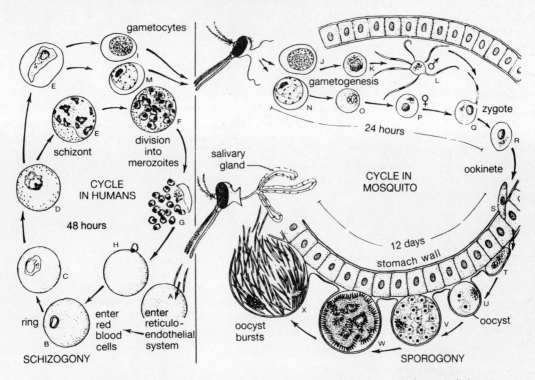

Fig. 6.6   Life cycle of *Plasmodium vivax*, a hemosporidian apicomplexan. *Left*, stages in human
host: sporozoites in mosquito's salivary gland move into human bloodstream; after initial
multiplication in liver cells, merozoites invade red blood corpuscles; each grows into a
schizont that produces 16 new merozoites by multiple fission (schizogony); these escape
by bursting the red blood cell and invade new corpuscles; some merozoites instead grow
into gametocytes that mature further only if ingested by a mosquito. *Right*, stages in
anopheline mosquito: gametocytes (gamonts) transform into gametes in the mosquito's
stomach; fertilized egg (ookinete) migrates into her body cavity, grows into large oocyst
within which multiple fission (sporogony) produces many sporozoites that break free
and migrate to the mosquito's salivary glands. (*From Storer et al.*)

6.45.   What characterizes the second type of sporozoans, the cnidosporans?

Phylum (or subphylum) *Cnidospora* is often divided into two separate phyla (or subphyla), the *myxosporidians* and
microsporidians, which are extracellular and intracellular parasites, respectively. Both are characterized by having a spore
with one or more *polar capsules*, each with a coiled filament that can be shot out to anchor the spore to the host's intestinal
lining. These parasites invade various tissues or body cavities, mostly of arthropods and fishes. Heavy myxosporidian
infections cause severe fish kills. Microsporidians cause serious diseases of economically valuable honeybees and silk-
worms.

6.46.   What are ciliophorans?

Characteristics that distinguish ciliophorans (ciliates) from other protists include *cilia* conjoined by a fibrous *infraciliature*, two nuclei (*macronucleus* and *micronucleus*), and often thread-ejecting *trichocysts*. The majority of ciliates are
free-living and abundant in both marine and freshwater habitats. They survive dry periods by encysting; in fact, cysts of
the common genus *Colpoda* have been found alive after 38 years' storage! In the rainy season, nearly any puddle contains
crowds of little speed demons, dashing about in all directions, leaving no doubt that these are not only among the most
successful, but also the *speediest* of all protozoans.

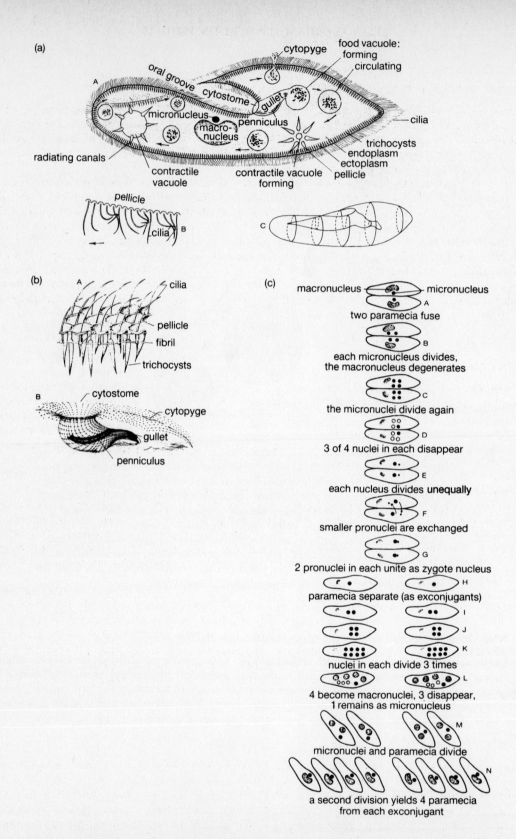

Fig. 6.7 *Paramecium caudatum.* (*a*) Structure: A, the entire ciliate; B, detail showing coordinated beating; C, three-dimensional form. (*b*) Structure of cell surface: A, pellicle with cilia, infraciliature, and trichocysts; B, infraciliature in the mouth region. (*c*) Sexual reproduction by conjugation followed by fission. [*From Storer et al.; (b) after Lund.*]

78

6.47.   What is the body architecture of ciliates?

Figure 6.7*a* depicts the structure of *Paramecium* as a representative ciliophoran. At least during some time of life, ciliates possess cilia that are interconnected by an infraciliature of microfilaments, including those known as kinetodesma that interconnect their basal bodies (Fig. 6.7*b*). Some cilia may be fused into leglike bristles *(cirri)* or membranelles surrounding the mouth. Although ciliates usually swim freely and rapidly by ciliary locomotion, certain creeping types move their fused, bristlelike cilia in coordination, like little legs.

Ciliates have a constant body form, imposed by their elastic *pellicle*, which still allows bending or temporary rounding-up by means of bundles of contractile filaments *(myonemes)* that lie within the pellicle. In sessile, stalked types such as *Vorticella*, which has a bell-shaped body like a tulip on its stem, the myonemes, twisted together into a large spiral fiber, extend the length of the stalk and can instantaneously coil it into a corkscrew. A ciliate's *cytostome* occupies a constant position, usually at the base of a depression, or *oral groove*, and opens into a *cytopharynx*, at the base of which food vacuoles form. Egestion takes places at a definite anal pore *(cytopyge)*, and the *contractile vacuoles* (typically two) are also fixed in position, each at a discharge pore opening through the pellicle. Ciliates are the only protozoans that consistently have two kinds of nuclei: a small, diploid *micronucleus* devoted to reproduction and a large, polyploid *macronucleus* with multiple chromosome sets that augment the cell's capacity for making vital proteins; in fact, the macronucleus may contain 500 times the DNA found in the micronucleus.

6.48.   How do ciliates reproduce?

Asexual reproduction typically consists of *transverse* binary fission across the rows of cilia, although budding and multiple fission occur in some. Sexual reproduction does not involve production of free-swimming gametes. Instead, meiosis is followed by a bidirectional exchange of haploid nuclei in an act of *conjugation* (Fig. 6.7*c*) between individuals, with disappearance of their separate plasma membranes and pellicles in a bridging region through which the nuclei will pass. By adherence of the oral regions, this conjugation bridge forms in a "kiss" that literally transforms both partners, since in the process of genetic exchange, each animal loses some alleles and gains others.

Genetic recombination can also be achieved without benefit of a sexual partner, by *autogamy*, in which the individual's micronucleus undergoes meiosis followed by nuclear mitosis, with disintegration of all the haploid nuclei but one. This one divides mitotically to form two genetically identical "gamete nuclei," which then fuse to form a new diploid nucleus. [If this seems ineffectual, review what you have learned about meiosis in Chapter 5. Only one allele of each gene pair passes into a gamete (in this case, a haploid nucleus), so if our ciliate has previously been heterozygous for some particular allelic pair (e.g., *Bb*), after autogamy and fusion of the remaining identical haploid nuclei it would not be heterozygous any more, but either *BB* or *bb*, which will alter its phenotype as the new genotype, multiplied within the new macronucleus, begins to affect the cell's metabolism.]

During both conjugation and autogamy, the macronucleus disintegrates and is later reformed by mitotic division of the new diploid nucleus. Conjugation seems to occur only between individuals of opposite mating type, which usually look just alike. However, in some ciliates, individuals that will engage in conjugation have a somewhat different appearance from those that will undergo only fission: they may be a bit smaller but still isomorphic, or some may be macroconjugants and others microconjugants. In this event, descendants of the macroconjugant get progessively smaller after each conjugation, while those of the microconjugant get larger, until all are of equal size. In the sessile *Vorticella*, the macroconjugant ("female") remains attached while the microconjugant ("male") breaks free of its stalk and swims off to find the female. After conjugation, the male simply disintegrates.

6.49.   What is the ciliates' mode of nutrition?

Most ciliates are holozoic, feeding on bacteria, algae, other protozoans, and even small metazoans. A number possess distinctive capsules known as *trichocysts* (Fig. 6.7*b*) from which threads are explosively discharged. The adaptive value of nontoxic trichocysts, such as those of paramecia, remains in question; they may serve defensively or help anchor individuals in feeding areas. The effect of toxic trichocysts *(toxocysts)* is dramatic, especially in the elephant-trunked *Dileptus:* a mere touch of its swaying "trunk" not only paralyzes the prey but may actually rupture its membrane and cause swift disintegration. When the trunk contacts a larger animal, such as the ciliate *Stentor* (Fig. 6.8*a*), it adheres and yanks out a piece of cytoplasm without killing the victim, which flinches as if hurt. The prey or morsel is then ingested by huge dilation of the cytostome, which lies at the base of the trunk, like an elephant's mouth. Watching *Dileptus* feed makes one very glad it is microscopic! Another raptorial ciliate with big ideas is *Didinium* (Fig. 6.8*a*), which can paralyze a paramecium larger than itself and then ingest it by prodigious mouth distention.

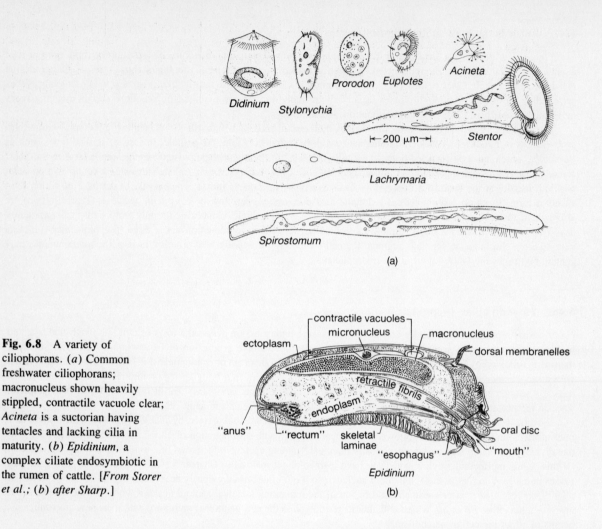

Fig. 6.8 A variety of ciliophorans. (*a*) Common freshwater ciliophorans; macronucleus shown heavily stippled, contractile vacuole clear; *Acineta* is a suctorian having tentacles and lacking cilia in maturity. (*b*) *Epidinium*, a complex ciliate endosymbiotic in the rumen of cattle. [*From Storer et al.; (b) after Sharp.*]

Not all ciliates are pint-sized predators; some are scavengers, invading the carcasses of small aquatic metazoans to disrupt the cells and ingest their contents; others are herbivorous and suck in entire strands of blue-green algae, such as *Oscillatoria*, distorting their bodies bizarrely in such gluttonous pursuit. Some ciliates are involved in symbiotic associations, either as host or as symbiont. *Paramecium bursaria* (green paramecium) plays host to endosymbiotic unicellular green algae (*zoochlorellae*); in experimental closed systems provided only with dissolved minerals and light, *P. bursaria* and its algae do well: the animal host produces enough carbon dioxide for the symbionts, and the latter synthesize enough organic materials to sustain both partners in this mutualistic association. Few ciliophorans are parasitic except for certain suctorians (which have tentacles that serve as mouths) that are endoparasitic in fish or other ciliates. A number of ciliates are ecto- or endocommensals, such as *Epidinium* (Fig. 6.8*b*), which lives in the rumen of cattle and has attained an epitome of internal structural complexity for a unicellular organism.

6.50. How are ciliophorans classified?

Variously considered to constitute a phylum or subphylum, ciliophorans are considered enough alike to be placed together in a single class, *Ciliata*, in which case the terms "ciliate" and "ciliophoran" become essentially synonymous. Some taxonomists abandon "Ciliata" as a taxon and simply recognize three or four classes of ciliophorans, while others retain the single class Ciliata to indicate the basic homogeneity of this group, despite differences in size, body form, and diet. The major subgroups are distinguished mainly on the basis of extent and type of ciliation, but since taxonomic schemes still differ, the following summary of major types represents a compromise:

1. *Kinetofragminophorans* bear *only simple cilia*, not compound ciliary organelles such as membranelles or bristles. *Dileptus* and *Didinium* are examples.

2. Cilia are present only on young *suctorians*, with just the infraciliature being retained throughout life. Adults are often sessile and stalked and have tentacles that adhere to passing prey (such as other ciliates). The prey are immobilized and their contents sucked out. A suctorian much smaller than a paramecium can hold several of the latter immobilized floating in the midst of its hapless victims like a miniature science fiction monster (see Fig. 6.8*a*, *Acineta*).

3. *Hymenostomatids* ("membrane mouth"; e.g., *Paramecium*) have uniform body ciliation, but possess *membranelles* along the oral groove.

4. *Peritrichs* (e.g., *Vorticella*) essentially lack body ciliation but have a row of cilia surrounding the mouth region *(peristome)*. They are typically bell-shaped and stalked and are often colonial.

5. *Spirotrichs* have a conspicuous set of membranelles *spiraling* around the mouth. They include *heterotrichs*, such as *Stentor*, which also have short cilia over the entire body; *hypotrichs*, such as *Stylonychia* and *Euplotes* (cf. Fig. 6.8*a*), with body ciliation reduced to ventral tufts forming bristles (cirri); and *entodiniomorphs*, such as *Epidinium*, which not only lack body cilia but also have quite restricted oral membranelles.

Chapter 7

The Lower Metazoans: Parazoa and Mesozoa

Sponges appear to have originated from a different protistan ancestor than that which gave rise to nearly all other multicellular animals, quite possibly a colonial choanoflagellate resembling today's *Proterospongia* (see Fig. 6.3*b*), since collar cells (*choanocytes*) with their distinctive ruff of extremely long microvilli are found only sporadically in animals other than these. *Phylum Porifera* ("pore bearers"), the sponges, is placed to one side of the kingdom Metazoa, as comprising members of *subkingdom Parazoa* ("beside animals"). Although their fossils indisputably occur in early Cambrian (and perhaps Precambrian) deposits, there is no evidence that in their 600 million-year history they have given rise to any line of descendants other than more modern sponges. A second phylum, *Placozoa* ("plate animals"), represented by a single species, has at least tentatively been placed with sponges in the Parazoa. Two enigmatic types of minute parasites, with very simple bodies but complex life cycles, are disputedly grouped in *subkingdom Mesozoa* and considered intermediate between ciliated protozoans and metazoans; however, some consider one or both types to be instead simplified, retrograde flatworms. All multicellular animals other than parazoans and mesozoans share common developmental patterns and have true tissues, and constitute *subkingdom Eumetazoa* ("true later animals").

7.1. What are the origins of multicellularity?

Well over 100,000 species of unicellular protists are still getting by quite contentedly, having forgone the temptations of multicellularity for a billion years or so. They occupy ecological niches that are unexploitable by larger organisms and can often survive adversity that would kill larger beings. In their smallness, protists represent great gene pools of countless individuals, whereas the total population of many metazoan species may not amount to even a few million individuals. Nevertheless, multicellularity has been profitable for the living world, for in the sense of diversity and number of *species*, multicellular organisms have dominated the scene, at least for the past 600 million years.

7.2. What mechanisms may produce multicellularity?

Two models have been developed to explain the development of multicellular organisms: the cohesion model and the syncytial model.

7.3. What is the cohesion model?

When a protozoan divides, the daughter cells say "goodbye" and amble away in different directions to seek their destinies. When a protozoan *colony* produces new zooids (individuals in a colony), the latter remain physically conjoined and ordinarily cannot wander off, even if they so wish. But when the fertilized egg of a metazoan starts to divide (into two, four, eight cells, etc.), these cells do not go off to pursue independent destinies. In fact, they resent the idea. This can be nicely demonstrated with sponges.

> **Example 1:** You can cut off a chunk of sponge and strain it through a finely woven cloth that separates the cells from one another and then watch the cells (in a small vessel of seawater) under the microscope. They soon become amoeboid, even if they were not previously so, and creep about restlessly, as if wondering what happened. They do not seem actually attracted to one other, but whenever they do come in contact, they stop moving and form little clusters, like sheep huddling against a cold, cruel world. This phenomenon is called *contact inhibition*.

Contact inhibition occurs widely during embryonic development in metazoans. It suggests that specific molecular "markers" exist either in the plasma membrane itself or in a coating secreted external to this, and that the cells stay together because of some interaction between markers, which are *species-specific*.

Example 2: Isolated sponge cells of two different species, of contrasting color (e.g., purple and orange) for ease of discrimination, are mixed together. When viewed under a microscope, the purple cells and orange cells brush by each other and never stop creeping—until each meets a cell of its own kind. Soon there are purple clumps and orange clumps, but no bicolored ones, which demonstrates that the surface markers are species-specific.

If reaggregating sponge cells form a sufficiently large mass, some of them seem to be reminded that they were once collar cells and belong on the inside, where they can do their job of filtering plankton from the water, so they go to the inside; other cells jostle about until they too are where they belong, and a tiny spongelet is thus reconstituted, ready to resume feeding and growth.

Example 3: A parallel behavior to the above can be seen in mouse embryonic cells, separated chemically (instead of by sieving): maintained in vitro in a nourishing culture medium, cells that were formerly stationary become mobile and creep about until they contact others of the same tissue type, at which point they stop moving. This is called *homoadhesion*. At the same time, the cells may also have certain affinities for some *different* kind of embryonic tissue and try to stick to it as well (*heteroadhesion*).

These apparently conflicting tendencies are important to the migrations of embryonic cells when the form of various organs and body structures is first shaping up (*morphogenesis*).

Example 4: In a culture of homogenized cells of embryonic mouse liver, muscle, and skin cells, the liver cells are very homoadhesive and hardly a bit heteroadhesive, so they end up as a solid mass in the center. The myoblasts are less devotedly homoadhesive and somewhat more heteroadhesive, so they form a layer around the core of liver cells. The epidermal cells are the least homoadhesive: they barely tolerate touching each other, even if only along the sides, and they seem to abhor touching other tissues even more. Thus *they* end up forming a one-cell-deep layer on the outside, as if wishing to get away from it all, if this could be managed.

This interaction of homoadhesive and heteroadhesive tendencies (which changes dynamically as embryogeny proceeds) allows embryonic cells to enter into new relationships, where they can influence each other's further destinies. It also explains how dissociated sponge cells can first reaggregate and then further shuffle about until they have reestablished the overall sponge body plan. But the next questions to be answered are what are these surface markers, what determines when and how they appear, and how do they permit varying degrees of homo- and heteroadhesivity to exist simultaneously?

Thus the evolution of surface marker molecules that cause cells to stick together instead of going their own ways may have been instrumental in the appearance first of protozoan colonies, then of metazoans. This need not imply that eumetazoans arose from colonial ancestors (although it seems likely that sponges did), but that production of surface markers would make possible *both* multicellular colonies *and* multicellular organisms, by keeping cells from going their separate ways following mitosis.

7.4. What is the syncytial model?

Another possibility regarding the origin of metazoans is that some inventive *multinucleate* protozoan one day decided to partition its cytoplasm, perhaps in response to a surface-to-volume problem, and thereby became multicellular. Such cytoplasmic division of a multinucleate mass *does* take place in protozoans that undergo multiple fission, but thereafter, the swarm of daughter cells disperses. Such dispersal, of course, could be circumvented by the production of the selfsame surface markers mentioned above, or some sticky binding material (such as the intercellular cement we find in metazoan tissues) could have been secreted to hold the cells together. Nevertheless, this *syncytial model* is not supported by most patterns of metazoan embryogeny: the early embryo does *not* become multinucleate and then divide up its cytoplasm; instead, the embryonic cells divide completely, but do not migrate away.

7.5. How is collagen important to multicellularity?

Multicellular animals synthesize the unique protein *collagen*, which makes up connective fibrils; *en masse*, these fibrils form sheets and cords of *collagenous connective tissue*. Even if the evolution of multicellularity did involve the development of cohesive cell behavior, the resulting multicellular bodies would have been held together too flimsily to be particularly sizable or motile, unless some binding substance were produced that could knit the tissues together strongly. This proved to be collagen, a protein so vital that it actually makes up 25 percent of the total protein in a human body.

Curiously, collagen contains plentiful amounts of an amino acid (hydroxyproline) that is not represented in the genetic code by any base triplet. This means it cannot be incorporated into protein molecules in the usual way involving the interaction of mRNA and tRNA; instead, its incorporation depends on special enzymes. These enzymes had to have been of wide enough occurrence among animal-like protists to account for the fact that collagen is synthesized by sponges as well as by eumetazoans; collagen is not synthesized by plants or fungi.

7.6.   What are the advantages of multicellularity?

Unicellularity has *its* advantages, for the very small can pit vast numbers against mortality and the threat of extinction. Protists and prokaryotes in the aggregate can probably outride a catastrophe that would strip the world bare of "higher" forms of life; they may endure to be the last living things on earth. But their highest cards are fecundity and encystment, and they do have their limitations. The advantages of multicellular organisms lie in their greater size, environmental control, cell specialization, and even their senescence and natural death.

7.7.   Of what advantage is greater size?

Size alone can be advantageous, for the dinner is usually smaller than the diner. Furthermore, if a protozoan is eaten, that is the end of it. Multicellular organisms—most plants and a number of animals, including sponges—often respond to being munched upon by *regenerating* the lost parts. In fact, commercial sponge collectors prudently replenish the supply by anchoring sponge fragments to weights and sinking these in favorable areas for future harvesting.

7.8.   How do multicellular organisms have greater control over their environment than do protozoans?

Environmental control can be accomplished more effectively by larger, more complex organisms. If an amoeba is caught in a drying water drop, it may lack time to encyst, and it certainly can't hop across to a larger puddle. When the tide goes out, a sponge cannot run away, but it *can* close its body orifices, trapping seawater within and maintaining its moisture until submerged again. Mobile metazoans simply go out with the tide and come back in again for feeding. In addition, multicellular organisms have greater potential for resisting predation: not only size, but greater mobility and special protective devices—including noxious secretions, stings, spines, and the like. Sponges guard their apertures from invasion with sharp needlelike *spicules* that somewhat limit access by animals seeking shelter. Although some protists live on land in moist, sheltered locations, none but multicellular organisms could develop the adaptive devices against dehydration that have let them enjoy an active life in dry habitats and thereby inherit the continents.

7.9.   Of what advantage is cell specialization?

Cell specialization becomes possible with multicellularity. Just so many goodies can be packaged in a single cell, in terms of specialized organelles. In metazoans, an entire cell may be geared for some particular job. The major types of specialized cells seen in most metazoans are epithelial, muscular, nervous, supportive-connective, and vascular (see Chap. 2). Sponge cell types will be dealt with below; sponges do not have true tissues, but they do possess cells specialized for food capture, contraction, food storage, reproduction, and secretion of skeletal elements. Death of any one cell can often be compensated for by division of another cell of the same type.

7.10.   Are senescence and natural death adaptive?

Only people worry about dying, and most animals' old age is cut short by disease or predation. Aging and death really enter the picture with multicellularity. Unicellular organisms *do* die—of dehydration or by predation, or from other external forces—but they are not ordinarily *programmed* to die of endogenous causes after some set life span. Unicellular organisms are essentially immortal: when a paramecium divides, which of the daughter cells is old and which is new? Both are both. Both are equally young, ready to grow again and divide again, *ad infinitum,* with sex thrown in only occasionally, as needed, to keep asexually reproducing paramecium populations from dying out. (On the other hand, sexual recombination is quite unknown in a number of sarcodines, flagellates, and cyanophytes, and they seem to get along just fine.) The daughter cells are also equally *old:* ancient, established genomes may go on and on through the immortality of unicellular life.

By contrast, multicellular *animals,* at least, grow old and eventually die from endogenous causes if not killed first.

Many examples exist of the programmed death of metazoans, often associated with reproduction, as when Pacific salmon cease to feed as they begin their single upstream breeding migration and are moribund by the time they arrive at the spawning grounds.

> **Example 5:** Corroborative of genetic limits to longevity is the behavior in tissue culture of non-tumorous metazoan cells, which divide only a certain number of times and then die out, even when regularly transferred to fresh nutrient media. This also applies to human tissues: significantly, cells explanted from older persons divide fewer times in culture before dying than those taken from younger individuals.

When programmed death enters the scene, the immortal bridge becomes restricted to the gametes that will form the zygote (or to the vegetative cell that becomes an asexual bud). Asexual reproduction still does exist in many metazoans as a conservative force, operating against change. But sexual reproduction, now a consistent rather than intermittent phenomenon, becomes the main propagative device of multicellular organisms, and this increases the novelty of individuals through the mechanism of genetic recombination. Each new sexually produced individual is to some degree unique. With multicellularity, regularity of sexual recombination, and internally generated aging and dying, the stage of life was set for change. Constancy has been the hallmark of unicellular life (although one-celled organisms necessarily have kept pace through mutation with environmental changes), but change has been the hallmark of multicellular life, with continuing production of new species capable of exploiting new habitats and innovative ways of life.

7.11.   What are the characteristics of sponges?

Sponges (Fig. 7.1a) are the simplest multicellular animals that have achieved size, variety, and abundance. They lack organs, true tissues, definite nervous and sensory cells (although all cells are irritable), and structures allowing extracellular digestion. As adults they are sessile in marine and freshwater habitats and rely on flagellated choanocytes to set up water currents that draw in minute organisms and organic particles through pores in the body wall, to be captured by the collar cells or by amoeboid phagocytes. Most sponges are brightly colored: red, pink, purple, or yellow; the adaptive significance of such pigmentation is unknown.

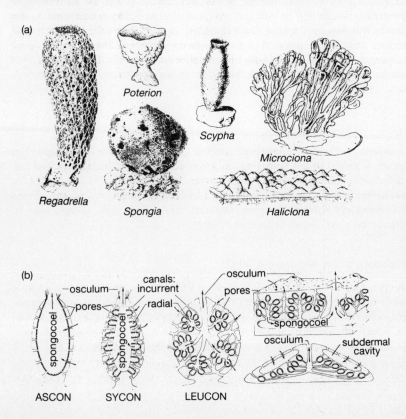

Fig. 7.1 Sponges. (*a*) Class representatives, external view: class Calcispongiae, *Scypha;* class Hyalospongiae, *Regadrella;* class Demospongiae, *Poterion, Spongia, Microciona,* and *Haliclona.* (*b*) Internal structure of canal systems: *from left to right,* asconoid, syconoid, and leuconoid body plans; *far right above,* part of an encrusting sponge; *below,* a freshwater sponge. (*From Storer et al.; Regadrella and Scypha after Lankester,* Treatise on Zoology, *A. & C. Black, Ltd.*)

7.12. What is the architecture of sponges?

Sponges may be individual or colonial. All have a body wall perforated by minute incurrent pores (dermal ostia) that strain the incoming water flow and one or more large excurrent openings (*oscula*) through which water exits as from a chimney or volcanic vent. Some dermal ostia are tubules running through the center of cylindrical cells called *porocytes,* while others are extracellular pores (*prosopyles*) penetrating the nonliving matrix of the body wall. Individual sponges are typically vase-shaped and range in size from limy sponges no larger than a rice grain to glass sponges 30 cm tall.

Colonial sponges may form clusters of individuals that arise by budding from horizontal runners, or they may form undivided, globose mounds up to a meter across. Such compact colonial forms are called *compound sponges* because the asexually produced zooids are so intimately conjoined that their individual limits defy definition. The shape of a sponge, maintained by skeletal fibers or crystalline spicules, is considerably affected by environment. The same species may rise from the substratum as delicate towers when growing in quiet waters or form a low, compact mass when exposed to surf or turbulence. Some sponges fit into crevices or under rocks, assuming a flat, encrusting form. Despite this variability, only three basic body plans exist (Fig. 7.1*b*): asconoid, syconoid, and leuconoid.

7.13. What is the body plan of asconoid sponges?

Asconoid sponges such as *Leucosolenia* possess the simplest body plan (Fig. 7.2*a*). They are small (10 cm tall or less), tubular, and radially symmetrical, with a thin body wall enclosing a large central cavity (*spongocoel*) that opens to the outside by way of a single, large osculum. Choanocytes line only the spongocoel. Most dermal ostia are pores through cylindrical porocytes that can contract to close the pore. Because the volume of water within the large spongocoel is too great to be expelled rapidly through the osculum, the rate of water flow is slower than in other types of sponges. Also, food particles entering the spongocoel are readily swept away uncaptured, so that this body plan is less efficient than the others.

7.14. What is the body plan of syconoid sponges?

Syconoid sponges such as *Scypha* benefit from increased internal surface area (Fig. 7.2*a* and *b*). The body wall extends out into numerous fingerlike projections, which may be externally visible as separate fingers or conjoined, so that the outer surface of the sponge is smooth with regularly spaced round openings. Each evagination of the body wall encloses a *radial canal,* which is continuous with the spongocoel. The dermal ostia open into these slender canals, which are lined with choanocytes, while the spongocoel proper is lined only by flat, epithelium-like cells. Food capture takes place in the radial canals. Syconoid sponges are larger than asconoids, but are still noncompound, radially symmetrical, and tubular or vase-shaped.

7.15. What is the body plan of leuconoid sponges?

Leuconoid sponges are solitary or compound and are the largest, most complex sponges. They include the commercial bath sponges. Additional feeding efficiency is gained in this body plan, in which numerous incurrent canals lead into clusters of tiny chambers lined with collar cells, from which water is discharged into excurrent canals leading to the osculum. Even a quite modest leuconoid sponge may possess millions of such flagellated chambers, and the rate of food capture is very high. The spongocoel is essentially reduced to narrow excurrent canals. Compound sponges may be irregular in shape or form a radially symmetrical mass shaped like a head or bowl.

7.16. What is the architecture of sponges at the microscopic level?

Microscopic organization includes cells, matrix, and skeletal elements (Fig. 7.2*a* and *b*). The matrix (mesohyl) is gelatinous and proteinaceous, penetrated by dispersed collagen fibrils. It contains scattered cells and skeletal components. The skeleton is composed of a meshwork of fibers of *spongin* (an insoluble, sulfur-containing protein similar to the keratin found in hair, feathers, and nails) and/or separate or conjoined spicules of many shapes (Fig. 7.2*c*), consisting of calcium carbonate or siliceous material ($H_2Si_3O_7$). Five major types of cells are morphologically distinguishable:

1. *Choanocytes* bear a single flagellum that beats spirally from base to tip, so that water is drawn toward the collar of long microvilli that are both interconnected and spaced apart by microfibrils. Water drains between the microvilli, while particles adhere to the outer surface of the collar and are carried down to the cytoplasm for ingestion. Captured food may be digested in food vacuoles within the choanocyte itself or transferred to an amoebocyte.

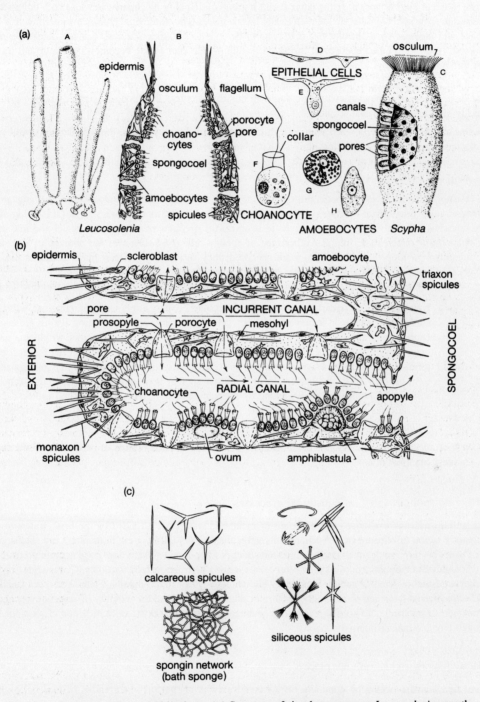

Fig. 7.2   Sponge structure and histology. (*a*) Structure of simple sponges: A, *Leucosolenia* growth habit; B, body wall section of an asconoid sponge, *Leucosolenia;* C, body form and section of a syconoid sponge, *Scypha;* D–H, cell types. (*b*) Detail of body wall of *Scypha* showing radial and incurrent canals, cell types, and spicules. (*c*) Sponge skeletal elements: spicules and spongin fibers. [*From Storer et al.;* (*a*) *and* [(*c*) *after Hyman.*]

Choanocytes also play roles in reproduction. They are reported to be capable of transforming into cells that will give rise to either eggs or sperm. In the latter event, a group of neighboring choanocytes lose both collars and flagella and turn into spermatogonia that divide meiotically to become sperm. Alternatively, when a sperm cell is drawn into a sponge and ingested by a choanocyte, if the sperm belongs to the same species, it is not digested but triggers a remarkable behavior by the choanocyte: losing its flagellum, the choanocyte becomes amoeboid and transports the sperm through the mesohyl to an egg cell.

2. *Pinacocytes* are flattened cells that form a single layer over the exterior and line the spongocoel where choanocytes are lacking. They are somewhat contractile and can slightly compress the sponge's surface and close its pores, an especially vital role in intertidal species.

3. *Porocytes* are cylindrical cells, each with a hole in the center that forms a pore through the body wall of asconoid and syconoid sponges. By contracting, they can close the pores.

4. *Myocytes* are slender, tapering cells that ring larger pores and oscula, contracting like a muscular sphincter to close those apertures.

5. *Amoebocytes* locomote by pseudopodia through the mesohyl. Besides "ordinary" amoebocytes that engage in feeding and food storage, five special functional types can also be discriminated morphologically, but it is not known to what extent amoebocytes can change from one type to another. (*a*) *Archaeocytes* have large nuclei and may give rise to eggs or sperm or differentiate into any other type of sponge cell. (*b*) *Collencytes* are nonmotile, star-shaped cells with threadlike pseudopods by which they are anchored; they secrete the collagen fibrils scattered throughout the mesohyl. (*c*) *Lophocytes* are highly mobile and secrete bundles of collagenous fibrils. (*d*) *Spongioblasts* secrete spongin fibers. (*e*) *Scleroblasts* secrete spicules. A complex spicule is produced by several scleroblasts working cooperatively.

> **Example 6:**   A remarkable instance of scleroblast cooperation is seen when internal buds (*gemmules*) are formed, usually in freshwater sponges, where they serve as an adaptation against seasonal changes. During gemmule formation, a number of archaeocytes first cluster together within the mesohyl and may then give off some chemical signal that prompts nearby scleroblasts each to secrete a unique, spool-shaped spicule. Having done so, the scleroblasts migrate toward the massed archaeocytes and position their little spools next to one another so that their outer and inner ends form the mosaic surfaces of a spherical shell making up the gemmule wall. The siliceous shell protects the enclosed archaeocytes through drought or winter; when favorable conditions prevail, the archaeocytes escape through a pore (micropyle) and develop into new sponges.

Despite such transitory interactions, sponge cells do not form the coordinated masses of cells of the same type that are known as "tissues" in eumetazoans.

7.17.   How do sponges carry on nutritional processes?

Food getting, gaseous exchange, and diffusion of metabolic wastes are all aided by the brisk water flow provided by the flagellary action of choanocytes. A leuconoid sponge only 10 cm tall by 1 cm in diameter can filter more than 20 liters (L) of water per day, while much larger compound sponges filter up to 1500 L a day. Food is captured phagocytically, either by choanocytes or by amoebocytes along the canals, and is digested in food vacuoles. Choanocytes ingest mainly particles that are bacteria-sized or submicroscopic, while amoebocytes capture dinoflagellates and other small plankton. Food can be transferred from cell to cell, either before or after digestion, and the mobility of amoebocytes accomplishes internal transport of nutrients. No special respiratory or excretory structures exist, since diffusion of gases or wastes can occur through any external or internal surface.

7.18.   How does reproduction take place?

1. *Asexual reproduction* occurs by gemmule formation (described above) or by formation of external buds that may arise directly from the parent sponge or from horizontal runners. The buds may detach and float away, or remain attached to form a colony or compound sponge. Sponges also enjoy a fantastic regenerative capacity. Even a small fragment can grow into an entire new sponge.

2. *Sexual reproduction* is initiated by maturation of ova and sperm from oogonial and spermatogonial cells representing transformed choanocytes or archaeocytes. Some species are *hermaphroditic* (bisexual, or *monoecious*), with each individual producing both eggs and sperm. Others are *dioecious* (unisexual), with separate male and female individuals. Sperm are shed into the excurrent passages and swept out the osculum, to be carried into the pores of another sponge.

There they are captured by amoebocytes or choanocytes and, if of the right species, are not digested, but transported by their captors to the egg cells, which are fertilized and begin development within the mesohyl.

7.19.   Is the embryonic development of sponges similar to that of "higher" animals?

Embryogenesis in sponges follows quite a different pathway from that in eumetazoans. Sponges have two types of larvae, *amphiblastula* and *parenchymula*, the former occurring in calcareous sponges and primitive demosponges, the latter in all other species.

(*a*)   *Amphiblastula* larvae form by a unique series of events (Fig. 7.3*a*). The zygote cleaves into a spherical mass of cells with smaller, flagellated *micromeres* at one pole and larger, unflagellated *macromeres* at the other. This mass reorganizes itself into a hollow ball similar to the blastula stage of eumetazoans (see Fig. 8.2), with flagella directed inward. The cells then *rotate* end for end so that their flagella point outward, a process known as *inversion*, and the hollow larva can then swim about actively. When it finds a place to settle, the amphiblastula invaginates its flagellated hemisphere so that the micromeres lie internal to the macromeres, which now form an outer layer; this double-walled cup is reminiscent of a eumetazoan gastrula and is often called such, but in eumetazoans the micromeres overgrow the macromeres, and not vice versa. The macromeres then differentiate into pinacocytes, and the micromeres into choanocytes and all other cell types.

(*b*)   *Parenchymula* larvae (Fig. 7.3*b*) are solid, not hollow, and are covered with an outer layer of flagellated cells. Some of the internal cells may secrete a few spicules. Emerging from the mesohyl of the parent, the parenchymula may swim about for up to 2 days before settling upon its anterior end. A drastic reorganization then takes place, in which the flagellated cells lose their flagella, migrate internally, and become amoebocytes or myocytes, or again grow flagella to become choanocytes, while the internal cells migrate outward to form pinacocytes.

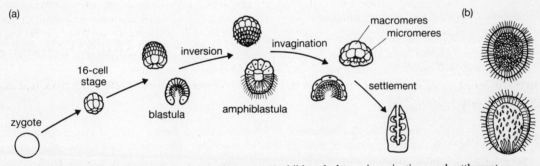

Fig. 7.3   Sponge development. (*a*) Cleavage, amphiblastula larva, invagination, and settlement in the calcareous sponge *Scypha*. (*b*) Parenchymula larvae.

7.20.   What happens after the larvae settle and the cells differentiate?

Postembryonic development often recapitulates less complex body plans. Leuconoid species frequently go through asconoid and syconoid stages before attaining the leuconoid level of body organization. After attaining their final form, some marine sponges may live for many years.

7.21.   What are the classes of sponges?

There are four classes of sponges: Calcispongiae, Hyalospongiae, Demospongiae, and Sclerospongiae.

7.22.   What characterizes the calcisponges?

Class Calcispongiae (= *Calcarea*) comprises sponges with calcareous spicules of *calcium carbonate*, which are needle-shaped (monaxon) or three- or four-pointed. They are mostly under 15 cm tall and include asconoid, syconoid, and leuconoid species (see *Scypha*, a syconoid sponge, Fig. 7.1*a*).

7.23. What characterizes the hyalosponges?

Class Hyalospongiae (= *Hexactinellida*), the glass sponges, comprises sponges with six-rayed siliceous spicules and glass fibers typically fused into a beautiful lattice that supports a tubular or funnel-like body (see *Regadrella*, Fig. 7.1*a*). Although most do not exceed a height of 30 cm, some glass sponges reach 90 cm. The body plan is syconoid, with choanocytes restricted to the radial canals, and the lining of the spongocoel is a syncytium (undivided multinucleate sheet of cytoplasm). These are deep-sea sponges, found at depths of 90–5000 m. The lovely Venus's-flower-basket often houses a mated pair of commensal shrimp that enter through the prosopyles, grow, and then cannot exit through the lattice-covered osculum. They live a protected but captive existence, feeding on plankton that elude capture by the host. Formerly, the sponge skeleton with its dessicated occupants was given to newlyweds in Japan, as a symbol of lifelong marital fidelity (or entrapment?).

7.24. What characterizes the demosponges?

Class Demospongiae contains 95 percent of the approximately 10,000 species of modern sponges. These are all of leuconoid body type, often compound, and occur from the intertidal zone to great depths, and also in fresh water. Many are brightly colored, with pigment-filled amoebocytes. They may entirely lack skeletal materials or have skeletons of siliceous spicules that are *not* six-pointed, or of anastomosing *spongin* fibers, or of both.

Commercial bath sponges have skeletons of spongin only (see Fig. 7.2*b*). One family of demosponges bores into corals and mollusk shells, promoting their decomposition. Boring is accomplished by amoebocytes that secrete a chemical that etches the limy substance until fragments come loose and are engulfed by the amoebocytes and transported to the excurrent canals for release into the water current. A number of demosponges have internal symbionts: blue-green algae or intracellular bacteria. (Such endosymbiotic bacteria make up a third of the mass of the sponge *Verongia!*) Although the cyanophytes can serve in carbohydrate production, like any photosynthesizing symbiont, the adaptive significance of the intracellular bacteria remains uncertain.

7.25. What characterizes the sclerosponges?

Class Sclerospongiae, the hard sponges, represents only a few species of dense, heavy sponges with an internal skeleton of siliceous spicules and spongin, and an outer covering of calcium carbonate penetrated by microscopic pores and elevated star-shaped oscula. They occur mainly in crevices and grottos in coral reefs.

7.26. What is phylum Placozoa?

The only known placozoan, *Trichoplax adhaerens*, is a flattened marine animal only 2–3 mm in diameter and of irregular shape (Fig. 7.4). Its body consists of a dorsal layer of flagellated cells interset with bright, shiny spheres of unknown function, a ventral layer of flagellated cells and gland cells, and a central cavity containing fluid and loosely spaced fibrous cells. Gliding on its flagella, this little creature creeps over its future meal, exudes enzymes from the gland cells, and then absorbs the digestive products. Although classified in subkingdom Parazoa for want of contradictory data, *Trichoplax* might actually be allied to some ancestral *eumetazoan*, for it lacks collar cells and would appear an unlikely candidate for descent from choanoflagellate protozoans. However, you may recall that the outer cells of sponge *larvae* are flagellated but lack collars; the collars do not develop until the flagellated cells have become internalized. Genetic and biochemical analysis, realizable because *Trichoplax* is willing to reproduce in aquaria, should eventually resolve the mystery of its affinities.

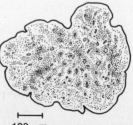

Fig. 7.4 *Trichoplax adhaerens.* About 500 μm long and extremely flattened, this simple marine organism seems to represent a distinct phylum, Placozoa.

100 μm

7.27.   What are the mesozoans?

Mesozoans are tiny internal parasites of marine invertebrates, blessed with very simple bodies but complicated life cycles. Their simplicity is suspect because of their parasitic habit, a mode of life that frequently induces retrograde evolutionary simplification from more complex, free-living ancestors. A number of instances can be seen in various phyla of endoparasitic species departing dramatically from nonparasitic species to which they are indubitably related. Although the affinities of mesozoans are unknown, some zoologists consider them retrograde flatworms (see Chap. 10). Nevertheless, mesozoans will be looked upon here as being very simple (rather than retrograde) metazoans. Although they are now placed together in a separate subkingdom, Mesozoa, the two groups of mesozoans are so very unlike that they probably deserve to be classified separately, perhaps as two different subkingdoms. Alternatively, one group may represent a valid subkingdom of very simple-bodied metazoans, while the other may stem from degenerate flatworms.

7.28.   What are the orthonectid mesozoans?

Orthonectids ("straight swimmers") are microscopic parasites within the tissue spaces of brittle stars, bivalves, polychaetes, and nemerteans. They are dioecious, with differently shaped males and females (Fig. 7.5a). The body consists of a single layer of ciliated cells surrounding an inner mass of either sperm or egg cells. Adults of both sexes escape simultaneously from the host, and sperm are released that penetrate the body of the female orthonectid, fertilizing the eggs within her. The zygotes grow into ciliated larvae that, released from the mother, seek out and enter the body of a new host. Within the host, the larva loses its cilia and becomes a syncytial mass of undivided cytoplasm that gives rise to asexual reproductive cells (agametes), each of which then develops into a new male or female orthonectid.

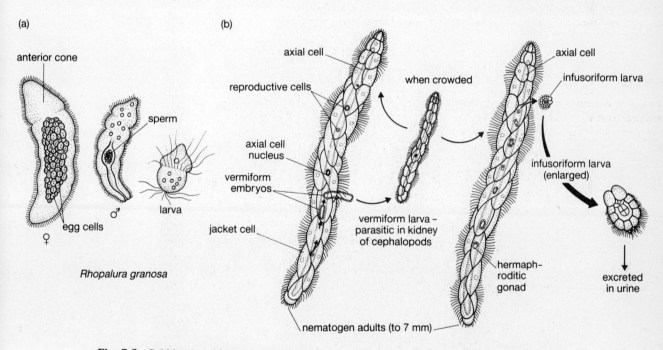

Fig. 7.5   Subkingdom Mesozoa. (*a*) An orthonectid mesozoan, *Rhopalura granosa*, parasitic in clams: adult female, male, and larva. (*b*) Life cycle of a dicyemid mesozoan parasitic in kidneys of cephalopod mollusks.

7.29.   What are the dicyemid mesozoans?

The life cycles and morphology of *dicyemids* ("two embryos") differ greatly from those of orthonectids, making their taxonomic grouping dubious. However, they too are anatomically very simple, yet have a complex life cycle. Dicyemids live only in the kidneys of bottom-dwelling cephalopods (octopuses, cuttlefish, and some squid). The adult form (*nematogen*) is 0.5–7 mm long and consists of a layer of 20–30 ciliated cells (the number is constant in any given species), which enclose a single, very long *axial cell* that contains minute reproductive cells, each of which can develop asexually into a *vermiform* (worm-shaped) larva that then exits the parent's body (Fig. 7.5*b*), but remains within the same host and develops into the adult dicyemid. Eventually, the kidneys become crowded with parasites, which stimulates an alternative life cycle. Now the reproductive cells of each nematogen develop into a gonadlike structure that produces both eggs and sperm simultaneously. Fertilization takes place within the "gonad," and the zygotes grow into *infusoriform larvae* (so called because they are only 0.04 mm long and resemble large ciliate protozoans, sometimes referred to as "infusorians"). These leave the host in the urinary flow and sink to the bottom, where they live for some time before another cephalopod host can be entered. Possibly they undergo further developmental stages within some unknown intermediate host.

Chapter 8

Eumetazoans: Development and Diversification

Aside from sponges and other minor but intriguing groups discussed in Chap. 7, all other multicellular animals belong to the subkingdom Eumetazoa and, despite much diversity of body patterns, share common pathways of embryonic development. Eumetazoans, whether radially or bilaterally symmetrical, differentiate true *tissues* from at least two and usually three layers of embryonic cells known as *germ layers*. Cell differentiation is more stable in eumetazoans than in the sponges, in which some cells change roles and appearance with impressive lability. The processes that orchestrate eumetazoan embryogenesis involve cell interactions that favor further development of complexity, so that integrated *organ systems* characterize most eumetazoans. Also, eumetazoans possess *nerve tissue,* specialized for irritability and the conduction and transmission of intercellular messages, and no such cells seem to exist in the lower metazoans.

8.1. What are the major aspects of eumetazoan embryogeny?

Embryonic development is studied from several basic angles, including description of developmental stages and investigation of the processes and mechanisms that bring them about.

8.2. What major processes are involved in embryonic development?

(*a*) *Mitosis* brings about multiplication of cells from one original cell, the zygote or asexual reproductive cell. Each daughter cell possesses sets of chromosomes identical with those of the original cell, so that all genes are ordinarily propagated to all descendant cells as the embryo becomes a multicellular system.

(*b*) *Morphogenesis* is the origin of form: the shape of a body and its parts, including internal organs. This involves *morphogenetic movements,* migrations of cells to other regions of the embryo, where they will come into contact with cells of different origin and enter into new relationships that influence the further development of both. These movements are accompanied by changes in homo- and heteroadhesive tendencies (see Question 7.3). Much more is known about how genes govern metabolic pathways by which, say, pigments are produced than how gene action brings about morphogenesis. But of course genes *do* govern the development of form. As a specific example, a single dominant mutant gene in humans determines that the hands and feet will bear one or more extra digits; thus, *polydactyly,* a major alteration in the morphology of the extremities, is a monogenic trait.

(*c*) *Cytodifferentiation,* specialization of cells in form and function, takes place gradually as morphogenesis proceeds. The zygote is *totipotent,* or fully competent, with unrestricted developmental capacity, but as mitosis and cell migrations proceed, the developmental potential of embryonic cells becomes progressively more restricted, a phenomenon known as *determination*. When determination advances to a certain point, the embryonic cells relinquish their undifferentiated appearance and transform into mature tissue cells.

(*d*) *Metabolism* is more intensive in the embryo than at any later stage of life. Most young organisms are totally dependent on the food reserves present in the egg cytoplasm, such as yolk, to provide energy and building blocks for synthesis. During cleavage, they are also dependent on a reserve of messenger RNAs and enzymes in the egg cytoplasm, present from before fertilization, to allow nucleic acids and proteins to be synthesized before the embryo's own chromosomes start to produce new RNAs.

8.3. What are the main stages in *early* embryogeny?

Activation, cleavage, blastulation, gastrulation, mesoderm formation, the formation of a body cavity (if present), neurulation, and origin of germ cells are the major stages of early embryogenesis.

8.4. What happens during activation?

Activation is the event that starts a cell on its way to becoming a whole new organism. Although some such event must take place when a vegetative cell starts to multiply to form an asexual bud, much more is known about the activation of egg cells, because they are comparatively large. *Fertilization,* the fusion of a sperm with the egg, is the most usual cause of activation, although *parthenogenesis* (development of an unfertilized egg) is a regular mode of development in a number of animal groups (e.g., aphids). Activation is marked by a *streaming of egg cytoplasm* that is ordinarily initiated by sperm contact and seems to result in a final reorganization of cytoplasmic constituents, which include a supply of mRNAs and regulatory chemicals as well as a certain amount of stored food. The egg may actually extend small pseudopods around the sperm and engulf it phagocytically. The sperm entry point can be very significant in setting future body axes before the egg even starts to divide: in the frog egg, for example, the sperm entry point sets the location of the future anus, for the plane of the first cell division passes through this point, dividing the embryo into left and right halves.

8.5. What happens during cleavage?

Cleavage is a series of mitotic divisions of the zygote or vegetative reproductive cell that take place without cell growth in between (Figs. 8.1, 8.2). Sea urchin eggs in the laboratory divide once an hour for 10 hours, transforming one large original zygote into more than a thousand tiny *blastomeres.* Throughout cleavage, the chromosomes are obsessed with only one activity: making more DNA (daughter chromosomes) to be transmitted to each new generation of cells, so that every blastomere possesses the same complement of chromosomes as the original cell. There are two types of cleavage: spiral and radial.

(*a*) *Spiral cleavage* (See Fig. 8.1E–G) occurs in most eumetazoans in which the first embryonic opening of the digestive tract is the mouth. These constitute the *protostome* ("first-mouth") phyla. The spiral effect begins when the embryo divides from four to eight cells. The first four blastomeres lie in one plane, and the next division produces a lower tier of four more cells. When cleavage is spiral, the mitotic spindle tilts obliquely, so that the two tiers of blastomeres do not lie one directly below the other, but in a staggered arrangement; thereafter, as each new tier of cells is added, the spindle tilts back and forth in alternation, so that the daughter cells do not lie directly below the previous tier.

(*b*) *Radial cleavage* (Fig. 8.1A–D) is characteristic of a more restricted group of phyla (which, however, includes all animals with backbones), in which the *anus* is the first opening of the embryo and the mouth develops later. These phyla are referred to as *deuterostomes* ("second mouth"). Since all known deuterostomes are of complex body organization, the deuterostome phyla probably arose as a side branch of the animal kingdom at a time when such complexity had already been attained. The nature of the genetic change that altered deuterostome development, from early cleavage onward, from the preexisting protostome pattern remains a mystery. In radial cleavage, the mitotic spindle does not tilt back and forth, but is oriented vertically starting with the third cleavage, so that the lower four blastomeres lie directly below the four above. As cleavage continues, each new tier of cells lies directly below the previous tier (see amphioxus, Fig. 8.2).

The body's basic symmetry and axes are established in just the first few cleavage divisions: (1) the first division, into two cells, separates the future left and right sides (at this stage, if one blastomere of a two-celled amphibian embryo is killed but left in place, the presence of the dead cell fools the surviving one, which goes on cleaving and eventually

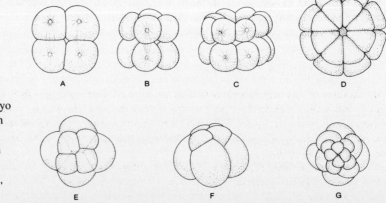

Fig. 8.1 Radial and spiral cleavage in eggs with very little yolk. A–D: radial cleavage, embryo viewed from side (A–C) and from top (D); spiral cleavage, embryo viewed from top (E, G) and from side (F). (*From Storer et al.;* D *after Gardiner,* G *after Hyman, others after Berrill and Karp.*)

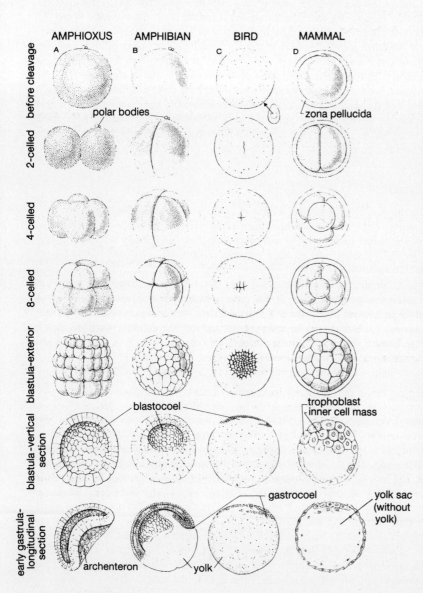

Fig. 8.2 Comparative development from zygote to early gastrula. A, amphioxus egg has very little yolk, cleaves holoblastically, has large central blastocoel, and gastrulates by invagination. B, amphibian egg has moderate yolk, resulting in unequal holoblastic cleavage and a reduced blastocoel; gastrulation involves involution and epiboly (overgrowth). C, bird egg, with much yolk, exhibits meroblastic (discoidal) cleavage of embryo on top of undivided yolk, viewed from top through blastulation, then in vertical section; gastrulation by delamination and inward migration. D, egg of viviparous mammal, essentially yolkless, cleaves holoblastically but thereafter adopts a developmental pattern reminiscent of reptile eggs, so that the apparent blastula consists of an inner cell mass (above) which will become the embryo proper, an outer trophoblast which will become the outermost extraembryonic membrane (chorion), and a cavity which represents an empty yolk sac rather than a true blastocoel; gastrulation is by migration of cells downward from inner cell mass. (*From Storer et al.; A after Hatschek, C after Blount and after Patten*, Early Embryology of the Chick. *McGraw-Hill Book Co., D, after Gregory and after Patten*, Embryology of the Pig, *McGraw-Hill Book Co.*)

produces a half-larva); (2) the second division, at right angles to the first, crosses the longitudinal body axis, creating two anterior and two posterior cells; (3) the third division, at right angles to *both* of the preceding, creates four dorsal and four ventral blastomeres and establishes the dorsoventral axis.

8.6. How does yolk affect cleavage?

Modification of cleavage by yolk is seen in the amphibian and bird eggs shown in Fig. 8.2. An "egg yolk" on our breakfast plate is the ovum itself, hugely swollen with stored food, the yellow, tasty complex of lipids and proteins that is the actual *yolk*. However, "yolk" is a generic term describing a variety of food reserves stored in the cytoplasm of *any* animal egg before fertilization. Yolk *impedes* cleavage, and the pattern of embryonic development can be greatly altered by the amount of yolk present.

When the new individual hatches quickly as an independent larva (as in amphioxus and most other marine invertebrates), yolk is scant and cleavage is complete (*holoblastic*), producing cells of equal size. A frog's egg has a moderate amount of yolk and can still cleave completely through its mass, but the yolk is concentrated in one hemisphere (i.e., the egg is *telolecithal*), and after the third division, the ventral-hemisphere cells, freighted with yolk, cleave more slowly than the yolk-free cells of the dorsal hemisphere, so that the latter become progressively smaller and more numerous; this is "unequal holoblastic" cleavage. The smaller blastomeres are called *micromeres,* and the larger *macromeres.* When still more yolk is present, the zygote cannot divide completely and so the embryo develops on the surface of the yolk. Insect eggs are *centrolecithal,* with a central yolk mass, but are small enough that the undivided yolk can be completely enclosed by a layer of embryonic cells; this is known as *superficial cleavage.* Bird, reptile, and certain fish (e.g., shark) eggs cleave *discoidally* (meroblastically), forming a flattened *embryonic plate* of micromeres on top of the huge yolk. An embryonic membrane, the *yolk sac,* then grows out from the embryo proper to surround the yolk, and, later, blood vessels that appear in this membrane absorb food and transport it to the embryo. Most mammalian eggs lack yolk and cleave holoblastically in the early stages, but soon the embryonic cells produce a yolk sac membrane around the nonexistent yolk and assume the shape of a flat disk on top of the yolk that isn't there. This is just one of many aspects of mammalian development that reflect embryogeny in ancestral forms.

8.7.   What are the events of blastulation?

Blastulation takes place after the embryo has cleaved to form a mass of cells, which in holoblastic embryos looks like a raspberry and is therefore called a *morula* ("raspberry"). The blastomeres then rearrange themselves into a *blastula,* a hollow ball enclosing a cavity (*blastocoel*) that is central in eggs with very little yolk and dorsally displaced in eggs with moderate yolk (Fig. 8.2). Some types of holoblastically cleaving embryos produce not a hollow blastula (*coeloblastula*), but a solid one (*stereoblastula*). Blastulation in discoidally cleaving embryos consists merely of the embryonic plate rising up a bit to come free of the yolk, forming a blastocoel that is a mere cleft between nonliving yolk below and the disk of embryonic cells above. At this time, the outer layer of blastomeres becomes a cohesive layer known as *ectoderm* ("outer skin"), which is called the first "germ layer" and will in time differentiate into skin cells and all tissue of the nervous system, along with the lining of the most anterior and posterior parts of the digestive tract and certain derivatives thereof. However, many of the blastula cells will not remain part of the ectoderm: they will move elsewhere through "morphogenetic movements," commencing with *gastrulation.*

8.8.   What happens during gastrulation?

Gastrulation produces a second germ layer, the *endoderm,* which lines the primitive gut (*archenteron*) of a two-layered embryonic stage known as a *gastrula* (Fig. 8.3). "Germ layer" denotes a layer of embryonic cells that have not even begun to transform into mature tissue cells, but have already undergone considerable restriction of developmental potential. The endoderm consists of cells that will form the lining of the gut and digestive glands (and, in land vertebrates, also the lining of the windpipe, bronchi, and lungs). During gastrulation, some cells of the blastula migrate internally to establish the archenteron. In embryos with little yolk, this is accomplished by *invagination* of surface cells, as seen in amphioxus (Fig. 8.2). Gastrulation of the frog egg involves active inward migration (*involution*) of some of the small, dorsal micromeres, while others continue to multiply and spread over the surface, overgrowing the large, ventral macromeres. This overgrowth (*epiboly*) eventually hides the macromeres from view. Gastrulation of discoidal embryos is accomplished by cells which simply *delaminate* (layer off) from the overlying ectoderm and/or by cells which migrate forward from a site equivalent to the blastopore of holoblastically cleaving embryos, pushing between the ectoderm and the yolk to form only a roof for the embryonic gut, which for now merely lies open to the yolk.

The opening into the archenteron (i.e., the primary site of inward cell movement during gastrulation) is the *blastopore.* Now a second major difference between protostomes and deuterostomes becomes manifest: the blastopore of protostomes marks the *anterior* pole of the embryo's longitudinal body axis and is the site of the future *mouth;* the blastopore of deuterostomes marks the *posterior* pole of that axis and is the site of the future *anus.* Whatever the mutational event that gave rise to the deuterostome branch of the animal kingdom, it must have entailed the reversal of the main longitudinal axis of embryonic development: the primordial deuterostome gastrula simply didn't know whether it was coming or going.

8.9.   Of what significance is mesoderm formation?

Mesoderm formation produces the third germ layer, or *mesoderm,* which fills the space between the ectoderm and endoderm. Body constituents derived from mesoderm include gonads and genital tracts, heart and blood vessels, musculature, connective tissues, excretory organs, and an internal skeleton where present.

> **Example 1:** *Further mesodermal development:* Most mesoderm cells are star-shaped, amoeboid *mesenchyme* that migrate actively and clump densely, delineating incipient organs, such as heart, bones, or moundlike limb buds. Early migrations and groupings of mesoderm cells in amphibian

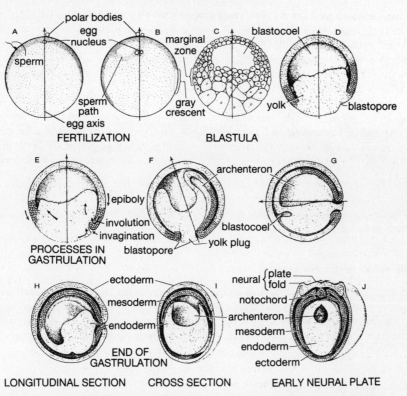

Fig. 8.3   Early development of the frog: A, arrow marks egg axis, denoted by polar body position; sperm contact activates egg cytoplasm and cortical rotation produces gray crescent (see B). B, entrance path of sperm (in plane of page and bisecting gray crescent) determines plane of first cleavage. C, blastula consists of upper micromeres and lower macromeres enclosing a reduced blastocoel. D, gastrulation commences with first appearance of blastopore (site of future anus in deuterostomes). E–I, gastrulation proceeds by involution of micromeres through blastopore, and by epiboly. F, notochord forms middorsally, mesoderm spreads lateroventrally from each side of the notochord, and J, thickening of dorsal ectoderm marks the beginning of neurulation. (*From Storer et al.*: A–C, E–G *adapted from Curtis and Guthrie,* Textbook of General Zoology, *John Wiley & Sons, Inc.;* D, H–J *adapted from Spemann,* Embryonic Development and Induction, *Yale University Press.*)

embryos are shown schematically in Fig. 8.4. First, mesoderm in the dorsal midline differentiates into a longitudinal supportive rod, the *notochord.* The remaining mesoderm becomes mesenchyme, which divides up into three major regions: (*a*) The *epimere,* the most dorsal portion, breaks up into *sclerotomes,* groups of mesenchyme that migrate medially to coalesce around the notochord as future vertebrae, *dermatome* cells that migrate to underlie the ectoderm as the future dermis of the skin, and *myotomes* that will become the segmental skeletal muscles of the body wall. (*b*) The *hypomere* sheets out ventrally between the endoderm and the dermatome, later becoming all organ muscles, throat muscles, the heart, and peritoneum. (*c*) The *mesomere* forms two longitudinal *urogenital ridges* just below the myotomes, in which kidneys and gonads will differentiate.

8.10.   Do all eumetazoans have a mesoderm?

A true mesoderm is not considered by most to be present in the radiate phyla (jellyfish, comb jellies, and the like), which are called *diploblastic* because they have only two germ layers, the ectoderm and endoderm. These differentiate into an outer *epidermis* and inner *gastrodermis,* usually separated by a gelatinous matrix that may or may not contain

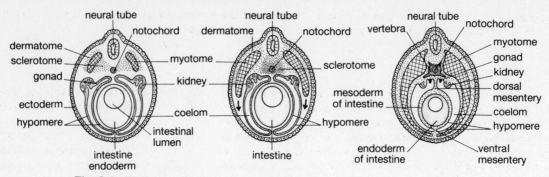

Fig. 8.4 Further development of the mesoderm in vertebrate embryos, generalized.

scattered cells. All other eumetazoans are *triploblastic,* having all three germ layers, including a true mesoderm, which greatly augments their developmental potential.

8.11.   What is the origin of the mesoderm?

The mesoderm arises differently in protostomes and deuterostomes. In the former, a single cell identifiable in early cleavage is the source of all mesoderm; after gastrulation, this cell, now located near the blastopore, begins to proliferate to form a mass of cells that soon fills the cavity between the ectoderm and endoderm. In deuterostomes, the mesoderm arises from endodermal cells in the roof of the archenteron and then sheets out in both directions between the ectoderm and endoderm until it meets midventrally, completely separating the other two germ layers.

8.12.   How does the body cavity form?

A *body cavity* allowing expansion of internal organs arises in nearly all bilaterally symmetrical metazoans. [The flatworms and nemerteans are exceptions and are referred to as *acoelomate* ("without-cavity") phyla because they have a solid mass of tissue between the lining of the gut and the epidermis (Fig. 8.5*a*).] Development of a body cavity takes place in several ways:

(*a*) *Pseudocoelomate* phyla have a body cavity, a *pseudocoel* ("false cavity"), that develops *between* the mesoderm and the endoderm (Fig. 8.5*b*). As a result, the gut lacks intrinsic muscle tissue and blood vessels and remains a rather flimsy epithelial tube, through the thin wall of which digestive products are absorbed directly into the fluid-filled pseudocoel. Although some remarkably successful animals such as nematodes (roundworms) have a pseudocoel, this restriction of mesoderm to the body wall has limited the evolution of more complex body patterns in the pseudocoelomate phyla.

(*b*) *Schizocoelomate* ("split-cavity") phyla, including mollusks and other complex protostomes, have a type of "true" body cavity (*coelom*) called a *schizocoel* because it develops as a split which opens up *within* the mesoderm, separating it into an outer layer (*parietal mesoderm*) adherent to the body wall and an inner layer (*visceral mesoderm*) that encloses the endoderm (Fig. 8.5*c, d*). This developmental event is fortuitously preadaptive for the evolution of complex body plans, for the visceral mesoderm becomes the musculature, connective tissue stroma (framework), and blood vessels of internal organs including the digestive tract and its derivatives.

(*c*) *Enterocoelomate* ("gut-cavity") phyla are deuterostomes in which the mesoderm pouches off from the sides of the archenteron (Fig. 8.5*d*), enclosing cavities that coalesce and expand to become another version of a "true" coelom that also comes to divide the mesoderm into parietal and visceral layers. This appears to be a separate evolutionary process from that seen in schizocoelous animals, but the same outcome is reached by convergence. An *enterocoel* is characteristic of lower chordates such as amphioxus, but vertebrates, which are also chordates and therefore deuterostomes, have simplified coelom formation to a splitting of the hypomeric portion of their mesoderm (see Fig. 8.4), in a manner convergent toward the pattern of coelom formation seen in higher protostomes.

The *lophophorate phyla* (brachiopods, bryozoans, and phoronids; see Fig. 12.3) seem to be intermediate between deuterostomes and protostomes in a number of respects; although they are traditionally considered protostome phyla, their manner of coelom development (Fig. 8.5*d*) is quite unique, resulting from the simple rearrangement of previously loose mesenchyme into parietal and visceral layers. Regardless of the developmental pathway by which a coelom is formed, this type of body cavity favors the evolution of structural complexity by its separation of the mesoderm into layers that contribute both to the viscera and to the body wall. All animals with a true coelom are called *eucoelomates,* and they represent the acme of complexity of both protostome and deuterostome lineages.

8.13.   What is neurulation?

Neurulation—the early development of the central nervous system—takes place while the mesenchyme is busily migrating hither and yon. In the bilaterally symmetrical eucoelomates, the *central nervous system* develops medially, down the length of the body, as an anterior brain and longitudinal nerve cord. In protostomes, most of the central nervous system develops *ventral* to the gut, whereas in deuterostomes such as vertebrates it develops *dorsal* to the gut. This suggests that the enigmatic mutational event that originated the deuterostome branch of the animal kingdom reversed not only the body's *longitudinal* developmental axis, but also its *dorsoventral* axis. The vertebrate central nervous system first appears as a thickening of the ectoderm (*neural plate*) in the midline, right above the notochord. The edges of this plate further thicken and rise, folding medially until they meet to form a hollow *neural tube,* which swells anteriorly to form the brain. The internal cavity persists throughout life as the brain ventricles and spinal canal, filled with nourishing cerebrospinal fluid. Only phylum Chordata has a hollow, instead of solid, central nervous system. The chordate "bubble brain" proved remarkably preadaptive for the evolution of a large, complex brain as seen in the vertebrates, because the

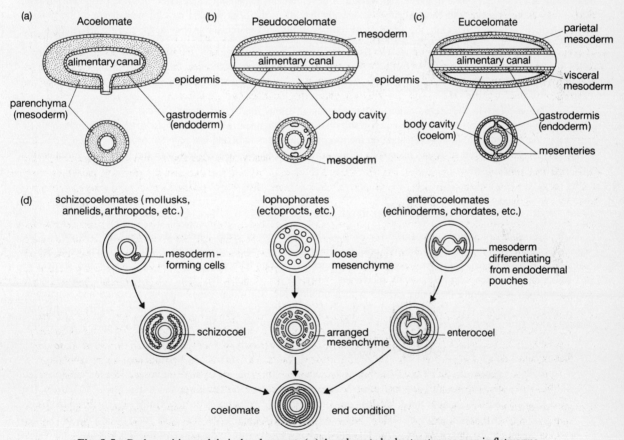

Fig. 8.5   Body cavities and their development. (*a*) Acoelomate body structure as seen in flatworms lacks a body cavity separating the gut from the body wall. (*b*) Pseudocoelomate body structure as in nematodes features a pseudocoel, or body cavity that develops between the mesoderm and endoderm, restricting musculature to the body wall. (*c*) Eucoelomate body structure features a body cavity (coelom) that develops within the mesoderm, dividing the latter into a parietal layer in the body wall and a visceral layer enclosing the endoderm (gastrodermis). (*d*) Pathways of development of a true coelom in schizocoelomates, enterocoelomates, and lophophorates; a schizocoel is characteristic of advanced protostomes, an enterocoel of deuterostomes; coelom development in lophophorates is unique.

living nerve cells can be nourished internally by the cerebrospinal fluid within the brain ventricles, as well as externally by blood vessels that cover the brain surface but cannot penetrate deeply into the interior. For some reason still unknown, blood vessels do not permeate masses of nervous tissue as they do other types of tissue, and this limits the capacity of the "solid" nonchordate brain for complex enlargement.

8.14.   What is the origin of germ cells?

The origin of germ cells, where known, presents a remarkable story. These spheroid, large-nucleated cells are the embryonic precursors of the spermatogonia and oogonia that will develop into sperm and eggs. Strange to say, evidence indicates that, in at least some animals, germ cells do *not* originate within the maturing gonad, but migrate into it from some distance away. For example, in the chick embryo, germ cells migrate by way of the bloodstream from a region in the yolk sac membrane considerably removed from the embryo itself, to lodge within the gonad, which only later takes on the appearance of an ovary or a testis. If this small portion of the yolk sac is experimentally removed, the chick is permanently sterilized, never maturing sex cells in its gonads. Microscopic inspection of the vestigial yolk sac of a mammalian embryo has revealed cells that look very much like germ cells, in the yolk sac membrane itself and also scattered along a route of migration leading toward the embryonic gonad. It has been suggested that germ cells are segregated at a very early time from the mainstream of development, remaining sequestered in some peripheral location where they escape contact with the regulatory chemicals that affect the behavior and developmental fate of the *somatic* (nonreproductive) body cells. Only the embryos of reptiles, birds, and mammals provide a site so *very* sequestered, but a similar principle may still operate in animals with less conspicuous protective devices.

8.15.   What mechanisms are involved in developmental regulation?

Regulation refers to the orchestration of developmental events that brings about coordinated morphogenesis and cytodifferentiation. Many fascinating research results have been obtained to date, but a sea of questions remain to be answered. At least some light has been shed on the mechanisms that control and integrate the processes of *ontogeny,* the development of the individual from conception to maturity. Questions 8.16–8.23 deal with the regulatory mechanisms that operate during ontogeny.

8.16   What is the role of gene regulation in ontogeny?

The overriding ontogenetic mechanism seems to be the regulation of genes, turning them "on" and "off" in specific sequences. Some genes may remain inactive until postembryonic life, being activated in due time to bring about sexual maturation, metamorphosis, and even senescence and programmed death. Other genes operate only in early life, sometimes just in the embryo, becoming permanently inactivated when they have done their job, like a drummer in an orchestra whose only job is to herald the rising curtain with a mighty roll. Still other genes may be activated and inactivated throughout life, as their products are needed for metabolism. A structural gene seems to be prevented from constantly making mRNA by a *repressor* protein that binds either to the chromosome directly at the locus of that gene or to another gene known as an *operator gene* that controls transcription by one or several structural genes. Such a sequence, of an operator plus a series of structural genes it controls, is termed an *operon.*

> **Example 2:**   The *lac* operon in the colon bacillus consists of three adjacent structural genes headed by an operator gene that must be derepressed before the structural genes can produce mRNA. The mRNA encodes three enzymes needed for the metabolic use of lactose (milk sugar). When lactose is *not* present in the medium, a *regulator* gene located some distance away from the operon produces a repressor mRNA that encodes a repressor protein specific for the *lac* operon. This repressor then binds to the operator gene, not the structural genes, a system that is economical since in that way one repressor can turn off the whole functionally related series of genes, like one switch controlling several lights in a room. The *lac* operon is turned on when lactose is available; the lactose molecule itself serves as the *inducer* that turns on the system, for when present, it binds to the repressor protein, which then cannot keep the operon turned off.

Little evidence exists to date as to whether the operon model applies to eukaryotes. Unlike that of prokaryotes, the genetic material of eukaryotes is fragmented into a number of individual chromosomes. Furthermore, chromosomal rearrangements, which involve inversion of gene sequences or translocation of blocks of genes from one chromosome to another, also characterize eukaryotes. As a result, over the passage of time operons would inevitably get broken up, so that an operator gene could no longer control a whole sequence of adjacent structural genes. Perhaps in eukaryotes, a repressor protein must attach directly to a structural gene. Alternatively, each individual structural gene may have its own adjacent operator locus to which a repressor binds. Cases are known in which activation of one gene apparently triggers activation of a number of other genes that are scattered among the chromosomes.

Example 3: The larvae of two-winged flies (Diptera) have few, but gigantic, chromosomes, each consisting of many DNA molecules grouped like strands in a cable. [Such *polynemic* ("many-threaded") chromosomes accomplish the same thing as polyploidy (multiplication of chromosome sets, as we saw in the macronuclei of ciliate protozoans)—namely, greater RNA-producing capacity.] Any gene actively producing RNA in such a giant chromosome produces a conspicuous puffed-out region in which the complementary strands of the many DNA molecules have separated, allowing RNA synthesis to proceed. Insect molting and metamorphosis are controlled by a hormone, *ecdysone*. When ecdysone is administered experimentally to a larva that is not ready to mature, within a few minutes a puff appears at a specific locus, and some minutes later, a few quite distant genes also puff up. In some manner ecdysone triggers the activation of those genes, perhaps by inactivating a repressor gene (or its product, the repressor protein) so that a number of genes can be derepressed in sequence, even though they do not lie in physical juxtaposition along a chromosome. The gene first activated seems to produce some factor that activates the rest.

8.17.   What is known about regulatory factors?

Regulatory factors of various kinds control ontogeny from zygote to adult. Although little is known of these factors, some have been found to be proteins, while others are smaller organic molecules (such as certain hormones and intraspecific communicative chemicals known as pheromones). Their precise mode of action is again usually unknown. Some seem to be highly specific, like a key fitting a lock; in other cases, a major morphogenetic event may be elicited by a bewildering range of experimentally applied agents, which in this event must serve merely as generalized facilitators.

8.18.   Is the egg cytoplasm involved in developmental regulation?

If the nucleus of an egg is removed, its cytoplasm will still be able to incorporate radioactively tagged amino acids into proteins; this shows that egg cytoplasm contains supplies of mRNAs even before fertilization, because no new RNA can be produced after removal of the nucleus. These preformed mRNAs and other cytoplasmic constituents are not randomly distributed within the egg, at least not after the sperm makes contact. Certain eggs with distinctly visible cytoplasmic granules show clearly that sperm penetration triggers a brisk streaming of egg cytoplasm, thereby concentrating these granules in localized regions of the egg, where they remain throughout cleavage (Fig. 8.6a). The following examples present evidence that such *regionalization of egg cytoplasm* is an important factor in developmental regulation.

Example 4: If an amphibian egg is cut in two along a plane that divides it from top to bottom, whichever half contains the nucleus will produce a normal (but undersized) larva; however, if the egg is bisected in a plane that separates the dorsal and ventral hemispheres, a nucleated dorsal half-egg can still produce a normal embryo, but a nucleated ventral half-egg produces only a disorganized mass of cells. Apparently the dorsal hemisphere contains vital regulatory chemicals, lacking in the ventral hemisphere, but symmetrically distributed between the left and right halves.

Example 5: Certain eggs are accessible to *fate mapping*, which is accomplished by daubing their surface with *vital dyes* before they begin to cleave. Vital dyes soak into the cytoplasm and remain there harmlessly. As the embryo develops, dyes of different colors betray the part of the egg cytoplasm that ends up forming a certain part of the embryo (Fig. 8.6b). Such fate maps are consistent for a given species, showing little individual variation.

Example 6: The cytoplasm of the frog egg is visibly regionalized even before fertilization, with yellowish yolk concentrated in the lower hemisphere and granular, grayish cytoplasm in the upper hemisphere. The outer cytoplasm (cortex) of the upper hemisphere is darker than the internal cytoplasm. When the sperm cell begins its penetration, at some point along the egg's equator, the cortex begins to rotate over the inner cytoplasm so that a crescent-shaped region of light-gray inner cytoplasm comes to mark the sperm entry point. This *gray crescent* is more permeable than other parts of the egg's surface and seems to be the primary site for exchange of materials between the egg and its environment. It also establishes the embryo's posterior end and the site of the future blastopore, for gastrulation

Regionalization of cytoplasmic granules during
cleavage of a tunicate egg

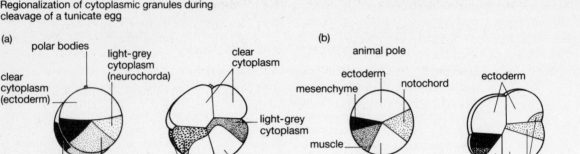

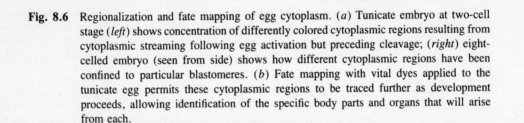

Fig. 8.6   Regionalization and fate mapping of egg cytoplasm. (*a*) Tunicate embryo at two-cell
stage (*left*) shows concentration of differently colored cytoplasmic regions resulting from
cytoplasmic streaming following egg activation but preceding cleavage; (*right*) eight-
celled embryo (seen from side) shows how different cytoplasmic regions have been
confined to particular blastomeres. (*b*) Fate mapping with vital dyes applied to the
tunicate egg permits these cytoplasmic regions to be traced further as development
proceeds, allowing identification of the specific body parts and organs that will arise
from each.

will later commence at the gray crescent. Also, the plane of the first cleavage bisects the gray crescent,
establishing the bilateral symmetry and dorsoventral axis of the future tadpole.

8.19.   What are (a) determinate (mosaic) and (b) indeterminate regulation of development?

As a result of cleavage, nuclei with chromosomes identical with those of the zygote become enclosed in packets of
cytoplasm that are quite unidentical and contain differing regulatory factors that will influence gene activity in different
ways in various embryonic cells. When blastomeres are separated at the four- or eight-celled stage, or when a nucleus is
taken from a later embryo and transplanted into an enucleated egg cell, remarkable differences show up between most
deuterostomes and most protostomes.

(*a*)   If a two-celled deuterostome embryo, such as a sea urchin, is divided into two separate blastomeres, *identical*
(*monozygotic*) *twins* result. Such blastomere separation at the four-cell stage produces identical quadruplets. In fact,
even at a much later stage, a mouse embryo consisting of many dozens of cells can be bisected along a plane that
leaves each half capable of growing into a complete, twin mouse. Similarly, a nucleus can be removed from a frog
blastula and transplanted into an enucleated frog egg, whereupon it acts as if it were a zygote nucleus and produces
a complete new tadpole, a younger twin ("clone") of the donor embryo. These data show that whatever gene
regulation may have taken place in these deuterostome embryos, the developmental clock can, so to speak, be set
back to zero, even using a nucleus from a cell as far along as the early gastrula stage. Thus, this pattern of regulation
is called *indeterminate*—it is reversible, at least up to a certain point. For the frog *Rana* this point is reached during
gastrulation. After this point has been passed, a nucleus taken from an ectodermal cell and placed into an enucleated
egg produces an abnormal embryo lacking a gut while a nucleus from an endodermal cell produces an embryo
lacking a nervous system; neither, of course, survives. However, another amphibian, the African clawed frog (*Xenopus
laevis*), can be cloned from nuclei of certain adult tissues (especially intestinal epithelium) transplanted into *Xenopus*
egg cytoplasm. So, in at least one vertebrate species determination can be reversed even in a fully differentiated
cell. Needless to say, this teases us with the possibility of cloning replacement organs for human patients, if we
could only figure out how to deregulate our own mature tissues.

(*b*)   Such reversibility of determination does *not* occur in protostome embryos. Separate their blastomeres at a two- or
four-celled stage, and you will not get twins or quadruplets, but inviable half- or quarter-embryos. Apparently, gene
regulation proceeds irreversibly in protostomes, even from the first cleavage division. Accordingly, this developmental
pattern is termed *determinate,* or "mosaic" because the early embryo is a mosaic of blastomeres whose individual
fates cannot be reversed.

8.20.   When do intercellular regulatory molecules affect ontogeny?

Intercellular regulatory influences become manifest during gastrulation, when the embryo's genes begin to produce new mRNAs and new proteins appear. Thus adjacent embryonic cells begin to communicate, partly by means of chemical factors that exert a localized effect. This localized communication is seen in the following examples of *induction*, where embryonic cells of one type influence the further development of cells of another type.

> **Example 7:**   In amphibian eggs, as gastrulation proceeds, dorsal micromeres pass into the interior through a region that forms the *dorsal lip* of the blastopore (see Fig. 8.3). While traversing this region, the micromeres become capable of producing one or several regulatory chemicals of profound effect. This can be shown by excising the dorsal lip tissue from one gastrula and inserting it under the ectoderm of the belly region of another gastrula. Here the graft will induce the host tissues to produce a well-formed secondary embryo! Apparently the dorsal lip tissue can both deregulate the host's ectoderm and stimulate morphogenesis of part or most of a new body.

> **Example 8:**   The vertebrate retina originates as cuplike outgrowths from the sides of the embryonic brain. When these *optic cups* contact the overlying epidermis, they stimulate those cells to grow inward and form a lens. If the optic cup is excised, no lens forms; if the cup is transplanted beneath the surface ectoderm in some bizarre location, it will influence the surrounding cells to form lens and eyeball, although the transplanted retina cannot function visually without proper connections to the brain.

> **Example 9:**   Morphogenesis of the mammalian kidney depends on reciprocal induction between two components: (*a*) a tubular outgrowth from the embryonic cloaca (terminal chamber of the intestine) and (*b*) the *nephrogenic plate* of the mesomere. When the two components come into contact, the cloacal bud branches to become the kidney's collecting ducts, and the nephrogenic tissue forms the rest of the kidney. If a cloacal bud is excised, no kidney at all will form on that side of the body.

> **Example 10:**   A proteinaceous *neurulizing factor* that induces brain formation has been identified in amphibian larvae. This factor can be suppressed by graduated doses of lithium, with the result that brain development is inhibited from the forebrain rearward. A small dose of lithium reduces only the cerebral hemispheres; a larger dose suppresses cerebral development totally, so that the eye-forming part of the brain (diencephalon) becomes anterior and produces a single, median eye (a condition known as *cyclopia*). With still further suppression of the neurulizing factor, no eyes at all are formed, and the entire forebrain is absent.

8.21.   When do hormones exert their regulatory effects on development?

Hormones are *systemic* regulatory chemicals; that is, they circulate throughout the organism by way of the bloodstream from the endocrine glands that produce them. They exert their effects later in embryogeny than the inductive factors described above, because they are secreted by fully differentiated glandular cells. Nevertheless, their developmental effects operate both prenatally and postnatally.

> **Example 11:**   In the absence of sufficient pituitary growth hormone, a human being will remain a midget. Untreated congenital thyroid deficiency results in cretinism: physical dwarfism and mental retardation. Thyroid hormone is also responsible for metamorphosis of amphibian larvae: a thyroidectomized tadpole never becomes a frog, while a young tadpole injected with thyroid hormone transforms prematurely into a minifrog. The bodily and behavioral transformations of puberty are mostly triggered by sex hormones; a baby chicken injected with testosterone (a male hormone) prematurely develops comb and wattles and precociously attempts to crow and mount other baby chicks. We have seen above that in insects the hormone ecdysone controls molting and metamorphosis. By dissemination through the bloodstream, a hormone can reach every body cell and coordinate simultaneously the responses of all cells susceptible to its action.

8.22.   What is intraspecific regulation?

Intraspecific regulation is seen wherever the development of certain individuals is affected by other members of the same species, usually through the agency of small organic molecules that act as *pheromones*.

Example 12:   Larvae of the marine spoonworm *Bonellia* will differentiate into females if they settle to the bottom in an area where other *Bonellia* are absent. These females then secrete a pheromone that causes nearby undifferentiated larvae to become males, thus assuring themselves of mates.

Example 13:   Caste differentiation in at least some termite species is controlled by pheromones produced by the adult king and queen of the colony. Except at a certain time of year, winged reproductives are not produced. Instead, a "king pheromone" suppresses maturation of male larvae (nymphs) and a "queen pheromone" similarly suppresses female nymphs, so that nymphs of both sexes develop into sterile workers that both look and act quite differently from the reproductives they might have become without such repressive parents.

8.23.   What is interspecific regulation of ontogeny?

Interspecific regulation, in which development of a member of one species is affected by some inductive influence exerted by another species, is probably more widespread than we know wherever symbionts develop within the bodies of hosts. Example 14 deals with such a symbiotic interaction, and Example 15 with a predator-prey relationship.

Example 14:   The tiny wasp *Trichogramma* inserts its eggs into the body of various host insects, within which the larvae develop to pupation. *Trichogramma* larvae that mature inside the body of a moth or butterfly turn into adult wasps with wings and plumose antennae. But *Trichogramma* larvae maturing within the body of an alderfly metamorphose into wingless adults with club-shaped antennae. Different nutritional environments provided by the two types of host are probably responsible for modified patterns of gene regulation in the parasite.

Example 15:   Rotifers are microscopic eumetazoans that live in fresh water. Development in the genus *Brachionus* is influenced by some chemical liberated into the water by another rotifer, *Asplanchna*, which preys upon *Brachionus*. When *Brachionus* is cultured in the presence of *Asplanchna*, its body form is modified, particularly with respect to a pair of lateral spines. The regulatory influence of the predator causes these spines to become extra large and movable instead of fixed, so that they can be directed straight outward. This makes *Brachionus* a thornier morsel for another rotifer to subdue.

8.24.   How are ontogeny and phylogeny interrelated?

Ontogeny is the developmental history of the *individual; phylogeny* is the *evolutionary history of the group* in question, up to and including life itself. Paleogenesis, ontogenetic extensions, and neoteny demonstrate interrelationships between ontogeny and phylogeny.

8.25.   What is paleogenesis?

Paleogenesis is the concept that embryonic development partly reflects evolutionary history and was first stated as the pithy but somewhat misleading aphorism "Ontogeny recapitulates phylogeny." If we took this verbatim, we could imagine a human uterus containing sequentially a fish, a reptile, and, finally, a human baby. It is not true that descendant species resemble *adult* ancestral forms, but it *does* seem true that descendant ontogenies recapitulate ancestral ontogenies— that is, developmental pathways are remarkably *conservative.* The cause of this conservatism is that evolutionary change rests on what has gone before. Descendant organisms cannot necessarily shed ancestral developmental pathways that seem no longer necessary to postnatal life: genes do *not* necessarily mutate just because their products are no longer needed, and a trait does not necessarily disappear because it seems no longer useful. If, in a cave-dwelling fish, a mutant form appears that lacks eyes, the eyeless mutant would probably survive better than those with large, useless eyes, simply because making useless eyes wastes a great deal of developmental energy in an environment where food is no doubt scarce. But being in the dark does not of itself *cause* eye-forming genes to mutate. Since genes do not mutate on demand, descendant organisms may have to pass through ancestral ontogenetic sequences, even when these are no longer useful. However, ancestral carryovers may be more necessary than they seem.

Example 16:   *Gill slits and aortic arches* go hand in hand in vertebrate evolution and ontogeny (Fig. 8.7). Aortic arches are arteries that carry blood from the heart dorsally past the gill slits, to join the branches of the great dorsal aorta above. All vertebrate embryos develop both pharyngeal gill slits and aortic arches at some point in development, yet only in fishes and larval amphibians will the gill slits become functional in respiration and the aortic arches subdivide to provide the gill circulation. The gill slits themselves could probably be dispensed with in air-breathing vertebrates, for the pattern of adjacent skeletal elements (gill arches) should suffice to guide aortic arch development, but these slits nonetheless develop and then regress. Persistent gill slits, mostly skin-covered, represent an embarrassment to a few folks, for saliva collects in these passages and may drip mortifyingly from an opening on the neck. Gill slits in mammals are just one of many examples of excess evolutionary baggage carried forward by descendant species.

Aortic arches are another story. Even though they first develop in the proper position to serve gills that will never exist, these arches remain essential even to air-breathing vertebrates for they establish an embryonic pattern that shapes the distribution of all major arteries in the anterior part of the body. Remnants of modified arches have become pulmonary arteries serving the lungs, portions of carotid arteries serving the head and subclavian arteries serving the arm, and an indispensable conduit to the dorsal aorta that carries blood posteriorly. The aortic arches provide a prime example of how evolutionary processes "make do" with what is available, modifying the old rather than scrapping it outright to plunge into some brave, new venture.

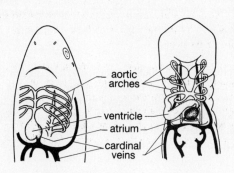

aortic arches

ventricle

atrium

cardinal veins

Fig. 8.7 Comparative ontogeny of circulatory system in shark and human. Embryonic shark (*left*) and human (*right*) both develop a series of aortic arches, which only in the shark will become the gill circulation. At this stage the hearts of both are tubular, but the human heart will later develop atrial and ventricular septa; both possess cardinal veins, which will persist in the shark but be mostly supplanted by the precavae and postcavae in the human.

8.26.   What are ontogenetic extensions?

Ontogenetic extension involves adding on new stages in descendant ontogenies, while ancestral pathways (phylogeny) persist, so that development becomes a lengthier process (this does not necessarily mean a longer prenatal period, for ontogeny also continues postnatally, sometimes through an entire series of larval stages).

Example 17:   Primitive crustaceans begin life as a *nauplius* larva with three pairs of head appendages and no thoracic or abdominal legs. The nauplius is free-swimming and independent in lower crustaceans. It metamorphoses directly into a number of rather different adult forms: brine shrimp, conchostracans, ostracods, cladocerans, etc. Most of these transformations involve addition of body segments posteriorly, with acquisition of additional legs. The nauplius of higher crustaceans molts to a *protozoea* larva, with legless abdomen but some thoracic appendages. This in turn metamorphoses to a similar but larger *zoea,* and that to a *mysis* stage, so called because it resembles the adult form of the opossum shrimp, *Mysis.* In decapod shrimps and lobsters, the mysis is the final planktonic stage, molting to adult form. Crabs add one additional subadult stage, the *megalops.* The nauplius and protozoea stages of lobsters and crabs take place within the egg. The first free-swimming stage of crabs is the zoea, but in lobsters that stage too is passed within the egg, and the lobster hatches as a mysis. Thus crustacean ontogeny both recapitulates the old and adds on the new.

Example 18:   At birth the mammalian heart has essentially attained its adult organization except for size, with right atrium and right ventricle receiving deoxygenated blood from the body and pumping it out to the lungs, and left atrium and ventricle receiving oxygenated blood from the lungs and pumping it out to the body. But the heart does not develop these four chambers straightforwardly. The embryonic heart is at first a tube that soon bends into a U, like a fish heart not yet divided into chambers; as in the latter, the posterior portion (sinus venosus, the first chamber of the fish heart) is the pacemaker.

This tube soon develops a definite atrium and ventricle, the two major pumping chambers of the piscine heart. Next, the atrium becomes subdivided by the growth of an atrial septum, while the ventricle remains undivided, a condition that persists into adulthood in amphibians. Next, a ventricular septum begins to grow anteriorly, partly separating the right and left ventricles, as in most adult reptiles. Finally, the ventricular septum is completed, an event that occurs only in mammals, birds, and crocodilians. Meanwhile, the sinus venosus, an actual but reduced heart chamber in amphibians and reptiles, finally dwindles to a mere node buried in the wall of the right atrium of birds and mammals, but still remains vital as the heart's pacemaker. The most anterior part of the fish heart, the truncus arteriosus, starts to subdivide lengthwise in amphibians, and completes a spiraling partition in reptiles, birds, and mammals to become the crisscrossing bases of the pulmonary and aortic trunks.

Human babies are occasionally born with a gross circulatory defect simply because the human truncus also starts out undivided and as the partition develops, it may fail to spiral properly. In this event the bases of the two major arterial trunks do *not* cross, and faulty arterial connections are made, so that blood entering the heart from the lungs goes right back to the lungs, while blood returning from the body goes straight back to the body without cycling through the lungs. That such babies even survive to be born (after which corrective surgery can be undertaken) depends partly on the ventricular septum's remaining incomplete and partly on the fact that the fetus receives oxygen by way of the placenta, not the lungs. So we see that from the basic linear sequence of fish heart chambers (sinus venosus, atrium, ventricle, truncus arteriosus), each group of vertebrates adapting to land life has added a further developmental stage while retaining preexisting stages in their own ontogenies.

8.27.   What is neoteny?

Neoteny is a developmental event with major evolutionary implications: a previously larval form becomes sexually mature and the former adult stage is simply deleted. Neotenic populations of modern species are known: the aquatic, gill-breathing salamander *axolotl* is a neotenic form of the tiger salamander, as shown by the fact that axolotls can respond to thyroid hormone or stress (e.g., captivity) by metamorphosing. It is possible that certain major evolutionary changes have involved neoteny. For example, tunicates are sessile, but start out life as tadpole-shaped larvae; one group of tunicates, larvaceans, retain that shape throughout life, and from such a presumably neotenic form the fishlike body of early vertebrates could have evolved.

Table 8.1   Eumetazoan Diversification

Subkingdom Eumetazoa
   Radiata (radially symmetrical, lacking central nervous system)
      Phyla: Cnidaria, Ctenophora
   Bilateria (bilaterally symmetrical, with central nervous system)
    Acoelomata (lacking body cavity)
      Phyla: Platyhelminthes, Rhynchocoela, Gnathostomulida
    Pseudocoelomata (having a pseudocoel)
      Phyla: Gastrotricha, Rotifera, Nematoda, Nematomorpha, Acanthocephala, Entoprocta, Kinorhyncha
    Eucoelomata (having a true coelom, developed within mesoderm)
     Protostomia (blastopore forms mouth; schizocoel present)
      Phyla: Brachiopoda, Ectoprocta, Phoronida, Annelida, Mollusca, Arthropoda, Onychophora,
        Sipuncula, Echiura, Pogonophora, Tardigrada, Priapulida
     Deuterostomia (blastopore forms anus; enterocoel present)
      Phyla: Echinodermata, Chordata, Chaetognatha, Hemichordata

8.28.   How have eumetazoans diversified and gained complexity?

Basic similarities in embryonic development suggest that all eumetazoans probably descended from a common protistan ancestor, perhaps a zooflagellate. Their evolutionary progress and diversification are summarized in Table 8.1.

Chapter 9

The Radiata:
Cnidarians and Ctenophores

Picture a colorful marine garden filled with more than 10,000 species of strange blossoms and drifting, transparent bells and globes. Arm these beauties with lethal stinging or adherent tentacles and endow them with predaceous appetites—and you have the Radiata, "flowers" with a difference!

9.1. What are the Radiata?

Radiates are eumetazoans with primary radial symmetry. They include two phyla, *Cnidaria* (coelenterates) and *Ctenophora* (comb jellies), which are considered the simplest living eumetazoans (and both have been blessed with names that start with a silent "c"). They are almost all marine and are thought to be the most numerous nonmicroscopic animals in the sea.

9.2. What characteristics are common to radiates in general?

(*a*) The radiate body plan consists of a centrally located mouth, around which the body parts are radially arranged. The main body axis is oral-aboral. The mouth opens into a *gastrovascular cavity* with a central stomach in which extracellular digestion occurs and from which lead systems of canals serving for internal distribution. A true anus is lacking, so that indigestible residues must be egested by way of the mouth (although two minute anal pores opposite the mouth of comb jellies may permit elimination of small particles).

(*b*) The body wall consists cross-sectionally of an inner tissue layer, the *gastrodermis*, derived from the embryonic endoderm, and an outer tissue layer, the *epidermis*, derived from ectoderm. Each cellular layer consists of several types of specialized cells described below. These two cell layers have the characteristics of true epithelial tissues: they adhere to each other firmly along the sides, and their activities are coordinated by the action of nerve cells. Between these two epithelial layers lies a nonliving gelatinous matrix, the *mesoglea*, which may be massive or paper-thin, contains few to many connective fibrils, and may or may not be penetrated by amoeboid wandering cells. If cells *are* present in the mesoglea, they have wandered in from the ectoderm and are not equivalent to mesoderm. The radiate phyla accordingly are considered *diploblastic*.

(*c*) The *nervous system* consists of one or two *nerve plexuses*, networks of neurons that form junctions with one another and with *sensory cells* and *effectors* (glandular and contractile cells that carry out the response) by way of junctional gaps known as *synapses*, at which neurotransmitter chemicals are secreted. Unlike synapses of higher animals, the interneuronal synapses are mostly *bidirectional*, so that excitation can be propagated in either direction across the junction. Although the fibers of certain neurons may group together to form *nerves*, there are no clusters of neuron cell bodies equivalent to the brains and ganglia of higher animals.

Free-swimming radiates possess gravity-sensing organs known as *statocysts*. These typically are fluid-filled sacs lined with "hair cells" (sensory cells bearing nonmotile cilia, of wide occurrence in the animal kingdom) and containing a solid crumb of calcium carbonate. Passive movements of this granule excite hair cells and keep the organism informed of its position in the earth's gravitational field; most mobile radiates tend to counteract downward drift by swimming upward periodically. [Cnidarians also may have simple eyes (ocelli), which will be described later.]

9.3. What advantages do radiates enjoy over sponges?

(*a*) *Contractility and mobility:* Sponges are motile only as larvae, and their powers of contraction are slight. Many radiates are mobile throughout life, and even sessile forms are highly contractile.

(*b*) *Presence of true tissues:* Sponges have specialized cells, but coordinated activity of these cells is sporadic and localized. Radiates have two coordinated epithelial layers.

(c) *Extracellular digestion:* Sponges are limited to filtering microscopic particles that must be digested intracellularly, in food vacuoles. Radiates possess a gastrovascular cavity into which glandular cells secrete digestive enzymes, so that digestion is wholly or partly extracellular.

(d) *Nervous tissue:* Sponges lack nervous tissue, but well-developed nerve networks, sensory cells, and even simple sense organs occur in radiates.

(e) *Prey-capturing devices:* Special mechanisms allow radiates to subdue and ingest large prey, permitting intermittent feeding, instead of the nearly constant filter feeding of sponges. Some radiates even wolf down animals larger than themselves.

9.4.   What are the characteristics of Cnidarians?

Cnidarians (Fig. 9.1a), also called *coelenterates* ("sac gut"), are much more numerous and diverse than ctenophores. They even include a few species that have successfully established themselves in fresh water. They are characterized by circumoral (surrounding the mouth) tentacles armed with batteries of stinging capsules. Most coelenterates develop from a ciliated, free-swimming planula larva, which is actually a solid gastrula (stereogastrula), formed when cells migrate inward during gastrulation to fill the blastula cavity completely, rather than forming the double-walled cup described in Chap. 8. When gastrulation proceeds primitively, the cells simply move inward from all sides to form the endoderm, but

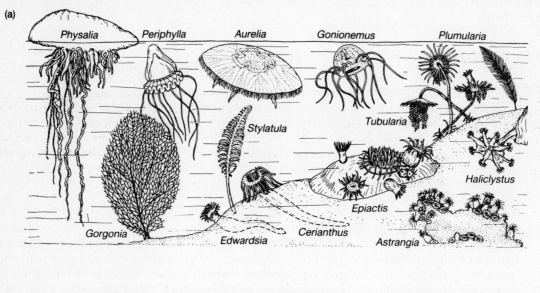

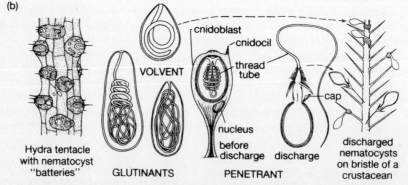

Fig. 9.1 Cnidaria. (*a*) Marine cnidarians in typical habitats (not to scale). Class Hydrozoa: *Tubularia, Plumularia, Gonionemus, Physalia;* Class Scyphozoa: *Aurelia, Haliclystus, Periphylla;* Class Anthozoa, subclass Alcyonaria: *Gorgonia, Stylatula;* Subclass Ceriantipatharia: *Cerianthus;* subclass Zoantharia: *Edwardsia, Epiactis, Astrangia.* (*b*) Cnidoblasts and nematocysts. (*From Storer et al.*)

even when cells do invaginate from the blastopore, the end product is still a stereogastrula lacking both mouth and archenteron. This suggests that the ancestral form that gave rise not only to coelenterates but perhaps to all other eumetazoans was a *planuloid*, much like a modern planula.

> **Example 1:** A planula develops definite anterior and posterior ends, but retains an ovoid, radially symmetrical body form. If some hypothetical planuloid became a creeping form, with ciliation confined to the undersurface, its symmetry would thereby have become bilateral; such a creeping planuloid is a logical candidate for the common ancestor of all Bilateria. Coelenterates that produce skeletons (mainly corals) occur in the fossil record from the beginning of the Paleozoic, but tiny, soft-bodied planuloids could leave no fossil trace. The argument for their existence must rest entirely upon embryological evidence.

9.5.   Why are coelenterates "cnidarians"?

With the exception of one species of comb jelly, coelenterates are the only metazoans that possess cells that produce organelles known as *cnidae* ("nettles"). A *cnida* (Fig. 9.1*b*) is a fluid-filled capsule that fills most of the interior of a cell known as a *cnidocyte*, or *cnidoblast*. The cnida contains an inverted thread that can be shot out by an explosive eversion (turning inside out).

9.6.   How do cnidae discharge?

The process of eversion depends on hydrostatic pressure. The fluid within the cnida is very high in solutes, so that its internal osmotic pressure is some 140 atmospheres. This should cause a massive inflow of water, except that the capsular membrane remains impervious to water until appropriate stimuli change its permeability. The sensory portion of cnidae varies: a single bristle derived from a nonmotile flagellum (*cnidocil*), or a cone consisting of one long and several short nonmotile sensory cilia, or a circlet of microvilli. The stimulus that affects the trigger region is probably more chemical than mechanical; particularly effective is the amino acid *glutathione*, which is found in animal bodies, the favorite food of most of our marine "flowers." As the permeability changes and water pours into the cnida, the thread is everted like a glove finger and, once everted, cannot be drawn in again. The entire cnida must be shed and regenerated by the cnidocyte.

9.7.   What types of cnidae are there?

(*a*) *Nematocysts* are of cosmopolitan occurrence, each cnidarian species having 1–7 of some 40 distinguishable varieties. These cnidae evert unbranched threads used in prey capture. Three main categories of nematocysts are (1) *glutinants*, which adhere to the body of the prey; (2) *volvents*, which wrap bola-like around bristles on the body of such prey as small crustaceans; and (3) *penetrants*, with open, often barbed, tips that inject a toxin into the prey's body. Nematocyst toxins may attack nerve or muscles, causing paralysis, or inflict tissue necrosis. Most cnidarian species lack nematocysts capable of injuring persons, but some notable exceptions are the Portuguese man-of-war (a hydrozoan colony), the sea nettle (a jellyfish), and the sea wasp (a cube jelly). In fact, sea wasp toxins can kill a person in only 3 minutes!

(*b*) *Spirocysts*, possessed by sea anemones and some corals, eject fine tubules that solubilize into an adhesive web used for enmeshing prey or adhering to the substratum.

(*c*) *Ptychocysts*, characteristic only of tube anemones (cerianthans), extrude sticky threads used in building the cerianthan's tube.

9.8.   What additional types of cells do cnidarians possess?

(*a*) *Epitheliomuscular cells* make up most of both the epidermal and the gastrodermal layers (Fig. 9.2). Commonly called "T cells" because of their columnar body and drawn-out base, these cells both make up the epithelial layer of each tissue and bring about body movements. The contractile bases of the T cells always face the mesoglea. The bases of the epidermal T cells are all oriented longitudinally, so that when they contract, the tentacle or body will shorten. The bases of the gastrodermal T cells are oriented at right angles to these, around the circumference of body or tentacle, so that their contraction lengthens that body part. Some gastrodermal T cells bear flagella that stir the contents of the gastrovascular cavity and promote digestion and internal distribution. They also may engulf

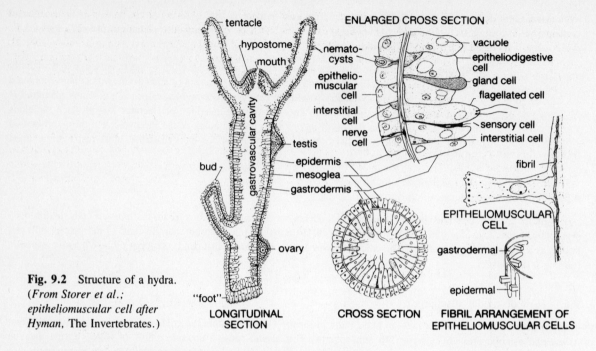

Fig. 9.2 Structure of a hydra. (*From Storer et al.; epitheliomuscular cell after Hyman*, The Invertebrates.)

microscopic particles from the partly digested food and complete the digestive process intracellularly, in food vacuoles. If development of the elongated base is suppressed, epitheliomuscular cells become simple columnar epithelium. By suppression of the epithelial portion, the elongated bases come to resemble smooth muscle fibers and may group to form muscular bands, as around the bell margin of scyphozoan jellyfish.

(*b*) *Gland cells* are columnar epithelial cells dedicated to secretion. Epidermal gland cells secrete mainly *mucus*, for lubrication, protection, and sometimes adhesion. Gastrodermal gland cells secrete *digestive enzymes* into the gastrovascular cavity.

(*c*) *Sensory cells* occur scattered among the epithelial cells of both epidermis and gastrodermis, their bases prolonged into nerve fibers that synapse with the underlying plexus. Ciliated "hair cells" line the gravity-sensing statocysts of jellyfish. Free-swimming coelenterates also usually have *ocelli* (simple eyes) consisting of an epidermal patch or cuplike depression lined with photosensitive cells. Cubomedusae possess four quite sophisticated ocelli: in each, a retinal cup backed with a screen of black pigment encloses a spherical lens that gathers and focuses light, so that these talented jellyfish can swim accurately toward mere points of illumination. Scyphozoan and cubozoan medusae have pendant, club-shaped sense organs (*rhopalia*), each containing an ocellus and a statocyst, guarded by an overhanging hood and lateral flaps (*lappets*).

(*d*) *Neurons* of coelenterates resemble those of higher animals in having a cell body from which extend *nerve fibers* of various lengths. Some are *multipolar*, with fibers extending in several directions, and are mainly responsible for slowly spreading excitation. *Bipolar* neurons have only two fibers, extending in opposite directions from the cell body; these mainly occur in rapid through-conduction tracts needed for the coordinated swimming movements of medusae. Unlike other metazoan nerve cells, many coelenterate neurons are not directionalized, secreting neurotransmitters from any fiber ending, so that impulses can pass either way across the synaptic junction.

(*e*) *Interstitial cells*, small round cells with large nuclei, are tucked between the bases of the epithelial cells of both layers. They prudently preserve an embryonic capacity for producing cells of any other type, including gametes. They are therefore responsible for reproduction and regeneration. Some cnidarians can regenerate from a mere fragment.

> **Example 2:** Hydra, a simple freshwater polyp, constantly renews its tissue cells, so that older cells pass downward and are shed from the base of the body column. Accordingly, hydras appear to be potentially immortal.

In preparation for sexual reproduction, one or more interstitial cells will give rise to temporary or permanent *gonads*. In the *testes*, the interstitial cells become *spermatogonia* that undergo meiotic divisions to become sperm. In hydras, eggs arise in the epidermis, singly, from interstitial cells that enlarge, using surrounding cells for food, and undergo meiosis with the unequal cytoplasmic division characteristic of oogenesis, wherein the extra set of chromosomes is discarded in a minute *polar body*. Most coelenterates have actual ovaries in which many eggs mature. An interstitial cell can also give rise to a bud, multiplying asexually to produce a complete new individual that may either break free to live independently or remain attached as a *zooid* in a colony. Strange to say, although parent and bud are genetically identical, developmental regulation in the growing bud may result in a new individual with a body form quite different from the parent.

9.9.   How do cnidarians exploit polymorphism?

The production of individuals of the same species that differ in appearance and function is known as *polymorphism*. Polymorphism is a very effective adaptive device in cnidarians and is responsible for much of their conspicuous success. In many coelenterates, two distinct types of individuals, *polyp* and *medusa* (Fig. 9.3), constitute *alternating generations*, with the polyp generation reproducing asexually and the medusa generation sexually. The polypoid and medusoid indi-

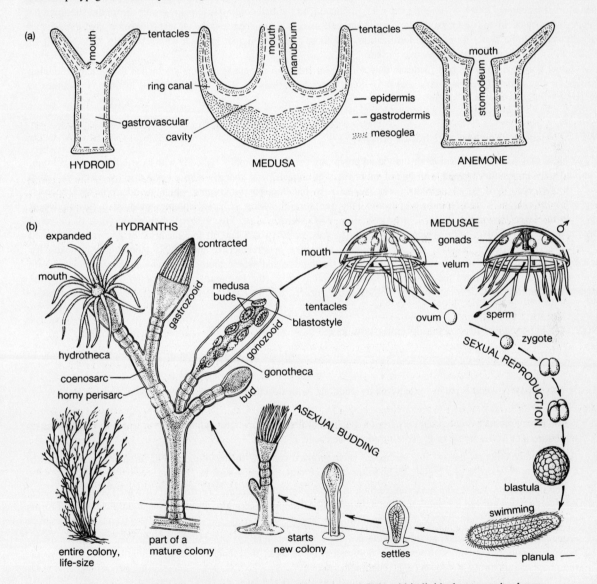

Fig. 9.3   Cnidarian body forms and typical life cycle. (*a*) Polypoid individuals as seen in classes Hydrozoa (*left*) and Anthozoa (*right*); medusa shown upside down to facilitate comparison. (*b*) Life cycle of the hydroid *Obelia*, showing characteristic alternation of asexual (polypoid) and sexual (medusoid) generations. (*From Storer et al.; modified from Wolcott*, Animal Biology, *McGraw-Hill Book Co.*)

viduals often differ greatly in size, diet, and habitat and therefore do not compete with one another. They also allow the species to exploit a wider habitat and range of food items. Alternatively, the polyp generation often buds off not only medusae but also other polyps that remain conjoined to form a *colony*. Frequently, the zooids in a colony are of two or more kinds, a diversity that permits a sophisticated *division of labor*.

9.10.   What are the characteristics of medusae?

Medusae are jellyfish of various types: (a) Hydromedusans (see *Gonionemus*, Fig. 9.4a), (b) scyphomedusans (see Fig. 9.7), and (c) cubomedusans (see Fig. 9.6). All have a somewhat bell- or umbrella-shaped body, with convex *exumbrellar* and concave *subumbrellar* surfaces, and a mouth at the end of a pendant *manubrium*. Epidermis and gastrodermis are separated by a great mass of jellylike mesoglea, penetrated by *radial canals* extending from the stomach. Marginal tentacles usually fringe the rim of the bell, and the manubrium may or may not be prolonged into streamerlike, nematocyst-laden *oral lobes*. Swimming is by jet propulsion, as the bell contracts, expelling water from the subumbrellar space. Sensory ocelli and statocysts occur around the bell margin. Permanent ovaries or testes develop, and gametes are usually shed into the water for external fertilization. The three types of jellyfish are dealt with separately (Questions 9.16, 9.21, 9.22), according to their respective classes.

Some polypoid colonies bud medusae that are not set free, but remain permanently attached as gamete-producing zooids or *gonophores* (see Figs. 9.4c and 9.5). Such gonophores may be more or less reduced in form, sometimes to mere sacs filled with eggs or sperm.

9.11.   What are the characteristics of polyps?

Polyps may be solitary or colonial. Individual polyps are sessile or creeping, typically with a cylindrical body column, ending orally in a mouth fringed with one or more rings of tentacles and aborally with a basal disc by which the polyp adheres or creeps. Many polyps reproduce only asexually, by budding, but *anthozoans*, which include some 6000 species, are polypoid cnidarians that reproduce both asexually and sexually. Most polypoid colonies produce zooids of two or more forms (see *Obelia*, Fig. 9.3b; hydrocorals, Fig. 9.4f; and *Physalia*, Fig. 9.5b). Polymorphic types of polypoid zooids include: (a) *gastrozooids*, which digest food; (b) *gonozooids*, which bud off free medusae or give rise to attached gonophores that produce gametes; (c) *dactylozooids*, fingerlike or streamerlike individuals (sometimes mistaken for tentacles), heavily armed with nematocysts, that serve for defense and prey capture; (d) *pneumatophores*, which inflate with a gas (e.g., air) to form the float of a planktonic colony (e.g., the Portuguese man-of-war); (e) *nektophores*, the muscular swimming bells of some floating colonies; and (f) *bracts*, leaflike individuals that protectively flank other zooids.

9.12.   What types of life cycles do cnidarians have?

Cnidarian life cycles are diverse, often, but not always, displaying the alternation of generations characteristic of this phylum.

1.   A polyp may produce a polyp sexually or by budding, with no intervening larva or medusoid form (e.g., hydras, see Fig. 9.4e).

2.   A polyp may produce a polyp sexually from a zygote that first develops into a *planula* larva, which eventually settles to mature into a new polyp (e.g., corals and sea anemones).

3.   A polypoid colony may bud off free medusae, which produce gametes; the resulting planula larva settles and becomes a polyp that develops a new colony by budding (e.g., colonial hydroids such as *Obelia*).

4.   A polypoid colony may bud gonozooids, which in turn bud attached medusae (gonophores), which produce gametes; the zygote develops into a planula larva that metamorphoses into an actinula, which swims about, settles, and forms the first polyp of the new colony (e.g., *Tubularia*, see Fig. 9.4d).

5.   A medusa produces gametes, and the zygote develops into a planula larva that settles and becomes a minute sessile polyp; the polyp may bud off a few other polyps that creep away independently, and eventually each polyp metamorphoses straight into a medusa (e.g., cubomedusae).

6.   A medusa produces gametes, and the zygote develops into a planula that settles and becomes a tiny polyp; the polyp (scyphistoma) eventually undergoes a transverse type of budding (*strobilation*) to produce young medusae (e.g., *Aurelia*, a scyphomedusan, see Fig. 9.7b).

7.   A medusa produces gametes, and the resultant planula larva metamorphoses into another planktonic larval form, the *actinula*, which develops directly into a medusa (e.g., *Aglaura*, a hydromedusan).

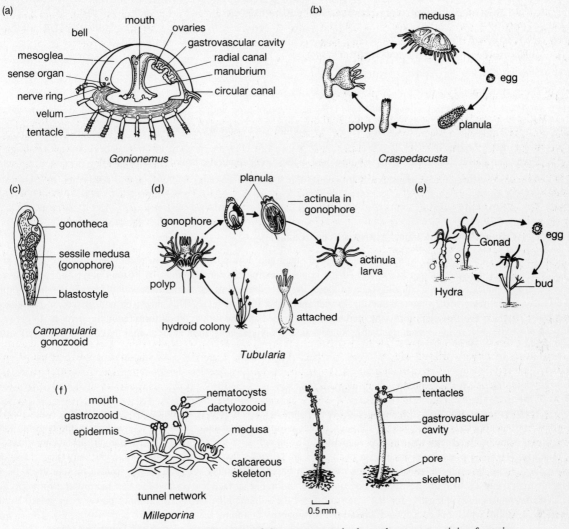

Fig. 9.4 Class Hydrozoa. (*a*) Structure of *Gonionemus,* a hydromedusan; note origin of ovaries from epidermis. (*b*) Life cycle of *Craspedacusta,* a hydromedusan. (*c*) Gonozooid of the hydroid *Campanularia* containing attached medusae (gonophores). (*d*) Life cycle of *Tubularia,* a colonial hydroid bearing permanently attached medusae (gonophores), within which the planula larvae develop to the actinula stage. (*e*) Life cycle of a freshwater hydra, which is solitary and has no medusa generation. (*f*) Hydrocorals: *left,* portion of colony; *center,* detail of dactylozooid (stinging individual); *right,* detail of gastrozooid (feeding individual).

This variety of life cycles shows that although alternation of generations is the general rule, one generation may be dominant and the other reduced, and in some cases one or the other generation is totally absent. Whether the medusoid or the polypoid body form occurred first in coelenterate evolution remains a matter for debate.

9.13.   What are the major taxonomic groups of cnidarians?

Phylum Cnidaria is divided into four classes: *class Hydrozoa,* considered to be the most primitive, includes hydras, colonial hydroids, floating colonial siphonophores, hydrocorals, and hydromedusae; *class Cubozoa,* the cube jellies, or cubomedusae; *class Scyphozoa,* with a dominant jellyfish generation, the *scyphomedusa,* and a reduced or absent polypoid generation; and *class Anthozoa,* colonial and solitary polyps with no medusoid forms at all.

9.14.   What are the distinguishing characteristics of hydrozoans?

1.   The mesoglea is acellular (lacking cells).

2.   Cnidocytes occur only in the epidermis.

3.   Gametes develop only in the epidermis.

4.   The gastrovascular cavity of polyps is not divided by radial septa.

5.   Hydrozoan medusae lack streamerlike oral arms and capture food with a ring of tentacles around the bell margin.

6.   Hydromedusae have a shelflike marginal *velum* that improves swimming efficiency by restricting the aperture through which water is expelled from the subumbrellar space.

7.   Hydrozoan life cycles include both polypoid and medusoid generations, or only one of these.

8.   The only freshwater cnidarians are hydrozoans: the hydras, a colonial hydroid, and a few types of freshwater hydromedusae.

9.15.   What is the evolutionary status of hydrozoans?

The trachyline medusae, which lack a polypoid stage and develop directly from planktonic actinula larvae, may represent the primitive cnidarian life cycle. At the opposite extreme we find solitary hydrozoan polyps of two different orders that completely lack a medusoid stage; of these, the actinulid polyps much resemble actinula larvae that have settled down to a benthonic life, without bothering to metamorphose into medusae. This might be considered an example of neoteny in which a former adult form (the medusa) is suppressed and the former larval form (actinula) persists. In between these two extremes are hydromedusae that produce a small polypoid generation and colonial hydroids that bud off small medusae.

9.16.   What are the characteristics of hydromedusae?

Hydromedusae are small (0.5–6 cm) and often shaped like small bells or thimbles. *Gonionemus* (Fig. 9.4a) may be considered anatomically representative, although its behavior is unusual since it prefers to crawl about the bottom, using adhesive pads on its tentacles to cling to vegetation. The mouth is at the end of a stalked manubrium lacking extended oral arms. Hanging down from the margin of the bell are tentacles that typically number four, but may range from one to many. Since their gastrodermal cells lack contractile bases, locomotion depends on the striated muscle fibers of the epidermal T cells, which are best developed on the subumbrellar surface and around the bell margin and form circular sheets that pulsate rhythmically, forcing water out from the subumbrellar cavity. Swimming power is increased by the shelflike velum, which projects inward around the bell margin, confining the water jet to a narrower orifice. The *subepidermal nerve plexus* projects two circumferential rings of nerve fibers, one above and one below the velum. The lower ring contains *pacemakers* that impose the rhythm of pulsation upon the entire nerve network. The bell margin is abundantly supplied with individual sensory cells, statocysts, and ocelli. Some hydromedusae are attracted to light (positively phototactic), while others display negative phototaxis and descend into deeper water by day (many types of marine plankton carry out such diurnal vertical migrations, which may protect them from harmful radiation). Medusae are either male or female; since the gonads are epidermal, the gametes are shed directly into the subumbrellar space.

The life cycle of medusae of *order Trachylina* includes a planula larva, which metamorphoses into an actinula, which matures into a medusa. Other hydrozoan medusae belong to various suborders of *order Hydroida* and have a polypoid generation that may be either larger or smaller than the medusa. The freshwater jellyfish, *Craspedacusta* (suborder Limnomedusae), grows to a diameter of 20 mm, but arises as a minute bud from a 2-mm, degenerate, tentacleless polyp that is solitary or forms a colony of only a few zooids (Fig. 9.4b). *Obelia* (suborder Leptomedusae) has well-developed polypoid and medusoid generations (see Fig. 9.3b). The medusae shed gametes that give rise to zygotes. Each zygote grows into a planula. This settles and transforms into a polyp that gives rise by budding to a hydroid colony 2–3 cm tall, bearing two kinds of nearly microscopic zooids: gastrozooids (*hydranths*) and gonozooids (*gonangia*). The latter bud off medusae that grow to a diameter of only 2 mm, but are still much larger than individual polyps in the colony.

9.17.   What are the characteristics of colonial hydroids?

Colonial hydroids (order Hydroida) are mostly only a few cm tall and usually have a finely branched external skeleton (*perisarc*), composed of chitin and protein and secreted by the epidermis. The perisarc may stop at the base of the zooids, leaving them naked (*athecate*), or may be expanded into protective sheaths: *hydrothecae* around the tentaculate feeding polyps and *gonothecae* around the reproductive polyps. Hydrothecae sometimes have lids that open only when the gastrozooids extend for feeding. All zooids are interconnected by a common gastrovascular cavity within a living tubular *coenosarc* that is partly lined with flagellated cells that carry food away from gastrozooids to young buds and gonozooids. Hydroid colonies may bud off free medusae (e.g., *Obelia*) or retain the medusoid generation as an attached *gonophore*. *Campanularia* (Fig. 9.4c) produces medusa buds along a thecate gonozooid, as in *Obelia*, but unlike the latter, these buds are never set free. The life cycle of *Tubularia* (Fig. 9.4d) also includes attached gonophores, but these are so simplified that they no longer bear much resemblance to a jellyfish. The egg is fertilized *within* the parent gonophore, becomes a planula, and is finally liberated as an actinula larva. This soon settles and becomes the first polyp of the new colony.

9.18.   What are the characteristics of hydras?

Hydras (order Hydroida) are solitary, freshwater polyps having body lengths of 3 cm or less and 6–10 hollow tentacles surrounding a mouth raised on a conical *hypostome* (see Figs. 9.2 and 9.4e). They are not attached and form no skeleton. A hydra can glide on its basal disc or somersault by bending over to attach its tentacles to the substratum, then detaching its disc and looping across to reattach beyond the tentacles. A hydra with grand ideas can secrete a gas bubble on its basal disc and float to the surface, where it can reattach to vegetation such as the underside of water lily pads.

Hydras are versatile at reproduction: they can produce asexual buds that eventually detach as small versions of the adult and, at the same time, develop temporary gonads. A conical testis liberates sperm into the water. Eggs mature singly and are retained in the parent's body wall while being fertilized and developing into an embryo. A planula stage is lacking. Instead, a hollow blastula is formed, and during gastrulation, the inner cells simply delaminate to form the endoderm, which becomes the gastrodermis. Only a thin membranelike layer of mesoglea is secreted between the two cell layers. Before the embryo is set free, it is enclosed in a protective cyst in which it can pass the winter, hatching in spring as a miniature hydra. The gastrodermis of green hydras (*Chlorohydra*) contains quantities of unicellular green algae as symbiotic *zoochlorellae*, which nourish their hosts when prey becomes scarce.

9.19.   What are the characteristics of hydrocorals?

Hydrocorals (order Hydrocorallina) are colonial, secreting an internal, epidermal skeleton of calcium carbonate, with pores through which the zooids emerge (Fig. 9.4f). Hydrocorals may be branching or encrusting, forming extensive masses that are often bright yellow or orange owing to symbiotic zooxanthellae. Gastrozooids are surrounded by fingerlike dactylozooids armed with powerful nematocysts. Although charitably called "defensive zooids," the dactylozooids of *stinging* or *fire corals* (*Milleporina*) are potently offensive! When the zooids are withdrawn, hydrocorals can be distinguished from small types of stony corals (anthozoans) because the latter have radiating septa in each skeletal cup.

9.20.   What hydrozoans form floating colonies?

Floating colonies occur in the orders Chondrophora and Siphonophora. *Chondrophores* such as *Velella*, the purple by-the-wind sailor, have a single, large, central gastrozooid, which not only does all the feeding for the colony, but also forms a chitinous float in the shape of a horizontal platform; from this platform hang a marginal ring of dactylozooids and a number of gonozooids bearing gonophores (Fig. 9.5a). *Velella*'s obliquely set sail usually keeps it from being driven ashore as it scuds lightly before the wind.

Siphonophores (Fig. 9.5b), such as *Stephalia* and *Physalia*, have a gas-filled float formed by the body of the original zooid, the pneumatophore, which in *Physalia* (Portuguese man-of-war) may reach a length of 30 cm with a median sail-like portion so structured that the colony sails about 45 degrees to the right or left of the wind direction. *Physalia*'s float contains gas of the same composition as air, which cannot be expelled to allow the colony to submerge. But many other siphonophores (such as *Nanomia*, which has a float filled 90 percent with carbon monoxide) rise by secreting gas into the float from a gas gland and sink by releasing it, thereby performing extensive vertical migrations. Below the float hang the other zooids of the siphonophore colony: (a) bottle-shaped gastrozooids that can plaster their expanded mouths over the body of fish too large to ingest whole; (b) gonozooids that bear gonophores or, in some species, liberate free medusae; and (c) dactylozooids, which in *Physalia* may reach a length of 10 m and bear nematocysts that can inflict painful stings upon swimmers. (The neurotoxin is somewhat neutralized by a liberal application of meat tenderizer.) Fish merely brushed by the streaming dactylozooids are promptly paralyzed, but it may take hours for the feebly contractile fishing zooids of a large *Physalia* to elevate the prey within reach of the waiting gastrozooids. Although *Physalia* merely sails serenely afloat, more ambitious siphonophores jet along by pulsations of muscular "swimming bells" (*nektophores*), which are modified attached medusae.

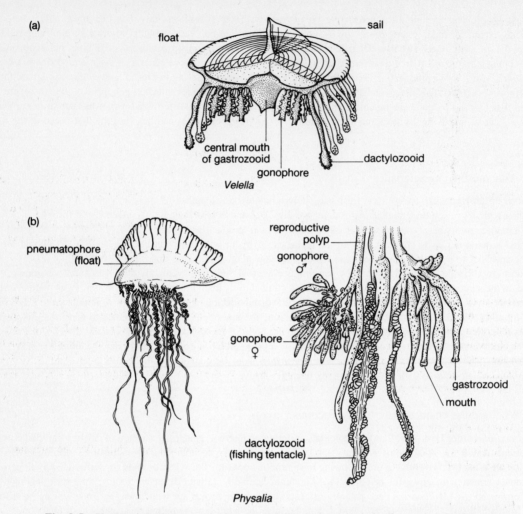

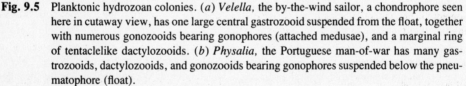

Fig. 9.5 Planktonic hydrozoan colonies. (*a*) *Velella,* the by-the-wind sailor, a chondrophore seen
here in cutaway view, has one large central gastrozooid suspended from the float, together
with numerous gonozooids bearing gonophores (attached medusae), and a marginal ring
of tentaclelike dactylozooids. (*b*) *Physalia,* the Portuguese man-of-war has many gastrozooids, dactylozooids, and gonozooids bearing gonophores suspended below the pneumatophore (float).

9.21. What are cubozoans?

Class Cubozoa includes the cubomedusae, curious jellyfish that are squarish in cross section (Fig. 9.6). The nearly
cuboidal bell is usually only 2–3 cm tall, but in some species reaches 25 cm. Each corner of the bell margin bears one
or a group of tentacles, which may be several meters long in large individuals. The base of each tentacle is a flattened
blade (*pedalium*). Sense organs (rhopalia) bearing ocelli and statocysts occur along the sides of the bell margin. The
cubomedusan ocellus is the most complicated eye found in any radiate animal. The round lens focuses light upon the
retina, and the underlying pigment layer screens the eye so that light cannot reach the retina except through the cornea,
which allows the positively phototactic cube jelly to tell quite accurately the direction from which light is coming.
Cubomedusae are especially strong swimmers, for the bell margin turns inward as a flaplike *velarium* that much concentrates
the water jet. This membrane functions like a hydromedusan velum, but differs structurally. Rather than having radial
canals extending from the stomach through the mesoglea, cube jellies have four *radial pouches* so extensive that the body
wall remains solid only at the four corners.

The cubozoan life cycle remains unknown except for one species, *Tripedalia cystophora*. Its planula transforms into
a solitary sessile polyp only 1 mm tall. This may bud off a few more polyps, which creep away. Each polyp eventually
metamorphoses directly into a cubomedusa.

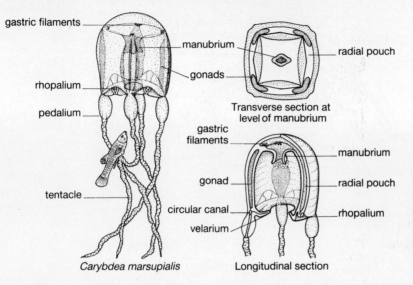

gastric filaments

manubrium

gonads

radial pouch

Transverse section at level of manubrium

rhopalium

pedalium

gastric filaments

manubrium

gonad

radial pouch

tentacle

circular canal

rhopalium

velarium

Carybdea marsupialis   Longitudinal section

Fig. 9.6 Class Cubozoa. The cubomedusan *Carybdea marsupialis*, a sea wasp. Note square cross section (*upper right*) and shelflike velarium (*lower right*) imparting speed in swimming.

The nematocyst toxins of the sea wasp, *Chironex*, inflict excruciating stings that can bring death to imprudent bathers, especially along the Queensland coast of Australia and other tropical inshore waters. If death does not occur within 20 minutes, the victim may survive the extensive tissue necrosis of the afflicted parts, but bear a web of lifelong scars as a reminder of having been enmeshed in a fiery tangle of sticky tentacles. Ingenious Aussies devise a complete nylon body suit from two pairs of panty hose, one put on the usual way, the over over hands, arms, and torso, with a slit for the head, which had better be kept above water unless another stocking is added as a mask. Strangely, the deadly nematocysts of *Chironex* do not sting through panty hose, but watch out for runs!

9.22. What are the scyphozoans?

Class Scyphozoa includes some 200 species of "true" jellyfish, or scyphomedusans, which often are quite large and have brilliantly colored internal organs.

9.23. How are scyphomedusae distinguished from hydromedusae?

1. Scyphomedusae are generally much larger than hydromedusae; in fact, the dark-blue *Cyanea* may have a bell 2 *meters* in diameter, with tentacles as much as 70 meters long!

2. Scyphomedusae usually have quite short marginal tentacles and fish by means of four long or highly frilled oral lobes suspended from the manubrium.

3. Scyphomedusae have no velum or velarium extending inward from the bell margin, but a powerful *coronal muscle* that rings the bell margin helps compensate for this lack.

4. The bell margin is usually scalloped, rather than entire, with a sensory rhopalium, flanked by a pair of lappets, lying within each indentation.

5. The stomach is not a simple sac, but extends into four gastric pouches bearing pendant *gastric filaments*, which give off digestive enzymes and are also armed with nematocysts that continue to subdue prey even after ingestion.

6. The gonads do not develop from the epidermis, but from the gastrodermal lining of the gastric pouches, so that the gametes are shed through the mouth.

7. The mesoglea is not acellular, but contains amoebocytes and connective fibers more massive than in hydromedusae. It is penetrated by a complex system of branching radial canals that eventually join a marginal ring canal. (Hydromedusae have only four radial canals plus the ring canal.)

8. The medusa is always the dominant generation, while the polyp generation is tiny or absent.

   Example 3: *Aurelia*, the plate-sized moon jelly, exemplifies a typical scyphozoan life cycle (Fig. 9.7). The male and female medusae shed sperm and eggs, respectively, the zygote develops into a *planula*, and this settles and becomes a hydralike polyp, the *scyphistoma*. The latter eventually undergoes strobilation (transverse budding), releasing young medusae (*ephyras*) serially from the outer

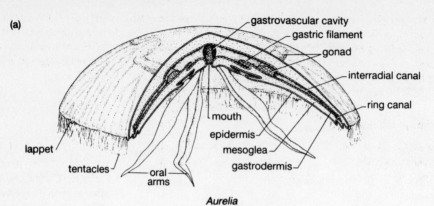

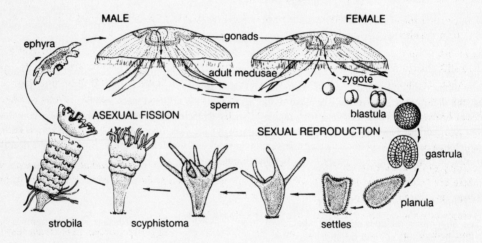

Fig. 9.7 Class Scyphozoa, *Aurelia*. (*a*) Structure; note long oral arms and short marginal tentacles, lappets marking the location of a sensory rhopalium, and gastrodermal origin of gonads. (*b*) Life cycle with reduced polypoid generation (not to scale); young medusae (ephyrae) arising from polyp by transverse budding (strobilation). (*From Storer et al.*)

end of the strobila. Various modifications of this representative life cycle occur. A few species have branching, colonial scyphistomas supported by a tubular skeleton. Some pelagic scyphozoans retain the scyphistomas in gastric brood pouches, or delete the polyp generation entirely. The single egg of *Pelagia* develops directly into a miniature adult.

9.24. What are the major types of scyphozoans?

(*a*) *Semaeostomids* (order Semaeostomae) are "typical" jellyfish with bowl- or saucer-shaped bells bearing scalloped margins. They are of worldwide distribution, from polar to tropical seas, and include the above-mentioned *Cyanea*, *Aurelia*, *Pelagia*, and the dreaded "sea nettle," *Chrysaora*.

(*b*) *Coronatids* (order Coronatae) are the only scyphomedusae to have a nerve ring for pulsation control. They are easily recognized by a deep constriction of the bell at a *coronal groove*. Many live in the deep sea.

(*c*) *Stauromedusans* (order Stauromedusae) such as *Haliclystus* (see Fig. 9.1*a*) mainly inhabit frigid coastal waters. They hang from algae and are attached by a stalk that projects from the aboral surface of the trumpet-shaped body, waiting serenely in ambush for crustaceans and young fish.

(*d*) *Rhizostomids* [order Rhizostomae ("root mouth")] lack marginal tentacles but have immensely frilled oral lobes bearing many *secondary mouths* that lead by way of canals into the stomach. The original mouth is lost by fusion of the oral arms.

Example 4: The rhizostome medusa *Cassiopeia* is particularly interesting. It lies upside down on the floor of shallow mangrove swamps, sunbathing the symbiotic algae (zooxanthellae) that crowd its mesoglea so densely that *Cassiopeia* can live entirely on the products of photosynthesis. Even so, the medusa is not averse to swallowing plankton and small animals that become trapped in mucus on its frilly oral arms and are carried by mucus cords into the secondary mouths.

9.25. What are the anthozoans?

Class Anthozoa includes over 6000 species of coelenterates that exist only as polyps, sometimes solitary, more frequently colonial. These polyps reproduce both asexually and sexually, and a medusoid stage is totally lacking. They include some of the most familiar radiates, as well as the most ecologically and economically important.

9.26. How do anthozoan polyps differ from hydrozoan polyps?

1. Although anthozoans range widely in size, the individual anthozoan polyp (Figs. 9.8 and 9.9) is usually considerably larger than a hydrozoan polyp.

2. The mouth of an anthozoan, centered within an *oral disc,* is not round but slitlike, often bearing at one or both ends a stiff ciliated groove (*siphonoglyph*) that serves to maintain water circulation through the gastrovascular cavity even when the mouth is closed. These features modify the primary radial symmetry of anthozoans.

3. The mouth does not open directly into the gastrovascular cavity, but into a flattened, tubular *pharynx* lined with ciliated epidermis, which extends inward at least half the depth of the gastrovascular cavity.

4. The mesoglea, though thin, is nevertheless much thicker than that of hydrozoan polyps and is densely fibrous, with many amoebocytes.

5. The lining of the anthozoan gastrovascular cavity is much increased by series of *radially arranged partitions;* some of these are "complete mesenteries" anchored to both pharynx and body wall, while others are "incomplete mesenteries" that reach only part way into the digestive cavity.

6. Nematocysts occur in the mesenteries, not just in the epidermis.

7. In most anthozoans the muscular part of the epidermal T cells has disappeared, except in the tentacles and oral disc, so that body contractility relies solely on the circularly oriented bases of gastrodermal cells, together with longitudinally oriented muscle fibers in the mesenteries. Where the body column meets the oral disc, the circular gastrodermal fibers form a ringlike *sphincter* that covers the oral disc when the tentacles retract, so that both mouth and tentacles are hidden from view.

8. Except for hydras, which form temporary epidermal gonads, hydrozoan polyps reproduce asexually. Anthozoans routinely develop gonads in their mesenteries, and gametes are shed through the mouth; some retain their eggs and brood them internally. A number of anthozoans are *protandrous hermaphrodites,* producing first sperm and later eggs. The zygote develops into a planula larva that settles to become a new polyp.

9. Colonial hydrozoans differentiate at least two types of polyps: feeding and reproductive. Colonial anthozoans are more frequently monomorphic.

10. The two major subclasses of class Anthozoa—Zoantharia and Alcyonaria—display, respectively, *hexamerous* symmetry (in which tentacles and mesenteries occur in multiples of six) and *octamerous* symmetry (having eight tentacles and septa). (Hydrozoan polyps lack mesenteries and have tentacles that vary in number but tend to be few.)

9.27. What are the alcyonarians?

Subclass Alcyonaria, the *octocorals,* have eight *pinnate* tentacles (with side branches like feathers); eight complete, *unpaired* mesenteries; and a *single* siphonoglyph (Fig. 9.8). Nearly all are colonial, with tiny zooids that are interconnected by a *coenenchyme,* consisting of a fleshy mesoglea penetrated by numerous gastrodermal tubes that are continuous with the stomachs of individual polyps. The coenenchyme is covered externally by a layer of epidermis that is continuous with that of the polyps. Amoebocytes in the coenenchyme secrete an *internal* skeleton of scattered or fused calcareous spicules, or of a horny material. The skeleton is sometimes richly colored, as in precious red coral (*Corallium*), used in making jewelry. Many octocorals contain symbiotic yellow-brown plantlike protists (zooxanthellae). Subclass Alcyonaria includes orders Stolonifera, Alcyonacea, Pennatulacea, and Gorgonacea.

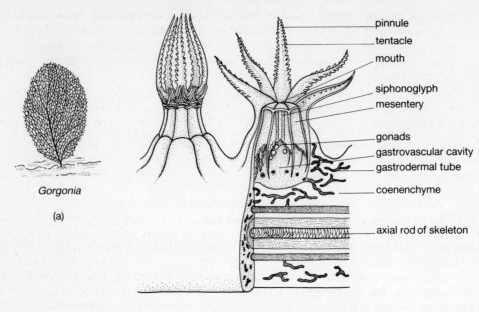

pinnule
tentacle
mouth
siphonoglyph
mesentery
gonads
gastrovascular cavity
gastrodermal tube
coenenchyme
axial rod of skeleton

Gorgonia

(a)

(b)

Fig. 9.8   Class Anthozoa, subclass Alcyonaria, a sea fan. (*a*) General growth form of the sea fan
Gorgonia. (*b*) Structural detail of colony: note eight pinnate tentacles per polyp, internal
radial mesenteries, and internal colonial skeleton.

9.28.   What are the characteristics of stoloniferans?

Stoloniferans, such as organ-pipe coral (*Tubipora*), lack a coenenchymal mass, but produce a creeping mat of *stolons*
from which the polyps arise; a skeleton may be lacking, or composed of calcareous tubes or separate limy spicules.

9.29.   What characterizes order Alcyonacea?

Alcyonaceans, the *soft corals,* are mostly tropical. Their polyps, extended for feeding during the daytime, wave
gracefully above a rubbery mass of coenenchyme into which they withdraw when disturbed. The skeleton consists only
of scattered calcareous spicules. The colony form is irregular or encrusting, or has fingerlike or lobate projections, and
may reach a diameter of a meter.

9.30.   What characterizes the pennatulaceans?

A *sea pen* has small secondary polyps borne on a large *primary polyp* that has a stemlike base that anchors in sand,
and either an elongate body giving it the look of a quill pen or a platformlike fleshy body, as in the purple sea pansy,
Renilla. Minute, scattered calcareous spicules represent the skeleton.

9.31.   What are the characteristics of gorgonians?

These are the *horny corals,* which secrete an upright, branching skeleton of organic material (*gorgonin*) and may
also produce separate or fused limy spicules. Order Gorgonacea includes the familiar sea whips and the sea fan, *Gorgonia*.

9.32.   What are the ceriantipatharians?

Subclass Ceriantipatharia includes *tube anemones* (order Ceriantharia) and *black corals* (order Antipatharia), which
seem very unlike but do possess in common a unique internal organization of numerous, unpaired radial septa, all of
them extending from body wall to pharynx. Cerianthids are solitary polyps with many tentacles and elongated bodies
several centimeters long, which live buried in sand in tubes made of the threads and capsules of discharged ptychocysts.
When extended for feeding, the oral disc lies flat upon the sand, with tentacles radiating in all directions. By contrast,
the black corals are colonial, producing erect, branching skeletons of a horny material bearing *thorns,* and their polyps

have only six simple and nonretractile tentacles. Although black corals live in relatively deep tropical waters, they are sought by divers for their skeletons, which are used in jewelry and are becoming rare.

9.33. What are the zoantharians?

Subclass Zoantharia includes hexamerously symmetrical polyps having more than eight tentacles that are rarely pinnate, six (or multiples of six) *pairs* of internal mesenteries, and nearly always *two* siphonoglyphs. They include both solitary and colonial forms, and if a skeleton is produced, it consists of solid calcium carbonate and is always *external* to, and secreted by, the epidermis. This subclass includes the sea anemones, corallimorphs, and stony corals.

9.34. What are the characteristics of sea anemones?

Sea anemones (order Actiniaria) are solitary polyps that attach by a flat *basal disc* to many solid objects—piers, rocks, shells—in coastal waters throughout the world. Both the body column and the numerous tentacles are often brilliantly colored. They range in size from diameters of 4 mm to over a meter at the oral disc. A few can inflict stings painful to humans. The arrangement of internal septa is complex, with six pairs each of primary (complete), secondary, and tertiary mesenteries (Fig. 9.9). In many species, the lower edge of their incomplete mesenteries is extended into threadlike *acontia* armed with nematocysts. The acontia can often be protruded through pores in the body wall as an extra means of defense or aggression. Sea anemones possess not only nematocysts, but also cnidae known as *spirocysts*, which have an unarmed, spirally coiled adhesive thread used in prey capture and attachment to the substratum. Some anemones cover their body columns with protective bits of shell. Despite their strong adhesion to the substratum, which can defy the most powerful surf, anemones can creep on the flattened basal disc, and a few are even known to break free and swim away from predators by rotational flexing movements. *Stomphia* responds in this manner to the approach of a species of starfish that preys on it (or to an extract of this predator) but does not swim away from starfish that do not attack it.

Sea anemones often reproduce asexually, by longitudinal fission, transverse fission, or *pedal laceration,* in which fragments of the pedal disc are pinched off. Some species form entire clones of asexually produced individuals that remain closely aggregated. Clear zones are maintained between adjacent clones, because individuals of more belligerent clones attack those of less aggressive clones, causing them to flinch away. When reproduction is sexual, some anemone species are dioecious ("two-house," i.e., each sex housed in a separate body), while others are monoecious ("one-house," i.e., both sexes in the same body) and protandrous ("first male"), producing first sperm and later eggs. Gametes may be shed through the mouth for external fertilization, the zygote producing a hollow blastula that gastrulates by invagination or ingression to form a typical planula larva. Some species produce planulae provided with enough yolk to sustain them until settlement, whereas the planulae of other species feed on plankton, gradually developing a pharynx and mesenteries, but not growing tentacles until after attachment.

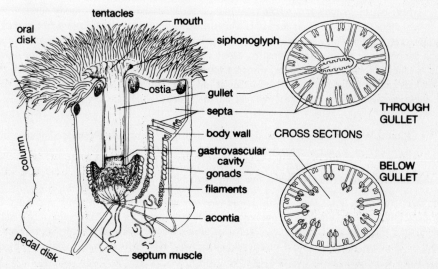

Fig. 9.9 Class Anthozoa, subclass Zoantharian, a sea anemone. *Left,* longitudinal cutaway view; *right,* cross sections showing arrangement of internal septa (mesenteries). (*From Storer et al.*)

Although most anemones capture fish and other large prey by means of their tentacles, some species bear only short tentacles and feed on plankton trapped on their surface by mucus and cnidae and transported to the mouth by ciliary action. Certain anemones are partners in interesting mutualistic symbiotic relationships. Many host symbiotic unicellular algae housed in their gastrodermal cells. Some are hosts for small, bright damselfish, known as "anemone fish" because they survive only in intimate contact with certain large anemones.

> **Example 5:** After the moments of first acquaintance, when the damselfish contacts the host tentatively and flinches as if stung, host and symbiont become adapted and the fish's mucus covering comes somehow to inhibit nematocyst discharge. The fish actually squirms down among the host's tentacles as if for contact comfort and, no doubt, gains protection from predators. The anemone benefits from its symbiont's movements, which dislodge silt, and sometimes from the snaring of fish that pursue anemone fish into their host's lethal thicket of tentacles. At least in aquaria anemone fish drop surplus food among their host's tentacles, probably as a version of food-caching behavior that inadvertently provides food for the host.

Symbiotic relationships also exist between certain anemones and hermit crabs. The anemone affords its host protection while getting a free ride on the snail shell inhabited by the crab. When the growing hermit crab switches to a new shell, it gently holds and massages its anemone until the latter glides over from the abandoned shell to the new one.

9.35.   What are the characteristics of corallimorphs?

Corallimorphs (order Corallimorpharia) form colonies of what appear to be pint-sized anemones, often with knobs on the ends of their radially arranged tentacles. However, the zooids are physically conjoined and resemble stony corals, except that no skeleton is formed. Beautiful *Corynactis*, the "beadlet anemone," is one of few red species that still look red in deep water: it is fluorescent, absorbing the shorter wavelengths that penetrate to those depths and reradiating them as longer, red wavelengths.

9.36.   What are the characteristics of stony corals?

Stony corals (order Scleractinia) are mostly colonial and secrete a heavy *external* skeleton of calcium carbonate. The zooids are interconnected laterally by a sheet of tissue that is actually a fold of body wall containing extensions of gastrovascular cavity. However, each has its own skeletal cup into which it can withdraw by squeezing its body down between the numerous radial partitions (*sclerosepta*) that project up into the polyp's body, pushing up the basal layers and fitting between each pair of mesenteries (Fig. 9.10*a*). Although the polyps are usually only 1–3 mm in diameter, reef-building scleractinians construct massive colonial skeletons, for the polyps actually ride on top of their skeletons, and their epidermal tissue continues to deposit new calcium carbonate for as long as the colony lives. Individual zooids die and are reabsorbed, but the colony as a whole may live on for thousands of years. Stony corals rely on symbiotic zooxanthellae to help them fix calcium from seawater. This the algae seem to do by using up free carbon dioxide, with the result that more carbon dioxide is set free by the conversion of soluble calcium bicarbonate to insoluble calcium carbonate, which the polyps secrete as skeletal matrix. Stony corals cannot live in water too deep for light to reach their symbiotic algae, so they colonize shallow water, growing toward the surface until at sea level the top of the reef dies off, since most corals cannot endure long exposure to air at low tide. However, sea levels have been rising ever since the peak of the latest ice age (15,000 years ago), and reef-building corals, which are restricted to tropical seas, have been keeping pace.

> **Example 6:** At Eniwetok atoll, which rises from a drowned volcanic seamount, the outer wall of the reef now plunges vertically more than 2 km to the original reef base. Some scleractinian colonies of the Great Barrier Reef of Australia may be the world's oldest living things, for they began to grow when the continental shelf was first inundated and have apparently been at work ever since.

Coral reefs take three main forms (Fig. 9.10*b*). A *barrier reef* usually lies well offshore, separated from land by a wide lagoon or even miles of open sea. A *fringing reef* grows directly out from the land, with only a narrow lagoon, if any, separating it from the shore. An *atoll* is a reef that encircles a central lagoon, usually as a result of the coral's building up from a subsiding volcanic cone. Coral islands form wherever drifts of coral sand pile up above sea level; then birds and currents bring plant seeds that root and build soil.

Coral reefs, added to by calcium-fixing (coralline) algae, form habitats for many forms of life that could not exist in the open sea.

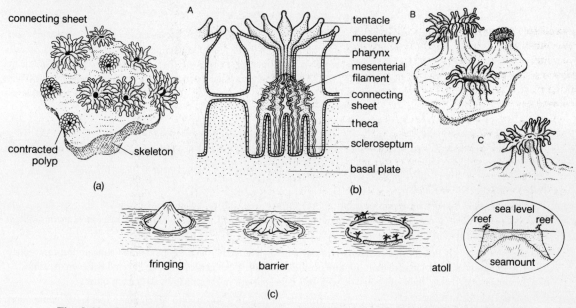

Fig. 9.10 Class Anthozoa, subclass Zoantharia, stony corals. (*a*) General growth form, as seen from above. (*b*) A, longitudinal section of a coral polyp, showing sclerosepta of its skeletal cup; B, expanded polyps in lateral view; C, new polyp arising by budding. (*c*) Types of coral reefs.

Example 7:   Coral reefs support entire animal communities that rely mainly on the photosynthetic activity of the unicellular zooxanthellae that live within the coral polyps. Some fish and invertebrates feed directly on coral, including the infamous crown-of-thorns starfish, which presses its everted stomach onto the living coral and soon moves on leaving a gleamingly bare skeleton behind. Parrot fish do not spare even the skeleton: they bite off entire chunks of coral, digesting the flesh and voiding clouds of coral sand. A host of small animals shelter among the corals, and predators hover in wait.

No ecosystem other than the tropical rain forest rivals coral reefs in complexity and productivity. Entire human cultures have long depended on the bounty of lagoons protected by coral reefs. Deplorably, valuable reefs are now dying from pollution and siltation wherever human populations and construction are on the rise.

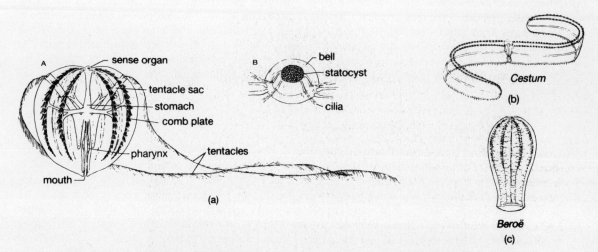

Fig. 9.11 Ctenophores. (*a*) *Pleurobrachia*, a tentaculate comb jelly: A, structure; B, detail of apical sense organ. (*b*) The aberrant tentaculate *Cestum* (Venus's girdle). (*c*) *Beroë*, a nude (tentacleless) ctenophore. (*From Storer et al.; after Hyman.*)

9.37.   What are comb jellies?

Comb jellies include about 60 species of marine planktonic animals belonging to *phylum Ctenophora* ("comb bearer"), so called for their eight meridional rows of *comb plates* formed from the fusion of long cilia. By day they flash prismatically as their ciliary plates refract light; at night they are often bioluminescent, glowing like little lamps. Most are about the size and shape of walnuts or gooseberries (Fig. 9.11*a*). Ctenophores share certain basic radiate characteristics with coelenterates (see Question 9.2), but although a large amount of mesoglea gives them some resemblance to a cnidarian medusa, there are many differences between medusae and ctenophores.

9.38.   How do ctenophores compare with medusae?

(*a*) Ctenophores are *biradially* symmetrical, because most have a single pair of oppositely placed *tentacles*, which may or may not retract into sheaths.

(*b*) Ctenophores do not swim by jet propulsion, aboral surface first, but propel themselves *mouth first*, by their *eight rows of ciliated comb plates*. They usually maintain a vertical position, *mouth upward*, whereas most medusae float mouth downward and swim with mouth and tentacles trailing.

(*c*) The nervous system is limited to a *subepidermal plexus*, concentrated around the mouth and beneath the comb-plate rows. A single statocyst is located aborally, and ocelli are absent. If the nerve fibers are severed between the statocyst and the comb-plate rows, the eight rows of cilia still beat, but not in coordination with each other.

(*d*) Each of the two long, *branching* tentacles arises from the bottom of a pouch, into which many species of ctenophores can retract their tentacles. Except in one species, nematocysts are lacking, and instead the tentacles bear unique "glue cells" (*colloblasts*). A colloblast is cone-shaped, with a helical thread that spirals outward from the cell's narrow inner end, then splays out as individual fibrils, each terminating in a granule of adhesive material. When caught by these colloblasts, even an active young fish cannot escape the voracious ctenophore, but the more usual prey is small crustaceans. When prey is captured, the tentacles are retracted to bring food to the mouth.

(*e*) The distribution of canals that arise from the stomach is biradial rather than radial, with two canals paralleling the long tubular pharynx, two serving the tentacle sheaths, and two pairs of meridional canals positioned on each side. An aboral canal extends from stomach to apical statocyst, where it branches to open to the surface as a pair of minute anal pores flanking the statocyst.

(*f*) The mesoglea contains connective fibers and amoebocytes of ectodermal origin, as in scyphomedusae, but the mesoglea of ctenophores also contains an anastomosing network of smooth muscle fibers. Within the tentacles these muscle fibers are organized into bundles, so that the tentacles are highly motile and retractile.

(*g*) All ctenophores are monoecious, simultaneously possessing a straplike ovary and testis in the wall of each meridional canal. Gametes are shed through the mouth and fertilized externally, except for species that brood their eggs.

(*h*) Whereas most medusae arise asexually from a polyp generation, ctenophores have no polypoid state, are never polymorphic, and reproduce only sexually. Cleavage is determinate (mosaic) in ctenophores, but indeterminate in coelenterates. The blastula is not hollow but solid, with outer micromeres enclosing a few macromeres. Gastrulation involves invagination and epiboly (i.e., extension of one part of the blastula over and around another part). No planula larva is formed, but instead the gastrula quickly develops into a free-swimming, spheroidal *cydippid larva* that closely resembles adult sea gooseberries (*Pleurobrachia*), although it may transform into quite a different adult shape.

9.39.   What are the important groups of ctenophores?

(*a*) *Class Tentaculata* includes comb jellies having a pair of tentacles, although these may be much reduced in some species. *Cydippids* such as *Pleurobrachia* (Fig. 9.11*a*) are most representative of their class and develop without much modification from the cydippid larva. *Cestids* such as Venus's girdle (*Cestum*) are much elongated in the plane of the tentacle sheaths; that is, the body extends laterally into two equal beltlike lobes on each side of the mouth, and the animal swims by eel-like muscular undulations, as well as by its ciliary combs (Fig. 9.11*b*). *Cestum* may reach a length of 1 m, and truly resembles a living belt. The *platyctenids* such as *Ctenoplana* are creeping benthonic forms, greatly flattened dorsoventrally by a profound shortening of the oral-aboral axis. Although the tentacles are retained on what has become the dorsal surface, the comb rows are much reduced or lacking in the adult.

(*b*) *Class Nuda* includes ctenophores such as *Beroë*, which lack tentacles (Fig. 9.11*c*). Swimming along with its huge mouth forward, *Beroë* simply engulfs any animal that absentmindedly neglects to get out of its way.

Chapter 10

The Acoelomate Bilateria

Of all bilaterally symmetrical animals, flatworms of *phylum Platyhelminthes* ("flat worm") are considered most primitive. The platyhelminthes, the tiny gnathostomulids (*phylum Gnathostomulida*), and the nemerteans, or ribbonworms, of *phylum Rhynchocoela* exhibit an *acoelomate* level of body organization: they have no body cavity separating the body wall from the gut and other internal organs. Many biological advances in organization and function first appear in these three phyla of acoelomate Bilateria, which occupy a transitional status between the radiates and the higher Bilateria.

10.1. What is the evolutionary status of acoelomates?

There is little doubt that nemerteans arose from free-living flatworms (turbellarians), but the origin of the latter from the Radiata remains more inferential than established. Like cnidarians, flatworms have a gastrovascular cavity with a mouth but no anus, and many complete the digestive process intracellularly, in food vacuoles within the gastrodermis. The position of the turbellarian mouth is primitive, being approximately at the middle of the underside, as is also the case in the dorsoventrally flattened, creeping platyctenoid comb jellies. Cleavage in the more primitive turbellarians is spiral, a characteristic shared with ctenophores, and a stereogastrula is produced by epiboly. The mouth opens near the position of the blastopore, and the internal mass of endoderm cells rearrange to form the wall of a saccular gut, which later becomes branched in the more sophisticated turbellarians. The turbellarian epidermis characteristically contains capsular *rhabdites* of complicated internal structure that may be evolutionarily related to the cnidae of cnidarians; produced by deeper-lying *rhabditogen* cells, rhabdites migrate outward to lie between the epidermal epithelial cells. They discharge, perhaps defensively, and disintegrate into a slimy coating.

Two orders of especially simple and presumably primitive turbellarians—*acoels* and *macrostomids*—have been proposed alternatively as resembling the body plan of an ancestral flatworm. Like modern acoels, an *acoeloid* ancestor would have lacked a gastrovascular cavity, the mouth opening upon a loose mass of digestive cells. The nervous system would have consisted merely of a subepidermal network, with a further plexus around an anterior statocyst. By contrast, as in modern macrostomids, the mouth of a *macrostomoid* ancestor would have opened into a simple, saccular digestive cavity (Fig. 10.1). Whereas macrostomids today have separate male and female openings (gonopores) and a copulatory apparatus, a macrostomoid ancestor may have lacked these features. Modern macrostomids have a pair of nerve cords that extend longitudinally from the anterior statocyst, an arrangement somewhat reminiscent of the way in which the radial nerves of ctenophores spring from their aboral statocyst. However, an ancestral macrostomoid might have possessed only a subepidermal nerve net, instead of longitudinal cords.

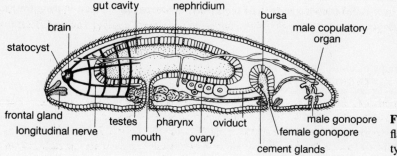

Fig. 10.1 Postulated ancestral flatworm of the macrostomoid type.

10.2. What biological advances appear in flatworms?

(*a*) *Bilateral symmetry* is characteristic of animals with a lively approach to the world: actively swimming or creeping in pursuit of food and other amenities. It is prerequisite to the evolutionary advancement of all higher metazoan phyla.

(*b*) *Cephalization*, the development of a head at the anterior end, bearing a concentration of sensory cells (receptors) and sense organs that meet the environment first as the animal moves along, appears in flatworms even before the relocation of the mouth from a ventral to an anterior position.

(*c*) *Centralization* of the nervous system involves a massing of neural cells into a brain and major longitudinal nerve trunks. The brains of more advanced, creeping turbellarians receive sensory input from cephalic receptors, including *chemoreceptors* which detect solutes, *rheoreceptors* which detect water currents, *thigmoreceptors* which detect tactile contacts, and *photoreceptors* which detect light. (Acoels and macrostomids are tiny forms that swim among bottom detritus; their major cephalic neural concentration surrounds a gravity-detecting statocyst.) The longitudinal nerve trunks are not simply bundles of fibers: they contain nerve cell bodies and synaptic junctions, as well as fibers; these nerve trunks or cords give off motor fibers to the periphery and receive sensory ones.

(*d*) *Mesoderm* as a true embryonic germ layer first appears in flatworms, which represent the most primitive indisputably triploblastic phylum. Some zoologists contend that radiates with cells in their mesoglea should be considered triploblastic, but such cells are derived from ectoderm and do not appear to be homologous with mesodermal tissues of higher metazoans. In any event, the true mesoderm that appears in flatworms provides a new developmental potential, for it differentiates into many structures that are either rudimentary or lacking in the radiates.

10.3. What structures differentiate from the flatworms' mesoderm?

(*a*) *Muscle layers*—from outside in: circular, diagonal, and longitudinal—allow wormlike lengthening, shortening, and flattening of the body and permit the flatworm to turn, twist, raise or lower its ventral surface, and sometimes to swim with a beautiful skirtlike movement brought about by metachronous waves of lateral contraction.

(*b*) *Protonephridia* ("first kidneys") carry out osmoregulation and excretion (see Question 10.10).

(*c*) *Accessory reproductive structures*, including genital ducts, penis, seminal glands, and yolk glands, enable flatworms to lay internally fertilized eggs within shells that protect them while embryonic development proceeds. The capacity of shelled flatworm eggs to survive exposure on land paved the way for the exploitation of terrestrial animals as hosts for many parasitic species of flukes and tapeworms.

10.4. What are turbellarian flatworms?

Turbellarians, the free-living flatworms, are both the most diverse and the most representative members of the phylum, for they have not undergone the structural and functional modifications selected for by a parasitic mode of life. Nearly all turbellarians are aquatic, some in fresh water, and many are found beneath rocks intertidally. However, a few species are terrestrial in moist habitats and reach impressive lengths of up to 60 cm! It is somewhat of a shock, during the rainy season of southern California, to find occasionally a brown, spatulate-head flatworm, as long as a snake, bumbling confusedly across one's lawn, probably lured by rain into forsaking the protection of a greenhouse. Most turbellarians glide eerily along a mucous track propelled by epidermal cilia, but larger species also employ muscular waves of contraction that, by sweeping lengthwise from anterior to posterior, lift and lower portions of the ventral surface, a type of creeping highly developed in snails. Nearly all turbellarians are predatory, although some feed on diatoms, especially when young. The predilection of planarians for liver-baited traps suggests some propensity for scavenging, at least in these freshwater forms.

10.5. What are the major groups of class Turbellaria?

The two major groups of turbellarians are the archoophorans and the neoophorans.

10.6. What are the archoophorans?

Archoophoran turbellarians include the more primitive orders, which lack yolk glands and have *endolecithal* eggs (i.e., eggs with internal yolk) that develop by spiral cleavage.

(*a*) *Acoels* (order Acoela) are marine, usually less than 2 mm in length, with a ventral mouth but no digestive cavity, a nervous system limited to plexuses beneath the epidermis and around the anterior statocyst, no protonephridia or genital ducts, and no definite gonads, with gametes arising instead from scattered *gametogonia* cells. Acoels swim along the bottom by means of cilia.

(b) *Macrostomids* (order Macrostomida) are tiny marine and freshwater species with a simple, saclike digestive cavity and no protrusible pharynx. Their primitive brain is centered upon the statocyst and gives off a pair of lateral nerve cords. Both ovaries and testes are present, with separate gonopores and a penis for sperm transfer.

(c) *Polyclads* (order Polycladida) are impressive marine turbellarians with greatly flattened ovoid bodies that seldom exceed 2 cm, although some grow larger than a saucer. Many are gaudily colored, especially tropical species that feed on coelenterates and store undischarged nematocysts in their skin as a borrowed means of self-defense. Their gastrovascular cavity is elongate, with many highly branched lateral pouches (*diverticula*). The extensible pharynx is either tubular or bell-shaped, draping over intended prey like a great hooped skirt. They have well-developed brains and numerous eyes.

10.7. What are the neoophorans?

Neoophoran turbellarians have yolk glands and *ectolecithal* eggs, i.e., the ovum is laid surrounded by a mass of *yolk cells* from which it derives nourishment while the embryo develops within the egg capsule. The presence of yolk cells has greatly modified the developmental pattern from that seen in archoophoran turbellarians. The neoophorans include a number of orders of organisms having greater structural complexity than the majority of archoophorans. Only two of these will be considered here.

(a) *Rhabdocoels* (order Neorhabdocoela) represent a large group of marine and freshwater flatworms that have a simple, unbranched digestive cavity and a bulbous pharynx that in many species can be protruded through the mouth.

(b) *Triclads* (order Seriata) are marine, freshwater, and terrestrial species that have a protrusible tubular pharynx opening at the conjunction of the three branches of the Y-shaped gastrovascular cavity. The freshwater *planarians* (Fig. 10.2) are the best-known turbellarians, since they take to being raised in captivity and have become the subjects of many investigations concerning regeneration and the chemical basis of learning.

10.8. How is the body of a representative turbellarian organized?

(a) The epidermis is a single layer of ectodermally derived epithelium anchored onto a cementing *basement membrane* equivalent to that of higher animals. On their free surface, the epidermal cells bear cilia and microvilli. Numerous gland cells lie within the epidermis, or below it with necks extending to the surface. *Duogland adhesive complexes* consist of (1) an *anchor cell,* which serves for attachment to the substratum and through which open the necks of a *viscid gland cell* secreting an adhesive material, and (2) a *releasing gland cell,* which secretes a material that destroys the adhesiveness of the viscid substance. Many turbellarians possess a *frontal gland,* an anterior glandular aggregation that also serves in anchoring the animal. *Rhabdites,* rod-shaped bodies secreted by deeper-lying rhabdogen cells, discharge, maybe in self-defense, and swell to form a mucous sheath about the body.

(b) The wall of the gastrovascular cavity is a single layer of epithelium derived from endoderm. In most turbellarians, some gastrodermal cells phagocytize particles for intracellular digestion in food vacuoles. Numerous diverticula along the tract increase the surface area of gastrodermis.

(c) Between the epidermis and gastrodermis lies a solid mass of mesodermally derived tissues: first a layer each of circular, diagonal, and longitudinal muscle cells, then a mass of *parenchyma* cells that serve for food storage and are cannibalized during starvation. Within the parenchyma lie the *protonephridial ducts* and *flame cells,* as well as the gonads and their ducts. At least one pair of lateral nerve trunks run posteriorly from the brain, typically with transverse connectives (*commissures*), so that the central nervous system is ladderlike. In ancestral turbellarians, the longitudinal nerve trunks are thought to have numbered eight and to have been radially arranged, ctenophore-style, but the number has been reduced, to the point that some flatworms have no more than the minimum of two.

10.9. How do turbellarians catch, ingest, and digest their food?

Turbellarians have no means of apprehending prey other than to glide up to or over it, wrap their bodies around it, entangle it in mucus, and pin it down by adhesive. They then ingest it whole or suck it up piecemeal by means of the muscular, usually extrusible, *pharynx.* The tubular pharynx of triclads can even puncture the cuticle of crustaceans at joints (where the cuticle is thin) and suck up the body contents by first exuding *proteolytic* (protein-digesting) enzymes from pores at the pharyngeal tip. Sucking is accomplished by peristaltic waves of contraction along the pharynx.

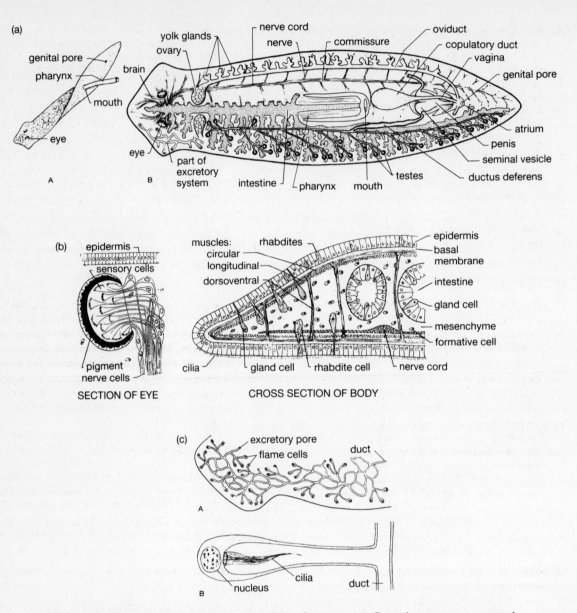

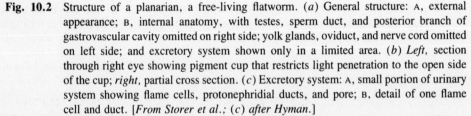

Fig. 10.2 Structure of a planarian, a free-living flatworm. (*a*) General structure: A, external appearance; B, internal anatomy, with testes, sperm duct, and posterior branch of gastrovascular cavity omitted on right side; yolk glands, oviduct, and nerve cord omitted on left side; and excretory system shown only in a limited area. (*b*) *Left,* section through right eye showing pigment cup that restricts light penetration to the open side of the cup; *right,* partial cross section. (*c*) Excretory system: A, small portion of urinary system showing flame cells, protonephridial ducts, and pore; B, detail of one flame cell and duct. [*From Storer et al.; (c) after Hyman.*]

Digestion is first extracellular, within the digestive cavity, then intracellular, in gastrodermal food vacuoles. The extensive branching of the gastrovascular cavity serves for internal distribution of nutrients (see Fig. 10.2*a*). Because their bodies are much flattened and are kept moist with water or mucus, turbellarians do not require special respiratory organs, gas exchange taking place at the body surface. Although most turbellarians are predatory, a few are commensal or parasitic on mollusks, crustaceans, and echinoderms.

10.10.   How do turbellarians conduct *osmoregulation and excretion?*

All flatworms except acoels possess primitive kidneys known as *protonephridia* (see Fig. 10.2*c*), consisting of *flame cells* and *renal tubules* leading to a *nephridiopore,* at which excretion occurs. The branching tubules terminate blindly at the inner end, in a *cap cell* that interdigitates with the adjacent tubule cell. Two to many flagella extend into the tubule from the cap cell, and the movement of these flagella (sometimes referred to as "cilia") can be seen as a flamelike flickering under the microscope, hence the name "flame cell."

Where the cap and tubule cells interdigitate, there are ultramicroscopic "windows" (*fenestrations*) covered by only a thin membrane, through which water and solutes pass from the tissue spaces into the tubule. The properties of this membrane, together with the possibility of reabsorption of valuable ions and other solutes by tubule cells as the urine flows along the network of ductules, may allow considerable regulation of the urinary components. Ion retention, in particular, may be vital for freshwater species, which may need the protonephridia more for ridding the body of excess water than of wastes. Ammonia, the main nitrogenous waste of most aquatic invertebrates, is easily excreted through the body surface, but the nephridia, deeply penetrating the parenchyma, may help with its removal from more internal tissues. The beating of the flame cell flagella not only propels fluid along the tubules toward the nephridiopores, but also sets up a negative pressure that causes fluid to be filtered across the fenestral membranes.

10.11.   What is the organization of the turbellarian nervous system?

The anterior brain is made up of a pair of *cerebral ganglia,* which receive sensory input from a variety of cephalic receptors, and two or more longitudinal *nerve trunks,* usually interconnected by transverse commissures, which give off sensory and motor nerve fibers to the periphery. Centralized nerve trunks, remember, are like our spinal cord: they contain entire nerve cells and synaptic junctions, as well as fibers, and are *not* equivalent to *nerves,* which are only bundles of nerve fibers. In the more primitive turbellarians, a statocyst is embedded in the brain, the longitudinal nerve cords number three or four pairs (dorsal, lateral, ventrolateral, and ventral), and a subepidermal nerve network is present. In higher turbellarians the statocyst, epidermal plexus, and all but two longitudinal nerve trunks have been deleted. The ventral pair—which is retained, although widely separated and conjoined by transverse connectives resembling ladder rungs—becomes enlarged and foreshadows the double ventral nerve cord of annelids and arthropods.

Fascinating experiments on learning and retention have been conducted on planarians, which can be trained by techniques of associative conditioning and will retain the memory despite bisection and regeneration.

> **Example 1:**   If a light flash and electric shock are coupled, a planarian will soon contract spasmodically when only the light is flashed. If such a trained animal is cut in two transversely, not only the original brain but the new brain grown from the separated tail end will retain the conditioning, as demonstrated by the fact that after regeneration both individuals will flinch to light alone. This shows that, in flatworms at least, whatever changes take place during the learning process are not stored in the brain alone, but throughout the nerve trunks as well.

Planarians also furnished early evidence that learning and remembering involve changes in the types of RNA produced by neurons.

> **Example 2:**   If a trained worm is bisected and the brainless posterior half is immersed in a solution containing an enzyme that destroys RNA at the cut surface, the new brain regenerated from that half lacks whatever information was stored posterior to the cut, for the regenerated worm does not flinch when exposed to a flash of light.

10.12.   What sense organs do the turbellarians have?

Ciliary receptors occur scattered throughout the epidermis and concentrated along the head and body margins. The ciliary beat maintains a water flow over chemoreceptive patches of cells. Other receptors are sensitive to touch and water currents. In addition turbellarians have from two to many ocelli, which detect light and its origin; most turbellarians are negatively phototactic. The planarian eye is an inverse cup, in which the distal ends of a group of modified sensory nerve cells fan out within a cup of black pigment (Fig. 10.2*b*). The pigment cup prevents light from reaching the photoreceptive endings, except for rays entering through the open side of the cup. Each photoreceptive neuron ending bears a number of *microvilli*, which spread like a crown from the end of the fiber or protrude at right angles along the distal portion of the fiber. This type of *rhabdomeric* photoreceptor characterizes all the protostome Bilateria. By contrast, the deuterostome photoreceptor bears an enlarged nonmotile *cilium,* the covering membrane of which is expanded into microvilli.

10.13. How do turbellarians reproduce?

Reproduction in turbellarians is both sexual and asexual. Planarians can regenerate experimentally from mere fragments. Each excised piece retains its original anteroposterior and dorsoventral polarities, and regeneration takes place more quickly from anterior than posterior fragments. Spontaneous fragmentation does not seem to constitute a natural means of reproduction, but a number of terrestrial and freshwater triclads and macrostomids do reproduce by *transverse fission:* the posterior end adheres to the substratum while the worm continues to crawl forward until it eventually snaps in two just behind the pharynx, producing a split personality. Each half then regenerates its missing parts. Some of the more primitive turbellarians fission transversely to form a longitudinal chain of small individuals that develop for a while before they detach. This is reminiscent of strobilation in scyphozoan cnidarians and appears to foreshadow the strobilation characteristic of tapeworms.

Sexual reproduction involves maturation of both eggs and sperm at the same time. Nearly all flatworms are simultaneous hermaphrodites. Sperm transfer is usually reciprocal, with each worm inserting its penis into the partner's vagina. During copulation, ejaculation is aided by seminal fluid from *prostatic vesicles.* Sperm differentiate from spermatogonial cells scattered within the parenchyma and migrate to the testes, where they pass into a *sperm duct* and collect in an expanded *seminal vesicle* just proximal to the penis. During mating the ejaculate is collected within the partner's *seminal receptacle* until fertilization.

The eggs mature within ovaries and are fertilized before being laid. In the archoophoran orders, no yolk glands are present, but the egg cell accumulates in its cytoplasm enough yolk for about a week of prenatal development. As the eggs are laid, their outer coating hardens into a shell. Marine polyclads lay eggs in strings of adhesive jelly secreted by *atrial glands,* which allow the eggs to be glued to a substratum rather than being washed away. The neoophoran turbellarians possess yolk glands, so that the egg itself lacks yolk but is surrounded by yolk-filled cells that nourish the developing embryo through several weeks of prenatal development within an egg capsule. Some freshwater turbellarians lay two kinds of eggs: thin-walled "summer eggs" which soon hatch and thick-walled "winter eggs" which remain dormant through the winter and survive freezing and drying before hatching in spring. Reproduction in planarians may be governed by day length as well as temperature, for these freshwater flatworms tend to reproduce by fission in summer and by sexual reproduction in autumn. Certain turbellarians reproduce by *parthenogenesis* (development of embryos from unfertilized eggs), and some of these species entirely lack males. Embryonic development in turbellarians is almost always direct, with tiny worms emerging from the egg capsule. However, certain polyclads have a planktonic ciliated larva with an apical sensory tuft that bears some resemblance to the pilidium larva of nemerteans and may constitute an evolutionary link between the two phyla.

10.14. What are flukes?

Picture yourself laid low by one imprudent swim in an inviting tropical waterhole or by eating a batch of raw or pickled fish spawned in the wrong part of the world, and you will realize that ignorance is *not* bliss and a knowledge of zoology may save your life or health. *Class Trematoda,* the *flukes,* represents a major group of parasitic metazoans that cause much death and debility among both human and nonhuman hosts. They lay enormous quantities of hard-shelled eggs (20,000 per day in the sheep liver fluke) that survive dehydration and ordinary sewage treatment processes. Such tremendous fecundity balances the mortality of the many fluke larvae that perish before finding their proper hosts. Over 6000 species of flukes have been described, with many more still undescribed.

10.15. What are the characteristics of flukes?

(*a*) *Somatic features:* A typical trematode (Fig. 10.3) has a flat, leaf-shaped or elongate body ranging in length from some less than 1 mm to an incredible 14-m monster that infests the muscles of the giant oceanic sunfish, *Mola mola.* An *oral sucker* rings the anteriorly located mouth, and in digenetic flukes (*order Digenea*), a second, *ventral sucker* aids attachment. The monogenetic flukes (*order Monogenea*), which are ectoparasites, have a large, posterior attachment structure, the *opisthaptor,* liberally equipped with hooks and anchors and usually a ring of suckers as well.

　　The outer covering of trematodes is quite unlike the turbellarian epidermis. It consists of a *syncytial* mass of undivided cytoplasm representing a fusion of processes extending outward from cells that lie in the mesenchyme (parenchyma) below the consecutive circular, longitudinal, and diagonal muscle layers. This *tegument* displays many pinocytic vesicles by which endoparasitic flukes absorb nutrients such as amino acids to supplement what they consume by mouth. The tegument also protects intestinal flukes against the host's digestive enzymes. Gaseous

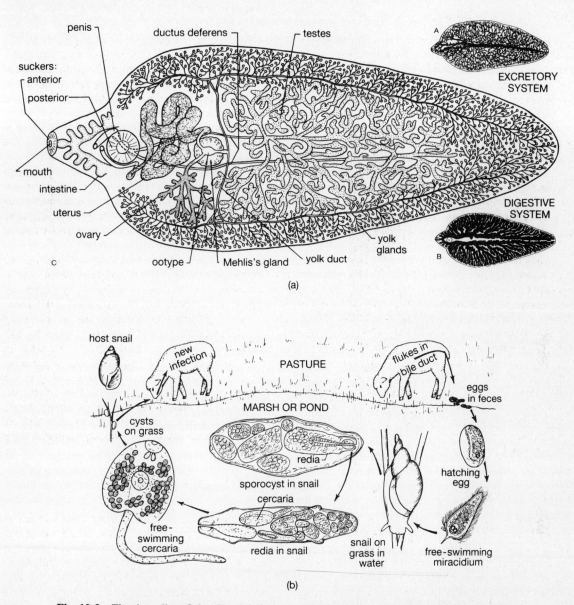

Fig. 10.3   The sheep liver fluke, *Fasciola hepatica*. (*a*) Structure: A, excretory system; B, digestive system (also see oral region of C); C, male and female reproductive systems. (*b*) Life cycle, showing penetration of snail by swimming miracidium; transformation of miracidium into sporocyst, which asexually produces rediae; asexual production of cercariae within a redia; and the free-swimming cercaria that leaves the snail and encysts on vegetation at the water's edge. [*From Storer et al.; (a) after Sommer and Landols, (b) after Thomas.*]

exchange and excretion of nitrogenous wastes take place through the tegument. Endoparasitic flukes are facultative anaerobes, surviving even in oxygen-poor locations within the host's body.

The mouth opens into a muscular pharynx, from which a short esophagus passes posteriorly, branching into two slender, cylindrical digestive tubes that run along each side of the body nearly to its posterior end. Secretory cells in the gastrodermis indicate that digestion is at least partly extracellular. Protonephridia are present, serving for osmoregulation in the ectoparasitic Monogenea, but being of uncertain function in the endoparasitic Digenea. A variable number of flame cells occur, which drain into two longitudinal collecting ducts that usually empty into a urinary bladder.

The nervous system is much like that of turbellarians, with an anterior pair of cerebral ganglia and usually three pairs of longitudinal nerve trunks. Sense organs are poorly developed in endoparasites, but ocelli occur in many ectoparasitic species.

(b) *Reproduction:* The most elaborate trematode body system is the reproductive system, which has both female and male components. The testes, usually two, send sperm ducts to a *copulatory organ* containing a seminal vesicle, prostate gland, and a cirrus or penis for inseminating the partner. (Technically, a *cirrus* is protrusible and a *penis* is not, but for simplicity only the latter term will be used here.) Sperm are stored in the seminal vesicle and ejaculated by way of the penis into the partner's vagina. Copulation is usually mutual, with each fluke transferring semen to the other. The sperm received at mating are stored in a seminal receptacle. Eggs from the single ovary, yolk cells from the vitelline glands, and sperm from the seminal receptacle all meet in a small chamber called the *ootype*. The ootype is surrounded by a number of secretory cells forming *Mehlis's gland,* which may lubricate the eggs for passage through the uterus. The egg is ectolecithal, and the surrounding yolk cells provide nutrients and a material that hardens to form the egg shell. The uterus may be long and convoluted, filled with eggs that are almost continuously being shed through the genital pore. The embryo develops within the egg capsule and hatches as a well-developed ciliated larva. The *onchomiracidium* larva of monogenetic flukes has ocelli, a digestive cavity, and a developing opisthaptor that helps it attach to a host, usually a fish, amphibian, or reptile. The larva matures directly into the adult fluke, which is usually an ectoparasite, although some species enter body cavities such as the mouth and urinary bladder. The life cycles of digenetic flukes, with stages in at least two different host species, are much more complex.

10.16.   What is the life cycle of a digenetic fluke?

Digenetic flukes are nearly all endoparasitic, infecting invertebrates (usually snails) as intermediate hosts and vertebrates of all types as definitive hosts. Reproduction in the intermediate host is asexual and produces two generations (referred to as "larvae"), while the hermaphroditic *sexual generation* (the "adult") inhabits many different organs of the vertebrate host, eggs being shed in feces, urine, or sputum. Although many variations occur, the following generalized scheme emphasizes the common aspects of the digenetic life cycle (Fig. 10.3b).

The larva formed within the egg capsule is a ciliated *miracidium,* which enters the body of certain species of freshwater snails, either when the egg is ingested or when the free-swimming larva penetrates the snail's skin. The miracidium migrates into the snail's digestive gland and transforms into a saccular *sporocyst.* Within the sporocyst, undifferentiated germ cells begin to develop asexually, maturing into a number of *secondary sporocysts* or *rediae,* that emerge from the body of the parent sporocyst but remain within the snail. Whereas the primary sporocyst must take in nourishment through its tegument, a redia has a mouth and saccular gut. This generation again reproduces asexually, producing tadpole-shaped *cercariae* from embryonic germ cells within the body of the redia or secondary sporocyst. The efficacy of these generations of asexual multiplication within the intermediate host is tremendous. For example, if a single miracidium of Manson's blood fluke (*Schistosoma mansoni*) enters the body of a snail, some 300,000 descendant cercarias may emerge over a period of some months.

Cercariae abandon the snail host and, according to their species, encyst on vegetation, encyst within a second species of intermediate host, or directly penetrate the skin of the definitive host. Species that encyst become pawns of fate: they can do no more unless ingested by a suitable definitive host. Those lucky few that do gain entry to the body of a definitive host change into *metacercariae* armed with suckers, ready to migrate to specific locations within the host's body, where they will grow to sexual maturity. Mating hermaphroditically, they produce enormous quantities of eggs, which leave the host by any suitable route.

10.17.   Which trematodes are dangerous human parasites?

(a) The *human liver fluke* (*Clonorchis* or *Opisthorchis sinensis*) is contracted by eating raw or pickled freshwater cyprinid fishes in which the metacercariae are encysted. The ingested worms excyst and migrate from the host's small intestine up the common bile duct into the liver, where they can survive for many years, producing eggs that pass down the biliary passages and exit with the feces. The flukes are small enough (1–2 cm) that a hundred or so may be harbored asymptomatically, but in the orient, where they are endemic, if waterways are contaminated with human sewage, repeated infection is the rule and eventually liver damage ensues. Treatment of sewage to kill the eggs and thorough cooking of fishes that could host the worm would be effective means of reducing the incidence of human infection.

(b) The *sheep liver fluke* (*Fasciola hepatica*) is primarily a parasite of sheep and other livestock, in which it causes "liver rot" but this species can prove a very damaging human parasite as well. It is mainly contracted by persons who eat watercress gathered from streamsides in pastures. The cercariae encyst on vegetation at the water's edge and, if eaten by an appropriate herbivore, will excyst and, totally lacking finesse, simply bore through intervening tissue into the liver, causing considerable distress during this migration. Since *Fasciola* becomes some 3 cm long and half as broad, it does more damage per worm than *Clonorchis*.

(c) The *giant intestinal fluke* (*Fasciolopsis buski*) is spread by human and pig feces. The cercariae encyst on water chestnuts, and human infection usually results from the practice of peeling these crisp delicacies with the teeth, accidentally swallowing the microscopic cysts. This fluke remains in the intestinal tract, where it is vulnerable to helminthcides taken by mouth, but when medication is not available, a number of these wide-bodied, 8-cm worms can cause intestinal obstruction. Infection can be avoided by dropping water chestnuts into boiling water before peeling.

(d) *Lung flukes* (*Paragonimus* spp.) are contracted by eating raw crab or crayfish containing the cysts. Infection is most likely in the orient and some parts of South America. The excysted metacercariae penetrate into the intestinal wall and are carried in the bloodstream to the capillaries of the lungs. Here they grow, living in walnut-sized capsules of connective tissue defensively laid down by the host; despite this containment eggs escape to be coughed up and expectorated, or swallowed and shed in feces. Lung flukes cause respiratory distress, including chronic cough and breathing difficulties. Infection is avoided by thoroughly cooking crab and crayfish.

(e) *Blood flukes* (*Schistosoma* spp.) are by far the most dangerous and prevalent of human trematode parasites, with an estimated 200 million persons infected today. Throughout much of Africa, the middle east, the orient, South America, and the West Indies, waterways polluted with human sewage become laden with schistosome cercariae ready and anxious to penetrate the skin of those who swim, wade, or do laundry. The cercariae burrow in actively, shedding their tails and producing a rash ("swimmer's itch"); if they are of schistosome species that normally infect birds, not mammals, they are soon overwhelmed by human immune responses and the rash soon subsides. But cercariae belonging to any one of three species, *Schistosoma manson, S. japonicum,* or *S. haematobium,* often overcome bodily defenses and are transported by the bloodstream to the liver, where they mature and then begin inching against the venous blood flow, aided by their suckers, until they settle down within small veins in the wall of the colon (*S. mansoni*), small intestine (*S. japonicum),* or urinary bladder (*S. haematobium*). These slender flukes are dioecious, and the larger males clasp one or more females between a pair of *gynecophoral* ("female-bearing") folds along their bodies. The females back into the tiniest veins (venules) to oviposit and then withdraw, leaving the eggs trapped in the venules. The eggs bear a spine by which they work their way out of the blood vessels into the tissues and eventually out into the lumen of gut or bladder, to escape in the fecal or urinary flow. This penetration causes pain and bleeding, but even worse results may ensue. As time goes by, the organ wall becomes scarified, and eggs trapped in the blood vessels are swept away, to be sieved out in the capillary beds of other organs, mainly the liver. Here the eggs cause inflammation and cirrhosis, as masses of scar tissue form around them, and the patient eventually may succumb to degenerative liver disease. Because of its prodigious egg output, *S. japonicum* infection bears the gravest risk. Avoiding skin contact with any water that may be infested with cercariae is the most effective way for a person to guard against infection, keeping in mind that thousands of cercariae can penetrate exposed skin in a few seconds. However, at a societal level, prevention of schistosomiasis may be most definitively approached by methods of sewage treatment effective enough to destroy helminth eggs.

10.18.　What are tapeworms?

Tapeworms are platyhelminths belonging to *class Cestoda*. Tapeworms can endanger human victims at two points in their life cycle: it is bad enough to contemplate harboring an adult tapeworm cosily ensconced within one's intestines, but this may be an inviting prospect compared with the results of swallowing tapeworm *eggs* and becoming host to their larvae! Cestodes occur throughout the world, although danger of human involvement is usually linked to unsanitary conditions. Nearly all species of vertebrates are subject to infection by at least one of the 1000 species of tapeworms in the adult or larval stage.

10.19.　What are the characteristics of a typical tapeworm?

Tapeworms are such highly modified flatworms that their turbellarian ancestry remains obscure. They are *not* thought to have evolved from flukes, but perhaps instead from the rhabdocoels. An adult tapeworm consists of an anterior anchoring device, the *scolex* (Fig. 10.4), which usually bears hooks and suckers, and a series of flattened segments (*proglottids*) that arise by strobilation from the neck of the scolex, so that the oldest proglottids are at the posterior end of the mature tapeworm, called a *strobila* (chain of proglottids). The cestode strobila comes close to being a colony instead of an organism, for no mouth or digestive cavity exists and each segment must independently absorb food from the host's

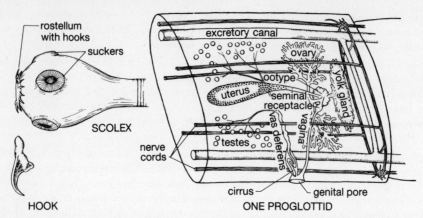

Fig. 10.4 The pork tapeworm, *Taenia solium. Left,* the attachment structure, or scolex, which
lacks a mouth; note young proglottids budding from neck of scolex. *Right,* a mature
proglottid containing both male and female reproductive organs. (*From Storer et al.*)

intestinal cavity. Furthermore, each proglottid matures a full set of male and female reproductive structures, after which
proglottids exchange sperm by copulating with one another. The most posterior proglottids, gravid with ripe eggs, usually
break off from the strobila, and although often soon to perish since they are carried out of the host's body, they display
enough coordinated muscular activity even after being shed to crawl away from the fecal mass and reach clean vegetation,
where they are more likely to be picked up by grazing animals. Perhaps the most cogent argument for considering the
strobila a single creature is that a common pair of nerve cords and four pairs of protonephridial collecting ducts run
throughout its length.

A proglottid is covered externally with a syncytial tegument reminiscent of that of trematodes, but, instead, having
a surface thrown into ultramicroscopic folds (*microtriches*) that perform the same function as microvilli, i.e., expanding
available surface area for absorption of food from the host's intestinal contents. Within the cytoplasm of the tegument
lie numerous *rhabdiform* organelles, conceivably derived from the rhabdites of turbellarians. Beneath the tegument the
proglottid is heavily muscled, first with a layer each of circular and longitudinal fibers, then with deeper transverse,
longitudinal, and dorsoventral layers enclosing the inner parenchyma. The hermaphroditic reproductive system includes
male testes, sperm duct, and eversible cirrus, which is inserted into the vagina of another proglottid during mating. The
vaginal canal dilates to form a seminal receptacle where sperm received during mating are stored to fertilize eggs in the
ootype. The eggs, provided with yolk from a vitelline gland, then accumulate in the uterus, which increases in size until
the gravid proglottid is little more than a sac of eggs. Usually, the proglottid first matures its male, and later its female,
organs.

10.20. What is the life cycle of a representative tapeworm?

Tapeworms exist as intestinal endoparasites in their definitive hosts, which are vertebrates, and as larvae in one or
several intermediate hosts, which may be vertebrates or invertebrates, according to species. Unlike digenetic flukes, few
cestodes reproduce at all within their intermediate hosts, using them merely as vehicles for dissemination. Instead, they
reproduce both sexually and asexually (by strobilation) within the definitive host (Fig. 10.5).

The embryonated egg is ordinarily freed into the environment by disintegration of the shed proglottid, although
certain cestodes such as the broad fish tapeworm keep all their proglottids, which simply liberate eggs through their genital
pores. The egg is resistant to drying and can survive for some time before being eaten by the intermediate host. Within
this host, a six-hooked *oncosphere* larva hatches out, uses its hooks to penetrate the gut wall, and eventually ends up in
the striated muscles, where it encysts. Within the cyst the larva grows into a *cysticercus* (bladderworm), having a spheroidal
body and containing an inverted scolex complete with attachment structures. The cysticercus remains dormant, unable to
develop further unless the flesh of the intermediate host is devoured by a carnivore. If eaten by an appropriate host species,
the bladderworm survives ingestion, and as the surrounding cyst is digested away, the larva everts its scolex and anchors
to the intestinal lining. The rounded body shrinks, and soon strobilation commences and proglottids are formed. As the
proglottids move posteriorly from the neck of the scolex, they grow in size, mature reproductively, mate, and become
gravid with eggs, as described above.

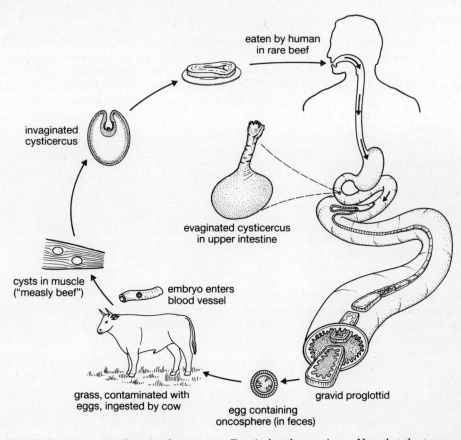

eaten by human
in rare beef

invaginated
cysticercus

evaginated cysticercus
in upper intestine

cysts in muscle
("measly beef")

embryo enters
blood vessel

grass, contaminated with
eggs, ingested by cow

egg containing
oncosphere (in feces)

gravid proglottid

Fig. 10.5 Life cycle of the beef tapeworm, *Taeniarhynchus saginata*. Note that the tapeworm uses the intermediate host (cow) only for dissemination (as cysts in beef) and carries out all reproduction, asexual (strobilation) and sexual, within the intestinal tract of the definitive (human) host.

10.21. How do tapeworms infect humans?

Tapeworms of several species can inhabit the human intestinal tract, mostly without causing severe symptoms. Once their presence is betrayed by the appearance of eggs or proglottids in the host's feces, they are vulnerable to oral medications that cause the scolex to lose hold and the entire strobila to be voided. However, in parts of the world where certain tapeworms are endemic, reinfection may be a persistent problem. In order to avoid becoming host to an adult cestode, it is only necessary to avoid eating meat that is not well cooked. In addition, most encysted cysticercus larvae are macroscopic and readily detected by meat inspection. At one time, tapeworm larvae enclosed in gelatin capsules were even sold to adventuresome persons wanting badly to keep their figures while indulging their appetites. This practice is now frowned upon, because some persons suffer side effects of nervousness, teeth grinding, and the like, and many lose no weight at all despite the food-absorbing activities of their pet parasite.

Much more serious than infection with the adult worm is *cysticerciasis*, which results from ingestion of tapeworm *eggs*. Risk of infection is greatest in rural parts of the world, where people are infected with adult cestodes and human sewage contaminates irrigation water or is even used as fertilizer for vegetable crops. If the eggs are swallowed, the oncosphere larvae migrate through the body, not only into the skeletal muscles, but also into the heart muscle, skeleton, brain, and other organs. The rice-sized, calcified cysts of the cysticerci of the pork tapeworm, *Taenia solium,* or the bean-sized cysts of the beef tapeworm, *Taeniarhynchus saginatus,* are revealed by x-radiation. Symptoms result both from toxins and mechanical pressure on adjacent tissues. Cysts in the brain may cause epileptiform seizures, and those in bones may lead to spontaneous fractures. Symptoms may not appear until years after the time of infection, and the prognosis is grave for victims carrying a heavy population of cysts lodged in vital areas.

10.22. Which tapeworms may afflict humans?

(*a*) *Hydatid cyst worm* (*Echinococcus granulosus*) is a thoroughly dangerous cestode, for it lives as an adult in the intestines of dogs, and its minute eggs, electrostatically clinging to dog hairs, can be readily swallowed, especially by children. Not to be confused with the dog tapeworm, *Dipylidium caninum*, which is larval in lice and fleas, *Echinococcus* is very tiny as an adult, with only about three proglottids in the strobila (Fig. 10.6). Accordingly, a dog may carry quite a number of these midget worms without acting ill. Domestic dogs are likely to become infected by eating rats, while wild canines would contract the parasite from their herbivorous prey, such as sheep, cattle, and wild ungulates.

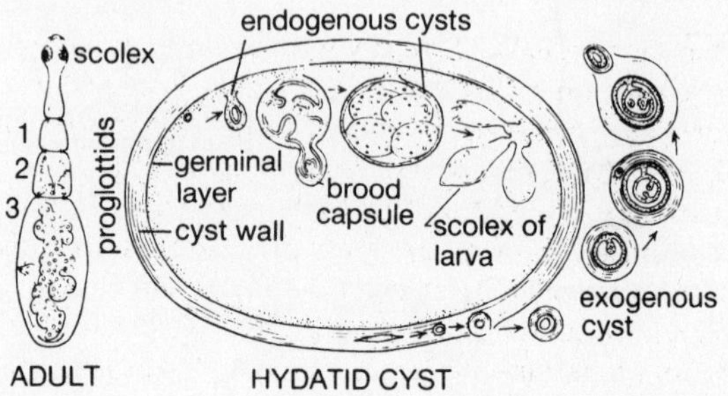

Fig. 10.6 Life cycle of the hydatid cyst worm *Echinococcus granulosus. Left,* the 3- to 6-mm adult, living in the intestinal tract of dogs or wild carnivores. *Right,* the hydatid cyst in an intermediate host (herbivorous mammals or humans), showing budding off of daughter cysts within the cyst formed by the original larva, or exogenously from its wall. (*From Storer et al.*)

The organism that becomes intermediate host by swallowing even a single *Echinococcus* egg is in for a rude surprise, for the cysticercus behaves quite atypically. Inexorably the encysted larva begins to swell into a *hydatid cyst* containing clear fluid; then daughter cysticerci commence to bud off the inner surface of the original larva, until the interior of the cyst is filled with "hydatid sand," each "grain" a minute infectious secondary larva. The hydatid cyst may gradually grow, at a rate of 1 cm per year, to a diameter of 20 cm or more. Lodged in the brain, one cyst can produce symptoms of a tumor and cause death if inoperable. When the cyst can be excised, great care must be taken to kill the hydatid sand by injecting the cyst with a chemical such as formaldehyde, after which the fluid is aspirated before removal of the bladderworm itself. If this precaution is not taken, rupturing the cyst could spread living "sand" throughout the body, sealing the patient's fate. Human infection, owing to contact with infected dogs, is on the increase, yet humans are actually only an incidental host. More usually, herbivores ingest the eggs, and when these eventually fall prey to packs of wild canines, the whole pack may become simultaneously infected by swallowing hydatid sand from a ruptured cyst in a single prey organism.

(*b*) *Broad fish tapeworm* (*Diphyllobothrium latum*) is endemic in the Great Lakes region of the United States and in many other parts of the world. It lives as an adult in the gut of fish-eating mammals, including people who relish raw fish, and is unusual in several respects. Eggs are shed in the feces of piscivorous animals and hatch in water into a free-swimming *coracidium* larva, which is in effect an oncosphere covered by a ciliated sphere. This swims about enticingly until eaten by a small crustacean (copepod), within which it develops into a hook-bearing *procercoid*. When the copepod is swallowed by a fish, the procercoid penetrates the host's gut and migrates into its striated muscles, where it does *not* become an encysted bladderworm, but simply grows into a slender, white, unsegmented juvenile worm (*plerocercoid*). When fish eat fish, the unharmed plerocercoid passes up the food chain from minnows to game fish such as pike. The plerocercoid transforms into the adult worm only when its fish host is eaten by a mammal. The adult's scolex lacks hooks and suckers, holding to the mammalian gut lining by means of a pair of folds. As the worm strobilates, ripe proglottids are not shed, and only eggs are expelled through the genital pores, so that in time the tapeworm attains an admirable length of 20 m and more than 3000 proglottids. This cestode selectively concentrates vitamin B12 (cobalamin) to such an extent that the host may fall ill with what appears to be pernicious anemia, even when eating a diet rich in vitamins.

(*c*) *Pork tapeworm* (*Taenia solium*) and *beef tapeworm* (*Taeniarhynchus saginatus*) are essentially restricted to humans for the definitive host and to domestic pigs, cattle, and even humans for intermediate hosts. *T. solium* grows to a length of 7 m, and *T. saginatus* to more than 10 m. For reasons we shall see in the next chapter, undercooked pork is seldom eaten, so *T. solium* infection is uncommon, but in some regions, cysticerciasis from the ingestion of *Taenia* eggs is both common and dangerous. Despite meat inspection, the cysts of *Taeniarhynchus* may escape discovery, so that even in the United States, where about 1 percent of beef cattle are infected, lovers of rare beef may find that an uninvited guest has come to stay. However, after two to three weeks, liberated noodlelike proglottids begin to crawl out of the host's anus, an event rarely met with equanimity, and a quick trip to the doctor allows the worm to be expelled.

(d) The *dwarf tapeworm* (*Vampirolepis nana*) is probably the commonest human cestode parasite, but its small size (2–4 cm) makes it of only moderate medical importance unless large numbers build up. An intermediate host (a variety of insects, including flour beetles) is *optional* in the life cycle. Human infection can begin either when an egg or a flour beetle carrying the cysticercus is swallowed. If an egg is ingested, the oncosphere hatches out and penetrates only into the gut lining, where it goes through an abbreviated *cysticercoid* stage, reentering the gut cavity in a few days to become an adult worm.

10.23. What are gnathostomulids?

Phylum Gnathostomulida ("little jaw mouth") includes some 80 species of slender, cylindrical acoelomate worms 1 mm or less in length, which live among sand grains in coastal marine shallows in many parts of the world, where they can survive anaerobic conditions. In spite of having escaped taxonomic description until the 1950s, they are surprisingly abundant. Like flatworms, they have a digestive cavity lacking an anus, and the mouth is ventral, behind the head. But these little chaps sport a pair of toothed lateral jaws with which they snap down bacteria and fungi that are scraped up by a comblike plate on the ventral lip. Their ciliated epidermis is also unusual: each cell bears a single cilium surrounded by a crown of eight microvilli. Unlike flatworms, the ciliary beat can be reversed, allowing these tiny creatures to back out of tight places. Gnathostomulids are hermaphroditic, and the male copulatory organ may be used in penetrating the partner's body wall to fertilize the large, solitary egg. The egg breaks through the body wall, adheres to the substratum, and undergoes direct development with spiral cleavage.

10.24. What are nemerteans?

The phylum *Rhynchocoela* ("beak cavity"), or *Nemertea*, contains only about 650 species of unsegmented worms, often found under rocks at the beach in shallow water, although some are pelagic, and one genus includes freshwater species, while another includes a few tropical land species. These worms are elongate and somewhat flattened (hence the appellation "ribbonworms") and are capable of great contraction and elongation and of spontaneous fragmentation in self-defense. Another name, "proboscis worms," refers to the phylum's major distinguishing feature: a *proboscis* that may be longer than the worm's own body. The proboscis is housed in a tubular cavity (*rhynchocoel*) that opens dorsal to the mouth and extends longitudinally nearly the entire body length (Fig. 10.7). This splendid device is shot out by hydrostatic pressure, everting in the process, and, according to the class of nemertean, either stabs the prey with a sharp barb (*stylet*) or enwraps it bola-fashion. A retractor muscle then returns the proboscis to its sheath while drawing the prey to the worm's mouth. "Nemertean" means "unerring," but doubtless the worms sometimes do miss long-distance shots, and some nemerteans must actually touch their prey before firing their proboscis. A captured nemertean frequently responds by everting and *autotomizing* (breaking off) its proboscis, which writhes about in a manner admirably suited for diverting a predator from the worm itself. Until the lost proboscis is regenerated, the unerring worm has little choice but to try to creep up on its prey and seize it at short range.

10.25. What advances over flatworms are seen in nemerteans?

The hotshot proboscis described above certainly provides an effective means of prey capture, but the nemerteans are of even greater interest by being the simplest living animals to display two innovations of great significance in the evolution of metazoan life: a *complete, tubular digestive tract* leading from mouth to anus, allowing ingestion, digestion, absorption, and elimination all to take place at one time, and a system of contractile *blood vessels* involved with internal transport. In addition, nemertean flame cells are intimately associated with the walls of these blood vessels, suggesting direct filtration of wastes from the blood into the protonephridia.

Whether or not these important biological advances were transmitted to descendant phyla by a nemerteanlike ancestor remains in doubt, but it is suggestive that some nemerteans have a planktonic *pilidium* larva that bears considerable resemblance to the trochophore larvae (see Fig. 12.1) of marine annelids and mollusks and of several other protostome phyla. Interestingly, the pilidium itself lacks an anus and somewhat resembles the planktonic larva of certain polyclads. Further developmental similarity to polyclads is that both display spiral cleavage and have a radially symmetrical gastrula with a tuft of long cilia at the dorsal pole.

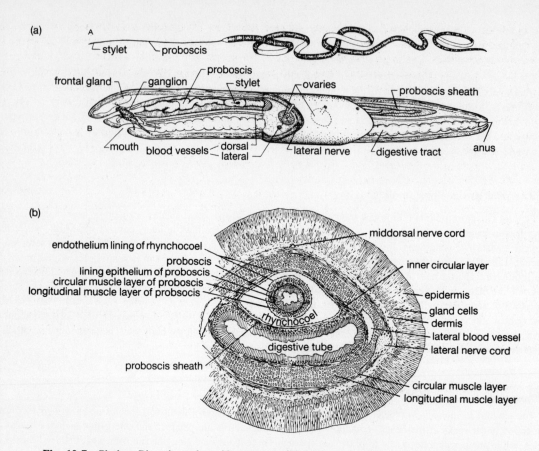

Fig. 10.7 Phylum Rhynchocoela or Nemertea, a ribbonworm. (*a*) General structure: A, external appearance with proboscis extended for prey capture; B, longitudinal section, showing digestive tract complete from mouth to anus, blood vessels, and proboscis retracted within its sheath (rhynchocoel). (*b*) Cross section showing proboscis within rhynchocoel, lying above gut, and muscle layers forming a solid body wall. (*From Storer et al.*)

10.26. What is the functional organization of a typical nemertean?

(*a*) The body wall of nemerteans is more highly organized than that of flatworms, with much less parenchyma tissue and superior musculature (Fig. 10.7*b*). The skin consists of an outer *epidermis,* a single layer of ciliated, mucus-secreting columnar epithelium, and an inner *dermis* with loose and dense connective tissue layers. Beneath the skin lies first a thick layer of longitudinal muscle fibers, then a layer of circular muscle, then an inner layer of longitudinal muscle. Within this lies a thin layer of parenchyma penetrated by bands of muscle fibers oriented vertically. Finally, a single layer of columnar epithelium forms the wall of the gut.

The well-developed musculature allows nemerteans to crawl by peristaltic waves of contraction that sweep down the worm's length; some pelagic species swim by dorsoventral undulations. The worms also can creep by cilia along a mucus trail.

(*b*) *Nutrition:* Nemerteans are exclusively carnivorous, capturing other invertebrates by means of the proboscis, which is shot out explosively and then retracted, bringing the prey to the mouth, where it is sucked in whole. Nemerteans of *class Anopla* have an unarmed proboscis that coils about the prey. Members of *class Enopla* have a proboscis armed with a stylet that repeatedly stabs the prey, apparently injecting a toxin by contractions of a muscular bulb at the stylet's base. Some nemerteans have rhabdites in the proboscis, further evidence of this phylum's descent from turbellarian flatworms. The complete digestive tract is tubular, consisting in linear sequence of mouth, buccal cavity, esophagus, glandular stomach, intestine with lateral diverticula, and anus.

Nemerteans possess a closed circulatory system, i.e., one in which blood is always contained within vessels, not disgorged into tissue spaces (as in the *open* circulatory systems of arthropods and most mollusks). The basic plan consists of two longitudinal vessels running on each side of the gut, interconnected anteriorly and posteriorly by expanded spaces (*lacunae*), but this circulatory pattern is often elaborated by addition of a middorsal vessel and transverse connectives. The larger vessels are contractile, but irregularly so, and the direction of blood flow often reverses. The blood contains amoebocytes and pigment-containing cells of uncertain function. Along the lateral blood vessels, a series of protonephridial flame cells protrude into the wall of the vessels, which in some species disappears so that the flame cells are literally bathed in blood. This suggests that wastes are being filtered directly from the blood into the protonephridia.

(*c*) The nervous system consists of a brain of four ganglia ringing the rhynchocoel and two major lateral nerve trunks running posteriorly, giving off a series of peripheral nerves and transverse connectives, a basic plan quite reminiscent of that seen in higher turbellarians. Additional smaller longitudinal nerve cords may also be present.

The epidermis contains both individual and clustered ciliated receptor cells and is also invaginated into richly innervated *cephalic grooves,* thought to function in tactile and chemical reception. In addition, most nemerteans have *cerebral organs* consisting of a pair of ciliated tubules in intimate contact with the brain; when the animal is hunting, a brisk current of water flows through these tubules, probably helping the worm locate prey by chemoreception. The eyes are inverse pigment cups, similar to those of planarians (see Fig. 10.2*b*), and number from two to hundreds. Nemerteans are negatively phototactic and hunt by night.

(*d*) *Reproduction* is basically sexual, although certain species also fragment spontaneously, with each piece becoming a new worm. Sexes are usually separate. Numerous gonads form along each side of the body, from clusters of parenchyma cells that develop into gametogonia. When the eggs or sperm have matured, each gonad develops a short duct and gonopore and the gametes are forced out by muscular contractions. Sometimes nemerteans aggregate before the gametes are shed, or a pair may come together, but no copulatory organs exist. The fertilized eggs may be laid in burrows or in strings of adhesive jelly, or set adrift in the plankton.

Chapter 11

The Pseudocoelomates

Pseudocoelomate animals represent a heterogeneous group of phyla having in common a type of body cavity known as a *pseudocoel* ("false cavity") that separates the endodermally derived intestinal epithelium from the body wall (see Fig. 8.5). This is in contrast to the *coelom* of higher phyla, which is a body cavity that arises *within* the embryonic mesoderm, separating it into parietal and visceral layers. Pseudocoelomates include the gastrotrichs, kinorhynchs, rotifers, acanthocephalans, nematomorphs, and nematodes.

11.1. What is the evolutionary significance of a pseudocoel?

By confining mesodermal derivatives to the body wall, a pseudocoel precludes the development of an intrinsic gut musculature, blood vessels, and connective tissues. For most of its length, the pseudocoelomate digestive tube consists of a single layer of cuboidal or columnar epithelium, through which digested food is absorbed into the body cavity. Despite this limitation and consequent lack of the type of evolutionary advancement seen in eucoelomate phyla, certain pseudocoelomates, such as rotifers and nematodes, are exceedingly successful, cosmopolitan animals.

Even a "false" body cavity can be advantageous in several ways. Its fluid provides a *hydrostatic skeleton*, making the body rigid when the muscles contract against the incompressibility of the liquid. It serves for internal distribution, and in fact no blood vessels occur in the pseudocoelomate phyla, which appear to have originated as an offshoot from the turbellarian flatworms and not from nemertean ancestry. Finally, it provides space for internal organs, especially reproductive structures, and is used as a brood chamber by some species.

11.2. What other features do pseudocoelomates have in common?

Other features that unite the pseudocoelomate phyla, although not of ubiquitous occurrence, include a collagenous cuticle, a pharynx made muscular by myoepithelial cells, a body wall musculature limited to longitudinal fibers, a syncytial epidermis, and a tendency for the number of cells or nuclei to be constant throughout a given species (*eutely*.) When nephridia are present, they are protonephridia, i.e., excretory structures that terminate internally in closed ends, so that materials are filtered across a membrane to enter the renal tubules. However, the structure of the terminal cells varies, there being flame cells (in acanthocephalans), flame bulbs (in rotifers), cyrtocytes (in gastrotrichs), and solenocytes (in kinorhynchs). *Flame bulbs* are multiple outgrowths of a multinucleate cell, each bulb bearing at least 30 flagella. A *cyrtocyte* is a cell that bears a single flagellum, which vibrates within a cylinder of stiff cytoplasmic rods. A *solenocyte* is similar to a cyrtocyte, but is multinucleate and bears two flagella, one long and one short.

11.3. What are gastrotrichs?

Phylum Gastrotricha ("belly hairs") includes about 400 species of microscopic marine and freshwater pseudocoelomates, mostly 50–1000 μm in length, with locomotory cilia often restricted to two lengthwise bands occurring only on the ventral surface (Fig. 11.1a). Primitive gastrotrichs have only one cilium per cell, a characteristic that may link them with the acoelomate gnathostomulids. The rest of the surface is covered with a cuticle that often bears spines, hooks, or scales. The head bears tufts of long cilia that are probably sensory. Duogland adhesive systems, similar to those seen in turbellarians, occur along the sides of the body or only at the forked tail end.

Beneath the cuticle lies a syncytial epidermis and thin bands of longitudinal and circular muscle fibers. The pseudocoel is nearly obliterated by the digestive, excretory, and reproductive structures it contains. The mouth is terminal and opens into a pharynx composed of myoepithelial cells that exert a pumping action to suck in bacteria, small protozoans, and the like. The noncontractile stomach-intestine, formed of a single epithelial layer including glandular cells, leads to the posterior anus. The protonephridia, typically a single pair, resemble those of gnathostomulids. Each consists of a terminal cyrtocyte and a long, convoluted tubule leading to a ventral nephridiopore. The protonephridia are especially large in freshwater species, which must rid their bodies of excess water. The convolutions of the nephridial tubule suggest that along its length either additional materials are absorbed into the urine or useful solutes are reabsorbed from the urine and transferred into the pseudocoel.

The brain consists of a cerebral ganglion on each side of the pharynx, connected by a dorsal commissure. Each

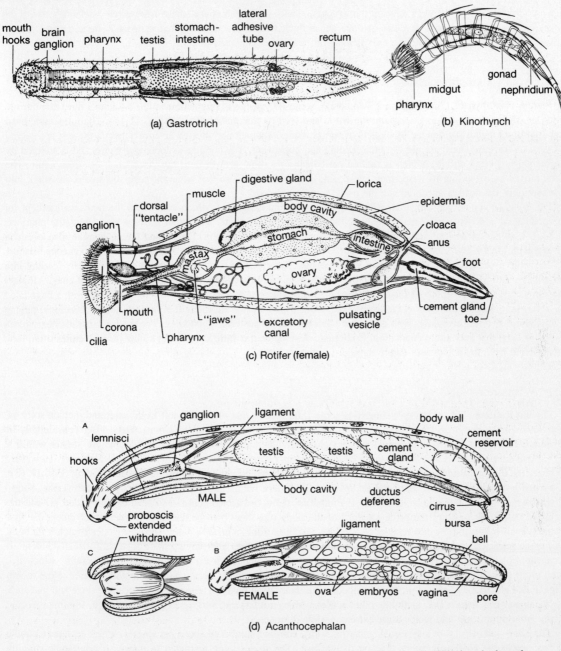

Fig. 11.1   Minor pseudocoelomate phyla. (*a*) Gastrotrich, dorsal view. (*b*) Kinorhynch, lateral
aspect. (*c*) Female rotifer, sagittal section. (*d*) Spiny-headed worm (acanthocephalan):
A, male; B, female; C, head with proboscis withdrawn. [*From Storer et al.; (c) after
Delage and Herouard, (d) after Lynch.*]

ganglion gives off a longitudinal nerve trunk that runs dorsolaterally. Sense organs are mostly limited to scattered sensory
cilia and a pair of ciliated pits on the head. A few species have photosensitive pigment granules within certain brain cells.

Unlike other pseudocoelomates, but like the acoelomate gnathostomulids, gastrotrichs are hermaphroditic. Only one
huge egg at a time is released, usually by rupturing the body wall. Sperm are usually transferred indirectly: the gastrotrich
flexes its body to gather sperm from its male genital pores by means of a *caudal organ* that is used in copulation and
then transfers the sperm through the female pore into a *seminal receptacle*. Freshwater gastrotrichs have degenerate male

organs, and their unfertilized eggs develop parthenogenetically. Like planarians, these freshwater species lay two types of eggs: a thin-walled egg that develops at once and a thick-walled dormant egg that withstands freezing and drying. Cleavage is determinate, but bilateral rather than spiral, and development is direct.

11.4.   What are kinorhynchs?

Phylum Kinorhyncha ("movable beak") comprises some 100 species of tiny marine animals (under 1 mm) that burrow in mud or silt in coastal waters throughout the world and feed on diatoms or detritus (Fig. 11.1b). The retractable head bears a terminal mouth surrounded by circlets of spines, some of which are directed anteriorly, while the rest are curved backward (recurved). These are used in both feeding and burrowing. The body lacks locomotory cilia and is covered by a segmentally divided cuticle, which is molted periodically to allow growth until the kinorhynch becomes adult. The animal moves by extending its head, then telescoping it into the trunk, so that when the recurved spines are stuck into the substratum, the head is anchored and the trunk is pulled forward.

The collagen cuticle is secreted by the underlying syncytial epidermis, which constitutes the body wall except for two dorsolateral and two ventrolateral bands of longitudinal muscle fibers. The brain forms a ring about the pharynx, giving off a single midventral nerve trunk with segmentally arranged ganglia. Dorsal and lateral ganglia also occur in each segment. Only a few species have ocelli, of the pigment-cup type, but all have scattered sensory cutaneous bristles.

The large pseudocoel contains fluid and amoebocytes. The digestive system consists of a buccal cavity leading into an esophagus into which salivary and "pancreatic" glands open, a stomach-intestine, hindgut, and terminal anus. All parts of the alimentary canal except the stomach-intestine are lined with thin cuticle. Two protonephridia, each with a flagellated *solenocyte*, lie bathed in the fluid of the pseudocoel, from which wastes are filtered. Kinorhynchs are dioecious, having one pair of either testes or ovaries with ducts that open by way of posterior gonopores. The male gonopores bear spines (*penial spicules*) that may aid in copulation. Embryonic development is little known, but kinorhynch juveniles often look quite different from the adults and mature through successive molts.

11.5.   What are rotifers?

Phylum Rotifera ("wheel bearer") comprises some 1800 species that mostly inhabit fresh water and include some of the *smallest known eumetazoans* (mostly from 40–500 μm, although a few grow to 3 mm). They differ considerably in body shape and mode of life. Some float and are globular, others creep or swim actively, some are sessile within a secreted cuticular tube, and a few form colonial aggregations (which are *not* produced by asexual budding). However, they share the distinguishing feature of an anterior *corona*, a ciliated crown used in locomotion and feeding (Fig. 11.1c). The corona is much reduced in endoparasitic species, which live within such invertebrate hosts as earthworms, slugs, and crustaceans. In the bdelloid rotifers such as *Philodina*, the two-wheeled corona and head can be pulled in, and the segmented cuticle also allows the posterior third of the body (the "foot") to be telescoped. The rotifer creeps by first anchoring itself by the toes, which receive an adhesive secretion from its *pedal glands*. Then the rotifer stretches its body with corona retracted and attaches its rostrum, finally releasing the toes and moving them forward for reattachment. It also anchors by its toes when using the ciliary current of the expanded corona to draw food to the mouth, which lies central to the corona. Some rotifers are suspension feeders, trapping minute organic particles that are carried to the mouth along ciliated grooves. Others are predatory, shooting out their jawed pharynx (*mastax*) like pincers to seize and macerate protozoans and tiny metazoans, including other rotifers. Many rotifers can survive in a dried-up state, like microscopic raisins, enduring drought and temperature extremes while dormant.

The body wall consists of a syncytial epidermis with a constant number of nuclei per species, which secretes a cuticle that lies *within* the epidermis rather than on the surface. The muscles are restricted to narrow bands, some running circumferentially, others running longitudinally to insert upon body parts that they serve to retract. The brain is a single dorsal ganglion, sending out nerves to the mastax, muscles, and various sense organs, including one to five very simple ocelli consisting of a small group of photoreceptors and a pigment cell.

The pseudocoel, filled with a branching network of amoebocytes, contains the gut, gonads, and a single pair of protonephridia with flame bulbs, the ducts of which often empty into a urinary bladder that voids into a *cloaca*, a terminal intestinal chamber that also receives the oviduct. The digestive tract begins anteriorly with the mouth, which often opens straight into the mastax. The mastax bears a pair of jaws (*trophi*) which work sideways, like those of gnathostomulids, and which are variously adapted for grinding or grasping. The ducts of a pair of salivary glands open just in front of the mastax, and a pair of *gastric glands* open into the gut where the esophagus (often ciliated) meets the stomach. A short intestine leads to the cloaca, which opens by way of an anus located anterior to the foot.

Rotifers are dioecious and reproduce only by gametes, but in many species unfertilized eggs develop *parthenogenetically*, in which case males are unknown or are only seldom produced. When males are produced, they are small and short-lived, since their gut is vestigial or absent. Since the males are already *haploid,* their testes are the site of sperm maturation, but not meiosis. The end of the sperm duct is modified into a copulatory organ that inseminates the female by puncturing her body wall (*hypodermic impregnation*). The eggs are produced in a syncytial mass of ovarian and yolk-producing tissue (*germovitellarium*), which allows yolk to pass directly into the developing ova. In species that have males, two types of eggs, amictic and mictic, are produced. *Amictic eggs* do not undergo meiosis and so remain diploid; these are thin-walled and develop without delay into females. *Mictic* eggs are haploid. If unfertilized, they remain thin-walled and develop into haploid males. If fertilized, they develop thick walls and remain dormant for months if necessary, capable of enduring dehydration and freezing. Cleavage is determinate and spiral, and development is direct except in sessile species, which have a motile larva resembling species of free-swimming rotifers.

11.6. What are acanthocephalans?

Phylum Acanthocephala ("spiny head") includes about 500 species of *spiny-headed worms* that range in size from 1.5 mm to over 1 m in length. Their distinguishing feature is a proboscis covered with recurved thorns, which can be protruded hydrostatically or pulled in by retractor muscles (Fig. 11.1*d*). The proboscis is used by the worm to anchor itself to the intestinal wall of a vertebrate host. A large acanthocephalan can cause acute pain and even death by completely perforating the intestinal wall, leading to peritonitis. The worms lack mouth and digestive tract and, like tapeworms, must absorb digested food through their body wall from the host's intestinal contents. The wall of the saccular, wormlike body consists of a tegument formed by a fibrous, extracellular cuticle overlying a syncytial epidermis permeated by fibers and containing lipid-filled spaces; branching *pore canals* open through the surface of the cuticle. Thin circular and longitudinal muscle layers of the body wall lie just within the tegument. The tegument takes in food by pinocytosis and through the pore canals; its cytoplasm contains enzymes that cleave dipeptides to amino acids and others that phosphorylate glucose and sequester it in such a manner that a steep concentration gradient is maintained for this sugar, expediting its speedy and continuous uptake.

The nervous system consists of a central ganglion lying within the proboscis sheath, which distributes nerves to the proboscis, body, and genital bursa. The organization of the nervous system and the two flame cell–type protonephridia (when present) are considered evidence of an evolutionary link to the turbellarian flatworms.

Sexes are separate, with the female being larger than the male, and multiple infection is necessary for reproduction to occur, since both sexes must be present in the same host. The gonads are suspended within the pseudocoel by a ligament reaching from the proboscis sheath to the posterior end. The gonopore is at the end of the trunk and in the male is furnished with a protrusible penis. The eggs are fertilized within the female's body, and as her ovary then disintegrates, her pseudocoel fills with developing embryos. Enclosed by a protective egg shell, the larva is expelled with the host's feces and is already armed with a crown of hooks for penetrating through the gut of the intermediate host, such as insects (certain cockroaches and beetles) or small aquatic crustaceans. Within the blood-filled body cavity of the arthropod host, the larva grows into the juvenile stage, or *acanthor,* bearing a circlet of hooks that it will use in attaching to the gut lining of its definitive (vertebrate) host until its spiny proboscis has developed. Fishes, birds, and various mammals that eat the invertebrate hosts receive the enclosed acanthors as a bonus. Humans are only an accidental host, and when the worms begin to inflict abdominal pain, the host can find relief through oral medications that make the worms release their hold.

11.7. Can horse hairs come to life?

People used to think so, since adult *horsehair worms* (phylum Nematomorpha) ("thread-shaped") resemble nothing more than long, dark hairs shed from a horse's tail at the watering hole, miraculously sprung to life. The lethargic females undulate slowly, as if barely animate, but the males swim and crawl actively, by whiplike motions. Over 200 species of nematomorphs have been described, all cylindrical and thin as a hair, but ranging in length from 10 mm to 70 cm. Their sudden appearance from "nowhere" is actually not proof of spontaneous generation, but results from their emergence, fully grown, from some inconspicuous invertebrate host—usually a beetle, cricket, centipede, or millipede that ingests a larva (or is actively invaded by one) while drinking at the water's edge. It is amazing that the parasite does not kill its host while growing, with repeated molts, to an adult length as much as 50 times that of the unfortunate arthropod. When the host comes to drink, the worm abandons it and enters the water, where it wriggles about like a living horse hair while quickly maturing sexually, mating, and ovipositing. Adult nematomorphs are doomed to early demise, for their gut is vestigial and they do not feed. They have only longitudinal muscles and so cannot lengthen their bodies, but are also unable to shorten them because of the thick outer cuticle, which, however, does permit lashing, whiplike movements. The cuticle bears plates, or *papillae,* and is secreted by the underlying epidermis, which is not syncytial, but cellular. The brain forms a ring around the esophagus, giving off a single midventral nerve cord.

Sexes are separate, and the gonadal ducts open into a cloaca (terminal intestinal chamber). A penis is lacking, but the worms still accomplish copulation as they coil their posterior ends together, bringing their gonopores in contact. The eggs are laid on aquatic vegetation, sometimes in gelatinous strings over 200 cm long. All but one nematomorph genus (which is marine and parasitizes crabs) belong to *class Gordioidea*, and their larvae bear some resemblance to kinorhynchs in having a segmented cuticle and protrusible proboscis armed with both anteriorly directed and recurved spines. When the eggs hatch, these formidably equipped larvae swim to the water's edge to await the coming of a thirsty potential host. Once inside the host, the larva must absorb nourishment through its body wall, for like the adult, it, too, has a vestigial gut.

11.8.   What are roundworms?

Roundworms, or *nematodes*, make up the phylum Nematoda ("thread") and may well be the most numerous of all metazoans, both as individuals and as species. Over 10,000 species have been described, but since it is suspected that many parasitic nematodes are host-specific (i.e., limited to a single species of host) and since almost every kind of animal and plant has its roundworm parasites, the actual number of nematode species may go into the millions! Morphologically, nematodes are fairly homogeneous except for size. They range from microscopic to over a meter long, yet nearly all are whitish, slender, cylindrical, elongated, smoothly tapering worms that move by a characteristic whiplike lashing, since they possess only longitudinal muscle fibers and have a cuticle that allows bending but prevents shortening of the body. The cuticle may be annulated (ringed), but the body is actually unsegmented.

11.9.   Why are nematodes so successful?

This is an interesting question. In addition to their notable success as parasites, free-living nematodes inhabit the sea from shore to abysses, fresh water, and soil. They flourish from tropical to polar regions and are successfully adapted to every kind of habitat, including deserts and hot springs. Ordinarily, any group of organisms that radiates so widely undergoes diversification of body form in the process of becoming adapted to many modes of existence. Even rotifers display much morphological differentiation, yet despite size differences, nematodes have not appreciably changed form during adaptive radiation. Their bodies are well-suited for living in interstices, and this they do admirably, finding living space in seemingly every niche and crevice, both literally and ecologically. Overall, nematodes are quite versatile, although individual species specialize in one of many different life styles. The minuteness of most nematodes also favors their abundance: 1 square meter of subtidal mud has been found to yield more than 4 million nematodes, while a single rotting apple has been reported to harbor 90,000.

11.10.   What type of specialization do nematodes exhibit?

As a phylum, roundworms display a prodigious range of physiological tolerances, but this range is divided among the different species. For example, the "vinegar eel" lives only in the acidic medium of homemade vinegar, and the "beer eel" revels in the felicitous habitat of mats soaked by generations of overfilled beer steins. Pig ascaris and human ascaris (Fig. 11.2*a–c*) are morphological twins: robust intestinal parasites up to 40 cm long, yet each appears strictly limited to its own host species and unable to endure the internal environment of the other. The evolutionary diversification of nematodes thus seems to have proceeded more along physiological and behavioral, rather than morphological, lines, other than size.

Nematodes are essential links in many food chains, for certain species represent the most cosmopolitan and abundant organisms that can feed directly on bacteria and fungi. Terrestrial nematodes actually live in the water film that coats soil particles. Like rotifers, they can survive dehydration and temperature extremes by *cryptobiosis:* dormancy in a dried-out state.

11.11.   What modes of life do nematodes exhibit?

Nematodes enjoy an impressive variety of possible life styles, as listed below. Multiply these options by a great range of potential hosts and specific food organisms and by an almost limitless diversity of geographic habitats, and you will realize how abundant and ubiquitous the roundworms are.

1.   Terrestrial, feeding on organic detritus and dung
2.   Scavenging on carcasses and/or on bacteria and fungi therein

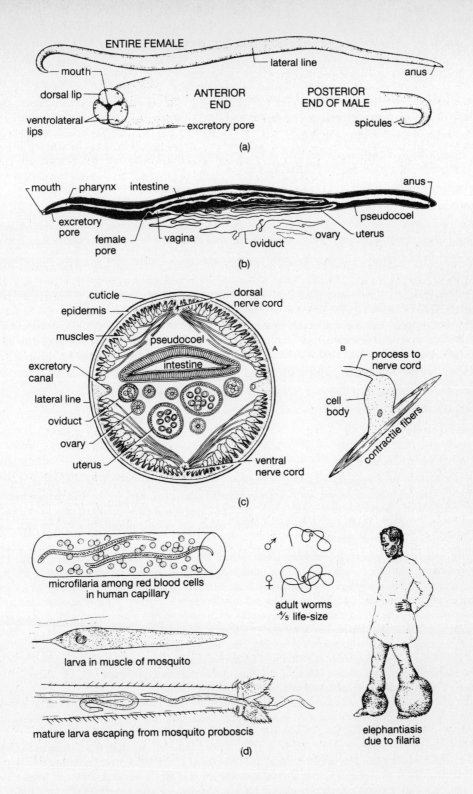

Fig. 11.2 Nematodes. (*a*) Pig ascaris, *Ascaris lumbricoides,* external features. (*b*) Internal structure of a female ascaris, lateral aspect, reproductive tract pulled out on left side. (*c*) A, transverse section of ascaris, showing pseudocoel, longitudinal muscle fibers (in cross section), and processes extending from muscle fibers to nerve cords; B, detail of one longitudinal muscle cell, shown lengthwise. (*d*) Life cycle of the filaria worm, *Wuchereria bancrofti.* [*From Storer et al.; (d) after Francis, after Fulleborn, and after Sambon.*]

3. Aquatic, feeding on bacteria in suspension or on detritus

4. Terrestrial or aquatic, ingesting fungi and yeasts whole

5. Aquatic, ingesting diatoms and other algae

6. Carnivorous on small soil and aquatic animals (including other nematodes), which are either masticated with cuticular teeth or pierced with a stylet and the contents sucked out

7. Ectoparasitic on roots, piercing root cells and sucking out the contents

8. Endoparasitic within various plant tissues, with juveniles migrating to other plants

9. Endoparasitic within plants until after mating, whereupon the female invades a sap-sucking insect larva; her offspring inhabit the larva through its metamorphosis to adult form, then abandon it for a plant pierced by the insect in feeding

10. Endoparasitic within plants, with only the early juveniles invading an insect host to mature partly therein before reentering a plant

11. Commensal, with the juvenile soil-dwelling worms entering an insect host and living commensally until the host dies, whereupon the worm eats the tissues of its deceased host

12. Free-living alternating with zooparasitic, as in the frog lungworm (*Rhabdias bufonis*), which lives and breeds in soil, producing larvae that penetrate frogs' skin, are carried to the lungs, mature and breed there, then lay eggs that are coughed up, swallowed, and shed with feces, to mature and live once more in the soil

13. Parasitic in only one animal host, being disseminated by eggs or a briefly free-living larva

14. Zooparasitic, with a vertebrate definitive host and an invertebrate intermediate host

15. Zooparasitic, with two intermediate hosts and a definitive host

11.12.   What are the characteristics of nematodes?

(*a*)   Nematodes develop from eggs that cleave determinately, but not spirally. Development is direct, and if a larval stage exists, it strongly resembles the adult except in size and sexual maturity. By the time the egg hatches, nearly all the young worm's body cells have already been produced, and it will grow almost entirely by cell enlargement. Like the majority of pseudocoelomates, nematodes are *eutelic:* the number of cells or nuclei in each body part is relatively constant in a given species.

(*b*)   The body wall consists of cuticle, epidermis, and muscle fibers that run only longitudinally (Fig. 11.2*c*). The collagenous cuticle is more complex than that of other pseudocoelomates. It is typically annulated and smooth, but may be striated, pitted, scaly, and bear patterns of *setae* (bristles) or papillae, at least some of which are sensory. Papillae on the head and lips are well-innervated and may be chemoreceptive. Sensory setae detect tactile stimuli. Some aquatic nematodes have a pair of simple eyespots, each made up of a pigment cup with a single sensory cell, but vision is certainly not this phylum's long suit. On each side of the head lies an *amphid*, an often complex cuticular invagination lined with nonmotile sensory cilia and thought to function in chemoreception. A posterior pair of unicellular glands, *phasmids*, which also may have a sensory function, are used taxonomically in dividing phylum Nematoda into two classes. Members of *class Aphasmida* lack phasmids and are nearly all free-living. Members of *class Phasmida* possess phasmids and are mostly parasitic.

The deeper layers of the nematode cuticle usually contain skeletal rods and fibers. The cuticle is toughened by a chemical process equivalent to that used to tan leather. During the worm's growth, the cuticle is molted four times, after which, molting ceases but growth still continues.

The epidermis, which secretes the cuticle, may be either cellular or syncytial. It is unique in having its nuclei confined to four longitudinal cords (dorsal, ventral, and lateral) that extend through the musculature to the pseudocoel, where nourishment is available. The longitudinal muscles are also highly unusual: each cell bears a cytoplasmic process that not only projects into the pseudocoel for nourishment, but also extends to rest upon either the dorsal or the ventral nerve cord. In other words, the muscles extend to reach the nerve cells, rather than the nerve cells sending fibers to reach the muscles.

The nematode central nervous system consists of a nerve ring that surrounds the pharynx, interconnecting four ganglia (dorsal, ventral, and lateral) and four longitudinal nerve trunks, which are also ganglionated (i.e., contain nerve cell bodies and synaptic junctions, as well as nerve fibers) and run within the four epidermal cords described above. The dorsal nerve cord appears to be entirely motor, while the other three have both motor and sensory functions.

(*c*) The digestive tract is long and straight, and complete from mouth to anus. The mouth is at the anterior end of the body, ringed by various radial arrangements of setae and three or six liplike lobes that may bear teeth, especially in carnivorous species. No eversible proboscis is present, but the mouth opens into a cuticle-lined buccal capsule that may bear a solid or hollow stylet used in stabbing prey or puncturing it and sucking out the contents. The buccal capsule opens into the tubular pharynx, the walls of which are composed of glandular and myoepithelial cells (as in gastrotrichs). In cross section, the pharyngeal lumen is three-branched (*triradiate*), reflecting the odd tripartite symmetry of the cephalic setae and lips. The long, straight intestine extends posteriorly and is made up of a single layer of columnar epithelium; it is guarded at each end by a valve that keeps food from being squeezed out of the gut by the pressure of fluid within the pseudocoel. Digestion is begun extracellularly by enzymes secreted by pharyngeal and intestinal cells, but is completed intracellularly, within the intestinal epithelium. Digestive products are transported through the gut lining into the pseudocoel, which serves for internal transport, no blood vessels being present. The alimentary tract terminates in a cuticle-lined chamber considered a cloaca in males, since the sperm ducts empty into it, but a rectum in females, since it is not joined by the female duct. The anus is located ventrally, just in front of the posterior tip of the body. Muscles simply open the anal valve, and the high internal pressure of the pseudocoelomic fluid causes an explosive defecation.

(*d*) Locomotion is accomplished entirely by contractions of longitudinal muscles, since motile cilia are absent throughout the phylum. Since the cuticle prevents the body from changing length, these muscles can bend the worm only from side to side in characteristic lashing movements. The strength of these movements is abetted by increasing the efficiency of the pseudocoelomic fluid as a hydrostatic skeleton. This increase is accomplished by a high osmotic pressure produced by solutes in that fluid. The body wall is permeable to water, which enters so long as the pseudocoelomic fluid is kept hypertonic, making the animal quite turgid. This fluid pressure is higher in nematodes than in any other animals having a body cavity.

(*e*) The excretory system of nematodes is unique. Although the presence of protonephridia in other pseudocoelomates suggests their presence ancestrally in roundworms as well, in modern nematodes they are lacking. Instead, one or two giant *renette cells* open to the exterior by a pore just behind the nerve ring. Alternatively, an H-shaped arrangement of two lateral tubes connected by a transverse canal, presumably evolved from the renette cells, opens at this pore. These structures are thought, but have not definitely been proved, to function in excretion, osmoregulation, or both.

(*f*) Reproduction is entirely sexual, and sexes are nearly always separate. The reproductive tract is tubular, often much coiled within the pseudocoel. In the female, the two tubular ovaries are continuous with oviducts that expand into a pair of uteri; the uteri unite at a common vagina, which opens by way of a gonopore that is usually about midway down the body. In some species the females secrete a sex-attractant pheromone.

   The male has one or two tubular testes leading into a long sperm duct that is dilated posteriorly to form a seminal vesicle. This opens into the cloaca by way of a muscular ejaculatory duct containing prostatic glands that secrete an adhesive material that may aid copulation. The cloacal wall bears sharp, curved penial spicules that can be thrust out of the male's anus into the female's genital pore to permit insemination. Internal fertilization is essential, even in aquatic species, since nematode sperm lack flagella (which is in keeping with the absence of motile cilia throughout the phylum). After fertilization, the egg is invested by a shell, the inner part of which is produced by the egg itself, the outer by the uterine wall. In some nematodes such as the vinegar eel (*Turbatrix aceti*) the eggs develop within the uterus, and the young are born *viviparously* (i.e., live).

11.13.   What parasitic nematodes afflict humans?

(*a*) *Human ascaris (Ascaris lumbricoides)* is contracted by swallowing eggs, usually on vegetation grown in soil contaminated with human sewage. The embryonated eggs promptly hatch, the young burrow through the intestinal lining and are carried by the bloodstream to the capillaries of the lungs. Here they grow for a few days, then break into the air sacs (alveoli), migrate up the air passages, and are swallowed again, now being large enough to hold their own against intestinal peristalsis, which they must do since they do not fasten onto the lining. They ingest the host's intestinal contents, mate, and lay great quantities of eggs that escape with the feces and are directly infectious to humans without need of an intermediate host. The worms may cause pneumonialike symptoms when in the lungs, allergic reactions to their secretions, and, in large numbers, intestinal obstruction. They are vulnerable to oral helminthcides while they remain in the intestine; however, they may migrate up the common bile duct and obstruct it, or perforate the intestinal wall, or migrate into the sinuses and burrow into the cranial cavity, or, less harmfully (except to one's sensibilities), come creeping out the nose. This is a large worm (to 40 cm), and a single female can lay 200,000 eggs a day, so it is not surprising that a great many people in rural areas become infected.

(*b*) *Hookworms* (*Necator americanus* and *Ancylostoma duodenale*) are exceedingly prevalent in moist subtropical and tropical regions and can pose a public health problem in children in rural areas of the southeastern United States. Their control depends on effective treatment of human sewage. The eggs hatch in moist, contaminated soil, and after feeding on detritus and molting twice, the minute *filariform larvae* await the barefoot passerby. They burrow through the thin skin between the toes, ride the bloodstream to the lungs, grow there, migrate into the alveoli and up the windpipe to be swallowed. Entering the small intestine, each worm grasps an intestinal villus in its mouth, bites off the tip and begins to suck blood, expelling the excess through the anus. These voracious little worms seldom exceed a length of 1 cm, but even a few can produce anemia, not only because they suck far more blood than they use, but also because they secrete an anticoagulant that causes the wound to bleed for some time after the hookworm has moved to another villus. Even though medication can free the host, reinfection is so likely that even today an estimated 60 million people suffer from hookworms in Asia alone.

(*c*) *Pinworms* (*Enterobius vermicularis*) afflict millions throughout the world, mostly children, often epidemically because their eggs are so minute they blow like dust and cling electrostatically to almost any surface. These tiny worms live in the gut, where heavy infestations cause intestinal upsets and generalized debility that may lower resistance to more serious disease. The infection is usually self-terminating unless reinfection occurs. The latter is very likely, for during the night the female pinworms migrate to the host's anus to oviposit, causing itching. Scratching the itch contaminates the fingers, and the child is quickly reinfected. Some 30 percent of the children in the United States experience pinworm infection, as do about 16 percent of the adult population.

(*d*) *Filariasis* and *elephantiasis* are caused by roundworms of the genus *Wuchereria* in tropical regions throughout the world (Fig. 11.2*d*). The adult worms grow to a diameter of 0.3 mm and a length of 8 cm for the female or 4 cm for the male. They live in the lymphatic vessels, where they can cause mechanical obstruction to lymphatic drainage as well as disturb the osmotic balance with their secretions, so that eventually the host's subcutaneous tissues swell with retained fluid and the edema stabilizes as a web of connective tissue forms. Since the swelling is most acute in the legs, these assume an elephantine appearance, hence the name *elephantiasis*. However, the scrotum, breasts, and lower arms may also become grossly swollen. The 0.2-mm-long *microfilaria larvae* are carried by the lymphatic flow into the bloodstream, where they migrate to the blood vessels of the skin, concentrating in the most superficial vessels at night, which coincides with mosquito activity. Ingested by a mosquito, the larvae grow to a length of about 1.4 mm, then migrate to her proboscis in readiness for reentering the vertebrate host. The initial infection may cause fever and even death, but even if the person survives and the initial symptoms subside, the edema of elephantiasis may eventually develop when infection is severe. Control requires mosquito abatement and avoidance of being bitten.

(*e*) *Trichinosis* is a potentially fatal disease contracted by eating undercooked meat, especially pork, bear, or walrus, containing the microscopic cysts of larval *Trichinella spiralis*. When swallowed, the dormant larvae are activated by digestion of the enclosing calcareous cyst and mature within 2 days to a length of 1.5 mm for males and 3 mm for females. The initial presence of the worms in the intestinal cavity causes nausea and vomiting, but if diagnosis is not made at this early time, the enteric symptoms subside as the mated females burrow into the intestinal wall. The females produce numerous larvae that bore through the host's tissues toward an encystment site in the skeletal muscles. The host may die during larval migrations, which cause muscular swelling and rigidity and difficulties in breathing and swallowing. If the patient survives until the larvae have safely encysted, these symptoms disappear, but the encysted worms continue to live in the muscles for some time. About 16 percent of the population of the United States carry encysted trichina worms, but because a mild infection may mimic the aches and fever of "flu," most never realize they have had a brush with trichinosis. The encysted worms cannot complete their maturation until their host's flesh is eaten, as by rats, dogs, cats, bears, and pigs. Pigs usually become infected by eating dead rats or pork scraps in garbage. Since pork is the main vehicle for human infection by this dangerous nematode, it must be cooked thoroughly. Not even storage in a domestic freezer is considered adequate to kill all encysted worms, and even if only a few remain to breed, serious symptoms may result. No wonder pork has been judged "unclean" and unfit for human consumption by several cultures.

(*f*) *Guinea worm* (*Dracunculus medinensis*) is prevalent in India, Africa, and the middle east, where people and other mammals become infected by drinking water that contains nearly microscopic copepod crustaceans (*Cyclops*) that host the larval worms. When the copepods are ingested, the larvae burrow out of the gut of their definitive host and migrate to the body cavities, connective tissues, and unfortunately sometimes to the brain meninges. The female grows to a diameter of 1 mm and a length of over a meter and, after mating, migrates to the host's subcutaneous tissue, usually in the foot, in preparation for egg laying. Here she produces an open ulcer, and whenever the foot is immersed in water, she liberates a cloud of larvae. These, after a brief free-living stage, are ingested by copepods and continue growth in the host's body cavity. Lying under the skin, the adult guinea worm resembles a varicose vein, but is not so easily removed surgically. Instead, it is still common practice to wind the worm out on a stick, no more than a centimeter a day, since breakage leads to severe inflammation and secondary infection.

(g) The *giant kidney worm* (*Dioctophyma renale*) is truly gigantic, the female attaining a diameter of more than 1 cm and a length of over a meter. This species only incidentally infects humans, but is worth consideration since it is endemic in the Great Lakes region of North America and is contracted by eating uncooked fish. Dogs, otters, and other piscivorous mammals are the usual definitive hosts, and the eggs are shed in urine. Eggs ingested by certain freshwater annelids develop into larvae in this first intermediate host. Fish that eat the annelids serve as the second intermediate host, with the larvae growing in their body cavities and mesenteries. When eaten by a mammal, the worms first inhabit the body cavity but eventually must migrate to the kidneys, where their eggs can escape with the urinary flow. Since a single worm can devour the tissue of an entire kidney, leaving only the capsule, *Dioctophyma* is a formidable parasite and may not be removable short of surgical extirpation of the infected kidney.

Chapter 12

Larval Links, Branch Points, and Minor Mysteries

Two major events in animal evolution—the development of a true body cavity (coelom) and the divergence of deuterostomes from protostomes—appear to center upon a group of phyla that differ greatly in adult body form but developmentally show evidence of relatedness. The phylogenetic links among these metazoan phyla are discussed in Questions 12.1–12.5, with the characteristics of each phylum being dealt with in subsequent questions.

12.1. How may pseudocoelomates and eucoelomates be linked?

We have already seen that a pseudocoel is a body cavity that lies between the endoderm and the mesoderm, whereas a coelom arises *within* the mesoderm, separating it into parietal and visceral layers and paving the way for the evolution of muscularized, vascularized internal organs. In its more primitive form, the coelom may lack visceral musculature and is distinguished from a pseudocoel by nothing more than a lining of a thin membranous *peritoneum*, which extends into the body cavity as *mesenteries* that support the viscera and coats the latter as a flattened epithelium known as the *serosa*, or visceral peritoneum. Several phyla exhibit curious blends of pseudocoelomate and eucoelomate characteristics.

(a) *Entoprocts* definitely have a pseudocoel, but unlike the body cavity of other pseudocoelomates, theirs is filled with gelatinous material enclosing parenchyma and amoebocytes, and muscle fibers are associated with the esophagus wall, as well as occurring in the tentacles. Entoprocts also differ from other pseudocoelomates by having a larval stage with an apical tuft of sensory cilia and a girdle of locomotory cilia, which appears to be a quite typical *trochophore* larva, a form characteristic of annelids, mollusks, and a number of other eucoelomate phyla.

(b) *Priapulids* bear a close resemblance to the pseudocoelomate kinorhynchs (see Fig 11.1*b*), but their body cavity is lined by a thin, cellular membrane that also forms mesenteries supporting the internal organs. Yet the gut posterior to the muscular pharynx is straight and thin-walled, and if the priapulid body cavity is indeed a true coelom, it qualifies as such only on the basis of this thin membranous lining. Furthermore, priapulid larvae resemble certain rotifers in which the cuticle of the posterior part of the body grows forward to form a *lorica* (case) into which the anterior part of the body may be withdrawn, and rotifers, you will recall, are pseudocoelomate.

(c) *Pogonophorans* are traditionally considered eucoelomate, although detailed study of their anatomy has failed to reveal a peritoneal lining that could qualify the body cavity for eucoelomate status. However, pogonophorans bear certain persuasive resemblances to annelid worms, which definitely are eucoelomate.

(d) *Tardigrades* are considered by some zoologists to display many similarities to the pseudocoelomate phyla, especially gastrotrichs (see Fig. 11.1*a*). They also resemble rotifers in being commonly parthenogenetic and in producing thin-shelled eggs under favorable conditions and thick-shelled eggs under unfavorable ones. As in most pseudocoelomates, the musculature consists only of longitudinally oriented contractile cells, which attach to the inner surface of the cuticle. Like gastrotrichs and rotifers, terrestrial tardigrades are masters of cryptobiosis (dormancy in a state of dehydration). On the other hand, during embryogeny, tardigrades do produce a series of five pairs of coelomic sacs as outpouchings from the gut (an *enterocoelous* mode of coelom development, which is actually characteristic of deuterostomes), but four pairs of these disintegrate, their walls differentiating into the muscle cells that form the subcuticular bands, while the fifth gives rise to the gonads, so that the cavity within the gonads (*gonocoel*) represents the only persistent remnant of a true coelom. The functional body cavity of tardigrades, therefore, is not a true coelom, and the limited musculature is confined to the body wall. The body cavity is termed a *hemocoel* for presumed similarity to the blood-filled tissue spaces of arthropods, to which tardigrades are sometimes thought related; however, tardigrades lack a heart and seem to have no real circulatory system.

12.2. Which phyla have a trochophore larva?

A *trochophore* ("wheel bearer") is a minute, free-swimming, planktonic larva that is characteristically top-shaped, with one or two girdles of cilia by which it spins merrily along (Fig. 12.1). The tip of the top bears an apical sensory plate with a tuft of long cilia. The trochophore has a complete gut, extending from a mouth located just below the main circumferential band of cilia (the prototroch) to an anus located at the bottom of the top, so to speak. This larva in typical or modified form occurs in the majority of eucoelomate protostome phyla and also links these with certain pseudocoelomate and acoelomate groups.

A typical trochophore larva occurs in polychaetes of phylum Annelida (the segmented worms), all marine members of phylum Mollusca except cephalopods (octopus, etc.), phylum Echiura, phylum Sipuncula, and the pseudocoelomate phylum Entoprocta. Larvae sometimes considered to be modified trochophores occur in phylum Brachiopoda, phylum Phoronida, and the bryozoans (phylum Ectoprocta). This represents *five* protostome phyla linked by a typical trochophore larva, and three more by larvae that may represent considerably modified trochophores, a relationship that suggests that modern trochophores preserve to the present day an ancestral larval form from which all these phyla may have diversified.

In addition, an extensive group of nemerteans (phylum Rhynchocoela) have a planktonic *pilidium* larva, which is very trochophore-like, with apical ciliary tuft and encircling bands of locomotory cilia, but interestingly has a flatwormlike saccular digestive cavity lacking an anus. This last feature enhances its resemblance to the larvae of certain turbellarian flatworms. Accordingly, a pilidiumlike turbellarian larva might well represent an ancestral link from flatworms to nemerteans, and the pilidium itself provides a link from the nemerteans to the protostome phyla that have a trochophore larva.

Since the annelids appear to be phylogenetically linked through their trochophore larvae to the other phyla mentioned above, it follows that those phyla that have no trochophore larva themselves but are in other ways undoubtedly related to the annelids are related to all the rest at one remove—kissing cousins, as it were. These would include the phyla Onychophora and Arthropoda and, quite probably, the Pogonophora and the parasitic Pentastomida as well.

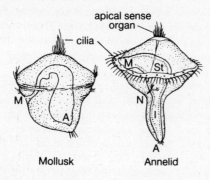

Fig. 12.1 Trochophore larvae. *Left,* larva of a mollusk (the limpet, *Patella*); *right,* larva of a marine annelid, *Polygordius.* M = mouth, A = anus, St = stomach, I = intestine, N = nephridium. (*From Storer et al.*)

12.3.   Which phyla possess a lophophore?

A *lophophore* is a spiral, U- or W-shaped extension of the body wall that bears hollow ciliated tentacles and encloses evaginations of the coelom. It is used for capturing food that adheres to mucus on the tentacles and is carried along ciliated grooves to the mouth. The lophophore can often be pulled in by a set of retractor muscles and protruded by the hydrostatic pressure of coelomic fluid. This novel device occurs in three phyla that are dissimilar in adult body form but share other features such as a U-shaped digestive tract, certain larvae that may be modified trochophores, and a unique manner of coelom development (see Fig. 8.5*d*). The three lophophorate phyla are Brachiopoda, Phoronida, and Ectoprocta (Bryozoa).

12.4.   Which phyla have metanephridia?

Up to this point, the type of excretory unit we have seen in acoelomates and most pseudocoelomate phyla has been the *protonephridium* (see Questions 10.10 and 11.2). In the eucoelomate protostomes, a new type of excretory unit emerges: the *metanephridium*, which collects and drains coelomic fluid. A metanephridium opens into the coelom by way of a ciliated funnel (*nephrostome*) that draws coelomic fluid into the excretory tubule, the first portion of which may be dilated into a sac, and finally expels it to the exterior via a *nephridiopore*. Unless unmodified coelomic fluid is to constitute the urine, the renal epithelium forming the sac and tubule must engage in active transport to reclaim useful solutes from the fluid before it is voided. Metanephridial kidneys are characteristic of mollusks, sipunculids, echiurans, phoronids, brachiopods, pogonophorans, and most annelids. (Some polychaete annelids have protonephridia, as do entoprocts and priapulids.) In the deuterostome line, vertebrate kidneys are derived from the metanephridial type, and functional metanephridial tubules are still found at least in embryonic life.

12.5. Which phyla may link protostomes and deuterostomes?

In Chap. 8 we saw that one branch of the animal kingdom, including echinoderms and vertebrates, consists of eucoelomate organisms known as *deuterostomes*, because the mouth is the second opening to appear during embryogeny, the anus developing earlier in the location of the blastopore. We saw that the deuterostomes also differ from other Bilateria, known as *protostomes*, in that (*a*) *cleavage* tends to be spiral and determinate in protostomes, radial and indeterminate in deuterostomes; (*b*) the *mesoderm* arises from the endoderm of the archenteron in deuterostomes, but from a single cell determined early in mosaic development in the case of protostomes; and (*c*) coelom development is *enterocoelous* (from gut outpouchings) in deuterostomes, but *schizocoelous* (by a split in a solid mass of mesoderm) in eucoelomate protostomes (see Fig. 8.5*d*). Some of the phyla considered in this chapter display a mixture of protostome and deuterostome developmental characteristics and so may stand near this great branch point in eumetazoan evolution. Inasmuch as all known deuterostomes are eucoelomate, the origin of this group must have taken place at a time when a true coelom was first evolving, along different developmental pathways. Although our knowledge of embryonic development is limited with respect to the majority of species in the phyla discussed here, the available data indicate that some animals living today still seem poised at that major evolutionary bifurcation.

The lophophorate phyla are protostomes that show such admixture of protostome and deuterostome traits that they might almost as well be classified as deuterostomes. Brachiopods display radial cleavage, and within class Articulata, coelom development is definitely enterocoelous. Radial cleavage is also characteristic of ectoprocts. Cleavage in phoronids shows an affinity for both radial and spiral patterns, with most species showing the radial type. Mesodermal and coelom development in phoronids shows intermediacy but is somewhat closer to the enterocoelous pattern. A number of lophophorates display a unique type of coelom development in which scattered mesenchyme simply coalesces into separate parietal and visceral mesodermal layers enclosing the body cavity. Despite variations in exactly how the coelom develops, in all three of the lophophorate phyla the coelom is partitioned in a manner characteristic of *deuterostomes*, which have three coelomic compartments: from anterior to posterior, the *protocoel*, *mesocoel*, and *metacoel*. In the lophophorates the protocoel is suppressed because these animals are essentially headless, but the mesocoel and metacoel occur in typical deuterostome fashion.

Knowledge of pogonophoran development has been limited to larvae and embryos collected while being brooded in their parents' tubes, usually at sea depths of 200 m or more, and even the adult worms have not been procured intact until fairly recently. Despite the difficulties of studying pogonophorans, it is known that cleavage is bilateral, a modification that could equally be derived from either the radial or the spiral pattern, and that the transitory blastopore occurs at the *posterior* end of the embryo, as in deuterostomes. On the other hand, mesoderm development takes place in two stages, the second of which definitely occurs as in protostomes. Pogonophorans were in fact once classified as deuterostomes, until it was discovered that they possessed a posterior body part (*opisthosoma*) that is segmented and bears setae that both chemically and morphologically resemble those of annelids. Anchored near the bottom of the tube by its setae, this opisthosoma had been broken off in specimens collected earlier. Furthermore, in earlier pogonophoran descriptions these animals were described as having a *middorsal* longitudinal nerve cord and a *ventral* heart, which are deuterostome traits, yet the worms live vertically within their tubes, and if removed therefrom, they curl up and do not crawl. Accordingly, the orientation of their dorsoventral axis is rendered problematic. The nerve cord could just as correctly be described as midventral, and the heart dorsal, as in annelids, and the direction of blood flow through the closed circulatory system would then conform to the annelid pattern.

12.6. What are entoprocts?

When large enough to be visible to the naked eye, an entoproct ("inner anus") looks like a tiny ovoid head sagely nodding atop a gooselike neck, the whole organism being no taller than 5 mm (Fig. 12.2*a*). All members of phylum Entoprocta are aquatic, and all but one of the approximately 60 species are marine, often in shallow water where they attach to algae. They are fairly common but often escape notice because of their minute size. Although some are solitary, the majority are colonial, with a number of these little nodding heads rising on their stalks from a common horizontal stolon or erect stem. The "head" is actually a body (*calyx*) shaped like an oval cup, encircled by a crown of 8 to 30 tentacles that surround both mouth and anus (hence the phylum name). Like those of lophophorates, the tentacles represent hollow extensions of the body wall, but they are not retractable in the manner of a bryozoan lophophore, nor do they lengthen or shorten. Instead, a band of longitudinal muscle fibers runs along only the medial wall of each tentacle, so that all can be folded inward together over the mouth and anus. When this occurs, a membrane that interconnects the tentacle bases contracts to cover the crown snugly.

The body is externally encased by a cuticle secreted by the underlying single-layered epidermis. Sensory bristles and pits occur on the body surface and are especially concentrated on the tentacles. A brain in the form of a single ganglion lies atop the stomach (actually it develops ventral to the stomach since the stalk arises from the original dorsal surface). From the brain three pairs of nerves extend to the tentacles, and three pairs to stalk and body. The musculature is limited to longitudinal fibers, mostly in the tentacles, tentacular membrane, and stalk. The large pseudocoel contains a gelatinous matrix enclosing both fixed parenchyma and amoebocytes. Two protonephridia lie within the pseudocoel and open by way of a common nephridiopore located just posterior to the mouth. The digestive tract is U-shaped, with a large stomach region, and its wall is made up of a single layer of ciliated epithelium. A thin layer of muscle fibers underlies the esophageal lining. Entoprocts are filter feeders, trapping in mucus on their tentacles microscopic plankton and organic particles that are drawn in by the ciliary current.

Asexual reproduction by budding occurs in both solitary and colonial species. In the former, the buds develop from the parent's calyx, break free, and move away before becoming sessile. In the latter, new zooids arise from stolons or erect branching stems. Under unfavorable conditions, entoprocts often autotomize the calyx portions, which are later regenerated from the stalks. Both dioecious and monoecious species occur, the latter being sometimes protandrous. The eggs are fertilized within the ovaries, shed through a gonopore that opens within the ring of tentacles, and brooded in a depression that lies between gonopore and anus. They undergo spiral, mosaic cleavage and hatch into typical trochophore larvae. The larvae enjoy a brief free-swimming existence before settlement. When settling, they first creep over the surface by means of a ciliated foot and, upon finding the substrate to their liking, attach by way of a frontal organ and metamorphose to adult form.

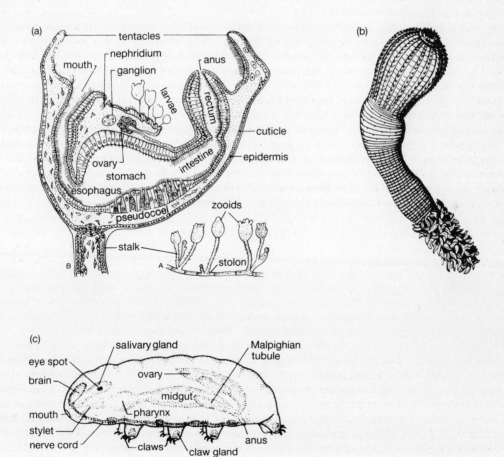

Fig. 12.2 Minor mysteries. (*a*) Phylum Entoprocta, *Pedicellina*: A, growth habit of colony; B, detail of calyx of one zooid. Note larvae being brooded. (*b*) Phylum Priapulida: external appearance of *Priapulus*. (*c*) Phylum Tardigrada, a water bear (to 1 mm). [*From Storer et al.; (a) modified from Becker, (b) from Wolcott, Animal Biology, McGraw-Hill Book Co.*]

12.7.   What are priapulids?

Members of phylum Priapulida ["phallus" (penis)] are named for their shape (Fig. 12.2*b*). They comprise nine known species of stoutly cylindrical marine worms that live in bottom mud from the shallows to considerable depths, mostly in colder Atlantic and Pacific waters of the northern hemisphere, and also from Patagonia to Antarctica. The small number of species does not indicate rarity, but simply that they inhabit a stable environment and have experienced little need for evolutionary diversification. Mainly, they differ in size, from 0.5 mm to 20 cm in length. Priapulids, especially the smaller species, bear a considerable resemblance to the pseudocoelomate kinorhynchs, which are less than 1 mm in length, but as mentioned above, the priapulid body cavity is lined by a thin membrane and thus technically qualifies as a coelom. Like kinorhynchs, the priapulid head and neck bear longitudinal rows of hooklike spines and can be retracted into the trunk as an *introvert*. This introvert makes up about a third the body length, the remainder consisting of the trunk and one or two caudal appendages. The latter, often finely divided, are hollow, containing extensions of the coelom, and are thought to have respiratory and sensory functions. Priapulids burrow by extending the introvert, using the hydrostatic skeleton of their coelomic fluid to achieve extension and rigidity, then pulling the body forward by invaginating the introvert, which remains anchored in the burrow by its hooks. Their activity is intermittent, for ordinarily they lie quietly, fully extended in a vertical position, with protruded introvert reaching to the surface. The pharynx and mouth are everted to allow slowly moving prey such as annelid worms to be seized by the circumoral hooks.

The priapulid body is covered by a thin, chitinous cuticle, which is molted periodically. This cuticle is secreted by the underlying one-layered cellular (not syncytial) epidermis, below which lie circular and longitudinal muscle layers. The central nervous system consists of a nerve ring around the pharynx and a single midventral ganglionated nerve cord. Surface papillae may be sensory in function.

The coelom is lined by a thin, cellular membrane that extends inward as mesenteries that support the internal organs. The coelomic fluid contains amoebocytes and, at least in *Priapulus*, a substantial number of corpuscles containing *hemerythrin*, a pinkish respiratory pigment that serves to bind and transport oxygen. Hemerythrin contains iron but does not in other respects resemble hemoglobin. The pharynx is muscular, toothed, and lined with cuticle; it leads into a straight, thin-walled intestine that terminates at the posterior anus. Paralleling the intestine on each side is a protonephridial excretory duct that bears great numbers of flagellated solenocytes along one side and has a tubular gonad attached along the opposite side, so that the duct serves to carry both urine and gametes.

Sexes are separate, and fertilization is external. In some species sperm release by males stimulates liberation of ova from nearby females. Cleavage is radial. The development of the body cavity remains obscure, and its eucoelomate nature is still subject to dispute. The larva burrows in mud and feeds on detritus. As mentioned above, it bears a certain resemblance to rotifers that develop a lorica (case) from the posterior part of the cuticle, into which the anterior part of the body can be withdrawn, but this resemblance may be due to convergence, not relatedness. Priapulids may indeed represent a link to the pseudocoelomates, but their relation to other eucoelomates remains a mystery.

12.8.   What are water bears?

Viewed under the microscope, these little animals (under 1 mm) have the appealing look of teddy bears toddling slowly along on four pairs of stumpy legs that bear terminal hooks or adhesive pads. Hence, members of phylum *Tardigrada* ("slow step") are popularly known as "water bears" (Fig. 12.2*c*). This resemblance is accentuated by a pair of appropriately placed red or black *eyespots*, each composed of a single pigmented cell. Although they are commonly treated as if allied with arthropods, the phylogenetic status of tardigrades is very debatable, and they remain a "minor mystery." "Minor" enough in size to escape the casual eye, the phylum constitutes some 400 species, many of which are quite cosmopolitan. They inhabit the water film covering mosses and lichens in damp terrestrial environments or live in bodies of fresh water, where they creep upon algae or among bottom detritus, for they cannot swim. A few are marine, inhabiting interstices among sand grains from the intertidal zone to considerable depths. Like rotifers, terrestrial tardigrades can survive hard times cryptobiotically, dehydrating from a water content of 85 percent to one of about 3 percent. In this state they can remain for years, resistant to heat or freezing (even to immersion in liquid helium at $-272°C$), oxygen deprivation, ionizing radiation, and even "pickling" in absolute alcohol. Their prime candidacy for survival of thermonuclear warfare and a "nuclear winter" brings some comfort in the thought of a ravaged world still populated by patiently plodding little water bears!

The plump, cylindrical tardigrade body is covered with a delicate, lamellar cuticle composed of the nitrogenous polysaccharide *chitin*, together with mucopolysaccharides. Chitin also makes up the exoskeleton of arthropods, but the ultrastructure of the tardigrade cuticle most closely resembles that of the pseudocoelomate gastrotrichs. The underlying epidermis is *eutelic*, but is cellular, not syncytial. The musculature is restricted to bands composed of single, longitudinal muscle cells, attached to the inside of the cuticle. The body cavity is large but is considered a *hemocoel* (although no circulatory system is present) since the true coelom, which develops in the enterocoelous fashion of deuterostomes, diminishes during development to only a cavity within the single, saccular gonad. Since the muscle cells develop from the other coelomic pouches, which form from the wall of the archenteron but do not persist, mesoderm formation also follows a deuterostome pattern. The anterior mouth leads into a buccal tube bearing a stylet apparatus, similar to that of herbivorous nematodes and rotifers, which is used to puncture plant cells. The contents are sucked out by means of a pharynx having myoepithelial walls, as in gastrotrichs, nematodes, and other pseudocoelomates. This bulbous pharynx leads to a tubular esophagus, which opens into a large midgut in which digestion and absorption take place. At the junction of the midgut with the short hindgut, some tardigrades have three large glands thought to be excretory in function and therefore named *Malpighian tubules*, after the excretory tubules characteristic of many arthropods. The central nervous system is similar to that of arthropods: a brain consisting of suprapharyngeal ganglia is connected by commissures that encircle the buccal tube to the subesophageal ganglia from which extends a *double* longitudinal midventral nerve cord with four segmentally arranged ganglia.

Molting occurs about 12 times during a lifespan of up to 30 months, and reproduction usually occurs only at such times, with the discarded cuticle being used to receive from 1 to 30 eggs laid by the female, to which are added sperm from the male. However, many species reproduce parthenogenetically, and males are rare or even unknown in some. Under favorable conditions, aquatic tardigrades lay thin-walled, rapidly hatching eggs and, under unfavorable conditions, thick-walled eggs that survive dehydration and overwintering. Moss-dwelling tardigrades lay only eggs having exceedingly thick, sculptured shells. Development is direct, with a unique pattern of holoblastic cleavage, and the juveniles use their stylets to break out of the egg capsule.

12.9.   What are bryozoans?

Bryozoans ("moss animals") belong to phylum *Ectoprocta*, meaning "outside anus," since the anus lies external to the ring of ciliated tentacles that surround the mouth (Fig. 12.3*a*). This ring is borne on a circular or U-shaped fold of the body wall called the *lophophore*, which can be pulled in by means of retractor muscles. Coelomic fluid is used to extend the lophophore.

Individual bryozoans are only about 0.5 mm in length, but nearly all the approximately 4000 species live in colonies assuming many forms, such as encrusting, foliose, branching, and globular. Each zooid lives in a tiny rectangular case (*zoecium*) secreted by its epidermis. The zoecium wall may be gelatinous or chitinous or may be made of calcium carbonate deposited beneath the cuticle. Certain calcareous bryozoan colonies, especially when dead, resemble delicate, miniaturized corals, but the zoecia lack the radial septa typical of stony corals. Other bryozoans resemble hydroid colonies, but the microscope reveals the beautiful whorl of the bryozoan lophophore and the cilia on its tentacles. Particles gathered by ciliary action are trapped in mucus on the tentacles and carried to the mouth, where they are pumped in by the muscular pharynx. Larger items are trapped as the tentacles close inward to form a basket, while outward flicks of the tentacles reject unwanted matter. Most of the U-shaped digestive tract is made up of a large stomach, in which extracellular digestion occurs, and a saccular cecum off the stomach, which is the main site of intracellular digestion (mainly of lipids). The gut lining is ciliated, but peristaltic muscular waves carry food through the stomach, which is separated by valves from the pharynx and from the posterior pylorus that leads to the rectum. Food reserves are stored in the gut epithelium. Nephridia are lacking, but amoeboid cells in the coelomic fluid collect and store wastes. The nervous system consists of a single ganglion that encircles the pharynx with a nerve ring and gives off nerves to the body and each tentacle.

Most individuals in a bryozoan colony are feeding zooids as described above, but modified zooids serve several other functions. Some, reduced essentially to a body wall without internal organs, form the stolons, attachment disks, and other parts of the colony structure. Others known as *vibracula* are modified into long bristles that flick away debris or small trespassers. In addition, because it is vital that the miniaturized bryozoan colony not be overgrown by other sessile creatures such as sponges, an additional type of protective zooid, the *avicularium* ("little bird"), is usually also present. This resembles a tiny bird's head, often on a flexible "neck," which has a pair of strong beaklike pincers that can repel or crush such intruders as larvae seeking a site for settlement.

Most ectoprocts are hermaphroditic, often protandrous. The gametes develop within the peritoneum and are set free into the coelom. Sperm are shed into the water through pores at the tips of two or more tentacles. The eggs are large, yolky, and few, and may pass out of the coelom by way of a special pore, but more usually are brooded, either in the coelom or in a modified zoecium known as an *ovicell*. In most marine species cleavage is radial, but the resulting larvae, which differ considerably in form but characteristically have a ciliated girdle and apical tuft of long cilia, may well represent modified trochophores. Larvae that are brooded lack a functional gut and, when set free, must settle quickly, budding off new feeding zooids before the young colony starves to death. Freshwater ectoprocts also produce hard-walled *statoblasts* containing asexual germinative cells. These statoblasts are liberated in fall as overwintering devices, directly giving rise to new zooids in spring.

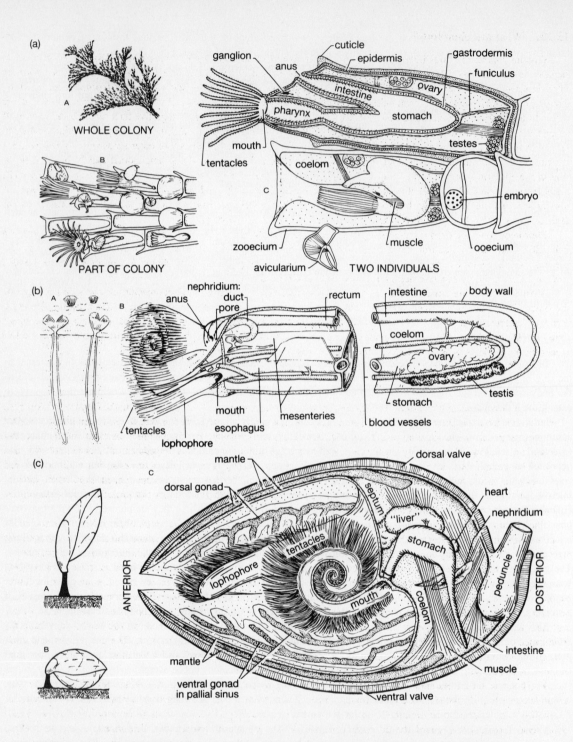

Fig. 12.3 The lophophorate phyla: Ectoprocta, Phoronida, and Brachiopoda. (*a*) *Bugula,* an ectoproct (bryozoan): A, colony, life-size; B, several zooids in zooecia; C, detail of zooids with lophophore expanded (*above*) and retracted (*below*); note defensive avicularium. (*b*) *Phoronis*: A, in habitat; B, internal structure with midpart of body, left half of lophophore, and most of lateral mesentery omitted; note U-shaped digestive tract with anus opening just below lophophore. (*c*) Articulate brachiopods: A and B, external appearance; C, internal structure of Magellania, with left half of lophophore and mantle removed. [*From Storer et al; (b) after Benahm, (c) after Shrock and Twenhofel.*]

12.10.   What are phoronids?

Phylum *Phoronida* constitutes only 10 species of wormlike lophophorates, which nonetheless are widespread in shallow temperate seas. Phoronids live in chitinous tubes buried in sand or attached to solid objects, often forming interwining aggregations, although they are not actually colonial. Even in those few species that reproduce asexually by budding and transverse fission, each animal is a separate individual, lying unattached within its own tube, which it never seems to leave. Since the chitinous tube is closed at the lower end, the phoronid's digestive tract bends back upon itself so that the anus opens anteriorly, just outside the U-shaped lophophore (Fig. 12.3*b*). The elongated, cylindrical body ranges from only a few millimeters to 30 cm in length. The body wall consists of an outer epidermis—which secretes the chitin tube—a thin superficial layer of circular muscle, and a thick, deeper-lying longitudinal muscle layer. A nerve ring in the epidermis encircles the base of the lophophore, giving off an epidermal nerve plexus serving the body wall, a single giant motor nerve fiber that runs along the left side of the body, and nerves distributed to the tentacles and the longitudinal muscle layer. Like bryozoans, phoronids are filter feeders. Food trapped by the ciliated tentacles of the lophophore is probably digested intracellularly at the bend of the U-shaped gut.

Phoronids have a closed circulatory system consisting mainly of a dorsal vessel that carries the blood anteriorly by peristaltic waves of contraction and a ventral vessel that propels it posteriorly. Red blood corpuscles (*erythrocytes*) are present, which contain a form of the oxygen-binding respiratory pigment hemoglobin. A pair of anteriorly located metanephridia open by separate nephridiopores flanking the anus.

Most phoronids are simultaneous hermaphrodites, maturing sperm within the peritoneum on one side and eggs on the other. The gametes are shed into the coelom and swept into the ciliated funnels of the metanephridia. The eggs, fertilized externally, are often brooded within the arms of the lophophore. Cleavage is radial, producing a planktonic *actinotroch* larva thought by some to represent a considerably modified trochophore. This swims about freely for a few weeks before it metamorphoses, sinks to the bottom, and secretes its tube.

12.11.   What are brachiopods?

Phylum *Brachiopoda* ("arm-foot") comprises the group of lophophorates best represented in the fossil record because of their sturdy bivalved shell. Today some 300 species exist, ranging in size from 5 mm to 8 cm, but more than 30,000 fossil species are known, some as large as 38 cm. The genus *Lingula* is considered a "living fossil" because, morphologically at least, it has remained essentially unchanged for the past 400 million years. Brachiopods are all marine, solitary, and sessile. Like clams, they have two calcareous shells (valves) covered by an organic *periostracum*, but while a clam has a left and right valve, the valves of a brachiopod are dorsal and ventral. When the valves are opened, most of the internal space is seen to be occupied by a huge, W-shaped lophophore with each of its two anterior projecting arms elaborately coiled. The body proper is confined to just the posterior region (Fig. 12.3*c*).

The phylum is divided into two classes. In *class Inarticulata* the two valves are much alike and are thin and made up of a chitinous material containing calcium phosphate. Posteriorly, the two valves are not articulated by a hinge, but are simply held together by muscles and are opened by a backward retraction of the body, which causes the pressure of coelomic fluid to force the valves apart. A long muscular *pedicle* extends posteriorly from between the valves, serving to anchor the brachiopod vertically in a sandy bottom. Brachiopods of the *class Articulata* are commonly called "lamp shells." Their two valves are dissimilar, the ventral one being more concave and having posteriorly a hole through which passes a stout, nonmuscular *peduncle* that anchors the brachiopod to a solid surface. In death the two valves often remain together, for they articulate by means of sturdy hinge teeth, and when found tossed up on the beach, they have the appearance of a miniature antique oil lamp, with the hole seemingly intended for the wick. The articulate brachiopods have strong valves formed of crystalline calcium carbonate, held together by powerful *adductor* muscles. Their lophophores are also furnished with a calcareous support lacking in the inarticulates.

The body wall of a brachiopod consists of an epithelial layer underlain by connective tissue. It extends anteriorly as a *mantle* consisting of two sheetlike lobes which are attached to the inner surface of each valve and which secrete material for shell growth. The edge of the mantle probably constitutes a brachiopod's main source of sensory information. It also bears setae (bristles) that guard the aperture when the valves are opened for feeding. The mantle lobes are hollow, containing channel-like extensions of the coelom, and where free of the valves, they also contain a layer of longitudinal muscle fibers. The coelom, as in all lophophorates, is divided into two compartments: a *mesocoel* that surrounds the esophagus and extends into the arms and tentacles of the lophophore and a *metacoel* that surrounds the other internal organs and extends into the mantle lobes. The coelomic fluid contains numerous cells, some containing hemerythrin. An open circulatory system is present: veins and capillaries are absent, and the dorsal heart simply pumps colorless blood into short anterior and posterior arteries that expand into hemocoel spaces that are separate from the channels of the true coelom. One or two pairs of metanephridia collect fluid from the metacoel by way of their ciliated nephrostomes, emptying through a nephridiopore on each side of the mouth.

Diatoms and other minute plankton are trapped by the cilia on the lophophore tentacles, carried to a ciliated groove along each arm of the lophophore, and thence to the mouth. The lophophore cannot be extended and retracted, so the feeding brachiopod simply allows its valves to gape, while the ciliary current moves water inward through an incurrent aperture and out through an excurrent aperture. The direction of ciliary beat can be reversed to drive off silt. As well as their role in food capture, the lophophore tentacles provide an extensive surface for gaseous exchange; uptake of oxygen from the water flowing by is promoted by the oxygen-binding capacity of the hemerythrin within some of the coelomocytes. The ingested food passes down a short esophagus to a large stomach into which a digestive gland opens. Digestion may take place intracellularly within the digestive gland, at least in *Lingula*. The intestine extends posteriorly, ending blindly in articulates, but opening by way of an anus in the inarticulates. The esophagus is encircled by a nerve ring that interconnects a small dorsal and a larger ventral ganglion, from which nerves are distributed to the mantle, lophophore, and muscles controlling the valves of the shell.

Most brachiopods are dioecious, shedding their gametes into the coelom to be carried out the metanephridial tubules. The embryo cleaves radially. As in deuterostomes, the mesoderm and coelomic pouches arise from the wall of the archenteron. The free-swimming larvae of inarticulates somewhat resemble the adults and after settlement simply mature without metamorphosis. The articulate larva, often considered a much modified trochophore, metamorphoses after a motile existence of only a few hours.

12.12.   What are sipunculids?

Members of phylum *Sipuncula* ("little siphon") are often called "peanut worms" because that is what some resemble when fully contracted in resentment of being gathered up by curious fingers. Since most occur in shallow water, actively burrowing in mud or sand or dwelling in mucus-lined retreats under rocks intertidally, they are reasonably familiar to beach explorers. Furthermore, they are sometimes exceedingly abundant in tunnels that they bore in coralline rock, reportedly up to 700 per square meter in some Hawaiian reefs. Sipunculids represent some 300 marine species ranging in length from 2 mm to an impressive 72 cm (Fig. 12.4*a*). The anterior part of the unsegmented cylindrical body is a slender, tubular introvert ending anteriorly in a mouth surrounded by short, ciliated tentacles or lobes, which are hollow but, unlike lophophorate tentacles, are contractile and do not contain extensions of the coelom. Instead, their internal cavities are confluent with a pair of tubular sacs that run alongside the esophagus, providing the fluid needed for extension of the tentacles and holding the fluid forced out of the tentacles when they contract. Deep ciliated grooves on the inner surface of the tentacles or lobes are used in collecting sediment or detritus, after which the introvert is pulled in by means of retractor muscles and the food swallowed. The introvert can then be extended again, by the hydrostatic action of coelomic fluid.

The trunk is wider than the introvert, is usually smooth, and never bears setae. The anus is not terminal, but located near the base of the introvert because the tubular digestive tract is U-shaped, spiraling back upon itself. The body is covered by a thin cuticle, similar to that of annelids, and the body wall consists of a single layer of epidermal epithelium underlain by a dermis containing sense organs and glands, followed by a layer each of circular, oblique, and longitudinal muscle fibers. Sensory cells are especially numerous at the tip of the introvert, and many species have a pair of pigment-cup ocelli embedded in the brain. The brain consists of a pair of ganglia located dorsally just behind the tentacles, with connectives around the esophagus joining midventrally to form a single, unsegmented longitudinal nerve cord that gives off many branches laterally.

The coelom is roomy, lined with a ciliated peritoneum that circulates the coelomic fluid, which has a large number of corpuscles containing *hemerythrin*. No blood-vascular system is present. Lying within the the coelom are the elongated sacs of one pair of metanephridia, which open near the anus. Excretion is aided by unique clusters of peritoneal cells, known as *urns* because each cluster projects into the coelom in the form of a microscopic vase capped with a ciliated cell. These *fixed urns* become detached and wander about in the coelom as *free urns*, each gathering a streamer of particulate wastes that can be removed by way of the metanephridial funnels.

Most sipunculids are dioecious, forming temporary gonads within the peritoneum at the base of the introvert retractor muscles. Immature gametes are shed into the coelom, ripen there, and pass out of the body by way of the metanephridia, to be fertilized externally. Cleavage is spiral, and in most species a typical trochophore larva develops, which may spend (according to species) from one day to a month in the plankton before metamorphosing and settling to the bottom. Similarities in embryonic development and in the structure of the body wall and arrangement of the nervous system suggest that sipunculids may have arisen as an early offshoot of the evolutionary line that produced annelids and arthropods, before body segmentation (*metamerism*) evolved.

12.13.   What are echiurids?

Phylum *Echiura* ["adder (snake) tail"] includes about 100 species of eucoelomate protostomes that are cosmopolitan in warm and temperate seas from the intertidal zone to depths of 2 km. In length they range from a few millimeters to

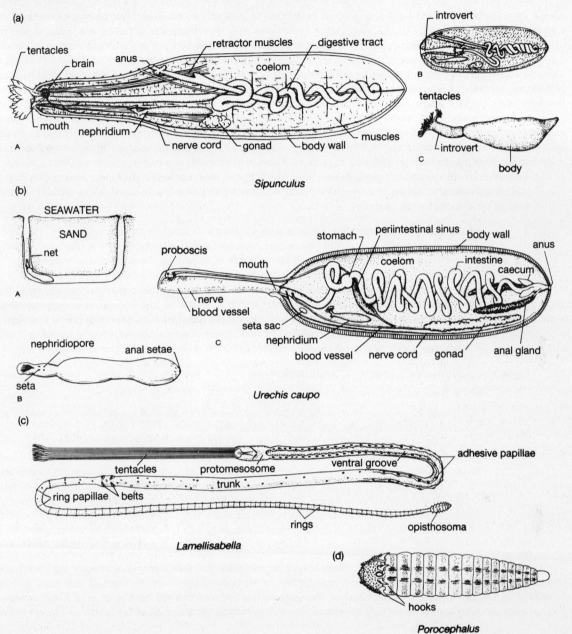

Fig. 12.4 A diversity of "worms." (*a*) Phylum Sipuncula, a peanut worm, *Sipunculus:* A, internal structure, with introvert and tentacles extended; B, internal structure, with introvert retracted; C, external appearance, fully extended. (*b*) Phylum Echiura, the spoon worm, *Urechis caupo:* A, in U-shaped burrow with slime net spread for food capture; B, external appearance; C, internal structure seen from left, with left side of proboscis removed. (*c*) Phylum Pogonophora, the beard worm, *Lamellisabella,* external aspect. (*d*) Phylum Pentastomida, the tongue worm or linguatulid, *Porocephalus,* external view. [*From Storer et al.; (a),* A and B *after Fisher and MacGinitie.*]

about 50 cm (Fig. 12.4*b*). Although the cylindrical, unsegmented trunk may remind one of a sipunculid and the organization of body wall and nervous system (with circumesophageal nerve ring and unsegmented ventral nerve cord) are much the same in the two phyla, echiurids and sipunculids are actually quite different in a number of respects. Echiurids do not have a U-shaped gut: the long, coiled digestive tube ends with a posterior, terminal anus. Echiurids do not have tentacles or an introvert, but instead have a trough-shaped or spoonlike proboscis (after which they are commonly called "spoon worms") that can be greatly extended or contracted, but cannot be retracted into the trunk. This proboscis is actually an anterior extension of the head. *Bonellia*, with only an 8-cm-long body, can stretch out its proboscis to a length of 2 m when feeding! The anterior ventral surface of the proboscis is characteristically flattened against the substratum, allowing detritus to be trapped in mucus and carried rearward to the mouth in a ciliated groove, through a "gutter" formed by the inwardly rolled edges of the more posterior part of the proboscis. Some echiurids cast off the proboscis defensively, later regenerating it. Whereas sipunculids lack setae, echiurids have a pair of hooked setae located ventrally just behind the proboscis and moved by setal muscles, as in annelids; additional circlets of setae may also occur near the posterior end of the trunk. Sipunculids lack blood vessels but have hemerythrin-containing coelomocytes, whereas echiurids have coelomic corpuscles containing hemoglobin and also usually possess a closed circulatory system, with contractile dorsal and ventral vessels and colorless blood. Echiuran metanephridia are very similar to those of sipunculids, each with a large sac occurring along the short urinary pathway, but whereas sipunculids have only one pair, echiurids may have hundreds. In addition, echiurids possess a pair of elongate *anal sacs*, which bear on their coelomic surface large numbers of ciliated funnels, which are similar to the nephrostomes of metanephridia and which probably also function in excretion, voiding wastes through the anus. Echiurids lack urns.

> **Example 1:** An interesting echiuran with a different mode of living is *Urechis caupo* ("fat inn-keeper"). Resembling a plump, pink wiener, *Urechis* inhabits a U-shaped tube it excavates in sandy bottoms along the California coast. From the mouth of its burrow, the worm exudes a mucous net in which plankton and minute organic particles become entrapped. This net is secreted as a tube by a ring of glands encircling the trunk just behind the ventral setae, and peristaltic contractions of the echiurid's body maintain a flow of water through both its burrow and its mucous snare. Eventually, *Urechis* detaches the net and devours both mucus and entrapped particles. *Urechis* is called "innkeeper" because its burrow is habitually occupied by commensals of several species, including a pea crab and a polychaete scale worm, which not only find shelter but relieve the host of some of its food supply.

Echiuran reproduction usually involves external fertilization, with the gametes, which mature in a single median gonad that develops within the peritoneum, being shed by way of the metanephridia. Sexes are separate.

> **Example 2:** Species of *Bonellia* exhibit a remarkable mechanism of sex differentiation, leading to extreme sexual dimorphism. Any larva that settles in an area where adult females are not present will mature into a female; however, any larva that contacts the proboscis of a mature female *Bonellia* is influenced by a pheromone she secretes to differentiate into a *male*, which remains fully ciliated and dwarfed (only 1–3 mm long). First attached to his mate's proboscis, the young male passes into her gut and eventually migrates into an expanded metanephridial sac that serves as a uterus and houses about 20 males in a sort of harem. The males lack proboscis, mouth, and anus and are nutritionally dependent on their gigantic mates. Sperm from the males fertilize the eggs while they are still within the nephridial sacs. The eggs develop by spiral cleavage into typical trochophore larvae, which live planktonically from a few days to 3 months before gradual metamorphosis to a wormlike form.

Interestingly, although echiurids lack all traces of body segmentation postembryonically, during the metamorphosis of some echiurid species, the longitudinal nerve cord displays transitory metamerism and the bands of mesoderm develop 10 pairs of segmentally arranged coelomic pouches. This suggests that echiurans evolved from an ancestral line in common with annelids but either lost, or never fully developed, the metamerism so characteristic of the latter.

12.14   What are beardworms?

Phylum *Pogonophora* ("beard bearer") comprises about 80 species of truly weird worms (Fig. 12.4*c*) that live in vertically opening, chitinous tubes embedded in the ooze of seafloors, usually from depths of over 200 m to the abysses of oceanic trenches. Most are threadlike, with a diameter of less than 1 mm and a body length of 5–85 cm, but more robust, gargantuan specimens exceeding a length of 2 m have been found living at great depths along rift zones some 9 km deep, where sulfurous hot waters sustain entire unlighted ecosystems based on the autotrophic capabilities of sulfur bacteria. If one had in hand only the expanded posterior portion (opisthosoma) of the beardworm's body, little fault could be found with the assumption that this fragment belonged to an annelid worm, for the opisthosoma is metameric, having its coelom internally divided by segmental septa, and it bears setae like those of annelids in both development and structure. Yet all the body anterior to the opisthosoma is unsegmented, although a longitudinal series of papillary rings encircles approximately its posterior third. About midway along the body occur two *girdles* of setae used in gripping the wall of the tube, so that the anterior and posterior halves of the body can contract independently. The cuticle, made of crisscrossing layers of mucopolysaccharides deposited between protruding epidermal microvilli, is like that of annelids.

Although the mesoderm is somewhat segmental, no septa divide the coelom metamerically, except in the opisthosomal region. Instead, coelomic partitions demarcate an anterior *protosome* (tentacles and head), a short *mesosome*, and an elongate *metasome* (all the trunk posteriorly to the opisthosoma). Except for the metameric opisthosoma, this tripartite division of the coelom is deuterostome-like. As mentioned above, the body cavity lacks a peritoneal lining, and its nature and ontogeny remain obscure.

Subepidermally, the body wall consists of a thin layer of circular muscles overlying a thick layer of longitudinal muscles. Mouth and digestive tract are totally absent, a remarkable lack not found in any other nonparasitic metazoan. However, the protosome bears anywhere from a single, spiral tentacle to a "beard" of over 260 long, hairlike *tentacles*, through which, it is thought, organic solutes are directly absorbed by active transport mechanisms. The tentacles are hollow and their surface area is tremendously expanded, not only by epidermal microvilli, but also by rows of minute *pinnules*, each of which is an outgrowth from an individual epithelial cell but which is large enough to contain a circulatory loop consisting of afferent and efferent blood vessels. This extensive surface area undoubtedly facilitates gaseous exchange as well as food absorption. The circulatory system is closed, with main longitudinal dorsal and ventral blood vessels, although which is which remains debatable because the worms are oriented vertically within their tubes and never crawl or exhibit righting responses when removed therefrom. If the single longitudinal nerve cord is considered to run mid-ventrally, as in annelids and related phyla, then the position of the heart, which lies within the protostome, should be described as middorsal, rather than (as has formerly been the case) midventral. A single pair of metanephridia draining the protocoel occur near the heart in the protosome.

Pogonophorans are dioecious, and the two cylindrical gonads lie within the metacoel, emptying separately through a pair of gonopores. The distal part of the sperm duct packages sperm into *spermatophores*, but nothing is known as to how fertilization occurs. Since many species brood the young within the adults' tubes, it has been possible to study cleavage, which follows a unique bilateral pattern, and although the transitory blastopore is located at the posterior end of the embryo as in deuterostomes, the later stages of mesoderm formation are characteristically protostome. The larva is wormlike, with an anterior girdle of locomotory cilia. Its life after quitting the parental tube remains a mystery.

12.15.   What are pentastomids?

Phylum *Pentastomida* ("five mouth") represents a group of some 90 species of wormlike parasites from 2–13 cm long, characterized by five short cephalic protuberances, four of which bear claws, with the fifth bearing the mouth plus two pairs of large hooks used in attachment to the host's tissues (Fig. 12.4*d*). The pentastomids (tongue worms) represent the only parasitic phylum of those considered in this chapter. Parasitism operates as a selective factor for modifying the parasite's body from the form typical of its free-living relatives; since the larva may be less modified than the adult, it is often useful to look at the larva for clues to evolutionary relationships. This brings a surprise, for pentastomid larvae resemble tardigrade larvae, having a plump, oval body, an anterior piercing apparatus, and two or three pairs of stumpy unsegmented legs ending in retractile claws. These larvae hatch from eggs shed with the feces or nasal exudates of the definitive host (usually a tropical reptile). When ingested by a suitable intermediate host (usually a fish or reptile), they penetrate the intestine and migrate about the body while growing and molting several times to become legless nymphs with a ringed abdomen that suggests segmentation. Finally, the nymphs become encapsulated and remain dormant until the intermediate host is devoured by a suitable definitive host. Within this host the pentastomid migrates to the lungs or nasal cavity and matures.

Despite tardigradish beginnings, the adult pentastomid does not look much like anything but itself. Although certain aspects of body structure suggest arthropodan affinities, it is a matter of dispute as to which group of arthropods might have given rise to pentastomids, so the latter are probably best classified in their own separate phylum. The thin, chitinous cuticle is molted several times during maturation (a fact equally true of tardigrades and arthropods). The body, though annulated, is actually unsegmented, and the digestive tract is a simple, straight tube with its anterior end adapted for sucking blood. The nervous system, as in annelids, arthropods, and tardigrades, has paired, segmentally arranged ganglia along a midventral longitudinal nerve cord. In annelid fashion the muscles of the body wall are arranged in circular and longitudinal layers, but are striated as in arthropods, instead of unstriated as in annelids. As in tardigrades, a blood-vascular system is totally lacking, although the body cavity is referred to as a hemocoel. Excretory and respiratory structures are also absent.

Chapter 13

Mollusks

The most enormous of all invertebrates (giant squid) and possibly the most intelligent (octopus?) are both mollusks. This tremendously successful and diversified protostome phylum includes more than 100,000 living and 35,000 extinct species, most bearing a shell of calcium carbonate that has contributed to a rich fossil record from the early Cambrian on. Although certain snails and slugs are terrestrial, mainly in moist habitats, and a number of freshwater snails and bivalves exist, the vast majority of mollusks inhabit the sea, from the intertidal zone to the abysses. Although abundant in the aggregate, many are being collected, without regulation, for food and the commercial value of their ornamental shells, and the increasing rarity of many species should serve as warning of potential extinction.

The history of humans and mollusks has been intertwined as far back as records exist, and no doubt prehistorically, ever since the first human tribe ventured upon an alien shore in search of food. Except for arthropods, no other invertebrate group has had such varied impact on human culture.

13.1. How have humans exploited mollusks?

(a) Humans have exploited mollusks for *food* since prehistoric times, as indicated by shell middens piled up around human encampments, which probably remained populated only until the mollusk supply was exhausted. Today considerable promise for an enduring yield rests with aquaculture technology, in which larval stages of commercially desirable mollusks, such as oysters and abalones, are cultured in aquarium systems and the young adults transferred to protected oceanic areas, such as estuaries; the annual harvest of cultured oysters numbers in the millions. Unfortunately, many species lacking commercial value are collected for food throughout the world, with no provision being made for their restoration.

(b) *Ornamental shells* have been valued as trade items for thousands of years, and ancient routes of commerce are often traceable by mollusk shells found far from their native habitats and often adorned with precious metals and gemstones. Because their superficial and deeper shell layers are differently colored, certain mollusks such as helmets (*Cassia*) have long been in demand for the carving of cameos. Shell collection was once limited to specimens cast up on beaches or collected by free divers, but with the advent of dredging and scuba (*s*elf-*c*ontained *u*nderwater *b*reathing *a*pparatus), only deep-water species remain unexploited.

(c) Certain human cultures formerly used shells as *money* in lieu of metal coins. Pacific native American tribes once used scaphopod shells as "wampum," and in the tropical Pacific the "money cowrie" long served as a medium of exchange.

(d) A *pearl* is often produced when some foreign object becomes lodged between the shell and outer flesh (mantle) of a bivalve. The mantle epidermis responds by encapsulating the object within thin concentric layers of calcium carbonate in a form known as *nacre* (mother-of-pearl), having light-refracting properties that produce a pleasing opalescence. Cultured pearls now augment the natural supply. A number of mollusks that do not produce commercially valuable pearls nevertheless have splendid nacreous layers lining their shells, which are used to make mother-of-pearl jewelry.

(e) Walkie-talkies require batteries, but anyone with strong lungs and a trumpeter's lip can make giant conch shells serve for *communication,* a use to which they have long been put by island peoples of the Indopacific. Far-reaching musical tones can be produced simply by blowing through a conch that is unmodified except for removal of the shell apex.

(f) Before there were plastics, the shells of clams enjoyed as chowder furnished the backbone of the *button* industry, for the crystalline calcium carbonate, although rigid, could be drilled without shattering.

(g) Perhaps the first source of nonfading *ink* for penmanship was procured from the ink sacs of cuttlefish taken in trawls for food. In addition, the bladelike internal *cuttlebone* has long been used as a source of calcium for canaries and other cage birds, since it is easily broken and digested.

(h) One of the most far-reaching ways in which mollusks have shaped human history resulted from the exploitation by a Mediterranean seafaring people, the Phoenicians, of a rock snail's purple exudate to dye cloth a permanent maroon hue that became known as *royal purple* because it was much prized and therefore quite expensive. Aided by their traffic in dyed cloth, the Phoenicians traded widely, carrying with them various elements of their culture, including the Phoenician *alphabet,* which even today is the basis of western linguistics.

13.2. How can mollusks get even?

(*a*) *Venomous* mollusks occasionally inflict pain and even death on the unwary. Of especial note are the small, but potent, blue-ringed octopus of the Great Barrier Reef and tropical cones (*Conus*), 4 cm or more in shell length. Owing to their sluggish nature, cones seldom sting people, but when they do so, they inject a venom far more potent per unit quantity than that of cobras. On the positive side, mollusk venoms are being investigated for potential medicinal applications.

(*b*) Certain snails are intermediate hosts for human parasites such as blood flukes, lung flukes, and human liver flukes.

(*c*) Bivalves cause food poisoning under certain circumstances. During summer months along the Pacific coast of North America, informed persons avoid consuming shellfish because of the possibility that filter-feeding species, mainly mussels and clams, have ingested quantities of toxic dinoflagellates and concentrated the poisons in their own tissues. In addition, mollusks can become infected with bacteria and viruses harmless to themselves but pathogenic in humans. Infectious hepatitis virus, for instance, remains virulent even through primary sewage treatment and can build up in the tissues of shellfish living near sewage outfalls.

(*d*) Herbivorous terrestrial snails and slugs commit acts of mayhem on garden and crop plants, resulting in a flourishing snail-bait industry. This is especially true of alien snails introduced as a possible food source.

> **Example 1:** The giant African snail was brought into Hawaii and then found unsafe for consumption because it is a vector of hepatitis. The European brown snail has ravaged California gardens since its importation as a source of escargot. In a masterpiece of turnabout biocontrol, both of these escapees are beginning to be brought under control by imported carnivorous snails.

(*e*) Chief destroyer of wooden hulls and pilings is the *shipworm, Teredo,* actually a bivalve with an attenuated, wormlike body. Another family of bivalves, the pholadids, contains both wood-boring and rock-boring species.

13.3. What are the features of mollusks in general?

Any phylum as diverse as Mollusca ("soft") requires that individual attention be given to the structural and functional adaptations seen in each of its classes. However, certain features characterize the phylum as a whole, and a hypothetical common ancestral body plan can be built up on the basis of these features (see Question 13.6).

13.4. What characteristics distinguish phylum Mollusca?

(*a*) The body is soft and unsegmented (Fig. 13.1*a* and *b*), consisting of a foot, a visceral hump, and a mantle: (1) The muscular *foot* is variously adapted for creeping or burrowing and in most classes is a part of a *head-foot complex.* In its most primitive version, the foot is a flat, ventral sheet on which the animal glides by ciliary action or creeps by waves of muscular contraction, secreting mucus for adhesion. It may also form a suction disc for anchorage against surf action. In bivalves and scaphopods, the foot is modified into a bladelike or conical digging structure and is extended by hydrostatic pressure but retracted by muscles. In cephalopods (octopus, squid, etc.), it is subdivided into a number of sucker-bearing tentacles and a funnel-like siphon for propulsion. The head lies anterior to the foot and is more or less well developed. It contains the mouth and oral structures and may bear sensory tentacles and eyes. (2) The *visceral hump* contains the internal organs and lies dorsal to the head-foot complex. (3) The *mantle* covers the visceral hump like a cloak and serves for respiration and shell production. The outer surface of the mantle secretes the shell, adding to its thickness and circumference throughout life. The inner surface is elaborated into gills (*ctenidia*) or a lung. By cilia or muscular pumping, the mantle maintains a flow of water through the mantle cavity, as needed for gaseous exchange and filter feeding, as well as flushing away of wastes. The gill cilia of bivalves are also adapted to carry food toward the mouth along special grooves. Jet propulsion by cephalopods is accomplished by muscular pulsations of the mantle that force water out of the mantle cavity through the siphon. The mantle also contains numerous sensory receptors, especially along its edge. Some mollusks can expand their mantle to cover the shell, exposing a slimy surface that deters predatory attacks by starfish.

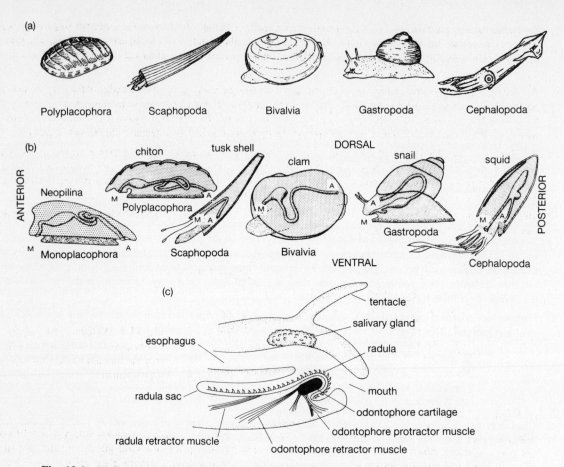

Fig. 13.1 Mollusks. *(a)* External features of the five major classes. *(b)* General features of body
organization in six classes of phylum Mollusca. Shell, heavily outlined; foot, darkly
stippled; digestive tract with mouth (M) and anus (A) depicted. *(c)* Mollusk head
(gastropod), longitudinal section showing radula mounted on odontophore with relevant
musculature. (*From Storer et al.*)

(*b*) A shell of calcium carbonate is usually present, ordinarily covered by an organic *periostracum* made of *conchiolin*,
protein "tanned" by quinones. The shell may consist of one, two, or eight valves. Its outer calcareous layer is made
up of vertical prismatic crystals of calcium carbonate (*calcite*) bound together by a thin proteinaceous matrix. The
inner layer consists of very thin lamellae of calcium carbonate in the form of calcite, *aragonite,* or both, deposited
over a thin organic matrix. Shells with the most spectacular nacreous linings may have up to 500 aragonite lamellae
per millimeter of thickness. The periostracum is particularly thick in freshwater mollusks, protecting the limy part
of the shell from erosion by acids, which may accumulate from decaying vegetation. Mollusks with external shells
can usually withdraw entirely into them and may have an organic or calcareous trapdoor (*operculum*) to seal the
aperture, protecting them from dehydration and certain predators. The snail that carries the blood fluke *Schistosoma
japonicum* is operculate and can survive the dry season nicely, making this fluke species particularly difficult to
control. Certain cephalopods have shells adapted for buoyancy regulation, such as the gas-filled chambered external
shells of nautiloids and extinct ammonites; the coiled, chambered internal shell of the small, deep-sea, squidlike
Spirula; and the cuttlebone of *Sepia,* which is divided by numerous thin septa into narrow spaces filled with fluid
and gas, mainly nitrogen.

(*c*) A *radula* is present in the buccal cavities of all mollusks except bivalves. This unique feeding device consists of a
long, chitinous ribbon set with many rows of recurved chitinous teeth; the ribbon passes over a cartilaginous
odontophore, which can be protruded from the mouth by muscular action or withdrawn into a ventral evagination
of the buccal cavity known as the *radula sac* (Fig. 13.1*c*). The radula itself is pulled forward and backward over
the odonotophore by special radulatory protractor and retractor muscles, producing a scraping action variously
employed for rasping vegetation or flesh, for drilling holes through the shells of other mollusks, and even for

excavating living spaces in rocks. Worn teeth are constantly replaced, at a rate of 1–5 rows per day, by growth from within the radula sac. In *Conus* a single venom-charged radula tooth is slid to the end of the proboscis, where it can be jabbed into the prey like a harpoon head and then left behind in the wound and replaced by another before the next attack.

(*d*) A patch of sensory epithelium known as an *osphradium* is sometimes found on the posterior edge of the gill membranes, where it is thought to monitor water quality, allowing the mollusk to stop circulating water through its mantle cavity when silt becomes excessive. Although osphradia occur only in bivalves and many gastropods, their occurrence in two such different classes of mollusks suggests that they represent an ancestral feature, unique to this phylum.

(*e*) A ciliated *style sac* that opens off the stomach of most bivalves and many gastropods serves to produce and rotate a mucous mass, sometimes solidified into a *crystalline style*, that acts like a reel to draw in through the esophagus a food-laden string of mucus. Although not present in all molluscan classes, its occurrence in these two diverse groups marks it as an ancestral feature unique to phylum Mollusca.

(*f*) Most molluscan classes have a unique posttrochophore larva known as a *veliger*, which possesses an expanded membrane (*velum*) used in swimming (Fig. 13.2).

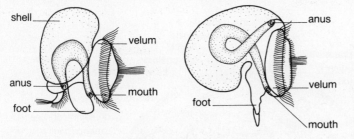

Fig. 13.2 Veliger larva. *Left*, before torsion; *right*, after torsion; only the rotation of the digestive tract is shown.

13.5. What are some additional characteristics of mollusks?

(*a*) The molluscan digestive tract is complete, muscular, and regionalized, with an expanded stomach and lengthy intestine containing ciliary tracts for sorting particles. It receives the ducts of several glands, usually including salivary glands that produce enzymes for extracellular digestion and a major digestive gland off the stomach that may secrete additional extracellular enzymes or digest particles intracellularly, or do both. The anus opens into the mantle cavity.

(*b*) Except in cephalopods, the circulatory system is open, and the heart has two posterior chambers (*atria*, or *auricles*) which collect blood from the gills and one anterior pumping chamber (*ventricle*) which expels blood into arteries. The arteries disgorge into expanded tissue spaces, so that the body cavity is a hemocoel. The true coelom is restricted to a *pericardial cavity* in which the heart lies. The blood contains amoebocytes, and a copper-based respiratory pigment, *hemocyanin*, is dissolved in the plasma. Hemocyanin is pale greenish blue when oxygenated and colorless when deoxygenated.

(*c*) The *pericardial coelom* that surrounds the heart collects wastes by filtration through the heart wall and from glandular cells in the pericardial membrane. These wastes are removed by way of metanephridia with ciliated funnels that primitively open directly into the pericardial space. Each metanephridium is expanded into a sac with greatly convoluted walls, providing much surface area for tubular secretion and reabsorption. From this sac a urinary duct (*ureter*) leads to a nephridiopore that primitively opens at the rear of the mantle cavity. A single metanephridium, or one, two, or six pairs, may be present.

(*d*) The basic pattern of the molluscan nervous system includes a ganglionic nerve ring around the esophagus; this nerve ring gives off a subepidermal plexus and two pairs of longitudinal nerve cords: a ventral (*pedal*) pair to the foot and a somewhat more dorsal (*visceral*) pair to the internal organs. Paired *cerebral, pedal, visceral,* and *pleural ganglia* typically occur, tending to be concentrated in the nerve ring of cephalopods and gastropods.

(*e*) Most mollusks are dioecious, the main exceptions being a number of gastropods that are simultaneous or protandrous hermaphrodites. In the more primitive mollusks, one pair of gonads occurs dorsolaterally on each side of the pericardial coelom, and the gametes are shed into the coelom and pass to the exterior by way of the nephridial ducts, but various modifications and accessory reproductive structures occur in more advanced mollusks. Fertilization may be internal or external, cleavage is spiral and determinate, and most marine species except cephalopods have a trochophore larva that develops into a free-swimming veliger.

13.6.   What was the body plan of the ancestral protomollusk?

Because of its soft nature, the molluscan body has provided no good fossil evidence about its organization. However, comparative and developmental studies suggest that the earliest, Precambrian mollusks may have been somewhat as shown in Fig. 13.3. The body was probably rather slender and wormlike, with a ventral, flat, creeping foot. The uncoiled, caplike, single shell probably consisted only of a thin layer of conchiolin, but before the evolution of this protoshell, the mantle may have been covered only by a thin chitinous cuticle, possibly with calcareous scales as seen in the few existing caudofoveates, which are considered to be the most primitive living members of the phylum. The gills of the ancestral mollusk probably occurred in the rear part of the mantle cavity, with anal and nephridial openings and osphradia also located posteriorly. The digestive system possessed a radula, style sac, and digestive gland opening off the stomach. The two pairs of gonads shed gametes into the pericardial coelom, which was drained by a single pair of metanephridia. The circulatory system was open, with paired auricles receiving blood from the gills and a single ventricle pumping blood forward through an anterior aorta into the hemocoel. The head was probably rudimentary and fused to the mantle, as in modern chitons. The nervous system may have consisted mainly of a circumesophageal nerve ring and longitudinal pedal and visceral nerve cords.

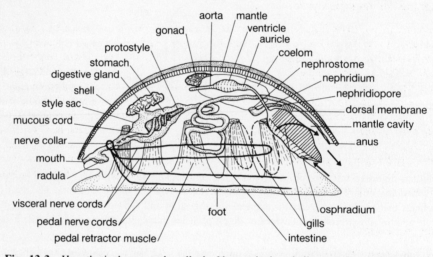

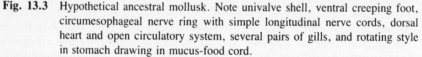

Fig. 13.3   Hypothetical ancestral mollusk. Note univalve shell, ventral creeping foot, circumesophageal nerve ring with simple longitudinal nerve cords, dorsal heart and open circulatory system, several pairs of gills, and rotating style in stomach drawing in mucus-food cord.

A few species still exist today of a class of one-shelled mollusks, the Monoplacophora, that is otherwise represented only by Paleozoic fossils. The body plan of this class shows certain traces of internal metamerism, including 5 segmentally arranged pairs of gills, 6 segmentally arranged pairs of metanephridia, and a ladderlike nervous system, somewhat as in chitons, giving off 10 segmentally arranged pairs of pedal nerves. This may constitute evidence that the ancestral protomollusk may have been incipiently metameric, as are the annelids to which mollusks are obviously developmentally related. However, we have no certain evidence that this was the case, for no trace of segmentation remains in the major classes living today, or in any known molluscan larval form.

13.7.   What are the major characteristics of the molluscan classes?

Although some contemporary mollusks seem to have remained in a relatively primitive condition since the early Cambrian, an impressive degree of diversification can be seen to exist, not only *among* the eight molluscan classes, but also *within* the three most advanced of these. The eight classes and their major characteristics are discussed in Questions 13.8–13.15.

13.8.   What are the caudofoveates?

Class Caudofoveata ("tail-little pit") includes some 70 species of marine, wormlike, burrowing forms from 2–140 mm long, lacking head, shell, and nephridia and having the foot reduced to a small oral *pedal shield* near the anterior mouth. A radula is usually present, used in feeding on detritus and microorganisms. The mantle is covered only with a chitinous cuticle and calcareous scales, and the mantle cavity is located posteriorly, enclosing a pair of gills. Caudofoveates are dioecious, but little is known of their reproductive habits.

13.9.   What are the solenogasters?

Class Solenogastres ("tube stomach") includes about 180 species of wormlike marine mollusks, usually under 5 cm long, that have only a narrow, midventral, longitudinal *pedal groove* on which they creep over corals and hydroids, upon which they prey. They lack a shell, having only calcareous spicules embedded in the integument. The head is poorly developed, and a radula is usually absent. Nephridia and gills are also lacking. They are hermaphroditic and probably copulate, though mating has not yet been observed.

13.10.   What are the monoplacophorans?

Class Monoplacophora ("one-plate bearer") is represented today by only about 10 species ranging from 3–300 mm long. These apparently are the descendants of a group that disappeared from the fossil record in the Devonian, when monoplacophorans seem to have abandoned the continental shelf for the deep-sea basins, where they survived to the present at depths of 2–7 km, in both the south Atlantic and east Pacific. The shell is single and cap-shaped or is in the form of a low cone and, in both modern and fossil species, is distinguishable from any gastropod shell by the presence of from three to eight muscle attachment scars, where a gastropod would have only one. *Neopilina*, which first turned up in a dredge haul in 1952, shows an interesting degree of bodily metamerism, having 8 pairs of pedal retractor muscles, 10 pairs of pedal nerves off a ladder-type nervous system with 10 commissures, 5 lateral pairs of gills, and 6 pairs of metanephridia (Fig. 13.4a). The mouth, which is anterior to the broad, flat foot, is flanked by a flaplike velum and postoral tentacles, and a radula is present. The heart is unique, being divided into two auricles and *two* ventricles. Each of the four gonads attaches by way of a gonoduct to one of the metanephridia. Monoplacophorans are dioecious, but nothing yet is known of their reproduction and development.

13.11.   What are chitons?

Class Polyplacophora ("many-plate bearer") includes mollusks with eight overlapping shell plates that allow them to conform snugly to irregular rock surfaces, to which many cling with their broad, creeping foot, even against powerful surf, although others hide beneath rocks or live subtidally. These are the *chitons*, which comprise some 600 living and 350 fossil species ranging from 2–30 cm in length (Fig. 13.4b). Most chitons are instantly recognizable by their armadillolike shell, but in some species the valves are mostly or totally covered by the fleshy mantle. The mantle of all chitons extends beyond the plates as a *girdle*, which helps the animal resist being pried loose; the chiton clamps down tightly against the substratum and then raises the inner margin of the mantle, creating a vacuum between girdle and foot. The shell plates are penetrated by large numbers of minute canals terminating at the surface in microscopic secretory and sensory organs known as *aesthetes*, which are unique to chitons. More than 1000 aesthetes may occur per square millimeter of shell. Some aesthetes have a lens and a group of photoreceptive cells serving as light detectors, for chitons have poorly developed heads and lack cephalic eyes.

The mouth is located ventrally, just in front of the foot, and is armed with a radula used in scraping the film of algae and other small organisms off the surface of rocks, as well as sometimes excavating a depression in the rock that precisely fits the shape of the chiton and to which it homes after foraging. Homing is assisted by a magnetic sense dependent on the presence of *magnetite*, which caps some of the radular teeth. Chitons have a unique sensory device, the *subradula sac*, which is protruded from the mouth and pressed against the substratum before feeding commences, seemingly testing for the presence of edibles before the radula itself is protruded. The dilated stomach lacks a style sac but receives the ducts of a large gland that secretes digestive enzymes. Digestion is almost entirely extracellular—in the stomach, channels of the digestive gland, and intestine. The lengthy, looped intestine terminates at a posterior anus.

The large pericardial cavity, a coelomic remnant located dorsally beneath the two posterior shell plates, contains a heart with two auricles that receive blood from the gills, which hang down from the roof of the mantle cavity along most of its length. The ventricle pumps the blood into a single, anterior aorta. A single pair of metanephridia, which are tubular

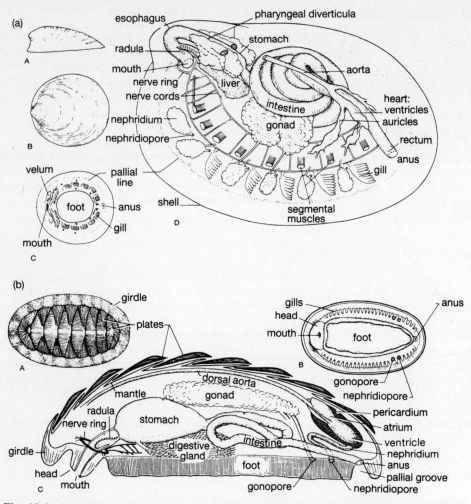

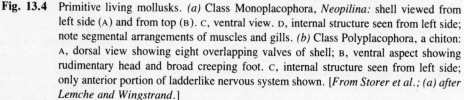

Fig. 13.4 Primitive living mollusks. *(a)* Class Monoplacophora, *Neopilina:* shell viewed from left side (A) and from top (B). C, ventral view. D, internal structure seen from left side; note segmental arrangements of muscles and gills. *(b)* Class Polyplacophora, a chiton: A, dorsal view showing eight overlapping valves of shell; B, ventral aspect showing rudimentary head and broad creeping foot. C, internal structure seen from left side; only anterior portion of ladderlike nervous system shown. [*From Storer et al.; (a) after Lemche and Wingstrand.*]

instead of saccular, opens into the pericardial coelom and voids urine through nephridiopores in the posterior part of the mantle cavity. The nervous system is primitive, lacking the paired ganglia characteristic of most mollusks, and consists mainly of a circumesophageal nerve ring that gives off nerves to the buccal cavity and two pairs of longitudinal cords: a ventral pair innervating the foot and a lateral *pallioviseral* pair serving the mantle and internal organs.

Sexes are separate in most chitons, with a single median gonad that gives off gametes by way of special gonoducts separate from the nephridia. Sperm are released into the sea, as are eggs also in some species, but in others the eggs are fertilized and brooded in the mantle cavity. The trochophore larva develops directly into the adult form, without an intervening veliger stage, suggesting that chitons are closer to the ancestral state than those molluscan groups in which the trochophore becomes a veliger larva.

13.12.   What are the characteristics of scaphopods?

Class Scaphopoda ("boat foot")—known as *tusk shells* from their single tapering shell, which is *open at both ends*—includes about 350 species of marine mollusks with a shell length of 4 mm to 15 cm. They burrow nearly vertically in mud or sand, at depths of 6 m to over 1800 m, leaving the narrow posterior end exposed for breathing. The reduced head and *conical* digging foot protrude from the buried anterior end of the shell, along with a large number of threadlike adhesive tentacles (*captacula*), which are unique to scaphopods (Fig. 13.5). Captacular cilia direct small particles to the mouth, but prey is caught by the adhesive tentacle tips and brought to the mouth by contraction of the captacula. The well-developed radula bears large, flattened teeth that crush and help ingest the prey. The stomach lacks a style sac, but a large digestive gland is present. Digestion is extracellular. The anus opens about midway along the mantle cavity, which runs the length of the shell. A single pair of metanephridia opens near the anus. Gills are absent, and gas exchange takes place through the mantle surface. The posterior shell opening serves for both inhalation and exhalation. Inhalation is by ciliary action and takes about 10 minutes, after which a violent exhalation (which also blows away silt) is accomplished muscularly, by foot retraction. A heart is lacking, and the circulatory system is reduced to *blood sinuses*. The nervous system possesses the typical molluscan pairs of cerebral, pedal, visceral, and pleural ganglia. Eyes and osphradia are absent.

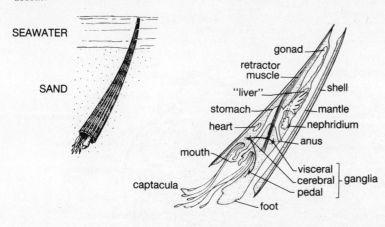

Fig. 13.5 Class Scaphopoda: *Dentalium*, a tusk shell. *Left.* position in habitat; *right*, internal structure seen from the left; note conical digging foot, threadlike adhesive captacula, and radula within mouth. *(From Storer et al.)*

Scaphopods are dioecious, and the single gonad opens by way of the right metanephridium. Eggs are fertilized internally and develop into trochophore and then veliger larvae, which are bilaterally symmetrical, as in bivalves, rather than twisted asymmetrically, as in gastropods. This symmetry, together with aspects of veliger metamorphosis, suggests that scaphopods and bivalves arose from the same branch of molluscan evolution.

13.13. What are the characteristics of gastropods as a whole?

Class Gastropoda ("stomach foot"), the most successful molluscan class, contains about 35,000 living and 15,000 fossil species, including marine, freshwater, and terrestrial species of one-shelled snails and shell-less slugs (Fig. 13.6*a*). Although basically bilaterally symmetrical, the gastropod body is typically both twisted by torsion that occurs during larval development and elongated and coiled in conformity with its single, coiled shell that ordinarily has only one aperture (Fig. 13.6*b*). The foot is large and flat for creeping, and the head is well developed, with eyes, sensory tentacles, and a radula that may be variously adapted for scraping, drilling, or piercing. The stomach typically has a style sac and also receives the ducts of a large digestive gland that functions in extracellular or intracellular digestion, or both, depending on the species. As a result of torsion, the heart has only *one auricle,* and usually only one nephridium and one gill are present. When gills are absent, the mantle cavity functions as a lung. Most gastropods have a sensory osphradium near the inhalant mantle opening.

Both monoecious and dioecious species occur. Although primitive gastropods shed their gametes into the water for external fertilization, in most species copulation and insemination take place, often preceded by courtship, and eggs are laid in protective capsules. Freshwater snails frequently brood their young within the oviduct. Marine gastropods have a trochophore stage that is often passed within the egg capsule, from which the young then emerge as free-swimming veligers. *Torsion* takes place quite abruptly, during the veliger stage, by uneven growth of the left and right muscles that attach the head-foot complex to the shell (see Fig. 13.2). This twists the internal organs through at least 180 degrees, allowing the head and velum to be withdrawn into the developing shell and the anus and gills to move anteriorly, to the shell aperture. *Coiling* is a separate phenomenon, which takes place only gradually, as the postlarval gastropod adds whorls to its shell and the body lengthens, like a conically tapering worm extending from shell apex to aperture. Coiling may be in one plane (*planospiral*, the more primitive type) or in two planes (*conispiral*), in which event the shell may taper conically (as in augers), or have a single large body whorl and depressed apex (as in *Helix*, the garden snail), or be intermediate between these extremes.

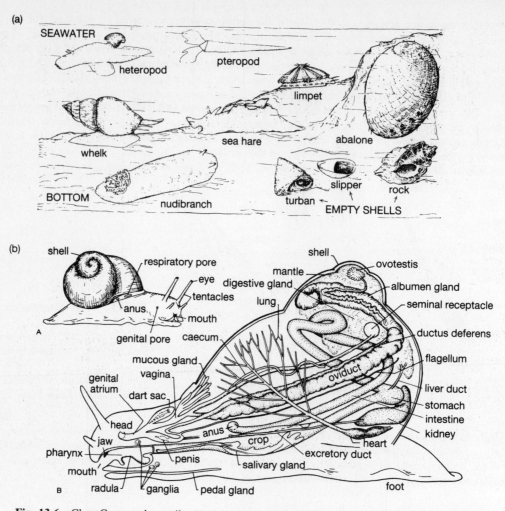

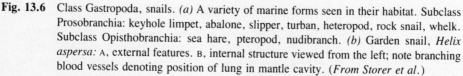

Fig. 13.6 Class Gastropoda, snails. *(a)* A variety of marine forms seen in their habitat. Subclass Prosobranchia: keyhole limpet, abalone, slipper, turban, heteropod, rock snail, whelk. Subclass Opisthobranchia: sea hare, pteropod, nudibranch. *(b)* Garden snail, *Helix aspersa:* A, external features. B, internal structure viewed from the left; note branching blood vessels denoting position of lung in mantle cavity. *(From Storer et al.)*

13.14. What are the characteristics of bivalves in general?

Class Bivalvia, or *Pelecypoda* ("hatchet foot"), includes some 20,000 living and 15,000 extinct species of mollusks having two lateral shell valves, *left and right,* which are hinged together *dorsally* (Fig. 13.7*a*). The foot is typically bladelike, adapted for burrowing, and operated by muscles that work against the hydrostatic pressure of blood within the hemocoel, although some species do not burrow, but live attached to a surface or even are occasionally free-swimming. The posterior mantle edges are modified into inhalant and exhalant siphons that can constrict or dilate to allow the ciliated gills to maintain a flow of water through the mantle cavity. The siphons may be long and tubular in some burrowing forms and may even be fused into a single pipelike structure, miscalled a "neck," since it protrudes from the clam's posterior end. Most bivalves are filter feeders, trapping small particles in mucus on the gills. Carried along ciliated grooves toward the mouth, the particles are sorted by flaplike *labial palps,* which reject some particles and direct others into the mouth (Fig. 13.7*b*). Bivalves are essentially headless, and cephalic eyes and a radula are lacking. However, the mantle edges may bear numerous tentacles containing tactile and chemical receptors and eyes that are usually simple pigment-spot ocelli, although in free-swimming scallops the eyes are remarkably complex and can probably resolve images rather well. A sensory patch in the lining of the exhalant mantle chamber is called an osphradium, although its position is poorly situated for detecting silt or other material about to be inhaled; unlikely to be the same as the osphradium of other mollusks, the function of this organ remains obscure.

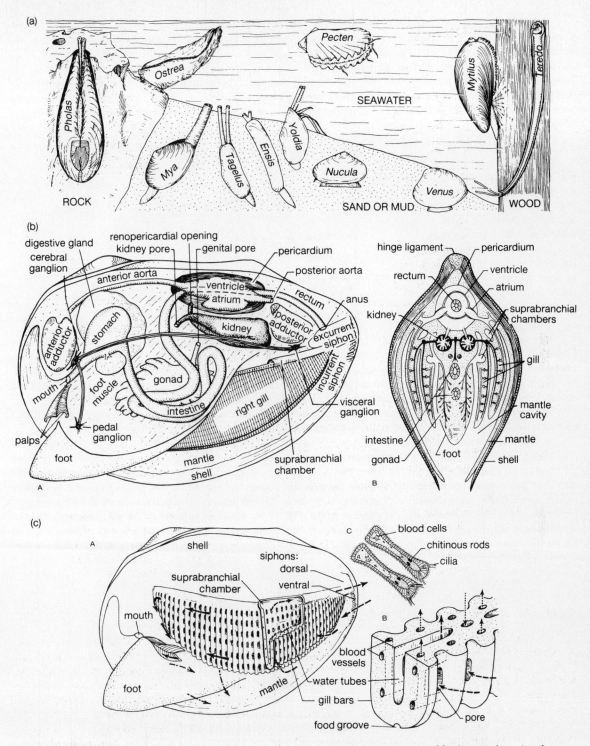

Fig. 13.7   Class Pelecypoda, bivalves. *(a)* Marine bivalves in their habitat. *(b)* Anatomy of freshwater clam, *Anodonta:* A, internal structure with left gills and mantle removed; B, schematic cross section in plane of heart. *(c)* Structure of gills in freshwater clam: A, part of left gill cut away to show internal structure; B, detail of part of gill; C, cross section of two gill bars (---→ = direction of water current, ·····→ = direction of blood flow, → = path of food particles to mouth, -·-·-→ = path of rejected particles). [*From Storer et al.; (b),* B *after Stempell.*]

Most bivalves are called *lamellibranchs* because they have *platelike* gills (Fig. 13.7c) composed of longitudinally arranged layers of W-shaped gill filaments. The lamellibranch stomach possesses a long style sac containing a crystalline style composed of solidified mucus and enzymes. As the style is rotated by the ciliated lining of the style sac, it reels in a food-laden mucous cord and the free end of the style, which projects into the stomach, is worn away by a chitinous *gastric shield*, liberating digestive enzymes; the style continues to grow from the base of the style sac (Fig. 13.8). The stomach contains a ciliated sorting area, where larger particles are directed toward the long, coiled intestine, while finer particles are diverted into the numerous openings of the digestive gland, in which their breakdown is completed intracellularly.

Fig. 13.8 Bivalve stomach. Arrows indicate food pathway, with mucus-food cord drawn in by rotations of crystalline style, smaller particles being diverted into ducts of digestive gland and larger particles being passed into intestine. Gastric shield erodes free end of style, liberating digestive enzyme.

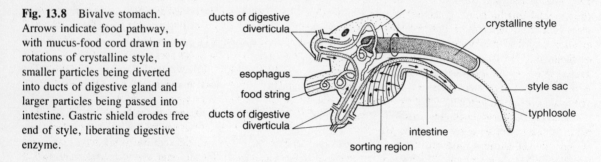

The dorsally located heart (see Fig. 13.7b) receives blood from the gills and pumps it out into an anterior, and sometimes also a posterior, aorta. The ventricle is unusual in being folded around the posterior part of the intestine (rectum), which passes right through the pericardial cavity. Most bivalves have no respiratory pigment in their blood. The single pair of metanephridia open into the pericardial coelom; each is complexly glandular at its proximal end and distally is expanded into a bladder that opens by way of a nephridiopore at the anterior end of the mantle cavity.

The nervous system includes a pair of *cerebropleural ganglia* interconnected around the esophagus and one pair each of pedal and visceral ganglia. In addition to the sensory equipment mentioned above, many bivalves have a gravity-sensing statocyst embedded in the foot.

Most bivalves are dioecious, with the gametes being shed from the large paired gonads into the mantle cavity above the gills (i.e., the suprabranchial cavity), where they are expelled with the exhalant current; however, some species brood their eggs in the suprabranchial cavity or within the gills. Hermaphroditic species may be protandrous or may produce eggs or sperm in reversible alternation. The embryos of marine species develop through trochophore and veliger stages, but freshwater species instead produce *glochidium* larvae (Fig. 13.9), which are parasitic on the gills or other body parts of fishes until old enough to resist being swept downstream. The freshwater mussel broods her eggs to larval status in her mantle cavity. Some species retain the glochidia until a fish comes close enough to inhale the larvae, which are then

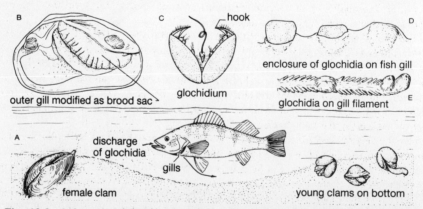

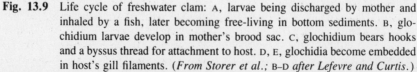

Fig. 13.9 Life cycle of freshwater clam: A, larvae being discharged by mother and inhaled by a fish, later becoming free-living in bottom sediments. B, glochidium larvae develop in mother's brood sac. C, glochidium bears hooks and a byssus thread for attachment to host. D, E, glochidia become embedded in host's gill filaments. (*From Storer et al.;* B–D *after Lefevre and Curtis.*)

forcibly expelled near the fish's mouth. This event is facilitated in certain species by elaboration of the mantle edge into a fishlike shape that may function as a lure. The glochidium anchors itself securely to the fish by means of a harpoonlike *byssus thread* (a long, tough filament) and a pair of hooked valves that close like tongs.

13.15. **What are the characteristics of cephalopods?**

Class Cephalopoda ("head foot") comprises about 650 living and 7500 vanished species, including nautiloids, ammonites, octopods, squids, and cuttlefish (Fig. 13.10). These represent the most highly mobile and advanced mollusks, although they are restricted to the sea and since the Mesozoic have declined greatly in total number of surviving species. They include the most massive of all invertebrates, the giant squid (*Architeuthis*), one specimen of which measured 16 m in total length, including the tentacles. Sperm whales, which look upon giant squid as a delicacy, sometimes bear scars some 25 cm in diameter, inflicted by chitinous claws on the tentacle suckers of their formidable prey. The cephalopod mouth is ringed with tentacles that number 8 in octopods, 10 in squid and cuttlefish, and more than 90 in contemporary nautiloids (Fig. 13.10*b*). The mouth is armed with both a radula and a pair of chitinous jaws, rather like a parrot's beak. The saliva is more or less venomous, though few species are harmful to humans. The eyes are large, and through parallel evolution and convergence, those of octopods, squid, and cuttlefish have come to closely resemble the structure of vertebrate eyes, even to the extent of being able to adjust the pupillary orifice and change focal length.

Although octopuses can creep upon their highly muscular tentacles, all cephalopods swim by jet propulsion. The muscular mantle forces water out of the mantle cavity through a siphon or funnel that can be bent in various directions to allow the animal to jet forward, backward, or vertically or to turn to the side, although backward movement, with tentacles trailing, is most rapid. The body of a cuttlefish or squid is elongated along its original dorsoventral axis, so that the former ventral region, with foot transformed into tentacles and siphon, now becomes the functional anterior end, while the former dorsal aspect becomes the functional posterior end, which may confuse us but does not in the least disorient the speedy cephalopods!

All cephalopods are predaceous. Benthonic species such as octopuses and cuttlefish prey mainly on crabs and shrimp, while pelagic squid are capable of catching even rapidly swimming fish, such as young mackerel. The prey, seized with the tentacles, is bitten and torn by the powerful beak and swallowed with the aid of the radula, which works in a tonguelike manner. However, octopuses are known also to use the radula to drill through the shells of gastropods as large as abalone, injecting through the drill hole a venomous saliva that paralyzes the shellfish so that it releases its hold on the substratum and can be flipped over and the flesh ripped out with the beak. The head also encloses a brain, which is larger and more highly developed than that of any other invertebrate. All the characteristic molluscan ganglia have been cephalized and fused together to form a brain with millions of nerve cells; one brain lobe is dedicated entirely to bringing about rapid color changes by contractions of the radial muscles that expand elastic pigment sacs (*chromatophores*), which are abundant in the integument. Additional ganglia occur in the gills and control the mantle musculature. A cartilaginous "skull" protects the brain and anchors the jaws and siphon. Octopuses have no other skeleton, but squids have an internal, horny *pen;* cuttlefish, the buoyant *cuttlebone;* and nautiloids and ammonites, an *external shell*, usually planicoiled, divided by septa into gas-filled chambers providing buoyancy. The only surviving genus of cephalopods with external shells is *Nautilus* [see Question 13.19(*a*)]. The small deep-sea *Spirula*, while related to cuttlefish instead of *Nautilus*, contains a planicoiled, chambered *internal* shell that serves for buoyancy regulation.

The internal structure of a squid, typical of cephalopods, is depicted in Fig. 13.10*c*. The digestive tract is *Y-shaped*, with a large cecum occupying the apex of the Y and both mouth and anus being located anteriorly. Digestion is extracellular within stomach and cecum and relies on enzymes secreted into the stomach by a "pancreas" and "liver." Absorption takes place through the lining of the cecum, which is elaborated into spiral, ciliated folds that sort out indigestible particles and conduct them back into the intestine, which terminates at an anus that opens into the mantle cavity close to the siphon. In most cephalopods an *ink gland* opens into the rectum; controlled amounts of ink are expelled defensively, either in blinding clouds or in small puffs that seem to distract predators.

Cephalopods are the only mollusks that have a *closed* circulatory system and *three* hearts. The median *systemic heart* pumps blood throughout the body by way of arteries into capillary beds, from which systemic veins deliver the deoxygenated blood to a pair of lateral *branchial hearts,* one located at the base of each gill. The branchial hearts pump the blood through the gill capillaries, and efferent branchial vessels then return the blood, now oxygenated, to the systemic heart. The respiratory pigment, hemocyanin, occurs in the blood plasma. The branchial hearts are also responsible for waste disposal by elevating blood pressure so that a plasma filtrate is forced out of the bloodstream either into the large renal sacs of the single pair of metanephridia or into the pericardial cavity itself, which drains to the metanephridia by a pair of *renopericardial ducts*. The nephridiopores open into the mantle cavity close to the gills. Cephalopod gills are unciliated and cannot set up a water current; instead, breathing is accomplished by the muscles of the mantle. During inhalation, water is drawn in all around the mantle edge; when the animal is not swimming, the water is exhaled by the same route,

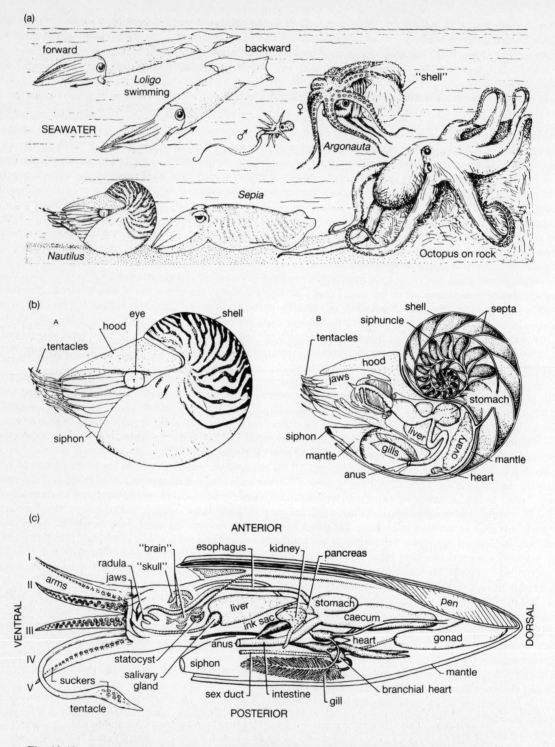

Fig. 13.10 Cephalopods. *(a)* Representative cephalopods. *(b)* Chambered nautilus, the only living four-gilled cephalopod: A, external features; B, sagittal section (two left gills removed); note that the globose body occupies only the newest chamber but is anchored by way of a ligamentous siphuncle to the wall of the original, smallest shell chamber. *(c)* Anatomy of a squid with mantle and arms removed on left side. *(From Storer et al.)*

but during swimming the mantle edge is snugged against a collar so that all exhaled water must pass out through the siphon, furnishing propulsion.

Most cephalopods are dioecious, and mating is preceded by courtship. No penis is present. Instead, sperm are packaged into hardened bundles (*spermatophores*) and stored in a large reservoir. During mating, the male reaches the tip of one modified (*hectocotylized*) tentacle into the reservoir, wraps it around several spermatophores, and then inserts these under the edge of the female's mantle or directly into her genital duct. The eggs contain much more yolk than those of any other mollusk, sustaining the embryo until it develops into a miniature adult a few millimeters long. At least some cephalopods die after mating for the first time.

Example 2: Female *Octopus vulgaris* lay only one batch of eggs, guard them until hatching, and then die. Certain squid species aggregate by the millions; when they have mated and the females have anchored the egg capsules securely to a substratum, all breeding adults of both sexes die. As a result of this pattern of programmed demise, few cephalopods may live longer than 1 to 4 years, which seems a pity since captive octopuses, at least, show considerable learning ability in two-choice operant conditioning tests involving discrimination of shapes and colors and have even been reported to solve detour problems requiring the animal to pull the cork from a long-necked bottle to reach the crab within. The very size that giant squid attain would seem to indicate considerably greater longevity for these huge animals.

13.16. What are the major subclasses of phylum Mollusca?

The classes Gastropoda, Bivalvia, and Cephalopoda each have diversified sufficiently that a brief résumé of their major subclasses is needed to complete our review of mollusks. Remember that each subclass possesses most of the characteristics of the class to which it belongs, so that only specializations and modifications will be considered here (Questions 13.17–13.19).

13.17. What features distinguish the gastropod subclasses?

Gastropods are divided into three subclasses based on the presence and location of their gills (Fig. 13.11). A great variety of forms adapted for different modes of life occur among these groups.

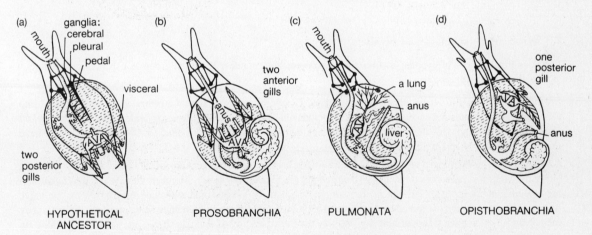

Fig. 13.11 Effects of torsion on adult gastropods. *(a)* Possible ancestral gastropod lacking torsion, showing posterior location of heart, gills, and anus, and untwisted nervous system. *(b)* A primitive prosobranch of order Archaeogastropoda, showing lifelong torsion with retention of both gills and both heart atria; note forward position of heart, gills, and anus, and twisting of nerve cords. *(c)* A pulmonate land snail showing retention of torsion with loss of one heart atrium and both gills, the latter being replaced by a lung formed from the mantle. *(d)* An opisthobranch, showing detorsion with return of the heart, anus, and remaining gill to a posterior position; note that the heart has a single atrium. *A* = atrium, *V* = ventricle. *(From Storer et al.; after Stempell.)*

(*a*) Members of *subclass Prosobranchia* ("forward gill") retain their larval torsion throughout life, so that the gill or gills and the anus remain in a forward position. This is true of the great majority of gastropods—marine, freshwater and terrestrial. Only one order, *Archeogastropoda,* including primitive forms such as limpets, keyhole limpets, abalones, tops, and turbans, has kept two gills, two auricles, and two nephridia. All other prosobranchs have only one gill, one auricle, and one nephridium. Order *Mesogastropoda* is characterized also by a radula with transverse rows of seven teeth; it includes littorines, turrets, wormshells, conchs, cowries, helmets, tuns, moon shells, violet sea snails, and heteropods. The last two deserve further mention, for both live in the open sea. Violet sea snails (*Janthina*) float beneath bubble rafts, preying especially upon the purple by-the-wind sailor *Velella,* a cnidarian (see Fig. 9.5*a*). Heteropods have a very reduced shell and a finlike foot by which they swim (see Fig. 13.6*a*). Order *Neogastropoda* is characterized by a radula with only three teeth per transverse row; it includes drills or murexes, whelks, volutes, olives, augers, and the venomous cones.

(*b*) Members of *subclass Opisthobranchia* ("rear gill") undergo *detorsion,* untwisting so that the anus and gill return to the posterior position; as a result of having undergone torsion as larvae, opisthobranchs have only one gill, one auricle, and one nephridium. Most opisthobranchs possess shells too small for their bodies to retreat into; some have only flimsy internalized shells; and others known as *nudibranchs* ("naked gill") have no shell at all, and their gills are exposed on the body surface, not tucked away under flaps of the mantle (see Fig. 13.6*a*). Species of the sea hare, *Aplysia* (so called for its rabbit-ear cephalic tentacles), may reach a chubby length of 1 m, thus being the largest known gastropods. Nudibranchs that feed on cnidarians are often brilliantly colored, warning predators that they can deliver secondhand stings by virtue of undischarged nematocysts stored in their integuments and in the plumose cerata that cover the backs of many. *Pteropods* ("wing foot"), or sea butterflies, are pelagic forms, with and without shells, that have armlike extensions (*parapodia*) used like wings to "fly" through the water (see Fig. 13.6*a*).

(*c*) *Subclass Pulmonata* includes terrestrial and aquatic gastropods that possess one auricle, one nephridium, and no gills at all. Instead, the mantle cavity is modified into a vascularized sac used for gaseous exchange on land and, secondarily, in water (see Fig. 13.6*b*). The intertidal *onchids* are slugs that have a posterior pulmonary sac and posterior anus. Pulmonates undergo torsion during development even though they pass through no trochophore or veliger larval stages, and most retain torsion throughout life, with anus and pulmonary opening located forward at the shell edge just behind the head. Those having one pair of cephalic tentacles are mostly freshwater snails, such as *Lymnea.* Pulmonates with two pairs of tentacles include terrestrial snails, such as *Helix,* and land slugs, such as *Limax.*

13.18.   What are the subclasses of bivalves?

The class Bivalvia, or Pelecypoda, is subdivided taxonomically on rather technical grounds, especially the nature of the hinge teeth and gill structure.

(*a*) *Protobranchs* ("first gill") have gill filaments arranged in rows along either side of a central axis and *not* folded into a W shape. Their foot is flattened ventrally. They include nut clams (*subclass Palaeotaxodonta*), which have *taxodont articulation* (i.e., a row of short teeth along the hinge margin), and solemyids (*subclass Cryptodonta*), which have thin, elongate valves lacking hinge teeth.

(*b*) *Lamellibranchs* have enlarged gills, with filaments folded back into a W shape. Adjacent filaments are interconnected either by cilia (*filibranchiate*) or by fleshy connections (*eulamellibranchiate*). Members of *subclass Pteriomorphia* are *epibenthonic* (i.e., living *on,* not *in,* the bottom of a body of water), often attaching themselves by one valve (e.g., oysters) or by byssus threads (e.g., mussels). Pteriomorphs have filibranchiate gills and lack siphons; they include arks, which have taxodont hinging, and also mussels, pens, oysters, jingle shells, and scallops. Scallops are particularly talented bivalves, for they can jet either forward or backward by expelling water from between their mantle flaps and are guided by two entire rows of remarkably well developed eyes along the mantle edges.

Members of *subclass Palaeoheterodonta* are mostly freshwater bivalves; these have eulamellibranchiate gills, but their siphons are very short and not tubular.

The great majority of burrowing clams, with tubular siphons, eulamellibranchiate gills, and shells lacking a nacreous lining, belong to *subclass Heterodonta,* which includes cockles, chamids, razor clams, giant clams, rock-boring pholadids, wood-boring shipworms, and several species of freshwater clams. Giant clams (*Tridacna*) at first bore into coral reefs, but as they grow, they eventually fall free to live epibenthonically, usually lying with valves agape, sunning the symbiotic algae that brilliantly color the much folded mantle. Giant clams may exceed a mass of 200 kg, and it is imprudent to thrust arm or leg between their gaping valves.

(c) *Septibranchs (subclass Anomalodesmata)* have one or no hinge teeth, and their gills are reduced to a muscular septum that divides the mantle cavity into dorsal and ventral chambers. These bivalves are predatory and employ pumping contractions of the septum to suck small worms and crustaceans into the mantle cavity, where they are seized by the muscular labial palps and shoved voraciously into the mouth.

13.19.   What are the subclasses of class Cephalopoda?

(a) *Subclass Nautiloidea* is a group of great antiquity, seen in Cambrian strata but represented today only by the *chambered nautilus (Nautilus)* (see Fig. 13.10b), which gives us some idea as to the body organization of the 2500 extinct forms of nautiloids. The chambered nautilus has two pairs of gills, two pairs of nephridia, and a large number (to more than 90) of tentacles that *lack suckers*. The eye is of the *pinhole-camera type,* lacking a lens for focusing, but instead having a small pupil through which seawater enters the eyeball. The siphon is not tubular but *bilobed.* An ink sac is lacking. *Nautilus* has a planicoiled external shell divided into chambers by septa, so that the globular body occupies only the newest and largest chamber. A cordlike ligament extends through the septa to the apex of the original chamber and anchors the nautilus within its shell, which it cannot leave. The gas-filled chambers serve as buoyancy regulators, allowing the nautiloid to hover at a given depth. They are not easy to observe alive in the wild, for by day they stay below the range of light penetration, rising to within scuba depth only at night. Extinct nautiloids probably bore close resemblance to the one surviving genus, but some had tapering, uncoiled shells.

(b) *Subclass Ammonoidea* is known only from fossil forms. The *ammonites* existed from the Silurian to the end of the Cretaceous, when they abruptly became extinct along with many other forms of marine and terrestrial animal life (see Chap. 4). That ammonites became extinct and nautiloids did not (quite) seems attributable to the fact that ammonite young lived in the surface plankton and were much tinier when hatched than nautiloid youngsters, which hatched at a larger size and lived in deeper water, where they would be better buffered from climatic crises. Ammonites had planicoiled, chambered, external shells that in some species grew to the size of a large truck tire. The suture lines marking septal attachments were often elaborately convoluted, rather than simple as in nautiloids.

(c) *Subclass Coleoidea* includes all *dibranchiate* cephalopods, which have only *one pair of gills and nephridia*. They lack an external shell, but have an *ink sac.* The eyeball is of the closed *vesicular* type, with anterior and posterior fluid-filled chambers, cornea, iris, and lens, and seems to focus images clearly albeit nearsightedly. The siphon is tubular, permitting swift jet propulsion and efficient maneuvering. The mantle is external and may bear lateral fins. The tentacles, or arms, number 8 or 10 and bear suckers. Coleoids include the extinct *belemnites,* known from their *internal,* straight, chambered shells. Living forms having eight shorter arms and two longer, extensible tentacles that are shot out to grab prey include: *Spirula,* with its internal, planicoiled, chambered shell; cuttlefish, with bladelike cuttlebone; and squid, with an internal horny pen or plate and a usually elongated body bearing posterior fins. The *vampire squid* are small, deep-sea forms that have two tiny retractile tentacles and eight large arms united in umbrella fashion by an interbrachial web reaching almost to the tips, allowing them to swim like a medusa and capture prey within the web. The octopods have only eight arms and a globular body lacking fins. Although most are benthonic, including the familiar *octopus,* the *argonaut* uses modified paddle-shaped arms to propel itself in the open sea. The female carries along, and can retreat into, her delicate, planicoiled egg case, for which she is often called a "paper nautilus," even though she is not a nautilus at all. The male often remains dwarfed and rides within his mate's egg case.

Chapter 14

Segmented Worms and Walking Worms

Annelids ("ringed") and *onychophorans* ("claw bearers") represent two apparently related phyla of soft-bodied, elongate protostomes that share the feature of *metamerism:* the body trunk consists of a longitudinal series of structural modules known as *metameres,* or segments, each of which, in the onychophorans and most polychaete annelids, bears a pair of fleshy, unsegmented appendages. Modern onychophorans are all nonburrowing, terrestrial species that stump along, caterpillar-fashion, on their ventrally directed legs. The majority of annelids are marine polychaetes, which use their laterally directed legs (*parapodia*) for swimming, creeping, or burrowing (see Fig. 14.2*a*). Fossil traces of annelids occur in Precambrian strata that may be as much as 700 million years old, while the only definite fossil onychophoran found to date occurs in marine sediments of the Cambrian Period, less than 600 million years of age. It is generally thought that onychophorans arose from annelidan ancestry and either were ancestral to, or shared a common ancestor with, the *uniramous* arthropods (so called for their unbranched legs; e.g., centipedes, millipedes, and insects). Annelids in turn appear to be evolutionarily linked with other protostome phyla that share a typical trochophore larval form, namely, the entoprocts, mollusks, sipunculids, and echiurids. A trochophore larva is characteristic of the polychaetes but lacking in the other two annelid classes (oligochaetes and leeches), which undergo direct development.

14.1. How does metamerism develop in segmented worms?

The development of body segmentation by maturing polychaete trochophores (Fig. 14.1) may to some extent recapitulate the evolutionary origin of metamerism in the protostomes. The annelid trochophore characteristically first develops its apical tuft of sensory cilia and one equatorial ciliary girdle, the *prototroch;* it then adds a second girdle of cilia, the *telotroch,* just above the anus and, finally, a third girdle, the *metatroch,* just below the mouth. When the trochophore begins to assume a vermiform (*vermi:* "worm") shape, it lengthens in a growth zone between the mouth and the telotroch. The region from mouth to apical sense organ becomes the head (*prostomium*), and the anal region posterior to the telotroch the *pygidium.* All the annelid body between prostomium and pygidium represents the *trunk.* This region, lying between the mouth and telotroch of the trochophore larva, first simply subdivides into a series of incipient segments, then elongates by the addition of new segments from a *germinal zone* located just in front of the pygidium. This is the *reverse* of

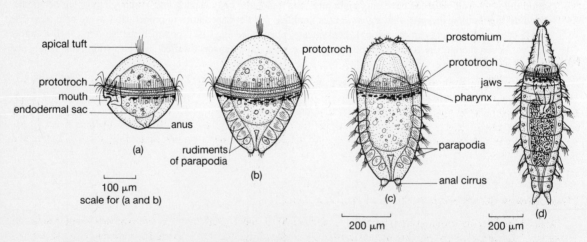

Fig. 14.1 Early stages in the life of the polychaete *Glycera.* *(a)* Trochophore larva (10 days). *(b)* Metatrochophore (4 weeks) showing developing parapodia. *(c)* Metatrochophore (7 weeks) with developing prostomium. *(d)* After metamorphosis, 8-week postlarva showing development of earliest body segments and protrusible pharynx with jaws.

178

strobilation as seen in tapeworms, in which the newest segments (proglottids) are produced in a germinal zone at the base of the scolex, with proglottids being progressively older toward the worm's posterior end. In all annelids, continued elongation of the body even in adult life is accomplished by the production of each new trunk segment just in front of the pygidium, so that the *oldest* segments lie just behind the head (unless the anterior end has been lost and regenerated). Polychaete larvae usually start to become metameric some time before settling to the bottom. The larvae of *Nereis fucata*, for example, have six or seven trunk segments before settlement.

14.2.   What is the evolutionary significance of metamerism?

Metamerism especially benefits burrowing annelids by subdividing the coelom crosswise into a series of compartments. As a result, individual segments or a short series of several segments can be made rigid by localized contractions of segmental muscles against the coelomic fluid, which cannot be displaced to another part of the body because it is trapped within each segment by transverse membranous partitions (*septa*). This much improves the efficiency of the annelidan *hydrostatic skeleton*, allowing terrestrial forms such as earthworms to penetrate even quite resistant soils.

The appearance of metamerism in planktonic larvae considerably before settlement suggests additional benefits of a metameric body plan. Six entire families of polychaetes are exclusively pelagic, swimming by serpentine lateral body undulations, a mode of locomotion that also benefits from a segmental anchorage of trunk muscles and a compartmentalized hydrostatic skeleton. During swimming, backsweeps of the oarlike parapodia are timed to coincide with peaks of lateral body undulations, which couples the effective strokes with localized trunk rigidity.

One further advantage of metamerism might be that during development of a metameric body plan, growth and complexity can be achieved by the serial repetition of a basic organizational "package." In annelids, each such module typically includes a pair of nephridia, a paired ganglion giving off segmental nerves, a set of segmental blood vessels, metameric muscles running from septum to septum, a set of chitinous bristles (*setae*), and usually a pair of parapodial appendages. The addition of segments by a process similar to budding could involve reiteration of a basic set of genetic instructions. In fact, some annelids do reproduce asexually by budding. The bristleworm, *Dodecaceria*, fragments the central region of its body into individual metameres, each of which buds off four young worms before dying.

Metamerism as seen in most annelids is *homonomous*, meaning that successive trunk segments and their appendages are essentially identical. This suggests that the same pattern of gene regulation and expression is followed in the maturation of each new trunk segment, except for the nonmetameric anterior portions of the digestive tube. Homonomous metamerism appears to be prerequisite to the evolutionary *diversification* of body segments and appendages, as occurs in certain polychaetes but is realized to a much greater extent in the arthropods. Such diversification of serially homologous segments, known as *heteronomous metamerism*, could have come about as a result of mutational or regulatory changes in the genetic packages controlling development of the body segments.

Heteronomous metamerism is fundamental to the tremendous success of the arthropods, which are thought to have evolved from annelidan ancestry. Metameric body architecture also arose in the deuterostome line, notably the chordates, but as an independent evolutionary event. The biological significance of metamerism is indicated by the fact that the most successful protostomes, the arthropods, and the most successful and advanced deuterostomes, the vertebrates, both have metameric body plans.

14.3.   What are annelids?

Annelids comprise some 10,000 kinds of segmented worms, which range in size from deep-sea species less than 1 mm long to giant tropical earthworms as much as 4 m in length. They constitute three major groups: *polychaetes* ("many hairs"), *oligochaetes* ("few hairs"), and *leeches*. The majority are aquatic, occurring in the sea as swimmers, crawlers, burrowers, and tube dwellers, or in fresh water, where they usually burrow in the bottom. Land-dwelling annelids are confined mostly to a subterranean existence as burrowers, except for certain leeches, which are fully terrestrial in humid environments.

14.4.   How is the annelidan body functionally organized?

Annelids usually have elongate, cylindrical or subcylindrical bodies made up of a *prostomium* (head), a series of similar trunk segments, and a terminal pygidium bearing the anus (Fig. 14.2*a*). The mouth opens ventrally between the prostomium and the *peristomium*, a postoral region consisting of the modified first trunk segment or several anterior trunk segments fused together. The prostomium primitively contains the brain and may also bear eyes and sensory tentacles and palps. In many annelids a "secondary head" is formed by a fusion of prostomium and peristomium, and often several

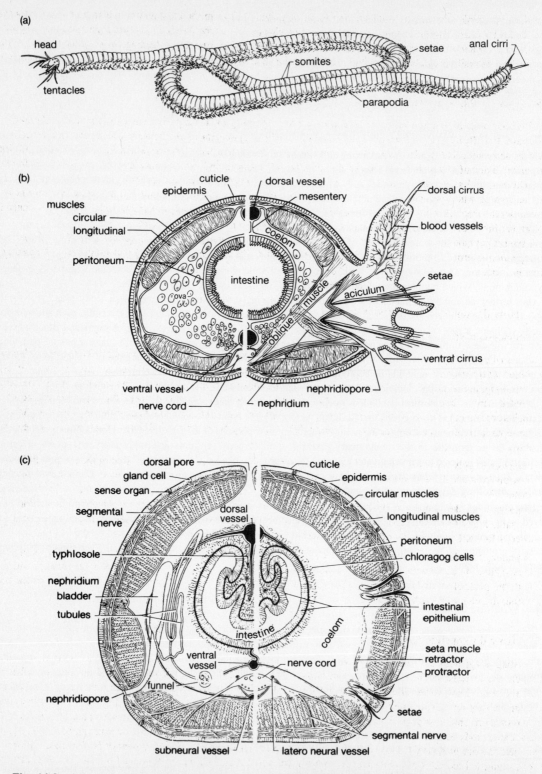

Fig. 14.2   Anatomy of representative annelids. *(a)* External features of the clamworm, *Neanthes virens*. *(b)* Transverse sections of a clamworm, showing eggs free in body cavity and parapodium on right side only. *(c)* Transverse sections of an earthworm, showing a metanephridium at left and setae at right. *(From Storer et al.)*

postperistomial trunk segments as well, and the brain may then shift from the prostomium to a more posterior location in the secondary head. The prostomium of *tubicolous* polychaetes (i.e., tube dwellers) is often elaborated into a spiral or funnel-shaped *crown* of feathery pinnate *radioles* that bear cilia used in suspension feeding. In polychaetes, most trunk segments bear laterally projecting appendages, or parapodia, that are best developed in creeping and pelagic species.

14.5.   How is the body wall of annelids organized?

The body wall of annelids (Fig. 14.2*b* and *c*) is lined internally by a thin peritoneum of squamous epithelium and is covered externally by a single layer of columnar epithelium bearing long microvilli, between which a mucopolysaccharide cuticle is deposited. The epidermis rests on a thin layer of connective tissue, below which lie thin circular and thick longitudinal layers of *obliquely* striated muscle fibers. Special protractor and retractor muscles move the parapodia forward and backward, and setal muscles project and pull in the setae, which are used for anchoring in a tube or burrow. The setae (and jaws, when present) are composed of the nitrogenous polysaccharide *chitin* and are ordinarily harmless to humans, but the tropical fireworms (amphinomids) have hollow, poison-containing, calcareous setae that readily penetrate the skin and break off, leaving the victim in acute pain and subject to secondary infection.

A typical parapodium is fleshy and divided into a dorsal lobe, the *notopodium*, and a ventral lobe, the *neuropodium*, each supported internally by chitinous rods (*acicula*). Each lobe bears a process (*cirrus*) that ranges from tentacular to leaflike in form.

14.6.   How do annelids move about?

Locomotion is accomplished in three main ways.

(*a*) *Burrowing* involves waves of peristaltic contraction that sweep down the length of the body. The segments encompassed within each wave first elongate by contraction of the circular muscles, then shorten and thicken by contraction of the longitudinal fibers. Forward progression takes place as the anterior end elongates and the prostomium, becoming pointed, drives into the substratum; then the anterior end becomes short and thick, the setae are projected for anchorage, and this wave of thickening passes backward while the anterior end lengthens again. The importance of the septa in limiting the movement of coelomic fluid, localizing the hydrostatic skeleton effect, has been mentioned above.

(*b*) *Crawling* in polychaetes is accomplished by the parapodia. Each parapodium is lifted free of the surface and moved forward (protracted), with bristles pulled in; then it is lowered onto the surface, with acicula and setae extended for anchorage; and finally it is pushed strongly backward (retracted). When a given parapodium starts its effective (backward) stroke, the one opposite it moves forward in a recovery stroke. Waves of parapodial activation move forward, rather than backward, because as each parapodium initiates its backward stroke, the one just anterior to it sweeps forward and starts its own effective stroke.

(*c*) *Swimming* in polychaetes involves lateral body undulations brought about by waves of contraction of the longitudinal musculature, in phase with the waves of parapodial movement, so that the effective stroke of a parapodium coincides with the peak of an undulatory wave. The metameric partitioning of the coelom enhances the rigidity of the body within the region of each undulation.

14.7.   How do annelids feed?

Feeding adaptations are varied. *Nonselective deposit feeders* include burrowing species, such as most oligochaetes, that ingest the material through which they are tunneling, digest the organic components, and eliminate the residue by way of the anus. Earthworms also reach out of their burrows to pull down surface vegetation. *Predaceous* polychaetes may hunt actively or wait in ambush in tube or burrow; they capture prey by shooting out an eversible pharynx armed with two or more chitinous jaws. The proboscis of the formidable bloodworm, *Glycera*, extends about 20 percent of the worm's entire body length and bears four hollow fanglike jaws that inject poison from venom glands. Some jawed polychaetes are not predatory but use their jaws for tearing off plant material. *Selective deposit feeders* lack a proboscis but have elongated tentacles or palps that are stretched out over the surface. Particles trapped in mucus on the surface of these are moved to the mouth along ciliated grooves. *Filter feeders* include most tubicolous polychaetes, which use their crowns of ciliated, pinnate radioles to set up a current that draws particles in suspension through the pinnules and carries them into a groove running down the length of each radiole to the mouth; here, larger particles are rejected and smaller ones swallowed.

Example 1: The parchment worm, *Chaetopterus,* lacks a radiolar crown, but traps plankton with a tubular slime net secreted from specialized anterior parapodia. Other wing-shaped parapodia maintain a strong water current through the U-shaped, parchmentlike tube, drawing particles into the slime net. As food is trapped, the posterior end of the constantly growing net is rolled up and cut free as a series of food balls, which are carried to the mouth along a ciliated groove.

Leeches range from predatory to ectoparasitic. Jawed leeches have three jaws arranged in a triangle, each bearing small teeth used in slicing through the tissues of the prey or host. Bloodsucking leeches that attack large animals slit the host's skin and pump blood from the host by means of a muscular pharynx, while exuding an anticoagulant, *hirudin,* into the wound. When small animals are attacked, all their soft tissues may be sucked in or they may be ingested whole. Jawless leeches instead have a tubular proboscis, covered and lined with cuticle, that can be extended from the mouth and forced into the tissues of the victim, often sucking up all soft parts.

14.8. How is the digestive tract of annelids organized?

The digestive tract is long, straight, muscularized, and differentiated into functional regions (Fig. 14.3*a*). The mouth opens into a muscular pharynx, which in a number of polychaetes and leeches forms an extensible proboscis used in obtaining food. Food is swallowed into a tubular esophagus that leads to a slightly expanded stomach. In earthworms, *calciferous glands* along the esophagus secrete excess calcium ions from the bloodstream into the gut for excretion. The stomach of earthworms is divided into an anterior, thin-walled storage region (*crop*) and a thick-walled *gizzard* in which the food is ground up. The crop of leeches is usually enlarged by a series of lateral pouches (*ceca*) that allow it to distend enormously since these animals feed only intermittently.

Digestion in annelids is entirely extracellular, by means of enzymes secreted by the epithelial lining of the stomach (or intestine, when a stomach is absent). The intestine is the site of absorption, and it often bears internal folds that increase the surface area. The peritoneum of the intestinal region produces yellow-green *chloragogen tissue,* which synthesizes glycogen and fats. When replete with these storage products, the choragogen cells are released into the coelom as free-floating *eleocytes.* Eleocytes concentrate particularly in regions of regeneration or healing, providing nourishment to the tissues by breaking down and releasing their contents into the coelomic fluid.

14.9. What is the organization of the annelid circulatory system?

Oligochaetes and polychaetes typically have a well-developed closed blood-vascular system, in which the blood is propelled by peristaltic contractions of the major dorsal and ventral blood vessels, particularly the former, which is considered the true heart. Blood passes forward in the dorsal vessel and posteriorly in the ventral vessel; these are interconnected by lateral vessels in every segment and also give off segmental branches serving the wall of the gut and the body wall (Fig. 14.3*a* and *b*). Many polychaetes also possess blind-ended contractile vessels, in which the blood simply flows to and fro. Oligochaetes often have one to five pairs of anterior commissures (called *aortic arches*) that do not pulsate rhythmically but do contract occasionally to assist the circulatory flow. In leeches the septa partitioning the coelom have largely disappeared, and the coelom is heavily invaded by connective tissue, reducing it to a network of channels and sinuses. Some leeches retain the basic circulatory pattern seen in other annelids, but the blood also flows through the coelomic spaces. However, in most leeches the original system of blood vessels has disappeared, and blood flows only through the coelomic channels, propelled by contractions of the channel walls.

14.10. How does gas exchange occur in annelids?

Gas exchange takes place anywhere through the mucus-moistened body surface, including parapodia and any other extensions of the body surface, such as radioles and tentacles, that can serve the function of gills. Oxygen transport is usually effected by respiratory pigments, but these are lacking in most leeches as well as certain polychaetes and small oligochaetes. Most oligochaetes (and some leeches) carry enormous molecules of extracellular hemoglobin in their blood plasma. Polychaetes possess a variety of respiratory pigments, found in the plasma or within blood corpuscles or coelomocytes: plasma and intracellular hemoglobins, plasma chlorocruorin, and corpuscular hemerythrin. *Hemoglobins* are proteins containing complex ring-shaped components called *porphyrins,* each with a central atom of iron to which the oxygen binds. *Chlorocruorin* is very similar to hemoglobin but appears green rather than red. *Hemerythrin* is an iron-containing nonporphyrin protein, chemically allied to the copper-containing hemocyanin found in a number of mollusks and crustaceans. More than one kind of respiratory pigment may be present simultaneously.

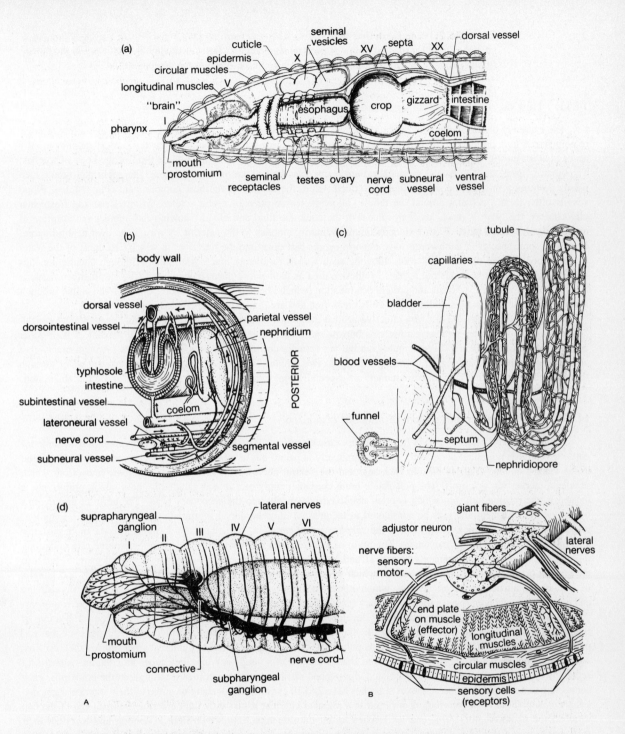

Fig. 14.3 Internal structure of an earthworm. *(a)* Anterior portion seen from left with nephridia omitted and three of the five pairs of aortic arches removed. The latter, erroneously called "hearts," contract sporadically, while the true heart, the dorsal vessel, pulsates regularly, propelling the blood forward. *(b)* Details of the circulatory system in one body segment. *(c)* An entire metanephridium with its associated capillaries. *(d)* Nervous system: A, lateral view of anterior portion; B, detail of ventral nerve cord and segmental nerves, arrows showing pathways of incoming sensory impulses from receptors and outgoing motor impulses to muscles. [*From Storer et al.; (d)*, A *after Hess.*]

Example 2:   *Amphitrite* has two kinds of hemoglobin: the one found within coelomocytes has a greater affinity for oxygen at low concentrations than the plasma hemoglobin circulating in the bloodstream. This facilitates the transfer of oxygen from the blood to the coelomic fluid.

14.11.   How do annelids excrete waste products?

The excretory organs of annelids are most typically paired metanephridia, but trochophore larvae and a few families of polychaetes have protonephridia with flagellated solenocytes. The metanephridia that characterize most annelids usually occur in all or most body segments of oligochaetes and polychaetes, but in only the middle third of the body of leeches. The ciliated funnel of each typical metanephridium opens into the coelom of one metamere; the nephridial tubule then passes posteriorly through the septum and lies coiled within the coelom of the next compartment (Fig. 14.3*c*). Blood vessels often form a network around the tubule, allowing reabsorption of useful solutes. In terrestrial and freshwater oligochaetes, the urine is much lower in salts than the coelomic fluid and blood, showing that tubular reabsorption of needed ions has taken place. Each metanephridium ordinarily empties to the exterior by way of its own nephridiopore. However, some species of earthworms have nephridiopores that open into the intestine, in which case most of the water is reabsorbed through the intestinal wall, allowing such worms to tolerate much drier soils. Although aquatic annelids excrete mainly ammonia, terrestrial species synthesize urea as their major nitrogenous waste. The chloragogen tissue is an important site of urea synthesis and protein deamination (which liberates ammonia made from the freed amino groups), and when replete with wastes, chloragogen cells break free and are drawn into the ciliated nephrostomes.

Because their cuticles are not very effective against evaporative water loss, terrestrial annelids must tolerate considerable dehydration. In dry weather earthworms burrow more deeply, but may still lose up to 70 percent of their water content, which causes them to become quiescent but not actually dormant. However, some species secrete a mucus cyst and undergo *estivation* (dry-season dormancy) for up to two months. When the rains come, they take up water through the skin and resume activity. Earthworm tunnels are especially beneficial in facilitating soil aeration and water absorption.

14.12.   How is the nervous system of annelids organized?

The brain receives sensory nerves from cephalic sense organs and is connected by commissures encircling the anterior gut to a ventral longitudinal nerve cord, which primitively is paired but in most annelids has become more or less fused in the ventral midline (Fig. 14.3*d*). In each segment the ventral nerve cord swells into a ganglionic mass from which segmental nerves are given off. The polychaete brain consists mainly of a dorsal ganglionic mass which primitively is located in the prostomium, but which is often displaced posteriorly into the secondary head. The brain of earthworms lies in the third segment and may be considered to include not only the dorsal *suprapharyngeal*, or *cerebral*, ganglion, responsible for sensory orientation, but also the *subpharyngeal ganglion*, since the latter dominates all the more posterior ganglia and is responsible for movement. The brain of leeches takes the form of a large ganglionic nerve ring encircling the pharynx or proboscis in segments 10 and 11 and includes not only the cerebral and subpharyngeal ganglia, but also the ganglia of other anterior body segments, which have migrated posteriorly to join the brain. The brains and ganglia of annelids contain neurosecretory neurons, which produce neurohormones that are disseminated by way of the bloodstream and regulate reproduction, regeneration, secondary sexual characteristics, and, in certain leeches, chromatophore responses.

Sense organs are most diverse and highly developed among polychaetes. Most polychaete eyes consist of a cup of retinal cells, with a spherical lens occupying the interior of the cup, but the pelagic *Vanadis* has vesicular eyes that can probably discriminate objects. Certain tubicolous polychaetes have simple clusters of photoreceptive cells on their crowns; when these photoreceptors are suddenly cast into shadow, the worms snap back into their tubes in a *shadow-withdrawal reflex*. Most oligochaetes lack eyes, but isolated lens-shaped photoreceptors are scattered throughout the epidermis, each connecting to a subepidermal nerve plexus. Leeches have 2 to 10 eyes located dorsally on some of their anterior segments. Each eye is a cluster of photoreceptors enclosed in a pigment cup that allows only light entering the open end of the cup to affect the photoreceptors. This enables leeches to tell the direction from which light is coming and to orient their movements accordingly. Leeches tend to avoid light, but some become positively phototactic when hungry, moving into exposed locations where they are more apt to contact a passing host.

Various types of ciliated cutaneous sense organs that detect tactile and chemical stimuli occur in all three annelidan classes. An entire ring of sensory papillae encircles each trunk segment of leeches; leeches that attack warm-blooded hosts have been found to be attracted to warm objects, indicating the presence of thermoreceptive cells, perhaps in these papillae. Leeches are also attracted to test objects smeared with blood or body secretions (including mucus, sweat, and skin oils) characteristic of their particular host species. Ciliated sensory pits (*nuchal organs*) occur in the heads of many polychaetes, being best developed in predatory species. These pits are apparently chemoreceptive and vital to hunting, for when they are destroyed, the worm ceases to feed.

Statocysts occur in certain families of burrowing or tubicolous polychaetes and serve to direct the worms' penetration of the substratum.

> **Example 3:**   The marine lugworm *Arenicola* burrows vertically downward in mud. In an aquarium, it will alter its angle of tunneling to compensate for any angle at which the tank may be tilted—unless its statocysts are experimentally destroyed.

14.13. How do annelids reproduce?

The reproductive systems of annelids show considerable variability. Most polychaetes are dioecious. They lack distinct gonads, instead developing masses of gametes within the peritoneum of all or some body segments. The gametes are shed into the coelom before meiosis occurs, and maturation takes place mainly within the coelomic fluid, so that the coelom of gamete-bearing segments becomes packed with sperm or ripe eggs. Gametes may reach the exterior in a number of ways, depending on species: by way of special gonoducts, the metanephridia, or the anus, or by complete rupture of the body wall. Quite commonly, in a process called *epitoky*, polychaetes produce special reproductive individuals (*epitokes*) that are adapted for swimming. Epitokes can arise by budding from the sides of the nonreproductive worm (*atoke*), by complete transformation of the atoke, or by the modification of only the posterior segments of the atoke, which break free and, although headless, swim to the surface and writhe about vigorously, shedding gametes.

> **Example 4:**   On certain specific nights at the start of the last lunar quarter of October and November, huge numbers of Samoan palolo worms extrude their epitokous posterior regions from their burrows and snap them off. This event takes place with such precise synchrony that, for a short while, the ocean surface is roiled by masses of writhing half-worms. Waiting natives, as well as seabirds and fish, scoop them up for a relished feast.

Oligochaetes are monoecious and have permanent gonads and accessory sexual structures, which are confined to only a few segments in the anterior third of the body. The paired ovaries and testes arise separately, in the septa of different segments, and give off premeiotic gametes that mature within *seminal vesicles* (if sperm) or *ovisacs* (if eggs). Sperm ducts and oviducts open into the coelom by way of ciliated funnels, and to the exterior by way of separate male and female gonopores. Other pores lead into segmentally arranged pairs of *seminal receptacles*, which receive sperm from the partner during copulation and later expel the sperm over the eggs as these are laid. Some aquatic oligochaetes have an eversible penis at the opening of a single, common, male gonopore at which all sperm ducts converge. Earthworms lack copulatory organs but bind their bodies together with a mucous tube, allowing sperm to swim from the male pores of each worm to the openings of the partner's seminal receptacles. Oligochaetes and leeches bear a girdle of glandular tissue, the *clitellum,* that first secretes the mucous copulatory tube, and later a series of chitinous *cocoons* filled with nutritive albumin. As each cocoon is freed from the clitellum, the worm backs out of it, discharging eggs into the cocoon as the oviducal openings are reached and then sperm, as the cocoon passes over the pores of the seminal receptacles. The ends of the cocoon seal, and the embryos, feeding first on yolk and later on albumin, develop into young worms that hatch, depending on species and climatic conditions, from 1 to 13 weeks after the eggs were laid.

Leeches are also hermaphroditic and have permanent gonads and ducts, but these differ considerably from those of oligochaetes. Usually there is only a single pair of ovaries, each enclosed in an ovisac from which a short oviduct extends anteriorly; the oviducts join to form a median vagina that opens by way of a female gonopore. The testes are spherical, containing an internal cavity in which sperm mature. They number from four to many pairs, arranged segmentally and joined in tandem by a sperm duct (vas deferens) running longitudinally along each side of the body. Each vas deferens leads anteriorly into a much-coiled epididymis or an expanded vesicle, from which an ejaculatory duct empties into a common median *atrium,* a muscular chamber that opens to the exterior by way of a single male gonopore. The atrium acts either as a penis or as a device for formation and expulsion of sperm packets (*spermatophores*). Leeches possessed of a penis copulate with their heads oriented in opposite directions, so that the penis of each can be inserted in the vagina of the partner. Leeches that lack a penis show less finesse, each worm simply stabbing its partner with a forcibly ejected spermatophore, which penetrates the integument and liberates sperm into the coelomic channels, through which they swim to reach the ovaries. Whichever method is employed, leech eggs are internally fertilized and are eventually shed into a series of cocoons produced and filled with nourishing albumin by the clitellar glands. Some leeches brood their eggs. They either carry the cocoons attached to their own bodies or lie protectively on top of cocoons that are attached to the substratum, until the young worms emerge.

14.14. What are the major taxonomic subdivisions of phylum annelida?

Although oligochaetes and leeches are sometimes grouped together into one class, Clitellata, by virtue of possessing a clitellum, most zoologists consider them different enough to be placed in separate classes, Oligochaeta and Hirudinea (Table 14.1). Class Polychaeta, which is considered ancestral to the other annelidan classes, is also the most diverse.

Table 14.1 Some Major Taxonomic Subdivisions of Phylum Annelida

Phylum Annelida
  Class Polychaeta
    Subclass Errantia
    Subclass Sedentaria
    Subclass Archiannelida
  Class Oligochaeta
    Order Lumbriculida ⎫
    Order Tubificida   ⎬ differentiated on the basis of their reproductive systems
    Order Haplotaxida ⎭
  Class Hirudinea
    Order Acanthobdellida ⎫
    Order Rhynchobdellida ⎬ differentiated on the basis of their feeding apparatus
    Order Gnathobdellida   ⎬
    Order Pharyngobdellida ⎭

14.15. What are the general characteristics of polychaetes?

Class Polychaeta ("many hairs") includes over 8000 species of mostly marine annelids that are externally and internally segmented; they usually have a trochophore larva, parapodia, bundles of numerous, long setae, and well-developed heads bearing eyes and sensory tentacles, palps, or radioles (Fig. 14.4a). They are mainly dioecious, with transitory gonads and epitokous reproductive forms.

14.16. What characterizes members of subclass Errantia?

Subclass Errantia includes the errant (wandering) polychaetes that seldom inhabit tubes or permanent burrows but, instead, usually swim, crawl, or tunnel actively. Their heads and sensory structures are particularly well developed, and a protrusible pharynx armed with jaws or teeth is used for getting food (Fig. 14.4b). Most are predaceous. Their trunk segments are similar, and each is provided with a pair of well-developed locomotory parapodia. There are a number of families of errant polychaetes, including:

(a) *Sea mice* (family Aphroditidae), with heavy oval bodies covered dorsally by long setae forming a mouselike fur

(b) *Scale worms* (family Polynoidae and Sigalionidae) with a dorsal surface covered by overlapping plates (*elytra*)

(c) *Nereids* (family Nereidae), large worms with (1) an eversible proboscis bearing one pair of chitinous jaws and (2) a well-developed head bearing four eyes and eight cirri

(d) *Glycerids* (family Glyceridae), burrowing species with a very long proboscis armed with four jaws

14.17. What characterizes members of subclass Sedentaria?

Subclass Sedentaria includes most polychaetes that live in tubes or permanent burrows. Sedentary polychaetes lack jaws, teeth, and a protrusible proboscis but often have long tentacles or prostomial radiolar crowns used in collecting particulate food. Their parapodia are often reduced or modified regionally for various applications. The many families of Sedentaria include:

(a)

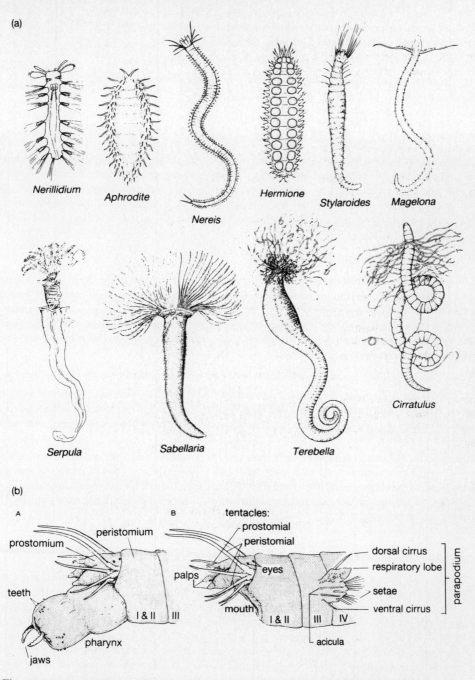

Fig. 14.4 Polychaetes. *(a)* Polychaete diversity; *upper row, left, Nerillidium,* an archiannelid; others represent subclass Errantia; *lower row,* members of subclass Sedentaria. *(b)* External features of anterior end of the clamworm *Neanthes,* with pharynx extended (A) and retracted (B). [*From Storer et al.; (a) after Grassé.*]

(a) *Spionids,* which live in tubes from which they project two long, armlike prostomial palps

(b) *Parchment worms* (e.g., *Chaetopterus*), which have fan-shaped parapodia for ventilating their U-shaped organic tubes

(c) *Cirratulids,* with many long, threadlike segmental gills

(d) *Lugworms* (family Arenicolidae), burrowers lacking head appendages

(e) *Sabellariids*, which cement sand grains into honeycomblike aggregations of tubes and have certain anterior setae modified into an operculum (trapdoor) for sealing the tube

(f) *Terebellids*, which project long, nonretractile tentacles from their tubes or burrows

(g) *Sabellids*, which have crowns of pinnate radioles and live in noncalcareous tubes

(h) *Serpulids*, which have radiolar crowns and live in calcareous tubes

14.18. What characterizes members of subclass Archiannelida?

Archiannelids, traditionally considered a separate and especially primitive class of annelids, probably instead represent a heterogeneous assemblage of small, aberrant polychaetes that have undergone convergent, secondary simplifications in the process of adapting to an interstitial or parasitic existence. They are well segmented internally but lack external segmentation; have a nervous system embedded in the epidermis, which is often ciliated; often have solenocyte protonephridia; are commonly dioecious; and usually lack parapodia and setae.

14.19. What are the characteristics of oligochaetes?

Class Oligochaeta includes 3100 species of terrestrial and freshwater annelids that undergo direct development, are externally and internally segmented, have reduced heads lacking cephalic appendages and eyes, lack parapodia, have short and relatively few setae, are hermaphroditic, and fertilize their eggs internally, laying them into a chitinous cocoon secreted by a clitellum (Fig. 14.5). They lack gills but have well-developed vascular networks in the skin. They are mainly herbivorous, especially on decomposing vegetation. Taxonomic division of the class is mostly based on technical differences in the reproductive system:

(a) *Order Lumbriculida* includes freshwater oligochaetes that have the male gonopores in the same segment as the testes.

(b) *Order Tubificida* includes marine, freshwater, and a few terrestrial forms that have one pair each of testes and ovaries in adjacent segments, with male gonopores in the segment immediately anterior or posterior to the segment containing the testes. Examples include the tube-dwelling *Tubifex*, often used to feed aquarium fishes, and the ubiquitous beachworm *Enchytraeus*, found in vast numbers in sand intertidally.

(c) *Order Haplotaxida* includes six families of earthworms and two families of freshwater and semiterrestrial oligochaetes. These have either two segmental pairs of testes followed by two segmental pairs of ovaries or one pair of each separated by an intervening segment. The sperm ducts extend through one or more segments from funnel to gonopore. This order includes the major family of temperate earthworms, Lumbricidae, which has male gonopores in segment 15, anterior to the clitellum.

14.20. What are the characteristics of class Hirudinea?

Class Hirudinea includes some 500 species of leeches, of which a few are marine or terrestrial, while the majority inhabit fresh water. Leeches range in size from 1–30 cm and are predatory or suck blood ectoparasitically. The body, consisting of 34 externally visible segments ringed with annular grooves, is dorsoventrally flattened, usually leaf-shaped, and has a large posterior sucker and often a smaller anterior sucker ventral to or surrounding the mouth (Fig. 14.6). Setae are lacking (except in one species), and parapodia and cephalic appendages are absent. Locomotion is accomplished inchworm-style—with the hind end anchored by the posterior sucker, the body elongates; the anterior end attaches to the substratum, and the ventral musculature contracts to bring the posterior end forward; the posterior sucker then reattaches just behind the head, after which the body stretches forward once more. Chromatophores occur in the skin, which is often brilliantly colored and patterned in stripes or spots of olive green, red, and black, and color changes may occur under neurohormonal control. A clitellum is present, and the directly developing eggs are laid in cocoons. All leeches are hermaphroditic, and only one male and one female gonopore are present. Internally, septa are lacking (except in the one species *Acanthobdella*, which also has a few segments bearing setae), and the coelom—invaded by connective tissue, muscle fibers, and chloragogen tissue—is reduced to a mere system of channels. Other deviations of leech internal organization from the typical annelidan pattern have been summarized in Questions 14.7–14.13 above.

Division of the class into orders is mainly on the basis of feeding apparatus:

(a) *Order Acanthobdellida* includes only one species, which parasitizes salmonid fishes and is considered primitive because two pairs of setae are borne on segments 2, 3, and 4 and the first five metameres are internally septate.

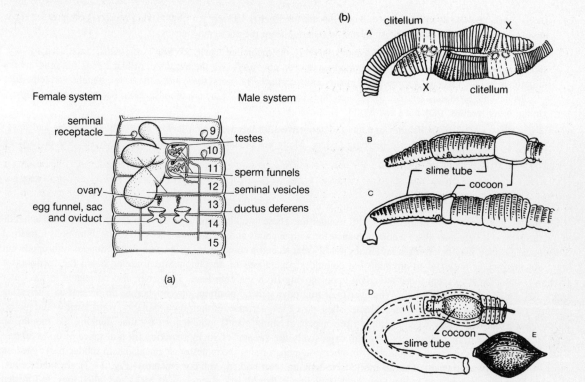

(a)

Fig. 14.5 Earthworm reproduction. *(a)* Reproductive system, dorsal view, with seminal vesicles removed on the right. *(b)* Mating and oviposition: A, earthworms in copulation (X = tenth segment); B, worm after mating, secreting slime tube and cocoon; C, worm crawling backward out of slime tube and cocoon, laying eggs into cocoon; D, freed slime tube containing cocoon; E, cocoon. [*From Storer et al.; (b),* A *after Grove,* B–D *after Foot and Strobell.*]

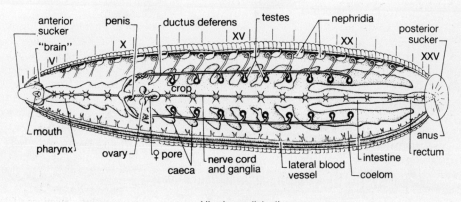

Hirudo medicinalis

Fig. 14.6 Internal structure of *Hirudo medicinalis,* the medicinal leech, ventral aspect. Note absence of internal partitions, and the highly distensible crop with numerous lateral pouches (caeca). (*From Storer et al.*)

(*b*) *Order Rhynchobdellida* includes freshwater and marine leeches lacking jaws but having a protrusible proboscis for blood sucking and a circulatory system that is separate from the coelomic sinuses.

(*c*) *Order Gnathobdellida* includes most leeches, notably the medicinal leech, *Hirudo medicinalis,* once used by physicians to drain "bad blood" from their patients and even now useful for the treatment of black eyes and other marks of combat and for improving circulation in surgically reattached fingers, toes, and ears. These aquatic and terrestrial species have no proboscis but usually possess three chitinous jaws. Many are bloodsuckers, but others are predatory. Unlike other leeches, they have red blood.

(*d*) *Order Pharyngobdellida* includes aquatic and semiterrestrial leeches which lack a protrusible proboscis and jaws, but which may have one or two piercing stylets. Many are predaceous rather than parasitic, sucking in food by pumping actions of the pharynx.

14.21.   What are onychophorans?

This is a good question. Zoologists are still wondering. These "walking worms" may represent a link between annelids and arthropods. Or they may share a common ancestor with annelids and, in turn, be ancestral to the uniramous terrestrial arthropods. Or they may be on offshoot "living fossil" that shared a common ancestor with the uniramous arthropods but were not themselves ancestral to anything but countless generations of onychophorans trudging down the corridors of time with few changes except those attendant upon moving from sea onto land. They may also be related to the water bears and perhaps the pentastomids. Whatever their true phylogenetic position, onychophorans do present an interesting mélange of annelidan and arthropodan characteristics, plus a few surprises of their own. Notably, whosoever would molest an onychophoran had better wear goggles: from a pair of slime glands mounted on papillae flanking the mouth, an onychophoran only a few centimeters long can explosively fire streams of sticky secretion for a distance of up to 50 cm! This secretion hardens promptly, entangling or otherwise embarrassing the molester. It can also help subdue prey, because most of the 70 living species of these inoffensive-looking, soft-bodied, velvety creatures (Fig. 14.7) are predaceous, mainly on worms, insects, and snails. Onychophorans live in damp, dark places, under rocks and fallen logs, in forested regions scattered throughout the West Indies and parts of Mexico and the southern hemisphere, including Australia, New Zealand, South America, and Africa. They mainly come forth to hunt at night.

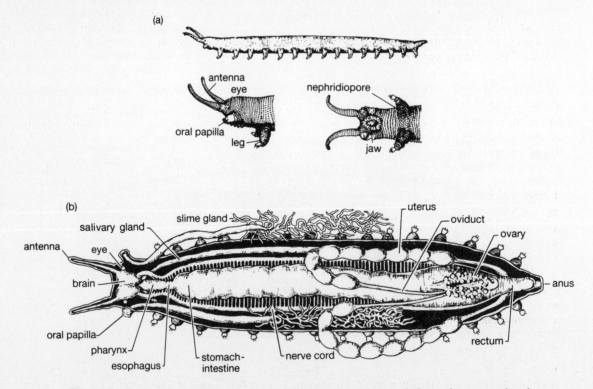

Fig. 14.7   Phylum Onychophora. *(a)* External features of a peripatus. *(b)* Internal anatomy, dorsal view. *(From Storer et al.; after Snodgrass.)*

14.22.   How are onychophorans like annelids and unlike arthropods?

(*a*)   The body is soft, and the body wall is typically annelidan, with a single-layered epidermis, thin dermis of connective tissue, and circular, diagonal, and longitudinal layers of muscle fibers. In contrast, arthropods have a chitinous cuticle sturdy enough to serve as an exoskeleton, and their body musculature consists not of solid muscle layers but of opposing sets of discrete muscles that attach to, and move, portions of the exoskeleton.

(*b*)   Each trunk segment bears a pair of fleshy, conical, unsegmented legs terminating in several ventral footpads and two chitinous claws; the legs are generally considered to be homologous with the parapodia of polychaetes, and the claws with annelidan setae. During locomotion, waves of contraction sweep down the body from anterior to posterior. As each segment is extended, the legs are lifted clear of the ground and swung forward, then are firmly planted and exert a pushing force by a backward stroke. Arthropod appendages are jointed and strengthened by their exoskeleton, and their movement does not involve bodily extensions and contractions.

(*c*)   The head is a simple structure, more like that of polychaetes than arthropods. It bears a single pair of unsegmented but annulated antennae and a pair of minute eyes that resemble the superior versions of annelid eyes seen in certain polychaete species, having a large chitinous lens and well-developed retina.

(*d*)   Each segment contains a pair of metanephridia, with ciliated funnels opening into coelomic sacs and tubules that terminate in tiny muscular urinary bladders. Each urinary bladder opens by way of a nephridiopore usually located at the base of each leg. Motile cilia are entirely lacking in arthropods, as are segmentally arranged metanephridia. The arthropodan classes considered most closely related to onychophorans have blind-ending Malpighian tubules that open into the hindgut and are not homologous with metanephridia.

(*e*)   The genital ducts are ciliated, as in annelids. As noted above, arthropods lack motile cilia.

14.23.   How are onychophorans like arthropods and unlike annelids?

(*a*)   Onychophorans are covered by a thin cuticle composed of chitin, as in arthropods, rather than of mucopolysaccharides, as in annelids. This cuticle is molted periodically, like that of arthropods, but it is much more permeable and is not divided into segmental plates.

(*b*)   The onychophoran body is not divided internally by segmental septa, as is the case with most annelids, but instead, as in arthropods, the true coelom is reduced to small sacs and the functional body cavity is a hemocoel. Also as in arthropods, the onychophoran hemocoel is partitioned into a middorsal pericardial sinus around the heart, a midventral perineural sinus around the nerve cord, a sinus around the gut, and a pair of ventrolateral sinuses.

(*c*)   Each leg of such onychophorans as *Peripatus* possesses a thin-walled basal vesicle, whose role seemingly is to take up moisture by everting—a function served by the coxal glands in the basal leg segments of some of the myriapodous ("many-legged") arthropods considered most closely related to onychophorans.

(*d*)   Onychophorans have a pair of chitinous jaws that, as in arthropods, are derived from appendages and come together medially like tongs; however, unlike arthropods, no other buccal appendages are present. The clawlike mandibles cut and macerate the prey, allowing saliva to enter the body of the prey, thus effecting external digestion so that only the liquefied tissues are sucked up. This procedure is unlike that seen either in annelids or in the myriapodous arthropods. The onychophoran digestive tract is simple and tubular, with a chitin-lined pharynx and esophagus and a straight stomach-intestine leading to a terminal rectum and anus. It is reasonably similar to that of centipedes.

(*e*)   The onychophoran heart and circulatory system are distinctly arthropodan and not at all like those of annelids. The heart is dorsal and tubular, with a pair of lateral openings (ostia) in each segment. The colorless blood, containing only amoeboid phagocytes, is propelled forward and ejected into the hemocoel just in back of the head, returning eventually to the pericardial sinus, from which it is sucked back into the heart by way of the ostia.

(*f*)   As in the myriapodous arthropods, gaseous exchange occurs through minute openings known as *spiracles,* from which microscopic air tubules (*tracheae*) extend directly to the tissues served. No such tracheal systems occur in terrestrial annelids, yet the onychophoran respiratory system differs enough from that of arthropods to suggest that it is simply convergent and not a true indicator of relatedness. For one thing, the spiracles occur scattered all over the onychophoran body, rather than in segmentally arranged lateral pairs; furthermore, arthropod tracheae branch, but those of onychophorans do not.

(*g*)   The nervous system is considerably more arthropodan than annelidan, with a large bilobed brain connected by circumpharyngeal commissures to a pair of longitudinal nerve cords; the latter bear segmentally arranged ganglionic swellings from which lateral commissures and segmental nerves arise. Arthropods also have a pair of ventral nerve cords, but these lie much closer together than in onychophorans, so that the segmental ganglia are fused instead of being conjoined by commissures. In contrast, the ventral nerve cord of annelids is single, and the brain is much smaller.

(*h*) Onychophoran reproductive systems generally resemble those of arthropods, with a single pair of large gonads, complicated genital tracts, and a single ventral, posterior gonopore. Male onychophorans have one pair of elongated testes and a complex genital tract including seminal vesicles and accessory glands; the two sperm ducts join to form a single ejaculatory duct in which the sperm are packaged into *spermatophores,* as is also the case in most centipedes and other myriapodous arthropods. In some species of onychophorans, the females possess seminal receptacles in which sperm are stored, but nothing is known of how the sperm or spermatophores are taken into the receptacles. In species that lack seminal receptacles, insemination is achieved in a manner no arthropod can manage: spermatophores simply deposited on the female's skin stimulate the breakdown of the underlying body wall, allowing sperm to pass directly from the spermatophore into the female's hemocoel.

Females have a pair of posteriorly located ovaries, and oviducts that are frequently expanded into a pair of tubular uteri. The end of each uterus joins with the other to open through a common genital pore. Australian onychophorans lay large eggs that resemble those of myriapods, having chitinous shells and a great quantity of yolk, so that cleavage can be only superficial. However, the majority of onychophorans are quite unlike both annelids and most arthropods in being live-bearing (*viviparous*). Both *ovoviviparous* and *euviviparous* species are known. In the former, the eggs are retained in the uteri until they hatch and the young are born by way of the genital pore. In the latter, the eggs have less yolk and may even cleave holoblastically, and the embryos are nourished by uterine secretions or even by a placentalike connection to the uterine wall. Viviparous onychophorans seem to be pregnant nearly all the time, producing up to 40 babies a year, which are born as pink, miniature versions of the adult worms.

Chapter 15

Joint-Legged Protostomes: The Arthropods

At least a million species of eucoelomate invertebrates enjoy the benefits of an external chitinous skeleton that protects the body and forms a series of jointed, segmentally arranged appendages. These highly successful animals, representing a peak of protostome evolution, are known as *arthropods* ("joint foot") for their jointed appendages. At present all joint-footed animals are placed in a single phylum, *Arthropoda,* although it is uncertain whether the sturdy chitinous *exoskeleton,* the greatest single innovation underlying the tremendous success of arthropods, evolved once or several times. Whether they should be considered separate phyla, subphyla, or classes, the major groups of living arthropods (Fig. 15.1) are *chelicerates* (spiders, scorpions, etc.), *crustaceans* (crabs, shrimp, etc.), and *uniramous mandibulates* (centipedes, millipedes, insects, etc.).

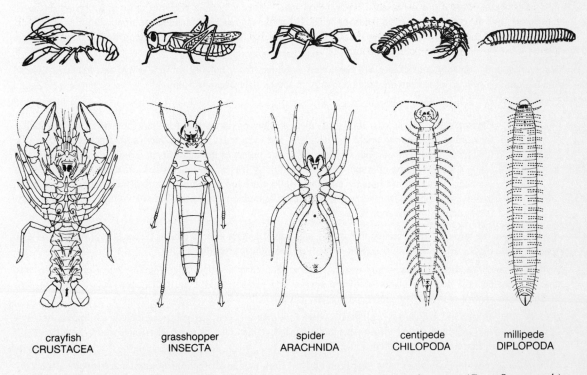

| crayfish | grasshopper | spider | centipede | millipede |
| CRUSTACEA | INSECTA | ARACHNIDA | CHILOPODA | DIPLOPODA |

Fig. 15.1 Arthropods. Representatives of five major groups, lateral and ventral aspects. (*From Storer et al.*)

15.1. How do arthropods affect humanity?

Without arthropods the world we live in would be a very different place, and it is doubtful that we *would* be living in it! We focus much energy and expense upon the control of arthropods harmful to humans, notably certain insect species, but in their countless numbers and diversity, arthropods are of overwhelming importance ecologically; for our own long-term benefit, we would be wise to guard arthropods from the effects of environmental pollution and the wholesale application of broad-spectrum pesticides that target a few species but strike down many.

15.2. How are arthropods beneficial to humans?

(*a*) Arthropods form vital links in marine, freshwater and many terrestrial *food chains.*

Example 1: In the sea, minute crustaceans such as copepods and euphausids are among the most abundant organisms that can feed directly upon *nanoplankton* (unicellular plankton, notably diatoms and dinoflagellates that convert solar energy to biomass). Without these *primary consumers,* all other levels of the food chain would collapse, since *secondary consumers* (larger invertebrates, fishes, birds, and marine mammals) are not adapted to feed directly upon nanoplankton and so could starve in the midst of plenty. The same applies to freshwater ecosystems, in which tiny crustaceans and aquatic insect larvae such as mosquito "wrigglers" constitute the primary consumers of phytoplankton and in turn are fed upon by infant fishes. On land, insects form the menu of so many insectivorous birds, reptiles, amphibians, and even mammals such as shrews, bats, and anteaters that countless vertebrates starve when insects become scarce.

(*b*) Imagine yourself trying to find food in a world where land plants are restricted to mosses, ferns, and conifers because flowering seed plants (angiosperms) do not exist. Arthropods, chiefly insects, are responsible for the success of modern seed plants, for most of these depend on insects to disseminate their pollen. Wind-pollinated grasses and grains are latecomers on the scene of plant evolution and would never have come into being if insects had not perpetuated ancestral angiosperms that were *not* wind-pollinated. Considering the wealth of animal life that relies for food upon the nectar, foliage, fruits, and seeds of flowering plants, an alien picture emerges of the earth that might have been, were it not for insects. Even today, most of our vegetable and fruit crops still depend on insect pollinators. But the costs of biological catastrophe should not be couched in human terms alone: consider the impact on animal life if most species of undomesticated flowering plants, from annuals to towering trees, were to become extinct for want of insects to pollinate them.

(*c*) Many arthropods, notably crustaceans, are consumed as human *food,* notably shrimp, crabs, and lobsters. Several hundred million pounds of shrimp are taken annually in the Gulf of Mexico alone.

(*d*) Commerce in *silk* helped shape human history.

Example 2: For at least 2000 years the secretions of silkworm caterpillars furnished a major impetus for trading between the orient and occident, and even today synthetic fabrics have not supplanted silk for fine textiles. The cocoons of some 2000 caterpillars of the moth *Bombyx mori* are needed to make 1 kg of silk at a cost of some 150 kg of mulberry leaves, the caterpillars' food supply. Silk culture has been carried on in China for more than 4500 years and when finally introduced to Europe stimulated massive plantings of mulberry trees. Today, the mulberries remain, but silk culture is practiced only in the orient, where labor is less expensive.

(*e*) *Spider silk* can be stronger than steel of the same gauge, and of more even diameter; accordingly, the silk of such spiders as the black widow is used in certain modern technological procedures.

(*f*) Insectivorous arthropods—spiders, many insects, centipedes, scorpions, and others—help *control insects* that otherwise might become pests on plants. Ladybird beetles and parasitic wasps are propagated on a large scale and distributed to control outbreaks of such economically detrimental insects as thrips and aphids.

(*g*) Soil-dwelling arthropods, which are ecologically vital as *scavengers and decomposers* include millipedes and sow bugs (crustaceans) that feed on leaf refuse, ants and other insects that demolish carcasses, and wood-chewing insects such as termites that help recycle fallen timber.

(*h*) *Esthetic and psychological values* have long been derived from human perceptions of certain arthropods, especially ornamental butterflies, moths, and beetles, which have inspired many works of art and literature. Scarab beetles were revered as sacred in ancient Egypt, bees and ants have long served as paragons of industry, and spiders are viewed as models of patient craftsmanship. Even today, fireflies, singing katydids, land hermit crabs, and other arthropods are kept as diverting pets in various parts of the world.

15.3. How can arthropods harm humans?

Exceedingly few arthropod species have negative impacts on human health and economy, but without doubt, those few impose dreadful costs.

(*a*) Ticks, mites, and bloodsucking insects such as mosquitos, fleas, lice, and bedbugs look upon people and their pets and livestock as just another meal. Screwworm flies and botflies have parasitic larvae that produce painful sores and abscesses.

(*b*) Other arthropods inflict painful and sometimes dangerous bites or stings in self-defense.

Example 3:   In the United States far more people die of bee and scorpion stings than of snakebite. The only potentially lethal spiders in North America are the female black widow and the brown recluse. Fire ants may kill livestock by occasional mass attacks. The poison claws of tropical centipedes, while seldom fatal, can deal quite agonizing nips.

(*c*) Some insects, ticks, and crustaceans are *vectors* of diseases that afflict human beings and domestic animals and plants.

Example 4:   Bloodsucking mosquitos transmit malaria, yellow fever, viral encephalitis, and filaria worms. Tsetse flies transmit the trypanosomes that cause African sleeping sickness in man and ngana in cattle, and reduviid bugs carry the trypanosomes causing Chagas's disease. Rat fleas are a major carrier of bubonic plague. Ticks transmit Rocky Mountain spotted fever and Q fever. Copepods are vectors for larval stages of the guinea worm *Dracunculus* and the broad fish tapeworm. Crabs and crayfish may harbor encysted lung fluke larvae.

(*d*) Wooden structures may be destroyed on land by termites and other wood-chewing insects, and in the sea by boring crustaceans such as gribbles (isopods).

(*e*) Crop and ornamental plants may be demolished by herbivorous and parasitic arthropods, notably plant mites and such insects as boll weevils, Japanese beetles, Mediterranean fruit flies, locusts, aphids, and scale insects. Stored grains and flour are attacked by certain beetles. Significantly, the most damaging species are often exotics, freed from the predators or parasites that control them in their native lands.

15.4.   Do all arthropods stem from a common ancestor?

Whether or not all joint-legged protostomes stem from a common ancestor, they share many features that otherwise would have to be explained in terms of convergence or parallelism. A number of their features suggest that arthropods arose, directly or indirectly, from annelids.

15.5.   How do arthropods resemble annelids?

(*a*) Both phyla are metameric. Metamerism is pronounced in the embryonic development of all arthropods and is also conspicuous in adults of primitive species.

(*b*) Primitively, both arthropods and polychaete annelids have appendages on every body segment.

(*c*) The nervous systems of both are constructed along the same lines, with a dorsal brain, connectives ringing the digestive tract, and a ventral longitudinal nerve trunk with segmental ganglia and nerves.

(*d*) The arthropod blood-vascular system seems to be derived from the annelid pattern, even though arthropods lack capillaries and have an open circulatory system. The hearts appear to be homologous: in both phyla, the true heart (as opposed to accessory or booster hearts) is a dorsal tube that propels the blood forward.

(*e*) The excretory coxal glands of chelicerates and the antennal or maxillary glands of crustaceans seem to represent modified metanephridia, the basic kidney type seen in other eucoelomate protostomes such as annelids and mollusks.

(*f*) Although most arthropod eggs are too yolky to cleave completely as annelid eggs do, in those few arthropods that do show holoblastic determinate cleavage, the mesoderm arises from the very same blastomere ("4d") as it does in polychaetes.

15.6.   How do arthropods differ from annelids?

(*a*) Arthropods have a protective and supportive exoskeleton formed from the chitinous cuticle. This exoskeleton consists of from 40 to 80 percent of the nitrogenous polysaccharide *chitin,* the balance being proteins, hardened and stabilized by tanning (formation of cross bridges between molecules) and sometimes calcium carbonate as well. The cuticle is often covered by a waxy *epicuticle* that reduces evaporative water loss. The exoskeleton is thin along lines of articulation, allowing flexibility. The cuticle must be molted periodically to permit growth. Its weight and the trials of molting restrict the size attained by arthropods. In contrast, the annelid cuticle is very thin, composed not of chitin but of mucopolysaccharides, is unsupportive and inefficient at preventing water loss, and is not molted. However, annelids *do* synthesize chitin for their jaws and setae.

(*b*) Annelids typically show homonomous metamerism: the body segments are commonly alike. Arthropods are characterized by *heteronomous metamerism* (modification of segments in different body regions), together with *tagmatization,* the fusion of embryonic body segments to form compound body sections, such as head, thorax and abdomen.

(*c*) Arthropods have a *fixed number* of body segments, but within most annelid species the number of segments is variable.

(*d*) The serially homologous parapodia of polychaetes are ordinarily just alike. The serially homologous segmental appendages of arthropods are typically *specialized* for such functions as stimulus detection, food getting, mastication, defense, locomotion, egg carrying, and copulation. Through tagmatization, appendages of successive segments may become closely grouped, as in the mouthparts of insects and crustaceans.

(*e*) Arthropods have skeletomuscular systems, in which striated muscles are arranged in opposing sets and attached to two or more different exoskeletal plates of the trunk or limbs. The arthropod body is supported upon its locomotory appendages, which can be moved in various planes by the attached muscle sets: forward by *protractors* and backward by *retractors,* to the side by *abductors* or toward the body by *adductors;* they can be bent by *flexors* or straightened out by *extensors*. Arthropods are accordingly the most agile of all invertebrates (with the possible exception of some cephalopod mollusks). By contrast, the body wall musculature of annelids forms solid circular and longitudinal layers that shorten and lengthen the body, together with protractors and retractors that move the parapodia forward and back.

(*f*) Arthropods are remarkable in having multiply innervated muscle fibers controlled by (1) *phasic* neurons that bring about rapid but brief contractions, (2) *tonic* neurons that evoke slow but sustained and powerful contractions, and (3) *inhibitory* neurons that prevent contraction. This diversity permits arthropods to execute sophisticated modulations of rate and duration while possessing relatively few muscle fibers, since each fiber can contract slowly or rapidly.

(*g*) Arthropods demonstrate patterns of instinctive behavior far more complex than those of annelids: courtship rituals, elaborate web building by spiders, caring for eggs and larvae, and some remarkable patterns of prey capture, described later. Social insects build complicated shelters and have complex communication systems, such as dance patterns by which worker honeybees guide other foragers to distant food sources.

(*h*) Many arthropods have keen image-resolving eyes. Also, experiments with honeybees and hermit crabs have shown that these arthropods (and probably many others) have color vision. Large, complex *compound eyes* (see Fig. 24.7) evolved in trilobites, crustaceans, and insects, but even the simpler eyes of spiders permit those that hunt to leap upon prey with a fine degree of accuracy.

(*i*) Terrestrial arthropods have not only impervious waxy epicuticles that protect them from dehydration, but also specialized respiratory and excretory organs that let them lead active lives under even exceedingly arid circumstances. As a result, arthropods have colonized the land with a success enjoyed by no other invertebrates. The internalized respiratory organs of land arthropods include *book lungs* and *tracheal* systems. Excretion of urine with little water loss is accomplished by the combined action of Malpighian tubules that end blindly in the hemocoel and the water-reabsorbing *hindgut,* into which the Malpighian tubules open.

(*j*) Unlike annelids, arthropods have an open circulatory system in which the coelom is much reduced and the functional body cavity is a hemocoel into which blood spills from the open ends of arteries. In some crustaceans venous channels are present, but capillaries are lacking. The dorsal heart, usually tubular with segmental openings (ostia), sucks in blood from the surrounding pericardial cavity, which is a sinus of the hemocoel, rather than a coelomic vestige as in mollusks. The open circulatory system is more likely a limitation of arthropods than an advance, since blood flow can be accomplished more efficiently in a well-developed closed circulatory system. Most terrestrial arthropods do not rely exclusively on the bloodstream for transporting gases, but have branching systems of air tubules reaching from the body surface to the tissues. This helps limit the size that terrestrial arthropods can attain.

(*k*) Unlike annelids, arthropods lack cilia.

(*l*) In all species having a complete digestive tract, although most of the alimentary tube is lined with endoderm, the most anterior and posterior portions are lined with ingrown ectoderm. The former, or *stomodeum,* ordinarily forms only the buccal cavity, while the latter, or *proctodeum,* constitutes only a short rectum. The stomodeum and proctodeum of arthropods are remarkably long and are lined by an inward extension of the chitinous *cuticle*. The stomodeum makes up the entire *foregut,* including buccal cavity, esophagus, and stomach, which function in ingestion, mechanical digestion (trituration), and food storage. The proctodeum makes up a hindgut of variable length, which absorbs water and compacts the feces. Between the chitin-lined foregut and hindgut lies the *midgut,* which secretes digestive enzymes (that are sometimes passed forward into the foregut to allow chemical digestion to commence there) and carries out most digestion and absorption of solutes. The midgut is often outpocketed to form cecal pouches or large digestive glands.

15.7.   What are the subdivisions of phylum Arthropoda?

Subphylum Trilobitomorpha: Extinct trilobites, with body divided by longitudinal furrows into three lobes (an axial lobe flanked by lateral lobes); head, thorax, and abdomen distinct; appendages biramous (two-branched).

Subphylum Chelicerata: Appendages include *chelicerae* and commonly *pedipalps* and four pairs of walking legs; body usually divided into *cephalothorax* (fused head and thorax) and abdomen; mandibles and antennae lacking.

  Class Merostomata: Aquatic, with abdominal appendages modified as external gills.

    Subclass Eurypterida: Extinct "water scorpions."

    Subclass Xiphosurida: Horseshoe crabs.

  Class Pycnogonida: Sea "spiders."

  Class Arachnida: Scorpions, whip scorpions, pseudoscorpions, solpugids, opilionids, spiders, ticks, and mites.

Subphylum Crustacea   (Chap. 16): Mostly aquatic, with gills; exoskeleton hardened with calcium carbonate; appendages biramous; two pairs of antennae; mandibulate; primitive forms have nauplius larva. Branchiopods, copepods, ostracods, barnacles, isopods, amphipods, euphausids, shrimp, lobsters, hermit crabs, crabs, etc.

Subphylum Uniramia   (Chap. 16): Primitively terrestrial with tracheae and Malpighian tubules; one pair of antennae; mandibulate; appendages uniramous. (May be a separate phylum, of onychophoran ancestry.)

  Class Chilopoda: Centipedes.

  Class Symphyla: Garden centipedes.

  Class Diplopoda: Millipedes.

  Class Pauropoda: Pauropods.

  Class Insecta: Insects.

Note: subphylum Mandibulata traditionally has included all arthropods having antennae and mandibles, but appears to be a polyphyletic grouping and so is replaced here by subphylum Crustacea and subphylum Uniramia.

15.8.   What were trilobites?

  Paleozoic seas were home to over 4000 species of trilobites ("three-lobed"), which flourished particularly in the Cambrian and Ordovician Periods and dwindled into extinction by the end of the Paleozoic, 200 million years ago. These most ancient and primitive of known arthropods ranged in length from 0.5 mm to nearly a meter and mostly crept along the sea bottom, although some were adapted for pelagic swimming, while others burrowed. The flattened trilobite body (Fig. 15.2) consisted of an anterior *cephalon* (head), a thorax, and a posterior *pygidium,* each subdivided by longitudinal

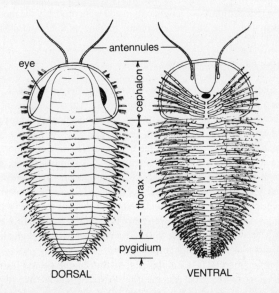

Fig. 15.2   A trilobite, *Triarthrus becki,* of the Ordovician Period. (*From Storer et al.; after Beecher.*)

furrows into three lobes. The cephalon, composed of at least five fused segments, was covered by a shield *(carapace)*. A pair of widely set compound eyes consisted of many cylindrical *ommatidia* remarkably like those of living arthropods. The mouth was located centrally on the underside of the cephalon, behind a liplike *labrum* from which extended one pair of long, slender antennae. Posterior to the mouth were four pairs of biramous (Y-shaped) cephalic appendages that in some species had their basal segments (coxae) thickened medially into toothed processes (similar to the gnathobases of horseshoe crabs) that probably served as food-crushing "jaws."

The thorax consisted of separate segments, which allowed trilobites to roll up. In pelagic species, the lateral margins of these segments were prolonged into spines that aided flotation. Each segment bore a pair of biramous appendages, each consisting of a medial branch adapted for walking and a lateral branch bearing a row of filamentous projections that may have served as gills or, when more robust, as spinous devices for digging, swimming, or filtering food. The posterior pygidium consisted of fused segments covered by a solid shield; each pygidial segment bore a pair of appendages like those of the rest of the body, but progressively diminishing in size toward the rear.

A trilobite hatched out as a planktonic *protapsis larva* 0.5–1 mm long, made up essentially of just the cephalon; metameres were added through successive molts until the larva became a *merapsis,* consisting of cephalon and pygidium. During further molts, thoracic segments were gradually added from the anterior border of the pygidium, until the full number of metameres had appeared and the final *holapsis* stage resembled a tiny adult.

15.9. What are chelicerates?

Chelicerates ("claw horn") are arthropods that lack antennae and mandibles and have as their first appendages a pair of *chelicerae* that are adapted for piercing or grasping, but not chewing. This subphylum, *Chelicerata,* includes merostomates, pycnogonids, and arachnids.

15.10. What are the features of class Merostomata?

Merostomates include the fossil *eurypterids* and modern *xiphosurids,* aquatic chelicerates characterized by a swordlike telson at the end of the body, five to six pairs of abdominal appendages modified as gills, and a cephalothorax bearing one pair of widely spaced compound eyes.

15.11. What were the eurypterids like?

The extinct eurypterids ("broad wing") had a wide carapace over the cephalothorax, and a long jointed abdomen consisting of a *preabdomen* and a *postabdomen.* The preabdomen had seven appendage-bearing segments, and the *postabdomen* had five segments lacking appendages and ended in a spiked telson. This arrangement resembles that of scorpions, so that eurypterids are called "water scorpions" (Fig. 15.3). The cephalothoracic appendages included a pair of small chelicerae and four pairs of walking legs, of which the last pair was paddlelike and was probably used in swimming. The abdominal appendages were modified as gills. Eurypterids grew to the impressive length of 3 m and were probably the most formidable predators inhabiting the estuaries and river systems in which the earliest vertebrates are thought to have evolved.

15.12. What are the characteristics of xiphosurids?

Xiphosurids ("sword tail") are mainly represented today by the *horseshoe crab, Limulus,* which occurs commonly in shallow water in the Gulf of Mexico and northwest Atlantic (Fig. 15.3). This living fossil has remained essentially unchanged since the Triassic Period, nearly 200 million years ago. Its horseshoe-shaped carapace arches beyond its legs, which are usually hidden from view when the animal is seen from above. The abdominal segments are fused and covered dorsally by a single skeletal plate, which is movably jointed to the carapace. The telson, as long as the rest of the body, is a mobile spike that the animal uses in righting itself when overturned. The cephalothorax bears the chelicerae, five pairs of walking legs, and a posterior pair of chilaria. The chelicerae are *chelate,* i.e., terminate in pinchers used in handing food into the mouth. The first four pairs of legs are also chelate and serve not only for walking but also for picking up food (bottom-dwelling invertebrates and algae), which they push between the *gnathobases,* spinous food-crushing processes on the medial surfaces of the coxae (basal segments) of these legs. The fifth pair of walking legs bears a cluster of tarsal processes used like a broom during burrowing. The *chilaria* resemble gnathobases minus the rest of the legs and are probably the coxal portions of a pair of degenerate legs.

The first abdominal appendages are broad flaps fused at the midline to form the *genital operculum,* on the underside of which are two genital pores. The remaining five pairs of appendages are also flaplike and fused together medially, but the underside of each pair is elaborated into a series of about 150 thin, membranous, parallel lamellae forming the "pages"

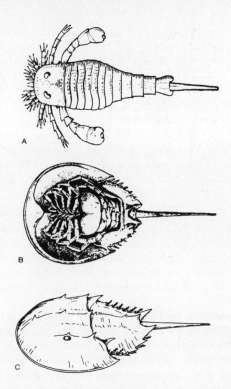

Fig. 15.3 Subphylum Chelicerata, merostomates: A, an extinct water scorpion, or eurypterid. B, C, the living xiphosuran (horseshoe "crab"), *Limulus polyphemus*. *(From Storer et al.)*

of a *book gill*. The lamellae are irrigated by oarlike movements of the flaps, which actually are used as oars in upside-down swimming by young *Limulus*. Blood containing hemocyanin fills the hollow lamellae of the book gills.

Limulus has two conspicuous *lateral eyes* that are compound but unlike those of insects or crustaceans, since the ommatidia are not compactly grouped and are probably too few to permit image resolution, although they may detect movement. The leading edge of the carapace bears a *frontal organ* (probably chemosensory) and a pair of tiny *median eyes*, which may discriminate only light and shadow; each consists of a cup formed by retinal cells, with the cavity of the cup occupied by a lenslike thickening of the overlying cornea.

Food crushed by the gnathobases is transferred forward to the mouth by the chelicerae. It passes down the chitin-lined esophagus to an expanded *crop* and thence into a muscular *gizzard*, the chitinous lining of which forms longitudinal, toothed ridges that pulverize the food. The particles then pass through a valve into a midgut consisting of an anterior stomach, into which open the ducts of two large digestive glands, and a posterior intestine. Enzymatic digestion takes place in the midgut and digestive glands. Fecal residues empty through the short rectum to the anus, located ventrally just in front of the telson base.

The circulatory system consists of a dorsal, tubular heart, which draws in blood from the pericardial cavity through its lateral series of ostia and pumps it out into anterior and lateral arteries that branch extensively before terminating in tissue sinuses. These drain ventrally into two main sinuses from which the blood flows into the lamellae of the book gills, thereafter returning in vascular channels to the pericardium. Four pairs of excretory coxal glands filter wastes from the blood and regulate water balance producing a more dilute urine when the animals are in brackish rather than salt water. These glands join a common renal tubule on each side, and the tubule opens at an excretory pore in the coxa of the fifth walking leg.

In spring, American horseshoe crabs migrate into bays and estuaries and come to shore at night, where the females burrow in sand at the high-tide level and lay up to 300 eggs. The eggs are fertilized externally by semen shed by the male, which has ridden ashore on top of the female by grasping her abdominal carapace with his modified hooklike first walking legs. The 3-mm eggs are centrolecithal and cleave holoblastically, producing a 1-cm-long *trilobite larva*, so called for its resemblance to those extinct arthropods, which are often considered ancestral to the merostomes and other chelicerates. The trilobite larva adds book gills and a longer telson with each molt until it gradually assumes the adult form.

15.13. What are "sea spiders"?

Pycnogonids ("compact knee") have lots of knees, but they seem more gangly than compact, since these animals are nearly all multijointed legs, with very little body (Fig. 15.4). Superficially, they resemble spiders and are commonly called "sea spiders," but they are actually very different. The 500 or so species of pycnogonids range from polar to tropical seas. Although common, most escape notice because they are only a few mm long. A few deep-sea species sport impressive

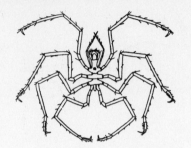

Fig. 15.4 Subphylum Chelicerata:
a pycnogonid or "sea spider,"
Nymphon. (From Storer et al.)

leg spans of up to 75 cm. The body is slender, with an anterior head that is fused with the first trunk segment. The head terminates in a tubular proboscis bearing a mouth armed with three rasping teeth. Some pycnogonids use the proboscis to penetrate the soft tissues of cnidarian polyps and sponges, which are sucked up and macerated by the muscular pharynx; others tear off bits of flesh with the chelicerae, which flank the mouth. The chelicerae are not chelate; each is a jointed hook. Additional cephalic appendages include a pair of segmented *palps,* well furnished with sensory bristles, and a pair of unique *ovigerous legs,* used in grooming and for egg carrying by the male. Toward the rear of the head is a median tubercle bearing four simple eyes.

The trunk consists of four to six segments. On both sides of each trunk segment, a large process projects laterally, articulating with an *eight-segmented* walking leg that ends in a terminal claw. The very reduced abdomen lacks appendages and seems little more than a place to put the anus. So hard put are pycnogonids for body space that their intestinal ceca extend into the legs for nearly their entire length, and the U-shaped gonad also sends lateral branches deep into the legs. Ripe eggs move upward within the leg to its base and are shed through a gonopore on the coxa. The male, facing oppositely and holding onto the female's underside, fertilizes the eggs as they are laid and collects them onto his ovigerous legs in spherical masses of about 1000 eggs apiece, glued together with secretions from cement glands on the femoral segment of his legs. At hatching, the larva is a *protonymphon* with only three pairs of trisegmented appendages, representing the chelicerae, palps, and ovigerous legs. It may continue to ride on father's legs or sequester itself within the tissue of a host polyp. New segments and appendages are added with each molt until the full complement is realized.

Digestion is intracellular, occurring in the epithelial cells lining both ceca and intestine. Cells reportedly detach from the lining and wander about the lumen, engulfing particles and later becoming reattached. Such lining cells are also said to concentrate waste, detach, and exit with the feces, thus taking the place of excretory organs, which pycnogonids lack. Blood vessels are also absent. The hemocoel is partitioned by a horizontal membrane into dorsal and ventral compartments. The heart pumps blood anteriorly in the dorsal compartment, where it passes through an opening in the partition into the ventral hemocoel, which extends into the legs. Eventually, the suction produced by dilations of the heart draws the blood dorsally through another opening at the rear of the hemocoel. No gills or other respiratory organs are present.

15.14.   What are the arachnids?

Class Arachnida ("spider") includes not only spiders but a diverse assemblage of chelicerates that evolved from aquatic ancestors (probably eurypterids) but today are nearly all terrestrial and comprise one of the most successful and ancient groups of land-dwelling invertebrates (Fig. 15.5*a*). True (but aquatic) scorpions occur in Silurian deposits, and all arachnid orders had appeared by the Carboniferous period.

15.15.   How did arachnids adapt to life on land?

(*a*)   They developed a water-impervious waxy epicuticle.

(*b*)   Internalization of book gills produced book lungs, which are protected from dehydration because they lie recessed within abdominal air chambers that open only by way of small, closable spiracles.

(*c*)   Some arachnids developed abdominal air passageways in the form of branched or unbranched tracheal tubules.

(*d*)   Water conservation was improved by synthesis of nitrogenous wastes into relatively nontoxic purines, mainly guanine, which can be excreted in a solid urine.

(*e*)   Malpighian tubules lying bathed in blood within the hemocoel allowed for selective removal of wastes, while the hindgut epithelium specialized for reabsorbing water; thus urine formation came to involve very little water loss.

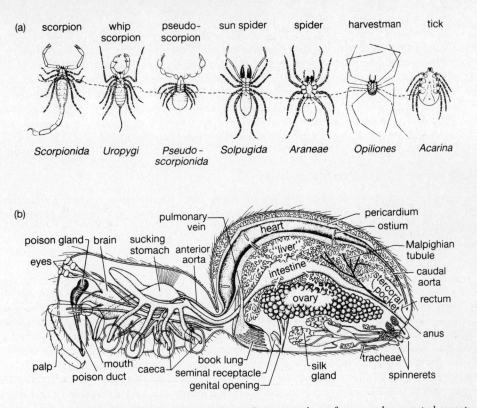

Fig. 15.5   Subphylum Chelicerata: arachnids. *(a)* Representatives of seven orders, ventral aspect: chelicerae black, pedipalps unshaded, walking legs stippled; dashed line denoting division between the anterior cephalothorax and posterior abdomen. *(b)* Internal structure of a spider, lateral view, with only bases of tracheae (air tubules) shown. [*From Storer et al.; (b) after Leuckart.*]

(*f*) Primitively, *spermatophores* were produced to allow sperm to be transferred to the female's genital tract without danger of dehydration; in male spiders, the pedipalps become modified into syringelike organs for injecting semen into the female's gonopore.

(*g*) Reproduction on land was achieved through viviparity or by the laying of eggs protected by a water-retaining envelope (*chorion*).

(*h*) An aquatic larval stage was deleted from the life history.

15.16.   What is the body organization of arachnids?

The arachnid body consists of an unsegmented cephalothorax (*prosoma*) and an abdomen that primitively is segmented and divided into a wider preabdomen and narrower postabdomen but appears unsegmented in spiders (Fig. 15.5*b*). The appendages, which arise from the cephalothorax, include a pair of chelate or fanglike chelicerae; a pair of pedipalps, which may be chelate or leglike; and four pairs of walking legs.

15.17.   How do arachnids feed?

Most arachnids are predatory, killing and rending their prey with their chelicerae and pedipalps while disgorging quantities of digestive enzymes from the midgut so that predigested liquids are pumped in by muscles that dilate the pharynx and sometimes also the esophagus, or "pumping stomach." The tubular midgut bears many lateral, branching diverticula that become filled with partially digested food. The midgut lining contains glandular and absorptive cells, secretes all digestive juices, and absorbs the digested food, which is mainly stored in interstitial tissues surrounding the diverticula.

15.18.   How do arachnids excrete wastes?

At the point where the midgut joins the short hindgut, one or two pairs of slender, blind-ended *Malpighian tubules* commonly arise. These branch within the hemocoel and absorb wastes through their thin syncytial walls, carrying the filtrate to the hindgut, where water reabsorption takes place. Arachnids that lack Malpighian tubules carry out excretion by means of *coxal glands,* thin-walled, spherical vestiges of the coelom that take up wastes from blood in the surrounding hemocoel and conduct them by way of a long, convoluted renal tubule to an excretory pore located in the coxal segment of the appendage of that metamere. Apart from the absence of cilia (which all arthropods lack), coxal glands appear to be modified metanephridia. Some arachnids have both coxal glands and Malpighian tubules.

15.19.   What specialized glands do arachnids have?

Spiders, scorpions, and pseudoscorpions have poison glands. Silk glands occur in spiders, pseudoscorpions, and certain mites. Arachnid silk is a protein similar to the silk produced by caterpillars, and is used in constructing shelters, egg cases, and snares.

15.20.   How does respiration occur in arachnids?

The slitlike spiracular openings of up to four pairs of book lungs occur on the ventral surface of the arachnid abdomen. Tracheal spiracles may also be seen on the abdomens of certain arachnids, especially tiny ones that cannot accommodate the more massive book lungs. A *book lung* consists of a series of thin, blood-filled, chitinous lamellae sequestered within a chamber formed by invagination of the ventral abdominal wall (Fig. 15.6). A muscle attached to the dorsal wall of this chamber opens the *pulmonary spiracle* and draws in air by expanding the chamber. Except when this muscle contracts, the spiracular slit remains closed, conserving the high humidity necessary for gas exchange. The "pages" of the book lung are spaced slightly apart by short bars, so that air circulates freely over their surfaces. The blood filling the lamellae contains oxygen-binding hemocyanin. By contrast, the blood of arachnids that have no book lungs and rely instead on tracheal systems lacks respiratory pigments. When present, *tracheae* are chitinous tubules, branching or unbranched, occurring singly or in clusters, that terminate directly on the tissues served and contain fluid at their inner ends where gases diffuse between the cells and the terminal *tracheoles.*

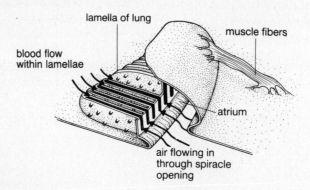

Fig. 15.6 Book lung, diagrammatic section.

15.21.   How is the arachnid circulatory system arranged?

In most arachnids the circulatory system is well developed. The dorsal, tubular heart lies in the anterior half of the abdomen and bears as many pairs of ostia as the original number of abdominal segments occupied by the heart (seven in scorpions). The heart gives off one pair of small abdominal arteries per segment, a posterior aorta to the rear part of the abdomen, and a large anterior aorta that extends into the cephalothorax, sending branches into the appendages and to the head region. Veins carry oxygenated blood from the book lungs back to the pericardial cavity.

15.22.   What are the components of the arachnid nervous system?

The nervous system of scorpions (Fig. 15.7) is more primitive (i.e., closer to the ancestral condition) than that of most arachnids: only the cephalothoracic ganglia are fused into a ring surrounding the esophagus, and the ventral longitudinal

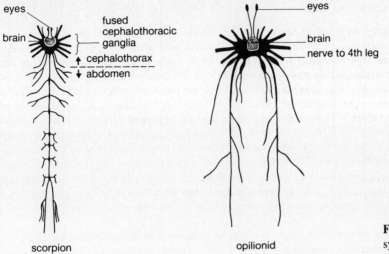

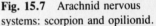

Fig. 15.7   Arachnid nervous systems: scorpion and opilionid.

nerve cord runs posteriorly, with a segmental ganglion giving off nerves in each abdominal metamere. In other arachnids, all the segmental ganglia are incorporated in the circumesophageal ring. The brain itself lies dorsal to the esophagus, receives the optic nerves, and gives off nerves to the chelicerae. The ventral mass of fused ganglia gives off nerves to all the other appendages and a single pair of nerves to the abdomen (see the opilionid in Fig. 15.7).

15.23.   What sensory organs do arachnids have?

Sense organs include eyes, slit sense organs, and sensory bristles. The eyes are simple, consisting of a cuticular thickening forming a combined cornea and lens, a layer of pellucid cells forming a vitreous body, and, finally, a retinal layer of photoreceptors backed by a membrane that is *reflective* in nocturnal arthropods. Such nocturnal species have *indirect eyes* in which the light-sensitive ends of the photoreceptors are oriented toward the reflective layer (*tapetum*), whereas diurnal species have *direct eyes* with photoreceptors oriented toward the outside. Many arachnids have both direct and indirect eyes for day and night vision. The eyes of most arachnids detect mainly movement, but those of hunting spiders probably can resolve images.

Slit sense organs occur by the thousands on the appendages and bodies of most arachnids. Each is a slit covered by a thin membrane that bows inward to contact the dendrites of sensory neurons. Slight changes in membrane tension excite the sensory endings, giving the arachnid information on body position and posture (proprioception), and allowing it to detect sound vibrations. *Sensory bristles*—simple setae or long, fine hairs—are abundant over the body surface and especially on the appendages. The bristles are connected with the sensory endings of nerve cells that are stimulated when the bristles are moved by touch or air currents.

15.24.   How do arachnids reproduce?

Arachnids are dioecious and often show considerable sexual dimorphism, the male being usually the smaller. The gonads are single or paired, and genital ducts lead to a single ventral orifice near the anterior end of the abdomen. The female tract contains seminal receptacles that hold sperm received during mating. In primitive arachnids, males produce spermatophores, which are deposited on the substratum and taken into the female orifice, but male spiders fill the swollen terminal segment of their pedipalps with semen from their own orifice and inject it into the female pore. Since most arachnids are predatory, elaborate courtship rituals have evolved that allow prospective mates to come in contact without one mistaking the other for prey. This sometimes happens anyway.

15.25.   What are the major orders of arachnids?

More than 100,000 species of arachnids are known to date, including the most ancient truly terrestrial animals. Eleven orders of arachnids have living representatives; seven of these are described in Questions 15.26–15.32.

15.26.   What are the characteristics of scorpions?

Scorpions (order Scorpionida) comprise 800 species from 13 mm to 18 cm long (some fossil species reached 86 cm). Mostly nocturnal, they hide by day, often in burrows that they excavate and defend territorially from other scorpions. The body consists of an undivided cephalothorax, covered dorsally by a carapace, and jointed abdomen made up of a wider preabdomen of seven segments and a narrow postabdomen ("tail") of five segments plus a terminal *stinging* apparatus (see Fig. 15.5*a*). The venom is usually no more toxic than a bee sting, but the North African *Androctonus australis* has a neurotoxic venom as powerful as cobra venom, which can kill a person in 6 hours. Also, species of *Centruroides* occurring in Mexico, Arizona, and New Mexico have caused hundreds of known deaths, especially among children.

A scorpion's chelicerae are stubby and chelate, while its pedipalps are long and bear large pinchers resembling a crab's. Prey is seized with the pedipalps and held while the scorpion arches its postabdomen over its back until the poison barb can be driven home. The envenomed prey is then manipulated so that the chelicerae can first belabor the head end, working in alternation like two little hands while the tissues are soaked with digestive juices. Thus the scorpion sucks in an externally digested broth. Scorpions drink water as available, but desert species can tolerate extreme aridity because their cuticle is highly impervious to evaporative water loss. Even so, they can survive water loss equivalent to 40 percent of their body weight.

Scorpions have four segmental pairs of book lungs, and oxygen is taken up by hemocyanin in the blood. Excretion is accomplished by two pairs of Malpighian tubules and a pair of coxal glands. The nervous system shows less fusion than that of other arachnids, since it retains a longitudinal ventral nerve cord with seven segmental ganglia. Despite scorpions' nocturnal habits, the poorly developed eyes are of the direct type. When walking, a scorpion holds against the ground the tips of a pair of comblike *pectines,* which are attached to the second abdominal segment and seem to be used to detect vibrations and sense the nature of the substratum.

Scorpion courtship is a lengthy affair. The smaller male grasps the female's pedipalps between the fingers of his own, and the two arch their postabdomens over their backs and rub them together, with venom sacs folded down so that stinging cannot occur. The male then pulls his mate along while he walks backward, until his pectines sense a surface suitable for spermatophore deposition. A male from which the pectines have been removed never deposits spermatophores. The male seems to be rather choosy, because he may lead his mate about on this courtship promenade for hours or even days. Once the male deposits his spermatophore, he draws the female over it, then pushes her backward so that the tip of the spermatophore penetrates her gonopore. After insemination, the partners separate. From a few months to more than a year, the female retains her internally fertilized eggs within her genital tract. All scorpions are viviparous, but some species are ovoviviparous and others euviviparous. Ovoviviparous species produce large, yolky eggs that develop within the ovarian tubules; cleavage is meroblastic, and the newborn young weigh less than the original eggs. The eggs of euviviparous species have little yolk, display holoblastic, equal cleavage, and develop within tubular diverticula of the ovary. These diverticula develop terminal clusters of *absorptive cells,* which rest against the mother's digestive ceca and absorb nutrients that are transferred to the embryos in a manner analogous to the placental nourishment of mammalian fetuses. At parturition, depending on species, some 6–90 infant scorpions only a few mm long are born; they crawl onto the mother's back, descending at the end of a week for their first molt, an arduous task that may take several hours, while the mother waits until the molted youngsters get on board once more. Over a period of several weeks the juveniles gradually forsake their mother, but until then she lets them suck up parts of her liquefied meals.

15.27.   What are whip scorpions?

Whip scorpions (order Uropygi) of about 85 species are widely distributed in tropical and subtropical Asia and America, including the southern United States, which is home to the impressive 6.5-cm *Mastigoproctus giganteus* (see Fig. 15.5*a*). Uropygids are often called "vinegaroons" because a whip scorpion can shoot a burning spray consisting of 84 percent acetic acid and 15 percent caprylic acid from large *anal glands* to each side of its whiplike (but harmless) tail. As sour as vinegar tastes, it usually contains no more than 5 percent acetic acid, so you can see that it is imprudent to handle a vinegaroon bare-handed.

Whip scorpions hunt by night, walking on only the last three pairs of walking legs and holding aloft the slender, elongate first pair, which play the role of sensory antennae. Prey is grasped with the robust pinchers of the pedipalps and thoroughly macerated by the fanglike chelicerae. During whip scorpions' elaborate courtship ritual, the male holds the tips of his mate's antenniform legs with his chelicerae, the female picks up the deposited spermatophores with her genital region, and the male then employs his pedipalps to thrust them into her gonopore. The female is oviparous, laying 7–35 eggs, which she glues to her body. She remains within a shelter that serves as a brood chamber until the young have hatched and undergone several molts, and dies soon after the young have departed.

15.28. What are pseudoscorpions?

Pseudoscorpions (order Pseudoscorpionida) make up some 2000 species of beguiling little beasts that, although common, escape notice because they are usually less than 8 mm long and hide away under bark and leaf mold (see Fig. 15.5a). Their name means "false scorpion," and they do look like miniature scorpions that forgot to grow tails. Nevertheless, they are actually efficient predators upon other small arthropods, for their long chelate pedipalps contain poison glands that open at the tips of their fingers. The chelicerae form pinchers that break open the body wall of the prey, allowing the pseudoscorpion to regurgitate digestive juices into the cavity and to suck up all the prey's internal tissues, leaving behind only the empty exoskeleton.

Pseudoscorpions lack book lungs, but two pairs of tracheal spiracles open on their third and fourth abdominal segments. Their excretory organs are coxal glands. They may be eyeless or have one or two pairs of eyes of the indirect type.

The chelicerae of pseudoscorpions are unusual in that the ducts of *silk glands* open at the end of the movable fingers. These glands are used to make a silk-lined chamber in which molting occurs. The silk glands also assist reproduction. Courtship tactics vary. In some species, the male merely deposits a spermatophore on the ground, and if a female comes close, she is attracted by its odor and straddles it, taking the sperm mass into her genital orifice. In other species, the male lays down silk threads around the spermatophore, and these guide the female to the spermatophore. Still other pseudoscorpions have a courtship ritual similar to that of true scorpions, in which the male grasps the female with his pedipalps and by dancelike maneuvers positions her over a deposited spermatophore. Finally, there are species in which the male evaginates from his genital region two long tubes which emit a sex attractant. When a female approaches, the pair dance without touching until a spermatophore is deposited, then the male seizes his partner's pedipalps and pushes her down upon it. The inseminated female constructs an igloo-shaped brood chamber of debris snugly lined with silk and holds her eggs within a sac attached to the underside of her body, nourishing the embryos and hatchlings with a secretion from her ovaries until they emerge from the brood sac after two molts.

15.29. What are solpugids?

Solpugids, or solifugids, (order Solpugida or Solifugae) are husky, *diurnal* arachnids that reach a length of 7 cm and can run very swiftly. Because they are often abroad by day, they are nicknamed "sun spiders," although they are not spiders at all (see Fig. 15.5a). Of the 800 known species, 100 occur in the southwestern United States, frequently in hot deserts. where they seek shelter in burrows or under rocks.

The carapace of the solpugid cephalothorax is crossed by a transverse seam, giving the effect of a separately movable head. Two conspicuous eyes are closely spaced together at the front of the carapace, just behind a pair of enormous chelicerae. Each chelicera forms a pair of vertically articulating pinchers that are sharply pointed for grasping and killing prey, even lizards. The pedipalps are leglike, but are held aloft in a threatening attitude, which may intimidate the curious, since solpugids are more apt to stand their ground, with chelicerae spread and pedipalps aloft, than to run away. Such spunk should not be rewarded with a boot sole, for solpugids do not have a venomous bite. When not being brandished defensively, the pedipalps are useful in capturing prey and climbing vertical surfaces (even glass), for each is tipped by an adhesive organ. Although solpugids have 10 leglike appendages (counting the pedipalps), they run on only the last 3 pairs, for the first walking legs, smaller than the rest, serve as tactile organs.

Solpugids have a well-developed tracheal system with three pairs of ventral spiracles. They possess both Malpighian tubules and coxal glands for excretion.

Compared with other arachnids, male solpugids are rather brusque lovers. The male simply seizes a female, briefly pats and strokes her into quiescence, then flips her over and thrusts the tips of one chelicera into her genital orifice. Then he ejaculates upon the ground, picks up the semen with his other chelicera, and pushes it into her orifice. He then swiftly leaps away before her appetite or affronted dignity imperils him, for solpugids also eat solpugids. The female is lightly burdened with maternal duties; she simply digs burrows, lays 50–200 eggs, and abandons them.

15.30. What are daddy longlegs?

Opilionids (order Opiliones or Phalangida) are commonly called "harvestmen," or "daddy longlegs," because in most the walking legs are exceptionally long and thin (see Fig. 15.5a). Opilionids readily snap off any leg by which they may be caught, but a lost leg is never regenerated. Certain spiders also have long, skinny legs, but spiders always have a "wasp waist" that sharply demarcates cephalothorax and abdomen, whereas opilionids have *no waist*: the body is ovoid and the abdomen is externally segmented. The chelicerae are small and chelate, and the pedipalps resemble short legs.

Opilionids include 3200 species with leg spans from 1 mm to 16 cm. Most favor humid woodlands. Unlike most other arachnids, they cannot survive long without food and water. Many are omnivorous, seizing snails and small invertebrates with their pedipalps and also scavenging on carcasses, fallen fruit, and vegetable matter. Unlike other arachnids, they do not digest food externally but suck it up in a particulate state, to be digested in the midgut.

Opilionids have one pair of direct eyes, elevated on a single median tubercle in the center of the cephalothorax. For excretion, they have a pair of coxal glands, and for respiration a tracheal system with a pair of abdominal spiracles (plus, in very active species, secondary spiracles opening into the legs).

The reproductive systems of opilionids are unique among arachnids. The male has a long tubular *penis*, and the female a protrusible *ovipositor* several times the length of her body. During copulation the partners face each other, while the male's penis is protruded from his genital orifice and passed between the female's chelicerae and under her body to reach her gonopore. The female later uses her ovipositor to penetrate deeply into humus or decaying wood, in which she lays several hundred eggs.

15.31.   What characterizes spiders?

Spiders (order Araneae) are the most familiar and diverse order of arachnids, with some 32,000 described species (see Fig. 15.5*a* and *b*). They are also probably the most maligned, for although all have a venomous bite, only a few are dangerous to humans and all help keep insect numbers down. In the United States only the black widow, *Latrodectus mactans*, and the brown recluse, *Loxosceles reclusa* (also called "violin spider" for the violin-shaped marking on its cephalothorax), inflict bites that are occasionally fatal. The venom of *Loxosceles* is hemolytic, often causing ulceration at the bitten site. Black widow venom is neurotoxic and causes about 5 deaths per 1000 bites. Since the shiny black female *Latrodectus* often builds her irregular web in sheds and lumber piles, under garden furniture, and in ground covers, we need to be alert for this common spider, which is usually recognizable by a red hourglass mark on the underside of her abdomen. The male *L. mactans* is a tiny, harmless, black-and-white spider, which not infrequently makes a tidy snack for his enormous mate, if he fails to run fast enough after inseminating her within her web.

Spiders are the only arachnids with chelicerae in the form of hollow fangs, which inject venom from large poison glands found anterior to the brain. The chelicerae of hunting spiders are often toothed for chewing the prey while it is digested by regurgitated enzymes. Spiders with untoothed chelicerae simply inject digestive juices into the puncture wound, and the liquefied tissues are sucked out by the powerful action of the "pumping stomach" (actually part of the esophagus).

Unique to spiders are the nozzle-shaped *spinnerets* that deploy several kinds of silk from large abdominal glands. Many spiders use silk to make webs that ensnare prey: orb webs, horizontal sheet webs, funnel webs, tube webs, and irregular webs.

> **Example 5:**   The most admired specimens of arachnid weaving, orb webs, are suspended between shrubs in the open. The frame of an orb web is first established with nonsticky silk, after which sticky silk is laid down in a concentric spiral. After a few days the adhesive threads become dry and ineffectual, and the spider usually devours its entire web, regaining valuable protein for the silk glands. Although most spiders prefer the out-of-doors, the spindly-legged house spiders (pholcids)—often mistaken for daddy longlegs (which are not spiders at all)—are possessed of an ungovernable urge to colonize human habitations and complicate the lives of persons who find spiders and webs disconcerting.

Silk is used in many different ways.

> **Example 6:**   As they walk about, most spiders play out a thread of nonsticky silk that serves as a safety line from which they dangle if suddenly swept off their support. A similar dragline glued to the substratum at intervals safeguards hunting spiders when they leap upon prey. Silk cocoons enclose the eggs, which may then be hung in the mother's web or carried about by female hunting spiders. After the spiderlings hatch, they often disperse by playing out a floating lifeline on which they sail away, sometimes being transported at high altitudes for hundreds of kilometers. Some spiders use silk to line lairs; there, they lie in ambush for passing insects and also spend the winter. A snug tubular burrow some 2 cm in diameter is excavated, lined, and sealed with a silken cap by trap-door spiders, which lurk with trapdoor barely ajar until prey comes by, seize the victim in one swift pounce, and drag it into the tube.

The marvelously diverse prey-capturing strategies employed by spiders of various families exemplify complex innate behaviors that owe much to genetic programming and little to learning.

Example 7: Unlike the web spinners, which construct webs according to species-specific patterns, many spiders are active predators. The keen-sighted wolf spiders (lycosids) and jumping spiders (salticids) hunt actively and pounce accurately upon prey, as do many species of the large, heavy tarantulas that make up a separate taxon of aranids (infraorder Mygalomorphae, with nearly all other spiders belonging to infraorder Araneomorphae). Crab spiders (thomasids) lurk in ambush upon plants. The nocturnal net-casting spider, *Dinopsis,* hangs head down, supported by a few silk strands, holding with its front legs a small, dense net with which it scoops insects out of midair. The bolas spider, *Dichrosticus,* dangles a single line bearing a droplet of sticky secretion that mimics the sex-attractant pheromone of certain female noctuid moths. When the male moth approaches this lure, the bolas spider briskly whirls its line and the sticky droplet strikes and ensnares the unfortunate suitor. Water spiders (agelenids) live underwater, within a bubble of air trapped in a thimble-shaped bell of silk attached to aquatic vegetation. As the air in the bell becomes depleted, the spider returns to the surface and, hanging upside down from the surface film, suddenly jerks its abdomen and hindlegs downward, pinching off an air-filled droplet, which it carries down and releases into the bell. Within its bell the water spider lies in ambush for any small aquatic animal, or it may dash to the surface to grab insects that fall into the water, retreating into the bell to feed. In winter, water spiders go deeper and build a new air-stocked bell in which they hibernate.

Primitive spiders respire by means of two pairs of book lungs, but in more advanced spiders, the posterior pair is replaced by tracheal systems that complement the remaining book lungs. As blood courses through the lamellae of the book lungs, plasma hemocyanin takes up oxygen. The heart is longest and has up to five pairs of ostia in spiders that rely only on book lungs for gas exchange, but becomes smaller and may have as few as two pairs of ostia in spiders having elaborate tracheal systems, since the latter rely much less on blood for oxygen transport. The heart beats only a few times a minute in spiders that lie quietly in wait, but over 100 times a minute in actively hunting spiders. Excretion is mainly accomplished by a pair of Malpighian tubules, since coxal glands are more or less reduced in all but primitive spiders, which have two pairs. The main nitrogenous waste is the purine *guanine,* which is excreted by cells in the cloacal wall, as well as the Malpighian tubules.

Most spiders have eight eyes, of which the anterior median pair are indirect eyes that aid nocturnal hunting, while the rest are direct eyes. Except in cursorial (running) species, the photoreceptors per eye are too few for image discrimination. Slit sense organs in the joint between the most distal of the longer leg segments (tarsus and metatarsus) serve web builders in detection of web vibrations that signal the capture of prey or the approach of a cautious male with amatory intentions, who may pluck the threads of the web like guitar strings. Hunting spiders rely on slit organs grouped into *lyriform organs* for proprioceptive information on both leg position and joint movements; this information allows them to backtrack to finish a meal from which they have been scared away.

In his abdomen just below the gut, the male spider has a pair of tubular testes that extend into convoluted sperm ducts; these ducts pass anteriorly to unite at a seminal vesicle that opens by way of a single median gonopore. The female's ovaries, which resemble bunches of minute grapes, are similarly located, and as the ripe eggs rupture into the cavity of each ovary, they pass forward in a pair of oviducts that unite at a median vagina flanked by a pair of sperm receptacles. In preparation for mating, a male spider constructs a special, tiny *sperm web* into which he ejaculates. He then draws the semen into a complicated *palpar organ,* which differentiates from the terminal segment of each of his pedipalps and bears a sperm reservoir and a copulatory projection. Thus charged, he goes off in search of a female, toward which he directs species-specific signals (such as a patterned waving of legs or pedipalps) that render her quiescent and allow him to approach with some degree of safety; he injects semen into her sperm receptacles with first one pedipalp and then the other. Certain mistrustful suitors first immobilize their mates with lines of silk before attempting to copulate.

15.32.   What are ticks and mites?

Order Acarina includes around 25,000 species of parasitic *ticks* and free-living and parasitic *mites,* a number of which damage animals and plants and also carry pathogenic microbes that afflict humans and livestock.

Example 8: Mites of the genus *Sarcoptes* tunnel into mammalian skin, causing *scabies* (seven-year itch) in humans and *mange* in animals. Some free-living mites have parasitic larvae, such as harvest mites, the larvae of which are *chiggers* that bite terrestrial vertebrates, exuding enzymes that digest the deeper skin layers, causing intense itching; chiggers also transmit such maladies as Asian scrub typhus. *Water mites,* conspicuously bright red, are benthonic in fresh water and shallow inshore marine waters. Their larvae parasitize such invertebrates as clams and aquatic insects.

The large acarines known as ticks can be grouped into a single suborder (Ixodides or Metastigmata), but the much tinier mites (averaging 1 mm long) are so diverse as to make up five different suborders, which may well represent an artificial assemblage (see Figs. 15.5*a* and 15.8*a* and *b*). Similar in form to ticks are the *spider mites* that attack plants: both have an ovoid body undivided into cephalothorax and abdomen, from which protrudes a cluster of mouthparts (erroneously called the "head"), which is the part embedded in the host's integument or used to pierce plant cells. The chelicerae and short pedipalps form this head (*capitulum*), while the other appendages [six in the larva, eight in the juvenile (nymph) and adult] are used in creeping and climbing. In ticks, the terminal segment of each walking leg bears a pair of prehensile claws. Mites of other suborders display modified body forms. *Beetle mites* resemble miniaturized beetles. *Gall mites* [that stimulate plants to produce enclosing structures (galls)] and *follicle mites* (that inhabit the hair follicles of mammals) are worm-shaped.

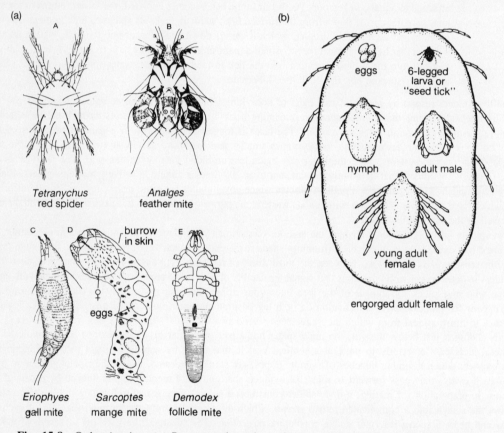

Fig. 15.8 Order Acarina. *(a)* Representative mites. *(b) Boophilus annulatus*, the Texas cattle fever tick, immature stages shown to scale within outlines of an engorged adult female. (*From Storer et al.; after U.S. Department of Agriculture.*)

Ticks, which may reach a length of 2 cm, suck the blood of reptiles, birds, and mammals. The six-legged larvae feed and drop off when engorged to molt to an eight-legged sexually immature nymph. This in turn takes a blood meal, drops off, and molts to the adult stage. Ticks can go without food for many months, but feeding must precede each episode of egg laying.

Example 9: Soft (*argasine*) ticks feed only briefly, at night, and drop off the host as soon as satiated. However, they inhabit their hosts' nests or lairs, and have only to climb aboard when hungry again. Hard (*ixodine*) ticks remain attached to their hosts for several days, during which time copulation occurs. After a single meal as an adult, the female ixodine lays her eggs and dies. Ixodines are negatively geotactic: they crawl up vegetation to where they can fall upon the bodies of passing animals, being attracted by odor to appropriate host species.

Hard ticks occasionally cause paralysis or death by secreting neurotoxic saliva. In addition they transmit Rocky Mountain spotted fever, tularemia, hemorrhagic fever, Q fever, anaplasmosis, encephalitis, and Texas cattle fever. Soft ticks may carry relapsing fever and other diseases.

Acarines lack book lungs and respiratory pigments and breathe by tracheal systems. The circulatory system commonly lacks a heart and is a mere network of tissue spaces. Excretion is usually accomplished by one to four pairs of coxal glands or a pair of Malpighian tubules, or both. Eyes are poorly developed and often lacking, and sensory setae are the principal sense organs. Female acarines have a single ovary with accessory glands and a seminal receptacle. Males possess a pair of lobate testes with sperm ducts that join at the single gonopore, and often have a chitinous penis. When a penis is lacking, semen may be placed on the ground and taken up into the female's genital orifice, or a spermatophore may be deposited and taken up by the female alone or with the aid of the male's chelicerae.

Chapter 16

Crustaceans and Uniramian Mandibulates

Although crustaceans and uniramian arthropods are alike in having antennae, mouthparts known as mandibles and maxillae, and compound eyes (in most), they are now considered different enough to represent separate subphyla. The conventional subphylum Mandibulata is being replaced by the subphyla Crustacea and Uniramia. In fact, if the uniramians—which are all primitively terrestrial—evolved from a terrestrial ony-chophoran ancestor, they may not even belong in the same phylum with arthropods that shared a common aquatic ancestor, possibly of the trilobite type. However, for now we shall treat crustaceans and uniramians as subphyla within phylum Arthropoda.

16.1. What are crustaceans?

From hatchling fishes gorging on copepods to giant whales placidly licking euphausids and mysids off their baleen whiskers, to humans dipping lobster tails in drawn butter, the vote is unanimous: crustaceans are a tasty bunch! Menaced by the appetite of a world of hungry people and animals, crustaceans (Fig. 16.1) respond with such unbridled fecundity that as larvae and adults they overwhelmingly outnumber any comparable group of aquatic metazoans.

Crustaceans are mainly aquatic arthropods with two pairs of antennae and skeletons often hardened by calcification. Their characteristics and body organization are well represented by the common freshwater crayfish (Fig. 16.2).

16.2. What is the organization of the crustacean head?

The cephalic organization of crustaceans is unique in including *two pairs of antennae*. These, along with one pair of *mandibles*, constitute the only appendages of the *nauplius*, the first larval stage of most crustaceans (see Fig. 16.3). By adulthood, the appendages of two more segments have been added to the head, as food-handling *maxillae*. The nauplius has a single, median eye composed of three to four pigment-cup ocelli, which permit the larva to orient by detecting the direction of light; this "nauplius eye" persists as the adult eye of certain small crustaceans such as copepods, but in most species, adults have a pair of sessile or stalked compound eyes (see Fig. 24.7). The eye of a lobster (*Homarus*), with some 14,000 ommatidia, can probably resolve images, at least crudely, but most crustaceans may detect little more than light and movement.

The excretory organs are also located in the head, as a pair of blind sacs often called *green glands*, or, more specifically, *antennal glands* if they open onto the base of the second antennae or *maxillary glands* when they open on the second maxillae. This arrangement is reminiscent of the coxal glands of chelicerates. Most larval crustaceans have both antennal and maxillary glands. Each green gland, like a coxal gland, originates internally as a small *end sac* derived from an embryonic compartment of the coelom, so that it too may be homologous to a metanephridium lacking ciliation. Fluid from the surrounding hemocoel passes into the end sac by pressure filtration and is subsequently modified by resorption of useful solutes as it drains through a spongy mass (labyrinth) and along a convoluted *renal tubule* to a urinary bladder that opens by way of a nephridiopore in the basal (coxal) segment of the appendage. These excretory organs mainly regulate ion balances, since nitrogenous wastes (mostly ammonia) diffuse through the gills and wherever else the cuticle is thin. In freshwater crustaceans such as crayfish, the green glands produce a copious, dilute urine low in salts, ridding the body of excess water while conserving vital ions.

16.3. How is the trunk of the crustacean body organized?

The trunk is composed primitively of a series of similar segments terminating in a *telson*, at the base of which the anus opens, but considerable departures from this arrangement have taken place in the course of adaptation toward different modes of life. Usually the trunk is subdivided into thorax and abdomen. Tagmatization is common, and head and thorax may be fused as a cephalothorax. Postcephalic appendages may be restricted to the thorax or borne on all trunk segments.

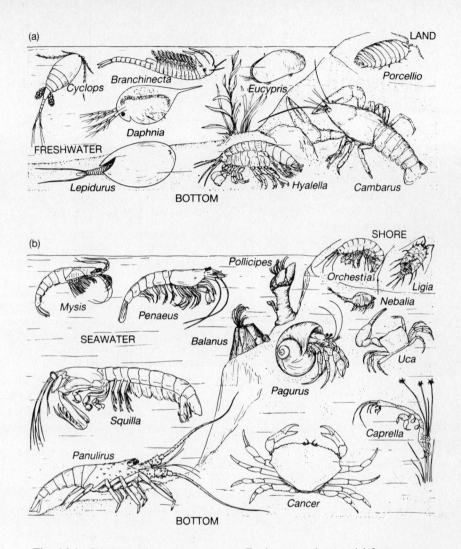

Fig. 16.1 Representative crustaceans. *(a)* Freshwater and terrestrial forms, not to scale: Copepoda, *Cyclops;* Branchiopoda, *Branchinecta* (fairy shrimp), *Daphnia* (cladoceran), and *Lepidurus* (tadpole shrimp); Ostracoda, *Eucypris* (bean shrimp); Malacostraca, *Porcellio* (sow bug, an isopod), *Hyalella* (amphipod), *Cambarus* (crayfish). *(b)* Marine forms, not to scale: Cirripedia, *Pollicipes* (goose barnacle), *Balanus* (acorn barnacle); Malacostraca, *Nebalia* (phyllocaridan), *Mysis* (opossum shrimp, a peracaridan), *Ligia* (rock slater, an isopod), *Orchestia* (beach hopper, an amphipod), *Caprella* (skeleton shrimp, an amphipod), *Squilla* (mantis shrimp), *Penaeus* (decapod shrimp), *Panulirus* (spiny lobster, a macuran decapod), *Pagurus* (hermit crab, an anomuran decapod), *Uca* (fiddler crab, a brachyuran decapod), *Cancer* (edible crab). *(From Storer et al.)*

Crustacean appendages are primitively *biramous* (Y-shaped), having a basal component, the *protopodite,* which bears a medial branch, the *endopodite,* and a lateral branch, the *exopodite.* These appendages may be modified for many different uses (see Fig. 16.2*b*). Most crustaceans are benthonic and have several pairs of appendages modified as sturdy walking legs, which are secondarily uniramous. Pelagic species possess swimming legs, usually fringed with stiff setae that increase their surface area.

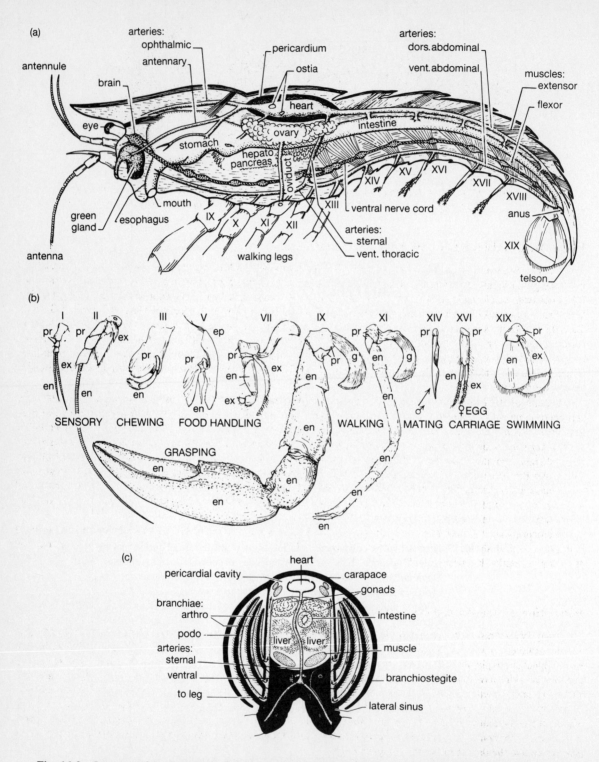

Fig. 16.2 Structure of a representative crustacean, the crayfish. *(a)* Internal anatomy of a female, lateral view; I–XIX = the 19 body segments characteristic of malacostracans. *(b)* Structural diversification of serially homologous appendages with respect to various functions. *(c)* Schematic transverse section at the level of the heart showing channels of the open circulatory system. Arrows show direction of blood flow; dark areas, deoxygenated blood; light areas, oxygenated blood. *(From Storer et al.)*

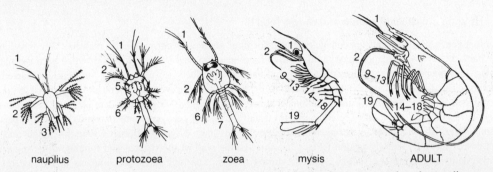

nauplius   protozoea   zoea   mysis   ADULT

Fig. 16.3 Development of the decapod shrimp, *Penaeus*. Appendages are numbered according to body segments to which they attach. (*From Storer et al.; after Muller and Huxley.*)

16.4. What is the nature of the crustacean exoskeleton?

The *exoskeleton* of larger crustaceans is heavily impregnated with calcareous deposits, forming a rigid "crust" that gives the group its name. It consists of (1) a thin *epicuticle* of lipid-impregnated protein that may also be calcified and (2) a thick *endocuticle* made up of an outer pigmented layer containing tanned proteins, chitin, and calcium salts, a middle layer that is most heavily calcified and contains more chitin and less protein, and a thin inner layer of protein and chitin. Each distinct trunk segment is enclosed by a dorsal plate, the *tergum,* and a ventral plate, the *sternum.* A cloaklike *carapace* may cover the thorax, cephalothorax, or entire body. When present, gills usually lie in a *branchial chamber* beneath the carapace. The *gills,* which are actually finely divided extensions of the cuticle that enclose blood-filled channels, are often attached to the bases of the thoracic appendages, so that movements of these appendages circulate water over the gills.

Molting of the exoskeleton (*ecdysis*) is controlled by hormones. A *molt-inhibiting neurohormone* is secreted by clusters of brain cells (X organs), whose axons extend into a pair of *sinus glands* where the hormone is released into the hemocoel. When secretion of this factor is inhibited by such environmental factors as temperature or photoperiod, a *molting hormone* is then secreted by Y organs located beneath the muscles that adduct the mandibles. In ecdysis, first the epidermis separates from the overlying cuticle and secretes a new epicuticle. Next, enzymes are secreted through canals into the space above the new epicuticle, and these gradually dissolve the old endocuticle, so that most of its components are reabsorbed and little remains to be shed but the old epicuticle. The crustacean now imbibes water and its body swells, splitting the old cuticle along some particular seam; the animal then pulls itself out. Further swelling takes place before the new epicuticle hardens, a new endocuticle is secreted, and calcification ensues.

If an appendage has been lost or cast off (autotomized), a *limb bud* forms soon after the amputation; the replacement limb grows rapidly during the premolt period and is unfolded from a sac during molting. Ecdysis is delayed until the premolt regeneration has been completed.

16.5. How are chromatophores important to crustaceans?

Many larger crustaceans are pigmented, brightly or cryptically, by means of *chromatophores* with branching cytoplasmic processes; chromatophores occur in the subcutaneous connective tissue and may contain white, black, brown, yellow, blue, or red pigments. *Monochromatic* chromatophores contain only one pigment, and *dichromatic* chromatophores two. Shrimp alone have *polychromatic* chromatophores containing three or more pigments. Movement of pigment granules within the noncontractile processes of the chromatophores is governed by hormones. Each pigment seems to be controlled separately, by a pair of opposing hormones that cause the pigment granules either to concentrate around the nucleus, making the coloration inconspicuous, or to disperse evenly throughout the cell, maximizing coloration. These *chromatophorotropic* hormones are produced by neurosecretory cells in the eyestalks, which also control the migration of black pigment around the ommatidia of the compound eyes.

When the compound eyes are light-adapted, a *pigment sleeve* surrounds each ommatidium so that only light directly entering the ommatidium reaches the sensory cells clustered at its base. As the eye becomes dark-adapted, the pigment sleeve migrates away, leaving the sides of the ommatidium transparent to light rays passing through at an angle; this increases the eye's light-gathering capacity and allows the animal to detect movement in dim light, but results in a probably indistinct *superposition image* as opposed to the mosaic *apposition image* perceived when the eyes are light-adapted.

16.6.   How do crustaceans obtain and digest food?

Food is gotten using various appendages adapted for filter feeding or grasping solid material. The most anterior one to three pairs of trunk appendages may be adapted as food-handling *maxillipeds* closely grouped with the maxillae and mandibles. Each mandible bears a fingerlike palp, and its medial surface is usually heavy and serrate, adapted for biting and grinding. The digestive tract is ordinarily straight, extending from the ventral mouth into a tubular esophagus that is often expanded into a triturating (grinding) stomach lined with chitinous ridges and denticles and calcareous ossicles, which serve to pulverize the food. The short midgut usually gives off ceca, one pair of which is often expanded into a large, solid *digestive gland* that secretes most of the digestive enzymes. Digestion occurs within the midgut and triturating stomach (when present), and absorption takes place through the wall of the midgut and tubules of the digestive gland, which also contains calcium-, glycogen-, and fat-storing tissues. The long intestine compacts undigested residues into feces, which are eliminated by way of the terminal anus. Except for the midgut, the entire digestive tract is lined by a delicate layer of chitin, which is shed at each molt.

16.7.   How is the crustacean circulatory system arranged?

The open circulatory system includes a dorsal heart that may be tubular and extend the length of the trunk or may be a compact vesicle restricted to the thorax. Ostia in the heart wall allow blood to be sucked in from the surrounding pericardial sinus of the hemocoel. In simpler crustaceans, the heart and even arteries may be absent, and the circulatory system may be represented only by blood sinuses. In more advanced species, the heart gives off a number of arteries that branch progressively into vessels of capillary size, which then open into sinuses. In crayfish, the heart gives off an anterior ophthalmic artery, a pair of antennary arteries, a pair of hepatic arteries (to the digestive gland), a posterior abdominal artery, and a ventral sternal artery (see Fig. 16.2*a* and *c*). Blood returning from the tissues collects in a large sternal sinus, passes through channels in the gills, and returns oxygenated to the pericardial sinus.

16.8.   What are the components of the crustacean nervous system?

The crustacean nervous system displays varying degrees of concentration and fusion. In the most primitive state—as seen in fairy shrimp and other branchiopodans—the dorsal brain gives off lateral connectives that pass ventrally around the gut and continue posteriorly as two separate ventral cords, each with segmental ganglia that are connected transversely to the ganglia of the opposite cord, in a *ladderlike* arrangement. In crayfish, the ventral cords and their ganglia have fused medially, giving the appearance of a single cord and ganglionic series, as in annelids. In crabs, all the ventral ganglia have fused into a single mass.

In addition to the median and compound eyes described above, crustacean sense organs include sensory hairs, especially chemosensitive *esthetascs*, which are concentrated on the first antennae. Certain higher crustaceans have gravity-sensing statocysts that usually contain sand grains.

16.9.   How do crustaceans reproduce?

Reproduction is always sexual, except for certain ostracods and branchiopods that reproduce parthenogenetically. Excepting cephalocaridans and some barnacles, crustaceans are dioecious. The paired gonads are located dorsally in the thorax or abdomen, with separate ducts opening ventrally. Many crustaceans produce nonmotile sperm that lack flagella, and they may transmit these by spermatophore. The male often has an appendage specialized for copulation, and in a few groups an actual penis is present. The eggs are often carried by the female, attached to certain appendages or held within a brood chamber.

16.10.   How does development proceed?

Primitive crustaceans produce small eggs that cleave holoblastically, but most higher crustaceans have centrolecithal eggs that cleave superficially. Crustaceans pass through at least one larval stage before attaining the adult complement of segments and appendages (Fig. 16.3). The first larval stage, the *nauplius*, is free-swimming except in most decapods (e.g., crabs and lobsters), where this stage takes place before hatching. In branchiopods and ostracods, the naupilus metamorphoses directly to adult form. In barnacles, it molts to a *cypris larva* (so called because it resembles the ostracod *Cypris* in having a bivalved carapace); the cypris then metamorphoses into the adult form. In all other crustaceans, the cypris stage is missing, and the nauplius transforms into a *protozoea*, which is also passed within the egg in most decapods. The protozoea molts to a *zoea*, the stage at which most decapods hatch. Later developmental stages are called *postlarvae*, since they have acquired all appendages requisite for adult life, but may still undergo further metamorphic changes before the adult body form is reached. Decapod shrimp and lobsters pass through a postlarval *mysis* stage, which molts to adult form. Crabs molt from a mysis to a *megalops,* which metamorphoses to the adult. As noted in Chap. 8, example 17, this crustacean tendency to lengthen ontogeny by adding new stages shows how evolution may have proceeded within the group. Each planktonic larval stage, however, has its own special adaptations for flotation and food getting. Being freshwater creatures, crayfish lack planktonic larval stages and simply hatch as miniatures of the adult.

16.11. What are the important subgroups of crustaceans?

Considering Crustacea to be a subphylum (instead of a class within obsolescent subphylum Mandibulata), the major taxonomic subdivisons dealt with here are as shown in Table 16.1.

Table 16.1 Major Taxonomic Subdivisions of Crustaceans

Subphylum Crustacea
  Class Cephalocarida: primitive; 7 species
  Class Branchiopoda: leaflike legs, mostly freshwater; 800 species
    Order Anostraca: fairy and brine shrimp
    Order Notostraca: tadpole shrimp
    Order Diplostraca: hingeless, bivalved carapace
      Suborder Cladocera: water fleas
      Suborder Conchostraca: clam shrimp
  Class Ostracoda: hinged bivalved carapace; bean shrimps; 2000 species
  Class Copepoda: most tiny; single eye; 7500 species, some parasitic
  Class Mystacocarida: one-eyed "moustache shrimp"; 3 species
  Class Branchiura: "fish lice"; 75 species, parasitic
  Class Cirripedia: barnacles; 900 species; one-third are parasitic
    Order Thoracica: goose and acorn barnacles; nonparasitic
    Order Acrothoracica: boring barnacles; nonparasitic
    Order Ascothoracica: endoparasitic and ectoparasitic species
    Order Rhizocephala: parasitize decapod crustaceans
  Class Malacostraca: over 20,000 species; 19 body segments, all bearing appendages
    Subclass Phyllocarida: bivalved carapace, foliaceous legs
      Order Leptostraca: marine filter feeders; 25 species
    Subclass Eumalacostraca: larger malacostracans
      Superorder Hoplocarida: mantis shrimp
      Superorder Peracarida: 7 orders with brood pouch formed by thoracic leg bases; 40% of all crustacean species
        Order Mysidacea: opossum shrimp; 450 species
        Order Isopoda: 14 similar legs; some parasitic; 4000 species
        Order Amphipoda: body laterally compressed; 5500 species
      Superorder Eucarida: eyes stalked, gills thoracic; all thoracic segments fused to carapace
        Order Euphausiacea: planktonic krill; 90 species
        Order Decapoda: 10 walking legs; shrimp, lobsters, anomurans, crabs; 8500 species
          Suborder Dendrobranchiata: shrimp hatching as nauplius
          Suborder Pleocyemata: hatch as zoea; mostly creep, not swim
            Infraorder Caridea: 2 pairs of chelae; most decapod shrimps
            Infraorder Stenopodidea: 3 pairs of chelae; coral shrimps
            Infraorder Astacidea: big-clawed macrurans
            Infraorder Palinura: nonchelate macrurans; spiny lobsters
            Infraorder Anomura: 5th legs reduced; hermit crabs, lithode crabs, porcelain crabs, mole crabs
            Infraorder Brachyura: abdomen tightly reflexed; true crabs

16.12.   What are cephalocaridans?

Class Cephalocarida ("head shrimp") contains only seven species but is significant as representing probably the most primitive of living crustaceans. Cephalocaridans occur in sand or mud from the intertidal to 300 m, along both coasts of North America and in the West Indies, Japan, and New Zealand. Less than 4 mm long, the body consists of an eyeless, horseshoe-shaped head, followed by a trunk of 19 segments, which are all similar except that the first 9 bear appendages that resemble the second maxillae. The protopodite of each bears not only an endopodite and exopodite, but an additional lateral lobe (*pseudepipodite*) that gives the limb a three-branched appearance. Cephalocaridans are hermaphroditic and are unique in discharging both sperm and eggs through a common duct. One species is known to brood its single egg in an ovisac attached to the first trunk segment.

16.13.   What are branchiopods?

Class Branchiopoda ("gill foot"), so called because the legs are flattened, leaflike *phyllopodia* used in gas exchange as well as swimming, includes a number of commonly abundant small crustaceans that inhabit fresh water (except for some marine cladocerans, and brine shrimp, which flourish in saline lakes). Their first antennae and second maxillae are vestigial, and the last abdominal segment bears slender, paired processes (*cercopods.*) Most branchiopods are filter feeders, using setae along the margins of their legs to sweep phytoplankton, bacteria, or organic particles into a ventral food groove between the legs, where adhesive glands glue them into masses that are worked forward to the mouth as swimming proceeds.

(a)   *Anostracans* (fairy and brine shrimp) lack a carapace, have a head with stalked eyes, and have an elongated trunk of 20 or more segments, up to 19 of which bear phyllopodia. They include the largest (10-cm-long) branchiopod species. Anostracans swim merrily about, upside down, the females towing egg sacs (see *Branchinecta*, Fig. 16.1*a*). This curious posture relates to light orientation: if an aquarium is illuminated from below, these crustaceans turn over and swim right side up; they just seem to like the sun on their bellies!

(b)   *Notostracans* (tadpole shrimp) have a broad carapace that covers the head and half the trunk like a shield through which the closely spaced, sessile compound eyes can be seen (see *Lepidurus*, Fig. 16.1*a*). The slender, flexible abdomen bears up to 70 pairs of appendages, and the telson has two long spinelike processes.

(c)   *Diplostracans* are laterally compressed, with a bivalved carapace. The two halves of the carapace are not hinged, but simply folded middorsally. They can be pulled together by an adductor muscle in the *conchostracans,* or clam

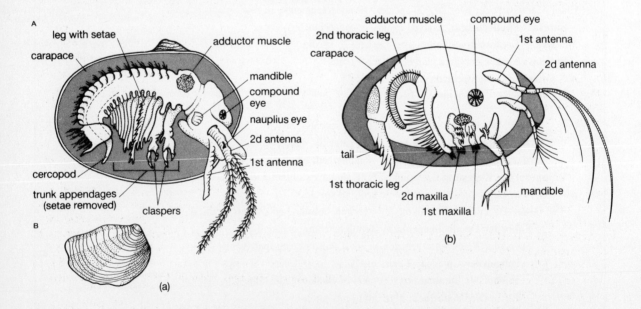

Fig. 16.4   Bivalved crustaceans. *(a)* Conchostracan (clam shrimp): *Cyzicus mexicanus* with right valve of the shell (carapace) removed; B, external aspect of right valve. *(b)* Ostracod: *Skogsbergia* with right valve and right appendages removed.

shrimps, which do look remarkably like infant clams when the shell is closed with the entire body tucked inside (Fig. 16.4*a*). The bivalved carapace of *cladocerans* (water fleas such as *Daphnia*) does not enclose the head, which is prolonged ventrally into a beaklike process that gives the cladoceran the look of a tiny bird (see *Daphnia*, Fig. 16.1*a*). Cladocerans retain their eggs in a brood chamber formed by the carapace, dorsal to the abdomen. In summer, cladocerans often produce only female offspring, parthenogenetically, from diploid eggs. As winter approaches, the eggs undergo meiosis and a few males are produced; these fertilize the haploid eggs, which develop hard walls that help them withstand cold and dehydration. Most branchiopods, in fact, inhabit temporary ponds, and their eggs may be blown about like dust for months or years before hatching.

16.14.　What are ostracods?

Class Ostracoda ("shelled") includes about 2000 living species of marine and freshwater bean shrimp (Fig. 16.4*b*; see also *Eucypris*, Fig. 16.1*a*), characterized by a bivalved carapace that has a distinct hinge line and is impregnated with calcium carbonate. The calcified shells provide this group of small crustaceans, mostly only a few mm long, with an excellent fossil record, continuous from the Cambrian, and more than 10,000 fossil species are known. The valves are apposed by an adductor muscle and may lock together by hinge teeth, clam-fashion. The entire body is contained within the closed shell, but some species have a notch allowing just the antennae to poke out. The ostracod body is nearly all head. The very reduced trunk bears only one to two pairs of legs, and the robust antennae are the main locomotory appendages, being fringed for swimming in planktonic species, while in benthonic types the first antennae are stout and spinous and are used for digging in mud and detritus. Most ostracods root around in the bottom ooze, but terrestrial species of Africa and New Zealand burrow through the humus of damp forests. Some ostracods are predatory, grasping worms, snails, and small arthropods with their antennae or mandibles and rending the prey with their maxillae. Others are scavengers, herbivores, or filter feeders. Certain marine species are photogenic, expelling clouds of luminescent secretion from a gland in the labrum (upper lip). Most ostracods lack gills, heart, and blood vessels. Some retain both antennal and maxillary glands into adulthood, but the majority keep only one or the other. All retain the *nauplius eye*, and one family also possesses compound eyes. Freshwater ostracods often reproduce parthenogenetically, and males are unknown in some species. While most ostracods simply shed their eggs into the water or attach them to a solid surface, some brood their eggs in the dorsal part of the shell cavity. Their nauplius larvae are unique in having a bivalved carapace like that of the adult.

16.15.　What are copepods?

Class Copepoda ("oar foot") represents an important group of crustaceans, including more than 7500 species, most of which are planktonic in the sea, although many live in fresh water, some even inhabit the film of water on mosses and soil particles, and a number are parasitic. Free-living copepods (e.g., *Cyclops*, Fig. 16.1*a*) are rarely more than a few millimeters long, but parasitic forms become larger, exceeding 32 cm in *Penella*, which parasitizes whales and fish. Parasitic species are often so highly modified that they are scarcely recognizable as arthropods. Free-living copepods have only the median nauplius eye (hence the name *"Cyclops"* for one major genus), and their first antennae are uniramous and often impressively oversized, like the horns of a longhorn steer. The main locomotory appendages of planktonic copepods are the short, biramous *second antennae,* which work either like rotors or like oars, propelling the copepod in smooth or jerky fashion. Most planktonic copepods are filter feeders, and their mandibular palps or maxillae are variously modified for producing feeding currents and trapping particles as small as 1.5 µm. The head is fused with one or two thoracic segments, as a cephalothorax, and the appendages of the first thoracic segment are modified as uniramous maxillipeds. The remaining two to five thoracic segments bear flattened, biramous appendages that aid swimming or are used for crawling by benthonic species. The narrow cylindrical abdominal segments are limbless, but the anal segment bears a pair of conspicuous, often elaborate, *caudal rami*. Free-living copepods lack gills, and most also lack heart and blood vessels; their excretory organs are maxillary glands. Many are luminescent, having *photophores* composed of epidermal glandular cells. Female copepods of certain species secrete pheromones that serve as species-specific sex attractants. The male clasps the female with one appendage while using another to transfer a spermatophore to the gonopore on her first abdominal segment. The eggs are usually retained in an ovisac that serves as a brood chamber. The nauplius larva molts five to six times before attaining a *copepodid* form, which molts five times more before adulthood. Some freshwater species lay both thin-shelled eggs and thick-shelled overwintering eggs, and many freshwater copepods survive desiccation by encysting within an organic capsule.

Over 1000 copepod species are ecto- or endoparasitic. In most, the larvae are free-swimming and attach to the host at either a nauplius or a copepodid stage, and the adult is more or less modified for parasitic life. However, the larvae of monstrilloids are not free-swimming, but endoparasitic in polychaetes and gastropods, and dispersal is by way of a free-swimming adult form that lacks mouth and gut and cannot feed, surviving only long enough to breed.

16.16.   What are mystacocaridans?

Class Mystacocarida ("moustache shrimp") includes 10 species of inconspicuous, but widely distributed, crustaceans only 0.5 mm long, which live among sand grains intertidally. Only the nauplius eye is present, both pairs of antennae and maxillae are well developed and setose, the first thoracic appendages form maxillipeds while the remaining are reduced to simple lamellae, and abdominal legs are absent.

16.17.   What are branchiurans?

Class Branchiura ("gill tail") includes 75 species of bloodsucking ectoparasites that attack marine and freshwater fish and amphibians. Branchiurans have an oval body 5–10 mm long, covered dorsally by a broad carapace from which only the unsegmented, legless abdomen protrudes. A pair of sessile compound eyes are present. The first maxillae are modified into attachment suckers, and the uniramous second maxillae form stout claws. A *sucking cone* is formed from plates (*labrum* and *labium*) around the mouth, and a piercing spine is used to puncture the host's skin and possibly inject venom. The four pairs of thoracic legs, biramous and setose, are employed in swimming from host to host. The eggs are deposited on the bottom and hatch at a postnauplius larval stage, which is parasitic at once.

16.18.   What characterizes barnacles?

Class Cirripedia ("curl foot") includes some 600 free-living or commensal and 300 parasitic species of widely distributed marine crustaceans that are unique in that all species are sessile in adult life (see, e.g., *Balanus* and *Pollicipes*, Fig. 16.1*b*). The life cycle (Fig. 16.5) commences with a planktonic nauplius that molts to a unique larval stage, the *cypris*. The cypris too is planktonic, but soon seeks a place for permanent settlement. The cypris larvae of nonparasitic barnacles are pheromonally attracted to adults of their kind, so that dense aggregations are formed, providing the propinquity needed for cross-copulation. Barnacles reproduce only sexually, are mostly hermaphroditic, and copulate by means of a tubular penis often longer than the barnacle's entire body. All barnacles brood their eggs in an ovisac within the cavity of the mantle, which is derived from the bivalved carapace of the cypris. Rather than being attracted to other barnacles, the cypris of parasitic barnacles is chemically attracted to appropriate host species. Those of commensal species seem to follow two different chemical signals for settlement: they are attracted to specific hosts, such as whales, and, on the host, aggregate by their attraction to previously settled conspecifics.

16.19.   What are the characteristics of nonparasitic barnacles?

Leading a comparatively blameless existence, nonparasitic barnacles heavily colonize rocky shores and wharf pilings, are sometimes commensal on other marine animals, and misguidedly settle upon the hulls of boats where their accumulation

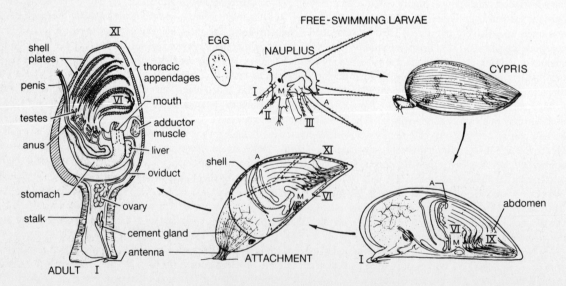

Fig. 16.5   Goose barnacle, *Lepas*. Life cycle and adult structure. M = mouth, A = anus, appendages I and II = first and second antennae, III = mandibles. (*From Storer et al.*)

creates such drag that eventually the unfortunate settlers have to be scraped off and are thereby doomed, since they are unable to reattach anywhere.

(a) *Order Thoracica* includes the familiar pedunculate *goose barnacles* and stalkless *acorn barnacles*. The thoracican cypris glues itself to the substratum by cement glands at the base of it first antennae, and its body rotates so that its life as an adult will be spent upside down, with the carapace opening upward. This allows the six pairs of curled, biramous thoracic appendages (*cirri*) to be thrust out and retracted, often rhythmically, capturing plankton in a basket formed by long setae that fringe the cirri. The bivalved carapace of the cypris persists as the mantle, which covers the adult body and becomes invested with calcified plates. In the acorn barnacles, an additional volcanolike cone of calcareous plates surrounds the entire organism. Both goose and acorn barnacles adhere so strongly to the substratum that they can withstand powerful surf, and both can close their carapace aperture tightly enough to withstand prolonged exposure to air at low tide. But acorn barnacles, also protected by their outer cone of plates, are better adapted to the high-tide zone, for they can endure even longer periods of exposure. Some thoracican species, although monoecious, also produce dwarf *supplemental males*, which live attached to the hermaphroditic individuals.

(b) Naked barnacles of *order Acrothoracica* secrete no plates, but seek shelter by boring into limy materials such as corals and mollusk shells; they are dioecious.

16.20.   What are the characteristics of parasitic barnacles?

(a) *Ascothoracicans* are ectoparasitic on brittle stars and crinoids or endoparasitic in stony corals, starfish, and sea urchins. They are not greatly specialized for parasitism, other than having prehensile abdomens and first antennae, plus sucking mouthparts if ectoparasitic or absorptive mantle lobes if endoparasitic. In fact, they are considered the most primitive living barnacles, being dioecious and having a chitinous, bivalved carapace covering the adult body, and six or fewer pairs of cirri.

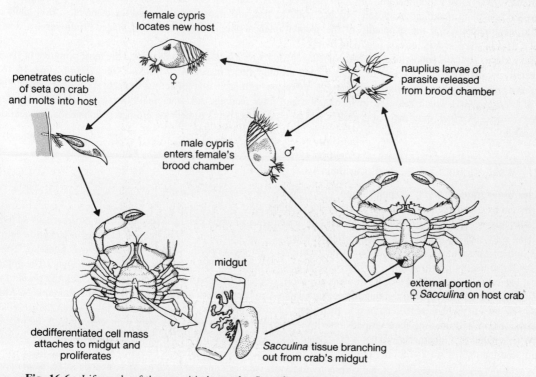

Fig. 16.6   Life cycle of the parasitic barnacle, *Sacculina*. Nauplius larva molts to cypris form. Female cypris settles on host crab, molts into the host's body as dedifferentiated cell mass that attaches to crab's intestine, invades its body with rootlike processes, then erupts to the surface as a bulbous mass containing a brood chamber. Male cypris is attracted to female, enters her brood chamber, and releases dedifferentiated cells that become a testis within her seminal receptacle.

(*b*) In contrast, *rhizocephalans* ("root heads") are so dramatically modified that their barnacle nature would be unsuspected were it not for their typical nauplius and cypris larval stages (Fig. 16.6). The female cypris settles on a suitable host, usually a crab or hermit crab, attaches by her first antennae to the base of a seta, perforates the integument, and molts into the host's body as a mass of dedifferentiated cells that settle on the host's midgut and send absorptive, rootlike processes throughout the host's body. This massive parasitic invasion sterilizes the host and prevents further molting. Eventually, the parasite erupts to the surface at the base of the host's abdomen, forming a bulbous brood chamber. A male rhizocephalan cypris, attracted to a female of his kind, attaches to the opening of her brood chamber and ejects a mass of dedifferentiated cells that migrate into a seminal receptacle, where they redifferentiate into a testis. This certainly represents the ultimate reduction of a husband to his biological function!

16.21.   What are malacostracans?

Class Malacostraca ("soft shell") seems misnamed, for it includes the largest and hardest-shelled of all crustaceans. But anyone who has enjoyed munching soft-shelled crabs, which can be eaten shell and all, has a practical understanding of the name "malacostracan," since the larger species of this class are compromised by the softness of their exoskeletons for some hours after molting, when their weight cannot be borne by the thin new cuticle and they hide away in helplessness. However, malacostracans include nearly 75 percent of all crustacean species, many of which are quite small and not so helpless after molting.

Malacostracans are characterized by a body plan known as the *caridoid facies* (as in the crayfish, Fig. 16.2*a*), with a head composed of five fused segments, a thorax of eight somites, and an abdomen of six somites plus a terminal telson that is simply a median extension of the nineteenth segment. Every one of the 19 body segments bears a pair of appendages; those of the nineteenth are *uropods*, often fan-shaped, that flank the telson and, with it, form a *tail fan* used in backward swimming. The other abdominal appendages are small, biramous *pleopods*. Primitively, the thoracic appendages are all similar and biramous, but the endopodite is better developed and persists as the only branch in the walking legs of benthonic species. The gills typically develop from basal projections (epipodites) of the thoracic legs; however, mantis shrimp and isopods have abdominal gills. In most malacostracans, one to three pairs of anterior thoracic appendages have become food-handling maxillipeds, closely grouped with the first and second maxillae and the mandibles. Both pairs of antennae are biramous, but the exopodite of the second antenna often forms a flattened blade. Primitively, a dorsal carapace, lost in some orders, covers head and thorax.

Class Malacostraca is subdivided into two subclasses. *Subclass Phyllocarida*, considered the most primitive of living malacostracans, includes only 25 species of small, but cosmopolitan, marine *leptostracans* (see *Nebalia*, Fig. 16.1*b*), which have eight rather than six abdominal segments, a bivalved carapace, and foliaceous thoracic appendages resembling those of branchiopods, which are used in locomotion and filter feeding. All other malacostracans belong to *subclass Eumalacostraca*, of which only the major groups will be considered here, namely superorders Hoplocarida, Peracarida, and Eucarida.

16.22.   What characterizes the hoplocaridans?

Superorder Hoplocarida ("weapon shrimp") includes the *stomatopods*, or *mantis shrimp*, which can exceed 36 cm in length and have unusually long tubular hearts, which extend most of the length of the trunk and bear 13 pairs of ostia. Mantis shrimp (see *Squilla*, Fig. 16.1*b*) are formidably armed with a hugely enlarged second pair of thoracic legs adapted for seizing prey such as fishes. The terminal finger, often bearing long spines, fits like the blade of a jackknife into a groove in the next segment; it can be shot out and retracted with great speed, trapping the victim as it folds up. These fingers are also handy in self-defense, for they end in a sharp point and can be driven home like a dagger. The folded raptorial legs are also used as bludgeons, to break the shells of crabs and gastropods; they can even crack the glass of an aquarium! During ritualized social fighting, these pugnacious creatures club one other with the folded raptorial legs. Most stomatopods wait in ambush at the entrance to their deep burrows, but some leave to hunt, swimming with their oarlike, gill-bearing pleopods and using their uropods only as rudders. They accurately spot prey with their stalked, compound eyes. When creeping, a mantis shrimp has a six-legged look, for the raptorial and reduced second thoracic legs are held forward, and the animal walks on only the last three pairs. The female compresses her 50,000 eggs into a ball, which she carries about with her thoracic legs and constantly turns and cleans. The eggs hatch as zoea, which live in the plankton for 3 months.

16.23.   What characterizes peracaridans?

Superorder Peracarida includes seven orders with 40 percent of all crustacean species, all characterized by having in the female a ventral brood pouch formed by large, overlapping horizontal plates that project from the coxae of certain thoracic legs. Most peracaridans are only a few centimeters in length. The three major orders are Mysidacea, Isopoda, and Amphipoda.

(a) *Mysids,* or *opossum shrimp,* look like small shrimp from 1.5–3 cm long and are often pelagic, swimming in great swarms that form an important food source for fishes and baleen whales (see *Mysis,* Fig. 16.1b). Their large carapace covers the cephalothorax but is not attached to the last four thoracic segments. Gas exchange is by thoracic gills or through the inner surface of the carapace. Mysids are considered primitive malacostracans because both antennal and maxillary glands are present and the heart is long and tubular.

(b) *Isopods* ("equal foot") are characterized by a dorsoventrally flattened body lacking a carapace, sessile compound eyes, one pair of maxillipeds and seven pairs of uniramous thoracic legs that usually look alike and are adapted for crawling (see *Ligia,* Fig. 16.1b). Their abdominal pleopods are adapted as gills, and the uropods are sometimes fan-shaped. Most of the 4000 species are marine, but some inhabit fresh water, and the pill bugs and sow bugs (see *Porcellio,* Fig. 16.1a) are truly terrestrial and lack any aquatic larval stage. In addition to gills, they have tubelike cuticular invaginations (*pseudotracheae*) that aid in breathing air. However, terrestrial isopods lack a waxy epicuticle and avoid dehydration by hiding under stones, emerging by night to feed on decaying vegetation. Wood-boring isopods ("gribbles") such as *Limnoria* commit mayhem on wooden pilings, docks, and boat hulls. Parasitic isopods include both bloodsucking species that attach to fishes and more highly modified types that parasitize crabs and other crustaceans.

(c) *Amphipods* ("both foot") include more than 5500 species of marine, freshwater, and terrestrial peracaridans that are characterized by a laterally compressed body lacking a carapace and by sessile compound eyes, one pair of maxillipeds, and gills borne on the bases of the seven pairs of uniramous thoracic legs (see *Hyalella,* Fig. 16.1a). The abdominal appendages include three anterior pairs of pleopods and three posterior pairs of uropods and are often adapted for jumping [as in beach hoppers or "sand fleas," (see *Orchestia,* Fig. 16.1b)]. Two or more pairs of thoracic legs may be specialized for grasping, as *gnathopods,* which have a terminal jackknifing finger (see *Caprella,* the skeleton shrimp, Fig. 16.1b). Relying for respiration only on gills and integument, terrestrial amphipods are restricted to damp beach sands and humid forests.

Ectoparasitic amphipods are dorsoventrally flattened, like isopods; the whale louse eats its way into its host's skin until it lies sheltered in the cavity formed.

16.24.   What are eucaridans?

Superorder Eucarida ("true shrimp") is characterized by stalked, compound eyes, a carapace to which all thoracic segments are fused, and thoracic gills. The females usually carry their eggs on the pleopods. The two eucaridan orders are Euphauslacea and Decapoda.

(a) *Order Euphausiacea* ("brightly shining") includes 90 species of widely distributed, pelagic, shrimplike eucaridans up to 3 cm long. Many have light-producing photophores and are brilliantly bioluminescent. Most euphausids filter-feed on zooplankton or phytoplankton, depending on the fineness of the mesh of the setae on the endopodites of their biramous thoracic appendages, none of which are specialized as maxillipeds. Long bristles on their abdominal legs aid swimming. Euphausids swim in dense swarms of over 60,000 per m^3 and constitute most of the "krill" on which baleen whales feed. The largest of the baleen whales, blue whales, reach a length of 30 m and consume up to 4 tons of krill a day. Euphausids cannot copulate: the male attaches a vase-shaped spermatophore next to the female's genital pore. The eggs hatch as nauplii and go through nine stages to reach maturity in 2 years.

(b) *Order Decapoda* ("ten-footed") includes some 8500 species of shrimp, crabs, lobsters, crayfish, and anomurans, which typically have 10 conspicuous uniramous thoracic walking legs, at least one pair of which is usually chelate. The other, more anterior, three pairs of thoracic appendages are biramous maxillipeds. Unlike that of euphausids, the carapace snugly encloses the gills in a *branchial chamber.* Nearly all decapods have a statocyst located in the base of each first antenna. The statoliths are commonly sand grains that the decapod inserts into its statocysts after each molt. Some statocysts are of complex form and have channels within which the movement of fluid provides rotational as well as gravitational information.

16.25.   What are natant decapods?

Decapod shrimps or prawns are often called *natantians* ("swimming") because some do swim expertly and are even pelagic, but the majority are benthonic and swim in only short backward darts. Shrimp usually have laterally compressed bodies with the carapace projecting forward as a pointed *rostrum.* The first two to three pairs of legs are chelate, and the abdomen is large and well developed for backward swimming with the aid of the tail fan formed by telson and uropods. Natant decapods actually fall into two different suborders. The commercially important *penaeid* shrimps make up *suborder Dendrobranchiata* (see *Penaeus*, Fig. 16.1*b*), the only decapods to hatch from the egg at the nauplius stage. All other shrimps, along with crabs, lobsters, and anomurans, belong to *suborder Pleocyemata*, in which the eggs are borne on the female's pleopods until they hatch as zoeas. Pleocyemate shrimps include *caridids* (*infraorder Caridea*), in which only the first two pairs of thoracic legs are chelate, and *stenopodids* (*infraorder Stenopodidea*), which have pinchers on the first three pairs of legs, the third pair being much enlarged.

16.26.   What are reptant decapods?

Heavier-bodied decapods are often called *reptantians* ("creeping") because most do creep and few are adapted for sustained swimming. They include astacids, palinurids, anomurans, and brachyurans.

(*a*)   *Macrurans* ("big tail") are reptant decapods with large, muscular, delicious abdomens bearing tail fans; when pressed, they shoot backward by abdominal flexion. Actually, macrurans belong to two separate infraorders: (1) *Astacidea* includes freshwater crayfishes (see *Cambarus*, Fig. 16.1*a*) and large-clawed lobsters such as the tasty *Homarus americanus*, in which the first walking legs are enlarged into heavy chelipeds; (2) *Palinura* includes the spiny lobsters (see *Panulirus*, Fig. 16.1*b*), which have no chelate walking legs at all.

(*b*)   *Infraorder Anomura* ("odd tail") includes a diversity of reptant decapods in which the abdomen is usually somewhat reduced and tucked under, and the fifth pair of walking legs is reduced or turned upward (so that at a casual glance anomurans look 8-legged instead of 10-legged).

> **Example 1:**   The burrowing *thalassinoid* shrimps—ghost and mud shrimps—have one pair of chelipeds of unequal size and a large, flattened abdomen that is quite loose-jointed for making U-turns within the burrow. *Paguroids* have a pair of chelipeds and an asymmetric abdomen with pleopods only on the female's left side. The most familiar paguroids are *hermit crabs*, with large, twisted abdomens that conform to the whorls of the gastropod shells that these "crabs" inhabit (see *Pagurus*, Fig. 16.1*b*). The large *coconut crab* wears a mollusk shell when young and aquatic but abandons this when it moves ashore. *Lithode crabs*, including the commercial king crab (*Paralithodes*), never wear a mollusk shell but carry their asymmetrical abdomens well tucked under. *Galatheoids* have one pair of large chelipeds and symmetrical abdomens with well-developed tail fans, kept more or less flexed under the cephalothorax. They include the pelagic *Galathea* (known as "red crab" or "squat lobster"), which swims by flexing its abdomen, and the *porcelain crabs*, which much resemble true crabs except for having whiplike antennae and only eight large legs. *Hippoids*—sand, or mole, crabs—lack chelipeds and have a symmetrical abdomen flexed under the cephalothorax. They occur intertidally on sandy beaches, buried with only eyes and antennae exposed.

(*c*)   *Infraorder Brachyura* ("short tail") includes the true *crabs*, in which the reduced abdomen, which lacks a tail fan, is tightly flexed into a depression on the underside of the cephalothorax (see *Cancer*, Fig. 16.1*b*). The male's abdomen is narrow and V-shaped, while the female's forms a broad U adapted for carrying eggs. The eyes, usually widely spaced, are lateral to the antennae, which are short. The first legs typically form heavy chelipeds. Crabs can be very agile and can move in all directions, but they run sideways. Most are marine or inhabit brackish water, but a number are freshwater, amphibious, or terrestrial. Amphibious species, such as fiddler crabs (see *Uca*, Fig. 16.1*b*) and ghost crabs, remain in air for extended periods, needing only an occasional dip to replenish the water in their branchial chambers. Land crabs are largely restricted to coastal regions, because they must return to water to breed. Some cheat dehydration by digging burrows that reach all the way to a shallow water table, like little wells. Others have a lunglike adaptation of the branchial chamber, into which protrudes an enlarged pericardial sac; the membrane of this sac not only aids gaseous exchange, but provides a surface along which rain water or dew is pulled in by capillary action, keeping the gill chamber moist.

16.27.   What are the uniramian arthropods?

A separate subphylum, *Uniramia*, has been proposed for five classes of terrestrial arthropods—centipedes, symphylans, pauropods, millipedes, and insects—which have certain features in common and appear to have evolved from a *terrestrial*, perhaps onychophoran, ancestor. All have appendages that are primitively unbranched (uniramous), and the cephalic appendages include a single pair of antennae, a pair of mandibles, and one to two pairs of maxillae. Primitively, food is picked up directly *under the mouth* rather than being carried forward from behind, as in crustaceans. Respiration is by *tracheal systems*, and excretion by *Malpighian tubules*. The eyes, whether simple or compound, are sessile. The head is enclosed in a sclerotized *head capsule*, and the trunk is externally segmented and multilegged. In insects, also called *hexapods*, the trunk is subdivided into a thorax of 3 segments, each bearing a pair of legs, and an abdomen of 9–11 segments, which is legless in adults but may bear a variety of appendages in larvae. The other uniramian classes, often called *myriapods*, have a greater number of trunk segments and appendages (Fig. 16.7). Although myriapods are different enough to warrant being placed in four separate classes, they do share a number of features including an elongated trunk composed of a series of similar, leg-bearing segments; a ventral nerve cord with a ganglion in each segment; a tubular heart, with segmental ostia, that runs the length of the trunk; and tracheal systems with spiracles that cannot be closed. Most myriapods lack a waxy epicuticle and are thus restricted to humid habitats.

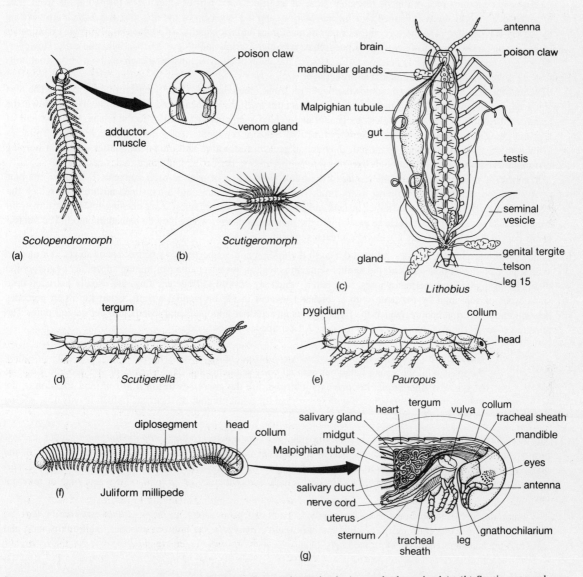

Fig. 16.7 Myriapodous uniramians. *(a)* Scolopendromorph centipede (a standard centipede). *(b)* Scutigeromorph centipede. *(c)* Internal anatomy of a centipede, seen from above. *(d)* Female symphylan, *Scutigerella*. *(e)* The pauropod, *Pauropus*. *(f)* Juliform millipede. *(g)* Anterior portion of the millipede shown in *(f)*.

16.28. What are the characteristics of class Chilopoda?

(a) Chilopods ("lip foot"), or *centipedes*, are distinguished by having their first pair of trunk appendages modified into curving *poison claws* that overlap medially and inject venom from glands that open at their tips (Fig. 16.7*a*). All, some 3000 species, are predaceous.

Example 2: The largest centipede, the 26-cm-long tropical American *Scolopendra gigantea*, has a bite occasionally fatal to humans. Other centipedes inflict bites which are painful but not lethal even to children, but which thoroughly immobilize their prey. When a centipede has subdued its dinner, it holds on to the meal with its claws and second maxillae, manipulates the food with its first maxillae, and crushes it with the mandibles. The ingested prey passes into a long, straight digestive tract, which receives the ducts of salivary glands near its anterior end, while a pair of Malpighian tubules open into the short hindgut (Fig. 16.7*c*).

The head structures of a centipede resemble those of an insect. One pair of antennae is located at the front. The upper lip of a preoral cavity is formed by a median labrum, and the lower lip by the first maxillae, which are overlain by the second maxillae. The mouth contains a pair of mandibles and a tonguelike *hypopharynx*. Burrowing centipedes lack eyes, and most others have clusters of ocelli that probably discriminate little more than light and dark. However, the long-legged house centipede *Scutigera* (Fig. 16.7*b*) has compound eyes consisting each of about 200 visual units that are quite similar to the ommatidia of insect eyes.

Behind the head is an elongated, somewhat flattened, trunk, composed of 15–170 leg-bearing segments. The legs are strong and clawed, allowing most centipedes to scuttle swiftly. Segments and legs are much alike down the length of the trunk to the small, legless pregenital and genital segments just anterior to the telson. The last pair of legs are often elongated and are often of sensory function. Because centipedes lack a waxy epicuticle and the spiracles of their tracheal systems cannot be closed, they are subject to desiccation and must retreat after nocturnal hunting into resting chambers reached through burrow systems that they construct in soil under stones and logs.

The heart is tubular, extending the length of the trunk, with a pair of ostia in each segment. Typically, one pair of spiracles occurs laterally in almost every trunk segment, each opening into a hair-lined atrium from which the tracheae arise. House centipedes have, instead, a series of middorsal spiracles, each opening into an atrium from which two large fans of short tracheae extend into the pericardial cavity, where they lie bathed in blood. The ventral nerve cord is double and bears a ganglion in each segment.

(*b*) Male centipedes have 1–24 testes, located dorsal to the midgut and connected by a pair of sperm ducts to a median ventral pore on the genital segment; the genital segment is legless but bears a pair of minute appendages (*gonopods*) that help manipulate the spermatophores. The latter, which are several millimeters long, are usually extruded onto small webs of silk spun by the male with a spinneret located inside his genital atrium. Courtship often precedes deposition of a spermatophore, which the female picks up with her gonopods and inserts into her genital pore. The unpaired ovary is tubular, lying dorsal to the gut. All centipedes are oviparous.

Example 3: Females of *subclass Epimorpha* wrap their bodies around a ball of 15–35 eggs, brooding the mass until the young hatch out with all body segments and 21–170 pairs of legs present. Females of *subclass Anamorpha* lay one egg at a time, use the gonopods to roll it in soil, and abandon it; the young of this subclass hatch with fewer than the adult number of segments (which is only 15), adding more at each molt.

(*c*) The *brooding centipedes* of subclass Epimorpha fall into two orders. *Order Geophilomorpha* includes eyeless burrowing centipedes that tunnel through loose soil by means of earthwormlike extensions and contractions of the body, while their 31–170 pairs of short legs serve mainly to anchor the centipede during burrowing. *Order Scolopendromorpha* includes the most typical centipedes, which have either 21 or 23 pairs of legs and may or may not have eyes.

(*d*) The nonbrooding anamorphic centipedes have only 15 pairs of legs as adults. In *order Lithobiomorpha* the legs are proportioned as in scolopendromorphs, while in *order Scutigeromorpha*, the house centipedes, the trunk is short and stout and the antennae and legs are exceedingly long.

16.29. What are the characteristics of class Symphyla?

Symphylans comprise about 160 species of widely distributed uniramians that resemble tiny centipedes only 2–10 mm long, with an elongated trunk of 14 segments that bears 12 pairs of legs and one pair of spinnerets (Fig. 16.7*d*). Although eyeless, they have a well-developed sensory pit organ at the base of each antenna and run rapidly and twist flexibly when maneuvering through leaf mold. Their cephalic appendages are similar to those of centipedes, except that the second maxillae are fused medially to form a labium. Poison claws are lacking, and symphylans feed only on plant materials. Although most consume decaying vegetation, some species attack living plants and can be serious horticultural pests, especially in greenhouses.

Only the first three trunk segments are furnished with tracheae, which open by way of a single pair of spiracles on the head. An eversible *coxal sac* at the base of each leg (also a feature of onychophorans and primitive insects) is used in taking up moisture.

The genital pores open ventrally on the fourth trunk segment. The reproductive behavior of *Scutigerella* is probably representative. The male simply deposits several hundred stalked spermatophores and ambles away, his duties done. When a gravid female encounters a spermatophore, she takes it into her mouth and stores it in her buccal pouches. As she lays each egg, she takes it with her mouth from her unpaired gonopore, attaches it to the ground or a plant, and smears it with sperm from her mouth. Parthenogenetic development of unfertilized eggs is also common among symphylans.

16.30.   What are the characteristics of class Pauropoda?

Pauropods ("small foot") represent about 500 species of midget myriapods less than 2 mm long; they are eyeless and inhabit soil, humus, and rotting vegetation (Fig. 16.7*e*). The head bears a lateral pair of discoidal sensory organs and a pair of unique Y-shaped antennae. One branch of the Y subdivides again, into two flagella and a club-shaped sensory structure. The trunk consists of 11 segments plus a telson, and each segment—except the first, last, and the telson—bears a pair of short legs. Like millipedes, to which pauropods may be related, *a single tergal plate covers each two segments*. A heart is lacking, and tracheae are also absent in most. As in millipedes, the gonopores open on the third trunk segment. The single ovary lies ventral to the gut, whereas the testes lie above the gut. The male roams about like Johnny Appleseed, depositing spermatophores here and there, to be taken up by any passing females. The eggs are laid in humus, and the young, like those of millipedes, hatch with only three pairs of legs.

16.31.   What are the characteristics of class Diplopoda?

(*a*)   Despite their common name, *millipedes* do not possess a thousand legs, but some of the more than 7500 species manage better than half that number. The name *diplopod* ("double foot") refers to the distinguishing characteristic of millipedes, to wit, that they have two pairs of legs on nearly every conspicuous body segment (Fig. 16.7*f*). This peculiarity derives from the fact that the embryonic somites fuse in pairs, forming *diplosegments* covered by a common tergal plate. Each diplosegment not only bears two pairs of legs, but has two more or less distinct ventral (sternal) skeletal plates, two pairs of spiracles opening into atria that give off tracheal tubes, two pairs of ventral nerve ganglia, and two pairs of heart ostia. Only a few trunk segments are apparently not diplosegments: the first segment (collum) is legless and forms a collar around the head, and the second, third, and fourth segments bear only one pair of legs apiece. These four segments are often referred to as the millipede's thorax, and the remainder, which may consist of more than 100 diplosegments, as the abdomen. In most familiar millipedes, the trunk is essentially cylindrical and is covered by an integument that lacks a waxy epicuticle but is hardened by calcium salts. The calcification of the exoskeleton along with the fusion in pairs of the abdominal somites and their dorsal (tergal) skeletal plates seems to strengthen the trunk as the millipede plows sedately forward through leaf mold and loose soil, propelled by the pushing force of its numerous legs and using its smoothly rounded head as a sort of ram. Despite a plethora of appendages, millipedes lack speed, and when molested, many do not try to run away, but simply coil up, exposing only their more heavily calcified dorsal and lateral surfaces while emitting noxious chemicals from *repugnatory glands* that occur one pair per most body segments. A variety of chemicals, including phenols, quinones, and hydrogen cyanide, are secreted by different species, some of which can discharge jets as far as 30 cm, giving the curious naturalist a rude surprise.

The millipede head (Fig. 16.7*g*) bears a pair of short antennae and a pair of very large mandibles with serrate and rasping surfaces, used in masticating decaying vegetation, which is the primary diet of most millipedes (some rock-dwelling species are predatory or omnivorous). As in other uniramians, the upper lip is formed by a labrum, and the lower lip by the first maxillae; however, in millipedes, the latter is enlarged into a broad, flat plate, the *gnathochilarium*, which forms the floor of the preoral chamber, and the second maxillae are absent. In certain tropical species, the labrum and gnathochilarium are prolonged into a piercing beak used in obtaining plant juices. Some millipedes are eyeless (but have scattered integumentary photoreceptors), but the heads of most bear rows or lateral clusters of up to 80 ocelli. Most millipedes are negatively phototactic.

(*b*)   The digestive tract is a straight tube with salivary glands that open into the preoral cavity, and a foregut leading to a long midgut. As in insects, the midgut produces a thin, chitinous *peritrophic membrane* that surrounds the food while permitting enzymes and digestive products to diffuse through readily. Two pairs of long Malpighian tubules empty into the gut at the junction of the midgut with the long hindgut. The heart extends the length of the trunk, and a short aorta carries blood forward into the head. One pair of ostia occurs in each thoracic segment, and two pairs in each abdominal diplosegment. Every diplosegment bears four spiracles, each opening into a chamber that gives off tracheal tubules.

(c) The female's long, tubular ovaries lie fused together just below the midgut, but each oviduct continues separately into its own pouch (vulva) that leads into a seminal receptacle and opens by way of a gonopore on the third segment. In males, the similarly located testes pass into sperm ducts that open on the third segment, either separately through a penis at the base of each leg or through a single median penis. The penis is not actually used in copulation. Instead, males of most species have one or both legs of the seventh segment modified into complexly shaped *gonopods* that form a reservoir into which the male, by looping his body, can insert the penis and ejaculate the semen. Once his gonopods are charged with sperm, the male courts the female using a variety of signals, according to his species: pheromones, stridulations, and tactile stimuli. During copulation the male holds the female and twists his body about hers, bringing his gonopods into contact with her genital pores, then employing them to transfer sperm into her vulvae. The tiny millipedes of subclass Pselaphognatha differ from all other millipedes (which belong to subclass Chilognatha) in that the males lack gonopods and instead deposit spermatophores on little webs from which two signal threads extend; these threads guide the female to the spermatophores, which she then takes up into her vulvae. Parthenogenesis is common in the pselaphognathans.

The eggs are fertilized with sperm from the seminal receptacles as they are being laid, often into a nest constructed by the female. Some millipedes build up a dome-shaped chamber of fecal matter into which the eggs are dropped through a central chimney, which is later sealed. Members of order Chordeumida enclose their eggs in cocoons of silk secreted by glands on the preanal segment. In some species, one or both parents remain coiled about the nest until the eggs hatch at the end of several weeks. The juveniles have only the first three pairs of legs upon hatching and add more legs and trunk segments with each molt.

16.32.   Why are insects so successful?

Members of class Insecta are the only uniramians to have only *three pairs of walking legs* as adults (Fig. 16.8*a*), which gives them a comparatively short, compact body, preadaptive to taking to the air. Insects are the only invertebrates that have managed to evolve *wings,* which has given them a very unfair advantage. By active flight and wind transport, winged insects have penetrated essentially every type of terrestrial habitat capable of sustaining life. If a single gravid female is storm-blown to some new environment where genetic isolation and new selective factors encourage speciation, she may turn out to be the founding mother of a whole new adaptive radiation. Insects are quite prolific and appear to enjoy a lively mutation rate, for they certainly multiply and speciate with unbridled enthusiasm.

Example 4: During only the 5–6 million years that the volcanic Hawaiian islands have been in existence, flies of a single genus, *Drosophila*, have evolved into 238 endemic species. Some 800,000 species of insects already have been taxonomically described, but the total number is probably more than twice as great.

Insect evolution is also promoted by their short generation times: a new generation of *Drosophila* can be produced every 10 days under laboratory conditions. Insects also tend to be small.

Example 5: Although insects range from less than 1 mm to 20 cm in length, few are longer than 2.5 cm. This means that large breeding populations, representing substantial gene pools, can be accommodated in limited areas. An acre of grazing land that could support a single steer provides food for countless millions of insects.

Insect mouthparts have become widely diversified from the basic handling-and-chewing type, specializing for many different diets, so that insects in the aggregate can utilize many food resources.

Herbivorous insects have proved to be good little biochemists: they have coevolved with land plants for several hundred million years, and many are very adept at tolerating toxic plant products (e.g., alkaloids) and may even accumulate these in their tissues, making themselves inedible. No wonder that insecticide-resistant populations of nearly every insect species detrimental to humans continue to spring up within a few years of the introduction of each new chemical product: variability exists in their gene pools, and the better biochemists (i.e., those better equipped metabolically to tolerate the poison) survive to reproduce, and do so abundantly. Fortunately, few species *are* detrimental from the human standpoint, and many are beneficial and ecologically essential, especially as disseminators of pollen and as food for many other creatures.

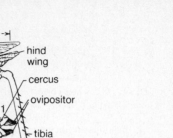

Fig. 16.8   Insect structure. *(a)* External features of a female grasshopper. I, II, III = thoracic segments, 1–11 = abdominal segments. *(b)* Grasshopper head, front view, and mouthparts. *(c)* Wing movement during flight: A, contraction of tergosternal (vertical thoracic) muscles depresses the tergum and displaces the wing bases so that the wingtips are raised; B, wing rotation, brought about by subsidiary muscles; C, downstroke produced by contraction of longitudinal muscles that shorten the thorax, causing the tergum to bulge upward and the wingtips to move downward. [*From Storer et al.; (c) after Snodgrass.*]

16.33.   What is the external anatomy of insects?

(a)   The insect head (Fig. 16.8*b*) bears one pair of antennae; these vary greatly in length and morphology, but are always richly provided with sensory bristles, pegs, pits, and plates that are mainly concerned with olfaction. So delicate are these sensors that a male moth may track an upwind female for at least a kilometer, while a male deprived of his antennae looks dumbly at a female without a notion of what to do with her. Two large, convex, compound eyes usually cover the sides and much of the top of the head.

Example 6: Each compound eye of a dragonfly—an arthropodan pursuit plane that literally scoops up other insects in midair—contains up to 30,000 ommatidia, allowing excellent mosaic vision. Conditioning tests with honeybees have demonstrated that these insects have color vision, although insensitive to red, and that they can readily discriminate between solid and broken abstract figures when correct discriminations are rewarded with sugar water.

Ocelli are lacking in many adult insects, but when present, usually number three and are clustered on the forehead; their functions are incompletely known, but they are sensitive to light of very low intensity.

Insect mouthparts are primitively designed for manipulation and mastication, but have often become specialized for various diets. The upper lip (labrum) covers the mandibles; these are just anterior to the maxillae and the labium, which is actually a fused pair of second maxillae. A tonguelike hypopharynx arises from the floor of the prebuccal cavity. Not only are the mouthparts of many insects radically modified for piercing, sucking, and the like, but larvae and adults of the same species may have mouthparts adapted for quite different diets. For instance, caterpillars have primitive chewing mouthparts, whereas in adult butterflies the maxillae are fused into a long, coiled proboscis used in probing the nectaries of flowers.

(*b*)   The insect's trunk is divided into a thorax and an abdomen. The thorax consists of three segments: *prothorax, mesothorax,* and *metathorax.* The abdomen has 9–11 segments, the last bearing sensory cerci and reproductive structures. The exoskeleton of each segment consists of a dorsal *tergum* (or *notum* in the thoracic region), a ventral *sternum,* and a *pleuron* on each side. Each of these regions contains thickened areas, or *sclerites,* separated by areas of thinner cuticle at sutures and points of infolding. Insects stop molting when the adult stage (*imago*) is reached, but in certain instances bodily expansion continues (as in queen termites) by extension of the areas of thin cuticle, so that eventually the segmental sclerites become widely separated. The insect epicuticle, richly impregnated with waxy compounds, resists evaporative water loss more effectively than that of other uniramians. The integument contains numerous cutaneous sense organs (*sensilla*) that register chemical and mechanical stimuli, including sound, which may be detected either by sensory bristles or by more complex *tympanal organs* composed of sensory cells within an air-filled space covered by a thin tympanic membrane that serves as an eardrum.

In adults, legs are confined to the thorax, with one pair articulating with the pleural sclerites of each thoracic segment. Although the legs are primitively adapted for walking, many are modified for special uses such as grasping prey, clasping, jumping, swimming, burrowing, and carrying balls of pollen.

(*c*)   Insects are the only winged invertebrates. They appeared on land 350 million years ago and primitively were wingless, like modern silverfish. Insect wings are outgrowths of the exoskeleton. It has long been speculated that insect wings first evolved as volplaning surfaces; however, recent experiments with models have shown that even tiny incipient winglets serve as excellent solar panels, and the more surface area they present, the more effectively they absorb solar heat. This may then have been the initial selective factor encouraging development of lateral cuticular outgrowths, with gliding, and eventually flight, being bonuses added when wing expansion had reached the point of aerodynamic competence.

Typically, insects have four wings. One pair (*forewings*) develops as an outgrowth of the cuticle of the mesothorax, and a second pair (*hindwings*) develops on the metathorax (see Fig. 16.8*a*).

Example 7: In two-winged flies (dipterans) the hindwings are reduced to *halteres,* small knobs that act as gyroscopes, correcting for flight instabilities (roll, pitch, and yaw). In other flying insects such corrections are made by way of a *dorsal light reflex,* to wit, postural adjustments that keep the dorsal ommatidia of the eyes most brightly illuminated from above.

Each wing consists of two layers of cuticle supported by hollow *veins,* which are conduits for blood; the wing also contains air tracheoles and sensory nerve fibers. When an immature insect molts to the adult stage, the wings emerge much crumpled, and the imago pumps blood into them, causing them to expand before they harden.

16.34.   How do insects fly?

Flying is made possible by special anatomical features. Each wing articulates with the lateral edge of a tergum and with a pleural process that serves as a fulcrum. Because this fulcrum is so close to the inner end of the wing, very slight displacements of the wing base medial to the fulcrum cause the wing tips to move through wide arcs (Fig. 16.8*c*). Contractions of vertical muscles running from sternum to tergum depress the tergum, which pushes down on the wing base and causes the wingtips to move upward. The downstroke is accomplished in different ways. In primitive winged insects, such as dragonflies and cockroaches, the muscles that move the wings downward are *directly* attached to the wing bases; in more advanced insects, such as bees and dipterans, that have very rapid wingbeats, the downward stroke is brought about *indirectly,* by the contraction of muscles that run longitudinally within the thorax, shortening the thorax and making the tergum bulge upward. Both direct and indirect sets of flight muscles operate in producing the downstroke in beetles and grasshoppers. The direct flight muscles also serve to tilt the wings as they move up and down, so that the wingtip pathway describes an ellipse or a figure eight, providing forward thrust. Some insects can modify the angle of tilt to allow hovering and even backward darts.

Wingbeat frequency varies from about 4 per second in butterflies to 1000 per second in midges. At frequencies of less than 30 per second, each wingbeat is initiated by a single burst of impulses from the nerve ganglion of that thoracic segment, but at higher frequencies, wingbeats result from alternating tremors of the antagonistic vertical and longitudinal muscle sets; these tremors are *myogenic* (caused in response to the stretching of each muscle set when its antagonist contracts) and are simply kept going by much less frequent rhythmic bursts from the central nervous system. The speediest flying insects appear to be dragonflies (40 km/h) and horseflies and sphinx moths (50 km/h).

16.35.   What is the internal anatomy of insects?

The internal organization of insects (Fig. 16.9) bears a general resemblance to that of other uniramians (see Questions 16.36–16.40).

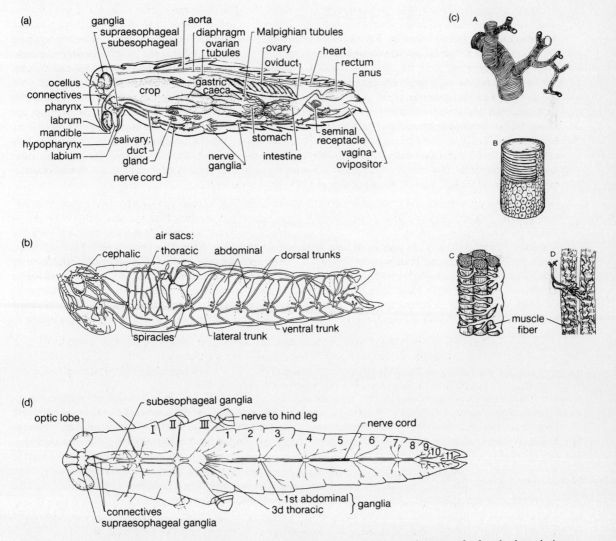

Fig. 16.9   Internal structure of the grasshopper, a representative insect. *(a)* Internal organs of a female, lateral view. *(b)* Respiratory system, showing spiracles, tracheal trunks, and air sacs on the left side only. *(c)* Detail of tracheae: A, larger trunks, supported by chitinous rings; B, tracheal epithelium and spiraling chitin thread; C, terminal branches around muscle cells; D, tracheoles ending on muscle cells. *(d)* Nervous system, dorsal view. [*From Storer et al.; (b) after Albrecht, (c) after Snodgrass, (d) after Riley.*]

16.36.   What are the components of the insect digestive tract?

The chitin-lined foregut consists of a buccal cavity into which salivary glands open, an esophagus that expands posteriorly into a food-storing crop, and a muscular *proventiculus,* or gizzard, equipped with toothlike structures for grinding and macerating food. The saliva sometimes contains digestive enzymes and, in bloodsucking insects, may have anticoagulant properties. In certain insects (moths, bees, and wasps) the salivary glands secrete silk used in spinning a cocoon in which pupation occurs. The entrance to the midgut is guarded by a stomodeal valve; food in the midgut is commonly enclosed within a thin, permeable chitinous *peritrophic membrane* secreted by the epithelial lining of the midgut or posterior end of the foregut. The midgut is the chief site of enzyme production, chemical digestion, and absorption. Many insects have clusters of midgut ceca of uncertain function; these may house symbiotic gut bacteria. From 2–250 Malpighian tubules open at the midgut-hindgut junction; these take up nitrogenous wastes (mainly in the form of uric acid) and other materials from the blood in the hemocoel. The chitin-lined hindgut (intestine and rectum) is the site of reabsorption of useful solutes from the urine, and often has *rectal glands* that function in water reabsorption, so that a nearly waterless urine is often voided. Many wood-eating insects such as termites rely on the cellulose-digesting capabilities of protozoans that live symbiotically in their hindguts.

16.37.   How is the insect respiratory system organized?

The insect tracheal system opens by way of one to two pairs of thoracic spiracles and a pair of spiracles occurring laterally on each of the first seven to eight abdominal segments (Fig. 16.9*b* and *c*). Except in primitive, wingless insects, *spiracular muscles* open and close the spiracles, reducing evaporative water loss. Thickened rings of chitin support the tracheae, which form two main longitudinal tubes with cross-connections and also expand into air sacs. Internally, the tracheae branch progressively into terminal, fluid-filled tracheoles as small as 0.1 μm in diameter and so numerous that most cells are intimately served with tracheoles that may even push the cell's membrane inward so that the tracheole tips are surrounded by cytoplasmic components such as mitochondria. Aquatic insects also rely on tracheae for respiration. Adults breathe the film of air that adheres to their bodies when they dive. Aquatic larvae such as those of mayflies have abdominal *tracheal gills,* which are outgrowths of the body wall containing a rich supply of tracheae that exchange gases with the water through the thin cuticle of the gills. Mosquito larvae instead breathe air at the surface through snorkel-like breathing tubes.

16.38.   What are the components of the insect circulatory system?

The insect heart lies dorsally in a pericardial sinus and is tubular, extending through the first nine abdominal segments. After each pulsation the relaxing heart may be stretched by fan-shaped muscles attached to its exterior, facilitating the heart's refilling. Blood is propelled forward into a single aorta that carries it into the head. Rapidly flying insects have a thoracic *booster heart* that sucks blood from the wings into the aorta. Insect blood is usually colorless and is unique in containing an unusual sugar, *trihalose,* and many free amino acids that maintain osmotic balance in lieu of the inorganic plasma ions on which most other animals depend. Various amoebocytes are present, some of which can agglutinate to form plugs that seal wounds.

16.39.   How is the nervous system of insects organized?

The insect nervous system resembles that of other arthropods. The brain consists of a fusion of cephalic (supra-esophageal) ganglia that are associated with the eyes, antennae, and mouthparts and are connected by commissures with the subesophageal ganglia (Fig. 16.9*d*). The paired ventral nerve cord forms a longitudinal chain of median segmental ganglia that are often fused to varying degrees. The three thoracic ganglia are fused into a single mass in adult dipterans, and the eight abdominal ganglia of primitive, wingless insects have become fused into a lesser number in winged insects. Insects also have an autonomic nervous system corresponding to the parasympathetic nerves of vertebrates, which governs physiological functions. Insects often display complex instinctive behaviors including courtship and parental care, together with communication and division of labor in social species such as ants, bees, and termites.

16.40.   How do insects reproduce?

The female reproductive system consists of one pair of ovaries, each made up of a cluster of tubules (*ovarioles*) within which the eggs mature and are stored. The two lateral oviducts unite to form a median oviduct that opens into a vagina. The vagina receives the ducts of accessory glands and bears a pouchlike diverticulum, the *spermotheca,* that stores sperm received at mating. The male's paired testes consist of *sperm tubules* in which sperm mature and then pass down sperm ducts to be stored in the seminal vesicles. These open, along with accessory glands, into a median *ejaculatory duct* that leads to the penis. The gonopores of both sexes are usually on the eighth abdominal segment, which is essentially at the posterior end of the male's abdomen, but in the female lies anterior to additional segments that form an ovipositor used in depositing eggs. In most insects, the penis is inserted into the female gonopore and extrudes a spermatophore directly into her vagina, but certain primitive, wingless insects (along with many myriapods) simply deposit the spermatophore on the ground, to be taken up by the female.

16.41.   What roles do hormones play in insect development?

A system of endocrine glands produces hormones that regulate molting and metamorphosis (Fig. 16.10). First, neurosecretory cells in the brain (*pars intercerebralis*) secrete a *prothoracicotropic hormone* (ecdysiotropin) that stimulates *prothoracic glands* in the thorax to secrete *ecdysone*. In the presence of *juvenile hormone* (JH) secreted by the *corpora allata*, ecdysone can bring about molting only to a larval stage, but when the secretion of JH diminishes to a certain point, the larva molts to a pupa, and as the secretion of JH ceases entirely, the pupa molts to the adult form. Premature treatment of midge larvae with ecdysone results in the prompt appearance of localized puffs on their chromosomes, a visible indication that this hormone is involved with the activation of specific genes.

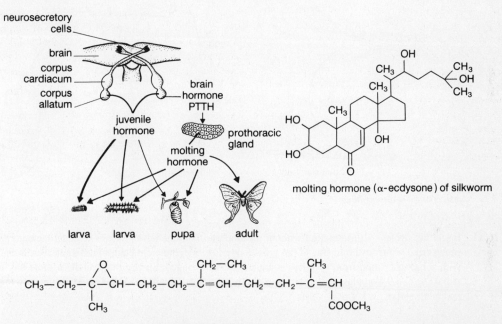

Fig. 16.10   Hormonal control of insect molting and metamorphosis. Certain brain cells secrete prothoracicotropic hormone (PTTH), which prompts the prothoracic gland to release ecdysone, a steroid hormone that triggers both molting and metamorphosis. However, in the presence of juvenile hormone from the corpora allata, metamorphosis is suppressed and the larva simply molts to another larval stage. In time juvenile hormone ceases to be produced, and in its absence, ecdysone also initiates metamorphosis.

16.42.   How does insect development proceed?

Insects lay land-adapted (*cleidoic*) eggs. The chitinous shell (*chorion*) is several layers thick, protecting the developing embryos from dehydration. It is secreted by ovarian follicle cells in the ovarioles and is penetrated by one or more pores (*micropyles*) through which sperm from the spermatheca enter as the eggs are laid. Except for the eggs of wingless collembolans, which cleave holoblastically, cleavage is superficial, on the surface of a central yolk mass. At hatching, the young of primitive, wingless insects resemble the adults and undergo no metamorphosis, simply increasing in size with each molt and finally attaining sexual maturity. All other insects undergo some type of metamorphosis (Fig. 16.11).

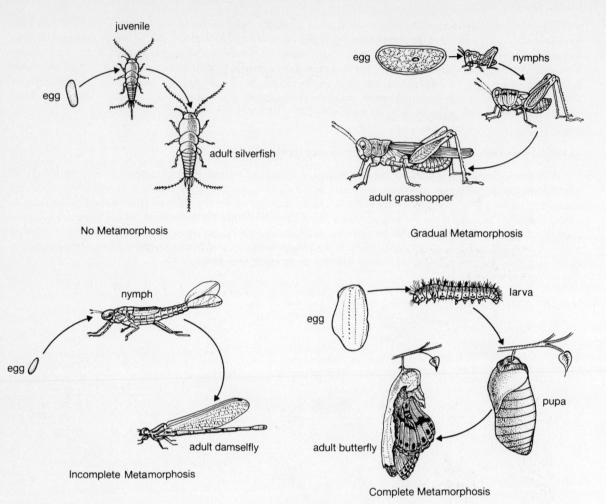

Fig. 16.11   Insect life cycles. *(a)* Maturation without metamorphosis in the silverfish. *(b)* Gradual metamorphosis, with progressive wing development, in the grasshopper. *(c)* Incomplete metamorphosis, as seen in the damselfly with abrupt transition from wingless aquatic nymph to winged, air-breathing adult, without any intervening pupal stage. *(d)* Complete metamorphosis, as in butterflies, in which a pupal stage occurs between the larval and adult stages.

> **Example 8:**   *Gradual metamorphosis* (as seen in grasshoppers) involves the progressive transformation, through several molts, of the wingless *nymph* to the winged adult. *Incomplete metamorphosis* (as seen in some terrestrial insects that have aquatic larvae, e.g., dragonflies) involves an abrupt molt from the wingless, often gilled, *naiad* to the winged adult; this molt takes place in air after the naiad has crawled above the waterline. Life cycles showing *complete metamorphosis* (as seen in flies, beetles, and butterflies) are interrupted by a quiescent stage, the *pupa*, that intervenes between larva and adult. The larva—often called a maggot, grub, or caterpillar—grows through several molts, then molts to the pupal state, which in turn molts to the adult, sometimes after a prolonged period of winter or dry-season dormancy. Pupal dormancy often is not *caused* by adverse external conditions, but sets in before the going gets bad. This type of genetically programmed dormancy, known as *diapause*, is usually initiated and terminated by *photoperiod*, but chilling may also have to intervene before diapause can be broken.

Fig. 16.12   Representative apterygote and exopterygote (hemimetabolous) insects. ♀ = female, ♂ = male. Note secondarily wingless condition of lice *(g)-(h)*. *(From Storer et al.)*

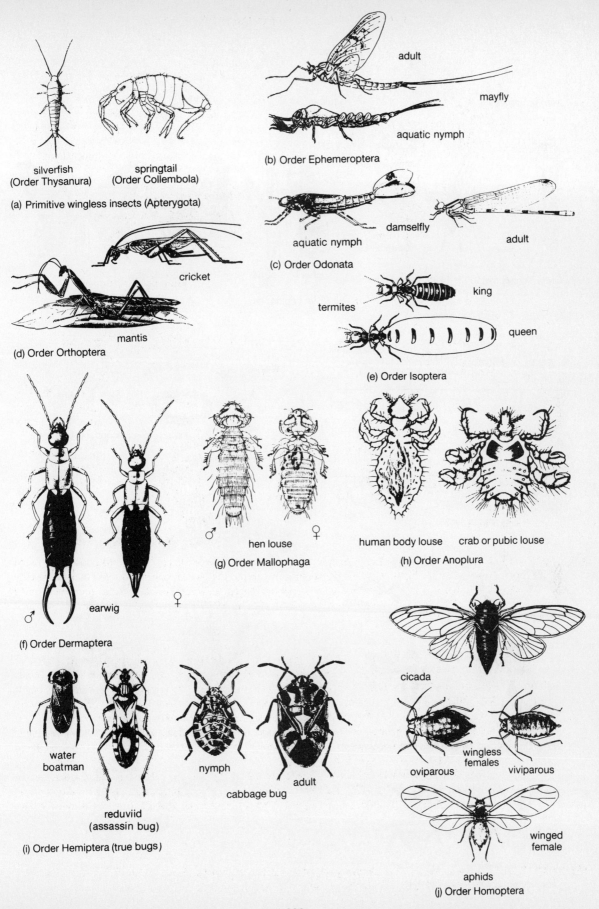

silverfish
(Order Thysanura)

springtail
(Order Collembola)

(a) Primitive wingless insects (Apterygota)

adult

mayfly

aquatic nymph

(b) Order Ephemeroptera

damselfly

aquatic nymph

adult

(c) Order Odonata

cricket

mantis

(d) Order Orthoptera

termites

king

queen

(e) Order Isoptera

♂

earwig

♀

(f) Order Dermaptera

hen louse

♂   ♀

(g) Order Mallophaga

human body louse   crab or pubic louse

(h) Order Anoplura

cicada

water
boatman

nymph

adult

reduviid
(assassin bug)

cabbage bug

(i) Order Hemiptera (true bugs)

wingless
females

oviparous   viviparous

winged
female

aphids
(j) Order Homoptera

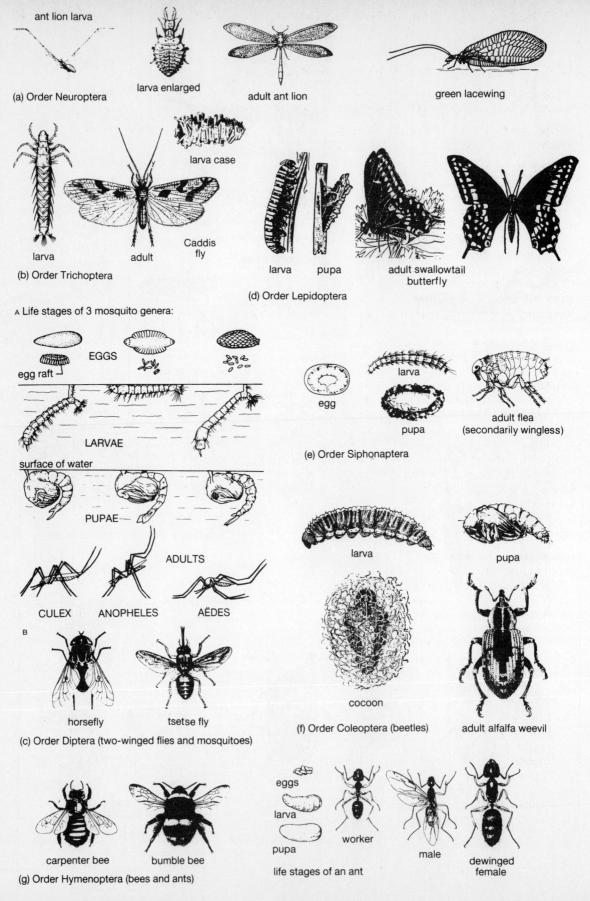

ant lion larva

(a) Order Neuroptera

larva enlarged

adult ant lion

green lacewing

larva case

larva

adult

Caddis fly

(b) Order Trichoptera

larva   pupa

adult swallowtail butterfly

(d) Order Lepidoptera

A Life stages of 3 mosquito genera:

egg raft

EGGS

LARVAE

surface of water

PUPAE

ADULTS

CULEX   ANOPHELES   AËDES

B

horsefly   tsetse fly

(c) Order Diptera (two-winged flies and mosquitoes)

egg

larva

pupa

adult flea (secondarily wingless)

(e) Order Siphonaptera

larva   pupa

cocoon

(f) Order Coleoptera (beetles)   adult alfalfa weevil

carpenter bee   bumble bee

(g) Order Hymenoptera (bees and ants)

eggs

larva

pupa

worker

male

dewinged female

life stages of an ant

234

16.43. What are some important groups of insects?

Insects fall into 2 subclasses: *Apterygota* ("wingless"), with only 2 orders, and *Pterygota* ("winged"), with 24 orders. Of these 26 orders of living insects, only some of the most familiar will be included here:

Subclass Apterygota: silverfish (order Thysanura) and springtails (order Collembola); primitively wingless; undergo *no* metamorphosis (Fig. 16.12*a*).

Subclass Pterygota: with wings, or secondarily wingless; undergo metamorphosis; includes nearly all insects.

Superorder Exopterygota ("outside-winged"): metamorphosis gradual or incomplete, with no pupal stage; wings develop externally from small pads present on the body of the earliest larval (nymph or naiad) stage (Fig. 16.12*b–j*).

Order Ephemeroptera ("ephemeral wing"): mayflies; larvae aquatic, with tracheal gills; short-lived adults cannot feed. The four long, net-veined wings are held vertically at rest.

Order Odonata ("toothed"): dragonflies, which hold their four slender, membranous wings straight out to the sides when at rest, and damselflies, which fold their wings vertically over their backs; predatory as adults on flying insects; as aquatic larvae, prey on other aquatic animals.

Order Orthoptera ("straight wing"): grasshoppers, katydids, mantises, walking sticks, crickets, cockroaches; with chewing mouthparts and hind legs often adapted for jumping. Winged species have straight, leathery forewings beneath which the membranous hindwings are folded up like fans. Many *stridulate*, producing sounds by scraping together portions of the exoskeleton.

Order Isoptera ("equal wing"): termites; social, with sterile worker and soldier castes; reproductives are briefly winged, with two pairs of long, net-veined wings of equal length that beat independently of each other and are shed after a short dispersal flight.

Order Dermaptera ("skin wing"): earwigs; very short, horny forewings under which the long membranous hindwings are folded; a pair of abdominal cerci used as tongs in apprehending prey.

Order Mallophaga ("wool eater"): biting lice; parasitic on birds and mammals; wingless; feed mostly on feathers, hair, and skin.

Order Anoplura ("unarmed tail"): sucking lice, e.g., head louse and crab louse of humans; flattened; wingless; parasitic on birds and mammals, with mouthparts adapted for piercing and sucking blood.

Order Hemiptera ("half wing"): true "bugs"; piercing-sucking beak that folds under the body; large, membranous hindwings covered at rest by forewings with leathery basal and membranous apical portions, the latter folded under the former when at rest.

Order Homoptera ("same wing"): leafhoppers, aphids, cicadas; mouthparts adapted for piercing and sap-sucking; wingless, or with membranous wings held rooflike over the abdomen; aphids usually parthenogenetic.

Superorder Endopterygota ("inside-winged"): larvae lack compound eyes; undergo complete metamorphosis with a pupal stage; wings develop internally (Fig. 16.13).

Order Neuroptera ("nerve wing"): lacewings and dobsonflies; predatory; long, membranous wings; dobsonflies have aquatic larvae; lacewings often have larvae known as ant lions that snare ants in funnel-like excavations in sand.

Order Trichoptera ("bristle wing"): caddis flies; mothlike adults and aquatic larvae that often live in movable cases covered by affixed bits of stone and plant matter.

Order Lepidoptera ("scale wing"): butterflies and moths; wings, bodies, and appendages coated with overlapping, pigmented scales; adult mouthparts form a coiled proboscis used in probing for nectar; larval mouthparts adapted for chewing, usually foliage; butterflies fold wings vertically over back and have knobbed antennae; moths fold wings horizontally and have pinnate antennae and larvae with silk glands for spinning cocoons.

Order Diptera ("two wing"): two-winged flies, e.g., houseflies, gnats, and mosquitos; forewings membranous; hindwings reduced to halteres; mouthparts are variable, adapted for lapping, chewing, piercing, sucking.

Order Siphonaptera ("siphon, wingless"): fleas; secondarily wingless; laterally compressed bodies; legs adapted for leaping; mouthparts adapted for piercing and bloodsucking; parasitic on birds and mammals.

Order Coleoptera ("shield wing"): 300,000 species of beetles; hard-bodied; heavy, horny forewings (elytra), unmoving in flight, cover the membranous hindwings at rest; mouthparts adapted for chewing, mostly on vegetation, although some (e.g., ladybirds and fireflies) are predaceous.

Order Hymenoptera ("membrane wing"): bees, wasps, ants; clear, membranous wings having few veins; chewing mouthparts; many with ovipositor modified into stinger; some wingless; most are social, with sterile worker caste.

Fig. 16.13 Representative endopterygote (holometabolous) insects. All of these undergo complete metamorphosis and pass through a pupal stage. (*From Storer et al.*)

Chapter 17

Echinoderms and Lesser Deuterostomes

Our survey of eumetazoans up to now has dealt with phyla in which the mouth is the first, and perhaps the only, major bodily orifice, opening in the location of the embryonic blastopore. All such phyla may be referred to as "protostomes," although this term as used often excludes the Radiata, and even Bilateria below eucoelomate status. The phyla we have yet to examine are those called "deuterostomes," so named because the mouth is the second orifice to appear in the embryo, and it is the anus that develops in the region of the blastopore. All known deuterostomes are eucoelomate, which suggests that they did not diverge from the original mainstream of eumetazoan evolution until a true coelom had begun to appear. An admixture of protostome and deuterostome developmental traits characterizes the lophophorates, suggesting that the branch point may have occurred during lophophorate evolution. Nevertheless, this remains obscure, and the profound differences seen today between developmental processes in the two groups provide little guidance in even hypothesizing the causal factors operative in this great divergence.

17.1. How do deuterostomes and protostomes differ?

The major differences are outlined in Table 17.1.

Table 17.1 Some Major Differences between Deuterostomes and Protostomes

Feature	Deuterostomes	Protostomes
Cleavage	Radial (see Fig. 8.1)	Spiral (typically)
Developmental regulation	Indeterminate: fate of blastomeres is not fixed irreversibly during first few cleavages; thus, monozygotic twinning can occur	Determinate, with mosaic cleavage: fate of blastomeres is irreversibly determined from the onset of cleavage, which rules out monozygotic twinning
Blastopore	Site of anus: marks *posterior* end of longitudinal body axis at gastrula stage; mouth opens later at opposite end of this axis	Site of mouth: marks *anterior* pole of longitudinal body axis
Mesoderm	Develops from wall of primitive gut or the archenteron and thus is derived from endoderm	Develops from single blastomere distinguishable in early cleavage, and not from the endoderm of the archenteron
Body cavity	True coelom always present, its development being characteristically *enterocoelous*, i.e., it first appears as hollow evaginations from the archenteron wall	If present, may be either a pseudocoel or a true coelom with *schizocoelous* development, i.e., it arises as a cleft in a formerly solid block of mesodermal tissue
High-energy storage compound used in muscle contractions	Creatine phosphate	Arginine phosphate
Major longitudinal nerve cord	If present, lies *dorsal* to the gut	If present, lies *ventral* to the gut
True heart	If present, usually lies *ventral* to the gut	If present, lies *dorsal* to the gut

17.2. What are echinoderms?

From the familiar starfishes that have delighted us since our childhood tidepool-exploring days, to the long-spined sea urchins that menacingly wave their toxic armament at skin divers, to leathery sea cucumbers relished in oriental cuisine as trepang, the echinoderms are a strange lot (Fig. 17.1). Because of their slow-moving habits, conspicuous size, and relative abundance in shallow inshore waters, most of us become thoroughly acquainted with these exclusively marine animals without ever realizing what misfits they are in the animal kingdom. Although developmentally they show indubitable relatedness to other deuterostome phyla, their larval stages are unique, giving little clue to the origin of the phylum Echinodermata ("hedgehog skin"). Furthermore, they appear to have sailed blithely off on their own evolutionary tangent,

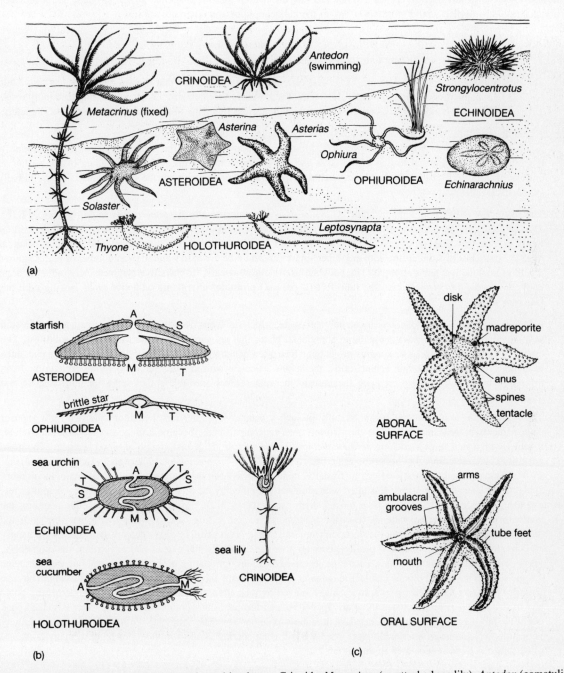

Fig. 17.1 Echinoderms. *(a)* Representative echinoderms. Crinoids: *Metacrinus* (an attached sea lily), *Antedon* (comatulid or feather star). Echinoids: *Strongylocentrotus* (sea urchin), *Echinarachnius* (sand dollar). Holothuroids: *Thyone* and *Leptosynapta* (sea cucumbers). Asteroids: *Solaster*, *Asterina*, and *Asterias* (starfishes). Ophiuroids: *Ophiura*, a brittle star. *(b)* Diagrammatic sections of the five major types of living echinoderms: M = mouth, A = anus, S = spines; T = tube feet. *(c)* External features of a starfish. *(From Storer et al.)*

so that their postembryonic stages and adult features represent an assemblage of characteristics shared not even with other deuterostomes. Those characteristics that lend themselves to fossilization display this uniqueness as far back as the lower (early) Cambrian, when the first echinoderm skeletons appear in the sedimentary record. This rich fossil record includes around 20,000 extinct species (compared with about 6000 living ones) and some 14 extinct classes (with 4 or 5 additional classes persisting to the present day). Even the oldest known echinoderm skeletal remains lack evidence of forms transitional with any other known phylum, except for a single recently discovered fossil species that appears to have had a series of pharyngeal gill slits, which would represent a link to the hemichordates and chordates.

17.3.   What characteristics distinguish the echinoderms?

(*a*)   The bilaterally symmetrical larvae are of several different and unique types that remain of disputed affinity, but come in two major versions: *pluteus* and *auricularia*. These undergo a clockwise 90° rotation during metamorphosis, so that the former left side becomes the oral aspect and the former right side becomes the aboral aspect of the adult.

(*b*)   The adults therefore exhibit secondary radial symmetry, adapted to their sessile or slowly creeping benthonic existence; in those few that show some degree of bilateral symmetry as adults (notably the heart urchins, or sea porcupines), this bilateralism is unrelated to that of the larva and derives from an anteroposterior lengthening of the postlarval oral-aboral axis, so that the new dorsal and ventral aspects are homologous with the *lateral* aspects of the radially symmetrical postlarval echinoderms.

(*c*)   The secondary radial symmetry of adult echinoderms is characteristically *pentamerous* (five-partite; Fig. 17.1*c*). This derives from the suturing of the skeletal ossicles to form a solid ovoid *test (theca)*, as seen in even the earliest fossil classes that display any symmetry at all: when the skeletal plates represent an odd number, such as five, no suture lies in a direct line with any other suture (as would be the case were an even number of plates sutured together), and the test gains strength. The body surface therefore usually displays five symmetrically radiating areas known as *ambulacra*, from which the tube feet (if present) protrude, alternating with five areas lacking tube feet (*interambulacra*).

(*d*)   Echinoderms have an *endoskeleton* of calcium carbonate, which develops from the *mesoderm* and underlies the skin as a lattice of separate *ossicles*, or as large ambulacral plates that are sutured together to produce a solid test. The ossicles or plates bear knobs or spinous projections that often penetrate the skin, justifying the phylum name since hedgehogs are spiny. In certain echinoderms, the spines articulate with knobs on the underlying skeleton and can be moved by muscles that insert upon the base of the spines. Such movable spines serve for both defense and locomotion.

(*e*)   The *tripartite coelom* develops asymmetrically in such a manner that after larval rotation, the two right anterior compartments become reduced while the two left anterior compartments become a unique *water-vascular system*, together with a system of branching *perihemal channels* that enclose the hemal system of blood lacunae. (The more posterior coelomic compartments become the body cavity proper, or *somatocoel*.) The water-vascular system (Fig. 17.2), originally used only for food-getting by early, nonmotile fossil echinoderms, evolved into the major locomotory apparatus of later, mobile forms. It is filled with seawater (modified by the presence of some proteins, coelomocytes, and a high concentration of potassium ions), which can be taken in or expelled through a sievelike opening, the *madreporite*. A *stone canal* extends from the madreporite to join a *ring canal* ("water ring") that surrounds the mouth. Excess fluid for the system is stored in from one to five *polian vesicles* that open off the water ring in nearly all echinoderms. The water ring typically gives off five *radial canals* that extend aborally within the ambulacra. From each radial canal, short *lateral canals* project in alternation, terminating in structures resembling tiny medicine droppers, each having a bulbous contractile *ampulla* at the inner end and a closed-ended *tube foot* distally. The tube feet, or *podia*, are extended by contraction and retracted by relaxation of ampullae. In most echinoderms the podia end in suction disks, so that when these are applied to a surface and the ampullae then relax, water is sucked back out of the tube feet and their disks adhere like miniature "plumber's friends," allowing the animals to cling indefinitely with little expenditure of energy. In fact, when we pull echinoderms loose from their attachments, many podia are often torn off, to be regenerated in due time.

(*f*)   *Pedicellariae*—present in starfish and sea urchins, but lacking in crinoids, sea cucumbers, and ophiuroids—are unique pinching devices. They consist of two or more claws that can be separated or apposed by muscular action and serve mainly to protect the echinoderm against predators and also against being settled upon by larvae of sponges and other encrusting organisms that eventually could smother the echinoderm. Pedicellariae are borne on long or short muscular stalks and are sometimes venomous (see Fig. 17.4*d*).

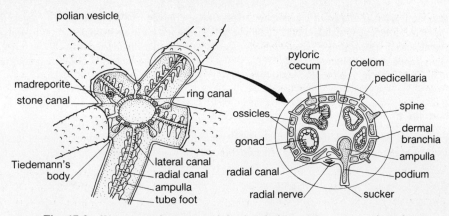

Fig. 17.2   Water vascular system: *left,* aboral view; *right,* cross section of arm.

17.4.   What are additional characteristics of echinoderms in general?

(*a*)   Cephalization is lacking; echinoderms display no tendency for nerve tissue and sense organs to be concentrated in a head. There is no true brain, but *ring nerves,* usually three, surround the mouth. *Radial nerves* (usually five) extend from the ring nerves and run parallel to the water-vascular system; they give off networks that serve the body at different levels. Few special sense organs occur, and most reception is by generalized sensory cells in the epidermis.

(*b*)   Excretory organs are lacking, but coelomic amoebocytes may concentrate wastes and remove them by simply migrating out of the echinoderm's body.

(*c*)   The *blood-vascular (hemal)* system is poorly developed, except in sea cucumbers, and consists as a rule of small sinuses surrounded by coelomic extensions (*perihemal channels*). The *hemal sinuses* form oral and aboral rings (and in starfish, a third, gastric ring), from which radial sinuses extend along each ambulacral area, paralleling the radial canals of the water-vascular system. The channel interconnecting the ring sinuses passes through a spongy *axial gland* of uncertain function, at the aboral end of which is a *dorsal sac* that is thought to pulsate, bringing about some feeble degree of circulation. Amoeboid coelomocytes occur in the blood as well as in the coelomic fluid and may serve in food distribution and waste concentration. Only sea cucumbers possess hemoglobin-containing coelomocytes employed in gas transport.

(*d*)   The coelom is large, with a ciliated peritoneum that circulates the coelomic fluid, which contains coelomocytes. *Tiedemann's bodies,* folded pouches in the wall of the water ring, are thought to produce the coelomocytes.

(*e*)   The digestive tract is complete and regionalized into esophagus, stomach, and intestine. These regions vary proportionally, and distinct digestive glands may or may not be present. Ophiuroids lack intestine and anus.

(*f*)   Gas exchange takes place through the epidermis of the body wall and podia. The epidermis is invaginated as the lining of *bursal sacs* in ophiuroids and is evaginated as minute *papulae* or *dermal branchiae* on the body surface of starfish and sea urchins. Each thin-walled papula covers an extension of the coelom, so that gases are readily exchanged between the surrounding water and the coelomic fluid. In many sea cucumbers, diffusion of gases through the integument is secondary to gaseous exchange across the membranes of internal *respiratory trees* that open off the cloaca and are filled and emptied by the pumping action of the cloaca.

(*g*)   Except for a few hermaphroditic species, sexes are separate and look alike. The *gonads,* multiple in all but sea cucumbers, are large, with simple ducts, and gametes are ordinarily shed into the water, where external fertilization occurs.

(*h*)   *Embryonic development* can be direct or indirect. Direct development without production of a planktonic larva takes place only in echinoderms that *brood* their eggs. In the great majority, the zygotes develop into free-swimming, transparent planktonic larvae, which may be either planktotrophic or lecithotrophic. In about 50 percent of echinoderm species, the eggs develop quickly into gastrulae that differentiate into a variety of larval forms; the larvae locomote by means of a number of circumferential bands of cilia, have a complete digestive tract consisting of stomodeum, esophagus, stomach, intestine, and anus, and are *planktotrophic,* utilizing the large, ciliated stomodeum for food collection. A variety of projections from the body wall may develop and later disappear, serving as larval adaptations for flotation and having no role in postlarval life. Most of the remaining species of nonbrooding echinoderms produce planktonic larvae that are nourished only by the egg yolk (i.e., are *lecithotrophic*) rather than by holozoic feeding.

(*i*) Echinoderms are masters of the art of regeneration. Ophiuroids, crinoids, and starfishes readily regenerate lost arms and even portions of their central disk. In fact, a number of these defensively autotomize their arms with consummate abandon, while certain sea cucumbers, secure in their capacity to regenerate organs, explosively eviscerate themselves, expelling gut, gonad, and respiratory organs. The coelomocytes play important roles in regeneration, phagocytizing necrotic tissue and bringing food to the multiplying cells.

17.5. What are the major subdivisions of phylum Echinodermata?

Echinoderms should be viewed with an eye to the fossil record, because so many entire classes of this phylum have vanished into extinction except for their petrified remains (Fig. 17.3).

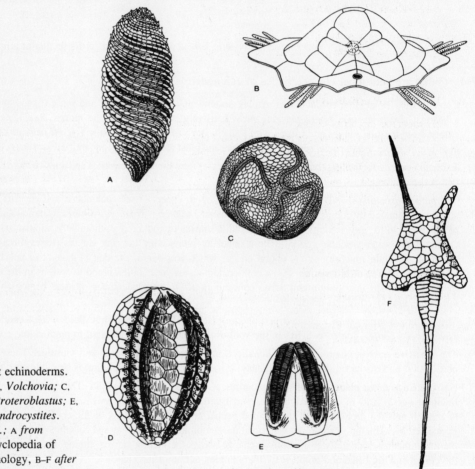

Fig. 17.3 Extinct echinoderms. A, *Helicoplacus;* B, *Volchovia;* C, *Lepidodiscus;* D, *Proteroblastus;* E, *Pentremites;* F, *Dendrocystites. (From Storer et al.;* A *from* McGraw-Hill Encyclopedia of Science and Technology, B–F *after Hyman.)*

Subphylum Homalozoa ("level animal"): extinct *carpoids* with asymmetrical, flattened bodies encased in fused test; Cambrian to Devonian
    Class Homostela: stalked, spoon-shaped test; no feeding arm
    Class Homoiostela: stalked, asymmetrical test with one feeding arm (brachiole), shown extending from upper left of *Dendrocystites* in Fig. 17.3F.
    Class Stylophora: asymmetrical test with feeding arm but no stalk

Subphylum Crinozoa ("lily animal"): body in a cup-shaped or globose theca (test); usually attached, often by an aboral stalk, with mouth upward and surrounded by arms or slender *brachioles*
    Class Eocrinoidea: extinct; oldest known echinoderms
    Class Rhombifera: extinct; pores perforating theca restricted to particular areas

Class Diploporita: extinct; pores perforating theca dispersed over entire surface
Class Blastoidea: extinct; theca with 5 ambulacra fringed by brachioles
Class Crinoidea: early Cambrian to present; 630 living species; most primitive living echinoderms
    Subclass Articulata: only subclass of Crinoidea (out of 4) having living representatives
Subphylum Echinozoa: lack arms and brachioles; modern species unattached
    Class Edrioasteroidea: extinct; 5 ambulacra bearing pores for tube feet; mouth upward
    Class Helicoplacoidea: extinct; body elongated with spiraling rows of ossicles
    Class Holothuroidea: Ordovician to present; 900 living species of sea cucumbers
    Class Echinoidea: Ordovician to present; 900 living species of sea urchins, sea porcupines, and sand dollars
    Class Ophiocistioidea: extinct; theca dome-shaped, 5-jawed mouth downward
Subphylum Asterozoa: 3600 species of starfishes and ophiuroids
    Class Stelleroidea: star-shaped, arms free and movable
        Subclass Asteroidea: starfishes
        Subclass Ophiuroidea: brittle stars, basket stars
        Subclass Somasteroidea: extinct Paleozoic sea stars

17.6.   What are the crinozoans?

Except for some species of class Crinoidea, all members of this subphylum are extinct. All the extinct classes were sessile, stalked or stalkless, attached to the bottom mouth upward, with the body enclosed in a complete test that extended orally to surround the mouth. The brachioles were pinnulelike, food-collecting projections of the body surface, found in most extinct classes of this subphylum but not to be confused with the heavier *arms* of crinoids. Since attachment precluded an anus directly opposite the mouth, the anus opened to one side through a pore in one of the interambulacral plates. Most of these extinct forms also possessed movable cover plates that could close protectively over the mouth and ambulacral food grooves. Only in crinoids did the theca become reduced to the form of a cup, with just a thin membrane covering the upper surface, on which both mouth and anus are located.

17.7.   What were the eocrinoids?

Class Eocrinoidea includes the oldest known echinoderms. Eocrinoids appeared in the lower Cambrian and became extinct in the Ordovician. They were stalked or stalkless, with a complete, pentamerous theca having five ambulacra and from five to many brachioles surrounding the mouth. Although they bore a superficial resemblance to crinoids, they were actually closer to the cystoids.

17.8.   What were the cystoids?

Cystoids (classes Rhombifera and Diploporita), which ranged from the Ordovician to the Permian, had complete, oval tests bearing five (less frequently, three) radially arranged ambulacra that extended aborally from the upwardly directed mouth (see *Proteroblastus*, Fig. 17.3D). Short brachioles either circled the mouth or extended down each side of the ambulacral grooves, their individual food grooves joining the latter. The theca of cystoids was perforated by pores that were either dispersed over the entire surface (*diplopores*) or restricted to particular areas (*rhombopores*); these pores seemed to have nothing to do with podia and instead probably allowed coelomic fluid and seawater to be exchanged for respiration.

17.9.   What were the blastoids?

Class Blastoidea was abundant from the Ordovician into the Permian. The ovoid theca, anchored aborally directly or by way of a short stalk, consisted of 13 plates arranged regularly into five elevated ambulacra and five depressed interambulacra, giving the whole blastoid somewhat the appearance of a five-rayed starfish bent backward with mouth uppermost and arms held captive by the interambulacral plates (see *Pentremites*, Fig. 17.3E). A row of densely placed brachioles completely edged the ambulacra, outlining the star shape. Unique folds and pores (*hydrospires*) paralleled the edge of each ambulacrum, probably allowing water to circulate inward between the thecal plates, abetting gas exchange.

17.10.   What are the crinoids?

Class Crinoidea is represented today by only about 630 species of *sea lilies* and *feather stars,* a pitiful remnant of the more than 5000 Paleozoic species (see *Metacrinus* and *Antedon,* Fig. 17.1a). Crinoids are an ancient group, dating from the lower Cambrian, and were once so abundant that many limestone beds throughout the world consist mainly of their skeletons.

Sea lilies are attached by a flexible stalk to the bottom, from 100 m to abyssal depths; only 80 species remain today. Some fossil sea lilies had stalks 25 m long! The stalks of modern species seldom exceed 1 m. The stalk is made up of a beadlike series of discoidal ossicles that permit limited bending and may bear whorls of short cirri.

Feather stars (*comatulids*) are not sessile, but swim by undulations of their arms, finding temporary anchorage, mainly among reef corals, by means of short tendril-like cirri that project aborally. Some of the approximately 550 comatulid species occur intertidally, while others inhabit the deeps.

17.11.   How is the crinoid body organized?

Aside from the stalk, when present, the crinoid body consists of the following:

(a)   A pentamerous "crown" is enclosed in a cup-shaped calyx with both mouth and anus located on the oral (upper) surface (see Fig. 17.1b). Although some crinoids possess more than 50 arms, typically 5 ambulacral grooves extend radially from the central mouth onto the oral surface of 5 main arms, which fork at once to give the appearance of 10 arms, all of which are *pinnate* (i.e., a row of pinnules is borne on each side of the ambulacral groove, so that each arm resembles a feather). As in the stalk, arms and pinnules are supported by articulated series of beadlike ossicles that permit flexion and extension. The ambulacral grooves are lined with ciliated columnar epithelium, the ciliary beat being directed toward the mouth. The grooves are bordered by rows of tentaclelike podia lacking suction cups but bearing mucus-secreting papillae. When small planktonic organisms contact the extended podia, they are trapped by the mucus and flipped into the ambulacral groove by a whiplike flick of the podia.

(b)   The crinoid water-vascular system is rather peculiar. A madreporite is lacking, but up to 1500 minute ciliated funnels perforate the membrane that covers the oral surface of the crown and probably allow hydrostatic pressure adjustments to be made. The water ring gives off not only radial canals, which branch into all forks and pinnules, but also a cluster of short stone canals at each interambulacrum which open into the coelom. Ampullae are absent, and extension of the tube feet is effected by muscular contractions of the ring canal itself.

(c)   Because crinoids are the most primitive living echinoderms, the organization of their nervous system may reflect the ancestral pattern for the phylum. It consists of three interconnected divisions: (1) the *aboral,* or *entoneural, motor system,* which innervates the stalk and sends five brachial motor nerves to the muscles of the arms and pinnules; (2) the *oral,* or *ectoneural, system,* which is sensory and consists of a circumoral nerve ring that connects with subepidermal radial nerves running just beneath the ambulacral groove in each arm; (3) a *hyponeural sensory system,* with a nerve ring that lies deeper than the oral ring and receives a pair of lateral sensory brachial nerves from the pinnules and podia of each arm.

(d)   Crinoids lack definite gonads, and gametes mature from the peritoneum in coelomic extensions (*genital canals*) within the pinnules. Mature gametes are shed by rupture of the pinnule walls. The eggs may be released into the water or glued to the exterior of the pinnule until hatching, or (in cold-water species) may be retained in brood chambers formed by saccular invaginations of the walls of arms or pinnules. The larva is a nonfeeding *vitellaria,* similar to larvae of sea cucumbers, having an apical tuft of sensory cilia and five girdles of locomotory cilia; this settles and attaches before undergoing metamorphosis. It is interesting that in the case of the single comatulid species for which development has been well studied, the feather star larva first changes into an attached, stalked form resembling a tiny sea lily and spends several months in this *pentacrinoid* state before the cirri develop and the crown breaks off and swims away, leaving the stalk behind.

17.12.   What are the characteristics of echinozoans?

Subphylum Echinozoa ("hedgehog animal") includes echinoderms that lack arms and brachioles and, except for certain extinct species, are unattached.

17.13.   What were the edrioasteroids?

Class Edrioasteroidea, which existed from the lower Cambrian into the Pennsylvanian, included echinozoan species having oval tests and aboral attachment stalks and species that had discoidal, flattened tests and rested on the bottom unattached, mouth upward—an orientation opposite that of modern sea urchins. Five straight or sinuous ambulacra extended aborally over the thecal surface, giving the edrioasteroid the appearance of a starfish or ophiuroid doing a backbend over a ball (see *Lepidodiscus*, Fig. 17.3c). Pores in the ambulacra permitted extension of tube feet. Movable plates could protectively cover the podia and ambulacral food grooves. Tubercles on the thecal plates suggest that some species had articulating spines.

17.14.   What were the helicoplacoids?

Class Helicoplacoidea was discovered in lower Cambrian strata around 1963. The helicoplacoid body was elongately ovoid, tapering at each end like a sweet potato (see *Helicoplacus*, Fig. 17.3a). The mouth was at one end, with the anus opposite. The body was enclosed in spiraling series of ossicles that allowed the animal to lengthen or contract by twisting the series together or apart. A single ambulacrum branched between the rows of ossicles. Although the ancestry of sea cucumbers remains debatable, the helicoplacoids seem possible candidates, for they show a cucumberish elongation of the oral-aboral axis and their reduction of thecal plates to spiraling series of ossicles could have been an early step in the progression by which holothuroidean ossicles have become reduced to microscopic dimensions, leaving their body wall leathery and flexible.

17.15.   What are sea cucumbers?

Class Holothuroidea ("sea cucumber form") seems to have originated in the Ordovician Period, but the extreme reduction of skeletal parts in this class of echinozoans has resulted in a scanty fossil record. Today, there are around 900 kinds of sea cucumbers, some of which are exceedingly abundant (see Fig. 17.1a and b). Species dredged for food yield around 10,000 tons per annum of dried body wall, marketed as trepang in the orient. The holothuroid body is elongated along the oral-aboral axis to resemble a cucumber or even a worm or snake.

17.16.   What are the organizational parts of a sea cucumber?

(a)   The body wall is soft or leathery, containing scattered microscopic ossicles of distinctive shapes, which are useful taxonomically. The animal rests on one side, which becomes its new ventral surface, on which are three ambulacra with well-developed tube feet, while the two ambulacra on the new dorsal aspect typically bear reduced podia. In some holothuroids the podia have become randomly scattered over the body surface and may be much reduced. In fact, the *apodid* ("footless") cucumbers have no tube feet except for those ringing the mouth.

The mouth, located at or near the new anterior end, is surrounded by 10–30 large, branching *buccal podia*, which are in fact greatly enlarged, modified tube feet adapted for getting food (Fig. 17.4a). Holothuroids are suspension or deposit feeders; plankton or particulate matter adheres to mucus on the extended buccal podia, which are then folded inward and stuffed into the mouth, one at a time, the food being sucked off as the podium is pulled out again, rather like a child sucking candy off its fingers. The crown of buccal podia can be entirely pulled in by the action of retractor muscles running lengthwise in the coelom. The madreporite of the water vascular system of holothuroids is unusual in that it lies free in the coelom rather than opening through the body surface.

(b)   The holothuroidean nervous system consists mainly of a circumoral nerve ring and five radial nerves extending the length of the ambulacra. In other echinoderms, the nerve ring dominates the rest of the nervous system in a perfunctorily brainlike fashion, but its removal from a sea cucumber does not prevent the animal from breathing, moving about, righting itself, and responding to light. Sensory reception depends mainly on scattered epidermal cells, although the burrowing apodids also have a statocyst near the base of each radial nerve, and a few possess eyespots in the form of clusters of photoreceptive pigmented cells at the bases of the buccal podia. The defensive behavior of molested sea cucumbers is initially to withdraw the crown of buccal podia and contract the entire body, shooting a stream of water from the anus; upon more severe disturbance, they often respond dramatically, by massive evisceration, expelling the respiratory organs, gonad, and even digestive tract, which of course must later be regenerated. Two major genera (*Holothuria* and *Actinopyga*) may defer such extreme measures by first expelling from the anus masses of sticky *Cuvierian tubules* that develop from the base of the respiratory organs; as well as entangling the molester, the Cuvierian tubules may also liberate toxic *holothurin,* which can render a good many innocent bystanders, such as fishes, utterly hors de combat.

(c)   Most holothuroids (with the notable exception of apodids, which breathe through their skins) have a pair of large, finely branched *respiratory trees* that lie in the coelom and open into the cloaca. A series of muscular contractions of the cloaca pump water into the respiratory trees, after which the cloacal suspensor muscles dilate the cloaca, causing the water to be sucked back out of the respiratory trees. The coelomic fluid contains coelomocytes called

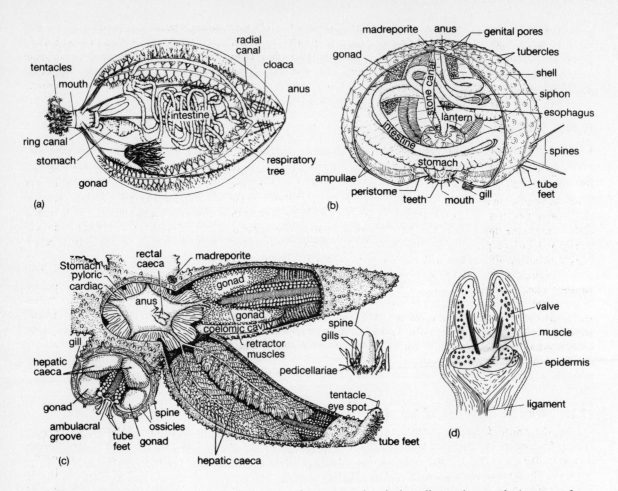

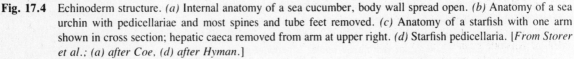

Fig. 17.4 Echinoderm structure. *(a)* Internal anatomy of a sea cucumber, body wall spread open. *(b)* Anatomy of a sea urchin with pedicellariae and most spines and tube feet removed. *(c)* Anatomy of a starfish with one arm shown in cross section; hepatic caeca removed from arm at upper right. *(d)* Starfish pedicellaria. [*From Storer et al.; (a) after Coe, (d) after Hyman.*]

hemocytes, which are filled with hemoglobin and take up oxygen that diffuses through the thin membrane of the respiratory trees.

In addition to the standard echinoderm hemal ring and radial hemal sinuses, certain holothuroids have remarkably well developed blood-vascular systems, including major dorsal and ventral blood vessels that run the length of the intestine. The dorsal vessel itself contracts peristaltically, maintaining a slow circulation, but in addition as many as 150 tubular hearts branch off the dorsal vessel and pump the blood through a series of lamellae that project inward from the intestinal wall, permitting absorbed nutrients to enter the blood, which then drains into the ventral vessel. A network of small blood vessels surrounds the left respiratory tree and assists in gas transport.

Dissolved wastes such as ammonia diffuse into the respiratory trees and are excreted with the exhalant water flow. Particulate wastes are ingested by coelomocytes, transported into the cavities of the respiratory trees, intestine, and gonadal tubules, and then removed from the body. Lacking respiratory trees, the apodids instead possess unique ciliated funnels through which waste-laden coelomocytes leave the body cavity and penetrate the body wall.

(d) As a rule holothuroids are dioecious and are unique among living echinoderms in having only one gonad; this is thought to be the primitive condition for the phylum. The gonad is made up of a large moplike cluster of tubules that join at a single gonoduct, which leads anteriorly to a middorsal gonopore located within or near the crown of buccal podia. Although about 30 species of sea cucumbers brood their young on the body surface or in chambers invaginated from the body surface, or even within the coelom itself, in the remaining approximately 870 species, the gametes are simply shed into the sea for external fertilization and development into planktonic larvae.

The *auricularia* larval stage, attained on the third day of development, strongly resembles the bipinnaria larva of starfishes and may constitute a link with this group, which actually belongs to another subphylum. The auricularia develops into a barrel-shaped *doliolaria* larva encircled by three to five ciliary girdles, and gradual metamorphic changes progressively convert this larva into the adult form. Certain holothuroids pass through a nonfeeding, barrel-shaped vitellaria larval stage much resembling that of crinoids and a few ophiuroids, constituting another possible evolutionary link among echinoderms of different subphyla.

17.17.   What are the echinoids?

Class Echinoidea first appears in Ordovician deposits and includes the spiniest of the spiny: some 7200 fossil and 900 living species of *sea urchins,* sea porcupines (heart urchins), and *sand dollars,* which have movable spines that articulate by concave bases with tubercles on the solidly sutured test (see Fig. 17.1*a* and *b*). Pores through the test allow extension of tube feet, which sea urchins use for creeping, along with pushing actions of the spines, but which in sand dollars and sea porcupines are ineffectual in locomotion but useful in respiration and food collection. All echinoids possess three-jawed, stalked pedicellariae, which can be extended and directed toward certain stimuli. Some pedicellariae are poisonous, giving off a venom that paralyzes small animals and repels larger ones.

The long digestive tract (Fig. 17.4*b*) commences with a buccal cavity and pharynx surrounded by a masticatory Aristotle's lantern, from which a slender esophagus leads into a small intestine; this loops once completely around the inside of the test, then leads into the large intestine, which performs a complete loop in the opposite direction before leading aborally to a rectum that opens by way of the anus. A slender ciliated tube, the *siphon,* which parallels the small intestine and opens into it at each end, appears to serve as a bypass for diverting excess water from the material in the small intestine. The echinoid hemal and nervous systems conform to characteristic echinoderm patterns.

There are both regular and irregular echinoids.

17.18.   What are regular echinoids?

Regular echinoids are sea urchins having globular tests with five interambulacra and five ambulacra extending all the way from the oral to the aboral pole. Through the pores of the ambulacra project rows of podia that can reach beyond the spines. The spines vary greatly in length among different species and range from club-shaped to exceedingly needlelike and toxic. The mouth is surrounded by a *peristomial membrane* shaped into a circular lip, five pairs of buccal podia, and five pairs of bushy gills. The gills are hollow extensions of the body wall, filled with coelomic fluid. The coelomic fluid is pumped in and out of the gills by the muscles of specialized ossicles that form the inner part of a buccal apparatus unique to echinoids and known for its shape as *Aristotle's lantern.* The urchin's mouth can be dilated to allow the five teeth of Aristotle's lantern to be protruded. These teeth form the oral ends of five calcareous plates (pyramids), which are manipulated by muscles that protract and retract the entire lantern and open and close the teeth. The teeth grow constantly, counteracting the erosive toll of masticating corals, kelps, and marine grasses and scraping encrusting algae, bryozoans, and the like off rocks.

> **Example 1:**   Along the coast of southern California, population explosions of sea urchins—quite possibly nourished by organic particles in the millions of gallons of daily sewage outfall poured into the sea in the most heavily ravaged areas—have destroyed vast tracts of giant kelp by gnawing through the plants at their bases, thus devastating the "kelp forest" community of fishes and invertebrates. In earlier times and without the sewage factor, sea urchins were kept in check by sea otters, to which they represent an especially favored delicacy.

17.19.   What are irregular echinoids?

The *irregular echinoids* are mostly adapted for burrowing in sand and have either flattened, discoidal tests (sand dollars) or ovoid tests with swept-back spines; echinoderms of the latter type (heart urchins) exhibit secondary bilateral symmetry, having the mouth at one end and anus at the other. The ambulacra of the upper (or aboral) surface resemble five flower petals and so are called *petalloids*. The podia of these petalloids are specialized for respiration, peristomial gills being absent. The podia of the ventral or oral ambulacra are modified for capturing food in the form of tiny organic particles and transporting these along food grooves to the mouth. Aristotle's lantern is present, but not protrusible, in sand dollars and is lacking in heart urchins.

17.20. How do echinoids reproduce?

All echinoids are dioecious. Regular echinoids have five, and irregular ones four, *gonads;* these are attached along the inner surface of the ambulacra and have short gonoducts that open by separate gonopores located on the upper surface of the test. In regular urchins, the gonopores are located in the five genital plates that encircle the *periproct,* a membrane-covered aboral opening through the test that accommodates the anus. In certain human cultures the large, ripe gonads are eaten raw as a gustatory treat. At present some 8 million kg of sea urchins is being collected off the coast of southern California, and the frozen gonads shipped mainly to Japan. If the urchin escapes this fate, muscular contractions of its gonads expel great streams of gametes into the sea, where external fertilization takes place. A few sea urchins brood their eggs on the membranes surrounding mouth or anus, using adjacent spines to keep them in place. Brooding species of irregular echinoids have deeply concave petalloids in which the eggs are held. The eggs on nonbrooding species develop into *echinopluteus* larvae, which have six long larval arms that increase buoyancy and are lost in the metamorphosis to adult form.

17.21. What were the ophiocistioids?

Class Ophiocistioidea, found from the Ordovician to the Devonian, consisted of curious creatures with dome-shaped aboral and flattened oral aspects (see *Volchovia,* Fig. 17.3в). These fossil echinoderms were unattached, and as in sea urchins, the oral surface was oriented downward. This surface bore five jawlike plates flanking the centrally located mouth and six gigantic podia in each of the five ambulacra, which were restricted to the oral surface.

17.22. What are the asterozoans?

Subphylum Asterozoa ("star animal") outweighs all other echinoderms in terms of living species, with some 3600 starfishes and ophiuroids (brittle, or serpent, stars). By contrast, the fossil record of the group, which begins in the Cambrian, has yielded only about 480 extinct species, perhaps because the skeleton of separate ossicles readily disassembles after death. The asterozoan body is star-shaped and consists of a flattened central disk and five or more radiating arms. The two main groups—asteroids and ophiuroids—were until recently granted the status of separate classes (Asteroidea and Ophiuroidea), but currently they have been demoted, without change of name, to the rank of subclasses within a single class, *Stelleroidea* (the only class in the subphylum). Whether this loss of dignity is warranted remains debatable, and the answer awaits biochemical analysis of DNA concordance between these two groups, for despite their free-moving arms and stellate form, asteroids and ophiuroids (see Fig. 17.1*a* and *b*) differ in many important respects.

17.23. How do asteroids and ophiuroids differ?

Class Asteroidea includes the *starfishes* (Fig. 17.4*c*; see also Figs. 17.1*c* and 17.2). *Class Ophiuroidea* includes the brittle, or serpent, stars and the basket stars. Table 17.2 lists the major ways in which these two stellate echinoderms may be distinguished from each other.

17.24. How do asteroids and ophiuroids *resemble* each other?

(*a*) In both, the body consists of a disk and five or more arms.

(*b*) The nervous systems of both conform to the typical echinoderm pattern described earlier, consisting of a circumoral nerve ring and superficial and deep radial nerves extending into the arms.

(*c*) The hemal systems are poorly developed and essentially alike, with oral and aboral hemal rings and radial hemal sinuses extending into each arm, and the body spaces contain unpigmented coelomocytes that appear to serve in waste collection.

17.25. What are arrowworms?

Arrowworms make up around 65 species of planktonic marine animals that range from 2.5–10 cm in length, with straight, torpedo-shaped bodies having lateral and caudal fins that somewhat resemble the feathers of an arrow (Fig. 17.5*a*). They belong to *phylum Chaetognatha* ("hair jaw"), named for the 4–14 incurved, nonchitinous spines that flank the mouth.

Table 17.2   Major Differences between Asteroids and Ophiuroids

	Asteroidea	Ophiuroidea
1. Central disk	Not sharply set off from arms.	Flattened and sharply set off from arm bases.
2. Arms	Limited flexibility; comparatively wide (sometimes to such an extreme that the starfish resembles a cushion or hexagon); usually number 5, but certain species have 6 or more [sun stars (*Heliaster*) may sport more than 40].	Flexible; slender; nearly always number 5 (although these 5 branch repeatedly in basket stars).
3. Regeneration of parts	Central disk can regenerate lost arms; any arms bearing a fragment of the disk can regenerate an entire new starfish.	Lost arms readily regenerate from their bases; disk itself has limited regenerative capacity, although *fissiparity* (asexual reproduction by division in two across the disk) occurs in some species.
4. Autotomy	Arms rarely autotomized defensively, but a number of starfish exhibit fissiparity, and one genus (*Linckia*) freely sheds its arms, which although lacking any part of the disk, can each regenerate a complete starfish.	Arms autotomize readily at any point along their length but cannot regenerate entire new animals.
5. Ambulacral grooves	Open.	Closed, i.e., no open ambulacral grooves run along the arms' oral surface.
6. Lateral canals serving tube feet	Come off the radial canal in a staggered arrangement (as in all echinoderms except ophiuroids; see Fig. 17.2).	Come off the radial canals in pairs directly opposite one another, a unique characteristic among echinoderms.
7. Podia	Ordinarily terminate in suckers.	Suckerless and tentaclelike; extend from openings between the oral and lateral shields covering each arm; used in food gathering but play no significant role in locomotion or anchorage.
8. Locomotion	Effected by *podia:* extended by contractions of the ampullae, podia attach to the substratum; their subsequent shortening (by relaxation of the ampullae) pulls the animal forward (in contrast, sand-dwelling sea stars use their suckerless podia like tiny legs).	Effected by *arms:* a beadlike series of articulated ossicles (*vertebrae*) supporting the arms are moved by intervertebral muscles, resulting in serpentine and oarlike arm movements that pull the animal forward in rapid jerks, with the spines providing traction; ophiuroids are the swiftest of the echinoderms.
9. Internal organs	Extend into the arms [e.g., digestive glands (*pyloric ceca*) and gonads].	Do not extend into the arms, which are mainly occupied by the vertebral ossicles.
10. Jaws	Lacking.	Five serrate, triangular jaws, useful in macerating food, converge at the center of the oral surface, forming the floor of a prebuccal cavity that encloses the mouth.
11. Stomach	Pouched; typically can be protruded from the mouth during feeding.	Large, saccular; cannot be protruded from the mouth.
12. Food and feeding	Many prey on bivalves by attaching their podia and exerting a steady outward pull until the shell gapes enough (as little as 0.1 mm) to allow insertion of the starfish's stomach between the valves.	Some consume algae or carrion; most are filter and/or deposit feeders that trap minute plankton and/or detritus in mucus and use their podia to compact the collected particles into food balls, transporting these to the mouth.

(*Table continues on page 248*)

Table 17.2　Major Differences between Asteroids and Ophiuroids *(continued)*

	Asteroidea	Ophiuroidea
13. Digestion	Extracellular in the stomach, and probably intracellular in the pyloric ceca, where most absorption takes place.	Extracellular in the stomach, where absorption also takes place, for pyloric ceca are lacking.
14. Intestine and anus	Intestine is short and leads to the aboral anus.	Both absent; mouth serves for egestion of indigestible residues.
15. Madreporite	Forms a conspicuous *sieve plate* on the *aboral* surface of the disk.	A simple pore on the *oral* surface.
16. Epidermis	Composed of columnar epithelium bearing *one cilium* per cell.	Except in basket stars, epidermis is syncytial, and surface cilia are confined to only a few regions.
17. Gas exchange	*Dermal branchiae* (papulae), which are thin evaginations of the body wall, serve for respiration.	Papulae absent; gas exchange occurs mainly through lining of 10 bursae that invaginate from the oral body surface and open by slits on the oral aspect of the disk; breathing is by bursal cilia or by pumping movements that expand and contract the bursae.
18. Pedicellariae	Present (Fig. 17.4*d*).	Absent.
19. Specialized sensory organs	Absent except for an eyespot at each arm's tip; eyespot consists of a cluster of up to 200 pigment-cup ocelli containing a photolabile red pigment.	Absent; only generalized epidermal sensory cells present.
20. Coelom	Large; fluid-filled.	Comparatively meager; crammed with organs within the disk and confined by the vertebrae to only the aboral parts of the arms.
21. Gonads	Ten in all; located in the arms, nearly filling them when ripe; open by gonopores located between the bases of the arms.	From 5 to many; develop on the coelomic side of the bursae; ripe gametes are shed into the bursae and expelled via bursal slits.
22. Hermaphroditic species	Rare.	Protandrous ones fairly common.
23. Brooding of eggs	Uncommon and restricted to circumpolar species, which hold the eggs by various means, including aboral depressions or baskets formed by spines between the arm bases.	Common in many species; eggs are held within bursae for direct development; young escape through bursal slits or, in some species, by the mother's popping off her entire aboral shield (later regrown); Some species are euviviparous, nourishing young via the bursal wall.
24. Larva	Nonbrooding species usually pass through two feeding planktonic larval stages: *bipinnaria* and *brachiolaria*.	If present, planktonic larva is an *ophiopluteus,* distinguished by 8 long larval arms and quite dissimilar to asteroid larvae.

The chaetognath body is unsegmented and consists of head, trunk, and *postanal tail,* the tail being a feature shared only with chordates. The body is covered by a thin cuticle secreted by an epidermis, which is single-layered except along the sides, where it forms a thick layer of *stratified squamous epithelium*. Chaetognaths are the only invertebrates to have a multilayered epidermis, a feature characteristic of vertebrates.

Except for their unmistakably deuterostome manner of development, chaetognaths might well have been considered pseudocoelomates, for their body cavity lacks a peritoneum, and the muscles of the body wall run only longitudinally, in lateral bundles. Rapid contraction of these muscles causes body flexion and dorsoventral movement of the caudal fin,

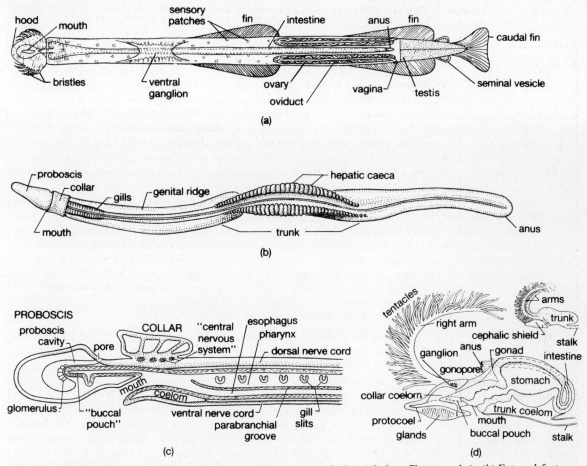

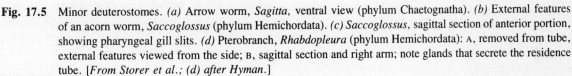

Fig. 17.5 Minor deuterostomes. *(a)* Arrow worm, *Sagitta,* ventral view (phylum Chaetognatha). *(b)* External features of an acorn worm, *Saccoglossus* (phylum Hemichordata). *(c) Saccoglossus,* sagittal section of anterior portion, showing pharyngeal gill slits. *(d)* Pterobranch, *Rhabdopleura* (phylum Hemichordata): A, removed from tube, external features viewed from the side; B, sagittal section and right arm; note glands that secrete the residence tube. [*From Storer et al.; (d) after Hyman.*]

and the animal darts forward, then floats and glides with the aid of its stabilizing lateral fins. During swimming, the rounded head is covered smoothly with a hood that is retracted when the arrowworm, a voracious predator, spreads its spines to capture prey, which ranges from copepods to small fishes and other chaetognaths. The prey is then pierced with anterior and posterior rows of teeth and possibly envenomed with a toxin from pits in the wall of the *vestibule*—a large chamber on the underside of the head—which leads to the mouth. The mouth opens into a bulbous, muscular pharynx, which passes through a septum separating the cephalic coelomic spaces from those of the trunk and leads into a straight intestine terminating at an anus; the anus is located just in front of the trunk-tail partition that separates the caudal coelomic compartments from those of the trunk. The tail includes all the body posterior to the anus, and not just the spatulate horizontal caudal fin. No heart or blood vessels are present, and the coelomic fluid serves for internal transport. Special excretory and respiratory organs are also lacking.

The nervous system includes a circumpharyngeal ganglionic ring featuring large cerebral and smaller lateral ganglia, from which nerves extend peripherally. Special sense organs include a pair of dorsal eyes, each made up of several fused pigment-cup ocelli, a cephalic *ciliary loop* of uncertain function, and longitudinal rows of *hair-fans,* clusters of long cilia that detect water vibrations.

Chaetognaths are simultaneous hermaphrodites, possessing a pair of ovaries in the trunk coelom and a pair of testes in the caudal coelom. The sperm are bundled into a single spermatophore. During mating each animal transfers its spermatophore to its partner, although self-fertilization also occurs. The spermatophore is simply stuck into the outside of the partner's body in the neck region; it then breaks open, allowing sperm to flow posteriorly to the two oviducal pores found just in front of the trunk-tail septum. The jelly-coated eggs are either shed into the plankton or carried attached to the parent's body for a while. Development is direct and no metamorphosis occurs.

17.26.   What are hemichordates?

Phylum Hemichordata ("half cord") includes a small group of marine invertebrates that were originally named for something they don't have: a *notochord*, the supportive axial rod that characterizes phylum Chordata; the supposed notochord found in hemichordates has proved to be no more than a slender diverticulum projecting into the proboscis from the tubular buccal cavity. On the other hand, hemichordates *do* share another feature otherwise limited to chordates: most are proud possessors of from two to many slits—referred to as *gill slits* or *pharyngeal clefts*—that extend from the pharyngeal cavity to the body surface. These clefts expel water that enters the mouth during feeding, while keeping food particles trapped within the pharynx to be swallowed. They are considered homologous with the gill slits of fishes, which mainly function in respiration as exhalant passages for water inhaled through the mouth.

Hemichordates may constitute an evolutionary link between echinoderms and chordates, for their embryonic development through gastrulation very closely resembles that of echinoderms. In addition, the hemichordate nervous system has components resembling the other two phyla: an *epidermal plexus* reminiscent of part of the echinoderm nervous system and (in some species) a *dorsal, hollow neural tube*, which is a feature otherwise seen only in chordates, but in hemichordates is confined to the collar that lies just behind the head.

The two classes of hemichordates are the *acorn worms* and the *pterobranchs*.

17.27.   What are the characteristics of acorn worms?

The approximately 70 species of *class Enteropneusta* ("gut-breathe," after the gill slits) are commonly known as acorn worms for the shape of their proboscis. The proboscis is used in excavating U-shaped burrows in sand or mud, usually in shallow waters where low tides may uncover the burrow openings, identifiable by a characteristic coiled-rope pile of feces deposited outside the rear orifice. The cylindrical, soft, worm-shaped animals range from 9–45 cm long, the body being clothed in cilia and divided into proboscis, collar, and trunk (Fig. 17.5b). The soft proboscis and collar can be made rigid for burrowing by seawater sucked through dorsal pores into their internal spaces; the coelom itself has become so filled with connective and muscle fibers that it cannot serve as part of a hydraulic skeleton. The mouth—located in the ventral, anterior margin of the collar—is permanently open and leads into a buccal cavity followed by the pharynx. The pharyngeal walls are perforated with lateral series of U-shaped pharyngeal clefts (Fig. 17.5c) that open into branchial chambers; these empty to the exterior by way of two longitudinal rows of *gill pores* along the anterior part of the trunk. Burrowing and other muscular activities cause water to enter the mouth and drain from the clefts, but organic particles are retained to be digested in the long, straight intestine, and sand, which is also retained, is voided from the terminal anus. In certain species, a series of sacs (*hepatic ceca*) occur dorsally along the midregion of the intestine. Although the esophagus is muscular and molds the food particles into a mucus-food cord, which it forces peristaltically into the intestine, the intestine has a poorly developed musculature, and the mucus-food cord continues to move posteriorly mainly by ciliary action.

The blood-vascular system is open. Blood flows anteriorly through the dorsal longitudinal vessel into a central sinus at the proboscis base. A closed, muscular, fluid-filled sac, the *heart vesicle*, lies against the central sinus, and its pulsations drive the blood anteriorly into a *glomerulus* that bulges into the proboscis coelom as a series of blood-filled evaginations, which possibly serve an excretory function. The blood is then gathered into a ventral longitudinal vessel and flows posteriorly, collecting digested food from blood sinuses in the gut wall.

The enteropneust nervous system consists mainly of an epidermal plexus that is thickened to form longitudinal nerve cords, chiefly the middorsal and midventral cords. The midventral cord runs anteriorly only as far as the collar. In the collar the middorsal cord separates from the epidermis and continues forward as the *collar cord*, which in some species is hollow, as is the central nervous system of chordates. Special sense organs are lacking, except for a ciliated groove, the *preoral ciliary organ*, which may test the quality of particles about to be carried into the mouth.

Sexes are separate, and two dorsolateral rows of multiple gonads open by separate gonopores along the anterior third of the trunk. Fertilization is external, and development is either direct or indirect, by way of a planktonic *tornaria* larval stage.

17.28.   What are the characteristics of pterobranchs?

Class Pterobranchia ("wing gill") is named for the two curved arms that extend from the collar and bear tentacles covered with cilia used for collecting minute plankton as food (Fig. 17.5d). Both arms and tentacles are hollow and occupied by extensions from the coelom, thus being similar to a lophophore. Pharyngeal clefts are absent, and the surface of the hollow arms and tentacles does provide the major site for gas exchange, since the rest of the body is enclosed in a secreted tube. Pterobranchs constitute only a few species of small hemichordates a few millimeters long, which mainly live in deep water. Most are sessile, tubicolous, and colonial: individual zooids are connected by their stalks to horizontal stolons, from which new zooids arise by budding. The proboscis is flattened like a shield and serves to secrete the tube and to allow the pterobranch to creep up to the mouth of the tube for feeding. The remainder of the body consists of the elongated stalk and a compact trunk containing the internal organs, including a U-shaped gut such as is found in a number of animals that inhabit blind-ending tubes.

Pterobranchs are dioecious. As with other hemichordates their early embryogeny suggests a strong affinity to echinoderms, but their planktonic larva, when known, is unlike that of enteropneusts. Some zoologists believe that archaic pterobranchs may have been the common ancestors of echinoderms and chordates.

Chapter 18

Phylum Chordata

The success story of chordates has a Cinderella quality: from humble beginnings this phylum has flourished to produce, within the past 475 million years, a radiation of backboned animals (vertebrates) that includes both the most gigantic and the most intelligent beings this planet has ever harbored. At present, phylum Chordata (Fig. 18.1) includes about 1300 species of tunicates (*subphylum Urochordata*), 30 species of amphioxus (*subphylum Cephalochordata*), and some 43,800 living species of animals with backbones (*subphylum Vertebrata*). Although far outnumbered by insect species alone, vertebrates tend to be much larger creatures than most invertebrates and they include many familiar animals that have accomplished a highly successful conquest of the land.

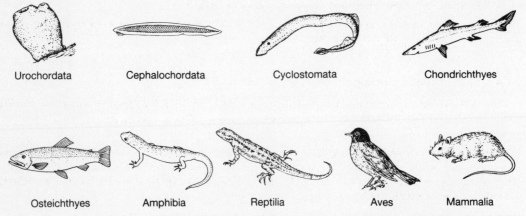

Fig. 18.1 Representatives of phylum Chordata (*left to right, top to bottom*): subphylum Urochordata (tunicate); subphylum Cephalochordata (amphioxus, lancelet); subphylum Vertebrata: lamprey (cyclostome), shark (chondrichthyan), salmon (osteichthyan), salamander (amphibian), lizard (reptile), robin (bird), rat (mammal). (*From Storer et al.*)

18.1. What characteristics distinguish chordates from other animals?

(*a*) The one most special characteristic of all chordates is their possession, at some stage of life, of a longitudinal skeletal rod, the *notochord*, which is both supple and supportive. The notochord is enclosed in a fibrous connective tissue sheath and consists of large vacuolated cells containing a gelatinous solution. The hydrostatic pressure of this solution makes the notochord rigid but still flexible. The notochord cannot be shortened, but in amphioxus and larval tunicates, muscles attach to its sheath and bend it from side to side during swimming. The backbone of vertebrates develops around the notochord and may largely obliterate it in postembryonic life, but in some fishes a persistent notochord runs right through the hollow bodies of the mature vertebrae.

(*b*) *Pharyngeal clefts* (gill slits), first used in filter feeding, represent a second feature that occurs in all chordates at some time of life and are shared only by the enteropneusts of phylum Hemichordata. Even reptiles, birds, and mammals, which are air breathers throughout life, develop pharyngeal clefts during embryogeny (see Chap. 8, Example 16).

(*c*) A *dorsal, hollow neural tube* extending almost the full body length is also unique to chordates and characteristic of them all at some stage of development. The neural tube is more or less expanded anteriorly to form a *hollow brain*. If any single feature is to be held responsible for the impressive evolutionary success of the chordate lineage, it is certainly this: ancestral chordates bequeathed a bubble brain, rather than a solid brain, to their descendants.

> **Example 1:** Since we all deplore vacant minds, what can be so marvelously preadaptive about a hollow central nervous system? The answer lies in a curious quality of nervous tissue: although nerve cells (*neurons*) have enormous metabolic needs and consume oxygen avariciously, masses of nerve

cells are not readily penetrated by blood capillary beds. Neuron cell bodies must be positioned close to surfaces at which nutritional exchanges can take place. In contrast with the solid brains of nonchordate metazoans, the hollow chordate brain is preadapted for great expansion, since nerve cell bodies can be located near either the inner or the outer surfaces of the central nervous system, close to a nourishing supply of what, in vertebrates, we call *cerebrospinal fluid*.

The reliance of brain cells on the cerebrospinal fluid is shown by the fact that in the course of the prodigious enlargement of the mammalian brain, regions of *gray matter*, to which neuron cell bodies are restricted, are limited to the walls of the brain cavities (*ventricles*) and to an external layer (*cortex*) that, even in humans, is only about 1 cm thick. At a less advanced stage, the gray matter of chordate brains lies only internally, along the inner surfaces of the neural tube. The rest of the brain tissues consists of *white matter*, interconnecting tracts of fibers covered with fatty protective sheaths.

The earliest version of the chordate brain may well have resembled the single, thin-walled vesicle that today makes up the brain of larval tunicates ("sea squirts"). Strange to consider that some 500 million years ago one such tiny bubble presaged the multitalented vesicular brain of the vertebrates.

(*d*) A *subpharyngeal gland*, or its homolog, the *thyroid gland*, is unique to chordates. The vertebrate thyroid binds iodine and regulates metabolism by way of such hormones as *thyroxin*.

(*e*) *Photosensitive cells* (including optic retinas, when present) develop from the *ectoderm of the central nervous system*, not from the epidermal ectoderm as in other phyla.

18.2. What other features characterize chordates in general?

(*a*) A *postanal tail* is found in all chordates during some stage of life, a feature shared with chaetognaths. Even human fetuses have short tails. The tail is muscular and serves for propulsion in tunicate larvae, cephalochordates, and vertebrate fishes.

(*b*) The chordate body is *metameric*, especially as regards the somatic musculature, which (except in tunicates) consists of segmentally arranged *myomeres* separated by thin connective-tissue *myosepta*. Segmentation persists throughout life in cephalochordates and vertebrates and in the latter is especially pronounced in the spine, the musculature, the gill slits, and the distribution of blood vessels and nerves serving the body wall.

(*c*) When a heart is present (in urochordates and vertebrates), it is located ventral to the gut.

(*d*) When an endoskeleton is present, it is derived from the *mesoderm* (as in echinoderms).

(*e*) Chordates are bilaterally symmetrical deuterostomes: the anus develops at the site of the blastopore, the coelom is enterocoelous in origin, and indeterminate cleavage admits the possibility of monozygotic multiple births.

18.3. What are tunicates?

Tunicates ("membrane") are marine chordates of subphylum Urochordata that have a unique, acellular, fibrous tunic enclosing the body.

18.4. What are the distinguishing features of tunicates?

(*a*) The *tunic*, secreted by the underlying *mantle*, is typically composed of a type of *cellulose (tunicin)*, which is remarkable, because cellulose is characteristically synthesized by plants, not animals. The tunic may be thin and delicate but is usually thick with a gelatinous, leathery, or even gristly texture. It may be transparent or opaque and is often brightly pigmented.

(*b*) The subphylum name, *Urochordata*, derives from the fact that the notochord is confined to the muscular tail, by which the tadpole-shaped larva swims.

(*c*) Tunicates are the only chordates that regularly reproduce asexually as well as sexually. Many colonial ascidians form fleshy masses in which individual zooids gradually die off and are reabsorbed while young ones continue to be produced by budding.

(*d*) No coelom ever develops, because during embryonic life, pouches fail to form along the sides of the primitive gut (archenteron). Tunicates are therefore the only chordates to have no body cavity other than irregular, unlined blood sinuses.

(*e*) The planktonic larval tunicate superficially resembles a tadpole only 1–5 mm long, with a long, muscular tail housing the notochord and neural tube and a pudgy body containing a complete digestive tract dominated by an enlarged, perforate pharynx used in filter feeding. The larva also possesses a heart and blood vessels, a trunk ganglion, and a single brain vesicle. The latter develops a *median eye* from its dorsal wall and also possesses sensory hair cells attached to a calcareous *otolith* that is subject to gravitational displacement, as in the statocysts of other phyla. Many elements of larval form are retained into adulthood in larvaceans (class Larvacea), but in other tunicates the larva undergoes rapid metamorphosis into a radically different adult form in which the pharynx hypertrophies into a great *pharyngeal basket*. During metamorphosis the tail is partly reabsorbed and partly shed, the notochord disappears, and the central nervous system becomes reduced to only the trunk ganglion (Fig. 18.2*a*).

(*f*) Feeding and respiration both occur as water is filtered through the pharyngeal slits, collected into a surrounding chamber (*atrium*), and expelled to the exterior via an *atriopore*. In the adult, the mouth is guarded by a muscular *incurrent siphon*, and a similar *excurrent siphon* guards the atriopore. Pharyngeal cilia propel the water current and direct food particles, trapped in mucus, along ciliated grooves to the esophagus. Most of the mucus is secreted by the trough-shaped *endostyle*, considered a forerunner of the thyroid gland because it concentrates iodine.

(*g*) The open circulatory system of tunicates is unique among chordates. The U-shaped heart of the adult expels blood directly into unlined spaces in the body mesenchyme, from which irregular channels carry it past the organs and back to the heart. As in vertebrates, the heartbeat is *myogenic*, i.e., the cardiac muscle fibers pulsate without nervous excitation. This is in marked contrast to the neurogenic heartbeats of nonchordates, which are dependent on motor nerve fibers. Each end of the tunicate heart has a *pacemaker* excitatory center from which the heartbeat is propagated. Curiously, the heart expels blood from one end for 2 or 3 minutes, then its beat *reverses*, and blood is pumped out the opposite end, as the two pacemakers function in alternation.

(*h*) *Vanadocytes* are unusual cells found in tunicate blood. These amoeboid cells transport surprising quantities of the element *vanadium*, which occurs at concentrations of 3 ppm or less in seawater but as much as 3700 ppm within the tunicate itself. Vanadium is taken up by vanadocytes in the walls of the pharynx, which then migrate into the tunic, where they disintegrate, liberating the vanadium, which acts as a reducing agent in the deposition of cellulose microfibrils. (In certain tunicates these cells, instead, transport iron, which serves the same reducing function.)

(*i*) Tunicates are the only chordates to lack excretory organs. Ammonia simply diffuses out by way of the vast pharyngeal surface area, while other wastes, such as uric acid, are picked up by amoeboid *nephrocytes* (*excretophores*) that wander through the blood and finally become fixed in clusters in certain body regions.

(*j*) *Hermaphroditism* is characteristic of tunicates but uncommon among other chordates.

18.5. What are the classes of subphylum Urochordata?

(*a*) *Class Ascidiacea* ("little sac") includes the familiar solitary and colonial *sea squirts*, renowned for the vigor with which large individuals respond to handling by shooting water jets from both siphons. Ascidians are the only chordates that are sessile as adults. Frequently they are colonial or compound. The zooids of compound ascidians have independent buccal siphons, but shared excurrent orifices. When the ascidian larva settles head down after some 36 hours of free-swimming life, its tail disappears and its organs rotate by 180° so that both mouth and atriopore face away from the attachment surface. As growth occurs, with pharyngeal hypertrophy, the tunic enlarges and thickens, covering the body of the sessile adult (Fig. 18.2*b*). A colonial or compound condition may then be achieved by asexual budding.

(*b*) *Class Thaliacea* ("luxuriant") has tailed larvae that metamorphose into barrel-shaped, planktonic adults known as *salps*, which have their incurrent and excurrent siphons facing in opposite directions (Fig. 18.2*c*). This positioning allows them to swim by jet propulsion, as water is taken in one siphon and forced out the other by contractions of the 6–10 muscle bands ringing the 10-cm-long body like barrel hoops. These transparent urochordates are often brilliantly bioluminescent. All salps bud asexually, producing individuals that either break free or remain attached in trains up to 20 m long. Some species show alternation of generations: a sexual generation liberates gametes, from which the tailed larvae arise; these mature into an asexual generation of adults lacking gonads, which bud off a new generation of salps having gonads and reproducing sexually. Salps are thought to have evolved from sessile, ascidian-type ancestors.

(*c*) *Class Larvacea* is of great interest because the planktonic adult stage is *neotenic*, meaning that the tadpole-shaped larval body is retained with only minor changes other than the development of gonads.

> **Example 2:** In *neoteny*, a phenomenon known elsewhere in the animal world, the former adult body form is simply deleted from the life history by sexual maturation of the larval stage. The Mexican

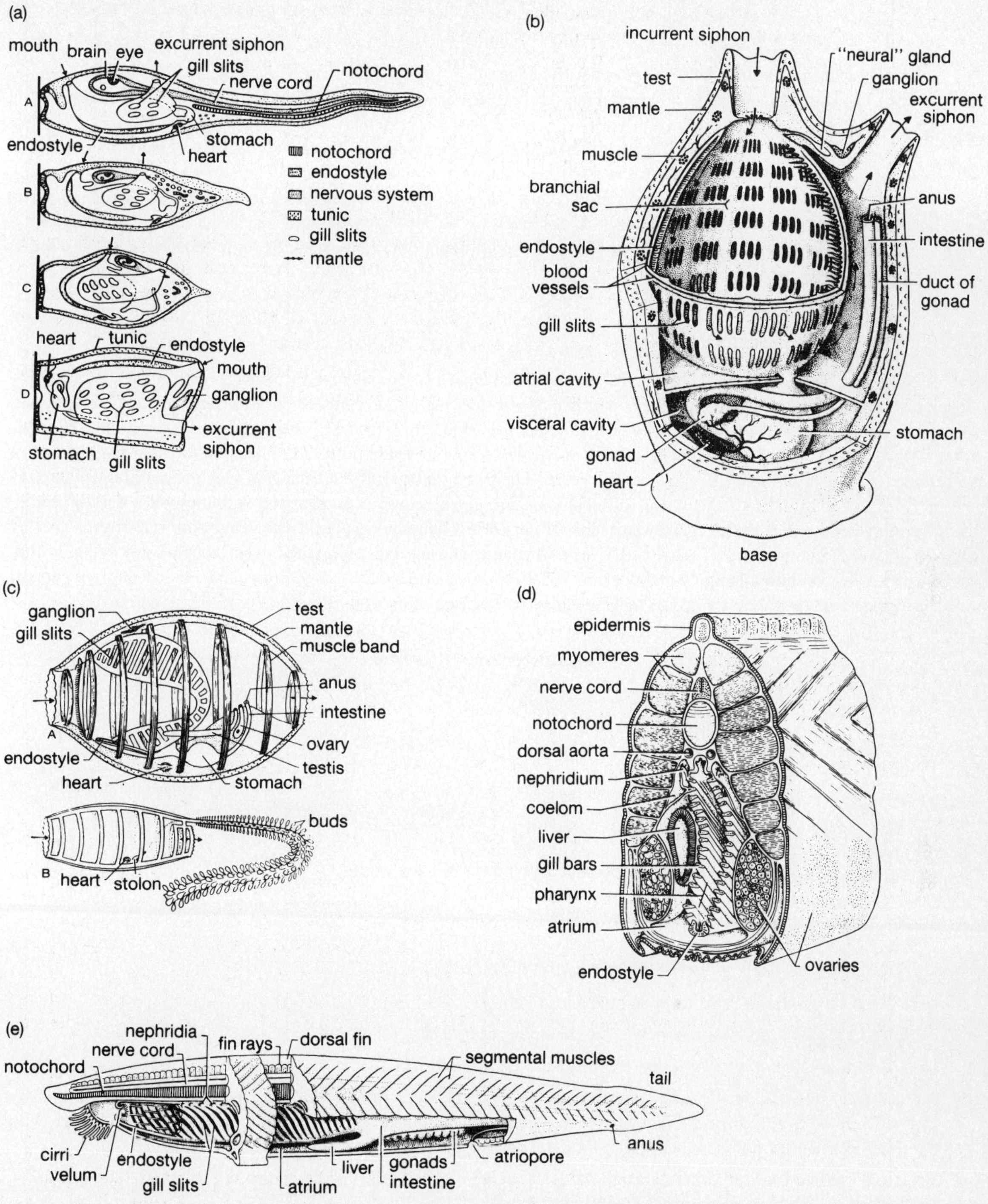

Fig. 18.2 Invertebrate chordates (arrows show path of water circulation). *(a)* Metamorphosis of an ascidian: A free-swimming larva settling on its head; B, reabsorption of tail with reduction of notochord and neural tube; C, disappearance of notochord and rotation of internal organs; D, enlargement of pharyngeal basket, deposition of tunic. *(b)* Structure of an adult solitary ascidian. *(c)* Salp: A, adult of sexual generation; B, adult of asexual generation trailing chain of buds. *(d)* Section through pharyngeal region of a female amphioxus. *(e)* Structure of amphioxus, body wall partly removed. [*From Storer et al.; (a) after Kowalewsky and after Herdman, (c) after Stempell, (d) after Kukenthal.*]

salamander, axolotl, reproduces in a gilled, larval condition, but under physiological stress may metamorphose into the familiar tiger salamander (*Ambystoma*).

The existence of a neotenic type of urochordate suggests a possible mechanism whereby the mobile, fishlike body form of cephalochordates and early vertebrates could have arisen within a phylum previously characterized by a sessile adult life.

Larvaceans are less than 5 mm long, but they often secrete gelatinous houses as large as walnuts. The house is actually a version of the tunic and is secreted by the epidermis. However, its architecture may be elaborate, featuring a protective sieve that keeps out particles too large to ingest, a funnel-shaped food net into which plankton is concentrated by the larvacean's beating tail, and even an escape hatch through which the larvacean can make a speedy exit if its house is seized by a predator. Splendid though the house may be, it is abandoned in only a few hours, and a new house is secreted.

18.6.   What are the cephalochordates?

Subphylum *Cephalochordata* ("head cord") includes the *lancelets*, or *amphioxus* ("sharp on both ends"), which seem the most vertebratelike of all invertebrates, whether or not they really do represent modern descendants of a common ancestor with vertebrates. These fishlike little marine animals (Fig. 18.2*d* and *e*), usually no more than 7 cm long, can swim by lateral undulations of body and tail, but prefer to burrow in sand with head protruding for filter feeding. They are reportedly tasty scrambled with eggs, but this seems a degrading prospect for creatures of such great zoological interest and probable antiquity. The characteristics of amphioxus are given below, in comparison with those of a very primitive living vertebrate, the filter-feeding, freshwater larva of the jawless lamprey eel (Fig. 18.3). This larva was given a species name of its own: "ammocoetes" ("sand bed," for its habitat), and the label has stuck even though we now know that ammocoetes grows up into a lamprey and does not represent a separate species. Amphioxus and ammocoetes are enough alike to suggest common ancestry and to make detailed comparison interesting.

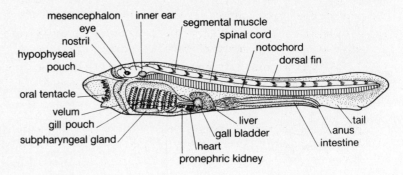

Fig. 18.3 Ammocoetes larva of a lamprey. Compare with Fig. 18.2*e*.

18.7.   How are amphioxus and ammocoetes similar?

(*a*) Both have a fish-shaped body form and expanded vertical caudal fin.

(*b*) Both lack the paired pectoral and pelvic fins found in most fishes.

(*c*) Both have a *persistent notochord* that extends from tail tip into head.

(*d*) The somatic musculature of both consists of a longitudinal series of narrow segmental myomeres separated by myosepta. In both, the dorsal musculature of the trunk is much thicker than the ventral layers, which contrasts with the usual invertebrate condition, in which the body wall is about the same thickness all around.

(*e*) Both lack jaws and feed by filtering small particles, using oral tentacles to gather and reject particles.

(*f*) The distribution of major blood vessels is similar and foreshadows the pattern seen in vertebrate fishes as a whole, and the primary direction of blood flow—posteriorly in a dorsal aorta, and anteriorly in a ventral aorta—is the same in both.

(*g*) Both have an enlarged pharynx with walls supported by a longitudinal series of cartilaginous gill bars.

(*h*) The mucous-secreting endostyle of amphioxus and subpharyngeal gland of ammocoetes both lie in the floor of the pharynx and seem to be homologous, but during lamprey metamorphosis the latter transforms from a mucus gland into the thyroid gland of the adult.

(*i*) The alimentary tract of both is straight and uncoiled and opens by way of an anus located just anterior to the caudal fin.

(*j*) In both, a *diverticulum* grows outward from the midgut during embryonic development. In amphioxus, this becomes the saccular *midgut cecum,* in which smaller food particles are digested and absorbed, whereas in ammocoetes it gives rise to the liver and gall bladder.

(*k*) Just in front of the anterior end of the neural tube, and embryonically attached to that structure, amphioxus has a dorsal median *flagellated pit* of chemosensory function; in a similar position, but opening into a *hypophyseal canal,* is the single median *nostril* of ammocoetes.

18.8.   In what major respects do ampioxus and ammocoetes differ?

(*a*) Amphioxus is marine; ammocoetes lives in fresh water.

(*b*) Cephalochordates attain a length of only 10 cm and then mature sexually without metamorphosis; when about the same size or even longer, and after a larval life of 5–7 years, ammocoetes *metamorphoses* into the adult lamprey, which often migrates to the sea.

(*c*) The notochord of amphioxus runs nearly the entire body length, whereas that of ammocoetes stops short at the level of the midbrain.

(*d*) The dorsal neural tube of amphioxus is expanded anteriorly into a single brain vesicle of great insignificance; that of ammocoetes is expanded into a conspicuous hollow brain displaying the three major divisions (*hindbrain, midbrain, forebrain*) characteristic of vertebrates.

(*e*) Scattered photoreceptive cells occur on the surface of the central nervous system of amphioxus, while ammocoetes sports a pair of prominent lateral eyes that develop from the brain wall.

(*f*) The amphioxus pharynx, which comes to occupy nearly half the entire body length and to have nearly 200 gill slits, opens into an atrium drained by an atriopore anterior to the anus; the considerably shorter pharynx of ammocoetes opens directly to the exterior by way of only 7 rounded gill slits.

(*g*) The respiratory function of the ammocoetes pharynx is enhanced by the development of *gill lamellae* in the walls of *pharyngeal pouches;* neither of these features occurs in amphioxus.

(*h*) The blood of amphioxus lacks corpuscles and is circulated by the pumping action of numerous muscularized bodies located along the major arteries in the pharyngeal region; ammocoetes blood contains corpuscles, some with hemoglobin, and is circulated by a single heart that develops along the ventral aorta below the esophagus.

(*i*) The cephalochordate excretory system consists of clusters of protonephridia with flagellated solenocytes, which project into a metamerically arranged series of vesicles that drain into the atrium. The excretory organs of the larval lamprey are *pronephric kidneys* with open coelomic funnels (*nephrostomes*) and, therefore, are of a metanephridial type.

18.9.   What are vertebrates?

Vertebrates are animals with backbones; i.e., they have a vertebral column. Their magnificent adaptive radiation across nearly 500 million years of geologic time is pictured in Fig. 18.4.

18.10.   What are the major subdivisions of subphylum Vertebrata?

Superclass Agnatha: jawless fishes
  Class Ostracodermi: extinct armored fishes
  Class Cephalaspidomorphi: lampreys
  Class Myxini: hagfishes

Superclass Gnathostomata: jawed vertebrates
  Class Acanthodia: extinct jawed fishes with spines for fins
  Class Placodermi: extinct armored fishes with jaws
  Class Chondrichthyes: cartilaginous fishes—sharks and rays
  Class Osteichthyes: ray-finned and lobe-finned bony fishes
  Class Amphibia: amphibians, usually with gill-breathing larva, terrestrial adult

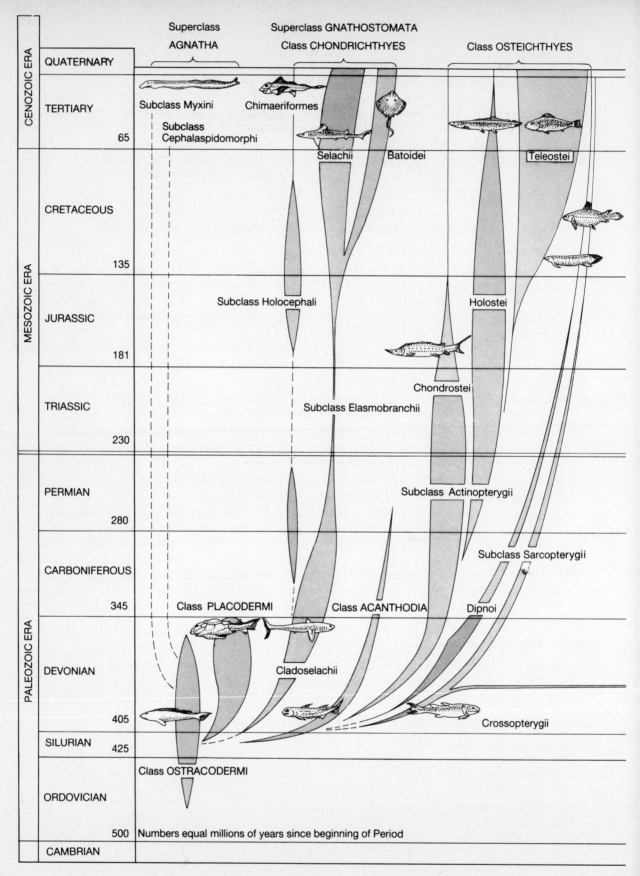

258

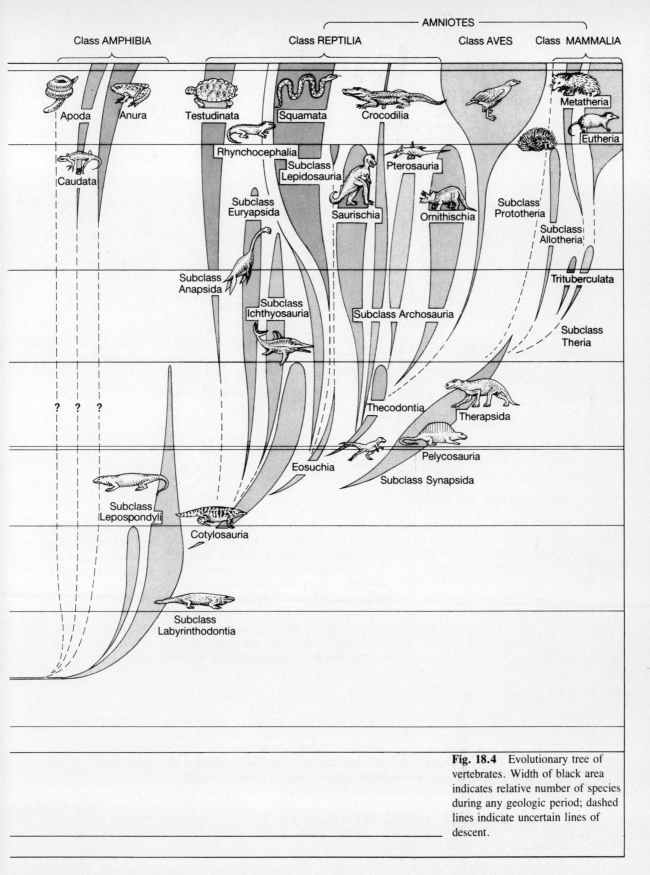

Class AMPHIBIA
Class REPLILIA
AMNIOTES
Class AVES
Class MAMMALIA

Apoda
Anura
Testudinata
Squamata
Crocodilia
Metatheria
Eutheria

Caudata
Rhynchocephalia
Subclass
Lepidosauria
Pterosauria
Subclass
Prototheria
Subclass
Allotheria

Subclass
Euryapsida
Saurischia
Ornithischia

Subclass
Anapsida
Subclass
Ichthyosauria
Subclass Archosauria
Trituberculata

Subclass
Theria

? ? ?
Thecodontia
Therapsida

Eosuchia
Pelycosauria
Subclass Synapsida

Subclass
Lepospondyli
Cotylosauria

Subclass
Labyrinthodontia

Fig. 18.4 Evolutionary tree of
vertebrates. Width of black area
indicates relative number of species
during any geologic period; dashed
lines indicate uncertain lines of
descent.

Class Reptilia: tetrapods with lungs and epidermal scales
Class Aves: birds, having feathers
Class Mammalia: tetrapods with hair, mammary glands

Amphibians, reptiles, birds, and mammals are sometimes grouped as *Tetrapoda* because of their land-adapted limbs, while *Pisces* includes finned gnathostomes (jawed fishes). Also, reptiles, birds, and mammals are known as *amniotes* because of the *extraembryonic membranes* (including the *amnion*) that protect their developing embryos, all other vertebrates by contrast being *anamniotes*.

18.11. What are the distinguishing characteristics of vertebrates?

A generalized vertebrate body plan is shown in Fig. 18.5. Specific distinguishing characteristics of vertebrates are discussed in Questions 18.12–18.46.

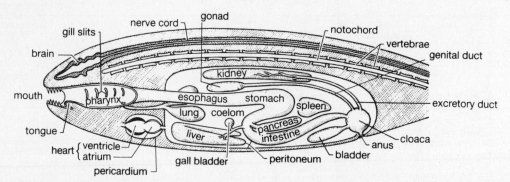

Fig. 18.5   Schematic representation of vertebrate structure. Both lungs and gills are shown, but these seldom occur together. (*From Storer et al.*)

18.12. How is vertebrate skin unique?

The vertebrate integument (Fig. 18.6) has two components: a deeper-lying *dermis*, consisting mainly of connective tissues, and an outer *epidermis*, which is unique in that it is not made up of a single layer of cells, but of a multilayered stratified squamous epithelium in which older cells die and are sloughed from the outer surface while new cells are replenished in a basal, mitotically active layer, the *stratum germinativum*. This innovative body covering preadapted vertebrates for life on land, since in terrestrial vertebrates the outermost epidermal layers (*stratum corneum*) consist of dead cells rendered impervious to water by intracellular concentrations of an insoluble protein, *keratin*. Dead, keratinized cells also make up the *scales* of reptiles, and such epidermal derivatives as feathers, hair, quills, claws, nails, hooves, rattles, epidermal armor (as in armadillos), whalebone (baleen), and the horny sheath (*rhampotheca*) that covers the jaws of turtles and birds in the absence of teeth. Epidermal-dermal developmental interactions produce such structures as horns and antlers, which have a bony core and a covering, transitory or permanent, of living or keratinized epidermis.

18.13. What glands occur in vertebrate skin?

Various *integumentary glands* occur in different vertebrates: *mucous* and *poison* glands in fishes and amphibians, *scent* glands in reptiles and mammals, *oil (sebaceous)* glands in birds and mammals, and *sweat, mammary, ceruminous (earwax),* and *eyelid* glands in mammals only. Multicellular integumentary glands arise from *epidermal* epithelium that grows down deeply into the underlying dermis, remaining open to the surface by way of a duct. These glands may be *tubular* (e.g., sweat glands) or *alveolar* (terminating in grapelike clusters of secretory cells, e.g., sebaceous and scent glands). Although located subcutaneously, *lacrimal* (tear) and *salivary* glands are also considered to be evolutionary derivatives of integumentary glands.

18.14. What structures occur in the dermis?

The vertebrate dermis contains such heavy concentrations of elastic and collagenous connective tissue that in some mammals and reptiles it can be tanned as *leather*. Among the connective fibers run blood and lymph vessels and nerves

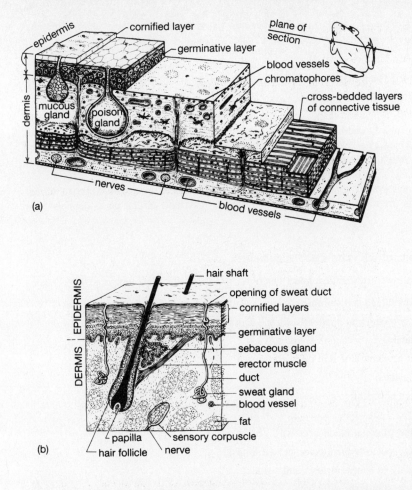

Fig. 18.6 Vertebrate skin. *(a)* Frog skin. *(b)* Mammalian (human) skin. *(From Storer et al.)*

serving tiny *cutaneous sense organs* and intrinsic *integumentary muscles* such as those that erect hairs or feathers (*arrectores pilorum* or *plumosum*). Cutaneous senses commonly include *touch, heat, cold, pressure, and pain.*

> **Example 3:** Mammalian tactile receptors include oval *Meissner's corpuscles*, deeper-lying *Merkel's discs*, and nerve fibers that branch around the bases of hairs. *Pacinian corpuscles* are pressure-sensitive organs found both in the dermis and in deeper body parts; they resemble oval onions the size of pinheads. Sharks and duck-billed platypuses have cutaneous *electrical receptors* of extreme sensitivity, which allow these animals to locate prey by the minute electrical fields that all living things generate. The keenest *heat* detectors known are found in facial pits in the cheeks of such pit vipers as rattlesnakes and along the jawline of pythons and boas.

18.15. What role has the dermis played in bone formation?

An important primitive property of the vertebrate dermis was its capacity for producing *bone*, a tissue unique to vertebrates (see Question 18.18). Dermal bony armor protected the earliest known vertebrates, the ostracoderms, probably from the onslaughts of large predatory invertebrates such as water scorpions. Bone is made up of an inorganic matrix of crystalline *calcium phosphate* $[Ca_{10}(PO_4)_6(OH)_2]$ deposited within an organic matrix consisting mainly of fibrils of the protein *collagen*. Bone cannot be digested by invertebrates, which lack the highly acid gastric juices of vertebrates and the protein-splitting enzyme *pepsin*, which operates in an acidic medium. Probably evolved first as a protective integumentary adaptation, bone later invaded the internal, originally cartilaginous, endoskeleton, with incalculable benefits for evolving vertebrates, for bone is a much stronger tissue than cartilage. Even today, most of the vertebrate skeleton is first laid down as cartilage, and as development proceeds, the cartilage is dissolved and replaced by bone; such bones are therefore called *replacement bones*, or *cartilage bones*. Bones that instead ossify directly within connective tissue membranes are called *membrane bones*, and most of these originated in the dermis. The membrane bones that form the face and roof of the cranium are referred to as *dermal bones,* even though they now lie subcutaneously, because they represent evolutionary derivatives of the thick bony scales that once guarded the brains of primitive bony fishes.

18.16.   What other structures evolved from the dermis?

Besides its lasting contributions to the vertebrate skeleton, dermal bone also gave rise to fish *scales*. Progressive thinning of the dentine-coated dermal bony armor and thick *cosmoid* scales of early fishes led to the types of scales seen in living fishes (see Fig. 20.1*b*), in which only one layer of bone is present: the rhomboidal *ganoid* scales of gars and sturgeons, in which the bone layer is covered by a layer of lustrous, highly resistant *ganoin,* and the even thinner *cycloid* and *ctenoid* scales of other bony fishes, in which the ganoin layer has been lost. Although we usually think of fish scales as covering the body surface, they actually lie in the dermis and are overlain by a thin, mucous epidermis.

Another debt owed this ancient dermal armor is that in placoderms its outer surface bore tiny, enamel-capped *dermal denticles;* vertebrate *teeth,* together with the toothlike *placoid* scales of sharks, may have originated as denticles that persisted after the armor was lost.

18.17.   What are the major functions of the vertebrate integument?

The complexity of vertebrate skin allows it to perform many functions, not all of which are unique to vertebrates:

(*a*)   Protection from predators (e.g., camouflage and warning coloration, quills, armor, poison and scent glands)

(*b*)   Protection from microbes (stratum corneum and phagocytes in the dermis)

(*c*)   Protection from ultraviolet light (tanning by melanin deposition)

(*d*)   Vitamin D synthesis (by ultraviolet irradiation of skin oils)

(*e*)   Nourishment of young (mammary glands)

(*f*)   Protection from dehydration (stratum corneum)

(*g*)   Heat conservation (hair, feathers)

(*h*)   Heat dissipation (sweat glands, spines, dermal blood vessels)

(*i*)   Social communication (scent glands, species-typical color and markings)

(*j*)   Light production (photophores, mainly in fishes, as lures and signals)

(*k*)   Locomotion (feathers, claws, hooves, skin webs as in bats)

(*l*)   Sensory information (see Question 18.14)

18.18.   How is the vertebrate skeleton formed?

The articulated, mesodermal endoskeleton (Fig. 18.7*a*) consists of *bones,* which are bound one to another by *ligaments* of collagenous connective tissue or are sometimes more firmly conjoined by *fibrocartilage* or even by immovable *sutures.* The two major parts of the endoskeleton are the axial skeleton and the appendicular skeleton.

(*a*)   In some vertebrates such as sharks, "bones" consist only of cartilage, but in most, the endoskeleton first *chondrifies,* as embryonic mesenchyme differentiates into *chondroblasts* that secrete cartilage, and later *ossifies,* as the cartilage is progressively eroded away by wandering cells called *chondroclasts,* while other cells known as *osteoblasts* lay down *bone matrix* in its place. A focal point of chondroclastic and osteoblastic activity is known as an *ossification center*. A long bone such as the human ulna (see Fig. 18.7*c*) has three ossification centers: an *epiphysis* at each end, and a *diaphysis* that makes up the shaft; when the epiphyseal cartilages that separate the ossification centers have been fully replaced, the bone can no longer grow in length.

(*b*)   In bone tissue, the inorganic matrix of calcium phosphate is deposited as crystalline *hydroxyapatite* within which embedded *collagen* fibrils provide tensile strength. Bone is a dynamic, adaptable tissue being continuously reabsorbed and redeposited along lines of stress in such a manner that strength is maximized while weight is minimized. It is unique to vertebrates, and its extraordinary weight-bearing potential preadapted vertebrates both for agile locomotion on land and for attainment of body bulk much greater than could be supported out of water by such alternative skeletal materials as cartilage, chitin, or calcium carbonate. Bone is deposited in microsopic cylindrical units known as *Haversian systems* (Fig. 18.7*b*), each of which in cross section consists of concentric lamellae of matrix surrounding

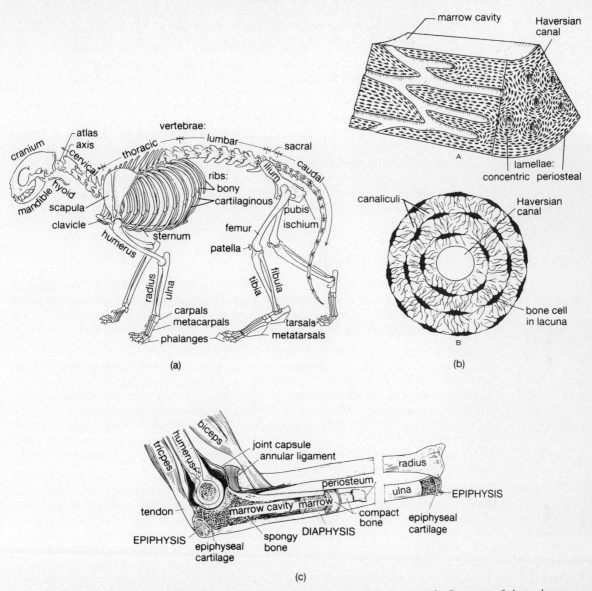

Fig. 18.7 The vertebrate endoskeleton. *(a)* Skeleton of a land vertebrate, the cat. *(b)* Structure of dense bone: A, Haversian systems (osteons) in transverse and longitudinal sections, in part of the shaft of a long bone; B, enlarged cross section of a single Haversian system. *(c)* Structure of a long bone and a synovial joint in the human arm, and tendons of insertion of triceps and biceps muscles; note epiphyseal cartilages at which bone can grow further in length. (*From Storer et al.*)

a central cavity (*Haversian canal*) in which run a tiny nerve and blood vessels. Trapped in small spaces (*lacunae*) within the matrix are the bone cells themselves, called *osteoblasts* when actively laying down matrix and *osteocytes* when quiescent. When matrix is eroded by the action of wandering cells called *osteoclasts* (as when the blood calcium level falls), the freed osteocytes revert to osteoblast status, laying down new matrix as soon as materials become available. Throughout life, bone repair and diametric growth occur by differentiation of new osteoblasts from a tough connective tissue membrane, the *periosteum*, that covers the exterior of each bone and is continuous with the ligaments. Even when a bone is fully ossified, if it articulates movably with another bone (Fig. 18.7c), the articulating surfaces are capped with *hyaline cartilage* and the joint cavity is lined with a *synovial membrane* that secretes a slick, lubricative fluid.

18.19.   What is the vertebrate axial skeleton?

The *axial skeleton* always includes a *cranium* that partially or fully encases the brain and a segmental series of *vertebrae* constituting a *spinal column* ("backbone" or "spine") that develops around the notochord and protects the neural tube (now called a *spinal cord*). In fishes, dorsal trunk muscles attached to the vertebral column flex it from side to side during swimming. In most tetrapods, the vertebrae of different regions are structurally specialized: *cervical* for neck flexibility and head mobility and support; *thoracic* for rib anchorage that allows respiratory movements of the ribs; *lumbar* for bearing much of the weight of the trunk; *sacral* for anchoring the pelvic girdle; and *caudal* for tail support and movement. The axial skeleton of bony fishes and tetrapods also includes metamerically arranged *ribs;* these develop in the connective tissue *myosepta* that segmentally interrupt the longitudinal trunk muscles. In tetrapods, short ribs fused onto the vertebrae form *transverse processes* that offer more surface for muscle attachment. In amniotes, the movable ribs extend ventrally to articulate with a *sternum* ("breastbone"), which usually consists of a segmental series of *sternebrae* that ossify where the myosepta intersect a midventral connective tissue seam (*linea alba*) running the length of the trunk. The articulation of ribs with sternum forms a complete "rib cage" that protects the heart and lungs and aids respiration (*costal breathing*).

18.20.   How is the appendicular skeleton organized?

The *appendicular skeleton* consists of the *limb bones* and the *girdles* that anchor them proximally. Vertebrates are unique in never having more than two pairs of limbs, *pectoral* and *pelvic*. This deplorable state of affairs results from a bit of costly economizing during the evolution of the vertebrate appendicular skeleton, which ruled out the possibility of earthly angels with arms, legs, and wings, and centaurs with four running legs and a pair of arms as well.

> **Example 4:**   The use of vertebrate forelimbs for flying has always been paid for by sacrificing manipulative capacity. Bipedal locomotion, typically less swift or stable than quadrupedal movement, is the price of freeing the forelimbs for grabbing or carrying things. Three pairs of limbs would have been *much* nicer to have, but the evolutionary process is woefully lacking in foresight. Certain primitive jawed fishes had entire lateroventral rows of fin spines and seem to have been courting the notion of evolving more than four limbs, but, unhappily, they settled for the economy of developing no more fins than were just needed to stabilize them against roll. So, although we can thank these early fishes for originating at least the two pairs of pectoral and pelvic appendages, we may wish they had thought about the needs of their terrestrial descendants.

The appendicular skeleton is lacking in lampreys and hagfishes, and ostracoderms either had no limbs at all or had simply a pair of stabilizers not homologous with true fins. In cartilaginous fishes each limb girdle consists only of a transverse bar that interconnects the two fins and lacks attachment to the axial skeleton. In bony fishes, the pelvic girdle is reduced, so that each pelvic fin is anchored by a separate pelvic bone, which is held in place only by the musculature, allowing great variability in the placement of these fins. On the other hand, in bony fishes the two primary bones of the pectoral girdle—the ventral *coracoid* and dorsal *scapula*—become anchored to the head by way of dermal bones, most notably the *clavicle*. In extinct bony fishes of the lobe-fin line (crossopterygians), the fins and their girdles contained bones seemingly homologous to those of tetrapods.

In tetrapods, the *pelvic girdle* consists of three bones (sometimes fused together): a ventral anterior *pubis*, ventral posterior *ischium*, and dorsally projecting *ilium*. A major departure from the piscine state is that the tetrapod ilium always articulates with the vertebral column in the sacral region, strengthening the support of the hindlimbs so that these assume the leading role in support and thrust. The salamander sacrum consists of only a single vertebra and its attached sacral ribs; that of humans represents a fusion of five vertebrae and their fused-on ribs, furnishing an extended articular surface for the ilia. In contrast to this sturdy pelvic anchorage, the *pectoral girdle* of land vertebrates, still consisting usually of the ventral coracoid and clavicle and the dorsally projecting scapula, is quite movable because it is *not* anchored to the skull as in fishes, but rather is merely attached by way of muscles to the vertebral column. Thus it is less well adapted for weight bearing and thrust than the pelvic girdle. In fact, in mammals the coracoid has become reduced and fused onto the scapula, and the clavicle may be almost gone.

The *limb bones* of tetrapods are serially homologous: the single proximal element is known as the *humerus* in the forelimb and the *femur* in the hind limb. The two distal limb bones are the *radius* and *ulna* in the forelimb and the corresponding *tibia* and *fibula* in the hind limb. The cluster of small wrist bones are *carpals,* with corresponding *tarsals* in the hind foot. As a rule, a series of five *metacarpals* or five *metatarsals* extends distally from the carpals or tarsals, respectively, each articulating with the two or three *phalanges* that make up the skeleton of each digit. Tetrapod vertebrates show many adaptive alterations of the appendicular skeleton, particularly its distal portion, which may even be modified into wings or finlike flippers (Fig. 18.8).

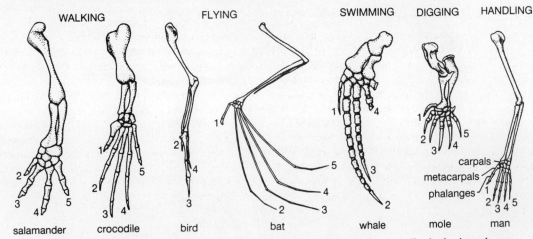

WALKING   FLYING   SWIMMING   DIGGING   HANDLING

salamander   crocodile   bird   bat   whale   mole   man

Fig. 18.8   Radiative adaptation of forelimb in various tetrapods. From proximal to distal, the homologous bones in each forelimb consist of humerus (upper arm), radius (lower arm bone on digit 1 side), ulna (lower arm bone on digit 5 side), carpals (wrist bones, typically eight), metacarpals (hand bones, typically five), phalanges (bones of digits, typically two in digit 1, three in others). (*From Storer et al.*)

18.21.   What is special about the vertebrate nervous system?

The vertebrate nervous system is by far the most advanced in the animal kingdom and is unique in a number of respects (see Questions 18.22–18.29). Its main anatomical divisions are the central nervous system (brain and spinal cord) and the peripheral nervous system (cranial and spinal nerves and peripheral ganglia).

18.22.   What is the overall organization of the central nervous system?

The vertebrate *central nervous system* (CNS), consisting of brain and spinal cord, is protectively enclosed in three membranes (*meninges*): an outer *dura mater*, central *arachnoid mater*, and an innermost *pia mater*. The dura mater and pia mater are rich in blood vessels. As in all chordates the vertebrate CNS is hollow, containing spaces filled with a liquid now called *cerebrospinal fluid*. However, the vertebrate brain (Fig. 18.9) is unique in consisting postembryonically of *five* vesicles: the *telencephalon* (cerebrum) and *diencephalon*, which make up the *forebrain;* the *mesencephalon (midbrain)*: and the *metencephalon* and *myelencephalon*, which constitute the *hindbrain*. The cerebral hemispheres each enclose a cavity known as a lateral *ventricle*, the diencephalon encloses the third ventricle, and the hindbrain the fourth ventricle, which opens into a fluid-filled subarachnoid space between the arachnoid and pia mater and continues posteriorly as the cerebrospinal canal that runs through the center of the spinal cord. The pia mater, a mere network of blood vessels held together by loose connective tissue, protrudes into each ventricle as a *choroid plexus* that facilitates the exchange of materials between the blood capillaries and the cerebrospinal fluid in the ventricles. Thus nourishment is made available both to the brain's outer surface and to cells lying in or near the walls of the ventricles.

The CNS contains *gray matter* and *white matter*. *Gray matter* consists of nerve cell bodies and fibers not coated with white, fatty myelin sheaths. *White matter* consists of fiber tracts that are myelinated. Communicative junctions between nerve cells occur only in the gray matter. Primitively, gray matter lies only deep within the CNS, nourished by cerebrospinal fluid, but in higher vertebrates it has increased so greatly that it has erupted to the outside and spread in a thin layer over the outer surfaces of such parts of the brain as the cerebrum and cerebellum.

18.23.   What are the functional regions of the cerebrum?

The gray matter areas of the cerebral hemispheres are the cortex, or *pallium,* and the *cerebral basal nuclei* (corpus striatum). In fishes, the pallium deals almost exclusively with olfaction, and it retains that function in tetrapods, but during the course of vertebrate evolution it has come to assume many additional roles, including sensory analysis of all types, initiation of voluntary movements, association, memory, and cognition. Three versions of the pallium are histologically and functionally distinguishable: the ancient *paleopallium,* concerned with olfaction; the *archipallium* of reptiles and birds, concerned with sensory analysis and association; and the *neopallium,* best developed in mammals, also involved with sensory analysis and association, but far more complex than the earlier versions (Question 22.6). The cerebral basal

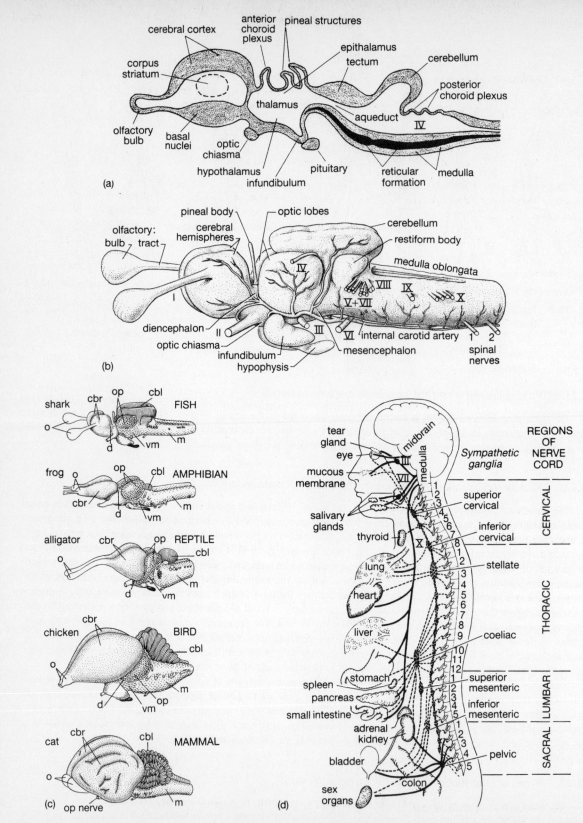

Fig. 18.9 Vertebrate nervous system. *(a)* Sagittal section of primitive vertebrate brain; cerebrospinal fluid in unshaded areas. *(b)* Shark brain, lateral view showing cranial nerves (I–X). *(c)* Representative vertebrate brains: O = olfactory lobe; cbr = cerebrum; d = diencephalon; op = optic lobe or tectum; vm = ventral midbrain; cbl = cerebellum; m = medulla oblongata. *(d)* Autonomic system: parasympathetic division, solid black lines; sympathetic division, heavy stippling and dashed lines. (*From Storer et al.*)

nuclei, known in the aggregate as the *corpus striatum*, are deep-lying gray matter areas that help control skeletal muscle action and coordination. In the bird brain, the corpus striatum forms a huge solid mass (see Fig. 21.15) and represents the highest brain center for organizing motor responses, especially complex instinctual behaviors.

18.24. What are the contributions of the diencephalon?

The diencephalon consists of the dorsal *epithalamus*, lateral *thalamus*, and ventral *hypothalamus*. During embryogeny, the *lateral eyes* develop as outgrowths from the walls of the diencephalon, and the nerve cells borne outward in these forming optic cups differentiate into the retina. The epithalamus primitively gives rise to a third, *median eye*, which in primitive vertebrates actually contains rods and cones and has a rudimentary lens and cornea (see Fig. 18.10c). In most vertebrates this organ is more glandular than eyelike and, even when furnished with lens and retina, now seems more involved with monitoring photoperiod than with actual vision. That this organ can arise from either of *two* median evaginations (*pineal* and *parietal*) of the diencephalon roof and that the retinas of the lateral eyes arise from its side walls point to a fundamental capacity of this part of the brain for differentiating photoreceptors. The thalamus contains a number of deep-lying nuclei involved in both sensory analysis and neural relays to the cerebrum. The hypothalamus is concerned with emotions, physiological regulation, reproductive and ingestive behaviors, and hormonal integration.

18.25. What are the functions of the mesencephalon?

In most vertebrates the midbrain is mainly concerned with control of eye movements and, in anamniotes, analysis of visual information as well. In mammals, the midbrain also becomes involved with acoustically oriented reflexes and muscle coordination; fibers from the midbrain to the cerebral basal nuclei secrete dopamine (*deoxyphenylalanine amine*), deficiency of which is a cause of Parkinsonism in humans.

18.26. What are the functions of the hindbrain and reticular formation?

The most conspicuous portion of the metencephalon, the *cerebellum*, is involved with muscle coordination and equilibrium. The myelencephalon, or *medulla oblongata*, relays impulses forward from the spinal cord to higher parts of the brain, regulates physiological activities, and receives or gives off most of the cranial nerves. From medulla to thalamus a fingerlike core of gray matter, the *reticular formation*, runs through the brain stem, receiving assorted sensory input. Its thalamic portion, the *reticular activating system*, maintains consciousness and alerts forebrain centers to meaningful stimuli. The nearby *raphe nuclei* produce *serotonin*, a neurotransmitter secreted from fibers extending into the activating system, which depresses the latter and so brings on *sleep*. Another closeby region (in mammals, the *locus coeruleus*) secretes *noradrenalin* onto the activating system, thus waking up the organism and (in mammals and birds only) being also somehow associated with the dreaming state.

18.27. What are the cranial nerves of vertebrates?

Nerves are bundles of nerve fibers that conduct messages to and from the CNS. *Spinal nerves* arise from the spinal cord between successive vertebrae, while *cranial nerves* (Table 18.1) arise from the brain itself. Cranial nerves number 10 pairs (I–X) in anamniotes (see Fig. 18.9b) and 12 (I–XII) pairs in amniotes (see Fig. 22.2B). *Sensory*, or *afferent*, nerves conduct messages only from the periphery into the CNS; *motor*, or *efferent*, nerves conduct impulses only from the CNS to the effectors (muscles and glands); *mixed nerves* carry both afferent and efferent fibers.

18.28. What are the functional divisions of the peripheral nervous system?

The *peripheral nervous system* (PNS) of vertebrates is both anatomically and functionally divided into two distinct parts: *somatic motor-sensory*, which controls bodily movements and is concerned with nonvisceral sensation, and *auto-nomic*, which regulates physiological processes. The PNS is made up of nerves, sense organs, and peripheral *ganglia*, the last being concentrations of nerve cells located outside the CNS. Somatic ganglia include *sensory ganglia* that contain the cell bodies of sensory neurons and lie close to the brain and spinal cord; the retina and inner ear are also ganglionic.

Table 18.1 The Cranial Nerves*

Olfactory nerve (I): actually a tract leading from neuron cell bodies located in *olfactory bulbs* to the olfactory area of the cerebral cortex; sensory, for detection of odors.†

Optic nerve (II): sensory; from retina to thalamus and midbrain.

Oculomotor nerve (III): motor; from midbrain to four of the six muscles that move the eyeball.

Trochlear nerve (IV): motor; from midbrain to *superior oblique* muscle of eyeball.

Trigeminal nerve (V): mixed (sensory and motor); from head (especially facial skin) to metencephalon and from metencephalon to jaw muscles.

Abducens nerve (VI): motor; from metencephalon to *lateral rectus* muscle of eyeball.

Facial nerve (VII): mixed; from metencephalon to head muscles and glands, from taste buds and visceral receptors to metencephalon.

Otic nerve (VIII): also known as "acoustic," "auditory," "vestibulocochlear," or "statoacoustic" nerve; sensory; from inner ear to metencephalon, for hearing, equilibrium, and gravitational orientation.

Glossopharyngeal nerve (IX): mixed; from taste buds to medulla oblongata, and from medulla to pharyngeal muscles and smooth muscles and glands of the head.

Vagus nerve (X): mixed; the main parasympathetic nerve serving internal organs including heart, lungs, digestive tract.

Accessory nerve (XI): also known as "spinal accessory" nerve; motor; from medulla to muscles that move the head (sternomastoid and trapezius).

Hypoglossal nerve (XII): motor; from medulla to tongue muscles.

*The names of the cranial nerves are easily recalled in correct order with the aid of a hoary mnemonic device that utilizes first letters to produce a memorable jingle, one version of which reads, "On Old Olympus's Towering Top, A Fat Old German Viewed A Hop!" ("Hop" meaning ski jump, one supposes.)

†The nerve fibers extending to the olfactory bulbs from the olfactory neuroepithelial cells occuring scattered within the nasal lining, are not bound together to form a single discrete pair of olfactory nerves; hence the tract from bulb to brain is commonly referred to as the olfactory "nerve."

The autonomic system (Fig. 18.9*d*) is uniquely dual, consisting of largely antagonistic *parasympathetic* and *sympathetic* divisions, the former facilitating normal organ functions, the latter mainly responsible for mobilizing the body for emergency "fight-or-flight" reactions. Parasympathetic fibers leave the brain in several cranial nerves, notably the vagus, and also emerge from the spinal cord in the sacral region, passing to ganglia located in or close to the organs served; short postganglionic fibers continue from these ganglia and end on glands or visceral muscle fibers. Sympathetic fibers exit from the middle (thoracolumbar) portion of the spinal cord, forming a chain of segmental ganglia that parallel the cord on each side and from which lengthy postganglionic fibers pass to the organs served.

18.29. What is special about vertebrate sensory reception?

Sensory reception in invertebrates is carried out by *sensory neurons* and their specialized endings; vertebrates possess not only sensory neurons but also *epithelial* and *neuroepithelial* cells that act as receptors. They also profit by some remarkably complex sense organs, notably eyes and ears, which contain not only receptors but also entire aggregations of neurons that not only are involved in sensory transmission but, at least in the case of the eye, also engage in some degree of primary sensory analysis.

18.30. What are the special features of the vertebrate eye?

The vertebrate eye (Fig. 18.10*a* and *b*) is of the *vesicular* pattern seen also in cephalopod mollusks (see Fig. 24.6). Although these eyes evolved independently, they are so highly convergent that both share many similar (analogous but not homologous) particulars, including an *iris* that changes pupillary dimensions and a *lens* with *ciliary muscles* for adjusting focal length. However, the vertebrate eye is unique in being a "brain eye": its retinal tissues grow outward from the diencephalon and are derived from the embryonic wall of this part of the forebrain. The photoreceptive *rods* and *cones* are in fact modified ciliated neuroepithelium that originated as part of the brain's inner lining. In addition to the two lateral eyes, the vertebrate diencephalon often develops photosensitive dorsal outgrowths that may form median eyes with tiny but identifiable retinas (Fig. 18.10*c*). Cyclostomes have both a median pineal eye and a less well developed parietal eye, tuatara and certain lizards have a functional parietal eye, and salamanders have a pineal eye. Even when much altered from an eyelike condition, the pineal appears to function as a seasonal timekeeper, tracking the relative duration of day and night and hormonally regulating annual reproductive cycles.

The retinas of the lateral eyes are complex neural structures within which the rods and cones communicate with

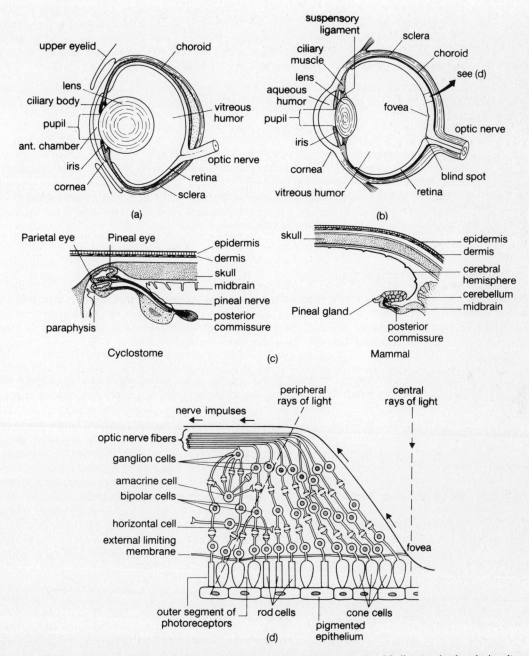

Fig. 18.10   Vertebrate photoreceptors. *(a)* Shark eye. *(b)* Human eye. *(c)* Median (parietal and pineal) eyes of a cyclostome, and their homolog, the pineal gland of mammals. *(d)* Structure of the vertebrate retina.

layers of neurons that lie entirely within the retina, and these in turn communicate with one another and with neurons called *ganglion cells*, the axons of which form the optic nerve (Fig. 18.10*d*). Whether or not a given ganglion cell will transmit impulses to the brain depends on the appropriate stimulation of one or more tiny patches of rods and cones that serve it and constitute its *receptor field*. The entire vertebrate retina is a patchwork quilt of minute receptor fields of variable shapes and orientations.

Example 5:   A given retinal receptor patch may be so shaped and oriented that it will be stimulated best by a bright, slender object presented vertically in the field of view. Another retinal patch may respond not when illuminated, but only when suddenly cast into shadow. Yet another may be excited by only an upward movement. This gives you some idea of the enormous complexity of the vertebrate retina's mosaic organization, for each example given implies a host of others (e.g., as well as upward-movement patches, there will also be retinal patches that respond to downward, leftward, rightward motion, and so forth).

Most vertebrates are capable of seeing in both dim light and bright light, and those that enjoy color vision—including bony fishes, lizards, turtles, birds, and primates—possess specialized photoreceptors that are excited only by light of certain wavelengths.

Example 6:   In the human eye the highly sensitive rods function only in dim light (because their light-absorptive pigment, *rhodopsin,* cannot regenerate in bright light) and provide nocturnal (*scotopic*) vision. *Cones* are insensitive to dim light, but since their light-absorbing pigment (*iodopsin*) regenerates in bright light, they furnish diurnal (*photopic*) vision. In addition, perception of color is possible because the human retina contains three types of cones that differ in the molecular structure of their light-sensitive pigment, so that one type of iodopsin most strongly absorbs light in the red portion of the spectrum, another most strongly absorbs green light, while a third type absorbs short, blue wavelengths.

18.31.   How does the vertebrate eye focus light?

The vertebrate eye focuses light by means of its curved *cornea* and a transparent *lens,* which is especially huge and spherical in nocturnal animals such as mice. The eye can also *accommodate* (change its focal length) for near or far vision. This is accomplished by eye muscles that change the corneal curvature (*corneal accommodation*), or change the shape of the lens (*intrinsic lenticular accommodation*), or move the lens forward or back (*extrinsic lenticular accommodation*).

Example 7:   The eye of a bony fish is "set" for near vision and accommodates for far vision when the *ciliary muscles* retract the lens toward the retina. In sharks, accommodation consists of pulling the lens forward. Fishes cannot use their corneas for focusing, because the refractive index of the cornea is the same as that of water. The eyes of tetrapods are normally adjusted for far vision. Near vision is accomplished in amphibians by moving the lens forward; in reptiles the ciliary muscles contract around the lens, squeezing it into a more spherical shape, while in mammals contraction of these muscles both bows the cornea and reduces tension on a suspensory ligament that holds the lens in place and keeps it stretched into a flatter shape. Released from tension, the mammalian lens rounds up, except that in humans, at least, loss of lens elasticity with age compromises lenticular accommodation, and corneal accommodation may not suffice to ward off the dread reading glasses. The bird eyeball is often nonspherical because of *scleroid bones* that ossify in its outer coat (*sclera*) and prevent distortion of the eyeball during flight. A bird's eye accommodates from far to near vision by means of modified ciliary muscles: *Brucke's muscle* compresses the lens into a rounded shape (as in reptiles), and *Crampton's muscle* increases the corneal bow.

18.32.   What is the functional organization of the vertebrate ear?

Small clumps of sensory cells (*neuromasts*), found widely scattered in the skin and mouth lining of primitive fishes, seem to have given rise to both the vertebrate *inner ear* and the *lateral line system* of aquatic anamniotes. Both of these are unique to vertebrates, and both are involved in detecting vibrations, i.e., *hearing.* However, the ear is also concerned with *proprioception,* the sense of bodily movement and position, for it monitors head movements and gravitational orientation. The inner ear is thought to have evolved from cephalic branches of the lateral line system (see Fig. 20.1*c*), a longitudinal series of tiny sense organs like neuromasts lying in a groove or tube along each side of the body, which are stimulated when water vibrations strike the animal's body. The inner ear first appears embryonically as an invagination from the surface which pinches off to form a hollow *otic vesicle* on each side of the hindbrain. Ancestrally, such invaginations could have borne neuromasts inward from the lateral line system, to become the sensory cells of the ear.

Each otic vesicle develops into a complicated *membranous labyrinth* (Fig. 18.11*a*) which lies within a bony or cartilaginous *otic capsule* which conforms reciprocally to its shape, becoming the *skeletal labyrinth* (Fig. 18.11*b*). The skeletal labyrinth contains a fluid (*perilymph*) that bathes the outer surface of the membranous labyrinth, which contains a separate fluid, *endolymph*. The acoustic function of the labyrinth (rudimentary in most fishes) depends on propagation of pressure waves through the *perilymph*, affecting sensory cells lying in the wall of the membranous labyrinth within a ribbonlike *organ of Corti*. In fishes, the rudiments of this organ occur in the wall of a gravity-sensing sac (*lagena*), which lengthens in tetrapods as its acoustic function develops, until in mammals it coils up like a snail and becomes known as the *cochlea*. Since the acoustic function of the ear is most highly developed in mammals, this will be considered further in Question 22.7.

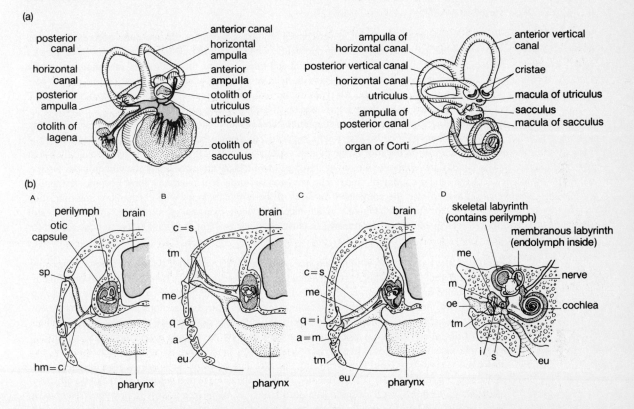

Fig. 18.11 The vertebrate ear. *(a)* Membranous labyrinth of inner ear, in fish (*left*) and mammal (*right*). Note relative size reduction for two gravity-detecting structures (sacculus and utriculus), and enlargement and coiling of sound-detecting lagena, which becomes the mammalian cochlea. The organ of Corti is a ribbon of sensory cells that monitor sound. *(b)* Position of the membranous labyrinth within the skeletal labyrinth and the relation of the inner ear to the middle and outer ears in shark (A), primitive amphibian (B), primitive reptile (C), and mammal (D); the black space shown between the membranous and skeletal labyrinths is filled with perilymph, while endolymph fills the membranous labyrinth. hm = hyomandibula, sp = spiracle, tm = tympanic membrane (eardrum), me = middle ear cavity, eu = eustachian tube, q = quadrate bone, a = articular bone, c = columella, i = incus, m = malleus, s = stapes, oe = outer ear canal.

The evolution of the lagena into an organ dedicated only to sound detection entailed no loss of gravitational sense, because the labyrinth has two other gravity sensors, the *sacculus* and *utriculus*. In bony fishes and amphibians, these sacs contain calcareous nodules (*otoliths*), sometimes handsome enough to polish for jewelry! In birds, mammals, and most living reptiles, otoliths have been replaced by microscopic calcium carbonate crystals embedded in a gelatinous blanket into which the "hairs" (nonmotile cilia) of the sensory cells are affixed; gravitational displacements shift the weighted blanket, exciting the sensory cells by pulling on their hairs. Sharks and rays lack calcareous nodules and instead take fine sand grains into the sacculus through a pore to the exterior.

Example 8: Labyrinthine gravity detectors are especially important to postural orientation when vision cannot serve. A labyrinthectomized fish successfully keeps itself right side up by directing its back toward the light, but when the tank is lit from below, the misguided creature will flip over and try to swim upside down.

The rest of the labyrinth consists of *semicircular canals*—one in hagfishes, two in lampreys, and three in gnathostomes. In the last group, the canals are oriented in three planes, like the seat, back, and one arm of a chair. This provides three coordinates for monitoring head movements that produce rotational displacement of endolymph within the canals. The receptive hair cells are clustered in a swelling (ampulla) along each canal, their sensory cilia being embedded in a gelatinous cone (*cupula*) that sways to one side or the other according to the direction of endolymph rotation.

18.33. How do vertebrates sense tastes and odors?

Taste and *olfaction* are chemical senses served in vertebrates by unique receptors that are quite different from one another. Taste receptors are epithelial cells bearing long microvilli, which are clustered into minute *taste buds* located mainly in the mouth and pharynx but also scattered over the head and trunk of some fishes; those of mammals are mostly restricted to the tongue. Only four types of taste buds have been identified, each being most sensitive to one of four primary taste acuities: *sweet, sour, salty* and *bitter*. All types are not universally present: cats, for instance, lack "sweet" taste buds. Additional flavor sensations result from a combination of gustatory and olfactory input.

Olfaction begins when molecules of certain shapes settle into complementary receptor sites in the membranes covering the sensory cilia of special *neuroepithelial* cells. These cells are scattered among the regular epithelial cells of the nasal lining and that of *vomeronasal organs* in the roof of the mouth, especially of birds and reptiles. When a snake flicks out its forked tongue, air-borne molecules adhere to the tongue, the tips of which are then thrust into the vomeronasal organs, increasing the keenness of the reptile's sense of smell. The vomeronasal organs of mammals have become vestigial as the inner nasal lining has expanded over the convoluted surfaces of the turbinate bones with concomitant multiplication of olfactory receptors. The olfactory sense of humans, while much less acute than that of most other mammals, is still incredibly sensitive, for odorous materials can be detected in concentrations as slight as 1/50,000 the amount needed for a substance to be *tasted*.

Example 9: Experiments with human subjects suggest that vertebrate olfactory receptor sites come in seven different shapes. This *stereochemical theory* is supported by the fact that the odors of synthetic organic compounds can often be predicted if their molecular architecture conforms to one of seven shapes, which produce seven primary olfactory sensations described as floral, ethereal, musky, pepperminty, camphoraceous, putrid, and pungent. An olfactory receptor cell represents an odd combination of neural and epithelial types: the columnar cell body lies among regular epithelial cells in the nasal mucous membrane, its free surface covered with nonmotile sensory cilia and its inner end prolonged into a fiber (axon) that passes into the olfactory bulbs of the brain. In anamniotes the roof of the mouth is also the floor of the cranium, so that in the absence of a separate nasal cavity, olfactory receptors are confined to the nostrils. The nostrils may be blind sacs useless in breathing (as in most cartilaginous and bony fishes) or tubules opening in the buccal cavity (as in lungfishes and most nonmammalian tetrapods). Most vertebrates have two external nostrils, but some fishes have four: two incurrent and two excurrent, through which water flows from anterior to posterior, without entering the buccal cavity. Living agnathans have a single nostril, served by two olfactory nerves.

18.34. Why is the vertebrate liver such a special organ?

The *liver* of vertebrates seems to have no real equivalent in invertebrates, although the term "liver" is often used promiscuously to designate any major digestive gland seen in eumetazoans. The vertebrate liver is not primarily a digestive gland at all, although it does secrete *bile salts* that aid fat digestion by emulsifying fat droplets and facilitating absorption of digested fats. It is the largest single organ and most versatile biochemical factory in the vertebrate body. It performs excretory functions, including the destruction of old red blood corpuscles and the removal, as *bile pigments*, of toxic organic products of hemoglobin breakdown. It also excretes in the bile food contaminants such as heavy metals and enzymatically inactivates other harmful chemicals such as drugs and insecticides. The liver converts the nitrogenous waste ammonia to less toxic urea. Its amoeboid Kupffer cells destroy bacteria by phagocytosis. It synthesizes a wide variety of *plasma proteins*, which are essential for clotting, osmotic balance, and antibody production. The liver utilizes blood sugar for the synthesis and storage of glycogen and breaks down this polysaccharide when blood sugar must be replenished. It stores vitamins such as A, D, E, and K and minerals such as iron. At times, the liver also produces red blood corpuscles.

18.35.   How do vertebrates breathe?

Vertebrates may breathe by means of lungs, internal gills, external gills, or the integument alone. Their most unique organs of gaseous exchange are the internal gills of fishes and larval amphibians and the lungs of tetrapods. Both internal gills and lungs are unusual in that they develop from part of the alimentary tract and are lined or covered by an epithelium that runs uninterruptedly throughout the entire alimentary and respiratory systems, although the specific differentiated state of this membrane varies regionally (e.g., most tetrapod air passageways are lined with ciliated epithelium, whereas the alimentary lining is not ciliated; the pulmonary air sacs are lined with squamous epithelium, whereas columnar epithelium lines most other parts of the tract).

Only vertebrate fishes and aquatic amphibians have pharyngeal gills that develop in the walls of gill pouches from which the pharyngeal clefts, now rightfully called "gill slits," open externally. While these complex internal gills and their support structures are developing, larval amphibians and unhatched sharks breathe through filamentous external gills that develop on the outside of the neck and usually disappear when the internal pharyngeal gills become functional. When amphibian metamorphosis commences, a pair of sacs grow posteriorly from the back of the pharynx, ballooning out into the coelom as simple lungs. Except for lungfishes and a few primitive ray-finned species, living bony fishes lack functional lungs, but a homolog, the *swim bladder*, develops from the gut (and sometimes even retains a duct into the gut); this sac serves mainly in buoyancy regulation but occasionally also in respiration.

Starting with amphibians, skeletal gill-support arches that became redundant as gills were abandoned were developmentally subsumed into the terrestrially adapted respiratory system as the first *laryngeal cartilages,* which guard the opening into the lungs (glottis). Amniotes never need gill arches, but after a fashion, some develop embryonically anyway and are transformed into additional cartilages of the larynx and trachea. As well as guarding the opening into the trachea, the larynx serves as the sound-producing "voice box" of all vocal tetrapods except birds (which have a special structure at the base of the trachea). Since the respiratory systems of various vertebrates differ considerably, they will be studied further as individual classes are discussed.

18.36.   What is unique about the vertebrate circulatory system?

That vertebrates have a closed circulatory system is nothing new, for such occurs in many invertebrates. But the vertebrate system is unique in having two distinct divisions: (1) a *blood-vascular system* that contains blood and includes heart, arteries, capillaries, and veins and (2) a *lymphatic system* that contains *lymph* and includes blind-ended *lymph capillaries, lymph vessels* (lymphatics), *lymph nodes,* and, in some anamniotes, *lymph hearts* for propulsion. The two systems communicate, so that the lymph eventually empties into the bloodstream.

18.37.   What is the adaptive value of the lymphatic system?

Lymph must be drained from the tissues, or an abnormal waterlogged state (*edema*) will develop. Most lymph originates as a protein-free filtrate of blood plasma that is exuded from the blood capillaries as a result of hydrostatic pressure; it serves as an *interstitial fluid* that bathes the tissues, but cannot be allowed to accumulate. Excess tissue fluid readily enters the lymph capillaries, since hydrostatic pressure within them is negligible. At the gut, lymph capillaries also transport lipids, since molecules so large cannot enter the blood capillaries. The lymph capillaries converge into lymph vessels, which have flaplike valves that prevent backflow.

If bacteria invade the tissues, they are apt to penetrate the lymphatics; thus, lymph nodes filled with defensive cells (mostly *lymphocytes*) cluster along the length of lymphatics, and as the lymph drains through these nodes, the cells detect such invaders and respond by secreting *antibodies*. Lymphocytes that multiply in the lymph nodes (and other lymphoid tissues) find their way into the bloodstream by way of the lymph flow.

18.38.   How is the vertebrate heart different from other hearts?

The vertebrate heart beats myogenically, and its beat is initiated by a single pacemaker. Nonchordate hearts beat neurogenically, but the nerves serving the vertebrate heart do no more than speed it up as needed, or slow it down again to a rate slightly under that at which it would beat if all nervous connections were severed. Myogenic pulsatility results from an inherent oscillatory mechanism in cardiac muscle fibers that causes them to contract rhythmically, even as isolated cells in tissue culture. The fibers are self-excitatory by way of a cyclic fluctuation in the cell membranes' permeability to sodium and potassium ions. When cultured cardiac cells come into contact with one another, they begin to beat synchronously, with the fastest-beating cell imposing its rate on the rest. In the intact heart, one specific chamber or region serves as pacemaker for the whole. The four primary chambers of the vertebrate heart, as seen in fishes, are, from posterior to anterior (the direction of blood flow), *sinus venosus*, *atrium*, *ventricle*, and *truncus arteriosus*. All four of these chambers are pulsatile throughout life in primitive fishes such as sharks, but in other vertebrates the sinus venosus and truncus arteriosus do not beat postembryonically.

The evolutionary and developmental histories of the vertebrate heart show the progressive loss of the sinus venosus as an actual chamber and its reduction to a *sinoatrial node* embedded in the wall of the right atrium, but even in this reduced state it still serves as the heart's pacemaker in all vertebrates. The truncus arteriosus also disappears as an actual pulsatile chamber, becoming subdivided into the bases of the great arteries (pulmonary and aorta) that spring directly from the heart. These changes are accompanied by partitioning of the atrium into separate right and left atria and, finally, the completion of a septum that divides the ventricle into separate right and left ventricles. The full spectrum of these changes in the vertebrate heart can be seen during embryonic development of higher vertebrates, such as birds and mammals, as well as by comparisons of living vertebrates (Fig. 18.12). These changes reflect the transition of tetrapod vertebrates from an aquatic to a terrestrial life. Fishes have a *single circulation* in which blood passes only once through the heart for each complete circuit of the body (Fig. 18.13). The progressive internal partitioning of the tetrapod heart into separate right and left pumps allows the "right heart" to collect deoxygenated blood from the body and pump it out to the lungs, and the "left heart" to receive oxygenated blood from the lungs and pump it out to the rest of the body. Since blood pressure and rate of flow decline greatly during passage through a capillary bed, this pattern of *double circulation*, with pulmonary and systemic circuits, increases the efficiency of circulation in land vertebrates by routing the blood twice through the heart for each complete journey through the body.

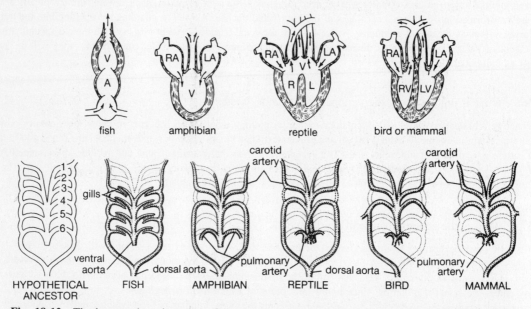

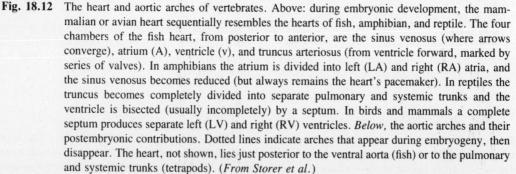

Fig. 18.12   The heart and aortic arches of vertebrates. Above: during embryonic development, the mammalian or avian heart sequentially resembles the hearts of fish, amphibian, and reptile. The four chambers of the fish heart, from posterior to anterior, are the sinus venosus (where arrows converge), atrium (A), ventricle (v), and truncus arteriosus (from ventricle forward, marked by series of valves). In amphibians the atrium is divided into left (LA) and right (RA) atria, and the sinus venosus becomes reduced (but always remains the heart's pacemaker). In reptiles the truncus becomes completely divided into separate pulmonary and systemic trunks and the ventricle is bisected (usually incompletely) by a septum. In birds and mammals a complete septum produces separate left (LV) and right (RV) ventricles. *Below,* the aortic arches and their postembryonic contributions. Dotted lines indicate arches that appear during embryogeny, then disappear. The heart, not shown, lies just posterior to the ventral aorta (fish) or to the pulmonary and systemic trunks (tetrapods). (*From Storer et al.*)

18.39.   Of what importance are the aortic arches?

The *aortic arches* (see Figs. 8.7 and 18.12) are another unique feature of the vertebrate circulatory system. These arches pass dorsally between the pharyngeal clefts, conducting blood from the ventral aorta to the dorsal aorta; in fishes and gill-breathing amphibians the central portion of each arch breaks up into a branchial capillary bed in which the blood

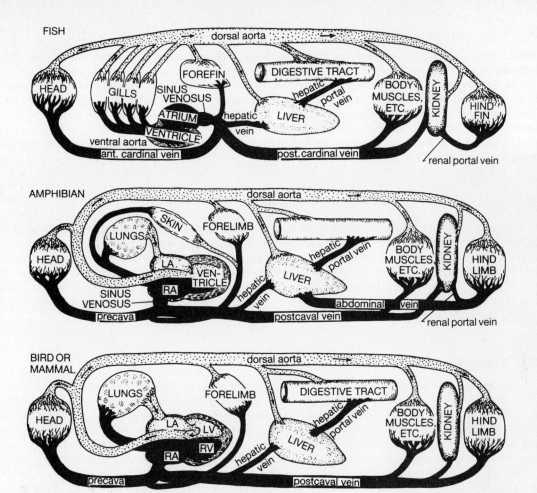

Fig. 18.13 Schematic lateral view of circulatory pathway in fish, amphibians, and birds and mammals. Arrows show direction of blood flow: oxygenated blood, white; deoxygenated blood, gray. Note that the crucial transition from a single circulation to a double circulation takes place between fishes and amphibians. In the single circulatory pathway of fishes, blood passes only once through the heart for each transit of the body; when the double circulatory pattern becomes established, blood passes twice through the heart for each complete transit of the body: that entering the right atrium is shunted to the lungs and returns oxygenated to the left atrium, to be pumped out to rest of the body. Despite the absence of a ventricular septum in amphibians, little mixing of oxygenated and deoxygenated blood takes place in the undivided ventricle. (*From Storer et al.*)

is oxygenated, but vertebrates that never have gills still develop aortic arches, which however do not proliferate into gill capillaries. Primitively, the aortic arches number six pairs, but even in gilled vertebrates only arches III–VI persist into postembryonic life. Certain of the aortic arches that form during the embryogeny of land vertebrates make essential contributions to the developing pattern of arterial distribution. Although arches I, II, and V make only a transitory appearance, arch III persists as the bases of the external carotid arteries serving the facial region and arch IV forms the paired or unpaired *systemic arch* that carries blood to the dorsal aorta from the aortic trunk at the heart. In amphibians and reptiles, the systemic arch persists on both sides, which must have struck ancestral birds and mammals as too much of a good thing, because in these two classes one systemic arch or the other has lost its connection with the dorsal aorta and remains simply as the base of the subclavian artery of that side (in birds the right systemic arch persists, and in mammals the left). Arch VI serves adult amphibians as part of the *pulmocutaneous artery* carrying deoxygenated blood to the lungs and skin; in amniotes, arch VI becomes merely the pulmonary arteries to the lungs, but its primitive attachment to the aorta remains useful in fetal mammals as the *ductus arteriosus,* which shunts blood from the pulmonary artery directly to the aorta, bypassing the lungs, to obvious benefit since fetal mammals exchange gases at the placenta.

18.40. What is a portal system?

Portal systems represent a unique and important feature of the vertebrate venous system: they allow materials entering the bloodstream at one capillary bed to leave the bloodstream promptly at a second capillary bed without being distributed throughout the entire body. Ordinarily, only arteries subdivide into capillaries, but in a portal system, a *vein* does so. Since blood in a portal vein has become deoxygenated during its transit of the first capillary bed, any organ that receives blood from a portal vein necessarily must be doubly vascularized: it must also receive oxygenated blood by way of an incoming artery. The portal and arterial blood flows mix by anastomosis at the capillary level, so that the tissues are in intimate contact with substances present in each. Vertebrates enjoy three such portal systems: hepatic, renal, and hypophyseal.

18.41. What does the hepatic portal system accomplish?

The *hepatic portal system* collects venous blood mainly from the wall of the gut and, by way of a huge hepatic portal vein, delivers this nutrient-rich blood into the liver, which thus is given first priority in taking up the contained amino acids, glucose, vitamins, etc. Considering the wide variety of vital functions performed by the liver (see Question 18.34), you can see the critical importance of the hepatic portal system in delivering useful food materials directly from gut to liver. But by the same token, the liver also receives greatest exposure to ingested toxins, and may be damaged thereby, as in the fatty degeneration and hepatic cirrhosis of alcoholism. Hepatic cirrhosis (hardening) impedes the portal blood flow, creating back pressure that may result in such niceties as varicose veins in the esophagus.

18.42. What is the significance of the renal portal system?

The *renal portal system* evolved in fishes as a means of detoxifying venous blood returning from the tail, since this blood is laden with wastes produced by the caudal muscles used in swimming. Throughout the life of a fish, its kidneys receive both oxygenated blood from the renal arteries and venous blood from the renal portal veins. In reptiles, birds, and mammals, the renal portal system appears only transitorily during embryogeny, but its legacy to the vertebrate kidney persists in the form of a double set of renal capillaries. Thus, arterial blood entering the kidney first passes through capillary clusters called *glomeruli*, where an especially high blood pressure causes water and small dissolved molecules to filter through the capillary wall into the adjacent kidney tubules, but the blood does not yet relinquish its oxygen. The still oxygenated blood then leaves the glomeruli by the way of efferent arterioles, which join the ancient renal portal capillary beds that surround the renal tubules along their length; here gaseous exchange takes place, additional materials are excreted from the bloodstream into the renal tubule, and a good deal of water and useful solutes are retrieved from the glomerular filtrate and transferred back into the bloodstream. Thus, the two anastomosing sets of renal capillaries provide the vertebrate kidney with two opportunities for exchanging materials between the bloodstream and the forming urine.

18.43. What is the importance of the hypophyseal portal system?

Although very localized, the *hypophyseal portal system* is vital to the coordination of a number of the scattered hormone-producing glands that make up the vertebrate *endocrine system*. It interconnects two capillary beds that lie quite close together. The first serves the *hypothalamus* of the brain, where it collects neurohormones secreted by certain neurosecretory brain cells, a short hypophyseal portal vein then delivers this blood directly into the capillary bed of the *pituitary* gland (*hypophysis*), so that only tiny amounts of these neurohormones need be secreted. Since hypothalamic neurohormones control the release of pituitary hormones, the brain thus exerts intimate control over the output of the vertebrate body's "master" endocrine gland.

18.44. How is vertebrate blood unusual?

Vertebrate blood contains a much greater proportion of cells to plasma than that of invertebrates; in humans, no less than 45 percent of the blood volume consists of cells. The cellular components of vertebrate blood are of three main types: (*a*) hemoglobin-containing *erythrocytes* involved in gas transport (see Question 22.11); (*b*) amoeboid *leukocytes* important in bodily defense; and (*c*) spindle-shaped *thrombocytes* (replaced in mammals by enucleated *platelets*), which help control hemorrhage.

Example 10: As seen in mammals, leukocytes make up several distinct types with different functions: *neutrophils* and *monocytes* are phagocytic; *basophils* secrete inflammation-promoting histamine; *eosinophils* multiply in the presence of parasitic worms; *B lymphocytes,* when activated, secrete specific antibodies; and activated *T lymphocytes* assist the B cells, produce antibodies that stay attached to their membrane and function only on contact, and retain a "memory" of the pathogen so that the immune response is accelerated upon repeated exposure.

18.45. How does the vertebrate urinary system differ from that of other animals?

The vertebrate excretory system is unique in several respects:

(*a*) Vertebrates are the only animals that possess, during their life history, *two* sets of functional kidneys. The embryonic kidneys, always located anterior to the definitive kidneys, meet the excretory needs of the developing fetus until the second set has matured enough to take over the job. Potential nephrogenic tissue lies dorsally all along the length of the embryonic coelom (Fig. 18.14*a*). In anamniotes, the most anterior part of this *nephrogenic plate* develops a pair of *pronephric kidneys* that function until the rest of the plate matures into a pair of *opisthonephric kidneys,* the definitive excretory organs (Fig. 18.14*b*). In amniotes, the pronephros briefly appears but is nonfunctional, and a pair of *mesonephric kidneys* quickly differentiate from the central portion of the plate to serve the needs of the embryo; after a pair of *metanephric kidneys* develop from the most posterior portion of the plate, the mesonephric kidneys cease to function.

(*b*) The *nephron,* or microscopic functional unit of vertebrate kidneys, is a unique evolutionary offshoot of the metanephridial type. Its history can be traced inferentially by examining the embryonic and postembryonic kidneys of living vertebrates. In pronephric kidneys (see Fig. 18.14*a*), each nephron has a ciliated funnel (*nephrostome*) and *coelomoduct* draining coelomic fluid in the same manner as seen in the *metanephridia* of the invertebrates. Near each nephrostome, a compact ball of capillaries (*glomerulus*) bulges into the coelom; the intraglomerular blood pressure forces a plasma filtrate through the capillary wall into the coelomic fluid, where it is drawn into the renal tubule by the nephrostomal cilia. In the next evolutionary stage, seen in the mesonephros, a cuplike expansion (*Bowman's capsule*) develops from the side of each nephron, enclosing a glomerulus; now fluid enters the renal tubule both by way of the nephrostome and by way of glomerular filtration directly into the capsule. This condition is also seen in the opisthonephric kidneys of adult anamniotes, in which coelomoducts persist but are narrow and vestigial. Finally, the nephrostomes and coelomoducts totally disappear (in metanephric kidneys), the double-walled cup forming the Bowman's capsule becomes the inner end of the nephron, and only glomerular filtrate, not coelomic fluid, is transported.

(*c*) The *ducts* that drain the embryonic (and ancestral) kidneys (see Fig. 18.14*b*) persist into postembryonic life as combined *urinary* and *sperm ducts* in male anamniotes, or only as sperm ducts in male amniotes, since the metanephric amniote kidney is drained medially by a new ureter (*metanephric duct*). In amniotes, therefore, passageways for sperm and urine remain separate until the point at which the two sperm ducts empty into the cloaca (of reptiles and birds) or join the urethra just below the urinary bladder (in mammals).

18.46. What are the unique features of the vertebrate reproductive system?

As in most eumetazoans, this system includes *gonads, genital ducts, glands,* and, in some vertebrates, *genitalia* such as copulatory organs. However, once more there are certain salient features found only in vertebrates:

(*a*) *Sexual indifference* is characteristic of the embryonic development of the vertebrate gonads, genital tracts, and genitalia. This sounds bad, but it simply means that up until a certain time, the gonads and genitalia are sexually indistinguishable and incipient genital tracts of each sex are present in both sexes. This condition predisposes lower vertebrates to a form of hermaphroditism involving sex reversals (even reversible reversals), especially in certain bony fishes (anemone fish, for instance, are reported to change sex back and forth according to which sex may be needed to complete a breeding pair, whereas female wrasses eventually end up as males if they live long enough). If the embryo is genetically male, the time comes when its gonads begin to secrete a tiny amount of testosterone; this is enough to masculinize the genital tract and genitalia. The opposite is not true of females: it is not the presence of estrogen, but the absence of testosterone, that allows the reproductive system to feminize.

Example 11: Cattle twins of opposite sex provide a startling illustration of the devastating effects of prenatal exposure of a genetic female to even the minute amounts of testosterone reaching her from her brother by way of anastomosing placental blood vessels. When the twins are born, *he* is normal,

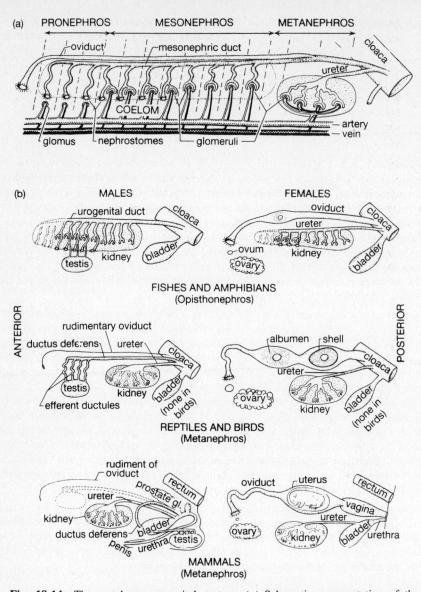

Fig. 18.14 The vertebrate urogenital system. *(a)* Schematic representation of the relationships among embryonic and postembryonic kidneys, urinary and genital ducts, and blood vessels. The *pronephros* (functional embryonic kidney of fishes and amphibians, vestigial in amniotes) consists of segmentally arranged tubules opening into the coelom by way of ciliated funnels, with a ball of capillaries (glomus) nearby; renal tubules drain laterally into a longitudinal pronephric duct (later known as mesonephric duct, finally becoming the male sperm duct in amniotes). The *mesonephros* (functional embryonic kidney of amniotes) develops posterior to the pronephros and is roughly equivalent to the postembryonic opisthonephros of anamniotes; note addition of cup-shaped Bowman's capsules enclosing capillary balls (glomeruli). The *metanephros* (postembryonic kidney of amniotes) develops posterior to the mesonephros with outgrowth of a new drainage duct (metanephric duct, here labeled "ureter"); the nephrons lack openings into the coelom and end blindly in Bowman's capsules. *(b)* Urogenital systems of male and female vertebrates, shown on one side only. Dotted lines indicate embryonic position of pronephros in fishes and amphibians; dashed lines show primitive position of testis during early mammalian development.

but she is ruined for life: she will grow up to be a steerlike "freemartin," behaviorally neuter, sterile, and with uncertainly bisexual or even masculinized genital tract and genitalia.

In the course of normal development, even after the vertebrate reproductive system has become genderized, genital structures of one sex persist at least vestigially in the opposite sex, a fact untrue of dioecious invertebrates.

> **Example 12:** In mammals the same embryonic structure that develops into the *glans penis* in males becomes the female *clitoris*, which is still important in sexual arousal and orgasm. Also, the embryonic structures that form the female *labia majora*, which enclose the more delicate genitalia and duct orifices, develop in males into the *scrotum* that holds the testes outside the body cavity.

(b) Vertebrate *gonads*, a single pair of ovaries or testes, are unique in being not only sites of *gametogenesis*, but also major *endocrine glands* that produce *sex hormones:* (1) *androgens* (chiefly testosterone) promote male traits and sex drive; (2) *estrogens* produce female traits; and (3) *gestagens* (chiefly progesterone) support pregnancy and promote parental behaviors. Both gametogenesis and sex hormone production are controlled by the hypothalamus of the brain, which secretes *gonadotrophin release factors* that prompt the pituitary gland to produce *follicle-stimulating hormone* (FSH) and *luteinizing hormone* (LH). Although their actions overlap, FSH chiefly promotes oogenesis and spermatogenesis, whereas LH stimulates sex hormone production in both sexes and triggers ovulation (egg release from a ripe ovarian follicle).

Chapter 19

Anamniotes I: Primitive and Cartilaginous Fishes

Fishes and amphibians are known as *anamniotes* because they do *not* lay amniote eggs in which the embryos themselves produce protective enclosing membranes, such as the amnion. These membranes are not to be confused with an egg shell or case, which is secreted by the mother's genital tract. Most oviparous anamniotes lay eggs provided at most with a gelatinous coating, but the eggs of hagfish and oviparous cartilaginous fishes are well protected by tough horny egg cases; still, these cannot develop on land, for the embryo would perish of dehydration. Furthermore, most anamniotes breathe by *gills* throughout their lives, although lungfishes have both gills and lungs simultaneously, and most amphibians start life with gills but become pulmonate upon metamorphosis.

The term "fishes" refers generally to vertebrates that have *fins* instead of land-adapted legs and possess gills throughout their lives. In actuality, fishes constitute some seven distinct classes of vertebrates: *Ostracodermi, Myxini* (hagfishes), and *Cephalaspidomorphi* (lampreys), of superclass Agnatha, and *Placodermi, Acanthodia, Chondrichthyes,* and *Osteichthyes,* of superclass Gnathostomata. Modern lampreys and hagfishes, and extinct placoderms and acanthodians, appear each to have arisen independently from ostracoderms, the earliest vertebrates represented in the fossil record. Chondrichthyans (cartilaginous fishes) quite possibly arose from early placoderms, and osteichthyans (bony fishes) from primitive acanthodians. Living fishes include some 30 species of lampreys, 21 species of hagfishes, 1000 species of chondrichthyans, and perhaps 30,000 species of bony fishes. Fossil evidence strongly suggests that the earliest tetrapod land vertebrates (labyrinthodont amphibians) evolved from lobe-finned fishes of class Osteichthyes, and even today most amphibians are restricted to moist habitats, lay their eggs in fresh water, and pass through a larval stage in which the limbs may either be lacking or feeble and augmented by a broad swimming tail. Despite their four-footed condition, in their developmental processes amphibians remain closer to the fishes than to their early offshoot, the reptiles. Reptiles did perfect the amniote egg, and a waterproof epidermis to boot, and passed on these traits to their own offshoots, the birds and mammals.

The story of vertebrate evolution from the ostracoderms on is spelled out in the fossil record in greater detail than the history of many invertebrate groups of similar magnitude, because vertebrates are fairly good-sized and their bones and teeth are so resistant that they have a good chance of becoming petrified. Dentition not only is useful in defining the diet of vanished animals, but also is valuable in determining relatedness among different vertebrate species.

19.1. What are agnathans?

Their fossilized armor plates tell us that primordial vertebrates had no jaws, but filtered food from water taken in through the mouth and strained through their pharyngeal clefts, in this respect resembling cephalochordates. However, whereas amphioxus draws in food only by ciliary action, even the earliest known vertebrates possessed a pharyngeal structure that enabled them to suck in water by muscular pumping, like little vacuum cleaners. These small armored fishes, dubbed "ostracoderms" ("shell skin"), existed as much as 500 million years ago and survived for over 150 million years before vanishing into history. Two highly modified types of jawless fishes, the hagfishes and lampreys, have persisted to the present day, and although they possess no armor at all, they still are of direct enough descent from the ostracoderms to warrant inclusion in the same vertebrate superclass, *Agnatha*.

19.2. What were ostracoderms?

Picture an odd little fish, its flattened trunk armored turtlewise with dense bony plates, grubbing along the bottom of some primordial bay or river, clumsily propelled by undulations of a sturdy tail clothed in the "chain mail" of smaller bony scales (Fig. 19.1). Not only does it search the dangerous shadows with wide-spaced lateral eyes, but its median pineal eye, centered on its forehead, gazes blandly upward, monitoring peril from above. What predators warrant such armoring and three-eyed watchfulness? This is an invertebrate world, in the Ordovician Period some 475 million years

280

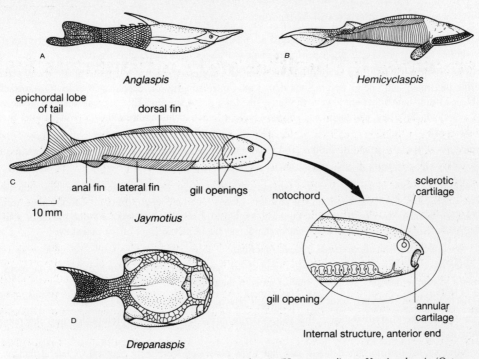

Fig. 19.1 Ostracoderms. Lateral views: A, *Anglaspis* (Heterostraci); B, *Hemicyclaspis* (Osteostraci); C, *Jaymotius* (Anaspida), with detail of anterior end. Dorsal view: D, *Drepanaspis* (Heterostraci).

ago. We may well owe our strong internal skeletons to the selective pressure exerted on these, our earliest timid ancestors, by a most formidable predaceous arthropod that lurked in the same habitats that have yielded ostracoderm remains: the giant eurypterid, or water scorpion. Up to 2 m in length, water scorpions at first may have relished these new munchables, but selective survival of individuals with armor too thick to munch would eventually have rendered ostracoderms essentially immune to predation by invertebrates, which cannot digest bone. Only vertebrates produce a protein-digesting enzyme that works under the intensely acidic gastric conditions needed to dissolve the mineral matrix of bone. Probably ostracoderms could not prey extensively upon each other, except for eggs or small fry, for they were (at least at first) filter feeders that sifted bottom mud with their jawless mouths.

Quite possibly, ostracoderms would be with us today if only they had refrained from evolving ungrateful, jawed descendants that could outswim them and no doubt could crunch them up, armor and all, like popcorn. As things turned out, jawed fishes first appeared in Silurian sediments laid down 425 million years ago, and ostracoderms vanished from the fossil record at the end of the Devonian, 345 million years ago. This means that they held out against the terror of jaws for a respectable 80 million years and during that time developed considerable diversity of body form with adaptation to many modes of life.

Example 1: Some ostracoderms came to have highly reduced armor and may have filter-fed on marine plankton. Another group (the coelepids, possible ancestors of hagfishes) had armor reduced to mere denticles, but apparently still sucked food from the bottom. One heavily armored type had a broad "saw" like that of a sawfish, but projecting forward *below* the mouth.

However, jaws did not evolve in these later, more specialized ostracoderms, but only in some earlier, more primitive type. This fact illustrates an evolutionary principle that seems to apply to the origin of each new vertebrate group: the innovative descendant lineage shows up quite *early* in the evolutionary history of the *ancestral* group, and the latter goes on to develop an extensive adaptive radiation while the new group is slowly getting under way.

Although some paleontologists would relegate ostracoderms to a side branch of the vertebrate evolutionary tree, it seems more likely that this group was ancestral to all other vertebrates, since bone is a very sophisticated tissue, and its deposition depends on a genetic and hormonal complex so involved that it would be highly improbable for both the tissue and its regulatory mechanisms to evolve independently in two or more different ancestral lineages. Absence of bony tissue from the skeletons of cyclostomes and chondrichthyans would therefore appear to represent a secondary loss of a very ancient vertebrate trait.

19.3. What were the characteristics of ostracoderms?

(*a*) Although some reached a length of 60 cm, most were only 20–30 cm long, but even this exceeded the size of most invertebrates (an important exception being the giant water scorpions mentioned above).

(*b*) The ostracoderm head and trunk were often dorsoventrally depressed, with a bony head shield, bony plates or spines covering the trunk, and one or two median dorsal fins. Interestingly, one cephalaspid order (Osteostraci) produced one-pieced head shields that are always the same size for a given species. This suggests that the larvae were naked, like lamprey ammocoetes, and their head shields developed upon metamorphosis, never growing after first becoming ossified.

(*c*) The mouth was jawless, round or slitlike, and usually rather small (which would increase the strength of suction).

(*d*) A pair of lateral eyes and a median dorsal pineal eye were present.

(*e*) The number of nostrils in the earliest ostracoderms remains in doubt, but since brain impressions show two separate olfactory bulbs and since a separate nostril probably served each bulb, these primordial forms have been placed in subclass Diplorhina ("two nostrils"). By contrast, members of subclass Monorhina ("one nostril"), also known as *cephalaspids*, had a single median nasal slit in the dorsal shield just in front of the pineal eye.

(*f*) Only two semicircular canals occurred in each inner ear.

(*g*) The single-piece head shield of cephalaspid ostracoderms has provided fossil evidence of a cartilaginous brain case enclosing a brain remarkably similar to that of modern lampreys.

(*h*) The pharyngeal wall contained some 10 pairs of gill pouches.

(*i*) The structure of the pharynx indicates that ostracoderms could suck water in forcibly, by muscular pumping, even drawing in fair-sized prey.

(*j*) Most ostracoderms had no paired fins at all, but some possessed stabilizing pectoral or lateral flanges, lobes, or spines, which are not considered homologous to the paired fins of jawed fishes.

(*k*) The tail was large and laterally compressed; it was also flexible, owing to the reduction of the bony armor to small scales, so that it could be undulated from side to side; the caudal fin was heterocercal (with unequal lobes), some species having an enlarged dorsal lobe, others an enlarged ventral lobe, probably as needed to counteract the shape and weight of the foreparts.

19.4. What are cyclostomes?

Present-day lampreys and hagfishes (Fig.19.2) have in common a scaleless, elongated, cylindrical, eel-like body; a round, jawless mouth; a tongue bearing horny "teeth"; spherical gill pouches; a single nostril; no paired fins; and poorly developed cartilaginous skeletons with persistent notochords and no bony tissue at all. The brain has very small cerebral hemispheres and an even more rudimentary cerebellum. The spinal cord is flattened, and unlike all other vertebrates, the dorsal (sensory) and ventral (motor) roots of the spinal nerves arise *alternately* instead of simultaneously and remain separate. The heart consists of sinus venosus, atrium, and ventricle, but no truncus arteriosus. Their kidneys lack the tubular capillary beds that arose in gnathostome fishes from branches of the renal portal veins. The gonad is single and lacks a duct; the gametes are shed into the coelom and pass through pores into the urogenital sinus from which they are shed. The eggs of both are externally fertilized.

Because of these similarities, lampreys and hagfishes were once classified together in class Cyclostomata ("round-mouthed"). Yet this is now abandoned as being a polyphyletic taxon. Lampreys seem definitely to have evolved from the osteostracans of subclass Monorhina. The relationships of hagfishes remain obscure, but some evidence suggests their separate origin from subclass Diplorhina. Although the term "cyclostome" remains useful in referring to the eel-like modern descendants of ostracoderms, lampreys are now placed in *class Cephalaspidomorphi* ("head shield form") and hagfishes in *class Myxini* ("slime"). Since ostracoderms were not eel-like in form, it might seem odd that modern descendants from two quite different ostracoderm groups would have converged in this respect. Actually, this is not as coincidental as it seems, for with the evolutionary loss of dermal bone and with no appendages to assist locomotion, swimming could require a great deal of lateral undulation, which becomes more effective as a body elongates to serpentine proportions.

19.5. What are the characteristics of lampreys?

Lampreys fall into three groups:

1. *Sea lampreys* migrate into fresh water to breed, and their filter-feeding ammocoetes larvae reside there for 3–5 years before migrating downstream and metamorphosing into parasitic adults. Adults suck blood from fishes by using the

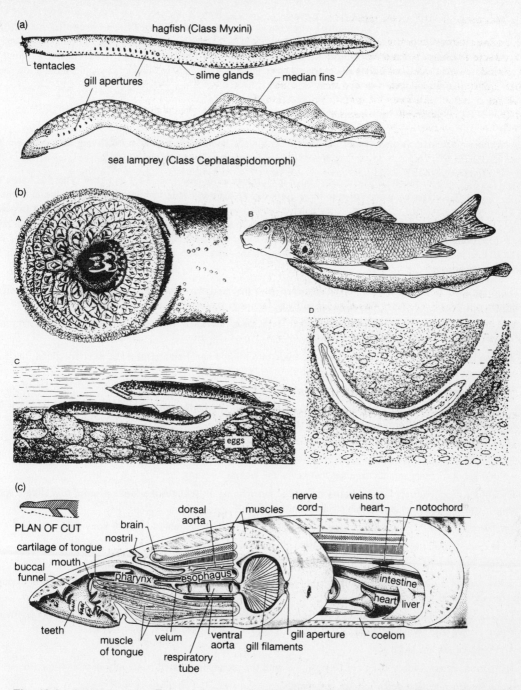

Fig. 19.2 Cyclostomes *(a)* External features, *(b)* Lampreys in habitat: A, detail of buccal funnel with horny teeth; B, lamprey clinging to host fish by way of buccal funnel; C, spawning, with male moving rocks to make a nest with downstream barricade; D, filter-feeding, nonparasitic ammocoetes larva in burrow on stream bottom. *(c)* Internal structure of anterior portion of adult lamprey, with left side of body removed in part. [*From Storer et al.; (a),* A *after Wolcott,* B *after Norman, (b) after Gage.*]

circular edge of the buccal funnel like a suction cup and using the horny, toothlike structures on the funnel and tongue to rasp through the the host's skin. The host may eventually be killed if the lamprey rasps entirely through its body wall and perforates the peritoneal cavity.

Example 2:   This occurred when sea lampreys invaded the upper Great Lakes after construction of the Welland Canal to carry shipping around Niagara Falls; this large species found the lakes to its liking, became resident, and demolished a lake trout industry that had yielded nearly 5 million kg a year. Only recently has a chemical pesticide poisonous only to lamprey larvae begun to allow the lake trout population to recover.

2.   *River lampreys* migrate into fresh water to breed and down to the sea when mature, but they do not feed parasitically, instead preying on crustaceans, mollusks, and worms.

3.   *Brook lampreys* remain in fresh water all their lives, but the adults do not feed at all, living only long enough to reproduce.

The characteristics of the ammocoetes larva have been reviewed in Questions 18.7 and 18.8. These lamprey larvae (see Fig. 18.3) may well have carried down to the present day an ancestral state that antedated the evolution of the bony armor by which ostracoderms are known. They hatch after two weeks' development, from 1-mm eggs laid in a nest of pebbles built mostly by the father, who uses his sucking mouth to lift and move pebbles to make a barricade that protects the eggs from being swept downstream. The larvae leave the nest in a week but remain filter feeders for up to 7 years.

The characteristics of *adult* lampreys are given below in the same order as those of hagfishes (see Question 19.6), to facilitate comparison:

1.   Lampreys are basically freshwater inhabitants; even those that go down to the sea as adults are *anadromous*, migrating upstream to breed and spend their long larval lives; lampreys never breed in the sea.

2.   The eyes and pineal body are well developed, with even the latter having a lens and pigmented retina (see Fig. 18.10*c*).

3.   The single nostril lies on top of the head and leads into a blind canal ending near the pituitary gland; the canal is *not* supported by cartilage rings.

4.   The skin is moistened by mucus, but great quantities of slime are not produced.

5.   Except during breeding, parasitic lampreys are usually found attached to the body of a host, ordinarily a fish, which it usually leaves still alive but much weakened.

6.   The underside of the head forms a *buccal funnel* studded with numerous horny "teeth"; a protrusible "tongue" bears more of the same, allowing the lamprey to attach firmly by suction and rasp into the host's flesh.

7.   An oral gland secretes an anticoagulant that keeps the wound bleeding.

8.   The simple digestive tract lacks a stomach, but the inner lining of the intestine forms a *spiral fold* (like a spiraling ramp), which increases the intestine's functional length.

9.   The blood salt level is much lower than that of seawater and is comparable to that of bony fishes.

10.   As in osteostracans, two semicircular canals occur in each ear.

11.   A segmental row of sensory pits, the lateral line system, extends down each side of trunk and tail.

12.   Seven pairs of gill pouches open to the exterior by separate pores; during breathing, water is sucked in through the pores and forced out again.

13.   The branchial skeleton is well developed, and an incipient vertebral column exists as paired cartilages that form *neural arches* over the spinal cord.

14.   Accessory hearts are lacking.

15.   The postembryonic kidney is an opisthonephros, with the pronephros functioning only in embryonic life.

16.   Sexes are separate, and death follows spawning.

19.6.   What are hagfishes?

Hagfishes inhabit the sea at depths from 24 m to over 1800 m, tending to stay near the bottom, where they live aggregatively, in burrows dug in mud. They both tunnel after buried worms and swim weakly to find distressed fish or even bait, which they swallow without second thoughts, hook and all. Little is known of their lives in the wild. The eggs, over 1 cm long, are laid in tough cases bearing hooks by which they attach to the bottom, and development is direct, with no larval stage.

The characteristics of hagfishes are as follows:

1. Hagfishes are exclusively marine.

2. The lateral eyes are degenerate and covered by thick skin, but the skin covering the median eye is transparent.

3. The single, large nostril opens at the very anterior end and is useful for inhalation since it leads posteriorly into the pharynx; the nasal canal is supported by cartilaginous rings, and a pair of nasal sacs open laterally into this canal (evidence of a diplorhinid origin of myxinoids).

4. Numerous large *mucous glands* along the sides of the coelom open through the body wall and can shoot out coiled threads of mucus and protein that unwind and trap the mucus against the body as a heavy coat of slime; a caught hagfish ties its body in a knot and works the knot forward until it pulls itself free of both predator and slime, then celebrates escape by sneezing to clear its nasal passage.

5. Hagfishes hunt actively for prey by olfaction, burrowing after buried invertebrates, and are also quickly attracted to bait or dead or dying fish.

6. The mouth is surrounded by six sensory barbels supported by cartilages; it lacks a buccal funnel but has one large dorsal horny "tooth," and the protrusible "tongue" bears on each side a toothed horny plate. These plates separate when the tongue is pushed out and lock together when the tongue is retracted, thereby either pulling in small prey whole or locking onto a pinch of flesh from larger prey. In the latter event, the hagfish pulls its knot trick again, slipping a knot forward along its body until it can push against the body of the prey, tearing off the bite. A healthy fish would not stand still for this, so snared and dying fish fall victim. Eventually, the hagfish succeeds in boring into the fish's interior, where it devours the soft body parts from the inside out, leaving only skin and bones.

7. No anticoagulant is secreted, for hagfish are not bloodsuckers.

8. The digestive tract lacks both stomach and spiral intestinal fold.

9. Hagfish blood is isotonic with seawater; sodium and chloride levels are much higher than in any other vertebrates, but in addition, high concentrations of potassium ion help maintain osmotic balance.

10. Each ear contains a single semicircular canal.

11. The lateral line system is very poorly developed.

12. Gill pouches number 5–14 pairs, according to species; each pouch empties into a tube that leads posteriorly before opening to the surface, so that the gill pores lie as far back as the middle of the body; these tubes may coalesce so that the pores may be fewer than the pouches; in some species all tubes on one side join to form a single excurrent opening on each side.

13. The branchial skeleton is poorly developed, and neural arches are lacking.

14. Accessory hearts occur as muscular bulbs along veins in the caudal region, but even so, blood pressure is extremely low and much blood is held in large blood sinuses.

15. Functional adult kidneys include both the opisthonephros and the segmental pronephros with open coelomic funnels (see Fig. 18.13).

16. Some species may be hermaphroditic, with the gonad producing sperm at one time and eggs at another.

19.7.   How did jaws evolve?

Jaws are excellent features: they let one bite. Biting is very useful in feeding, but jaws armed with teeth also constitute most vertebrates' main weapon. Vertebrate jaws represent a modification of the first pair of gill arches (together referred to as the *mandibular arch*), which gave up their original function of gill support and became braced against the skull. Gill arches are not one-pieced structures, but, rather, are made up of a string of cartilages, including dorsal *epibranchial* and lateral *ceratobranchial* elements (Fig. 19.3a). The ceratobranchial portion of the left and right first arches became the *mandibular cartilages* of the *primary lower jaw,* and the epibranchial portion the *palatoquadrate cartilages* of the *primary upper jaw*. The right and left halves of each jaw ordinarily fuse together in the midline, either directly or by way of encasing dermal bones.

Embryonically, the primary jaws of most vertebrates still are formed as cartilages and then become invested with a covering of dermal bones that make up the *secondary jaws*. In tetrapods these dermal bones usually knit the upper jaw firmly to the skull in such a way that the lower jaw alone remains movable, but in chondrichthyans and most bony fishes,

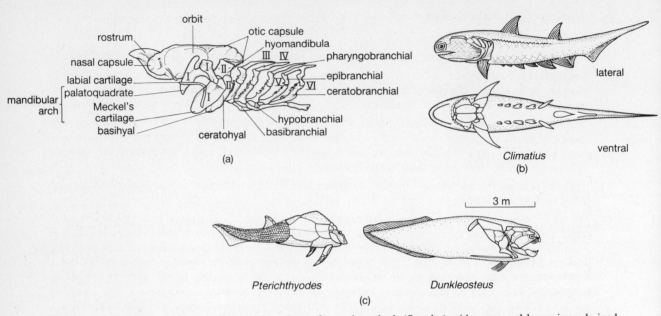

Fig. 19.3 The origin of jaws. *(a)* The primitive jaws of a modern shark (*Squalus*) with upper and lower jaws derived from gill arch I (mandibular arch) and upper jaw unfused to cranium; note upper portion of arch II (hyomandibula) involved in suspension of jaws from cranium. *(b)* *Climatius*, a very primitive jawed fish of the extinct class Acanthodia. *(c)* Placoderms: *left,* antiarch, *right,* arthrodire.

both jaws can be moved independently of the cranium. A dorsal portion of the *hyoid arch* (II), the *hyomandibular cartilage*, is often involved in jaw suspension in fishes. In placoderms the jaws articulated directly with the cranium (*euautostyly:* "true self-support") and the hyoid arch remained uninvolved; in primitive postplacoderm fishes the hyomandibula added its support so that the jaws articulated to the skull both directly and by way of the hyomandibula (*euamphistyly:* "true double-support"); finally, in more advanced fishes the hyomandibula came to intervene between the cranium and mandibular arch (*hyostyly*).

19.8. Who were the first "Jaws"?

Could gill arch I have developed into jaws in more than one line of ostracoderm descendants? This seems to be a distinct possibility, since two quite different groups of jawed fishes appeared during the Silurian. These were the *acanthodians* and the *placoderms*.

19.9. What were the acanthodians?

Acanthodians ("spinous") first appear fragmentarily in Silurian strata 425 million years old, with complete fossils being found from the Devonian into the Permian, when this group became extinct. During some 175 million years, they remained remarkably conservative in form, and most did not change appreciably from their earliest fossil appearance, which presents a surprising illusion of modernity (Fig. 19.3b). Acanthodians lived in both marine and freshwater habitats, and their fusiform bodies suggest that they were good swimmers. Their precise ostracoderm affiliation remains problematic. Formerly they were classified as placoderms, but the two groups appear to be distinct.

Acanthodian characteristics include the following:

1. The dorsal, anal, and paired fins consisted only of a sturdy bony spine forming the leading edge, with nothing more than a skin web extending posteriorly from spine to body. Although somewhat larger spines developed ventrolaterally in the pectoral and pelvic positions, an entire series of smaller paired spines occurred in between.

2. The caudal fin was heterocercal, with a larger upper lobe.

3. The rounded head terminated bluntly, with a large, often somewhat undershot, mouth.

4. The teeth, resembling those of sharks, were either fused onto the jaw cartilages or, as in sharks, attached to connective tissue bands allowing continuous tooth replacement.

5. Small squarish scales that grew as the body did typically covered the fish uniformly, but in many acanthodians certain scales were enlarged into gill covers, while in others the head scales became vestigial.

6. The cartilaginous cranium enclosed three semicircular canals.

7. The vertebral column consisted only of a segmental series of neural arches tenting the spinal cord above the notochord and a series of inverted *hemal arches* bracketing blood vessels below the notochord.

19.10.   What were the placoderms?

Placoderms ("plate skin") actually appeared later in the fossil record than acanthodians (late Silurian) and survived for a much shorter period, disappearing during the early Carboniferous (Fig. 19.3c). Primitive placoderms were benthonic and considerably resembled typical ostracoderms, especially in the extensiveness of their bony armor. Unlike the acanthodians, they radiated into many specialized types, including a gigantic predatory arthrodire (*Dunkleosteus*) 10 m long.
Major characteristics include the following:

1. Thick bony plates ordinarily armored the anterior half or third of the body, the posterior portion being naked or covered with small scales allowing mobility.

2. The scales and plates were of the *cosmoid* type, made up of a deeper layer of dense *lamellate bone*, a more superficial layer of *vascularized bone* (penetrated by blood vessels), and an outer coating of dentinelike *cosmine*, which in later placoderms formed toothlike dermal denticles. Retention of the denticle while the rest of the scale or plate becomes lost is thought to have been the origin of teeth, or, more specifically, the teeth of sharks and their relatives, if these descended from placoderms.

3. Placoderms themselves developed no true teeth, relying instead on dermal bones that covered the jaw cartilages and provided sharp slicing and serrate piercing edges, effective in predation but not renewable by replacement, as is often the case with teeth. In some, these "tooth plates" formed solid crushing surfaces for dealing with shellfish.

4. A *transverse seam* separating the head shield from the trunk shield allowed the head to be tilted upward, raising the upper jaw (which had become solidly attached to the skull by way of the dermal bones) while the lower jaw was dropped.

5. Paired movable pectoral and pelvic fins developed. The former underwent considerable adaptive diversification.

>   **Example 3:**   In some placoderms the pectoral fins became encased in bony tubes forming a jointed exoskeleton; in others they became wide and flattened, like those of rays (although less flexible). The pelvic fins became sexually dimorphic in certain groups, indicating their use in copulation, as the pelvic *claspers* serve chondrichthyans.

Class Placodermi includes three major orders (excluding the acanthodians, which were formerly considered a type of placoderm):

1. *Rhenanids* (order Rhenanida) included sharklike and raylike forms with reduced armor; although this resemblance may be convergent, rhenanids could have been ancestral to chondrichthyans.

2. *Arthrodires* (order Arthrodira: "joint neck") lived in both fresh water and the sea and had an especially wide gap between their head and trunk plates, which let them open their voracious mouths especially wide by tilting back their entire heads.

3. *Antiarchs* (order Antiarchi) could prop themselves up on long stiltlike pectoral fins encased in jointed armor. They lived on the bottom in fresh water, and the dorsal position of their eyes bespeaks alertness for prowling arthrodires.

19.11.   What are chondrichthyans?

The *class Chondrichthyes* ("cartilage fish") includes sharks and rays (*subclass Elasmobranchii*) and chimaeras (*subclass Holocephali*), fishes with jaws and paired fins, which entirely lack bony tissue (Fig. 19.4). This class appeared in the Devonian and first underwent an ambitious Paleozoic radiation of sharks with three-pronged (cladodont) teeth and terminal rather than undershot mouths. One group, the *xenacanths* (Fig. 19.5a), were benthonic in fresh water and had pointed (*diphycercal*) caudal fins and leglike paired fins with a median series of supportive skeletal elements; they died out without descendants in the early Mesozoic. The *edestoids*, represented by 2-m *Cladoselache* (Fig. 19.5b), had broad-based paired fins of limited mobility and five pairs of lateral gill slits. As in modern sharks, their vertebral column

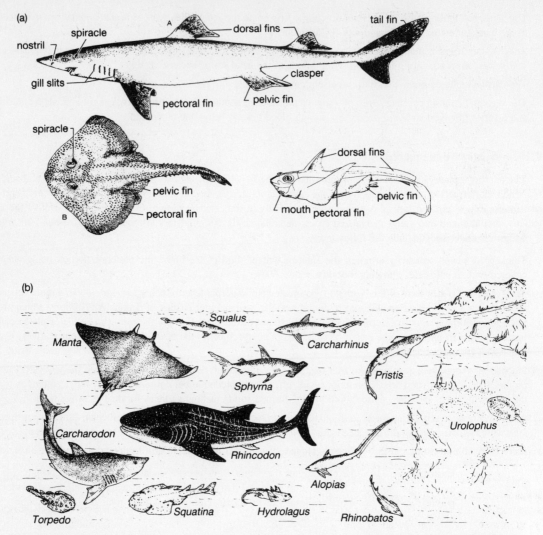

Fig. 19.4 Class Chondrichthyes. *(a)* Representatives of three major groups. Subclass Elasmobranchii: A, shark (*Squalus acanthias*), a selachian; B, skate (*Raja*), a batoid. Subclass Holocephali: C, chimaera (*Chimaera colliei*). *(b)* A variety of common cartilaginous fishes. Batoids: *Manta* (manta ray), *Torpedo* (electric ray), *Urolophus* (stingray), *Pristis* (sawfish), *Rhinobatos* (guitar-fish). Selachians: *Squalus* (spiny dogfish), *Carcharinus* (requiem shark), *Sphyrna* (hammerhead), *Carcharodon* (white shark), *Rhincodon* (whale shark), *Alopias* (thresher shark), *Squatina* (angel shark). Holocephalan: *Hydrolagus* (ratfish). [*From Storer et al.; (a)*, A *after Goode*, B *after General Biological Supply House, Inc.*, C *after Dean.*]

continued into the dorsal lobe of the forked caudal fin, but the two lobes were of equal size (*homocercal*) rather than the dorsal lobe being larger (*heterocercal*). Also, the vertebrae of *Cladoselache* consisted only of separate neural and hemal arches surrounding the large notochord, whereas modern chondrichthyans have complete vertebrae with centra that constrict the persistent notochord into a series of beadlike elements. *Cladoselache* had well-developed jaws with *amphistylic* suspension and had already developed the dental replacement apparatus of modern sharks: each functional tooth is the oldest member of a linear series of developing teeth riding on the same constantly growing ligamentous band; this band progressively moves each tooth in turn onto the jaw crest, after which the tooth is shed and replaced by the next in line. *Cladoselache* has been described as lacking claspers, but since claspers occur in other edestoids and in xenacanths (not to mention placoderms), all described specimens of *Cladoselache* simply may have been female.

By the end of the Paleozoic, the edestoids had developed a very successful adaptive radiation of sharks with three-cusped teeth, and the chimaeras had also appeared. But now the tremendous geologic and climatic changes that accompanied

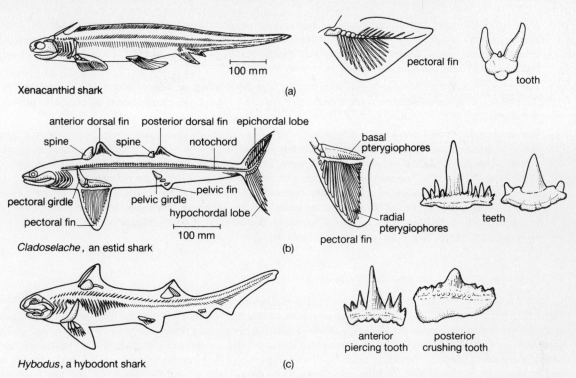

Fig. 19.5   Paleozoic sharks. Compare structure of pectoral fins and teeth of the *(a)* xenacanthid, *(b)* edestid, and *(c)* hybodont sharks.

the formation of Pangaea precipitated a mass extinction so overwhelming that nearly all the Paleozoic sharks died out except for a group represented by *Hybodus* (Fig. 19.5*c*). The hybodonts had forward rows of piercing teeth and rearward batteries of shellfish-crushing teeth, together with paired fins that articulated to the girdles by way of three large basal cartilages, allowing greater maneuverability than in sharks like *Cladoselache*. Their dentition preadapted the hybodonts to survive this time of crisis, since they enjoyed a diversified menu, and so they passed on to the next elasmobranch radiation a legacy of *tribasal limb articulation, variable dentition,* and certain adaptations related to benthonic life, such as a somewhat flattened body and heterocercal tail with enlarged dorsal and reduced ventral lobes. The characteristics and groups summarized in Questions 19.12–19.15 refer to post-Paleozoic elasmobranchs that radiated from hybodont ancestry. (Chimaeras, by contrast, have come down essentially unchanged from the Devonian. Happily stable in cold marine waters to depths of 1800 m, they are buffered from the uncertainties of life on the continental shelf.)

19.12.   What is the modern elasmobranch radiation?

Elasmobranch taxonomy is still in flux. Traditionally, subclass Elasmobranchii has been divided into just two orders: *Selachii,* the sharks, and *Batoidei,* the rays. More recent proposals subdivide living sharks into several different orders, but the lines along which these should be delineated remain in dispute. Whether or not Selachii is abandoned, the term "selachian" will remain widely in use to denote sharks in general. The major groups of living elasmobranchs are reviewed in Question 19.14.

19.13.   How do selachians and batoids differ?

(*a*)   The selachian body is typically fusiform, with dorsoventral flattening usually confined to the head; the entire batoid body is much depressed.

(*b*)   Selachians have five to seven lateral gill slits on each side of the neck, and a modified first gill slit, the roundish *spiracle,* is located close behind the eye; batoids have a large spiracle behind the eye and five pairs of gill slits located ventrally. The spiracle serves for both inhalation and exhalation when the elasmobranch is resting on the bottom, since oral inhalations at such times would lead to mouthfuls of sand; consequently, it is largest in benthonic species.

(*c*) Some selachians never rest on the bottom and rely on continuous swimming to force water in through the mouth and out through the gill slits; other selachians and most batoids do rest on the bottom, switching to spiracular breathing at such times. Batoids often nearly bury themselves by using their pectoral fins to flip sand over their backs until only eyes and spiracles remain uncovered.

(*d*) The pectoral fins of batoids are enlarged, wide and flat, and joined along the sides of both head and body; those of selachians, even when enlarged as in angel sharks (see *Squatina*, Fig. 19.4*b*), remain free of attachment to the sides of the head.

(*e*) The selachian tail is well developed and muscular, with a large caudal fin that usually has an enlarged dorsal lobe, and swimming is accomplished by S-shaped undulations of trunk and tail, sweeping the caudal fin from side to side. The tail of batoids is reduced and may lack fins entirely and be whiplike, sometimes with a defensive stinging barb at the base. The extent of tail reduction correlates with the size and flexibility of the pectoral fins: when the latter are smallish, as in guitarfishes, or inflexible and nonmuscular because they are occupied by large *electric organs* (in electric rays), the slender tail bears dorsal and caudal fins and serves for propulsion; when the pectoral fins can be moved in a flapping or undulatory manner, like wings, the tail either is a mere filament (in eagle rays, stingrays, and mantas) or lacks a caudal fin but bears two small dorsal fins (skates).

19.14.   How are living elasmobranchs classified?

Most taxonomic schemes recognize batoids as constituting one order (*Rajiformes*, or *Batoidei*) because all these highly flattened elasmobranchs resemble one another more closely than any resemble sharks. The ordinal classification of sharks into three orders recognizes the orders *Chlamydoselachiformes* (frilled sharks), *Hexanchiformes* (six- and seven-gilled sharks), and *Squaliformes* (all others, having two dorsal fins and five pairs of gill slits). Another widely accepted scheme recognizes eight major selachian orders: (*a*) *Hexanchiformes* (frilled and six- and seven-gilled sharks, having an anal fin and only one dorsal fin); (*b*) *Pristiophoriformes* (saw sharks, with no anal fin, two dorsal fins, and a rostrum forming a broad saw blade); (*c*) *Squaliformes* (dogfish sharks, having two dorsal fins and no anal fin); (*d*) *Squatiniformes* (angel sharks, with flattened body, no anal fin, large pectoral fins and terminal mouth); (*e*) *Heterodontiformes* (horned or bullhead sharks, with grasping and crushing teeth, anal fin, and a stout spine just in front of each of the two dorsal fins); (*f*) *Orectolobiformes* (carpet and whale sharks, with anal fin and two dorsal fins, and the mouth well in front of the eyes); (*g*) *Lamniformes* (white, thresher, basking, and mackerel sharks, with anal fin, two dorsal fins, and ring-type intestinal valve); (*h*) *Carchariniformes* (requiem, hammerhead, ground, cat, and hound sharks, with anal and two dorsal fins, and a movable lower eyelid that closes over the eyeball).

19.15.   What are some representative batoids?

Batoids include the following families:

(*a*) *Sawfishes* have sharklike bodies and toothed saws that they lash from side to side to disable prey (see *Pristis*, Fig. 19.4*b*).

(*b*) *Guitarfishes* have flattened bodies but round, muscular tails used for swimming.

(*c*) *Electric rays* grow to 2 m and produce shocks of more than 200 volts (V) with which they stun unwary fish (see *Torpedo*, Fig. 19.4*b*).

(*d*) *Skates* bear dorsal but not caudal fins on their tails and generate about 4 V from small caudal electric organs (see *Rhinobatos*, Fig. 19.4*b*).

(*e*) *Stingrays* have a single toxic barb on the whiplike tail, usually near its base (see *Urolophus*, Fig. 19.4*b*).

(*f*) *Eagle rays* often swim in midwater instead of resting on the bottom; their reduced tails bear three large, serrate spines with poison glands.

(*g*) *Mantas*, which reach a width of 6 m, have mouths that cannot close and lateral cephalic lobes that scoop plankton toward the mouth (see *Manta*, Fig. 19.4*b*).

(*h*) *Freshwater stingrays* live in tropical Asia and South America.

19.16.   What are some interesting sharks?

(*a*) *Frilled sharks* make up a single species that is widely distributed in deep water and preys on cephalopods. Named for the frilly margins of its six pairs of gill slits, this slender 2-m shark seems eel-like, for the single (dorsal) tail lobe trails in line with the body. As in the earliest known fossil sharks, the mouth is terminal, not ventral, and the lateral line system lies in an open groove; nevertheless, an anal fin, considered a more recent feature, is present. The eyes can be protruded and rotated upward for overhead vision. Frilled sharks are ovoviviparous, and up to 15 eggs are incubated internally for nearly 2 years.

(b) *Sand tiger sharks* experience an extreme in sibling rivalry: the female gives birth to only two young at a time because the first baby to hatch out in each of her two uteri eats up all its siblings before facing the outside world!

(c) The *mackerel sharks* are pelagic species including the harmless 9-m basking shark that swims with huge mouth agape, filtering plankton, and the far from harmless 6-m great white shark that comes up from directly below to attack sea lions (and sometimes divers and surfboarders). Great white sharks have also been known to chew on boat propellers, probably being attracted by the electromagnetic properties of metallic parts.

(d) *Whale sharks*, the largest of all fishes (to 18 m), mooch along sucking plankton into a terminal, slotlike mouth 3 m wide.

(e) *Hammerhead sharks* have their eyes and nostrils placed on laterally projecting cephalic lobes, possess very long dorsal tail lobes, and sometimes attack bathers in very shallow water.

(f) *Thresher sharks* have normally shaped heads but scythelike upper tail lobes, which they use to round up schools of fish and then disable them by slashing movements.

(g) *Saw sharks* resemble sawfish (which are batoids), but their pectoral fins are not attached to the head, their gills slits are lateral, and the "teeth" of their flattened saw are large and small in alternation.

19.17. What are the characteristics of elasmobranchs?

Elasmobranchs differ from bony fishes in the following respects:

1. Bone tissue is totally absent, and the skeleton consists of cartilage that may be further calcified for hardness and buoyancy reduction.

2. *Placoid scales*, like tiny recurved teeth, protrude through the epidermis (Fig. 19.6a) and reduce drag during swimming.

3. The skin is very tough and valuable as leather or, with denticles still attached, as *shagreen* for fine polishing operations.

4. The spiracles and five to seven pairs of gill slits are exposed, not covered by an operculum.

5. The paired fins are comparatively broad-based and cannot be maneuvered sufficiently to allow braking, hovering, or backing up.

6. The limb girdles consist of transverse bars united ventromedially, and the pectoral girdle lacks articulation to the cranium.

7. Dorsoventral flattening affects the head and often the body as well.

8. In sharks the dorsal lobe of the heterocercal tail fin, into which the vertebral column extends, is usually enlarged; this causes the tail to rise when swimming and tilts down the head, thus compensating for the upward-planing effect of the flattened head.

9. The mouth and paired nasal sacs are on the underside of the head.

10. Teeth are borne on the primary jaws (mandibular arch) and are constantly replaced by new teeth moving up into position on connective tissue bands.

11. No lung or swim bladder is ever present.

12. Elasmobranchs ("plate gills") are so called because their gill filaments protude from the walls of flattened partitions (*interbranchial septa*) that separate adjacent gill pouches (see Fig. 20.2a).

13. A unique *rectal salt gland*, just anterior to the cloaca, secretes excess salts from the bloodstream into the gut.

14. Except for freshwater rays, elasmobranchs osmoregulate by *urea retention*, storing up concentrations of this toxic nitrogenous waste that would knock most other animals cold. In fact, their body fluids contain so much solute (3.8 percent) that they are slightly hyperosmotic to seawater, so that they produce a copious, dilute urine. (A representative shark produces urine at a rate of over 1 ml/kg/h, a marine bony fish only 0.3 ml/kg/h.) This method of osmoregulation limits most elasmobranchs to the sea; however, the cosmopolitan bull shark ranges upriver for hundreds of kilometers, greatly lowering its body content of urea and NaCl during such incursions; and a population of this species even survives in a landlocked Nicaraguan lake.

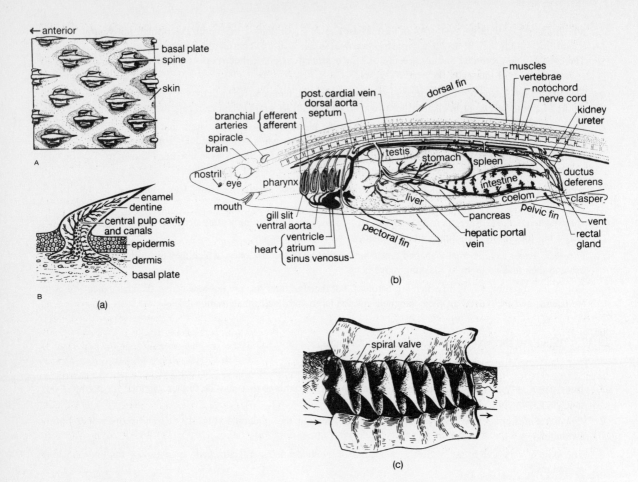

Fig. 19.6 Shark structure. *(a)* Placoid scales: A, view from above; B, median section through a scale, or dermal denticle, and skin. *(b)* Internal structure. *(c)* Intestine opened to show spiral valve. [*From Storer et al.; (a) after Klaatsch, (c) after Jammes.*]

15. All four heart chambers (sinus venosus, atrium, ventricle, conus arteriosus) are pulsatile.

16. The claspers of the male's modified pelvic fins can be flexed at right angles to the body and one or the other inserted into the female's cloaca while semen flows along a groove in the dorsal surface of the clasper.

17. The eggs are about 10 cm long, and in oviparous and ovoviviparous species, much yolk is present; in the former, eggs are laid within a tough keratinous case secreted by the anterior part of the oviduct; in the latter, they are held in utero until hatching, an incubation that may take as much as 2 years. Some sharks are actually euviviparous: the eggs of these species contain substantially less yolk, and since the dry weight of an egg is *less* than that of the newborn, it is apparent that the fetus receives nourishment from the uterine wall by way of a simple placenta. When they first face the world, young elasmobranchs are very large, compared with newly hatched bony fishes, and are better equipped to escape predation.

What are some additional characteristics of elasmobranchs?

1. The digestive tract (Fig. 19.6*b*) consists of mouth, buccal cavity, pharynx, short esophagus, J-shaped stomach with a pyloric sphincter closing its lower end, intestine typically with a spiraling partition (*spiral valve*, Fig. 19.6*c*) that increases its functional length, cloaca, and anus. The pancreatic and bile ducts enter separately just below the stomach. The bilobed liver is oil-rich (once a major commercial source of vitamins A and D), constitutes about 20 percent of the body weight, and in addition to its usual functions (Question 18.34) promotes buoyancy.

2. As in bony fishes, the gills are designed to allow countercurrent gas exchange. Since the flow of water past the filaments runs oppositely to the flow of blood in the branchial capillaries, an especially effective exchange of gases

is brought about through countercurrent exchange, because as the fluids move in opposite directions, the concentration gradients for their dissolved gases will not even out, as would be the case were the two fluid systems to flow in the same direction.

3. Certain sharks (e.g., great white), like certain bony fishes (e.g., tuna), are *warm-blooded*. Heat generated in deep-lying *red muscle* is retained within the body instead of being lost via the gills, because of countercurrent heat exchange between outgoing and incoming blood vessels in a *rete mirabile*. Accordingly, in water at 21°C, the deeper tissues may register a cozy 27°C.

4. Many sharks and rays (along with some bony fishes) locate prey, even when buried, by detecting the minute amounts of bioelectricity that emanate from any living body. Elasmobranch *electroreceptors* are located in cutaneous pits (*ampullae of Lorenzini*) concentrated on the snout. These may also aid navigation, by picking up voltage differences associated with ocean currents, or detecting the earth's magnetic field, or both.

5. Sexes are separate and sex reversal does not seem to occur. The single pair of gonads is associated with paired genital ducts. As in all gnathostome anamniotes, sperm are transported in the duct of the opisthonephric kidney. In the female a separate, tubular oviduct commences as an expanded ciliated funnel located close to the anterior ovaries and, in viviparous species, inflates posteriorly to form a uterus that opens into the cloaca.

19.18.   What are the holocephalans?

Chimaeras (*subclass Holocephali*) are aberrant deep-sea chondrichthyans that have descended little changed from Devonian times. They differ from elasmobranchs in lacking spiracles and a cloaca, having a fleshy operculum covering the laterally placed gill slits, bearing a robust poison spine on the first dorsal fin, and being scaleless when adult. The tail is slender and tapering, hence the common name "ratfishes." The body is not dorsoventrally compressed, and the mouth is terminal. In the short-nosed chimaeras, the face is blunt, and large *tooth plates* on each jaw impart a buck-toothed Bugs Bunny look. The eyes are large, and the forehead sports a hooklike cephalic *clasper* that holds onto the female while copulation is achieved by means of the actual pelvic claspers. Those species of known reproductive habits aggregate and move into shallow water for spawning, laying their eggs in sand or mud even up to the surf line. The eggs may be 25 cm long, and 15-cm young hatch out after 5–8 months. These primitive fish become vulnerable to commercial fishing during spawning time, which seems a pity. They were once considered transitional between cartilaginous and bony fishes, but of course this is not the case.

Chapter 20

Anamniotes II: Bony Fishes and Amphibians

The body fishes, or osteichthyans, are the dominant aquatic vertebrates of modern times, impressively diverse in appearance and way of life. They also include extinct forms ancestral to the earliest land vertebrates, the labyrinthodont amphibians. Amphibians ("double life") are appropriately named: the majority recapitulate through aquatic larvae the historical transition of the earliest tetrapod vertebrates from water to land. The fossil record bespeaks an impressive radiation of Paleozoic amphibians, followed by extinctions so extensive that only three main groups exist today. But early amphibians also included the ancestors of reptiles, the first fully terrestrial vertebrates.

20.1. How do osteichthyans differ from other bony fishes?

Osteichthyans ("bony fish") have at least some bone tissue, but so did ostracoderms, placoderms, and acanthodians. Certain osteichthyans have very little or no bone in their endoskeletons, but do have bony plates in the skin.

Any definition meant to distinguish osteichthyans from other bony fishes should include the following:

1. The paired fins, pectoral and pelvic, articulate movably with skeletal girdles that contain elements typical of four-limbed vertebrates as a whole, namely, two separate pectoral girdles, each consisting of endoskeletal coracoid and scapula plus dermal clavicle (together with additional dermal bones providing anchorage to the skull), and two separate pelvic girdles, each made up of a single endoskeletal element.

2. Their fin webs are supported by multiple bony rays.

3. The fin rays of their pectoral and pelvic fins either fan out from the base (in "ray-finned fishes") or fringe the edge of the fin's median lobe (in "fleshy-finned fishes").

(These criteria exclude acanthodians from class Osteichthyes, for they typically possessed more than two sets of paired fins, each fin was supported by no more than a single stout spine, and any anchorage provided the pectoral spines was by way of ventral dermal plates not homologous with the girdle elements of four-limbed vertebrates; even so, some taxonomists make Acanthodia a subclass of Osteichthyes.)

20.2. What characteristics distinguish osteichthyans from elasmobranchs?

Bony fish (Fig. 20.1a) differ from elasmobranchs (see, e.g., Fig. 19.6 and Question 19.17) in the following respects:

1. At least some bone is present in the dermis and/or endoskeleton.

2. Dermal bones roof and floor the cranium, encase the primary jaws, and make up part of the pectoral girdles.

3. Scales, derived from dermal bone (Fig. 20.1b), may be ganoid, cycloid, or ctenoid, but never placoid. Cycloid and ctenoid scales overlap like shingles. Whereas placoid scales erupt through the epidermis like teeth, the scales of osteichthyans are fully overlaid by a delicate, mucous epidermis. The vibration-detecting lateral line system opens to the surface by a row of pores that actually penetrate these scales (Fig. 20.1c).

4. The gills are covered by an operculum of bony plates, which can be flared out to increase the suction achieved by inhalation (allowing sizable prey to be "vacuumed" in).

5. The paired fins (either ray fins or lobe fins) are more maneuverable than those of sharks, and can be used to bring the fish to a dead halt in midwater, to hover or back up, or to execute tight turns; certain bony fish rely for locomotion more on their pectoral fins than on their tails.

6. The left and right limb girdles are separate; dermal bones anchor the primary elements (coracoid and scapula) of the pectoral girdle to the skull.

7. The body is usually laterally compressed, so that the transverse axis is shorter than the dorsoventral axis.

8. The tail is typically homocercal, with upper and lower lobes of equal size, and the vertebral column rarely extends into the caudal fin.

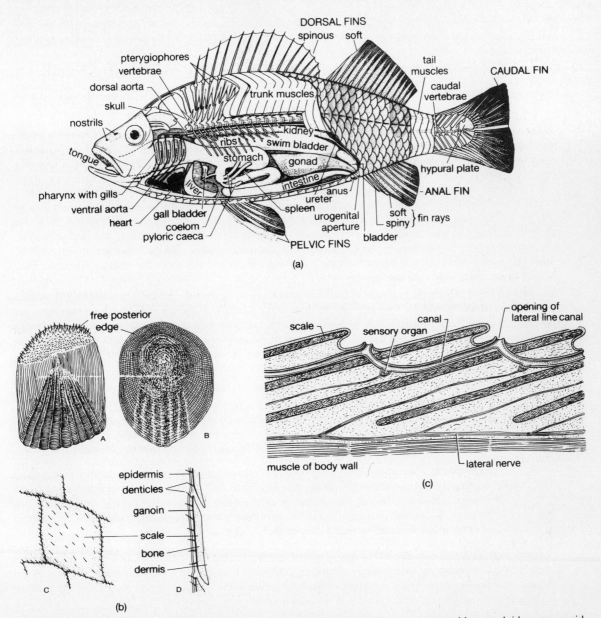

Fig. 20.1   Structure of bony fish. *(a)* Anatomy of a yellow perch. *(b)* Scale types: A, ctenoid; B, cycloid; C, D, ganoid, surface and section, respectively. *(c)* Section of skin showing scales lying below epidermis, and portion of sensory lateral line system, which detects vibration. [*(a) and (b) from Storer et al.*]

9. The mouth is usually terminal, and the nostrils are located dorsally on the snout; the olfactory sacs may or may not open into the buccal cavity.

10. Teeth are not borne on the primary jaws, but are fused onto the encasing dermal bones, from the rim of the mouth all the way back to the pharynx. Alternatively, the dermal bones themselves may be modified into shearing or crushing mouth edges. The teeth are usually attached to the bone by only a connective tissue hinge along the inner side, so that they fold down, allowing prey to slip off the teeth inwardly. Although teeth are repeatedly replaced, new ones do not continuously ride up into position on ligamentous conveyor belts as in elasmobranchs.

11. The stomach may be J-shaped, or the pyloric orifice may occur near the anterior end, with the body of the stomach forming a blind sac. Most bony fishes have numerous (up to 200) slender diverticula (*pyloric ceca*) that open into

the intestine just beyond the pyloric sphincter, secrete digestive enzymes, and carry on fat absorption. The intestine may be quite long and coiled, and a spiral valve is lacking except in lungfishes and the primitive chondrosteans and holosteans.

12. One or two lungs or an unpaired swim bladder is usually present.

> **Example 1:** One lung occurs in the Australian lungfish, and a pair in South American and African lungfishes and *Polypterus*, a ray-finned fish that dies if *not* allowed to breathe air.

The swim bladder, an evolutionary derivative of lungs, lies dorsal to the gut and primitively connects to the esophagus by a duct. When a duct persists, the swim bladder can function in air breathing, as in the ray-finned *Arapaima*, an enormous denizen of Brazilian rivers. The swim bladder serves principally as a means of adjusting buoyancy, letting the fish stay at a given depth without rising or sinking. Gas enters the bladder from a gas gland and is removed either by absorption into the bloodstream or by way of the duct, when present. When a fish is caught at some depth and brought quickly to the surface, the gas expands until the swollen bladder forces its way into the fish's throat, where the untutored mistake it for the tongue. The swim bladder also functions in sound production, as special muscles set its gas into vibration, producing hoots, bubblings, growls, sobs, and yelps. The gas can also be set into vibration by pressure waves in the surrounding water, and the largest group of freshwater fishes (*superorder Ostariophysi*) profits from this by having a chain of small bones that transmit vibrations from bladder to inner ear, improving hearing.

13. The gills are borne only on the four pairs of branchial arches, not along platelike partitions between the gill pouches (Fig. 20.2*a* and *b*). The double rows of filaments diverge from each arch like the arms of a V, expediting water flow through the lamellae; since this flow is opposite that of blood through the capillaries, a countercurrent exchange system operates that is so efficient that up to 80 percent of the dissolved oxygen is taken up (by contrast, chondrichthyans can absorb only about 50 percent, and we ourselves, lacking the benefits of countercurrent exchange, take up only about 20 percent.

14. Salt-regulating cells in the gills excrete excess salts from the blood streams of marine fishes and actively transport needed salts into the bodies of freshwater fishes.

15. The body fluids of bony fishes are hypotonic to seawater and require constant, active osmoregulation. Although the internal content of Na^+ and Cl^- in the body fluids is much less than that of seawater, bony fishes increase their solute content by concentrating K^+, and they do *not* retain urea. Instead, they rely on their gills and kidneys.

> **Example 2:** Marine osteichthyans produce a very scanty urine because the glomeruli in their kidneys (see Fig. 18.14*a*) are reduced or absent; thus their blood does not undergo pressure filtration, a process prodigal with water. Instead the renal tubule cells actively secrete wastes (mainly urea) directly into the urine. By contrast, freshwater bony fishes have kidneys with large glomeruli and void large quantities of dilute urine, thus ridding their bodies of excess water as well as wastes.

These mechanisms of osmoregulation can be modulated more effectively than those of chondrichthyans, and many bony fishes tolerate the taxing salinity fluctuations of estuaries while others freely migrate between rivers and the sea. *Anadromous* fishes (such as salmon) migrate upstream to breed, while *catadromous* fishes (such as the freshwater eel, *Anguilla*) move downstream to the sea for breeding.

16. Several families of bony fishes generate mild electrical fields that serve as "radar" in murky tropical waters and may also provide means for intraspecific communication. (By contrast, the electric eel and electric catfish generate stunning shocks of 600 and 300 V, respectively.)

17. Few bony fishes are live-bearers, but there are notable exceptions.

> **Example 3:** Guppies and surfperch inseminate the female by means of an extension of the anal fin which is analogous, but not homologous, to the claspers of chondrichthyans. The one surviving crossopterygian (lobefin), *Latimeria*, is ovoviviparous in the absence of any specialized sperm-transferring fin at all; since they breed in the deep sea, their success at viviparity remains a mystery. Surfperch are *euviviparous* to an amazing extent: the tiny eggs, which are fertilized and develop in the mother's *ovary*, have little yolk, and yet when the young are born, they are fully 3 cm long and almost sexually mature, for they mate when only 2 days old. However, the sperm remain dormant for months before fertilizing the eggs, giving the child brides a chance to grow up before experiencing motherhood.

18. Sex reversal is a common phenomenon in certain families. Most typical is the wrasse pattern, in which some males

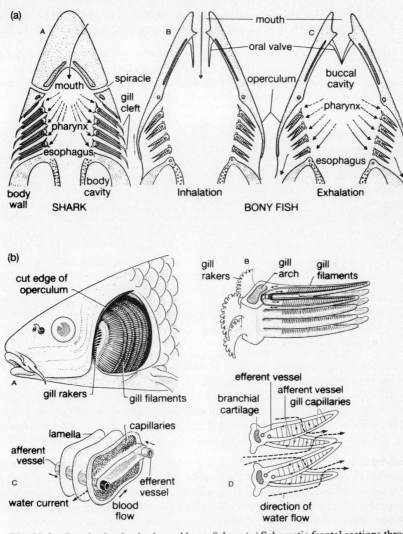

Fig. 20.2 Respiration in sharks and bony fishes. *(a)* Schematic frontal sections through
buccal cavity and pharynx: A, shark; note separate gill clefts, and gill filaments mounted on platelike septa extending from gill arches. B,C, bony
fish; note absence of septa, and opercula covering gills; inhalation: buccal
cavity expands with oral valve open and opercula closed; exhalation; buccal
cavity contracts with oral valve closed and opercula flared out. *(b)* Gill
structures in a carp: A, gills in gill chamber with operculum removed; B,
section of a gill arch with bony gill rakers that help retain food in mouth,
and several gill filaments. Arrows indicate direction of blood flow; deoxygenated blood in afferent vessels, black; oxygenated blood in efferent
vessels, gray. C, small portion of one gill filament, solid arrows showing
direction of blood flow and dashed arrow direction of water current. D, flow
of water (dashed lines) and blood in capillaries (solid lines) in opposite
directions provides countercurrent exchange. [*From Storer et al.; (a) after
Boas, (b) after Goldschmidt.*]

are born that way, while those born female eventually reverse sex and become secondary males, a type of hermaphroditism known as protogyny (as contrasted with the *protandrous* type seen in some invertebrates).

19. The eggs are quite small compared with those of chondrichthyans and are not laid in horny cases. Nevertheless,
yolk composition is such that even pinhead-sized eggs cleave discoidally, not holoblastically. The tiny young often

hatch in only a few days and are highly vulnerable, so that great quantities of eggs must be laid— up to 6 million per season by a cod. Certain species reduce predation by protective parental behaviors.

> **Example 4:**   Male sticklebacks build cylindrical nests to which they lure gravid females, then guard and fan the eggs until hatching. Father sea horses brood the eggs in an abdominal pouch. Males of several groups brood the eggs in their buccal cavities, and for some time after hatching, the fry flee for safety back into father's mouth.

20.3.   What are the main groups of osteichthyans?

In the mid-Devonian, when placoderms and acanthodians were still thriving but ostracoderms were dying out, three distinct types of osteichthyans appeared at much the same time, evincing a prompt onset of adaptive radiation, quite likely from acanthodian ancestry. These earliest osteichthyans included one type with radiating bony fin rays, representing *subclass Actinopterygii* ("ray-finned"), and two types with fleshy fins, once classified together as *subclass Sarcopterygii* ("fleshy-finned"), since they also had in common lungs and internal nostrils (*choanae*) opening into the mouth (thus being useful in air breathing). However, from their very first appearance in the fossil record, sarcopterygians constituted two separate groups, lungfishes and crossopterygians ("tassel-finned"), and since the development of the choanae has been shown to be totally different in these two groups and the skeletal support of the fleshy fins is also different, these are now recognized as separate subclasses: *Dipneusti*, the lungfishes, and *Crossopterygii*, the lobefins. "Sarcopterygian" is now merely a term of convenience.

20.4.   What are the characteristics of crossopterygians?

Subclass Crossopterygii is distinguishable from its start by division of the cranium into anterior (nasal) and posterior (otic) portions separated by a movable joint that may have allowed the mouth to gape wider by elevating the upper jaw. This jointing confined the brain to the otic half of the cranium.

Lobefins are best known from the fossil record, since all but one species, *Latimeria* (Fig. 20.3*a*), were extinct before the Cenozoic Era. *Latimeria* is representative of the *coelacanths* ("hollow spines," for the hollow cartilaginous fin rays of this group). Coelacanths are retrograde in ossification, and the group is considered an offshoot from the rhipidistians discussed below.

> **Example 5.**   *Latimeria*, a veritable living fossil, persists in deep water off southeast Africa, is occasionally taken on long fishing lines, grows to a pugnacious length of 1.6 m, and can survive for 2 hours out of water, despite the uselessness in respiration of its fat-filled lung. Fossils of extinct coelacanths show that the lung wall was curiously calcified, injecting some doubts as to how useful the lung was for freshwater coelacanths or even as a hydrostatic device for the many later coelacanths that moved to sea. Like sharks, *Latimeria* osmoregulates by urea retention. Fossil evidence had suggested that viviparity might have been an option for coelacanths despite their lack of copulatory appendages: *Latimeria* proved the point when pregnant females were caught, one with five 30-cm young in her single oviduct.

20.5.   Why are rhipidistians of particular interest?

The 1–4 m lobefins known as rhipidistians (*rhipid-:* "fan," for the arrangement of limb bones in some) died out by the end of the Paleozoic but were ancestral to the earliest amphibians. Their lungs seem to have been functional in respiration. The basal skeletal units of their lobed fins were three endoskeletal elements (*basal pterygiophores*) found also in chondrichthyans and primitive actinopterygians: the anterior (preaxial) *propterygium*, posterior (postaxial) *metapterygium*, and central *mesopterygium*. From these, several series of *radial pterygiophores* continue distally. Although in lungfishes and many crossopterygians the basals constitute part of a linear series of axial elements, with radials extending from the sides in a pinnate arrangement, in certain rhipidistians such as *Eusthenopteron*, the pterygiophores were so arranged that the mesopterygium appears to have been homologous with the tetrapod humerus and femur, the propterygium with the radius and tibia, the metapterygium with the ulna and fibula, the more proximal radials with carpals and tarsals, and the more distal ones with metacarpals, metatarsals, and phalanges (Fig. 20.3.*b*). Another striking point of similarity is the unique dentition of rhipidistians and the earliest known amphibians, or *labyrinthodonts* ("maze tooth"), since both had conical teeth with deep, complicated folds of dentine (Fig. 20.3*c*). Also, in both groups the attachment of the jaw muscles allowed the jaws to snap closed from a full gape, impaling prey on large palatine teeth. The leglike use of

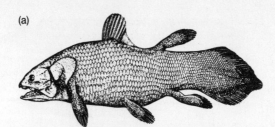

Latimeria, a deep-sea coelacanth

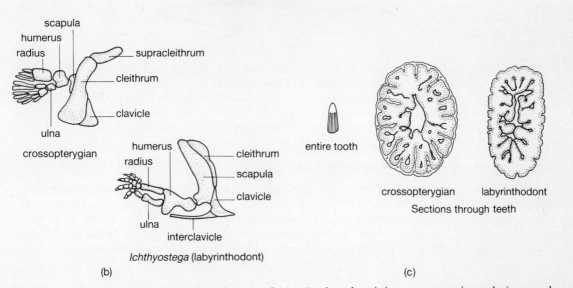

Ichthyostega (labyrinthodont)

Sections through teeth

Fig. 20.3 Crossopterygians and labyrinthodonts. *(a) Latimeria,* the only existing crossopterygian, a deep-sea coelacanth from the Indian Ocean. *(b)* Comparative structure of the forelimb and pectoral girdles of a rhipidistian crossopterygian and *Ichthyostega,* a labyrinthodont amphibian. *(c)* Sections through teeth of rhipidistian crossopterygian (*middle*) and labyrinthodont amphibian (*right*), showing complex folding of dentine; *left,* lateral view of entire tooth showing external striation. [*(a) From Storer et al.; after Romer.*]

crossopterygians' lobed fins in traveling overland from one water hole to the next is inferred from their sturdy skeletal support, but is further confirmed by the manner in which these fins are used by living lungfishes: although lungfishes cannot crawl on land, when walking across the bottom or swimming to the surface, they move their fins in a *diagonal* progression typical of walking tetrapods: left front, right rear, right front, left rear.

20.6. What are the characteristics of lungfishes?

Subclass Dipneusti ("two breath," i.e., air- and water-breathing) includes only three living genera (Fig. 20.4) found, respectively, in stagnant waterways of Australia, South America, and Africa—remnants of ancient Gondwanaland. Little is known of fossil forms, for the dipneust skeleton is poorly ossified, but lungfishes rapidly diverged from crossopterygians shortly after their appearance in the Devonian, quickly settling upon a somewhat eel-like shape. They also soon lost the tooth-bearing maxillae and premaxillae of the upper jaw and developed specialized crushing tooth plates.

Living lungfishes do not inhale air through their nostrils as their ancestors might have done: they gulp air by way of the mouth and swallow it into the lung or lungs.

Example 6: The 1.5-m Australian *Neoceratodus* mostly closely resembles fossil lungfish in having heavy fleshy fins by which it walks across the bottom tetrapod-fashion. It lives in oxygen-poor waters and augments gill respiration by occasionally swallowing air into its single lung, but it cannot survive if the water dries up. By contrast, the more closely related African *Protopterus* (which can exceed 2.2 m) and South American *Lepidosiren* have paired lungs and will suffocate in water if prevented from breathing air. Species of these two genera burrow into mud when the water dries up and estivate

Fig. 20.4 Lungfishes. The Australian *Neoceratodus* (A) uses its sturdy lobed fins to walk across the bottom of the stagnant waterways it inhabits. The South American *Lepidosiren* (B) and African *Protopterus* (C) have reduced fins that cannot bear their weight but are still moved on the diagonal, like legs, during swimming. (*From Storer et al.; after Norman,* Guide to Fish Gallery, *British Museum.*)

for months within a cocoon of hardened mucus and mud, their tails wrapped protectively over their vulnerable eyes. Estivating in burrows is a very ancient lungfish strategy: the spreading droughts of the Permian trapped some unfortunates to be fossilized inside their cocoons. The paired fins of *Protopterus* and *Lepidosiren* cannot support their bodies, for they have become reduced to slender tendrils, but during swimming they are still moved legwise in diagonal progression.

20.7.   What are the characteristics of ray-finned fishes?

Subclass Actinopterygii embodies the piscine success story: ray-finned fishes constitute some 24 extinct and 35 living orders including over 400 living families and 20,000 living species. Their pectoral and pelvic fins have no fleshy lobe, but consist of skin membranes supported by bony rays that radiate from a narrow base. Like fans, the fins can be spread out or folded together, and can also be raised, lowered, adducted, and abducted. Choanae are lacking in all but the benthonic marine stargazers. The three actinopterygian subclasses are Chondrostei, Holostei, and Teleostei.

20.8.   What are the chondrosteans?

Infraclass Chondrostei includes 10 extinct orders (mostly of paleoniscoids, the earliest known Devonian actinopterygians) and 2 surviving orders represented, respectively, by sturgeons and paddlefish, and by *Polypterus*, the African bichir (Fig. 20.5*a* and *b*). The endoskeleton of modern chondrosteans is cartilaginous, with little bone, and the short, straight intestine has a spiral valve.

20.9.   What characterized the paleoniscoids?

Paleoniscoids had fins supported by *both* closely packed radial pterygiophores and a greater number of bony fin rays. They had a single dorsal fin, a characteristic of actinopterygians as a whole, as opposed to the two dorsal fins of crossopterygians and primitive dipneusts (although in a number of teleosts the single dorsal fin has become divided into two visible parts). A large swim bladder counteracted the weight of heavy scales and bony plates. The jaws worked like mousetraps: the hyomandibula (of the hyoid arch) propped the jaws open, and contraction of the jaw adductor snapped the mouth closed forcibly, impaling prey on the maxillary and premaxillary teeth. Paleoniscoids were very successful for 200 million years, radiating into many forms convergent with those seen in modern bony fishes.

20.10.   What are the last surviving chondrosteans?

(*a*)   *Sturgeons* are (by convergence) quite sharklike in form, their heterocercal tails down-tilting the flattened snouts used for rooting prey out of bottom ooze. The underside of the snout bears sensory barbels that trail along the bottom, and the taste buds are external, on the rim of the protrusible, toothless mouth. These placid marine and freshwater fish reach lengths of 8 m. The endoskeleton is cartilaginous, and the vertebrae lack centra (i.e., are *aspondylous*), so that axial support is furnished mainly by the persistent notochord. Scales are lacking except for opercular shields and longitudinal rows of bony plates. Sturgeons' eggs are the classic caviar.

(*b*)   *Paddlefishes* live in fresh water in North America and China. They are essentially scaleless. A long, oar-shaped rostrum overhangs the huge mouth, which is kept wide open for filter feeding. The Chinese paddlefish reaches a length of 7 m.

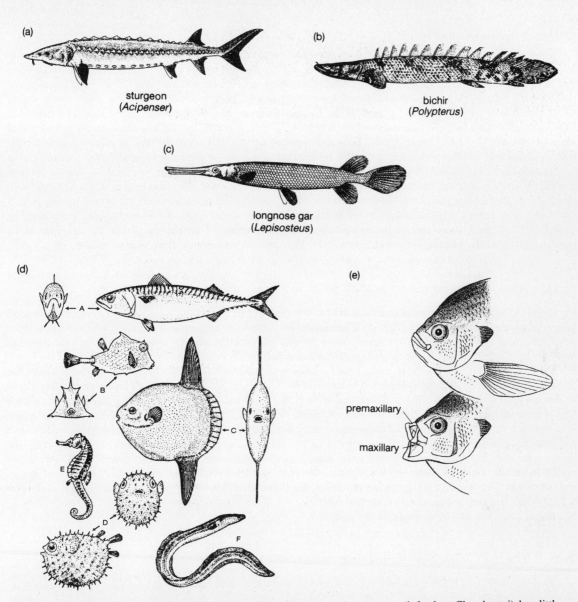

Fig. 20.5 Bony fishes (subclass Actinopterygii). *(a) Acipenser*, the sturgeon (infraclass Chondrostei) has little bone in the endoskeleton, but bony scales in the skin. *(b)* The bichir (infraclass Chondrostei) has a more fully ossified skeleton and paired lungs; air breathing is essential to survival. *(c)* The longnose gar (infraclass Holostei) can use its swim bladder as a lung; its body is armored in rigid ganoid scales. *(d)* A variety of teleosts (infraclass Teleostei): A, mackerel (*Scomber*), streamlined for rapid swimming; B, trunkfish (*Ostracion*), encased in a rigid carapace of dermal bone that permits mobility only of the fins and jaw; C, oceanic sunfish (*Mola*), a huge, leisurely swimmer that feeds mainly on jellyfishes; D, porcupine fish (*Chilomycterus*), shown defensively inflated with spinous scales erect; E, seahorse (*Hippocampus*), with tiny mouth at tip of snout, rigid trunk, and prehensile tail; F, common eel (*Anguilla*), with flexible, serpentine body. *(e)* Jaw specialization of teleosts, which permits mouth to be shot forward as prey is strongly inhaled. [*From Storer et al.; (a)–(d) after Norman.*]

(c) Bichirs are confined to tropical Africa and enjoy the distinction of being ray-finned fishes with paired lungs (of unequal size). To live, they must gulp air. A spiraclelike opening behind each eye may provide another breathing passage. Coated in hard, lustrous *ganoid* scales, bichirs use their stalked pectoral fins to walk along the bottom of lake and river edges, prowling for prey. Their endoskeletons are considerably more ossified than in sturgeons and paddlefish, and their vertebrae are *spondylous*, developing centra around the notochord.

20.11.   What are holosteans?

Infraclass Holostei is an erroneous but convenient taxon that takes in a polyphyletic assemblage of ray-finned fishes that arose from various paleoniscoids in the late Permian, mostly died out by the end of the Mesozoic, and today live only in North America, represented by 1 species of bowfin (*Amia*) and 10 species of gars (Fig. 20.5c). The latter, tightly clothed in a solid pavement of rhomboidal ganoid scales that make the body somewhat rigid, have long, slender snouts armed with conical fish-seizing teeth. Largest is the 3-m *alligator gar*. The swim bladder can be used as a lung. The *bowfin* also gulps air into its well-vascularized swim bladder, which is so efficient that this 1-m fish can survive out of water for 24 hours. *Amia* has a more flexible body than gars do, for its scales are of the thin, overlapping cycloid type. It preys voraciously on other fish and invertebrates alike, aided by a feature common to holosteans: the posterior end of each maxilla—which in chondrosteans is solidly attached to other cheekbones—now has come free, so that the cheek is covered by only a membrane, and the still-attached anterior part of each maxilla articulates by way of a kind of ball-and-socket joint with the front part of the neurocranium (the inner, cartilaginous braincase under the dermal roofing bones). These alterations provide great advantages in feeding: as the holostean mouth opens, the hyomandibulas, unrestricted by a solid shield of cheekbones, now swing outward, increasing the volume of the pharynx, while the free ends of the maxillae pull forward, stretching the mouth into an "O." Thus prey capture changes from snap to vacuum.

20.12.   What are teleosts?

Teleosts ("final bone"), of *infraclass Teleostei*, with bones fully ossified but of reduced density, represent over 95 percent of all living fish species (Fig. 20.5d). The tremendous success of this polyphyletic group, which stemmed from no less than four types of advanced holosteans during the mid-Mesozoic, began with improvements upon the holostean food-sucking mechanism. In teleosts the premaxillae and maxillae have become so mobile that they can shoot the mouth out into a cylinder (Fig. 20.5e). Next, the vertebral column no longer extends into the caudal fin, so that the latter has become a much more flexible and symmetrical homocercal fin, improving locomotion. Teleosts also have raised to a peak of efficiency the neural *Mauthnerian system* found in many bony fish and tailed amphibians: one giant neuron on each side of the hindbrain receives excitation from the otic nerve (VIII) and sends its giant axon posteriorly the entire length of the spinal cord, synapsing with spinal motor neurons all along that side of the body. This arrangement serves to trigger a simultaneous, strong contraction of all the muscles on one side, producing a powerful, evasive snap of the body and rapid acceleration. Bolstered by these fundamental improvements, teleosts have tremendously diversified in size, shape, coloration, dentition, diet, and behavior, making Osteichthyes the dominant vertebrate class in terms of numbers of species and individuals. Teleosts may not appreciate this status, for some 41 million metric tons of them are caught every year to feed people and pets and provide fertilizer and oils.

Major subgroups of teleosts are listed in Table 20.1.

20.13.   What are amphibians?

We know them today as frogs, toads, and salamanders, and a few of us may have glimpsed the reclusive caecilian; mostly we hear them singing through the spring nights in high-pitched trills, barks, and resonant honks. But our little modern friends, some 3500 species of insect and worm eaters, are only a remnant of the great Paleozoic radiation of labyrinthodonts, the first land vertebrates.

20.14.   Why did vertebrates come ashore?

In retrospect, invasion of the land seems a great advancement, but at the time, what good did it do? During the Devonian Period, when the first amphibians appeared, land plants already flourished, but no amphibian adapted to eat plants showed up in the fossil record until the Carboniferous, a few million years later, and even so, amphibians as a whole have remained staunch carnivores from their beginning to the present. On land there was precious little prey (except each other); although insects existed, they did not explode in abundance until the later Carboniferous. So going ashore was not a quest for food.

What other factors may have led toward development of a land-adapted skeleton on the part of fishes that already had internal nostrils, lungs, and muscular lobed fins? One may have been the need for better support of bulky bodies simply when moving about in shallow water, a selective factor that could operate long before either rhipidistians or their descendants ever crawled out on land. Another could be the need to escape drying water holes and to travel overland to others. A third factor could be relief from predation, for the waters teemed with voracious fishes, and the land was at first free of such perils. These factors defined a selective advantage for creatures with legs instead of fins.

Table 20.1   Major Subgroups of Infraclass Teleostei

Superorder Elopomorpha: have long-lived, leaf-shaped planktonic *leptocephalus* larvae; e.g., eels, tarpons, bonefish

Superorder Osteoglossomorpha: primitive; bony tongue, pressed against roof of mouth when biting; e.g., the electrogenic mormyrids

Superorder Clupeomorpha: pelvic fins in abdominal position, fin rays soft, swim bladder with connection to inner ear and open duct to pharynx; herrings

Superorder Protacanthopterygii: pelvic fins abdominal, fin rays soft, premaxillaries elongated; salmon and trout, lantern fishes

Superorder Ostariophysi: swim bladder connected to inner ear by *Weberian ossicles,* bladder usually with duct to pharynx, pelvic fins abdominal; most freshwater fishes (over 5000 species) including cyprinids (carps, minnows), piranhas, suckers, catfishes, electrogenic gymnotids

Superorder Paracanthopterygii: mostly marine, upper jaw not protrusible, pelvic fins far forward; e.g., toadfishes, anglerfishes, clingfishes, cods

Superorder Atherinomorpha: small surface feeders, many viviparous; needlefishes, flying fishes, cyprinodonts, poeciliids (mollies, mosquito fish, swordtails), four-eyed fish, grunions

Superorder Acanthopterygii: most marine teleosts, with spinous fin rays, pelvic fins forward, pectoral fins high on sides; e.g., squirrelfishes, sticklebacks, tube-mouthed fishes (sea horses, pipefishes), scorpionfishes, flatfishes (halibut, flounder, sole), triggerfishes, puffers, and *order Perciformes* (over 8000 species of perches, sunfishes, basses, remoras, surfperches, cichlids, surgeonfishes, parrotfishes, damselfishes, wrasses, mullets, gobies, mackerels, tunas, swordfish, marlin)

The labyrinthodonts did inherit a habitat previously unexploited by vertebrates, but they never fully conquered it: (*a*) their life cycle still included an aquatic larval stage (as known from small fossil forms with large external gills), and this is still true of most amphibians today, and (*b*) even now, after 350 million years of evolutionary opportunity, amphibian skin remains poorly adapted to retain water, and most species are still restricted to humid habitats.

20.15.   What were the early amphibians?

Extinct primitive amphibians are divided on the basis of vertebral structure into two subclasses: *Labyrinthodontia* (in which the body, or centrum, of each vertebra primitively consisted of two bony elements, *intercentrum* and *pleurocentrum*) and *Lepospondyli* (in which the vertebral body was a single, spool-shaped centrum). Both somewhat resembled big salamanders. Subclass Labyrinthodontia diversified into three orders: *Ichthyostegalia, Anthracosauria*, and *Temnospondyli*. Subclass Lepospondyli (a sideshoot from early labyrinthodonts) radiated into two orders: *Aistopoda* and *Nectridia*. The oldest tetrapod known to date is *Ichthyostega*, a primitive labyrinthodont.

20.16.   What were the ichthyostegans like?

Ichthyostegans ("fish roof," for their flattened heads) were heavy-bodied, long-tailed creatures superficially resembling giant salamanders but retaining certain fishy features (Fig. 20.6*a*). Found in Greenland sedimentaries of the late Devonian age, they were already well adapted to crawl on land. However, they were still clothed in overlapping fishlike scales, and the laterally compressed tail bore a caudal fin supported by dermal fin rays. They also possessed the labyrinthine tooth structure that links labyrinthodonts with rhipidistian fishes. Although the rhipidistian-like vertebrae still consisted of several separate elements, they already bore rudimentary intervertebral articular processes (*zygopophyses*) important in reducing torque during walking. The trunk *ribs*, slender in fishes, had become stout enough to keep the body weight from crushing the lungs when the animal was in shallow water or on land. The basic pattern of the tetrapod appendicular skeleton was already established, with *land-adapted legs* having one proximal bone (humerus and femur), two bones forming the lower limb (radius and ulna; tibia and fibula), a group of small wrist (carpal) or ankle (tarsal) bones, followed by five elongate bones of the hand (metacarpals) or foot (metatarsals), aligned with five series of phalanges supporting the digits (see Figs. 20.3*b* and 18.7*a*). (Curiously, modern amphibians have only four digits on their front feet, the thumb being vestigial.) The pectoral girdle was altered from the rhipidistian state by its detachment from the skull, with loss of two dermal bones and addition of a new median dermal bone, the *interclavicle*, which united the two girdles ventrally. A much greater change from fish to amphibian took place in the pelvic girdle, which now consisted of the three bones universal to the tetrapod pelvis: two ventral elements, a posterior *ischium* and anterior *pubis*, articulating medially with

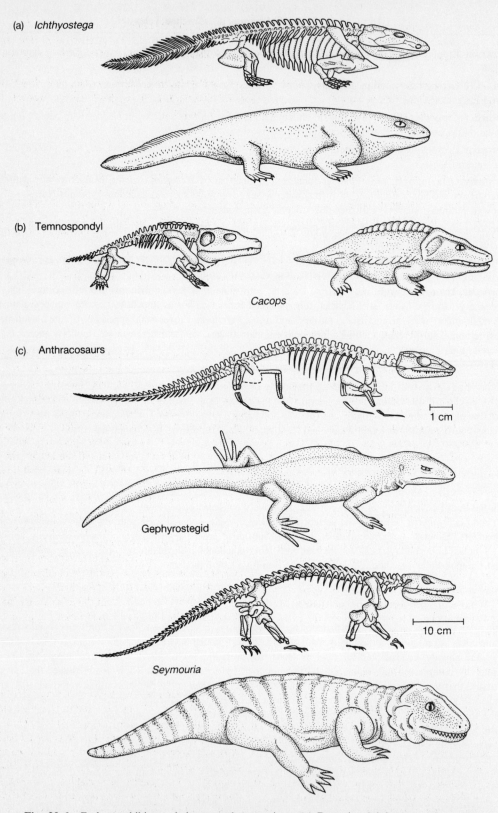

(a) *Ichthyostega*

(b) Temnospondyl

Cacops

(c) Anthracosaurs

1 cm

Gephyrostegid

10 cm

Seymouria

Fig. 20.6 Early amphibians: skeletons and restorations. (a) Devonian *Ichthyostega*, the earliest
known labyrinthodont, with fishlike scales and a caudal fin supported by fin rays. *(b)*
Cacops, a Permian temnospondylid labyrinthodont. *(c)* Reptilelike anthracosaurs: a
Carboniferous gephyrostegid and the Permian *Seymouria*.

its opposite number by way of a firm *pubic symphysis*, and one dorsal element, the *ilium*, anchored to the vertebral column by way of a *sacroiliac* articulation that for now involved only one *sacral vertebra* and its attached sacral ribs. Thus the separate pelvic bones of fishes, held in place only by the trunk musculature, came to be superseded by a complete bony pelvic ring.

As primitive as they were, ichthyostegans already had turned aside from the main pathway of labyrinthodont evolution, for they had lost a skull bone (intertemporal) still found in more recent groups. Even so, these animals have provided us with the oldest specimens found to date of vertebrates adapted for life on land.

20.17.   What were the lepospondyls?

Lepospondyls represented a sideshoot of amphibian evolution that appeared in Carboniferous times and died out without proven descendants at the end of the Paleozoic. They were aquatic or semiaquatic and somewhat resembled salamanders, but their vertebrae do not homologize convincingly with those of any other amphibians: when lepospondyls first appeared in the fossil record, they had already developed one-pieced vertebrae quite unlike those of rhipidistians and mainline labyrinthodonts. The vertebral centrum is not solid but spool-shaped, ossifying around the notochord with a hole persisting through the center for the notochord to pass through.

The *aistopod* lepospondyls were snakelike in form, with up to 200 vertebrae and with limbs reduced or absent. Certain *nectridians* were also snakelike, but most members of this order had broad, flat bodies and skulls bearing lateral, posteriorly curving "horns."

20.18.   What characterized the temnospondyls?

Temnospondyls were the flower and mainstream of the labyrinthodont radiation per se, for they flourished and diversified for a good 130 million years, from the late Devonian into the Permian, and still hung on into the Triassic Period of the Mesozoic before dwindling into extinction (Fig. 20.6*b*). These included both long-bodied, long-tailed forms and short-tailed types with stocky bodies, but all had very big heads—sometimes a third the entire body length—that became more and more flat and, in some species, a meter long. They waddled along so close to the ground that they could not open their mouths wide enough to seize food by simply lowering the mandible: instead, they evolved a unique muscular arrangement for opening the mouth by tilting up the entire cranium. Over time, their spraddling limbs tended to become reduced, and they probably led semiaquatic lives. The evolution of their vertebrae tells an interesting story.

In fishes each vertebra first appears as a number of separate cartilages associated with the notochord, and these cartilages (*arcualia*) do not fully coalesce to form a single complicated bone unless the vertebra is *spondylous*, i.e., develops a centrum around the notochord. This is also seen in embryonic development, even that of humans: the vertebrae first appear as arcualia, which enlarge and fuse to various degrees, and may subsequently ossify. Primitively, spondylous vertebrates have backbones in which *two* types of centrum, *intercentrum* and *pleurocentrum*, occur in alternation, and may be equal in size (as in the holostean *Amia*). In many fishes these two types of centra have simply fused in pairs to form one composite centrum. The more usual story, however, has been that in the evolution of a vertebra from separate elements (as in rhipidistians and early labyrinthodonts) to one complex bone, one or the other type of centrum has come to predominate while the other has diminished and disappeared. In the temnospondyls, the "winning" centrum was the intercentrum, and the pleurocentrum became small and eventually vanished. By contrast, in the other major branch of labyrinthodont evolution, the *anthracosaurs*, the pleurocentrum won out, and the intercentrum faded away, and that is why *our* vertebrae have pleurocentra today!

20.19.   What characterized the anthracosaurs?

The *anthracosaurs* appeared late in the Carboniferous and died out as such before the end of the Permian, but they had a great destiny: within this group arose a number of very reptilelike lineages, of which one did in fact make the great transition to reptilehood (Fig. 20.6*c*). The anthracosaurs did not succumb to the big, flat-headed trend that characterized most temnospondyls: the skulls were narrow and deep, favoring brain expansion; the heads were in better proportion to the body, favoring elevation; and in many, the legs became sturdy and capable of lifting the trunk well off the ground. Reptilelike species arose in several anthracosaur subgroups, such as the *seymouriamorphs* and the *diadectids*. *Diadectes* even developed chisel-like front teeth and crushing rear teeth suitable to a vegetarian diet, making it the first known terrestrial herbivorous vertebrate. On technical grounds the best candidate for ancestor of all reptiles (and through them, of all birds and mammals) seems to be yet another group of anthracosaurs, the *gephyrostegids*, which were smaller and had especially sturdy legs. The fact is, several different anthracosaurs are so intermediate in skeletal structure between amphibians and reptiles that they could be either, *but* the crucial difference lies in the structure of the developing egg, which, depending on whether it was amniote or not, would definitively identify an intermediate form as reptilian or amphibian, and for the groups mentioned above, the nature of the egg remains unknown.

20.20.   What is the history of modern amphibians?

Only three orders of amphibians—*Anura,* or *Salientia* (frogs and toads); *Caudata,* or *Urodela* (salamanders); and *Apoda,* or *Gymnophiona* (caecilians)—survive today, and the curious thing is that their antecedents remain obscure. Furthermore, they differ enough from one another to suggest that they may well each have stemmed from different labyrinthodont ancestry, although some taxonomists place all three in a single subclass, *Lissamphibia.* Unfortunately, the common features that link the Lissamphibia are mostly soft, not skeletal, and so in the absence of fossil evidence there is no way to be sure these traits were not in fact *labyrinthodont* features, too widely distributed in primitive amphibians to give us a good clue to the relatedness of the three modern orders.

From our foregoing discussion of the evolution of vertebrae, by which we saw that those of temnospondyls feature persistent intercentra, while those of anthracosaurs have pleurocentra, we might expect the vertebrae of modern amphibians to provide a clue to their origins, but this is not so. Surprisingly, they develop *neither* type of centrum. Furthermore, no convincing homologies seem to exist with the spool-shaped centra of lepospondyls.

> **Example 7:**   In the *anurans* a true centrum never forms; instead the arcualia of the neural arch grow down and around the notochord, producing a so-called *arch centrum.* Fossil frogs show vestiges of both intercentra and pleurocentra, indicating that they arose from a lineage that still had both, but apparently even those vestiges were lost in anuran evolution and a new type of centrum substituted. In nearly all vertebrates, the vertebrae develop from arcualia that are initially cartilaginous. This is not so in salamanders and caecilians: the centra form not from arcualia and not by replacement of cartilage, but by direct ossification within the connective tissue sheath of the notochord.

Left with such unresolved genealogies, all we can do is enjoy modern amphibians as they are (Fig. 20.7) and congratulate them for their ingenuity and persistence in sticking it out on land, through all the millions of years of competition and predation by vertebrates far better adapted to terrestrial life than amphibians can ever be.

Salient characteristics of modern amphibians are reviewed in Questions 20.21–20.29.

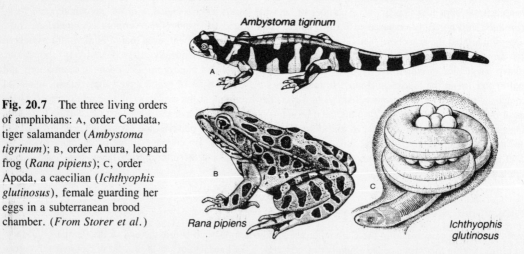

Fig. 20.7   The three living orders of amphibians: A, order Caudata, tiger salamander (*Ambystoma tigrinum*); B, order Anura, leopard frog (*Rana pipiens*); C, order Apoda, a caecilian (*Ichthyophis glutinosus*), female guarding her eggs in a subterranean brood chamber. (*From Storer et al.*)

20.21.   What is the amphibian life cycle?

(*a*)   A *typical* life cycle includes an aquatic larva that metamorphoses to a terrestrial adult form (Fig. 20.8). The eggs are usually laid in water, protected by only a gelatinous coating; cleavage is holoblastic unequal; and the larva is aquatic, hatching with filamentous external gills on each side of the neck, which normally disappear after internal gills have formed along the branchial arches in the pharynx (Fig. 20.9*a*). An operculum of skin covers the gills so that water leaves by way of a single excurrent pore on each side of the neck. The larva may be limbless or four-legged, but *never* has paired fins; its broad flat tail bears a caudal fin.

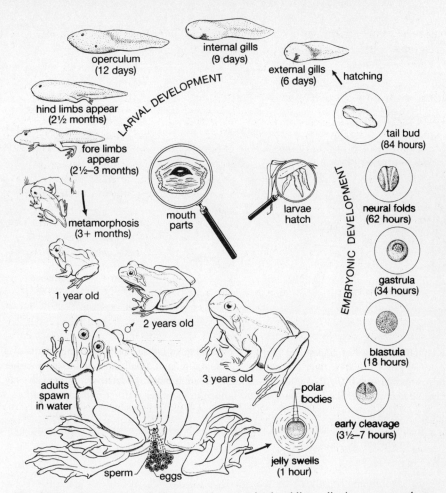

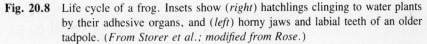

Fig. 20.8   Life cycle of a frog. Insets show (*right*) hatchlings clinging to water plants by their adhesive organs, and (*left*) horny jaws and labial teeth of an older tadpole. (*From Storer et al.; modified from Rose.*)

(*b*) High levels of the pituitary hormone *prolactin* (responsible for milk production in mammals) permit larval growth while suppressing metamorphosis; metamorphosis is triggered by an increased flow of *thyroid hormones*; insufficiency of dietary iodine or extirpation of the thyroid can prevent metamorphosis.

(*c*) A number of modern amphibians abridge or delete the aquatic stage of life, and/or provide parental care that reduces or eliminates the time their offspring must spend in the water.

> **Example 8:**   Plethodontid salamanders undergo direct development from eggs laid in moist humus; they lack both gills and lungs and exchange gases through their skin. Many caecilians, plus a few salamanders and anurans, are even viviparous, giving birth not to larvae but to postmetamorphic juveniles.

> **Example 9:**   Anurans exhibit an impressive array of protective parental behaviors: (1) The Australian gastric frog swallows her eggs and broods them in her stomach, regurgitating them after metamorphosis. (2) Arrow-poison frogs lay their eggs on land to be guarded by one parent until they hatch, whereupon the tadpoles attach by suckers to the parent's back and get a free ride to water. (3) The marsupial tree frog broods her eggs in a pit on her back, which she rips open when the tadpoles must escape to feed. (4) The male Darwin's frog broods one egg in each of his large vocal pouches, in which the embryos undergo metamorphosis to leave his mouth as tiny froglets. (5) The aquatic Surinam toad female

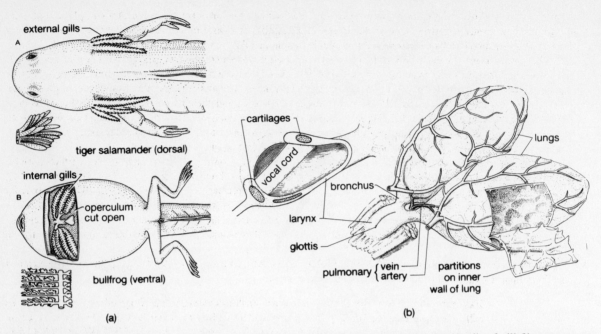

Fig. 20.9 Respiratory organs of amphibians. *(a)* Gills of amphibian larvae, insets showing details of gill filaments; A, external gills of a larval tiger salamander; B, internal gills of a bullfrog tadpole (at an earlier stage, tadpoles have external gills). *(b)* Respiratory system of the adult bullfrog: detail of larynx with only the right vocal cord shown, and saclike lungs with alveoli limited to the inner wall. (*From Storer et al.*)

broods her eggs in individual glandular pits on her back, where they remain through metamorphosis to emerge as young toads.

20.22. How do amphibians breathe?

Adult amphibians typically breathe by means of a pair of saccular lungs lacking bronchial trees (Fig. 20.9*b*). The lungs develop as posterior diverticula from the pharynx during metamorphosis, and open to the pharynx by way of a simple larynx with paired arytenoid cartilages. These close the laryngeal opening (*glottis*) during swallowing, since the glottis develops ventral to the esophageal opening. This lack of evolutionary foresight has predisposed land vertebrates to choking accidents, since material being swallowed must pass right over the top of the larynx. In anurans, the larynx contains two elastic bands, the *vocal cords*, and males produce calls that advertise territory and attract mates. Breathing in pulmonate amphibians is accomplished by the energetically inefficient method of *buccal pumping*, that is, using the throat muscles.

> **Example 10:** When a frog breathes, it first inhales through the nostrils, with mouth and glottis closed, by inflating its buccopharyngeal cavity; then it opens the glottis and contracts muscles of the body wall, while the elastic lungs recoil, expelling a stream of air that passes out through the open glottis and nostrils, over the top of the fresh air held in the throat sac. After this, the frog closes its nostrils and contracts its throat muscles, forcing air into the lungs through the still open glottis.

Throat pulsations also accomplish breathing in salamanders, with some gas exchange occurring even through the well-vascularized buccopharyngeal lining; this lining and the skin are the only sites of gas exchange in the lungless plethodontids. Some salamanders retain internal or external gills throughout life, with or without lungs.

20.23. What characterizes the skin of amphibians?

The *skin* of modern amphibians (see Fig. 18.6*a*) lacks the overlapping scales seen in fossilized skin impressions of some labyrinthodonts, although caecilians retain vestigial dermal scales. The digits lack claws or nails. The skin is kept moist with mucus from cutaneous glands and is richly vascularized, permitting *percutaneous respiration*. The epidermis is not very effective in preventing evaporative water loss, but a certain amount of keratinization does take place, especially in toads and newts when terrestrial. Only in amniotes does keratinization transform the outer skin layers into an "epidermal seal" of dead, squamous cells packed with the insoluble protein keratin.

When keratinization is too scanty to reduce evaporation effectively, one wonders how the process ever got started. The answer seems to be that depositing nitrogenous compounds in cells destined to be sloughed from the body surface provides an accessory route for excreting excess nitrogen with little water loss. The harmlessly sequestered protein can even be recycled digestively on molting, when certain amphibians devour their shed skin. Amphibians seem to have originated the mechanism of epidermal keratin synthesis so vital to amniotes, but they themselves have not been able to take full advantage of their innovation, for they are caught in an adaptive conflict: the hornier the skin, the less it can function in respiration, and amphibian lungs are nothing to write home about.

Amphibian skin not only loses water, it can *absorb it* from moist surfaces by mere contact, an especially useful talent for desert spadefoot toads, which live in deep burrows, estivating through the dry season but avoiding shriveling like prunes by soaking up even the meager moisture of desert soils. Percutaneous water uptake is promoted in some by increased urea retention, and also by the action of *antidiuretic hormone* from the posterior pituitary, which increases reabsorption of water in the kidney and even from urine stored in the urinary bladder. Amphibians can osmoregulate successfully in fresh water by producing copious dilute urine, but they would dehydrate in seawater, and so none live in the sea.

Venom glands occur commonly in the skins of salamanders and anurans. The venom can be irritating on contact and sometimes fatal if swallowed or introduced into the bloodstream.

> **Example 11:** Especially deadly are certain small arrow-poison tree frogs of tropical America, so called because their venom has long been used by natives to anoint dart tips. The most powerful known animal venom is *batrachotoxin* from the kokoi frog of Colombia: 1/100,000 oz (0.00028 g) can kill a person!

Anurans and salamanders with poisonous skin secretions often have brilliant skin pigmentation supplied by chromatophores and serving as *warning coloration*. Nonvenomous amphibians tend to be cryptically colored, and some can change color according to background.

20.24.   How do amphibians hear?

Hearing can be a problem on land when one has been used to doing it in the water. Amphibians retain the lateral line system of fishes only in their aquatic stages and lose it when terrestrial, thus relinquishing the major piscine device for vibratory reception. The ear alone remains to them, but primitively its labyrinth lies housed in the skeletal *otic capsule*, blanketed even further in bony fishes and amphibians by dermal roofing skull bones. In amphibians no cochlea is present, but as in fishes the gravity-detecting *lagena* contains a rudimentary organ of Corti and so can function in sound detection, provided that vibrations can reach it. A salamander lying prone is well positioned for ground vibrations to be transmitted through its jaw bones to the inner ear, but what about the more erect anthracosaurs and anurans? An excellent example of old structures taking on new functions is seen in the fate of the amphibian hyomandibula. This dorsal element of the hyoid arch (gill arch II) was seen earlier as having changed function in jawed fishes from gill support to jaw articulation with the skull. This positioned the inner end of the hyomandibula right against the cranium near the otic capsule, with the result that vibrations striking the overlying dermal skull bones would be apt to set the hyomandibula in motion, vibrating against the otic capsule (see Fig. 18.11*b*). In terrestrial amphibians, the second gill pouch, lying hard by the hyomandibula and no longer needed in gill breathing, furnished an opening to the exterior that could be covered by a thin *tympanic membrane* (eardrum) and also provided a cavity into which the hyomandibula could slip, now relinquishing its role in jaw suspension and forming a bony bridge from the tympanic membrane to the otic capsule. The capsule wall thinned down locally, leaving only a membrane-covered window (*fenestra*) against which the inner end of the hyomandibula (now called the *columella*) could fit with precision, transferring its vibrations to the perilymph fluid within the capsule (see Question 18.32). Thus the second gill pouch became the middle ear cavity, and its original connection with the pharynx persisted as the *eustachian canal*, which permits air to enter or leave the middle ear as needed to maintain equal pressure on both sides of the eardrum. The columella is the only *middle ear ossicle* in most nonmammalian tetrapods; in mammals it becomes the *stapes* (stirrup), the innermost of the three mammalian ear bones. However, amphibians developed a second ossicle all their own: the *operculum*, a fragment from the otic capsule itself that can be muscularly coupled to the inner end of the columella. Its use is best known in anurans: when it locks to the columella, the composite unit has greater inertia and so vibrates to tones of low frequencies; when it is uncoupled, the columella alone is set into vibration, by tones of higher frequencies.

20.25.   What advance in the digestive tract is seen in amphibians?

The digestive tract of amphibians shows greatest change from the piscine condition with respect to the structure and use of the *tongue*. The tongue of a fish or larval amphibian is a nonmuscular pad extending forward from the ventral part of the hyoid arch (*basihyal*) to lie on the floor of the buccal cavity. As metamorphosis takes place and the hyoid arch loses its role in gill support, its basihyal elements serve to anchor muscle fibers derived from the longitudinal throat musculature (*hypobranchials*) that now contribute to a highly mobile tongue. The amphibian tongue is often protrusible and can be flicked out to catch insects.

20.26.   What is the nature of the amphibian circulatory system?

The circulatory system changes metamorphically from the single circuit of fishes to the double circuit of tetrapods (see Fig. 18.12*b*). The heart of an amphibian larva is like that of a fish, and pumps only deoxygenated blood forward to the gills. During metamorphosis, the tetrapod pattern becomes established as the atrium divides longitudinally by the growth of an *atrial septum*, and a spiraling partition develops within the truncus arteriosus. Now blood passes *twice* through the heart during each complete circuit. Deoxygenated blood enters the sinus venosus, which now communicates only with the right atrium, while oxygenated blood returning from the lungs directly enters the left atrium; these two streams flow together into the single ventricle, but do not mix greatly, with the result that the spiral valve within the truncus can shunt mostly deoxygenated blood into the pulmocutaneous artery and oxygenated blood into the aorta.

20.27.   What are the characteristics of salamanders?

Salamanders (*order Caudata*) are modern amphibians with *tails*. Their limb girdles remain largely cartilaginous, and their legs are short and cannot be rotated under the trunk to elevate it. Instead, the legs spraddle to the sides, so that walking is laborious and involves not only the diagonal leg progression characteristic of quadrupeds, but also sinuous bending of trunk and tail from side to side, like a flexible fish. The largest species of the approximately 380 now living is a 1.5-m giant Japanese salamander. Some salamanders are *neotenic*, retaining larval gills and flattened tails throughout life and of course being entirely aquatic. Some aquatic species are eel-like with vestigial limbs. At the other extreme are the gill-less, lungless plethodontids, which have no larval stage at all and live entirely on land, in moist habitats.

Salamanders do not vocalize and lack eardrums, probably hearing poorly. Those called *newts* return to water to breed, where the male attracts the female by secreting a pheromone, then courts her with tail lashing and quivering, and leads her on an underwater nuptial parade in which she walks straddling his tail, and picks up with her cloacal lips a spermatophore that he deposits on the bottom. The eggs, laid singly or in clusters, hatch into slender, *legless tadpoles* that have external gills and metamorphose in a few months. Other salamanders court and mate on land, the female pressing her body down upon the deposited spermatophore. Later the female goes to water to deposit her eggs, which usually hatch out as *four-legged larvae* with external gills and tails bearing caudal fins for swimming. The viviparous fire salamander (*Salamandra*), brilliantly blotched with yellow and black, retains her eggs internally from summer to the following spring, when she goes to water and gives birth to 30–40 2-cm gilled larvae.

20.28.   What are caecilians?

Order Apoda ("footless"), or *Gymnophiona* ("naked snake"), includes some 165 species of *caecilians*. These elongate, 17-cm to 1.5-m tropical amphibians lack legs and limb girdles and are also essentially *tailless*, with the anus terminal or nearly so. Many burrow in moist soil, but others are aquatic throughout life, swimming eel-fashion. They are the only living amphibians with small dermal scales in their skins. Earthwormlike, their bodies are conspicuously metameric, because the well-developed segmental trunk muscles are their only means of propulsion. Their eyes are small, overgrown by skin in burrowing species, and a unique tentaculate sense organ lies on each side of the head in a groove extending from eye to tip of snout.

The male has a copulatory organ that he protrudes from his cloaca. Many species are viviparous: not ovoviviparous, but euviviparous in a very odd way. The unborn young, armed with special fetal teeth, gnaw away and eat the lining of their mother's oviduct, which keeps regenerating and producing a creamy "uterine milk." At birth these well-fed youngsters are 30–60 percent of their mother's length. Oviparous species lay their eggs in underground burrows, where the mother remains coiled about them until hatching (see Fig. 20.7). Some complete their larval stage in the egg, or undergo direct development. In other species, the burrow must lie close to water, since the young hatch as gilled larvae and will lead aquatic lives until metamorphosis. Aquatic caecilian larvae have internal gills, newtlike heads with well-developed eyes, and tails flattened for swimming.

20.29.   What are the characteristics of frogs and toads?

Order Anura ("tail-less") includes the most aberrant and successful modern amphibians: some 2900 species of frogs and toads. The larvae are legless tadpoles, first having external and then internal gills, and, what is remarkable, they are chiefly herbivorous. Except for the tailed frog (*Ascaphus trui*), which has a tail-like copulatory organ, anuran eggs are fertilized externally as they are laid in strings or clumps, with the male riding on the female's back, clutching her around the chest or waist, and voiding semen gradually as oviposition proceeds.

The conspicuous success of anurans relates to their agility on land, accomplished by drastic skeletal modifications that have made them famous leapers (Fig. 20.10). The skull has been made lighter by loss and thinning of bones, and the eye sockets (orbits) have become large skull openings that allow the eyes to be retracted; ribs are reduced or absent, and the vertebral column has been shortened to nine free vertebrae plus a slender *urostyle* composed of fused vertebrae, which makes the lower back rigid; the elongated ilia parallel the urostyle, articulating with the last movable vertebra (sacrum); the hind limbs are long and enlarged for jumping and swimming, and their musculature is so well developed that bullfrog legs often end up on folks' dinner plates. The hind toes are more or less webbed, depending on how amphibious the species may be, except in tree frogs, which have expanded suction pads at the tips of all digits.

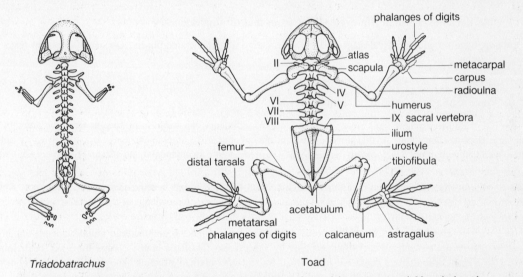

Triadobatrachus   Toad

Fig. 20.10   Skeletons of the anuranlike Triassic *Triadobatrachus* and the common toad. Note the broad, lightly constructed cranium, reduced tail, and shortened vertebral column in both, although *Triadobatrachus* had only slightly lengthened ilia and its posterior vertebrae were not fused into a urostyle. (*Toad skeleton from Storer et al.; after Jammes.*)

Anurans have large tympanic membranes on the sides of the head and are alert to air-borne sounds. Their prowess at vocalization is notable; sounds produced by vibration of the laryngeal vocal cords are amplified by the resonance of throat pouches, which in tree frogs inflate like balloons during singing.

A few of the most important anuran families are:

(*a*)   *Ranidae:* true frogs; amphibious; fully webbed toes and smooth, moist skin; e.g., bullfrog, leopard frog.

(*b*)   *Bufonidae:* true toads; pudgy, with rough skin bumpy with venom glands and a large ovoid *parotoid* venom gland behind the eye; males with undifferentiated gonads anterior to testes, which mature into ovaries if the testes are removed; e.g., *Bufo* and the viviparous *Nectophrynoides.*

(*c*)   *Hylidae:* tree frogs; small; toes unwebbed, having clinging pad on the tip of each toe and the terminal phalanx of each digit curved into a claw; e.g., *Hyla.*

(*d*)   *Dendrobatidae:* arrow-poison frogs; small, diurnal, toxic; transport tadpoles to water on parent's back.

(*e*)   *Pipidae:* tongueless frogs; feet hugely webbed; entirely aquatic, tadpoles filter-feed; e.g., Surinam toad (*Pipa*) and African clawed frog (*Xenopus*).

(*f*)   *Pelobatidae:* spadefoot toads; horny spade on hind feet used in digging deep burrows; nocturnal with vertical cat's-eye pupil; tadpoles voraciously omnivorous and cannibalistic, undergoing metamorphosis within 15–30 days.

Chapter 21

Amniotes I: Reptiles and Birds

Laying eggs on land can be a good strategy when laying them in water would expose them to a host of questing nostrils attached to voracious appetites. Laying eggs on land may become a necessity if habitats dry up and no other option exists. During the 65 million years of the Carboniferous Period, world climate was unusually warm and moist, and great forests of primitive trees flourished, often in swampy terrain. But the ensuing Permian Period became arid, and eggs laid on land could survive only if they had sufficient protection against water loss. This drying climatic trend heralded the downturn of amphibian dominance and the increasing ascendancy of their descendants, the reptiles.

21.1. How does the reptilian life cycle differ from that of amphibians?

Reptiles have no aquatic larval stage and do not undergo abrupt metamorphosis. Direct development takes place within an *amniote egg*.

21.2. Which came first, reptile or amniote egg?

The oldest known amniote eggs were found in Permian sediments, but we have no idea as to which animals found in the same deposits laid them. At that time several lineages of reptilelike amphibians coexisted, but probably only one of these achieved both direct development (as seen today in plethodontid salamanders) and also an effectively land-adapted egg.

During Carboniferous times several groups of anthracosaur labyrinthodonts became quite reptilelike in their skeletons and dentition, and for some of these, classification remains debatable without knowledge of their developmental processes.

> **Example 1:** The herbivorous *diadectids* are classified as anthracosaur amphibians by some taxonomists and as cotylosaur reptiles by others. No fossilized amniote eggs have yet been found in Carboniferous strata, yet animals of indubitably reptilian structure, known as *cotylosaurs*, did exist in the later part of that period. Definition of cotylosaurs as reptiles therefore has been based not on their eggs, but on structural comparisons with later reptiles that *were* known to lay amniote eggs.

What this points up is that evolution from one major animal group to another involves changes in entire constellations of traits: not only structural features that can be fossilized, but physiological and behavioral ones as well, and it is a combination of traits upon which taxonomic judgments usually have to be based. On the other hand, today we can say with confidence that if an ectothermal (so-called cold-blooded) tetrapod produces an amniote egg, it must be a reptile. Questions 21.4–21.16 deal with some of the major innovations that appeared in reptiles.

21.3. How did reptiles spread throughout the world?

Once vertebrates had become independent of water for reproduction, their dispersal on land was facilitated by the coalescence of the separate continents into the supercontinent of Pangaea, which remained a single land mass throughout the first half of the Mesozoic Era, known as the age of reptiles.

21.4. What changes made possible land-adapted eggs?

(*a*) The *maternal contribution* to such adaptation is investing the egg in a case or shell before it is laid. This helps appreciably: think how much faster a raw egg dehydrates after being broken into a bowl, compared with an egg with shell intact. An *eggshell* secreted by the oviducal lining could have been the first point of discrimination between a reptilelike amphibian and an amphibianlike reptile. Such a shell would protect against mechanical damage, scavenging insects, and the like, even before its water-conserving properties became crucial, and before the membranes characteristic of the amniote egg had evolved. Production of eggshells is not limited to tetrapods: consider the horny cases of chondrichthyan eggs and the chitinous cleidoic eggs of insects. But eggshells can be more than a simple casing.

When you crack the breakfast egg, notice that everything external to the yolk, which is the swollen ovum, constitutes protective investments secreted by the hen's oviduct: (1) a fibrous layer deposited directly on the vitelline membrane of the ovum, (2) the albumen (egg white), (3) two shell membranes of matted keratin fibers, and (4) the outer shell, of calcium carbonate with pores plugged by collagen. This provides a good blanketing, but eggshells must be pervious to gases and thus cannot avoid being somewhat porous to water.

(b) The *embryo* itself is responsible for the rest of the adaptive process. Only in the amniote eggs of reptiles, birds, and mammals do we find the embryo wrapping itself in *extraembryonic membranes* derived from its own proliferating cells. The potential for vertebrate embryos to grow extensive membranes outward from their bodies proper is already seen in fishes that lay large, yolky eggs. The embryo develops at first as a flattened disk on the surface of the yolk, at the same time sending out sheets of endoderm and mesoderm that end by completely enveloping the yolk, as the *yolk sac membrane*. The blood vessels that develop within this membrane absorb nutrients from the yolk and transport them back to the embryo proper. We know that the viviparous coelacanth *Latimeria* produces large eggs, so although the eggs of modern amphibians are small with little yolk, those of the large Paleozoic labyrinthodonts may have been as large and yolky as those of *Latimeria*. This size would have required their embryos to engage in yolk sac formation. This membrane-producing potential being once established, production of additional membranes would be a mere extension of this capability. The three new extraembryonic membranes, unique to the amniote egg, are the *amnion, chorion,* and *allantois.*

21.5. How do the extraembryonic membranes develop?

The development of these membranes as seen in the chick embryo (Fig. 21.1) is essentially the same as in reptiles and egg-laying mammals. The three germ layers (ectoderm, endoderm, and mesoderm) do not remain confined to the embryo proper, but sheet out over the surface of the yolk. The mesoderm forming the middle part of this sheet soon separates into two layers (an outer *somatic,* or *parietal,* layer and an inner *splanchnic,* or *visceral,* layer), separated by a fluid-filled space (the *extraembryonic coelom*). As a result of this separation, the membrane immediately against the yolk, the *splanchnopleure,* consists of endoderm covered by splanchnic mesoderm, while the outer membrane (*somatopleure*) consists of somatic mesoderm covered by ectoderm. While the splanchopleure goes on expanding over the yolk to become the yolk sac, the somatopleure begins to fold upward, arching over the embryo until the folds meet and fuse. The inner fold, which directly encloses the embryo, is the *amnion,* consisting of ectoderm to the inside and mesoderm to the outside. In the outer somatopleuric membrane, the *chorion,* the ectoderm is still on the outside and the mesoderm on the inside, as before folding took place. The chorion continues to expand until it fully encloses amnion, embryo, and yolk sac all together within the extraembryonic coelom. The final membrane of the amniote egg is the *allantois,* which grows outward as a pouch from the embryo's hindgut, lined with endoderm and coated with splanchnic mesoderm, like the gut itself. The allantois expands until it contacts the chorion on all sides and develops a rich supply of blood vessels. Both membranes together go on expanding until they directly underlie the inner shell membrane.

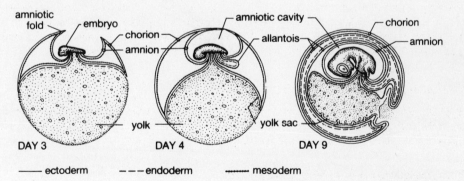

Fig. 21.1 Development of the extraembryonic membranes of the chick, a representative
amniote. These membranes, formed by the embryo but extrinsic to it, characterize
the eggs of amniotes: reptiles, birds, and mammals. Each membrane is made
up of two layers: the chorion and amnion of ectoderm and mesoderm, the yolk
sac and allantois of mesoderm and endoderm. The egg shell and albumen are
not shown. (*From Storer et al.*)

21.6. How do the extraembryonic membranes promote the embryo's survival?

Chorion and amnion together, expanding as the embryo grows, continue to envelop it within two fluid systems that prevent dehydration. The amnion also differentiates muscle fibers, which contract rhythmically, rocking the embryo to and fro. The yolk sac membrane develops vitelline blood vessels through which yolk is transported into the embryo's body. The allantois becomes essential as soon as the embryonic kidneys start to excrete urine, since in this neat little closed system urine has nowhere else to go. The allantois serves as a prenatal urinary bladder in birds, reptiles, and oviparous mammals, providing a site well removed from the embryo itself for sequestering toxic metabolic wastes such as urea. As development proceeds, gas exchange must become increasingly efficient. This function too is assumed by the allantois: gases diffuse to and from the allantoic circulation through the porous shell and thin overlying chorion. This surrogate lung serves the embryo (now a fetus) until a few days before hatching, when the chick employs its beak to break through into an *air sac* that lies between shell and shell membranes at the large end of the egg. Baby chickens hatch after only 3 weeks' incubation, but young alligators live within the egg for 2–3 months, and tortoises for 3–4 months, which attests to the efficiency of the microenvironment provided by the amniote egg. Even so, reptile eggs must often be buried, since over several months of development too much evaporative water loss would take place were the eggs laid in the open.

21.7. What adaptive changes occurred in reptilian skin?

Heavy keratinization of the epidermis created an *epidermal seal* that allowed reptiles full access to terrestrial habitats, even deserts. The outer layer of the epidermis, *stratum corneum*, now could form horny epidermal scales (as in snakes), plates (as in crocodilians and turtles), spines, and so forth. Dead keratinized epidermal cells furnished the digits with claws that provide traction during running and facilitate climbing, digging, grasping, and rending prey. Few cutaneous glands occur in reptiles.

21.8. What improvements occurred in the reptilian respiratory system?

Lobate lungs having increased alveolation and served by a lengthy *trachea* (permitting neck elongation) and a branching *bronchial tree* (Fig. 21.2) improved pulmonary efficiency, making cutaneous respiration unnecessary. Inhalation changed from the amphibian mode of *pushing* air into the lungs by compressing the throat, to *pulling* it in by contraction of special

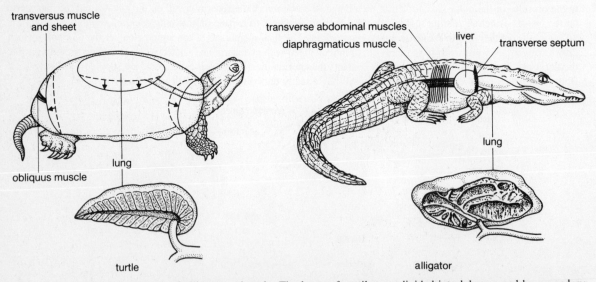

Fig. 21.2 Respiratory systems of alligator and turtle. The lungs of reptiles are divided into lobes served by secondary bronchi, providing a degree of alveolation much more extensive than seen in amphibian lungs. In the absence of a diaphragm (a mammalian feature), breathing is accomplished in the alligator by the action of the diaphragmaticus muscle (inhalation) and the transverse abdominal muscles (exhalation), while in turtles the body cavity is expanded, sucking air into the lungs, by forward rotation of the forelegs together with contraction of the obliquus muscle, which pulls the transversus backward.

muscles of inspiration that enlarge the body cavity, creating a negative pressure that draws air into the lungs. The *diaphragmaticus* muscle of crocodilians pulls the liver and attached transverse septum backward, expanding the thoracic cavity, while the *obliquus* muscle of turtles expands the entire coelom.

Gills are never present, so unneeded gill-support skeletal elements have found new use as additional laryngeal cartilages and cartilage rings that keep the trachea and bronchi from collapsing under negative pressures.

21.9.   How does the reptilian ear differ from an amphibian ear?

The ear structure is similar to that of amphibians, except that the eardrum, when present (it is absent in snakes and amphisbaenids), is often protectively recessed at the inner end of an *external auditory meatus*. A special middle ear bone, the *extracolumella*, runs from the tympanic membrane to the columella itself. In the inner ear, the lagena has enlarged and contains more acoustic receptors than in amphibians.

21.10.   What major advances are seen in the reptilian brain?

The cerebral hemispheres are much larger than those of amphibians, with great expansion of the archipallium. In consequence, the cerebral cortex attains dominance in sensory analysis and association. A rudimentary neopallium (the most complex version of the cortex, best developed in mammals) also originated in reptiles; although seen today only in crocodilians, if the reptilian neopallium is truly homologous with that of other amniotes, it must have appeared early enough in reptilian evolution to be shared by therapsids and archosaurs alike.

21.11.   How do reptiles thermoregulate?

Reptiles are the only ectothermal ("outside heat") amniotes, lacking physiological mechanisms for maintaining a stable body temperature. We call them "cold-blooded," but this is inaccurate: although their body temperature fluctuates with the environment, a number of reptiles are substantially more heat-tolerant and therefore more hot-blooded than the "warm-blooded" birds and mammals. In fact, the desert iguana (*Dipsosaurus dorsalis*) is most active at a body temperature of 42°C (compared with 38°C for humans), and tolerates up to 47°C (117°F). Besides being heat-tolerant, reptiles thermoregulate *behaviorally*.

> **Example 2:** Chilled reptiles sunbathe until body temperature is optimal, then avoid overheating by seeking shade or, if shade is lacking, face into the sun to reduce the surface area exposed. Many lift their bodies off of a hot substrate, or lie prostrate radiating body heat to cooler ground.

Some reptiles evolved specialized thermoregulatory structures.

> **Example 3:** Certain Paleozoic pelycosaurs (see Fig. 21.7) bore tall middorsal thermoregulatory sails of highly vascularized skin supported by elongated neural spines of the backbone: the pelycosaur could collect heat by presenting its flanks to the sun, or could dissipate excess body heat from the sail by facing the sun. Equivalent devices evolved in some dinosaurs, such as the series of erect bony plates that ran middorsally along the back of *Stegosaurus* (Fig. 21.5).

The larger Mesozoic dinosaurs were most likely *endothermal* ("inner heat"), because heat generated by muscular activity and other metabolic processes is conserved by a large body mass. In a warm climate, they probably retained this heat through the night with little energy cost and thus achieved a nearly *homeothermal* ("even-temperatured") state like that of birds and mammals, as opposed to the *poikilothermal* ("variable-temperatured") condition of modern amphibians and reptiles, which often become torpid at night and require a morning sunbath before perking up. However, even if dinosaurs *were* homeothermal through a combination of endothermy and climatic factors, they were still unequipped to deal with any prolonged drop in ambient temperature, for they lacked heat-retaining fur or feathers. The sudden extinction of dinosaur species that were still flourishing 65 million years ago may well have been precipitated by severe planetary cooling caused by earth's being struck by an asteroid or comet (see Questions 4.23–4.26).

21.12.   What new feeding devices did reptiles evolve?

Dietary innovations helped Paleozoic reptiles succeed, and varied feeding strategies persist in modern reptiles, in contrast with modern amphibians, which are all predatory, except for tadpoles.

Example 4:   Some turtles are carnivorous, others herbivorous. Most lizards are insectivorous, but iguanas are chiefly herbivorous, while the impressive monitors (including the outsized 3-m Komodo dragon) attack good-sized mammals. An extinct Australian monitor was over 6 m long and, given the monitor temperament, would have been a mean creature to meet on a lonely trail! Snakes and crocodilians are wholly predaceous, mostly enjoying a menu of other vertebrates.

The earliest known Carboniferous reptiles, small (10- to 20-cm), lizardlike *romeriid captorhinomorphs* (which looked very much like gephyrostegid amphibians, see Fig. 20.6c), were able to profit by the explosive proliferation of insects during the later Carboniferous, for their jaw mechanism became specifically adapted to an insectivorous diet. The chitinous arthropodan exoskeleton is indigestible, thus the soft internal tissues can be better digested if the insect can be crushed in the mouth before being swallowed. Paleozoic amphibians and rhipidistians could snap their jaws closed forcibly, but could not exert a static crushing force upon prey being held in the mouth. Subdivision of the jaw musculature into separate *pterygoideus* and *temporalis* muscles took place very early in reptile evolution and provides one means of distinguishing Paleozoic reptiles from amphibians: the pterygoideus continued to provide the quick closing snap, but the temporalis, oriented nearly at right angles to the closed jaws, could thus apply crushing force after the jaws were closed. (Place your fingers on your temples and clench your teeth; the contractions you feel are those of *your* temporalis, exerting static force; you can see how important this muscle is for crushing food during chewing.)

This new jaw mechanism also served to crush *plants,* and the herbivorous cotylosaurs soon evolved a correlated dentition adapted for shearing and mashing vegetation. Throughout the Permian Period and Mesozoic Era reptiles took increasing advantage of this great food source that few amphibians ever came to use. With an abundance of edible terrestrial plants, herbivorous reptiles could grow enormous, providing food in turn for carnivores from modest to awesome size.

21.13.   What changes are seen in the reptilian heart?

The reptilian heart has become further partitioned into right and left pumps by the growth of a *ventricular septum,* which is complete only in crocodilians. The truncus arteriosus is fully subdivided into separate pulmonary and systemic trunks. In modern reptiles, the systemic trunk also divides in two, all the way to the ventricle, so that one pulmonary and *two* systemic trunks spring directly from the ventricle.

21.14.   What innovations in the urinary system originated with reptiles?

Reptiles developed an entirely new type of kidney and urinary duct, which characterizes all amniotes. These *metanephric kidneys* and their ducts arise by the embryonic outgrowth of bilateral *cloacal buds* that interact inductively with mesoderm of the far posterior portion of the nephrogenic plate so that the cloacal buds form the *metanephric duct (ureter)* and *collecting tubules* (see Fig. 18.14), while the nephrogenic plate contributes the *nephrons* (Bowman's capsules and renal tubules, see Question 18.45). Thus reptiles established the amniote pattern of having a pair of functional embryonic *mesonephric* kidneys developed from the *central* portion of the nephrogenic plate and drained laterally by *mesonephric ducts,* followed postembryonically by a pair of metanephric kidneys not seen in anamniotes. The new, median ureters serve only to carry urine to the bladder, leaving the mesonephric ducts of the embryonic kidneys available solely for sperm transport. The renal portal system so important to fishes and amphibians that swim with their tails becomes nonfunctional in reptiles, but it still partly develops in the embryo and contributes its tubular capillary bed to the renal circulation (see Question 18.42).

21.15.   What advances do reptiles show in reproduction and rearing of young?

(a)   Internal fertilization is achieved in all reptiles by the development of a male copulatory organ. In crocodilians and turtles, a true penis develops from the cloacal wall. This cloacal penis is homologous with that of mammals and primitive birds. It is *hemotumescent,* i.e., becomes rigid and enlarged by blood engorgement. It contains two sinusoidal *corpora cavernosa* responsible for erection and is capped with a small, sensory *glans penis.* Instead of a true penis, lizards and snakes have paired *hemipenes,* eversible sacs on either side of the cloaca, which are housed in the base of the tail; only one hemipenis is used at a time.

(b)   The yolky eggs of oviparous reptiles are laid within protective shells and are usually buried by the mother. Female crocodilians cover the nest with mud and rotting vegetation, and the heat of decay incubates the eggs. Curiously, incubation temperature affects sex determination: low temperatures result in all hatchlings being female, high ones in only males. The mother stays nearby, guarding the nest and eventually helping the hatchlings escape by ripping away the hardened mud. Young crocodilians follow their mother about like peeping ducklings, as she defends them against assorted predators, including adults of their own kind. Dinosaur tracks showing babies walking along with adults suggest that some of these great reptiles lived in herds or family groups, with adults protecting the young. Crocodilians and turtles lay eggs, but many lizards and snakes are *ovoviviparous*.

21.16.   What skeletomuscular advances occurred in reptiles?

(*a*)   Skeletomuscular advances that were to improve the lot of amniotes as a whole began to appear even in the reptilelike amphibians. The gephyrostegid labyrinthodonts were small and agile and may well have been the direct ancestors of the earliest known reptiles (romeriid captorhinomorphs). The limb bones and girdles became progressively more sturdy, and a more complex appendicular musculature spread out over the surface of the trunk muscles. Compared with amphibians, which have only one sacral vertebra, reptiles developed a pelvic girdle which came to articulate with a *sacrum* made up of two to five fused vertebrae with their short *sacral ribs;* this superior pelvic anchorage allowed for improved support and greater thrust from the hind legs. Salamanders' legs spraddle to the sides, but reptilian limbs *rotated* so that the elbows came to point backward and the knees forward (Fig. 21.3). Thus the limbs can be brought under the trunk, elevating it and gaining mechanical advantage. Although cotylosaurs and mammal-like reptiles remained quadrupedal, early *archosaurs* were *bipedal,* as were a number of their dinosaur descendants, along with birds, which can seriously be thought of as "feathered dinosaurs." Even the quadrupedal dinosaurs carried the stamp of ancestral bipedality, for they had powerful hind legs and their backs were often highest at the pelvis, sloping down to small front quarters.

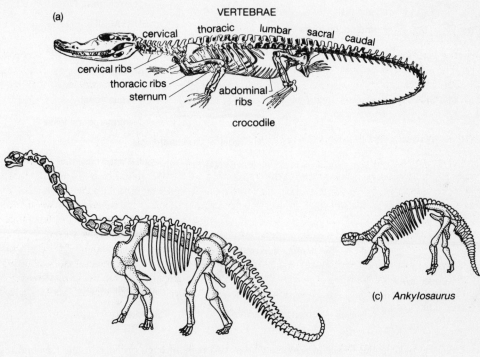

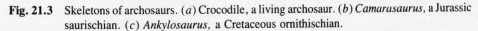

Fig. 21.3   Skeletons of archosaurs. (*a*) Crocodile, a living archosaur. (*b*) *Camarasaurus*, a Jurassic saurischian. (*c*) *Ankylosaurus*, a Cretaceous ornithischian.

Amphibians have only one cervical vertebra, but reptiles have a number, which glide upon each other rather freely, promoting neck flexibility. Head mobility also became enhanced as the first two vertebrae specialized into a ring-shaped *atlas*, which articulates with cranial processes called *occipital condyles*, and an *axis* bearing an anteriorly directed process that allows the head to be turned from side to side by rotation of the atlas upon the axis.

(*b*)   *Temporal fossae* arose in most reptiles and were passed along to the mammals and birds (Fig. 21.4). As mentioned in Question 21.12, one of the earliest distinctions between reptilclike amphibians and amphibianlike reptiles is that in the latter the jaw-closing musculature had subdivided into pterygoideus and temporalis portions, the latter extending

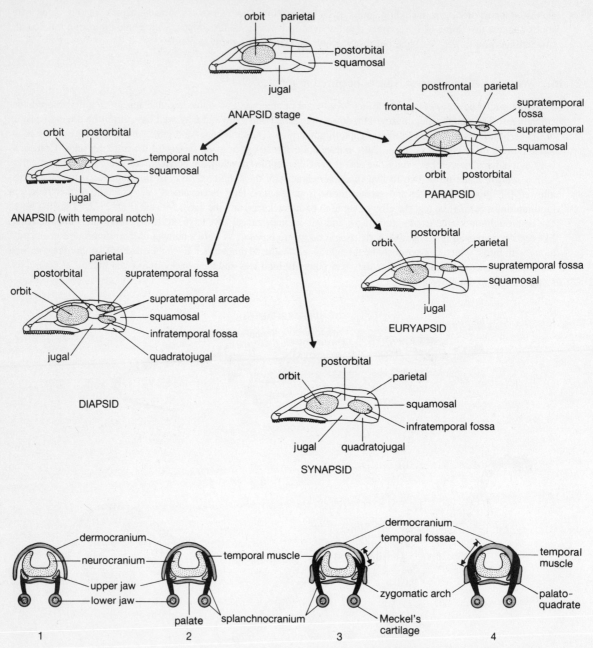

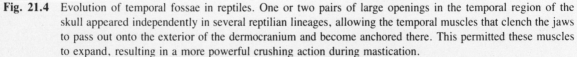

Fig. 21.4 Evolution of temporal fossae in reptiles. One or two pairs of large openings in the temporal region of the skull appeared independently in several reptilian lineages, allowing the temporal muscles that clench the jaws to pass out onto the exterior of the dermocranium and become anchored there. This permitted these muscles to expand, resulting in a more powerful crushing action during mastication.

upward from the rear of the jaw. In bony fishes and primitive amphibians and reptiles, the *dermocranium* of dermal bone forms a solid roof over the originally cartilaginous *neurocranium,* and the jaw musculature is confined to the space between the two. Development of one or two pairs of large openings (temporal fossae) through the dermocranium permitted the temporalis to exit onto the *outer* surface of the dermocranium, opening the way for its great expansion, with attendant improvement of chewing. Cotylosaurs and turtles lack these fossae, but these openings evolved *independently* in several other reptilian lineages, providing a basis for recognizing separate subclasses.

21.17. What are the major subdivisions of class Reptilia?

Reptiles are best understood in the light of the fossil record, for those living today are but a remnant of the great Mesozoic radiation (Fig. 21.5). The major subdivisions of this class are listed in Table 21.1.

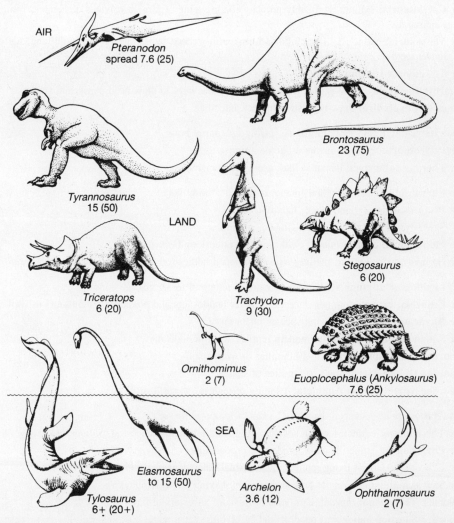

AIR

Pteranodon
spread 7.6 (25)

Brontosaurus
23 (75)

Tyrannosaurus
15 (50)

LAND

Triceratops
6 (20)

Trachydon
9 (30)

Stegosaurus
6 (20)

Ornithomimus
2 (7)

Euoplocephalus (Ankylosaurus)
7.6 (25)

SEA

Elasmosaurus
to 15 (50)

Tylosaurus
6+ (20+)

Archelon
3.6 (12)

Ophthalmosaurus
2 (7)

Fig. 21.5 Representative Mesozoic reptiles. All species shown lived during the Cretaceous Period, except for the Jurassic *Brontosaurus* and *Stegosaurus*. Length given in meters and (feet). *(From Storer et al.)*

21.18. What are the characteristics of turtles?

Turtles (order Testudinata) are the only anapsids alive today (Fig. 21.6). As a group they have existed essentially unchanged since they arose during the Triassic. Their major characteristics are as follows:

(*a*) In the absence of fossae, a large *temporal notch* at the rear of the skull gives the jaw muscles access to the outer surface of the dermocranium.

(*b*) Teeth are missing and the jaws are covered by a keratinous sheath, the *rhampotheca*, which may be edged for shearing vegetation or hooked for tearing flesh.

(*c*) The trunk is encased in armor composed of bone to the inside and epidermal keratinous plates to the outside. It consists of a dorsal *carapace* and ventral *plastron* connected only by rigid lateral bridges permitting mobility of

Table 21.1   Major Subdivisions of Class Reptilia

Subclass Anapsida: temporal fossae lacking

  Order Captorhinomorpha: extinct; includes romeriids, the earliest known reptiles

  Order Cotylosauria: extinct; a probably polyphyletic grouping of large herbivores descended from captorhinomorphs

  Order Mesosauria: extinct; 1-m amphibious forms with webbed feet, flattened tails, and long, narrow jaws with slender teeth

  Order Testudinata/Chelonia: turtles

    Suborder Pleurodira: side-neck turtles—bend neck sideways to draw head partly into shell

    Suborder Cryptodira: draw head straight back into shell

Subclass Synapsida: 1 pair temporal fossae, *below* squamosal bone

  Order Pelycosauria: extinct; mammal-like dentition

  Order Therapsida: extinct; mammal-like, ancestral to mammals

Subclass Euryapsida: 1 pair temporal fossae, *above* squamosal bone

  Order Protorosauria: extinct; small, lizardlike

  Order Sauropterygia: extinct; paddle-limbed, marine

  Order Plesiosauria: extinct; marine, with flippers; often long-necked

  Order Ichthyosauria: extinct; marine, with fish-shaped bodies (often listed as *subclass Ichthyosauria*)

Subclass Lepidosauria: primitive diapsids (2 pairs of temporal fossae); scaly skins

  Order Eosuchia: extinct; common ancestor to other lepidosaurs and possibly to archosaurs as well

  Order Rhynchocephalia: one living species (tuatara)

  Order Squamata: most successful modern reptiles; molt periodically

    Suborder Lacertilia/Sauria: 26–30 families of lizards

    Suborder Ophidia/Serpentes: 10 families of snakes

    Suborder Amphisbaenia/Annulata: 1 family of legless "worm lizards"

Subclass Archosauria: advanced diapsids (2 pairs of temporal fossae)

  Order Thecodontia: extinct; teeth in sockets; originated bipedality; ancestral to dinosaurs

  Order Phytosauria: extinct; aquatic, predatory, crocodilelike

  Order Crocodilia: largest living reptiles; quadrupedal, short-legged, amphibious, predatory

  Order Saurischia: extinct; reptile-hipped dinosaurs (having pubic symphysis)

    Suborder Theropoda: bipedal, carnivorous; includes ancestors of birds

    Suborder Sauropodomorpha: quadrupedal herbivores, often huge

  Order Ornithischia: extinct; bird-hipped dinosaurs (lacking pubic symphysis); mostly herbivorous and quadrupedal

    Suborder Hadrosauria: bipedal; mostly duck-billed and amphibious

    Suborder Stegosauria: with dorsal series of erect bony plates

    Suborder Ankylosauria: short-legged, heavily armored

    Suborder Ceratopsia: head with horns and bony shield

  Order Pterosauria: extinct; winged

neck, limbs, and tail. Turtles have only 10 trunk vertebrae, of which 8—along with their broad, attached ribs—are tightly fused to the overlying carapace.

(*d*)   The shell imposes restrictions on the limbs, so that the elbows must rotate forward and the knees protrude sideways.

(*e*)   Since the dimensions of the trunk cannot change, breathing is accomplished in a special way: inspiration results from expanding the entire body cavity within the shell by moving the pectoral girdle outward while contracting the

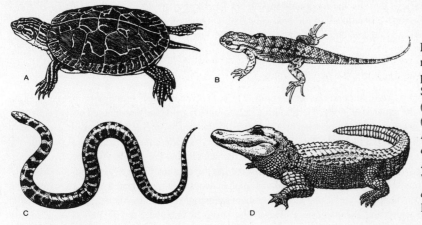

Fig. 21.6 Types of living reptiles. Order Testudinata: A, painted turtle *(Chrysemys)*. Order Squamata: B, spiny lizard *(Sceloporus)*; C, water snake *(Natrix)*. Order Crocodilia: D, Alligator *(Alligator)*. *(From Storer et al.: A–C from Stebbins,* Amphibians and Reptiles of Western North America; *D afterPalmer,* Field Book of Natural History.)

obliquus muscle at the rear of the trunk (see Fig. 21.2); expiration consists of pushing air out of the lungs by moving the pectoral girdle inward while the *transversus* muscle contracts at the rear of the coelom, shortening the body cavity.

(*f*) Copulation is by use of a true cloacal penis. Attendant reproductive behaviors include simple courtship (e.g., butting) and territorial fighting between males that may end in one overturning the other. The female finds a suitable place for excavating a deep nest with her hind feet. She lays 100 or so eggs, then buries and deserts them. Sea turtles must crawl inland above the high tide line for nest building, and the hatchlings instinctively find the sea by crawling toward the brightest horizon.

21.19. What are the major types of turtles?

Suborder *Pleurodira* ("side neck") includes only the side-necked turtles, so called because they bend the neck sideways to tuck the head partly into the shell. Suborder *Cryptodira* ("hidden neck") includes all other turtles, which withdraw the head straight back between carapace and plastron by flexing the neck ventrally; 5 of the 10 cryptodire families are:

(*a*) *Snapping turtles* (family Chelydridae), are heavy-bodied (to 80 kg) American freshwater turtles of wide dietary habits; they are known to waylay swimming snakes, mammals, waterfowl, and frogs, as well as fishes, which the alligator snapper attracts by lying with mouth agape, wriggling a wormlike tongue tip.

(*b*) *Terrapins*, or pond turtles, (family Emydidae) swim by paddling their webbed feet, but readily crawl ashore to sunbathe. They are omnivorous on aquatic plants and invertebrates. Female terrapins raised in captivity may lay viable eggs as much as 4 years after their last mating, from sperm held in their ovaries, a practice common to certain other turtles and snakes.

(*c*) *Tortoises* (family Testudinidae) are herbivorous, terrestrial turtles with high-domed shells. Giant species exceeding 100 kg are indigenous to the Galápagos archipelago and islands in the Indian Ocean.

> **Example 5:** Darwin first observed that races of Galápagos tortoises living on islands where their staple food, cacti, were treelike have especially long necks and carapaces with a high arch at the front, adaptations that allow them to reach up to browse, and that races living on islands where cacti are low-growing lack these specializations.

(*d*) *Sea turtles* (family Cheloniidae) have winglike front flippers and "fly" gracefully through the water. Some migrate great distances to return to the localized breeding areas where they began life. Because of the edibility of turtle flesh and eggs and the usefulness of their oil and shells (especially "tortoiseshell" from the hawksbill, used in making glasses frames and curios), they have become severely endangered.

> **Example 6:** Green turtles, *Chelonia mydas,* may weigh over 150 kg. They feed upon sea grass and sometimes crawl clumsily ashore to bask. Hawksbills may reach 100 kg and are omnivorous, with a craving for crabs and fish. On shore, they move their flippers in alternation, like legs, rather than together as green turtles do.

(*e*) *Leatherbacks* (family Dermochelyidae) are marine giants reaching a length of nearly 3 m and a weight of 800 kg, with a 3-m flipper span. The carapace is reduced to hundreds of small bony plates covered with leathery skin instead of keratinized plates. The skin is oily, and no barnacles encrust it. Leatherbacks relish jellyfishes, salps, and pteropods.

21.20. What were the synapsids?

Subclass Synapsida, the extinct *pelycosaurs* and *therapsids* (Fig. 21.7), evolved one pair of temporal fossae opening low on the side of the skull below the squamosal bone. Herbivorous pelycosaurs considerably resembled cotylosaurs, but the carnivorous ones developed a *heterodont* dentition foreshadowing that of mammals, including deep-rooted maxillary fangs like canines and premaxillary teeth like incisors. They also developed an *arched palate* that later made possible the separate nasal cavity of mammals. Their legs, though still spraddling to the sides, grew longer and slimmer. By the mid-Permian, pelycosaurs had given rise to the therapsids, a group of mammal-like reptiles including heavy-bodied herbivores and also carnivores (*theriodonts*: "mammal tooth") that developed mammalian dentition and graduated toward a mammalian skeletal condition. The synapsids were happily diversifying as the dominant terrestrial vertebrates when they were decimated by the mass extinction event that ended the Paleozoic Era. Those that survived into the Triassic Period were mostly carnivores, including those that gave rise to mammals.

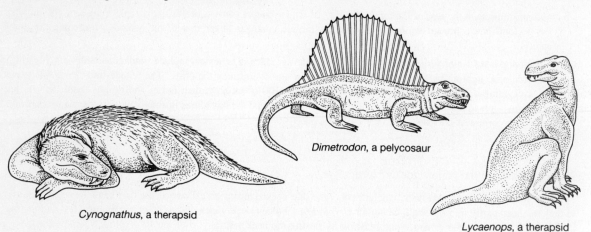

Dimetrodon, a pelycosaur

Cynognathus, a therapsid

Lycaenops, a therapsid

Fig. 21.7 Pelycosaurs and therapsids: mammalian forerunners. The pelycosaur *Dimetrodon* bore a tall thermoregulatory sail along its back. Pelycosaurs gave rise to such mammal-like therapsids as the fox-sized *Lycaenops* and wolf-sized *Cynognathus*.

21.21. What were the euryapsids?

Subclass Euryapsida developed one pair of fossae lying high on the side of the skull, above the squamosal bone. This extinct group included *marine* fish-eating Mesozoic reptiles, chiefly *ichthyosaurs* and *plesiosaurs* (see, e.g., *Ophthalmosaurus* and *Elasmosaurus*, Fig. 21.5). Both had front and rear flippers, but plesiosaurs had basically reptilian bodies and either long or short necks, while ichthyosaurs had very short necks and fish-shaped bodies with dorsal and caudal fins. Being quite unable to come ashore, ichthyosaurs were live-bearers. The vertically oriented caudal fin, moved from side to side like that of a fish, developed with the vertebral column extending down into the lower lobe, the reverse situation from that in sharks. Some ichthyosaurs attained lengths of over 7 m and were the most massive animals of their time. Ichthyosaurs are often placed in a separate subclass (Ichthyosauria), because intermediate forms are lacking between them and plesiosaurs and their temporal fossae open dorsal to a supratemporal bone that lies above their squamosal but is not present in plesiosaurs (a *parapsid* condition that may have evolved quite separately).

21.22. What are archosaurs?

Subclass Archosauria is characterized by two pairs of temporal fossae, a *diapsid* condition also found in *subclass Lepidosauria*. Because of this basic similarity, archosaurs and lepidosaurs (mainly lizards and snakes) perhaps should be grouped together in a subclass *Diapsida*. A common ancestor for both groups may be found in the *eosuchians* of the late Carboniferous and Permian; early eosuchians were lizardlike and quadrupedal, while Permian forms had modestly enlarged hind legs suggesting partial bipedality.

Besides their temporal fossae, archosaurs, the "ruling reptiles" of the Mesozoic, developed additional skull openings seen today in crocodilians: one or two *antorbital fenestrae* in front of each eye and a *mandibular fenestra* on each side of the lower jaw. These openings allowed space for muscle expansion, while preserving the strength of the skull. Subclass Archosauria included a stem group, *order Thecodontia* ("socketed teeth"), two orders of dinosaurs (*Ornithischia* and *Saurischia*, from which *birds* arose), the flying pterosaurs, and two independently evolved orders of amphibious fish eaters, the extinct phytosaurs and the crocodilians. Of all the archosaurs, only crocodilians and birds survive, the latter being sufficiently established in their own right to be assigned to a separate vertebrate class (Aves).

21.23.   What characterized the thecodonts?

Thecodonts started out in the Triassic as quadrupeds and ended up fully bipedal with hind legs much larger than forelegs. They gave rise to all other archosaurs. In contrast with lizards and snakes, their teeth were rooted in sockets, a general archosaur trait. Their most impressive descendants were the terrestrial dinosaurs.

21.24.   What were ornithischians like?

Bird-hipped dinosaurs (order Ornithischia) had a birdlike pelvic structure (Fig. 21.8*a*) in which the pubic bones did not meet midventrally, but instead swung backward along the lower edge of the elongate ischium. This provided a long surface for muscle attachment, which in later ornithischians was further lengthened by forward extension of the pubis. Although this pelvic structure might suggest that birds arose from ornithischians, other skeletal features make it more likely that they came from saurischian ancestors and developed their open pelvis later on.

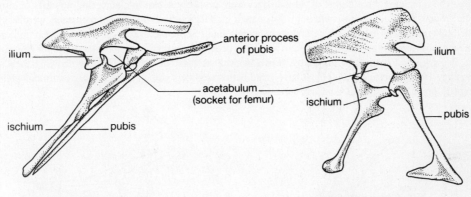

(a) *Tetraradiate* (Ornithischia)   (b) *Triradiate* (Saurischia)

Fig. 21.8 Dinosaur pelvic structure. (*a*) Right pelvis of a bird-hipped dinosaur (order Ornithischia). (*b*) Right pelvis of a reptile-hipped dinosaur (order Saurischia). In ornithischians
(and birds) the two pubic bones do not meet ventromedially at a pubic symphysis as
they do in saurischians (and most other reptiles, as well as mammals). This open pelvic
structure may facilitate laying large eggs or giving birth to large young.

Ornithischians were all herbivorous. Some were quadrupedal, including *stegosaurs, ankylosaurs*, and *ceratopsians* (see, respectively, *Stegosaurus, Euoplocephalus*, and *Triceratops*, Fig. 21.5). *Hadrosaurs* (see *Trachydon*, Fig. 21.5), were bipedal and included the amphibious *duck-billed dinosaurs*, which grew up to 10 m and 10,000 kg, and sported webbed forefeet and flattened tails for swimming.

21.25.   What were the saurischians like?

Reptile-hipped dinosaurs (order Saurischia) had a typical reptilian hip structure with pubic bones meeting midventrally (Fig. 21.8*b*). They fell into two extensive subgroups: the carnivorous, bipedal *theropods* and the herbivorous, quadrupedal *sauropods*, which became the largest animals ever to walk the earth (see *Brontosaurus*, Fig. 21.5). Sauropods had long, slender necks and tails and massive trunks that, like quadrupedal ornithischians, stood highest at the pelvis. At this point a large swelling in the spinal cord marked a *sacral plexus* several times larger than the brain, which probably controlled the hind limbs and tail. These enormous creatures reached lengths exceeding 25 m and weighed as much as 50,000 kg.

The predaceous theropods carried bipedality to such an extreme that their forelegs became markedly reduced in size and number of digits. Their long, powerful hind legs had four toes: three directed forward and one backward as in birds, with the rear toe sometimes vestigial. Their long bones were hollow, as in birds, even in such giants as *Tyrannosaurus rex* (see Fig. 21.5), which stood 6 m tall with a length of 15 m including a large head with 2-m jaws. More lightly built were the *coelurosaurs*, including the long-necked *ostrichlike dinosaurs* (ornithomimids: "bird mimics"), which also resembled birds in having three-fingered hands and toothless jaws sheathed with a keratinous beak. Despite their reptilelike hip structure, it is from this group that birds are thought to have come, a hypothesis strengthened by the recent finding that, upon reexamination, certain fossil specimens of these small, fleet-footed dinosaurs show unmistakable impressions of wing feathers or feather-attachment points. Feathers being the diagnostic hallmark of birds, these little dinosaurs should qualify as birds. The opening up of the avian pubic symphysis therefore seems to have been an event independent of that seen in ornithischians, although in both groups this may have been an adaptation for laying eggs that were exceptionally large with respect to body size.

21.26.   What were the first vertebrates to take to the air?

Pterosaurs (order Pterosauria) were winged archosaurs from a lineage quite different from that which gave rise to birds. Some became the most prodigious creatures ever to sail the skies. Although the skeletons of advanced pterosaurs show marked convergence with bird skeletons—eventually including loss of teeth, a keeled sternum to which the flight muscles attached, a shortened tail, elongated iliosacral articulation, and fused trunk vertebrae compensated by a long, flexible neck—the wing structure was totally different (Fig. 21.9). The wing was a membrane of skin, supported only along its leading edge by the arm, hand, and tremendously elongated fourth digit, and anchored medially along the sides of the trunk and legs. At the wrist joint, a digitlike carpal may have anchored an additional skin web stretching from hand to neck. The remaining three digits, of normal size, were probably useful in clambering up cliffs or simply getting around on the ground, since pterosaurs could not walk on their hind legs as birds do. Earlier pterosaurs were less convergent toward birds, having toothed jaws and long counterbalancing tails. Some pterosaur fossils show what appears to be the imprint of a furry body covering; these reptiles may well have been endothermal, with hairlike scales for heat conservation. The tiniest, sparrow-sized pterosaurs may have flown as well as bats, but larger forms with wingspreads of 7.6 m (*Pteranodon*) and 15 m (*Quetzalcoatlus*) must have fared better as gliders. Like condors, these enormous pterosaurs may have scavenged upon carrion such as dinosaur carcasses.

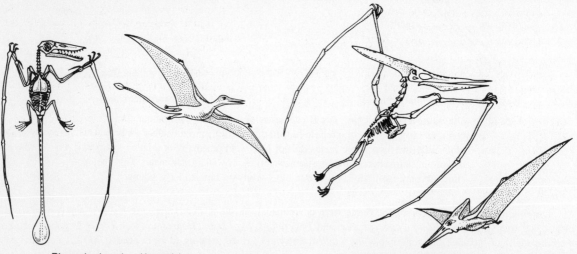

Rhamphorhynchus (Jurassic)   *Pteranodon* (Cretaceous)

Fig. 21.9   Pterosaurs. These flying reptiles had enormously extended fourth fingers that supported the skin membrane of the wing. Jurassic pterosaurs such as *Rhamphorhynchus* retained such primitive features as teeth and long tails, whereas Cretaceous species such as *Pteranodon* were toothless and nearly tailless and had developed a keel on the breastbone.

21.27.   What were the phytosaurs like?

Phytosaurs were aquatic, predatory archosaurs resembling crocodilians, but they arose quite separately and much earlier, becoming large and abundant during the Triassic. Their nostrils opened on top of their heads just in front of their eyes, not toward the tip of the elongate snout (as in crocodilians), suggesting that they lurked in ambush, submerged with only eyes and nostrils exposed. Their four short limbs were equal in size, suggesting that they may have branched off from early thecodonts that were still quadrupedal.

21.28.   What characterizes the crocodilians?

Crocodilians (order Crocodilia) are the largest of living reptiles. American crocodiles (*Crocodylus acutus*) reach lengths of 7 m and weigh 1000 kg. They have long trunks, snouts, and tails, and stubby legs. They swim with limbs trailing, propelled by lateral undulations of their powerful, flattened tails. When crocodilians first appeared in the Jurassic, their hind legs were substantially larger than their forelegs, indicating descent from bipedal thecodonts. This disparity progressively diminished, but even today the hindlegs remain somewhat the larger. Crocodilians' nostrils, found near the tip of their long snout, do not open into the mouth as in other reptiles; instead, they lead into an elongated nasal cavity separated from the buccal cavity by a *secondary palate*, formed by extensions of the premaxillae, maxillae, and pterygoids.

Both living and extinct crocodilians have been found with stones in their stomachs, apparently swallowed as a means of reducing buoyancy. Although larger crocodilians can kill large prey (even humans on occasion), insects, crustaceans, and fish form the bulk of their diet. When on land, crocodilians walk sedately with trunk well elevated, being given to only short bursts of speed; young crocodiles sometimes gallop like squirrels, moving both front legs together, then both hind legs, a gait familiar in mammals but most unusual in reptiles.

The four living families of order Crocodilia are alligators, crocodiles, caimans, and gharials.

21.29.   How do the four living families of crocodilians differ?

(*a*)   *Alligators* have broad, blunt snouts with widely separated nostrils (see Fig. 21.6). Their upper teeth overlap the lower. They have 17 or more teeth on each side of the lower jaw, the enlarged fourth tooth fitting into a pit in the upper jaw. Alligators are comparatively lethargic and may live 50 years. Much less aggressive than crocodiles, they can be grown on farms for their hides. Only two species exist today: the American alligator (*A. mississippiensis*) and the Chinese alligator (*A. sinensis*).

(*b*)   *Crocodiles* occur in Africa, Asia, Australia, and America. Unlike other members of this order, they are not confined to fresh water but inhabit estuaries and have even been sighted swimming far out at sea. Their snouts are narrow and markedly concave, with the external nostrils placed close together at the elevated tip; they open into the nasal cavity by way of a *single* orifice (bony naris). They have no more than 15 teeth on each side of the lower jaw, which *interlock* with the upper teeth; the enlarged fourth tooth fits into a *notch* in the upper jaw.

(*c*)   *Caimans* of tropical America are spared the fate of other crocodilians, which are often killed for their hides, because their belly skin (the part tanned for leather) is made commercially valueless by a reinforcement of large bony plates. They are very agile in water and surprisingly fast on land, and not reluctant to attack livestock. When threatened by hunters, they inflate their bodies, hiss, and jump about with jaws agape.

(*d*)   *Gharials* of India have extremely narrow, slender snouts armed with long, recurved, interlocking teeth. They grab fish with a sideways slash of the head. Although nearly 3 m long, they rarely harm people, which is fortunate since they are protected as sacred.

21.30.   What are today's most successful reptiles?

Subclass Lepidosauria ("scaly reptile") is by far the most successful group of modern reptiles, with some 3600 species of lizards, 2500 kinds of snakes, 130 species of amphisbaenids, and a single surviving species of rhynchocephalian (Fig. 21.10). By comparison, only about 250 species of turtles and 23 species of crocodilians remain alive, and all other great reptilian groups have long vanished. As in archosaurs, the skull of lepidosaurs is diapsid, but except for the rhynchocephalian *Sphenodon* (tuatara), they have developed other skull specializations entirely their own.

21.31.   Why is tuatara unique?

The New Zealand tuatara (see Fig. 21.10*a*) is the only living representative of *order Rhynchocephalia* ("beak head"). This 0.5-m relic shares the burrows of seabirds and feeds largely on insects that subsist on the guano. Its placid calflike eye gives no hint that it bites with bulldog tenacity. Tuatara is *not* a lizard: (*a*) its vertebrae are biconcave, not concave only in front; (*b*) its quadrate bone (involved in jaw suspension) is fixed, not movable; (*c*) its anus is a transverse, not a vertical, slit; (*d*) its diapsid skull reflects the archosaur state; (*e*) it has a conspicuous median eye.

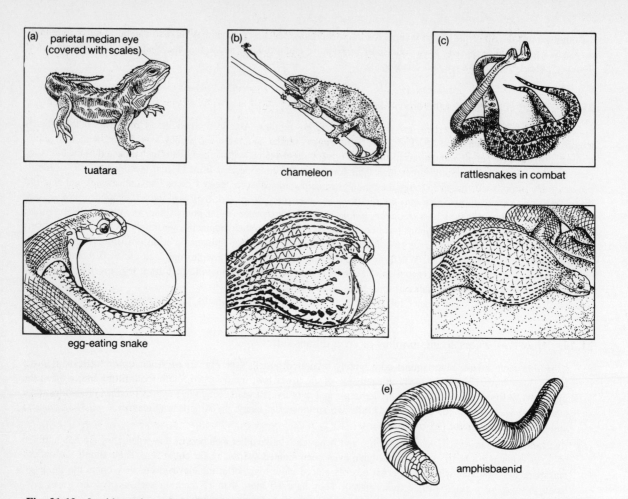

Fig. 21.10 Lepidosaurians. Order Rhynchocephalia: (*a*) the tuatara *(Sphenodon)* of New Zealand. Order Squamata: (*b*) African chameleon *(Chameleo)* capturing prey with extended tongue; (*c*) rattlesnake *(Crotalus)* males engaged in social fighting at the mating season; (*d*) egg-eating snake *(Dasypeltis)* showing extreme distensibility of jaws and throat; (*e*) a burrowing amphisbaenid.

21.32.   What characterizes order Squamata?

   Members of *order Squamata* ("scaly")—lizards, snakes, and amphisbaenids—not only have epidermal scales, but shed them as well: in periodic molts the outer part of the stratum corneum sloughs scruffily or is pulled off, inside out, in one piece (snakes and geckos). Males have unique paired copulatory organs (*hemipenes*). A semisolid urine is excreted, in which uric acid is the main nitrogenous waste; this conserves water, allowing many snakes and lizards to inhabit very dry habitats.

21.33.   How have lizards diversified?

   Lizards (*suborder Lacertilia,* or *Sauria*) range in length from 2 cm to 3 m, and have radiated into many habitats and modes of life. They run swiftly, sometimes bipedally, and many are arboreal. Smaller types are mainly insectivorous, larger ones herbivorous, except for the predaceous monitors. A few of the 26–30 living families are as follows:

(*a*)   *Gekkonidae,* geckos, are nocturnal, often bear adhesive pads of long setae or microvilli on their toes, usually lack eyelids and have a transparent scale (spectacle) covering the eye, and sometimes bark.

(*b*)   *Iguanidae,* the largest family of American lizards, includes true iguanas, which flaunt a dorsal longitudinal row of enlarged scales or spines; anoles (American "chameleons," for their rapid color changes); horned lizards; chuckwallas; fence lizards (see Fig. 21.6); etc.

(*c*) *Teiidae*, the whiptails, have squarish ventral scales and a forked tongue; they often reproduce parthenogenetically.

(*d*) *Scincidae*, skinks, have smooth, round-edged, shiny scales; some are limbless, and some are euviviparous, with well-developed placentas.

(*e*) *Anguidae*, alligator lizards, have rectangular scales over thin bony plates, a rather crocodilian form with a long, heavy tail that is readily shed and regenerated, small limbs or none at all, and a deep skin fold running lengthwise along each side.

(*f*) *Helodermatidae*, the Gila monster and Mexican beaded lizard, are the only venomous lizards, their grooved teeth and venom glands opening onto the outer gum of the lower jaw; heavy-bodied, with beadlike dorsal scales over dermal bony ossicles, they feed on small mammals, quail eggs, etc.

(*g*) *Chameleonidae*, chameleons, change color rapidly and have twig-grasping feet in which two digits oppose three, prehensile tails, independently movable eyes, and a tongue that can be shot out the length of the trunk to capture insects (see Fig. 21.10*b*).

(*h*) *Agamidae* include most Old World lizards such as the "flying dragon," which has elongated ribs that can be spread laterally, stretching the covering skin into a winglike, volplaning surface.

(*i*) *Varanidae*, the carnivorous monitors, including the 3-m "dragon" of Komodo island, have long, forked tongues and are most likely ancestral to snakes.

21.34. Why are amphisbaenids not legless lizards?

Suborder Amphisbaenia ("walking both ways") is named for amphisbaenids' two-headed look, with blunt head and blunt tail of the same size (see Fig. 21.10*e*). Excepting one species with forelimbs only, these 0.5-m burrowers cannot walk at all, being limbless and vermiform, with soft skin folded into a series of rings, giving them the look of stout earthworms. Their eyes and ears are overgrown by skin. Legless lizards do exist, but amphisbaenids are not lizards at all: the tail is very short and trunk elongate, they have only one lung, and a unique *median tooth* occupies the front of the upper jaw. This tooth fits neatly between the two front teeth below, forming a nipper that cuts bites of flesh out of prey too large to swallow whole. They eat insects mainly, and females usually lay their eggs in anthills, thus providing the hatchlings with ample food.

21.35. What are the special adaptations of snakes?

Snakes (*Suborder Ophidia*, or *Serpentes*) are limbless, which seems a poor way to be, but most are surprisingly fast, and their slender bodies let them get into really tight places. Their trunks are very long, and tails quite short. Most have wide transverse ventral scales (*scutes*) that can be moved forward and back by integumentary muscles, allowing a snake stalking prey to glide stealthily forward; when in a hurry, snakes throw their bodies into lateral curves. Snakes' eyes are lidless and covered by a transparent spectacle. External ear, eardrum, and middle ear cavity are lacking, but snakes are not deaf. Their forked tongue is flicked out, then thrust into the olfactory *vomeronasal organs* in the roof of the mouth. Snakes are entirely predatory, and many are armed with venom. Their upper and lower jaws are not firmly articulated, nor are left and right halves fused medially; this allows the mouth to be stretched hugely for swallowing large prey (see Fig. 21.10*d*). Dinner is ingested head first, with each half-jaw moved forward alternately, literally walking the prey in. The glottis of snakes does not lie at the back of the pharynx, but just behind the lower jaw, allowing snakes to breathe uninterruptedly during the laborious swallowing process. Recurved conical teeth are located on the jaws and roof of the mouth. In venomous snakes, certain teeth are modified into grooved or hollow fangs. Nonvenomous snakes either swallow their prey alive, or first kill it by constriction, snugging their coils around the victim until it suffocates. Since the sternum is missing, the ribs are unattached ventrally and the rib cage can be expanded to accommodate the body of the prey as it slips down the esophagus. One lung is vestigial, and no urinary bladder is present.

Snake venoms are loosely described as *neurotoxic*, i.e., killing by neuromuscular paralysis (e.g., cobras), or *hemotoxic*, breaking down tissues and causing hemorrhages, intravascular clotting, or both (e.g., vipers). This distinction is true only to a point, for all snake venoms contain digestive (hydrolytic) enzymes: *proteases* that digest tissue proteins, *phospholipases* that break down cell membranes, and *hyaluronidase* that makes tissues more permeable to other venom components. Cobra and sea snake venoms also contain *basic polypeptides* that block neuromuscular junctions, causing death by respiratory paralysis.

The major ophidian families are as follows:

(a) *Boidae*, viviparous American boas and oviparous Old World pythons, are the monsters of the serpent world; impressive but nonvenomous constrictors, they reach a length of 10 m, have vertical pupils, vestigial hind legs protruding as a spur on each side of the anus, and a row of thermoreceptive pits along the jawline.

(b) *Colubridae* comprise the largest family of nonvenomous snakes, including bull snakes, king snakes, garter snakes, racers, water snakes (see *Natrix*, Fig. 21.6), etc., but also include the venomous *rear-fanged* snakes, which have grooved fangs in the rear of the jaw and, except for the African boomslang, are not dangerous to humans.

(c) *Elapidae*, fixed-fang, venomous snakes with two or more short, erect, grooved or hollow fangs mounted on the maxillae, include cobras, mambas, all Australian poisonous snakes, sea kraits, and American coral snakes.

(d) *Hydrophiidae* include the viviparous sea snakes, which rarely or never come ashore, for they can barely crawl on land; they have flattened swimming tails, fixed fangs, and highly toxic venom.

(e) *Viperidae*, fold-fang, poisonous snakes, have a pair of long, hollow fangs that fold back against the roof of the mouth when not in use; *pit vipers* (American types being the rattlesnakes, water moccasin, copperhead, fer-de-lance, bushmaster) have a heat-sensitive facial pit on each cheek in front of the eye; *vipers* lack this pit; all have vertical eye pupils, spatulate heads, and slender necks (see Fig. 21.10*c*).

21.36. Are the dinosaurs really all dead?

Take a bird. Pluck off its distinctive plumage and discount recent flight-related adaptations, and you have a small, bipedal archosaur, warm-blooded (homeothermal) to be sure, but so might have been its ancestors, the coelurosaurs. More truth than quip is the observation that the dinosaurs didn't all go west: they are with us still, wearing feathers. However, since this connection was perceived only recently and because birds have acquired many specializations all their own, related to their fundamental commitment to flight, birds are already firmly ensconced in a separate class, *Aves*.

21.37. What were the first birds?

The two most ancient birds recognized to date are the Mesozoic *Archaeopteryx* and *Protoavis*.

21.38. Why is *Archaeopteryx* so interesting to zoologists?

Archaeopteryx ("ancient wing") is known from several very well preserved fossils found in Jurassic deposits 140 million years old (Fig. 21.11). Pigeon-sized, it was indeed a bird, because its arms, trunk and tail were well-clothed with *contour feathers*, but it displayed many reptilian features, some of which have been retained in more advanced birds,

Fig. 21.11 *Archaeopteryx* was a pigeon-sized Jurassic bird with teeth and a long reptilian tail. It may have caught prey such as insects by running and leaping with wing-assisted bounds, using its free fingers to seize and manipulate the prey.

Archaeopteryx

while others have been lost. The elevated position of its posteriorly directed toe—as in small bipedal dinosaurs—suggests that it too was a cursorial predator and used the curved talons of its free hand digits to seize prey.

The selective factors operating in the evolution of feathers remain problematic. These keratinous structures are considered homologous with reptilian scales, but transitional fossil forms have not yet been found. Perhaps they first arose as a heat-retaining fringing of ordinary epidermal scales, but the feather imprints of *Archaeopteryx* already show the central shaft (*rachis*) typical of body contour feathers and their larger versions, flight feathers. The rachis suggests that "flight feathers" may have evolved first, as an assist to locomotion, and that smaller editions of these later spread over the body surface. Aided by its large wing feathers, *Archaeopteryx* may have bounded along, even gliding at times, especially when running downhill or on irregular terrain.

(*a*) Reptilian features of *Archaeopteryx* that disappeared in later birds include teeth, three free hand digits bearing curved talons, and a long tail with 21 vertebrae.

(*b*) Traits shared by *Archaeopteryx* and coelurosaurs, which have been retained in modern birds include:

   1. Digitigrade bipedality, walking on the toes, usually with three toes directed forward, and a fourth (digit I, our "big toe") backward

   2. A fused metatarsal bone

   3. An intertarsal joint

   4. Hind limbs rotated under the trunk and oriented vertically

   5. A three-digit hand (digits II, III and IV further reduced)

   6. Articulation of the atlas with the skull by way of a single occipital condyle

   7. Epidermal scales on legs and feet

(*c*) Avian traits seen in *Archaeopteryx* but *not* in coelurosaurs include feathers, clavicles fused medially as a *furcula* ("wishbone"), and a hip structure lacking a pubic symphysis, with pubic bones paralleling the ischia.

21.39.   What was *Protoavis* like?

Protoavis, a crow-sized, toothed, long-tailed species found in Texas shales and originally misidentified as baby dinosaurs, predated *Archaeopteryx* by some 75 million years, yet in some respects was more birdlike. It lacked teeth in the back of the jaws, which therefore could be lightened, and its overall structure suggests that it could fly for short distances, rather than running as *Archaeopteryx* may have done. The original misidentification of *Protoavis* was due to the fact that no feathers were found with the two known specimens, but closer inspection has revealed small tubercles along the forearm and hand bones like those to which wing feathers attach in modern birds.

21.40.   What traits characterize modern birds?

As a group, birds are more homogeneous than other vertebrates, because their major adaptive complex is focused on the ability to fly. All possess the characteristics given in (*b*) and (*c*) of Question 21.38, and most also display the features discussed in Questions 21.41–21.56. Many of these features are flight-related in terms of reducing weight, increasing power, or compensating for other flight-related adaptations. Although not all birds *can* fly, in the aggregate, birds are the most efficient flying mechanisms ever evolved.

> **Example 7:** In ordinary flight great horned owls have been clocked at 64 km/h, starlings at 80 km/h, and hummingbirds at 112 km/h, while a peregrine falcon diving on a duck may plummet at over 160 km/h (100 mi/h). Soaring birds such as condors, that stay aloft with little wing movement by exploiting thermal updrafts, have wingspans as wide as 3 m (an extinct condor had a 5-m wingspan, a record for any bird).

21.41.   What is the foremost distinguishing characteristic of birds?

All birds have feathers (Fig. 21.12). Feathers develop from integumentary *feather follicles*, which are confined to specific *feather tracts*. Each feather remains rooted in its follicle by a basal quill, until it is molted and replaced. Feathers differ morphologically according to function: (*a*) flight, (*b*) streamlining, (*c*) insulation, (*d*) waterproofing, (*e*) visual communication (e.g., by species-specific coloration and displays), (*f*) acoustic communication (e.g., tail rattling by

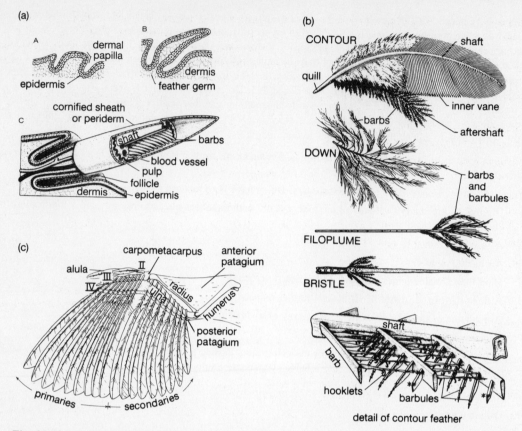

Fig. 21.12 Feather structure. (*a*) Development of a contour feather from a dermal papilla (A–C). (*b*) Four types of feathers, and detail of fine structure of a contour feather. The asterisk denotes barbules sectioned to show curved edge along which the hooklets slide, making the feather flexible. Preening resets disarticulated hooklets. (*c*) Arrangement of wing feathers (remiges), showing attachment of primaries to the hand and secondaries along the forearm (II, III, IV = digits). (*From Storer et al.*)

woodcocks and peacocks), (*g*) concealment by *cryptic coloration* and *countershading* (having a pale belly that counteracts shadow).

21.42.   What kinds of feathers do birds have?

(*a*) *Contour feathers* cover the body like shingles. They are unique in having the quill extended as a *shaft*, or *rachis*, that supports a *vane* made of closely spaced *barbs*. The barbs bear on each side a parallel series of small *barbules* provided with tiny hooks (*barbicels*), by which the barbules engage with those of adjacent barbs. The barbicels may become disengaged, but the bird sets this to rights by preening. Contour feathers on the body often bear a *secondary shaft* with fluffy, unhooked barbules.

(*b*) *Remiges* are large contour feathers that produce the overall shape of the *wing*, which may be (1) *narrow* and *tapering* in rapid flyers that catch insects on the wing (e.g., swifts) or make very long migratory flights (e.g., plovers); (2) *long, wide,* and *spreading* in birds that soar slowly over land (e.g., condors); (3) *very long* and *narrow* in oceanic soaring birds (e.g., albatrosses); and (4) *short, wide,* and *elliptical* in birds that maneuver through forests.

The remiges include *primaries* attached to hand (digits III and IV) and wrist, *secondaries* attached along the ulna, and *tertiaries* along the humerus. The *alula* is a tuft borne on digit II. In flight, the primaries, with their flexible rear margins, act like a propeller blade, accomplishing propulsion (forward pull). When turbulence is a problem, the primaries may be spread apart, reducing the turbulence by opening slots through which air can drain. The secondaries, with shafts quite asymmetrically set, provide power: on the upstroke they tilt like opened slats of a venetian blind, allowing air to drain between, then tilt in the opposite direction, like the slats of a closed blind, presenting a solid surface against the air during the downstroke. The tertiaries add to the surface area of the opened wing. The alula is elevated at slow flight speeds, avoiding stalling by increasing lift. Most lift is provided by the cross-sectional character of the spread wing: it resembles a section of an airplane wing, with the upper curved surface creating a longer pathway for air flowing over the top of the wing than for air flowing under the wing; the longer pathway rarefies the air passing over the wing, creating lift (Bernoulli's law).

(c) *Rectrices* are large contour feathers that form the *tail fan*. They attach to a stubby tail vestige and are controlled by muscles that can raise, lower, and spread them. They serve as a *rudder* in flight, a *brake* during landing, and a *counterbalance* during flight and perching. Rectrices can also be the key to a female's heart, as when they are spread into an erect fan during courtship.

(d) *Semiplumes* also have an elongate rachis, but their barbs and barbules are *plumulaceous,* i.e., do not hook together but fluff out like an ostrich plume; they help retain body heat.

(e) *Down feathers* are especially efficient at retaining body heat, enabling many birds, even small ones, to withstand extreme cold. The rachis is shorter than the longest barbs, which are highly plumulaceous.

(f) *Filoplumes* are hairlike, with only a few barbs at their tips; clustered around the bases of some contour feathers, they may serve a sensory function with respect to movement of those feathers.

(g) *Bristles* consist of the stiff rachis with barbs only at the base; they guard the nostrils from dust, serve as eyelashes, and fringe the mouths of birds that catch insects in midair.

(h) *Powder feathers*, often resembling down feathers, disintegrate at the tips into a fine, waterproof, keratinous powder that is spread throughout the plumage during preening. Absent in some birds, they are particularly abundant in herons and bitterns.

21.43. How is a bird's skeleton specialized?

The avian skeleton (Fig. 21.13a) displays many flight-related adaptations:

(a) Many bones are thin and hollow, with internal supportive trusses. A bird's entire skeleton usually weighs less than its plumage.

(b) The trunk vertebrae are *fused* together, minimizing distortion by the pull of the flight muscles.

(c) The ribs hook together by *uncinate processes* when in flight.

(d) The 8–24 cervical vertebrae have saddle-shaped (*heterocoelous*) centra that allow especially free neck mobility, in compensation for the rigidity of the trunk.

(e) The *sacrum* is long, articulating with an elongated ilium for a strong sacroiliac union aiding bipedal locomotion.

(f) The caudal skeleton—only four free vertebrae plus several fused into a terminal *pygostyle*—furnishes attachment for muscles that control the tail fan.

(g) Except in flightless ratites (e.g., ostriches) the sternum projects into a huge *keel* to which the flight muscles anchor (Fig. 21.13b).

(h) The *coracoids* of the pectoral girdle are braced against the sternum ventrally, while the clavicles are fused medially by way of an interclavicle element to form the V-shaped furcula.

(i) The jawbones are so narrow and light that teeth cannot form even though tooth-producing genes are still present.

(j) The hand is simplified into a rigid paddle that maneuvers the primary remiges: compared with coelurosaurs, the carpal and metacarpal elements of a bird's hand are further fused and reduced, digit III is large but immobile, and digit IV vestigial; only the remnant of II remains movable for raising the alula.

(k) In perching birds (passerines) a tendon passing over the heel and inserting on the terminal phalanges locks the toes in flexion around twigs even while the bird sleeps: as the knee and ankle joints bend when the bird settles down to rest, tension on the tendon makes the toes curl under (see Fig. 21.13b).

21.44. How do birds operate their wings?

Hugely enlarged *pectoral muscles* originate on the sternum and insert on the humerus. Contraction of the larger mass, the *pectoralis*, pulls the wings *down and backward* in the downstroke. The upstroke, which moves the wings *up and forward*, is accomplished in unique fashion, using a deeper breast muscle, the *supracoracoideus,* the tendon of which

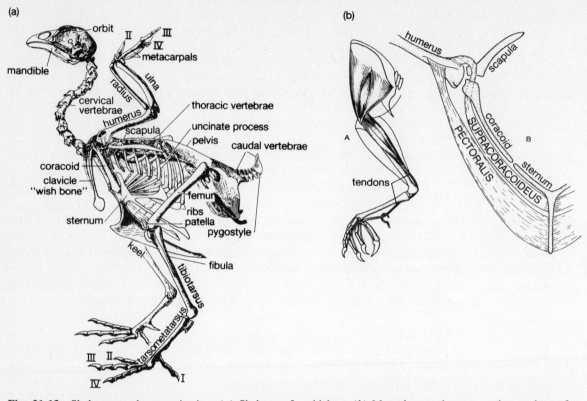

Fig. 21.13   Skeletomuscular organization. (*a*) Skeleton of a chicken. (*b*) Muscular attachments: A, leg tendons of a perching bird pass over the heel and through a ring-shaped tendon just before dividing into a separate tendon flexing each toe; when the leg is bent at the ankle as the bird settles to roost, tension on the tendons curls the toes around the perch and locks them in place. B, the muscles that raise the wings (supracoracoideus) insert on the humerus, but the supracoracoideus's tendon of insertion passes through a pulleylike ligament that changes its direction so that it inserts on the humerus opposite the insertion of the pectoralis. [*From Storer et al., (a) after Elenberger and Baum, (b) after Wolcott.*]

passes around onto the *upper* surface of the humerus, opposite the insertion of the pectoralis (see Fig. 21.13*b*). In strong flyers such as pigeons, the flight muscles constitute more than 30 percent of the body weight.

21.45.   What are the characteristics of birds' eyes?

They are *very good* eyes! The avian eye is larger relative to body size and its photoreceptors are denser per unit of retinal area than for any other vertebrate. Also, diurnal color vision is sharpened by *oil droplets* in the cones, making bird sight exceptionally acute.

Despite these improvements, the bird eye is basically reptilian:

(*a*)   A *sclerotic ring* of bones that ossify in the eyeball's outer coat (sclera) supports the front of the eyeball; in birds this may reduce distortion of the eyeball during flight.

(*b*)   Accommodation is as in reptiles: at rest the eye is set for distance vision, and contraction of the ciliary muscles compresses the lens, rounding it up for near vision. In birds, part of the ciliary musculature forms a separate *Crampton's muscle* that achieves further accommodation by bowing the cornea despite the constricting presence of the sclerotic bones.

(*c*)   The retina lacks blood vessels, but a vascular *pecten* projects from the back of the eye into the posterior chamber, allowing materials to be exchanged between the vitreous humor and the bloodstream.

(*d*)   A transparent third eyelid (*nictitating membrane*) can be extended over the eyeball from its nasal corner; in birds, this protects the cornea in flight as needed. (This separate eyelid should not be confused with the nictitating membrane of amphibians, which is simply a transparent upper portion of the lower eyelid.)

21.46.   Do birds hear well?

Birds have wretched senses of taste and olfaction, but they enjoy keen vision and hearing. They are much better than humans at discriminating separate notes in rapid cadence, as in many bird songs. The mammalian cochlea is much longer than its equivalent in birds, the *lagena*, but the density of its receptive hair cells is only 10 percent that of birds. In structure, a bird's ear is essentially reptilian: the eardrum is at the inner end of an external auditory canal, the columella serves as the middle ear ossicle, and the lagena is only a bit longer than in reptiles; but in number of sensory cells, the bird lagena is very far advanced indeed.

21.47.   What are the special features of birds' skin?

The epidermis is keratinized enough to protect against evaporative water loss, but scales are absent except on the feet, the skin elsewhere being rather thin and delicate. A keratinous beak (*rhampotheca*) covers the jaws, its shape being adapted for various diets. In the feather tracts, feather follicles arise by an interaction of epidermal epithelium with dermal papillae (see Fig. 21.12a). The feather itself is made up of dead, horny epidermal cells. The plumage can be erected by tiny dermal *feather erector muscles* (*arrectores plumarum*), which insert on each follicle. A bird may fluff out its plumage as a means of thermoregulation, or the plumage may be erected as a social signal, as in threat or courtship. Birds have few cutaneous glands, but they do have a special large *uropygeal gland* opening dorsally at the tail base, which produces oil that the bird spreads over its plumage by preening.

21.48.   How do birds thermoregulate?

In today's world, only birds and mammals are *homeothermal*, maintaining a stable body temperature *metabolically*, despite environmental fluctuations. A typical avian body temperature is 40–42°C (104–108°F), a life-threatening fever by our standards! This reflects birds' very high metabolic rate, as demanded by the energy costs of flapping flight. For cooling, birds may erect their plumage and pant, since they lack sweat glands. A few birds are *facultative homeotherms*: they can shut down their heat-generating metabolic systems and allow their body temperature to drop closer to ambient levels. Thus, poorwills become poikilothermal during hibernation, and hummingbirds nightly, since otherwise they would perish overnight from the demands of maintaining their diurnal temperature.

21.49.   What are special features of the avian digestive tract?

Birds are sharp-tongued, and love to eat and run. The small, pointed, horny tongue ordinarily does little good except in swallowing, but in woodpeckers it can be protruded to an astounding length to harpoon insect grubs. Fish-stabbing birds (e.g., herons) use the tongue to push impaled prey off the beak.

The highly distensible, thin-walled *crop* just above the stomach (Fig. 21.14a) allows birds to gobble piggishly when food is available. Later, food passes gradually from crop to stomach, which has two parts: an anterior, glandular *proventriculus* and posterior, heavily muscularized *gizzard*. Unable to chew, grain-eating birds swallow stones that grind up the seeds in the gizzard. In deference to weight reduction, many birds lack a gall bladder, and bile drains directly into the intestine by way of the bile duct. The intestine also receives three separate ducts from the pancreas. Where the intestine joins the short but dilated rectum, birds have two ceca in which symbiotic bacteria digest fibrous food. As in reptiles, the cloaca receives the rectum and openings of the urinary and reproductive ducts.

21.50.   What changes in circulation are seen in birds?

Compared with mammals, birds' hearts are *larger* relative to body size and beat more rapidly: the heart of a dog makes up around 1 percent of its body weight and beats 140 times per minute, whereas that of a rather large flying bird, the turkey vulture, constitutes 2 percent of its weight and beats 300 times per minute.

The avian heart is fully separated into left and right pumps by a complete ventricular septum. The heart now consists of left and right atria plus left and right ventricles. The sinus venosus has completely disappeared as a heart chamber, its vestige being embedded in the wall of the right atrium as the pacemaker. Where *two* systemic trunks leave the reptilian heart and bend to the left and right, respectively, as two aortic or systemic arches that join the dorsal aorta, in birds only the right systemic arch persists, while the left arch ceases to communicate with the dorsal aorta and remains only as the base of the left subclavian artery to the left shoulder and wing (the opposite situation from that in mammals).

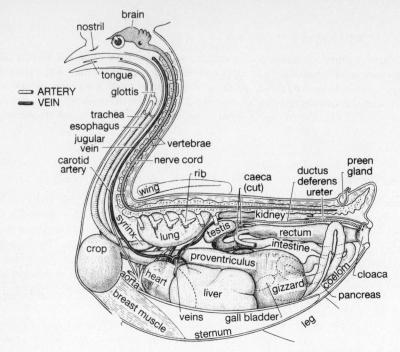

ARTERY
VEIN

(a)

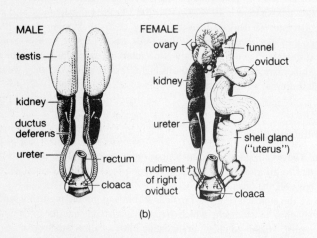

(b)

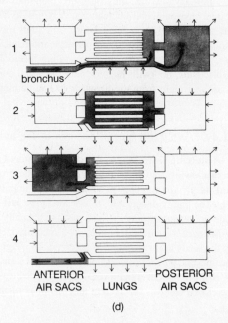

(d)

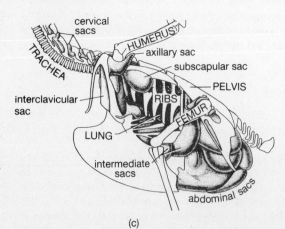

(c)

21.51.   How do birds excrete salts and wastes?

Bird kidneys are much like those of reptiles, and, like many reptiles, birds are *uricotelic*: i.e., they typically produce a semisolid urine in which the main nitrogenous waste is *uric acid*. This allows birds to conserve water, which is lost when a liquid urine is voided. Most birds lack a urinary bladder, another modification for weight reduction. Birds that drink seawater get rid of excess ions through *nasal salt glands* also found in certain squamates and turtles.

21.52.   How is the avian respiratory system remarkable?

The respiratory system of birds is quite unique and advanced (Fig. 21.14*a, c, d*). A long trachea leads from the larynx to the base of the neck, where it divides into two bronchi; at this junction, birds have a special sound-producing organ, the *syrinx*. The bronchi lead into the lungs, which change very little in size during breathing, because they are attached to the body wall and, instead of being saccular with alveolar clusters at the microscopic ends of the bronchial tree, are penetrated lengthwise by air passages (*parabronchi*) that open into thin-walled, distensible *air sacs*.

The manner of respiration is highly efficient: (*a*) no residual air remains in the lungs after exhalation (as is the case with all other pulmonate vertebrates); (*b*) fresh air passes *forward* through the lungs' microscopic *air capillaries* during each inspiration *and* expiration; and (*c*) *crosscurrent exchange* allows an especially thorough transfer of gases between air and blood capillaries. During inhalation the anterior and posterior air sacs become inflated, while the lungs are being compressed; compression of the lungs coupled with dilation of the anterior air sac pulls air forward through the lungs into those sacs, while about 75 percent of the air being inhaled passes straight into the posterior air sacs by way of the parabronchi. During exhalation, the air sacs are compressed while the lungs expand a bit; air in the anterior sacs is expelled from the body, and the air in the posterior sacs is forced forward through the lungs. Thus air flows unidirectionally and nearly continuously through the air capillaries of the lung, and any residual air remains in the air sacs. Gas exchange through the walls of the air capillaries is especially efficient because the blood flows crosswise to the air flow, maintaining a steep concentration gradient. So efficient is oxygen take-up that certain migratory birds zip along at altitudes as high as 6400 m, with good clearance even over the peak of Mt. McKinley!

21.53.   How do birds reproduce?

The reproductive system of birds (Fig. 21.14*b*) is unusual in several respects. The male tract resembles that of reptiles except that only ratites, ducks, and geese have a penis. In other birds, ejaculation takes place quickly, either when the cloacas of both partners are pressed together while the male stands on his mate's back, or when the female's oviduct is everted from her own cloaca and inserted into the male's. The female has only a single ovary (the left), but her undeveloped right gonad may develop into a *testis* if her ovary is destroyed by disease; this accounts for instances of sex reversal in domestic poultry: a former hen develops cock-feathering, acts like a rooster, and occasionally even sires chicks. The single (left) oviduct is greatly enlarged to accommodate the huge eggs. No more than one egg a day is laid, while younger eggs are gradually passing along the oviduct. collecting albumen, membranes, and shell.

> **Example 8:**   Birds' eggs are remarkably large for their body size, but that of the New Zealand kiwi is ridiculous: a 2-kg female lays a 0.5-kg egg! Gravid kiwis like to stand still in water, presumably resting their tired muscles by the buoyancy provided.

Reproduction is usually seasonal, with egg laying preceded by courtship, pair-bonding, and nest building. Clutch size ranges from one to more than a dozen in ground-nesting birds that have *precocial* chicks (i.e., ones able to walk and feed themselves soon after hatching). When chicks are *altricial* (helpless and unable to feed themselves), both parents usually cooperate in their care.

Fig. 21.14   Internal structure of a bird. (*a*) Lateral view of internal organs of a male, with intestinal caeca partly removed. (*b*) Urogenital systems; females have only a single functional ovary and oviduct. (*c*) Respiratory system: air sacs stippled lightly, lungs stippled darkly; also note syrinx in (*a*). (*d*) Avian breathing cycle tracing passage of one bolus of air (shaded), which is inhaled at (1) and exhaled at (4). The preceding bolus (one step ahead of the shaded bolus) would pass through the lungs into the anterior air sacs at (1), then be exhaled at (2). While the shaded bolus passes from the lungs into the anterior air sacs (3), a succeeding bolus is being inhaled into the posterior air sacs, and when the shaded bolus is exhaled (4), this succeeding bolus is forced forward into the lungs. Thus air moves forward through the lungs during inhalations (1,3) and exhalations (2,4) alike. [*From Storer et al.; (c) after Muller, (d) after Welty from Schmidt-Nielsen.*]

21.54.   How is a bird's brain unusual?

The avian forebrain differs from that of reptiles and mammals by being remarkably solid: the cerebral hemispheres are roofed by a thin archipallium, but most of their gray matter lies deep within, in the form of a rudimentary neopallium and huge corpus striatum (Fig. 21.15). Sensory analysis seems to take place chiefly in the large thalamus, which relays messages to the striatum. The latter organizes the complex instinctive behavior patterns characteristic of birds. Many patterns of bird behavior appear to have strong genetic determinants, even when modifiable by learning. Birds also exhibit a type of restricted learning known as *imprinting*, which takes place only at some sensitive period in early life but may have lasting effects (see Chap. 24).

Fig. 21.15   Bird brain, sectioned transversely through forebrain. The avian brain features a thin pallium (cerebral cortex) and large, solid corpus striatum and thalamus. The corpus striatum contains the highest centers for control of skeletal muscles; it is in intimate communication with the thalamus, a major sensory and relay area. The brain ventricles are filled with nourishing cerebrospinal fluid.

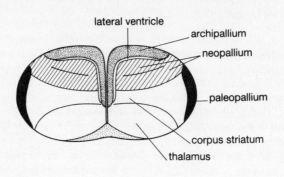

21.55.   How does migration benefit birds?

Many birds migrate very long distances, the champion being the Arctic tern, which breeds north of the Arctic circle and spends its nonbreeding life close to Antarctica. Most migratory birds fly north for breeding, profiting by the longer summer daylight of high latitudes. When a migratory pathway crosses the equator, the birds enjoy summer at *both* ends of their journey. As a result of their extraordinarily long migration, Arctic terns experience more daylight than any other creatures. The urge to migrate seems to be genetically determined.

> **Example 9:**   At the proper time of year, even inexperienced birds of migratory species exhibit a typical premigratory restlessness, jumping repeatedly in the correct migratory direction if they are allowed to see the night sky. Even when birds are kept under constant photoperiod throughout the year, they still exhibit two annual periods of premigratory restlessness. This indicates that such birds have a "biological clock" about a year long, a so-called *circennial rhythm*, which is merely adjusted and *not caused* by seasonal changes in photoperiod, since it persists in the absence of such change.

21.56.   How do migrants find their way?

In many species, first-year migrants must fly south without the help of older birds, which depart earlier.

> **Example 10:**   When golden plovers migrate from the Canadian Arctic to Argentina, the older birds follow an apparently learned, culturally transmitted route that arcs out over the Atlantic Ocean and exploits trade winds to curve westward to South America; the first-year migrants instead pursue a fairly straight trajectory SSE down the Mississippi valley, over the Caribbean and Andes, and down into Argentina.

> **Example 11:**   If birds that have never before migrated and therefore have learned no landmarks are banded and released 1000 km or so west or east of where they grew up, they typically fly along the correct bearing for about the right distance, but end up in the wrong place because they fail to correct for the east-west displacement. This indicates that genetic programming somehow affects recognition of migratory direction and the distance to be traveled, but there is no innate recognition of "goal."

Hand-raised birds of migratory species jump randomly in all directions if they cannot see the sky. In a planetarium, these inexperienced birds get no information from a stationary star map projected on the dome, but if the artificial sky is rotated even once, they *imprint* on the one star that does not move, which in nature would be Polaris, the North Star. After this, they jump in only one direction, which happens to be the correct migratory direction if the star *was* Polaris. However, if the planetarium sky is craftily shifted to rotate around some other bright star, the poor innocents fix upon the wrong bearing, and so may remain misguided for the rest of their lives. Thus, in most species studied, the migratory bearing itself seems to be genetically determined, but calculating this bearing requires the learning experience of fixating on the one unmoving star in the rotating heavens. Thereafter, the bird memorizes the star patterns and employs *sidereal navigation* by night, when most migration takes place.

Homing pigeons take their bearings from the sun's position, a *sun-compass reaction* that requires an accurate internal "clock" to compensate for the sun's apparent daily movement; this clock is dependent on an endogenous *circadian rhythm* about 24 hours long, which can be reset by keeping the birds under an artificial cycle of light and dark. With their "clock" reset, birds released far from home fly off in a predictably wrong direction and find their way back only when the effects have worn off. Pigeons trained to fly under total overcast can still home successfully by detecting the earth's *magnetic field*, and migratory birds too may employ a magnetic sense when visual cues are obscured. Landmarks, wind patterns, and good resting places are learned by experience.

21.57. What are the major subgroups of class Aves?

(*a*) *Subclass Archaeornithes* ("ancient birds") includes reptilelike birds, e.g., *Archaeopteryx* and *Protoavis* (see Questions 21.38–21.39).

(*b*) *Subclass Neornithes* ("new birds") includes all other birds, which have fused metacarpals, an elongate digit III, and 13 or fewer tail vertebrae. *Superorder Odontognathae* ("toothed jaw") includes only Cretaceous birds with teeth, such as the flightless, swimming *Hesperornis* and the ternlike *Ichthyornis*. All other birds—some 8700 species— are toothless and make up *superorder Neognathae* ("new jaw"), with 28 orders.

21.58. How have neognaths undergone radiative adaptation?

The radiative adaptation of neognath birds particularly involves diversification of feet and beaks with respect to various diets and modes of life (Fig. 21.16).

The entire foot may be long or short. Birds are *digitigrade* and stand on their toes with the ankle joint well elevated; we speak loosely of "long-legged" birds such as storks and cranes, when they are actually *long-footed*, and people mistake the ankle joint for a backward-bending knee, because most of the leg is held close to the trunk and hidden by feathers. Toes may be long and slender for wading, lobed or webbed for swimming, taloned for grasping prey, and so forth. Similarly, the horny beak (rhampotheca) may be serrate for catching fish, hooked for tearing flesh, spatulate for sieving, slender for probing, stout for cracking seeds, and so forth. Birds as a class exploit a tremendous variety of food resources, but since as a rule each species tends to be restrictive in its feeding strategy, many different species can coexist sympatrically with minimum competition. Such adaptive diversification has also produced numerous instances of *convergence* and *ecological equivalence*.

> **Example 12:** The indigenous songbirds of Australia were originally classified with families of Eurasian birds that they closely resemble, but recent DNA hybridization studies indicate that this diverse passerine avifauna represents an extensive adaptive radiation of a *single* family, the crows! Worse still, such DNA analyses indicate that certain other birds have even been placed in the wrong *order*! American vultures and condors closely resemble those of the Old World and have been grouped with them in order Falconiformes: carnivorous birds with flesh-tearing beaks; now it appears that the American species are really related to *storks* and belong with them in order Ciconiiformes.

21.59. What are the orders of superorder Neognathae?

(*a*) *Ratites*, or running birds that mostly cannot fly and have unkeeled sternums, make up the following five orders:

Tinamiformes: tinamous; running birds of Central and South America, with keeled sternum but weak wings.
Rheiformes: South American rheas; large, 3-toed, flightless birds with uncarinate (keelless) sternum, well-developed wings used in courtship dancing, plumulous remiges; females lay eggs in communal nest and male does all incubation.
Struthioniformes: African ostriches; to 2.2 m tall, 2-toed, uncarinate sternum, plumed wings deployed in courtship dancing; females lay eggs in communal nests and share incubation with male.
Casuariiformes: emus and cassowaries of Australia and New Guinea; to 1.5 m, uncarinate sternum, wings extremely reduced; male incubates eggs laid by one female.
Dinornithiformes: kiwis and extinct moas of New Zealand, the latter to over 3 m tall; wings exceedingly degenerate and remiges absent, unkeeled sternum, moas 3- or 4-toed, kiwis 4-toed with nostrils at tip of beak; male incubates eggs of one female.

Fig. 21.16 Radiative adaptation of birds. The evolution of birds into many forms adapted for different diets and modes of life has emphasized specialization of the legs and feet for wading, swimming, grasping, climbing, and perching, and of the beaks for cracking seeds, catching insects, cutting, probing, and sieving. (*a*) Representative types of feet. (*b*) Representative beak types. (*c*) A variety of birds. (*From Storer et al.; barn owl after Grinnell and Storer.*)

(b) Birds of the following orders have a keeled sternum and their wings are well developed and used for flying in air and/or water.

Podicipediformes: grebes; aquatic, diving after small animals and plant material; toes lobed, feet set far back on trunk, tail a mere tuft of down.

Sphenisciformes: penguins; marine, gregarious; use flipperlike wings to fly through water instead of air; southern hemisphere.

Procellariiformes: tube-nosed oceanic birds, albatrosses and petrels, mainly breeding on islands; rhampotheca composed of several plates, tubular nostrils carry nasal salt gland exudates toward the bill tip.

Pelecaniformes: pelicans, cormorants, boobies, frigate birds; web joins all 4 toes, gular throat pouch, fish-eating.

Anseriformes: ducks and geese; 3 toes in web, spatulate beaks covered by soft epidermis with many tactile sensory pits, except along horny edge.

Ciconiiformes: storks, herons, ibises; feet and toes usually long, often used in wading; bills mostly long, often used for stabbing fish. (Also, New World condors and vultures, soaring scavengers with flesh-tearing bills.)

Falconiformes: hawks, falcons, eagles, hook-beaked, diurnal predators that grab prey with talons; also, scavenging kites and Old World condors and vultures.

Galliformes: turkeys, pheasants, quail, grouse, chickens, megapodes; mostly ground-dwelling seed eaters, often flocking.

Gruiformes: long-legged cranes that inhabit grasslands; short-legged, marsh-dwelling rails; lobe-toed, aquatic coots.

Charadriiformes: shorebirds, 3 toes fully webbed or webbed just at base; sandpipers, stilts, avocets, oystercatchers, plovers, gulls, terns, auks, murres, puffins; bills and feeding strategies varied, diet of fish or aquatic invertebrates.

Gaviiformes: loons; aquatic divers to depths of 61 m, fish-eating; 3 toes webbed, feet so far to the rear that walking is nearly impossible.

Columbiformes: pigeons; young fed on "pigeon's milk" from crop.

Psittaciformes: parrots; bill hooked, upper jaw highly movable; mostly frugivorous; foot with 2 toes forward, 2 rearward, used prehensilely, with food grasped between front and rear lateral toes.

Cuculiformes: cuckoos, roadrunners; 2 toes forward, 2 behind with outer rear toe reversible; many cuckoos are nest parasites.

Strigiformes: owls; mostly nocturnal predators, large eyes directed forward, ringed by radially set feathers, asymmetrical ears aid sound location, toes feathered, plumage soft-edged for silent flight.

Caprimulgiformes: poorwills, nighthawks; catch nocturnal insects in midair with wide, bristle-fringed mouths.

Apodiformes: swifts and hummingbirds; unable to walk because of very short legs and tiny feet; wings stiff and pointed; swifts capture insects in midair by day; hummingbirds hover and probe tubular flowers for nectar.

Coliiformes: African mousebirds; small, first and fourth toes reversible, tail very long; scurry mouselike along branches.

Trogoniformes: trogons, quetzal; tropical; small, weak feet; brilliant plumage; short, stout beak with bristles at base.

Coraciiformes: kingfishers, hornbills, motmots, hoopoes, bee-eaters, rollers; mostly with beak disproportionately large; toes 3 and 4 fused at base.

Piciformes: woodpeckers, honeyguides, toucans; 2 toes forward, 2 or 1 behind; in woodpeckers, the rectrices are stiff for propping against tree trunks, and the protrusible tongue, which is barbed at the tip and used in probing for insects, is supported by a hyoid bone so lengthy that its sheath may encircle the eye socket.

Passeriformes: over 5000 species and 69 families of mostly small, terrestrial songbirds or perching birds; champion vocalizers; special tendon mechanism locks toes around slender perches, especially during sleep (see Fig. 21.13b); many feed on insects, others on seeds.

Important passerine families include: tyrant flycatchers; larks; swallows; shrikes; wrens; thrushes; mockingbirds and thrashers; crows, jays and magpies; chickadees, verdins, and tits; nuthatches; starlings and mynahs; tanagers; blackbirds, orioles, and meadowlarks; sparrows, finches, towhees, and buntings; vireos; wood warblers.

Chapter 22

Amniotes II: Mammals

Long before dinosaurs arose to dominate the land, the *therapsids*, or mammal-like reptiles, had begun a healthy expansion and were progressively accumulating constellations of traits we now consider mammalian. In fact, without knowledge of characteristics that do not fossilize, it remains a matter of dispute as to whether such forms as *Diarthrognathus* were reptiles (as presently classified) or mammals. As time went on, the therapsids fell under the shadow of emerging archosaurs that could outrun them and outgun them, so to speak, but as they slipped into extinction, the mammal-like reptiles left behind at least two different lineages that had crossed the boundary into mammalhood. How could these primitive mammals have survived in a world of dinosaurs? To begin with, they were *small*: mouse-sized, insignificant descendants of a reptilian group that had included some quite substantial types such as the herbivorous *Moschops* that stood 1.5 m tall at the shoulder and was 3 m long, although the actual ancestors of mammals were the less ponderous *carnivorous* therapsids, known as *theriodonts* for their mammal-like dentition. While the therapsids succumbed, their miniature descendants, the dawn mammals, could hide away in inconspicuous dens while flesh-eating archosaurs small enough themselves to relish such minor morsels were on the prowl. And when were such midget predators abroad? Probably only by day, when their body temperatures would be maximal. Homeothermal and insulated with fur, the tiny primitive mammals could fill ecological niches not exploited by reptiles: they could become creatures of the *night*. Legacies of nocturnal ancestry remain with mammals today: most are *color-blind* and have eyes well adapted to function in dim light, and many are of *crepuscular or nocturnal* habit, whereas most birds and reptiles are diurnal. Even while they remained small and secretive, the dentition of Jurassic mammals indicates that insectivorous, herbivorous, and flesh-eating types had already come into being. Some of the flesh eaters may have found unprotected dinosaur eggs so appetizing that they may well have contributed to the decline of certain species long before the mass extinction episode that ended the Mesozoic. Today, mammals fall into three categories: oviparous *monotremes*, *marsupials* that have rudimentary placentas and give birth to extremely immature young that are often housed in a *pouch*, and *eutherians* that have highly developed placentas.

22.1. What are the special characteristics of mammals?

Although mammals maintained a low profile for some 150 million years, they were busily evolving many innovative features both qualitatively and quantitatively superior to those of reptiles, so that when reptilian dominance was finally shattered and an abundance of ecological niches fell vacant, mammals were ready to move in, diversifying through the ensuing 65 million years to occupy and dominate every type of marine, freshwater, and terrestrial habitat accessible to air-breathing animals. These unique mammalian characteristics are discussed in Questions 22.2–22.16.

22.2. How are young mammals nourished?

Mammary glands occur in both male and female mammals, but are functional only in the latter. The pituitary hormone *prolactin* promotes milk secretion from these glands. The glands are located ventrolaterally along two *mammary lines* extending from the axial (armpit) to the inguinal (groin) region. The mammary ducts converge to open at separate pores clustered at the tip of a *nipple*. Milk contains protein, fat, milk sugar (lactose), and minerals, especially calcium, relative quantities varying according to species (e.g., human milk contains 3–5 percent fat, that of marine mammals 30–40 percent).

Newborns must find the nipple unassisted, through two innate behaviors: the *rooting reflex* and the *suckling reflex*. First the infant nuzzles about, seeking some protuberance to take into its mouth; next, it sucks vigorously upon this. Suckling stimulates sensory endings in the mother's nipple and, by way of a reflex arc, much increases the secretion of prolactin and another hormone, *oxytocin*, which makes the mammary ducts contract and eject the milk. Before they can begin to feed, newborn marsupials, little more than fetuses, must perform another vital instinctive response: unaided, they must crawl all the way from the vaginal orifice to the pouch, a formidable migration for an infant kangaroo no bigger than a jellybean.

22.3.   How do social relationships profit mammals?

Social relationships in mammals are influenced by the fact that the *mother* has the built-in food supply. Neonates range from *precocial* (well-furred, with eyes open, able to locomote; e.g., baby hares and hoofed mammals) to *altricial* (naked, with eyes sealed, unable to walk or thermoregulate adequately; e.g., newborn rabbits and mice), but in either case they spend some period of dependency with the mother. Unlike birds, in which mates commonly share parental duties equally, few mammals pair-bond, and few fathers care for their young. Exceptions are wolves, foxes, marmosets, and humans.

> **Example 1:**   The father wolf or fox brings food to his mate and offspring; fox kits are fed only by their own parents, but since wolves live socially in packs, older cubs beg food from any adult pack member and are rewarded with regurgitated meat. Marmosets are small arboreal monkeys that form nuclear families within a large troop; the father carries the infant, passing it to the mother only for nursing.

Many mammals live gregariously and mate promiscuously within the social group; in this event, males tend to be gentle toward all infants and actively protect the group against predators, as when male baboons hang back to threaten cheetahs while females and youngsters head for the trees. In some mammals maturation of the young takes several years, and affectional bonds to the mother may persist far beyond weaning (as in primates), unless the mother actively rejects her offspring at this point (as in bears).

The dependency of young and the vulnerability of females when pregnant or carrying infants are cardinal factors promoting social living, which is much more common in mammals than in other vertebrates. A mammalian *society* is not a simple aggregation such as a school of fishes, for mammalian social groupings demonstrate certain criteria by which a society can be defined: (*a*) individuals often occupy positions in a *social hierarchy*; (*b*) some *division of labor* exists, usually between sexes; (*c*) *communication* is complex and includes recognition of individuals; (*d*) the group is *cohesive*—its members tend to remain in visual contact, aggregating closely to rest, and individuals rarely strike out on their own; (*e*) the group is relatively *impervious* to strangers and may reject them aggressively. Mammals that live gregariously throughout their lives enjoy the protection of the group at all ages, but males in particular engage in ritualized *social fighting* that establishes their hierarchical position, which in turn can affect their future reproductive success.

> **Example 2:**   In a small wolf pack only the dominant male and his mate will produce cubs; in a larger pack the dominant male may register assent to mating by subordinates, by allowing courtship to proceed and then administering a gentle head pin or muzzle bite while the subordinate is locked *in copulo*, a compromising situation common to canids.

22.4.   What is special about mammalian skin?

(*a*) *Hair follicles* produce shafts of dead, keratinized cells known as *hairs* (see Fig. 18.6*b*). Long, lustrous *guard hairs* overlie insulative, wooly *underhairs*; sensory *vibrissae* (whiskers) flare from the snout; curving *eyelashes* protect the eyes; and *nostril bristles* filter air. Hair serves primarily in insulation, but secondary functions include advertisement of sex and maturational state (e.g., the mane of an adult male lion), recognition of species and individual (by markings), camouflage, feeding (e.g., "whalebone"), social fighting (e.g., keratinous rhinoceros horns), and defense (e.g., porcupine quills).

(*b*) *Hair erector muscles* (*arrectores pilorum*) in the dermis elevate the fur in threat or for heat retention.

(*c*) *Integumentary glands* are abundant and varied: (1) *Sebaceous glands* along the hair follicles secrete waterproofing oil that flows out along the hair shafts and does not have to be distributed by preening, as with the single uropygeal gland of birds. (2) *Sweat glands* cool the body by evaporation; sweat may also contain pheromonal substances.

> **Example 3:**   Double-blind tests indicate that men experimentally anointed with alpha-androstenol, a steroid component of axillary (armpit) sweat, generate positive feelings in women and aversive feelings in other men!

(3) *Scent glands* are not uncommon; they may serve defensively (e.g., skunk musk) or pheromonally, in marking territory and attracting mates. (4) *Ceruminous glands* secrete protective *ear wax* into the external auditory canal.

22.5.   What set of muscles is unique to mammals?

Muscles of facial expression (Fig. 22.1) enrich mammalian communication. Donald Duck notwithstanding, mammals are the only creatures that can scowl, sneer, and smile. We associate the contraction of specific facial muscles with particular emotional states: e.g., the *frontalis* raises the eyebrows as if in query; the *zygomaticus* pulls up the corners of the mouth in a smile; the *risorius* retracts the corners of the mouth, as in strain; the *caninus* deepens the angle of the mouth, as in exasperation; the *levator labii superioris* elevates the upper lip in a sneer; the *orbicularis oris* purses the lips as in doubt.

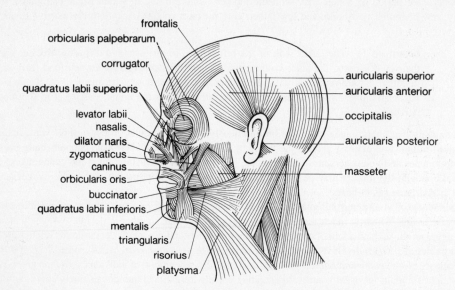

Fig. 22.1   A mammalian innovation: muscles of facial expression. See text for actions of some
of the superficial muscles of the human face.

22.6.   What major evolutionary advance is seen in the mammalian brain?

The brain of mammals (Fig. 22.2) has experienced tremendous expansion of the neopallium, the most highly organized version of the cerebral cortex, which appears rudimentarily in reptiles but now sheets out over the outer surface of both hemispheres as a thin cloak of gray matter only about 1 cm thick, displacing the old archipallium dorsomedially and the ancient olfactory paleopallium ventromedially (Fig. 22.3). In advanced mammals, the neopallium expands still more as the cerebral surface folds into *convolutions*. Eutherians have a large fiber tract, the *corpus callosum*, interconnecting the neopallium of the two hemispheres, allowing most information processed in one hemisphere to be transferred to the one opposite. The mammalian neopallium includes a new *somatic motor area* that initiates voluntary bodily movements. The actual output of signals from the motor cortex to the spinal cord is controlled by more ancient, subcortical motor centers: the *cerebral basal nuclei* (corpus striatum) mainly coordinate slow and precise movements, while the *cerebellar cortex* programs rapid and ballistic (pre-aimed) movements. Its new role in controlling the output of the cerebral motor cortex has led to great enlargement of the cerebellum and convolution of its surface.

22.7.   How does the mammalian ear function?

The mammalian ear (Fig. 22.4) is unique and advanced in several particulars:

1.   An external ear *pinna* (auricle) supported by elastic cartilage collects air-borne vibrations and, being movable in most mammals, helps detect the direction of the sound source.

2.   Three *middle ear ossicles*—hammer (*malleus*), anvil (*incus*), and stirrup (*stapes*)—occur only in mammals. These tiny bones not only transmit vibrations across the middle ear cavity but also, by concentrating them upon a tiny area (the *oval membrane*, which covers an aperture into the inner ear), amplify sound pressure some 30-fold, increasing hearing acuity. The stapes represents the inner portion of the old columella, while the malleus and incus are derived from bones (quadrate and articulare) involved in jaw suspension in nonmammalian tetrapods (see Fig. 18.11*b*).

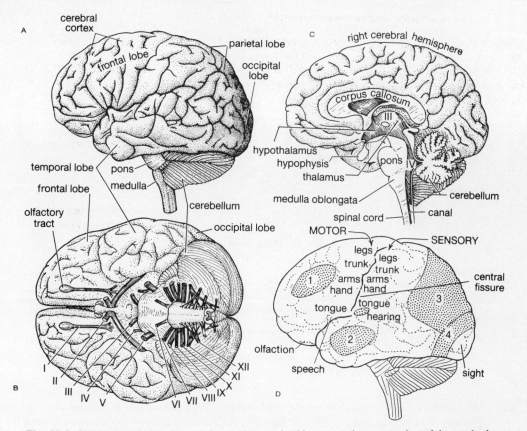

Fig. 22.2　The human brain. Mammals are characterized by tremendous expansion of the cerebral hemispheres and especially the neopallium, which is convoluted in the more advanced members of this class. A further mammalian innovation is development of a motor cortex that constitutes a new center for control of voluntary movements: A, viewed from left. B, ventral aspect showing roots of cranial nerves (I-XII). C, midsaggital section, III-IV = ventricles. D, localization of functions in the left hemisphere: association areas (stippled) are (1) frontal, (2) temporal (acoustic), (3) parietal, (4) visual. (*From Storer et al.*)

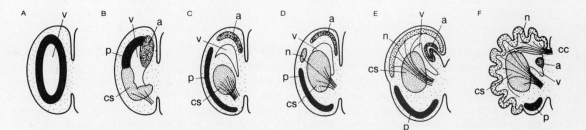

Fig. 22.3　Hypothetical construct of the evolution of the vertebrate cerebrum. White matter designated by scattered dots, gray matter by all other shading. Corpus striatum (b) finely stippled. Stippled regions designate the three versions of the cerebral cortex: paleopallium = p, archipallium = a, neopallium = n. Lateral ventricles = v; corpus callosum = cc. A, left hemisphere of primitive fish, with gray matter surrounding ventricle, probably nearly entirely olfactory in function. B, differentiation of gray matter into corpus striatum, paleopallium, and archipallium, as in modern salamanders. C, migration of pallial elements toward the outer surface while the corpus striatum remains internal, possibly as in primitive amniotes. D, appearance of the neopallium, as may have occurred in mammal-like reptiles. E, expansion of the neopallium in early mammals, with displacement of the archipallium and paleopallium. F, neopallium becomes convoluted in advanced mammals and a corpus callosum appears. (*Modified from Romer.*)

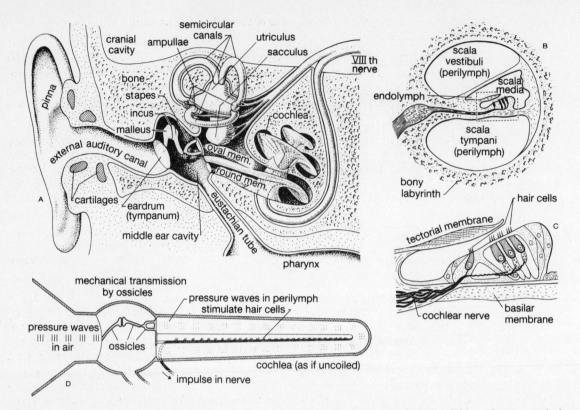

Fig. 22.4 The mammalian ear. Mammalian innovations as seen in the human ear include addition of an external pinna (movable in most mammals), addition of incus and malleus as middle ear ossicles, and coiling of the former lagena into a snail-like cochlea, with great lengthening of the sensory organ of Corti. A, overall structure of ear. B, cross section through one part of the cochlea; the scala media (cochlear duct) contains endolymph and represents a section through the membranous cochlea; dashed lines enclose part of the organ of Corti. C, detail of section of organ of Corti. The cilia of the hair cells are actually affixed onto the tectorial membrane and are mechanically stimulated when vibrations of the basilar membrane create displacements of the organ of Corti. D, diagrammatic longitudinal section of the ear, with cochlea uncoiled, showing translation of pressure waves in air to vibrations of eardrum and ossicles, to pressure waves that travel through the perilymph from base to apex of the cochlea (in the scala vestibuli), and back (by way of the scala tympani), to be dissipated into the air-filled middle ear cavity by vibrations of the membrane covering the round window. (*From Storer et al.*)

3. In the inner ear (see Question 18.32), the vesicle known as the "lagena" in nonmammalian vertebrates lengthens so greatly during mammalian embryonic development that it is obliged to coil up into the shape of a snail's shell, justifying the name *cochlea* ("snail"). As the *membranous cochlea* lengthens, so does the ribbonlike *organ of Corti* in its wall, with differentiation of many more sensory hair cells than in reptiles. The reciprocal shape of the *skeletal cochlea* that encloses the membranous cochlea produces a double fluid system: (1) *endolymph* fills the membranous cochlea (*scala media*), maintaining its cross-sectional shape and allowing room for vibratory displacements of the two membranes of the organ of Corti—the *tectorial membrane* lying above the sensory hair cells, and the *basilar membrane* lying below them; (2) *perilymph* fills the space in the temporal bone (i.e., the skeletal cochlea) that encloses the membranous cochlea, and serves to transmit vibratory pressure waves from the stapes to the organ of Corti. The membranous cochlea separates the skeletal cochlea into two canals, the *scala vestibuli* and *scala tympani*, which communicate by way of a pore (*helicotrema*) at the tip of the coil.

    The auditory pathway begins when pressure waves in air are translated into vibrations of the eardrum, which propagate through malleus, incus, and stapes, to the oval membrane. They are now transformed into pressure waves in the perilymph, which propagate up the cochlear coil through the vestibular canal and back down again by way of the tympanic canal, then dissipating into the air-filled middle ear cavity by way of a *round membrane*. Vibrations of the perilymph set the *basilar membrane* in motion at particular points along the organ of Corti where the dimensions of that membrane allow it to resonate to sound of some specific frequency.

> **Example 4:**   The human cochlea resonates to sounds with frequencies from about 16–20,000 Hz (herz = cycles/s): below 50 Hz the membrane vibrates as a whole instead of regionally and sounds are heard as bass rumbles instead of specific tones, but to higher frequencies such minutely localized parts of the basilar membrane resonate that a person with good hearing can discriminate some 300,000 different tones.

As the basilar membrane vibrates, it moves the adjacent layer of sensory hair cells up and down against the overhanging tectorial membrane. Since the latter is attached to the inner wall of the cochlear spiral, its own vibratory movements are translated into sideways displacements that tug upon the sensory cilia anchored to the outer part of the tectorial membrane. This stimulates the receptor cells, which transmit excitation to sensory neurons within the cochlea's *spiral ganglion*, their axons forming the acoustic branch of the otic nerve (VIII).

22.8.   What is different about the mammalian skeleton?

The skeleton of mammals is unique in a number of ways, some of which have improved mastication, hearing, cranial support, and pelvic support.

(*a*) Mammalian jaw articulation has shifted from quadrate-articulare to *temporal-dentary*. Where the primary jaws of fishes were each a single cartilage of gill arch I, in amphibians the posterior part of both upper and lower primary jaws separated off as the *quadrate* from the rear portion of the palatoquadrate, and the *articulare* from the rear part of the mandibular (Meckel's) cartilage (see Fig. 18.11*b*). Accordingly, in nonmammalian tetrapods the quadrate and articulare form the articulation between the jaws and the cranium, a condition known as *metautostyly*, which freed the hyomandibula from its role in jaw suspension and led to its becoming the columella of the middle ear (see Question 20.24). In mammals, the jaw has shortened, strengthening the bite, by a forward displacement of jaw articulation so that the dentary bone encasing Meckel's cartilage now articulates directly with the temporal region of the cranium (*cranioamphistyly*).

(*b*) When this innovation first took place in early mammals, it left the quadrate and articulare unemployed, and these bones had already become somewhat vestigial when rescued by their proximity to the columella and middle ear cavity: thus, the quadrate has come to be the mammalian *incus*, and the articulare (fused with a small dermal bone) the *malleus*. Once more we see how evolutionary opportunism can lead to adoption of new functions by preexisting structures, in this case improving mammalian hearing by the amplifying effect of these three ossicles.

(*c*) The mammalian mandible consists of only a single dermal bone on each side, the dentary, fused in the midline with its opposite number, forming a *chin*. This simplification, which strengthens the mandible, represents the ultimate reduction in the number of dermal bones encasing the Meckel's cartilage of the primary lower jaw.

(*d*) Mammals have a *secondary palate* that completely separates the nasal and buccal cavities and, posterior to this, a *soft palate* that continues this separation backward to the pharynx, where the *internal nares* open into the *nasopharynx* close to the glottis (the opening into the larynx). This lets mammals breathe while engaged in prolonged mastication. The secondary, or hard, palate that roofs the buccal cavity consists mostly of shelflike processes of the maxillae and palatine bones.

(*e*) Head support is improved in two ways: (1) The rear of the cranium is strengthened by the complete ankylosis of four bones (supraoccipital, basioccipital, left and right exoccipitals) that, in nonmammalian tetrapods, ring the foramen magnum through which the spinal cord passes. (2) The resulting single, complex *occipital bone* bears *two* occipital condyles, left and right, permitting a two-point anchorage between cranium and vertebral column, thus providing more stability than with the single occipital condyle of modern reptiles and birds.

(*f*) *Seven cervical vertebrae* occur in nearly all mammals, regardless of neck length. (Sloths have six or nine, and sirenians six.) While less numerous than in birds and certain reptiles, these provide ample head mobility.

(*g*) The mammalian larynx has gained components: a shieldlike *thyroid cartilage* protects it ventrally and an elastic cartilaginous flap, the *epiglottis*, folds down over the glottis during swallowing, reducing choking accidents.

(*h*) The mammalian pelvis is strengthened by the complete ankylosis of the three elements of each pelvic girdle (ilium, ischium, pubis) into a single *innominate* bone on each side; these articulate midventrally at the pubic symphysis. The extent of sacroiliac articulation varies among different mammals; the number of sacral vertebrae is usually 3–5 (3 in cats, 5 in humans), but up to 13 vertebrae form the sacrum of edentates.

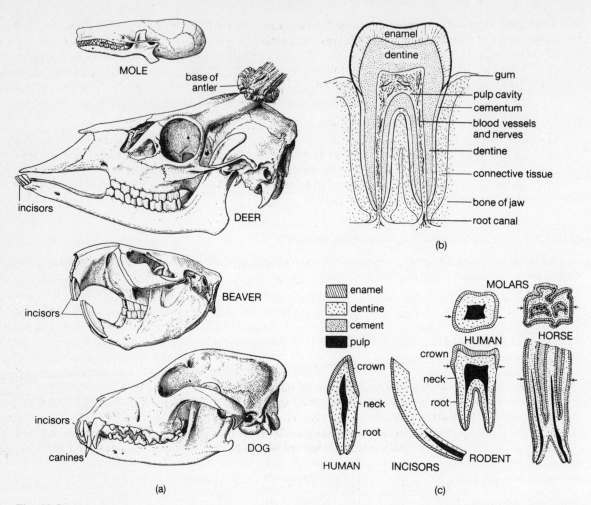

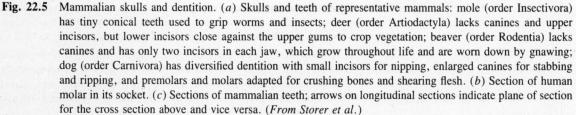

Fig. 22.5 Mammalian skulls and dentition. (*a*) Skulls and teeth of representative mammals: mole (order Insectivora) has tiny conical teeth used to grip worms and insects; deer (order Artiodactyla) lacks canines and upper incisors, but lower incisors close against the upper gums to crop vegetation; beaver (order Rodentia) lacks canines and has only two incisors in each jaw, which grow throughout life and are worn down by gnawing; dog (order Carnivora) has diversified dentition with small incisors for nipping, enlarged canines for stabbing and ripping, and premolars and molars adapted for crushing bones and shearing flesh. (*b*) Section of human molar in its socket. (*c*) Sections of mammalian teeth; arrows on longitudinal sections indicate plane of section for the cross section above and vice versa. (*From Storer et al.*)

22.9. How do their teeth help mammals come to grips with things?

Mammals enjoy the benefits of *heterodonty*, having *four distinct tooth types*: cutting or gnawing *incisors*, flesh-tearing *canines*, and crushing *premolars* and *molars* (Fig. 22.5). The crushing teeth have complex cusps and two and three roots, respectively, in contrast with the conical, single-rooted teeth of reptiles.

> **Example 5:** Humans, being omnivorous, possess all four tooth types: in each half-jaw 2 incisors, 1 canine, 2 premolars, and 3 molars, a dental formula of 2-1-2-3, for a total complement of 32. Mammals of more restricted diet have these tooth types variously represented; e.g., the dental formula of voles (meadow mice) is 1-0-0-3: canines and premolars are lacking, and the jaw is toothless in a wide gap (diastema) from the incisor to the posterior cluster of 3 molars. Deer, along with other ruminants, lack incisors and canines in only the upper jaw, so that their dental formula is 0-0-3-3/3-1-3-3.

Hemidiphyodont tooth succession is unique to mammals. This means that certain teeth come in only once, whereas others are preceded by one set of *deciduous* teeth. In humans, the deciduous set numbers 20 (2-1-2-0): incisors, canines, and premolars only. All of these are shed and replaced by teeth of the permanent set, to which are added the 12 molars, which actually belong to the first set but erupt later and are never shed. This is in contrast with the *diphyodont* succession of many reptiles and primitive mammals, in which the deciduous and permanent sets have the same number of teeth.

22.10.   What advances are seen in the mammalian respiratory system?

(*a*)   The fully alveolated lungs of mammals enclose a *bronchial tree*, which branches progressively to end in microscopic *bronchioles* that open into clusters of alveolar sacs (Fig. 22.6*a*). As a result, the lungs of a mammal provide much more internal surface for gaseous exchange than the partially alveolated lungs of a reptile of equivalent size. Human lungs provide some 90–150 square meters of alveolar surface area. A section of mammalian lung tissue looks quite

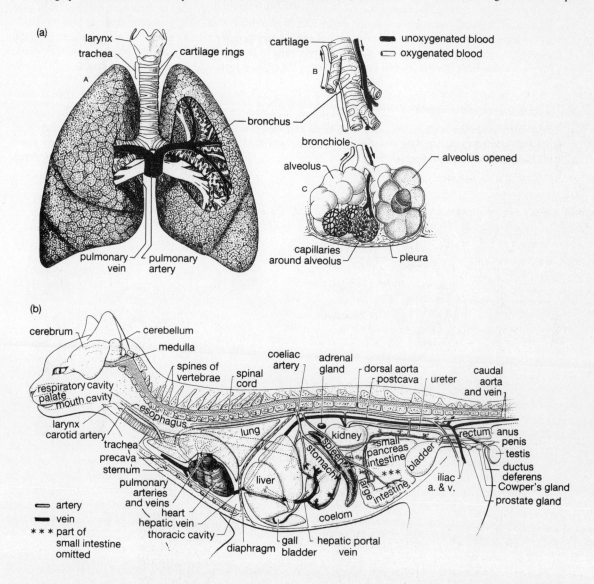

Fig. 22.6   Internal structure of mammals. (*a*) Human lower respiratory system: A, ventral view with heart removed; B, portion of bronchus and pulmonary vessels; C, microscopic representation of alveolar clusters with their surrounding capillaries. (*b*) Internal organs of a cat, lateral aspect. (*From Storer et al.*)

dense to the unaided eye, with openings visible only where bronchi or blood vessels have been transected; by contrast, the lung of a reptile such as an alligator appears loosely spongy in section because the bronchi terminate in macroscopic air spaces lined with alveoli.

(b)   The mammalian *diaphragm* (Fig. 22.6b) allows ordinary exhalation to be *energy-free*, for when it relaxes this muscle bows forward, forcing air out of the lungs by reducing the depth of the thoracic cavity. This new inspiratory muscle develops within the connective tissue *oblique septum* that separates the thoracic and abdominal cavities of reptiles and birds. When the dome-shaped diaphragm contracts, it flattens out and pushes the abdominal organs posteriorly, increasing the depth of the thoracic cavity with the result that air is sucked into the lungs by the lowered intrathoracic air pressure. This action is augmented by contraction of muscles that increase the circumference of the rib cage, notably the external intercostals. When the diaphragm relaxes, its passive forward movement causes the lungs to deflate. At the same time, relaxation of the external intercostals allows the ribs to fall inward, reducing circumference. Only during exercise (or when one exhales against resistance as in blowing up a balloon) must the force or volume of expiration be increased by contraction of muscles of the abdominal body wall together with muscles (such as the internal intercostals) that compress the rib cage. This is in contrast with other tetrapods, in which both inspiration and expiration are accomplished by muscular contraction and therefore require energy: in nonmammals, inhalation involves contraction of muscles that pull the oblique septum backward and expand the rib cage, while exhalation involves contraction of muscles that tense and flatten the oblique septum and compress the rib cage.

22.11.   How is the mammalian circulatory system advanced over that of reptiles?

(a)   In parallel evolution with crocodilians and birds (but independent of them), the postnatal mammalian heart is subdivided into separate right and left pumps by the growth of a complete ventricular septum. The sinus venosus disappears as a heart chamber but persists as the pacemaking sinoatrial node in the wall of the right atrium (Fig. 22.7).

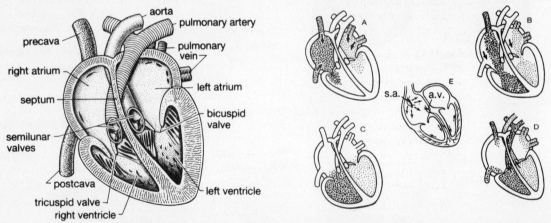

Fig. 22.7   The mammalian heart. *Left*, frontal section, ventral view. *Right*, heart action: A, atria filling, B, blood flows through expanded atria into relaxed ventricles. C, atria contract, further dilating ventricles. D, ventricles contract, forcing blood into pulmonary and systemic trunks, while atria relax and begin to fill again. E, spread of excitation (arrows), from the pacemaker, the sinoatrial node (s.a.), throughout the atrial walls and to the atrioventricular node (a.v.), from the latter through the interventricular septum, and thence ramifying throughout the ventricular musculature; the conduction pathway consists not of nerves but of modified cardiac muscle fibers. *From Storer et al.; modified from Best and Taylor,* The Human Body and Its Functions, *Henry Holt and Company, Inc.*)

(b)   Even as birds deleted the left systemic arch and retained the right, so mammals have deleted the right and retained only the left systemic arch, connecting the heart with the dorsal aorta. (Reptiles and amphibians carry forward from piscine ancestry the redundancy of *two* bilaterally symmetrical systemic arches that meet at the dorsal aorta; see e.g., Fig. 18.12.) The root of the mammalian right systemic arch remains as the base of the right subclavian artery to shoulder and arm.

(*c*) A chemoreceptive *aortic body* develops in the wall of the mammalian systemic ("aortic") arch, a small sense organ that mainly monitors blood oxygen level, supplementing the paired *carotid bodies* of all tetrapods, which occur in the walls of the carotid arteries serving the head and brain.

(*d*) *Pulmonary bypasses* improve fetal circulation in placental mammals, since before birth gases are exchanged in the placenta, not the lungs. (1) A flutter valve over an opening (*foramen ovale*) between the right and left atria allows blood to pass directly from right to left (but not in reverse), thus being shunted into the aorta rather than being pumped into the pulmonary trunk. (2) A short *ductus arteriosus* (the distal part of aortic arch VI) interconnects the pulmonary trunk with the systemic arch (IV), allowing most of the blood that does enter the pulmonary circuit to pass over into the aorta without going through the lung capillary beds. At birth the ductus arteriosus closes and becomes vestigial, and the interatrial flutter valve permanently seals shut.

(*e*) Mammalian blood is especially rich in *erythrocytes*: these hemoglobin-containing corpuscles make up at least 45 percent of the total blood volume. In mammals, these cells are unique in being discoidal and *enucleate*: the nucleus is extruded before the red corpuscle enters the blood flow; the corpuscle then survives (in humans) for about 120 days. All other vertebrates have ovoid, nucleated erythrocytes, even birds with their high metabolic demands. However, in mammals, enucleation increases the ratio of surface to volume and removes whatever impedance to gas exchange might be imposed by the presence of a nucleus. The *erythrocyte count* is unusually high in mammals.

> **Example 6:** Erythrocytes in turtle blood number 630,000/cc, whereas human blood contains 5,000,000/cc for males and 4,500,000/cc for females. Turtle erythrocytes measure 12.4 $\times$ 21.2 μm as compared with a diameter of 8.6 μm for fresh human erythrocytes, but size does not compensate for lower cell count, because larger corpuscles have a less favorable surface-to-volume ratio than smaller ones. Avian erythrocytes measure 9 $\times$ 18 μm and in small birds such as hummingbirds number 6,500,000/cc, but the ostrich, a large, flightless bird somewhat more comparable in activity and mass to a large man, has only 1,600,000/cc.

Erythrocytes contain hemoglobin, which binds O_2 to its iron-containing heme groups, and also binds CO_2 to one of its amino acids as *carbaminohemoglobin*. Erythrocytes also enable CO_2 to be transported as bicarbonate ion (HCO_3^-), because they contain an enzyme, *carbonic anhydrase*, which bonds CO_2 and H_2O to make H_2CO_3 (carbonic acid). This ionizes to H^+ and HCO_3^-. Placental mammals produce two different kinds of hemoglobin: adult and fetal.

> **Example 7:** Oxygen makes up 21 percent of the air at sea level, which has an atmospheric pressure of 760 mmHg [i.e., it displaces a column of mercury (Hg) by 760 mm in a mercury manometer]; thus, O_2 concentration (oxygen "tension") is expressed as 21 percent of 760, or 160 mmHg. Human adult hemoglobin is 80 percent saturated with O_2 at an oxygen tension of 80 mmHg, and the oxygen tension of fully oxygenated arterial blood is 100 mmHg. At the tissues, where O_2 is metabolically consumed, the oxygen tension drops to about 40 mmHg, and the hemoglobin "unloads" its oxygen, which diffuses into the tissues. Unborn mammals produce *fetal hemoglobin* with different properties: this molecule loads O_2 at the *unloading* threshold for *adult hemoglobin* and is 90 percent saturated at 40 mmHg, which is vital since otherwise oxygen could not be transmitted across the placenta from mother to fetus. As parturition nears, the fetus also begins to produce adult hemoglobin, which replaces the fetal type soon after birth.

22.12. How is mammalian thermoregulation superior to that of modern reptiles?

Modern reptiles are poikilothermal and can adjust their temperature only by *behavioral thermoregulation*. *Endothermy* and *homeothermy* characterize mammals, as well as birds and perhaps certain extinct archosaurs and therapsids. Mammals generate body heat by muscular contractions including shivering, and by metabolizing *brown fat*, leaving white fat for insulation. Brown fat contains large numbers of mitochondria packed with iron-containing cytochrome; when stimulated by sympathetic nerves activated by the hypothalamic temperature-regulating center, brown fat uses the energy of electrons to generate heat instead of building ATP.

> **Example 8:** Mammalian thermoregulation is so precise that body temperature is quite stable, except during *hibernation*. Normal body temperatures for monotremes are 28–30°C, for marsupials 33–36°C, and for eutherians 36–38°C. True hibernation is seen mainly in rodents of northern temperate regions: the body temperature drops to only a degree above ambient temperature and respiration and heart rate decrease drastically. A hibernating woodchuck's temperature declines to 4–14°C, its breathing slows from 262 to 14 breaths per min, and its heartbeat drops to only abut 10 per min; commencing hibernation

at a rotund 4.5 kg, it loses half its body weight while dormant, emerging famished in spring. Certain nonhibernatory mammals such as bears engage in "winter sleep," which is not true hibernation, because body temperature does not drop and the sleeping animal is readily aroused, even though in bears the heart rate may decrease from 40 to 10 beats per min.

Mammals retain body heat by using subcutaneous white fat, a covering of fur, and sometimes, countercurrent heat exchange systems in the extremities. Some mammals become *hyperphagic* as winter approaches, consuming excess food, which is converted to brown and white body fat. Most mammals molt in spring and fall, in autumn changing from a sparser summer coat to a winter coat that may be 50 percent denser (which is why fur trappers operate in winter). In addition, the hair erector muscles fluff out the fur, the better to retain body heat. Countercurrent heat exchange, best developed in polar mammals, involves transfer of heat from arteries running toward the extremities, to parallel-running veins carrying chilled blood toward the body; this exchange is so effective that the venous blood is rewarmed to full body temperature while the outbound arterial blood is chilled to almost the level of ambient temperatures. Thus, polar bears do not melt the ice on which they stand, because their paws are nearly as cold as the ice itself. The metabolism of icy extremities relies on enzymes geared to operate at such low temperatures.

Dissipation of excess body heat involves lassitude, evaporative cooling by sweating, and radiative cooling by dilation of dermal blood vessels. The outsized ears of desert jackrabbits, fennec and kit foxes, rabbit-eared bandicoots, and African elephants assist thermoregulation, since they expose a large, well-vascularized surface area for radiative heat loss.

22.13.   What is special about the mammalian excretory system?

The mammalian *metanephros* (see Question 18.45) is the only vertebrate kidney capable of such massive reabsorption of water from the glomerular filtrate that a liquid urine is formed that may be substantially hyperosmotic to blood plasma. This is because of the properties of the long *Henle's loop*, which is highly developed in the mammalian nephron but is quite rudimentary in reptiles and birds. Mammals have two types of nephrons (Fig. 22.8): *cortical nephrons* which lie entirely in the outer region (cortex) of the kidney and *medullary nephrons* which begin and end in the cortex, but extend a long hairpin loop far down into the medulla, and even into the renal pelvis and ureter of mammals adapted for arid habitats. The longer this loop, the more concentrated the urine can be. Urine formation by a mammalian medullary nephron involves the following processes:

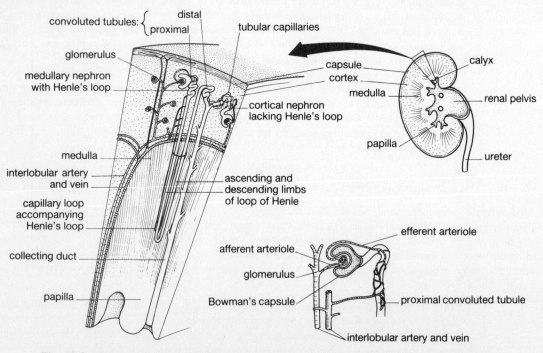

Fig. 22.8   The human kidney. Section through kidney and ureter, wedge section and large arrow designate the portion enlarged to show a cortical nephron and a medullary nephron with their glomeruli; the medullary nephron features a long Henle's loop accompanied by a capillary loop.

(*a*) *Ultrafiltration* of blood plasma at the glomeruli of the mammalian kidney produces a much greater quantity of glomerular filtrate than in other land vertebrates, owing to the high blood pressure and rapid circulation characteristic of mammals.

(*b*) *Reabsorption of useful solutes* takes place by active transport in the *proximal convoluted tubule*.

(*c*) *Tubular secretion* adds certain substances (notably creatine, ammonia, organic acids, H^+ and K^+) to the urine by active transport from the tubular capillary bed through the walls of the renal tubule and the collecting duct into which nephrons drain.

(*d*) *Water reabsorption* takes place osmotically, about 75 percent from the proximal convoluted tubule and the rest from the descending arm of Henle's loop, the distal convoluted tubule, and the collecting ducts.

(*e*) *Chloride ions* are actively transported out of the urine into the interstitial fluid by the cells that make up the *ascending* arm of Henle's loop, with Na^+ passively following. Some Na^+ and Cl^- then return to the urine by diffusion through the wall of the *descending* arm, to be transported out once more as they pass along the ascending arm. This multiplicative cycle acts as a salt trap, building up a gradient of NaCl in the urine within Henle's loop, in the surrounding tissue fluid, and in the accompanying capillary loops. In the human kidney this gradient normally amounts to about a fourfold increase in saltiness from the cortex to the bottom of Henle's loop in the medulla (Fig. 22.9). The wall of the ascending arm is impervious to water, but water *can* diffuse out of the descending arm and the collecting duct. The salt concentration built up in the interstitial fluid by the Cl^--transporting action of Henle's loop creates a steep osmotic gradient that draws water out of the urine, especially as it flows through the collecting duct toward the renal pelvis. As a result, the urine that passes from the mammalian kidney into the ureter can be saltier than blood plasma. Kangaroo rats are desert rodents that survive on seeds, with no drinking water or fleshy food at all. This is made possible by the extreme length of their Henle's loops and the accompanying capillary loops, which protrude out of the medulla as a tuft that extends right down into the ureter, creating along its exaggerated length such a massive salt gradient that only scanty quantities of highly concentrated urine remain to be voided.

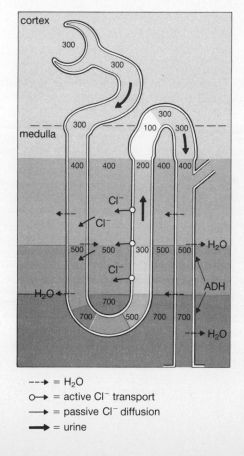

--→ = H_2O
o→ = active Cl^- transport
→ = passive Cl^- diffusion
➤ = urine

Fig. 22.9 Means by which the mammalian kidney produces a liquid urine that is hypertonic to blood plasma. Background shading indicates increasing salinity of the tissue fluid from cortex to medulla, paralleling the concentration of sodium chloride within the nephron and its collecting duct. Numerals indicate osmolarity values. (*Modified from Storer et al.*)

22.14.   What are the components of the male reproductive system in mammals?

The male reproductive system (Fig. 22.10) includes paired *testes* made up of convoluted *seminiferous tubules* separated by clusters of hormone-secreting *interstitial cells of Leydig*. The tubules exit the testes as *efferent ductules* that join the convoluted *epididymis* in which sperm maturation is completed. A *ductus (vas) deferens* extends from each epididymis to enter the urethra just below the bladder, so that a common tube, the *cavernous urethra*, enclosed by erectile tissue (*corpus spongiosum*), carries both urine and sperm from that point on. The bulk of the mammalian penis is made up of the corpus spongiosum together with paired erectile *corpora cavernosa*; erection is by hemotumescence (blood engorgement), but in some mammals (e.g., dogs) a bone, the *baculum*, ossifies between the corpora cavernosa, making the penis permanently rigid. As sperm enter the urethra in preparation for ejaculation, *seminal fluid* is added from several glands: milky, alkaline fluid from the *prostate gland* enhances sperm mobility; secretions of the paired *seminal vesicles* contain mucus and nourishing fructose, amino acids, and vitamins; and thick, mucinous fluid from the *bulbourethral glands* lubricates and protects the sperm. The urethra opens at the tip of the penis, which is expanded into a sensitive *glans penis* covered by a protective *prepuce* (foreskin).

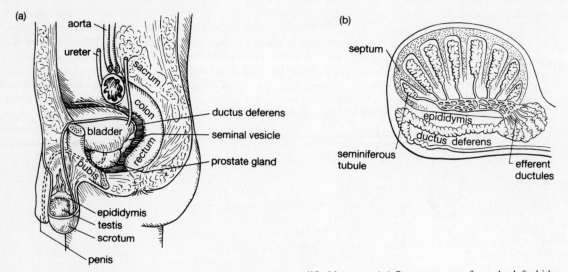

Fig. 22.10.   The urogenital system of male mammals as exemplified by man. (*a*) Structures seen from the left, kidneys omitted. (*b*) Section of testis and its ducts. *(From Storer et al.)*

The testes of most mammals are displaced out of the body cavity into a saclike *scrotum* suspended from the inguinal region. During embryonic development the testes first lie well forward in the abdominal cavity, as in other vertebrates, but they are anchored to the posterior body wall by the *gubernaculum*, a ligament that ceases to grow as the body lengthens, with the result that it eventually pulls the testes through an *inguinal canal* into the scrotum. In some mammals testicular descent occurs only during the mating season, but in others it is permanent. This peculiarity is necessitated by the high body temperature of mammals, which is ordinarily lethal to sperm, although some mammals (e.g., cetaceans) manage to remain fertile without benefit of a scrotum. (Avian sperm benefit by large abdominal air sacs that cool the testes by air-conditioning.)

22.15.   How does the female reproductive system function in mammals?

The female reproductive system (Fig. 22.11*a* and *b*) resembles that of other vertebrates in having paired *ovaries*, which release eggs into the coelom by the bursting of *ovarian (Graafian) follicles* (Fig. 22.11*b*) in the moment of *ovulation* triggered by a sudden, massive output of *luteinizing hormone* (LH) from the pituitary gland. Female mammals in reproductive state go through a series of *estrus cycles* that are suspended when pregnancy occurs. At the midpoint of each cycle they are typically "in heat"—compulsively seeking a mate—and ovulation takes place. However, certain mammals (e.g., cats and rabbits) ovulate only while actually copulating. The human *menstrual cycle* (Fig.22.11*c*) is modified from a typical estrus cycle in that sexual receptivity is not compulsive or limited to midcycle, and part of the uterine lining is sloughed if pregnancy does not occur.

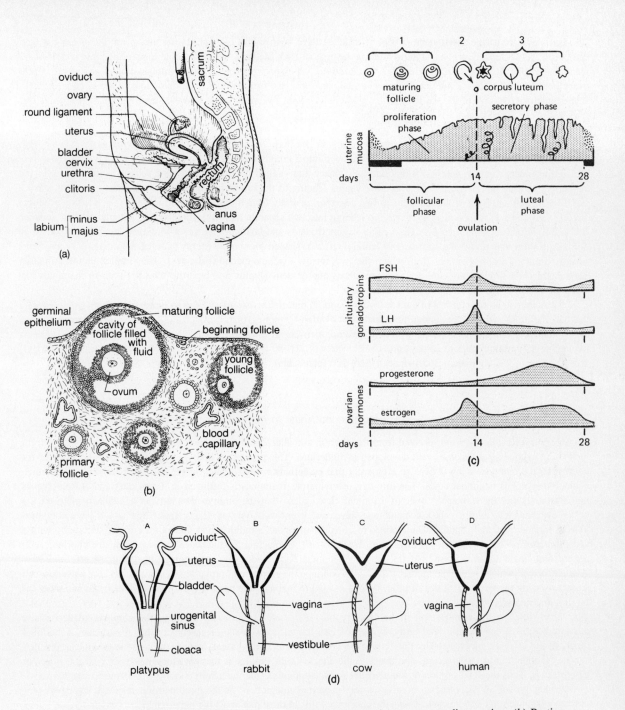

Fig. 22.11 Female reproductive system. (*a*) Human reproductive system, median section. (*b*) Portion of ovary showing eggs in ripening follicles. (*c*) Correlation of ovarian, uterine, and hormonal events in the human menstrual cycle. Horizontal black bar indicates duration of menstruation. (*d*) Types of uteri: A, duck-billed platypus, an egg layer, has rudimentary uteri that open separately into a urogenital sinus leading to a cloaca (lacking in viviparous mammals). B, rabbit has duplex uterus consisting of two uteri opening into a common vagina. C, cow has a bicornuate, Y-shaped uterus with separate horns and a common uterine body. D, human has simplex uterus with no horns. (*From Storer et al.*)

Until ovulation, the cells of the ripening follicles have been doing two jobs: transferring nutrients into the egg cytoplasm and secreting *estrogen*. After ovulation the cells of each ruptured follicle become transformed into a yellowish, glandular *corpus luteum* that mainly secretes *progesterone*, which prepares the uterus for implantation of an embryo. Each ovary is suspended within a mesentery (*mesovarium*, or *round* and *broad ligaments*), held close to the ciliated, funnel-shaped opening (ostium) of the *Mullerian duct*, and ciliary action sweeps the eggs into the slender *oviduct*. If fertilization does not occur as the eggs pass along the oviduct, the corpora lutea soon shrink, the uterine lining regresses or sloughs, and new follicles begin to ripen.

Posteriorly, the Mullerian duct widens into a *uterus*, the base of which bulges into the *vagina* as the *cervix* (Fig. 22.11*d*). In most mammals, the vagina and urethra open into a common *urogenital sinus*, but there is a trend—most fully realized in rodents and primates—for the external urinary and vaginal openings to become separated. In the egg-laying monotremes, each uterus opens into a separate vagina leading to the urogenital sinus, but viviparous mammals show a trend for the left and right Mullerian ducts to fuse together medially from the posterior end forward. Such fusion produces successively (*a*) a Y-shaped *bicornuate vagina* leading into two separate uteri (marsupials); (*b*) a *duplex uterus*, separate left and right uteri opening into a single median vagina (rodents and lagomorphs); (*c*) a V-shaped *bipartite uterus*, fused for only a short way above the vagina (carnivores); (*d*) a Y-shaped *bicornuate uterus* (hoofed mammals); (*e*) a *simplex uterus* forming a single chamber from which the two narrow oviducts extend (primates). The simplex uterus ordinarily accommodates a single fetus, and a bicornuate uterus one or two. Duplex and bipartite uteri are seen in mammals that have litters: the fetuses grow like peas in two pods.

The eutherian uterus consists of a thick layer of smooth muscle (*myometrium*) and an epithelial lining (*endometrium*) that becomes thick and secretory under the influence of ovarian hormones (first, estrogen and, after ovulation, progesterone) as the female's estrus cycle proceeds. When the endometrium is thickest, pregnancy becomes possible. Representative eutherian gestation periods in days are 13–21 (shrews), 21–30 (rats), 30–35 (rabbits), 52–69 (cats), 53–71 (dogs), 101–130 (pigs), 267 (humans), 365 (whales), 510–720 (elephants).

22.16. How does the chorioallantoic placenta contribute to mammalian success?

A *chorioallantoic placenta*, derived from the chorion and allantois of the amniote egg, appears in rudimentary form in some marsupials and becomes well developed in eutherians. This placenta is both an endocrine gland and a device for nourishing the embryo, a job it does so efficiently that eutherians are born much larger and more fully developed than neonate marsupials or monotremes. The tiny egg of viviparous mammals is fertilized in the oviduct and at first cleaves holoblastically but then switches over to discoidal cleavage so that the embryo develops as a flattened plate on top of nonexistent yolk; as the yolk sac membrane develops, it encloses only an empty space that would have contained yolk in monotremes and reptiles. The embryo reaches the uterus as a multicellular *blastocyst* covered externally with a syncytial *chorion* bearing numerous fingerlike *chorionic villi*. Contacting the thickened endometrium, the chorionic villi adhere and begin to sink in so that the embryo becomes *implanted*. Now the chorion begins to secrete *chorionic gonadotrophin*, which signals the mother's ovaries to continue the secretion of progesterone. Eventually, the placenta itself starts to secrete the large quantities of progesterone needed to maintain the proper state of the endometrium and quiet the uterine musculature.

Meanwhile, as the embryo and its extraembryonic membranes continue to develop, the *allantois* (which serves monotremes, as it does reptiles and birds, as a repository for urine and a site of gaseous exchange) assumes a modified function: the allantoic blood vessels, now known as *umbilical* arteries and veins, extend through the umbilical cord that attaches the embryo to the placenta, and underlie the chorion with capillaries that exchange materials with the maternal bloodstream. Thus the fetal portion of the placenta is *chorioallantoic*, derived from chorion and allantois together.

Further growth of the placenta involves a fetal-maternal interaction, as the endometrium responds by developing large blood sinuses in intimate contact with the chorionic villi. In most mammals the maternal blood vessels remain intact, but in primates, insectivores, bats, and carnivores, the villi erode right through their walls and lie directly bathed in maternal blood. In the former instance, when parturition takes place, the villi separate smoothly from the endometrium, which remains intact; in the latter case, the outer layers of the endometrium, including the placental portion, slough and are shed along with the fetal membranes so that some bleeding occurs.

The *form* of the placenta varies according to species: horses and swine have a *diffuse* placenta, with villi scattered over the entire surface of the chorion; in carnivores the placenta is *zonary*, with villi forming one or more belts around the circumference of the chorion; in humans villi persist on only one side of the chorion so that the placenta is *discoidal* and saucer-sized. When a young eutherian is born, its mother usually devours the placenta and fetal membranes and the umbilical cord soon shrivels and detaches, leaving a depression known as the *umbilicus* ("belly button").

22.17. How have mammals diversified?

Mammals range in body size from the 2-g, 4-cm pygmy shrew to the 100-ton, 30-m blue whale. They have diversified into forms adapted for running, leaping, burrowing, swimming, climbing, and even flying. Most are quadrupedal and *plantigrade* (walking on the full foot), *digitigrade* (walking on their toes), or *unguligrade* (walking on hooves sheathing the toe tips). Mammals also employ a wide range of feeding strategies and include omnivorous, herbivorous, and carnivorous species. Being homeothermal and insulated with fur or subcutaneous fat, they can occupy essentially any habitat where there is air to breathe.

This splendid mammalian radiation (Fig. 22.12) is almost entirely a product of the Cenozoic Era. As long ago as 200 million years, the two mammalian subclasses, *Prototheria* and *Theria*, arose, probably independently, from reptilian ancestry, but mammalian diversification did not get properly under way until 65 million years ago when reptilian dominance collapsed. Unhappily, of 34 mammalian orders all but 19 have already become extinct, and many more are becoming increasingly endangered by the expansion of human activities. The approximately 4500 species living today fall into the subgroups summarized below.

22.18. What are prototherians?

Subclass Prototheria ("first beast") includes three extinct orders (the Mesozoic triconodonts, docodonts, and multituberculates, named for the shape of their teeth) and one living order, *Monotremata* ("one opening").

22.19. What characteristics distinguish monotremes?

(*a*) Monotremes are the only surviving oviparous mammals. When laid, each soft-shelled egg is pea-sized (in echidnas) or 1.3 × 2.0 cm (in platypuses).

(*b*) The order name comes from the fact that, as in nonmammals, the terminal intestinal chamber (*cloaca*) receives both urinary and reproductive ducts, so that there is only one posterior body orifice.

(*c*) The female has two vaginas that enter the cloaca separately.

(*d*) The male's testes do not descend into a scrotum, and the urethra does not run through the penis, which carries only sperm.

(*e*) Nipples are lacking, and the young simply lick milk from pores along the mammary line.

(*f*) *Coracoids* and *procoracoids* (anterior coracoids) are present, as in therapsids, as separate bones of the pectoral girdle. Other mammals are different in that the coracoids have fused onto the enlarged scapulae and the procoracoids have vanished.

(*g*) Male monotremes have a *poison spur* on each ankle that delivers a toxic but nonfatal venom from a gland located under the skin of the thigh.

(*h*) Teeth are absent, except in juvenile platypuses.

(*i*) An ear pinna is lacking in platypuses and rudimentary in echidnas.

(*j*) Living monotremes are of moderate size, to about 3 kg, and 50 cm long.

> **Example 9:** *Ornithorhynchus anatinus*, the duck-billed platypus, inhabits fresh water in Australia and Tasmania, from tropical rivers to icy mountain streams. The platypus swims by its feet, which form outsized fans. The webbing on the forefeet extends so far beyond the digits that it must be folded backward to free the claws for walking or digging. When a platypus submerges, its eyes and ears are sealed by a facial furrow. Hunting invertebrates by means of electrical receptors on its leathery snout, the platypus grabs its prey with horny ridges along the margin of its "bill" and tucks tasty morsels into its cheek pouches for later consumption. The platypus is nocturnal, sleeping by day in a deep burrow dug in the riverbank. The female excavates a burrow 10–20 m long ending in a nest chamber lined with wet vegetation, which she transports in bundles wrapped around by her flat, beaverlike tail. After blocking her burrow with earth tamped by her tail, she incubates 2 eggs for only 7–10 days. The infants' eyes remain closed for 11 weeks, and they are nursed in the burrow for 4 months before being weaned.

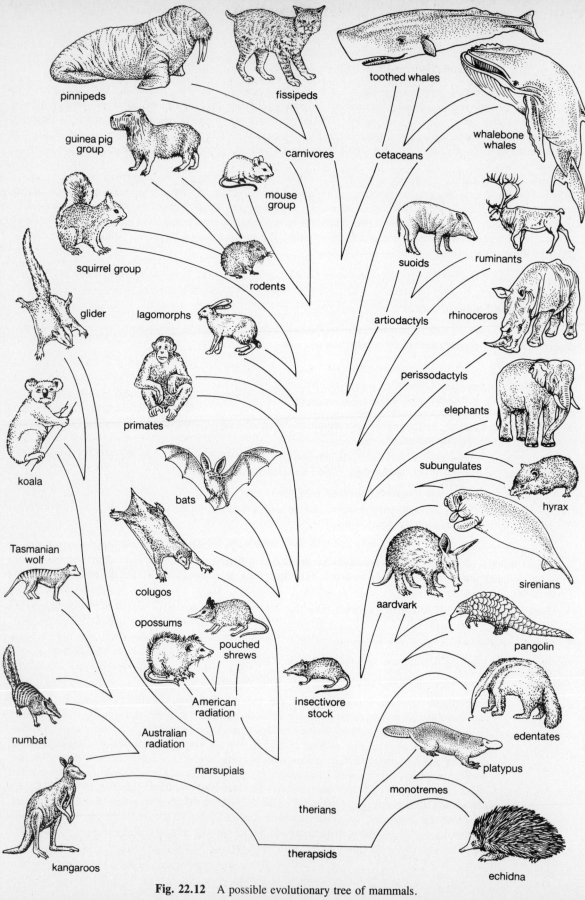

Fig. 22.12 A possible evolutionary tree of mammals.

Example 10: *Echidnas*, or "spiny anteaters" (*Tachyglossus* and *Zaglossus* spp.), inhabit Australia, New Guinea, and Tasmania. Clothed in stout 5-cm spines, these sturdy long-lived (up to 50 years) animals use their powerful claws to sink straight down into the ground at top speed until only the spiny back remains exposed. Their curved front claws also make short work of termite hills and infested logs, and the long, protrusible tongue is extended from the small mouth at the tip of the slender, toothless snout to gather the insects and crush them between horny ridges on the palate and back of the tongue. At the start of the breeding season a *brood pouch* develops on the female's abdomen, enclosing her mammary glands. A single, spherical egg is incubated in the pouch for 10–11 days, and the baby stays in the pouch until its growing quills become too painful for even a mother to bear. She then digs a burrow to conceal her baby and comes to nurse it every day or two until it is 3 months old.

22.20. What are therians?

All viviparous mammals belong to *subclass Theria* ("beast") and share most of the mammalian characteristics reviewed in Questions 22.2–22.16 above. Therians fall into two infraclasses: *Metatheria* ("after beast"), the marsupials, and *Eutheria* ("true beast"), the placental mammals.

22.21. What are marsupials?

Marsupialia ("little pouch") is the solitary order in infraclass Metatheria and includes live-bearing mammals that lack a placenta, or have only the rudiments of one. Today marsupials are restricted to the Australian region and South America, with a single species, the common opossum (*Didelphis marsupialis*) having become established in North America. South American marsupials evolved an adaptive radiation fully as impressive as that of Australian marsupials, even including large predators bearing a striking resemblance to wolves and saber-toothed tigers, but most of these died out after North and South America became connected and eutherians swarmed southward. In Australia marsupials range from animals the size of a shrew to the great gray kangaroo, but still larger species are known, though only as fossils: a buck-toothed marsupial lion (*Thylacoleo*) as big as an African lion, a wombat (*Phascolomis*) as large as a bear, and a slothlike *Diprotodon* larger than a rhinoceros.

22.22. What features distinguish marsupials from eutherians?

(*a*) Marsupials have either a rudimentary placenta (bandicoot and dasyurids) or no placenta at all, with the fetus relying mostly on its own yolk sac but also acquiring oxygen and some nourishment by way of close contact between its chorion and the uterine wall. Birth occurs after a short gestation period (opossum, 13 days; koala, 25–35 days; red kangaroo, 33 days; the marsupial "mouse" *Antechinus*, 30–33 days). The gestation period of *Antechinus* is close to that of a true mouse, but the inferiority of the marsupial intrauterine environment is apparent in contrasting sizes of the newborns: those of *Antechinus* are no larger than grains of rice, while newborn mice resemble plump jellybeans.

(*b*) The female reproductive tract is odd: the vagina branches into two lateral vaginae that extend anteriorly and then come together again medially to form a *vaginal sinus* into which the two uteri open separately. The vaginal sinus often extends posteriorly to open into the urogenital sinus, producing three vaginal passageways. The adaptive value of this redundant arrangement remains obscure.

(*c*) Female marsupials usually have a pouch (*marsupium*) that encloses their nipples and opens either forward (e.g., kangaroos) or backward (e.g., opossums, Tasmanian wolf), but some species have only a pair of skin folds flanking the nipples, while others lack even these.

(*d*) The young are born in a fetal state except for having large, clawed forelegs with which they clutch the mother's fur as they make the perilous journey to the pouch.

Example 11: Common opossums are larger than cats, but their neonates are smaller than honeybees, and only 60 percent ever reach the pouch. The squirrel-sized numbat's young are no larger than kernels of wheat.

Suckling is continuous since the nipple swells within the baby's mouth so that the youngster is unable to let go until mature enough to leave the pouch for short intervals. Family size is therefore limited by the number of nipples: the common opossum has 13 nipples but may produce 22 young, so that those arriving in the pouch after all nipples are occupied must starve. Roughly speaking, marsupial maturation from conception to weaning takes about twice the time needed for the maturation of eutherians of equivalent size, so when in competition with eutherians, marsupials are at a reproductive disadvantage.

(*e*) *Epipubic*, or marsupial, bones extend anteriorly from the pelvis in both sexes, possibly helping support the pouch or abdominal wall in females carrying young, although their presence in monotremes and pouchless marsupials, and their vestigial state in the pouched Tasmanian wolf, casts doubt on this hypothesis.

(*f*) Marsupial dentition is unique in several respects: (1) marsupials are more *monophyodont* than hemidiphyodont; i.e., their teeth come in only once, except for the last upper premolar, which is shed and replaced by a molar; (2) marsupials thus end up with 3 premolars and 4 molars per half-jaw, which is the reverse of the condition in primitive eutherians; (3) most marsupials have more than 3 incisors in each jaw half, whereas eutherians never have more than 3; (4) eutherians usually have the same number of incisors above and below, but in marsupials the number is rarely the same.

(*g*) A marsupial's brain is much smaller than that of a eutherian of equivalent size, and the unconvoluted cerebral hemispheres lack the corpus callosum of eutherians. Nevertheless, many marsupials, in particular the predatory dasyurids, are lively, alert, and speedy.

22.23.   What are the major living families of marsupials?

Except for the American didelphids and caenolestids, all living marsupials are restricted to Australia and neighboring islands.

(*a*) *American opossums* (Didelphidae) still enjoy considerable diversity in South America. The predatory water opossum has webbed feet and keeps its young from drowning by sealing its pouch tightly while it dives after small fish and aquatic invertebrates. The attractive murine opossums eat fruits and insects and are pouchless. The nocturnal common opossum is omnivorous, with a bare snout and scaly prehensile tail. Electroencephalography has shown that when an opossum "plays dead," it is wide-awake and alert, not unconscious or in shock; it is simply "playing possum."

(*b*) *Pouched shrews* (Caenolestidae) of South America are tiny and shrewlike, feeding on insects and spiders in moist Andean forests.

(*c*) *Dasyurids* (Dasyuridae) are intelligent, voracious Australian predators that are mostly mouse- to cat-sized, except for the medium-dog-sized Tasmanian devil and "tiger cat." The tiniest marsupial, the 7-cm common planigale, will nimbly tackle a grasshopper its own size, riding the insect like a cowboy on a bronco, until it can gnaw through the cuticle. Although the smaller dasyurids feed mainly on insects, they eat any prey they can subdue and they fearlessly assault animals as big as themselves.

(*d*) The *Tasmanian wolf* (Thylacinidae), the size of a large dog, has not been sighted since 1932 and may be extinct. It was persecuted for killing sheep in lieu of kangaroos.

(*e*) The *numbat*, or banded anteater (Myrmecobiidae), houses a 10-cm tongue in a 20-cm squirrel-sized body; this shy, bushy-tailed marsupial is close to extinction because of human intrusion and the growing scarcity of termite-ridden logs.

(*f*) *Marsupial moles* (Notoryctidae) of Australia's sandy deserts are remarkably convergent with eutherian moles: both burrow underground after worms and insects, both are nearly eyeless, and both have short tails and wide digging paws.

(*g*) *Bandicoots* (Peramelidae) are rat- to rabbit-sized and have pointed snouts, somewhat enlarged hind legs, and omnivorous appetites; desert species have long ears like rabbits.

(*h*) *Phalangers* (Phalangeridae), confusingly called "possums," occur throughout Australasia and are arboreal and mostly herbivorous, with bare or well-furred prehensile tails and an opposable thumb. The volplaning sugar glider looks remarkably like a flying squirrel; it feeds on insects, fruits, and flowers and reputedly bites into trees and licks up sap.

(*i*) *Koalas* and *great flying phalangers* (Phascolarctidae) are arboreal with molars adapted for chewing foliage; their first and second hand digits oppose the rest like a double thumb. The rotund, short-tailed koalas feed only on the leaves of certain eucalyptus species. Great flying phalangers are magnificent gliders measuring nearly a meter from nose to tail tip; they eat leaves and blossoms and travel by gliding as far as 100 m from the top of one tree to the base of the next.

(*j*) *Wombats* (Phascolomyidae) are stocky, short-tailed animals much like giant woodchucks or fat teddy bears, which munch ground vegetation and dig enormous tunnels with bark-lined colonial nesting chambers. They are unique among marsupials in having a single pair of upper and lower incisors that grow throughout life like those of rodents.

(k) *Macropodidae* ("big foot") includes kangaroos, wallabies, and quokkas, which hop nimbly on the toes of their enlarged hind legs, using the long tail as a counterbalance. The long feet have a single enlarged fourth digit flanked laterally by a smaller fifth digit, and medially by what appears to be a single small toe but actually consists of digits II and III grown tightly together. All are ground herbivores feeding on low vegetation, except for the arboreal tree kangaroos. Macropods have a fascinating reproductive strategy best known in the red kangaroo.

> **Example 12:** As soon as one fetus moves from uterus to pouch, the female red kangaroo mates again, but the presence of the suckling infant arrests development of the new embryo at about the 100-cell stage for some 235 days until the older sibling is mature enough to leave the pouch now and then and nurse only intermittently. The mother has two nipples, one of which has gradually enlarged as the first baby has grown; 33 days after resumption of development, the second offspring is born and attaches itself to the unoccupied, still tiny nipple. The composition of milk differs between the two mammary glands, for it changes as the youngsters grow. With a new infant in the pouch and the older joey still seeking the pouch for rest, safety, and occasional meals, the female mates once again. Thus at nearly all times, the female kangaroo is carrying three offspring at different stages of maturation.

22.24.   What are the major eutherian orders?

Eutherians represent the most successful of all tetrapod vertebrates in terms of habitat occupancy, variety of diet, and modes of life. A planetary tragedy of our time is that so many mammals in our legacy of wildlife, including those most familiar to us from childhood—rhinoceroses, lions, elephants, giraffes, tigers, zebras, gorillas, whales, and many others—stand close to extinction because of human expansion and habitat modification, as well as direct and often illegal slaughter for their pelts, horns, and tusks. Endangerment stalks many species of animal life, but because of our psychological ties with mammals, their plight touches us most profoundly. Captive breeding programs have preserved some species for the moment, but many cannot safely be repatriated, and the human population continues to burgeon. Reports of wild-born albinos, as among tigers, attest the critical reduction of breeding populations to a size at which inbreeding is bringing recessive genes into homozygous state. The orders summarized in Table 22.1 and Questions 22.25–22.52 contain many threatened species that can still be saved if enough persons become convinced of the importance of wildlife to the quality of human existence and the perpetuation of earth's biosphere.

22.25.   What are insectivores?

Order Insectivora includes the most primitive living eutherians. Insectivores (Fig. 22.13) have pointed snouts; mostly are plantigrade with 5 clawed toes per foot; have abdominal testes; and tend to be monophyodont with 44 teeth, including molars with conical, pointed cusps, and have the primitive dental formula 3-1-4-3. This order includes shrews, moles, and hedgehogs.

> **Example 13:** *Shrews*, with species ranging from 5 cm and the weight of a dime to a robust 65 cm, have mobile snouts and long vibrissae constantly aquiver after prey, since the smaller species of these voracious and largely carnivorous mammals perish of hypothermia if unable to devour some 75 percent of their body weight each day. Some shrews have a venomous saliva that helps them subdue large prey such as mice, although insects are a dietary staple. The water shrew has fur so dense it traps a layer of air that is carried underwater like a silvery overcoat, keeping the animal warm and dry even in icy mountain streams.

> **Example 14:** *Moles*, which live underground burrowing after worms and grubs, have velvety fur to which soil cannot cling; short, sturdy, forelimb bones; and wide spatulate hands that cannot be turned palm down, so that it is hard for moles to walk aboveground. The tiny eyes are nonfunctional, and the pectoral girdle has moved so far forward that these mouse-sized mammals appear neckless.

> **Example 15:** *Hedgehogs* are about 25 cm long and are covered on back and sides with sharp 2-cm spines that protect the animals more fully when they roll into tight defensive balls. Their diet includes fruit, nuts, insects, and small vertebrates including snakes, to the venom of which they are immune. Unlike other insectivores, hedgehogs hibernate up to 6 months a year.

> **Example 16:** *Elephant shrews* of Africa have a body length to 30 cm excluding a long, naked tail, but they are named not for their size but for their long, slender, wiggly snouts, reminiscent of an elephant's trunk. Although most are insectivorous, some species eat large prey and plant material as

Table 22.1   Major Groups of Living Eutherians

Order Insectivora: most primitive living eutherians; shrews, moles, hedgehogs

Order Dermoptera: arboreal gliders; 2 species of colugos; southeast Asia

Order Chiroptera: bats

   Suborder Megachiroptera: large, fruit-eating

   Suborder Microchiroptera: small, mostly insect eaters, orienting by echolocation

Order Primates: mostly arboreal, eyes placed for depth perception

   Suborder Prosimii: tree shrews, lemurs, lorises, tarsiers

   Suborder Anthropoidea: monkeys, apes, hominids

      Superfamily Ceboidea/Platyrrhina: American monkeys

      Superfamily Cercopithecidae/Catarrhini: Old World monkeys

      Superfamily Hominoidea: apes and hominids

Order Edentata: teeth reduced; anteaters, sloths, armadillos

Order Pholidota: covered with keratinous scales; pangolins

Order Lagomorpha: top incisors 2 behind 2; hares, rabbits, pikas

Order Rodentia: gnawing incisors 2 above, 2 below; mice, squirrels, beavers, capybaras, muskrats, porcupines, etc.

Order Cetacea: tail with horizontal fluke, hind limbs missing, front limbs flipperlike; whales and dolphins

   Suborder Odontoceti: toothed, single blowhole; sperm whales, beaked whales, killer whales, dolphins

   Suborder Mysticeti: lack functional teeth, have baleen plates, two blowholes; blue, finback, humpback, right whales, etc.

Order Carnivora: teeth adapted for carnivorous diet

   Suborder Fissipedia: usually terrestrial, with paws, not flippers; dogs, bears, raccoons, weasels, civets, hyenas, cats

   Suborder Pinnipedia: marine, with flippers; seals, walruses (sometimes given as order Pinnipedia)

Order Tubulidentata: African aardvark

Order Proboscidea: elongate trunk, tusks; elephants

Order Hyracoidae: tailless, hooflike toenails; Afroasian hyraxes

Order Sirenia: tail with horizontal fluke, hind limbs missing, front limbs flipperlike, herbivorous; manatees, dugongs

Order Perissodactyla: odd-toed hoofed mammals; horses, tapirs, rhinoceroses

Order Artiodactyla: even-toed hoofed mammals

   Suborder Suiformes: simple stomach, 4-toed; pigs, peccaries, hippopotamuses

   Suborder Tylopoda: 3-chambered stomach, 2-toed; camels, vicunas, llamas, guanacos, alpacas

   Suborder Ruminantia: 4-chambered stomach (3 in chevrotains); deer, giraffes, pronghorns, bovids

---

well. Elephant shrews hop on their elongated hind legs. Endangered mothers wait for their babies to grab securely onto the teats, then bound away, bouncing the youngsters along the ground.

22.26.   What are colugos?

The *colugos* of *order Dermoptera* are arboreal, cat-sized gliders. The two living species inhabit forests of southeast Asia. Miscalled "flying lemurs," colugos are noteworthy for having a skin web that reaches from the chin to the hand and fingers, then along the sides of the trunk to the hind legs and toes, and finally from toes to tail tip, giving the soaring animal the look of an oblong kite as it glides for up to 120 m so efficiently that it loses only about 10 m in height. Even the elongated fingers are webbed. In the treetops the colugo eats leaves and blossoms, creeping upside down along the underside of branches like a sloth, the female carrying her single baby securely in the hammock of her web. The teeth are unique in that the incisors and canines have two roots instead of one, and the lower incisors are comblike.

(a) Orders Insectivora (shrew, mole) and Chiroptera (bat)

shrew

mole

bat

flying squirrel

red squirrel

American porcupine

kangaroo rat

beaver

(b) Order Rodentia

long-tailed weasel

badger

gray fox

bobcat

(c) Order Carnivora

striped skunk

Fig. 22.13 Representatives of four eutherian orders. *(From Storer et al.)*

22.27. What are the characteristics of bats?

Bats make up the order *Chiroptera*, the second largest order of mammals. They enjoy cosmopolitan distribution because they can fly and even migrate long distances. The sternum is often moderately keeled for the attachment of enlarged pectoralis muscles as in birds. A skin membrane stretches between four exceedingly attenuated fingers, leaving the thumb free, and extends posteriorly to the hind legs. Many bats also have a web (*uropatagium*) extending from hind leg to tail tip, with which they scoop up insects in midair. The hind legs of bats have rotated so that the knees bend backward and the toes curl forward, which helps in both deploying the tail web and catching hold with the claws for hanging upside down, the roosting posture of bats in trees and caves. Some bats go dormant by day, allowing their body temperature to drop to ambient levels. Bats fall into two suborders: the larger Megachiroptera (fruit bats) and the smaller Microchiroptera.

Example 17: *Fruit bats* (suborder *Megachiroptera*) have doglike muzzles and a wing span to 1.5 m; they feed on fruit by night and by day sleep in flocks, hanging head down from tree branches like pendant fruits themselves, wrapped round by the cloak of the wings.

Example 18: Bats of suborder *Microchiroptera* (see Fig. 22.13*a*) are much smaller and orient in flight by *echolocation*, emitting ultrasonic yelps at frequencies of up to 130,000 Hz at a rate of 30–50 per s. *Leaf-nosed bats* have oddly shaped nostrils that may help focus the sonar beam, *vesper bats* have simple lips and muzzles, and *free-tailed bats* have the uropatagium confined to the tail base; all have large pinnae with a fleshy *tragus* protruding into the ear opening, possibly a device for better directionalization of the sonar echo. Although most microchiropterans capture insects, *fishing bats* pluck small fish from surface waters, *long-tongued bats* lap nectar from flowers, and tropical American *vampire bats* nip sleeping mammals and lap blood. Unfortunately, vampires and certain insectivorous bats can transmit rabies without being ill themselves.

22.28. What are the characteristics of primates?

Order Primates includes tree shrews, lemurs, monkeys, apes, and hominids. They are rather primitive mammals, in the sense of being generalized and closer to shrewlike ancestral mammals than are such mammals as horses, which are more advanced in the sense of being more highly specialized and further removed from the ancestral state. Primates have unusually long digits. Except for tree shrews, only the second hind toes of lemurs and the second and third hind toes of tarsiers bear claws; otherwise, primate digits bear flattened *nails*. The thumb and big toe are usually opposable for climbing; in chimpanzees and humans, the thumb is long enough to oppose the tip of the bent first finger in a "precision grip," used by chimpanzees when poking twigs into termite hills. Primate eyes are directed forward, allowing good *depth perception*, useful when leaping from tree to tree. This eye placement is stabilized by fully *cup-shaped eye sockets* (orbits). Order Primates is divided into the suborders Prosimii and Anthropoidea.

22.29. What are the prosimians?

Suborder Prosimii includes tree shrews, lemurs, lorises, and tarsiers. Aside from their pointed snouts, the south Asian *tree shrews* much resemble squirrels. They feed omnivorously by day and sleep in nests of leaves and twigs. The mother gives birth to two young, which she leaves alone in the nest except for a short visit every other day to let them nurse. Tree shrews form a link between shrews and primates but are classified with the latter for their well-developed brain and complete bony eye sockets. *Lemurs* are restricted to the island of Madagascar, where deforestation is imperiling them. They have woolly fur, long tails, and foxy muzzles; the strange *aye-aye* has a rodentlike skull with no canines and with incisors that grow throughout life. *Lorises* have a short muzzle and large, owlish eyes, and include the tailless slow loris of southeast Asia and the bushy-tailed bush babies of Africa. Bush babies live in scrub, feeding nocturnally on insects, birds' eggs, flowers, seeds, and fruit. *Tarsiers* are nocturnal, with enormous, protuberant eyes and large, thin ears. Their second and third toes are clawed, and the foot is elongated, enabling them to leap agilely after insects and lizards.

22.30. What are the characteristics of anthropoids?

Monkeys, apes, and hominids (suborder Anthropoidea) have especially large brains, producing a large cranial vault above a flattened face that is hairless upon and around the nose. The facial muscles are highly developed, and the lips freely protrusible. The ear pinnae are somewhat reduced and immobile, lying nearly flat to the head. The arms are usually longer than the hind legs. Anthropoids include the superfamilies Ceboidea/Platyrrhina, Cercopithecidae/Catarrhini, and Hominoidea.

22.31.   What are monkeys?

Monkeys are primates with tails. Monkeys of tropical America [*superfamily Ceboidea* or *Platyrrhina* ("flat nose")] lack cheek pouches and have nostrils that open laterally, to opposite sides of the nose; they include *cebids* (squirrel, spider, capuchin, and howler monkeys) that have prehensile tails and squirrel-size *marmosets* that do not. Asian and African monkeys, including macaques, baboons, guenons, and langurs [*superfamily Cercopithecoidea* or *Catarrhini* ("downward nostrils")], have parallel nostrils directed downward, often possess internal cheek pouches, and have calloused, hairless buttocks (*ischial callosities*) that swell up and redden in estrus females, advertising their readiness to mate. These monkeys lack prehensile tails.

22.32.   What are hominoids?

Hominoids [*superfamily Hominoidea* ("manlike")] include large, tailless primates: *apes* (family Pongidae) and *hominids* (humans and extinct humanlike species of family Hominidae). Like the catarrhinids, these Old World primates have downward-directed nostrils and hairless buttocks, but the latter do not swell up during estrus. In trees they clamber or *brachiate* (swing hand over hand); on the ground some walk bipedally, while others walk on the soles of the feet and the knuckles of the hands. The *pongids*, or apes, include the African chimpanzees and gorillas and the Asian orangutan, gibbon, and siamang.

> **Example 19:**   Gibbons are creamy and siamangs black, but otherwise these lesser apes look and act much alike: in trees they brachiate; on the ground they walk bipedally, long arms extended to the sides for balance. When they stand fully erect, the extreme elongation of their arms is evident, for their fingers touch the ground. Surprisingly, DNA analyses indicate that gibbons and siamangs are convergent rather than closely related. On the other hand, DNA concordance between chimpanzees and humans is so close that it is doubtful they belong in separate families.

> **Example 20:**   Orangutans are shaggy, red-haired, arboreal brachiators that live unusually solitary lives, except that juveniles remain with their mothers for 5 years. Gorillas and chimpanzees live on the ground in small social groups. Despite their formidable size and strength, the great apes are gentle and peaceful vegetarians, even becoming socialized to field investigators patient enough to outwear their shyness. However, chimpanzees are occasionally carnivorous, killing monkeys and baboons for food, and sometimes chimpanzee males even slay those of another group, using sticks as bludgeons, thus proving themselves even more humanlike than we once believed.

22.33.   What is the evolutionary history of hominids?

Some 25 million years ago, climatic changes caused forests to diminish and be replaced by *grasslands*: treeless prairies or tree-studded savannas. This new habitat encouraged mammals to venture into the open, where herbivores engaged in grazing and both they and their predators evolved toward fleet-footedness in open terrain. At this time early hominids began to abandon the arboreal life of forests, perhaps seeking carcasses on which to scavenge. The oldest known definitely hominid fossils are of the bipedal *Australopithecus africanus* and the stockier *A. robustus*, which inhabited south Africa 4 million years ago. The more robust *A. boisei* lived in east Africa from 1–3 million years BP (Before Present). The bipedal, fully erect *A. afarensis*, with a chimpanzee-sized (400-cc) brain, inhabited Ethiopia from at least 2.9–4 million years ago.

In 1972 Richard Leakey discovered in east Africa a nearly complete skull of a type known earlier from fragments collected at Olduvai and named *Homo habilis* by Louis and Mary Leakey. The complete skull, with a cranial capacity of 776 cc and characteristics that strongly link it to the later *Homo erectus*, has been radioactively dated to an age of 1.6 million years, and the discovery of stone tools at the Olduvai site suggest that *H. habilis* was a toolmaker, probably using chipped stones to cut flesh off carcasses. Regular meat eating represented another major divergence from other anthropoids, which are largely vegetarians. Probably first as a scavenger on fresh lion and cheetah kills, and later as a social hunter, *Homo* moved out upon the grasslands.

A skull of *Homo erectus*, previously known from China and Java, was discovered in east Africa in 1976 and radioactively dated at 1.5 million years old. *H. erectus*, with a cranial capacity of 1000 cc, stood 1.5–1.7 m tall. Evidence of group encampments, together with animal remains found along with tools, suggests that these hominids lived socially and killed prey as large as elephants and rhinoceroses. *H. erectus* disappeared about 300,000 BP, concurrent with the appearance of skulls with a 1200–1300 cc braincase (a 250,000-year-old specimen from England, and one 300,000 years old from Germany). With a cranial capacity within the range of modern humans (1000–2000 cc) and a more human facial skeleton and dentition, these skulls constitute the oldest known relics of *Homo sapiens*. The more recent remains known as *Neanderthal man* (from 130,000 BP) and *Cro-Magnon man* (from 40,000 BP) represent variations of a species that had already become polytypic.

22.34. What are the edentates?

Members of *order Edentata*—anteaters, sloths, and armadillos—either lack teeth or have degenerate, peglike teeth uncapped with enamel. Edentates originated in South America, and some grew to giant size, including armored 3-m glyptodonts and a ground sloth as big as an elephant. Some migrated into North America, but today only one species, the nine-banded armadillo, still occurs there.

Giant ground sloths became extinct in North America only some 10,000 years ago, and their bones are found in the La Brea tar pits of Los Angeles. The two living species of *sloths* are arboreal, hanging upside down from great hooked claws that number two or three per foot and creeping slothfully in very slow motion. They are so fully adapted to an inverted existence that their fur slants oppositely to that of other mammals, downward from belly to back, allowing rainwater to be shed; their necks are so flexible that the head can be rotated to be right side up while the body is upside down. Sloths still have nine grinding teeth on each side, which grow throughout life, so that even though all their front teeth are missing, they can adequately masticate foliage and twigs, digesting them in a chambered stomach like that of ruminants. Sloth hairs are pitted, the pits harboring green algae that camouflage their host. Small moths also live in the fur, their presence a mystery until it was discovered that whenever the sloth laboriously descends to the ground to defecate, the moths then lay their eggs on the droppings.

Armadillos have up to 25 cheek teeth adapted for an omnivorous diet. They are the only mammals armored in rows of bony dermal plates covered by plates of horny epidermis, commonly with soft skin between that allows them to roll up. *Anteaters* lack teeth and have long slender snouts housing even longer tongues. Great curved claws rip open termite hills, and the sticky tongue is then protruded from the tiny, round mouth at the tip of the snout and retracted when coated with insects. Arboreal anteaters have prehensile tails.

22.35. What are pangolins?

The scaly anteaters, or *pangolins*, of Afroasia represent *order Pholidota*. Pangolins are shingled with overlapping, keratinous scales except on the softly furred belly. The long, heavy tail is used prehensilely by tree pangolins, which feed mainly on tree ants, or as a brace by large (to 2 m) ground-dwelling species when they rear up to rip open termitaries with their stout claws. The snout is narrow and short, with a small mouth at the end, from which the tongue can be thrust to a distance of 30 cm. Pangolins are toothless, but swallow small pebbles that help grind up hard-bodied ants. Little is known of their private affairs, since they quickly die in captivity.

22.36. What are the characteristics of lagomorphs?

Order Lagomorpha includes hares, rabbits, and pikas. Lagomorphs have chisel-shaped gnawing incisors that grow throughout life. Once they were classified as rodents, but these two groups of gnawing mammals are simply convergent and have many important differences:

1. Rodents have one upper and one lower incisor in each half-jaw (1/1), while lagomorphs have 2 upper incisors and 1 below on each side (2/1), and their upper incisors are uniquely arranged, with a smaller pair directly behind the larger pair, their combined edges forming a broad chisel blade.

2. Rodent incisors are enamel-coated only in front; lagomorph incisors are enameled on the posterior surface as well. Next to the incisors both have a toothless gap (*diastema*) followed by the cheek teeth.

3. The exact number of premolars and molars differs within each order, but lagomorph molars are rootless. The cheek teeth of rodents are of equal size in both jaws, while the upper teeth of lagomorphs are larger than the opposing lower ones.

4. Lagomorphs have a broad palate and chew only with lateral movements; rodents have a narrow palate and, when chewing, move their jaws both laterally and forward and back.

5. Rodents have a rotating elbow joint that allows the often handlike forepaws to be pronated and supinated, making them useful in holding and manipulating objects; lagomorphs lack this feature and cannot use their paws as hands.

6. Lagomorphs, aside from pikas, have remarkably long ears, whereas rodent ears are ordinarily short and rounded.

7. Lagomorphs are tailless or have very short, furry tails, such as the white puff of cottontails; rodent tails are usually long and may be scaly or coated with short or long fur.

8. Lagomorphs commonly have enlarged hind legs and run by hopping, actually a bounding gallop; when walking they also favor a slow gallop-gait (first moving both forelegs forward, then advancing the hind legs together). Many rodents gallop when pressed, but lack the enlarged hind legs needed to accomplish great bounds; at leisure, they walk with the typical tetrapod diagonal leg progression.

9. Lagomorphs feed on green vegetation and bark; although largely vegetarian, rodents are a varied group with considerable diversity of diet.

> **Example 21:** *Hares* (including desert jackrabbits) are larger and rangier than rabbits, with longer ears; they make no burrows but live on the surface and, when panicked, spring away in great 6-m bounds. Diurnal, they radiate excess heat from their huge ears. Their young are precocial, born well-furred with eyes open and able to toddle, although they may remain in the fur-lined surface nest for up to 4 weeks. Rabbits burrow, and their young are *altricial*, born naked, with eyes sealed closed, in a subterranean nest chamber.

> **Example 22:** The small, tailless *pikas* live above timberline in the Rockies and Himalayas, but also frequent less rigorous habitats. Unlike the normally silent hares and rabbits, pikas whistle and bark loquaciously. Despite living where winters are cold, they do not hibernate but spend the summer cutting vegetation, spreading it to dry, then storing the cured hay.

Lagomorphs are successful, cosmopolitan, and fecund, providing abundant meat for many kinds of predators.

22.37. What are rodents?

In species (about 1687) and individual numbers, the members of order Rodentia outnumber all other mammals (Fig. 22.13*b*). Their characteristics have been given above in contrast with those of lagomorphs, but in addition it can be added that most rodents are nocturnal or crepuscular (active during twilight and predawn hours), and a number are true hibernators. They range from 5-cm mice to 1.2-m capybaras, most being small and secretive. Many dig burrows for shelter but feed on the surface (e.g., prairie dogs), while some (e.g., gophers) lead subterranean lives; others (e.g., tree squirrels) are arboreal, and some (e.g., beavers, nutrias, capybaras, muskrats) are amphibious in fresh water. Prolific, abundant, and diversified, rodents are of great value in many terrestrial food chains as primary consumers fed upon by carnivores. We ourselves, often viewing rodents as pests and carriers of diseases such as bubonic plague, should note that some bear valuable fur (chinchillas, nutrias, beavers, muskrats), and we also owe a debt of gratitude to the generations of laboratory-raised mice, guinea pigs, and rats on which medical and genetic research has long relied.

A few important families (out of the existing 34) are *Sciuridae* (tree, ground, and flying squirrels, chipmunks, marmots), *Geomyidae* (pocket gophers), *Heteromyidae* (hopping mice and kangaroo rats with fur-lined external cheek pouches), *Cricetidae* (white-footed mice, pack rats, lemmings, voles, muskrats, hamsters, gerbils), *Muridae* (Eurasian rats and mice, now cosmopolitan: house mouse, Norway rat, roof rat), *Erethizontidae* (porcupines), and *Caviidae* (guinea pigs).

22.38. What are the characteristics of cetaceans?

Order Cetacea includes whales and dolphins, the most perfectly adapted of water-dwelling mammals. They range throughout all oceans, and a few species inhabit rivers. Their streamlined, fish-shaped bodies make them excellent swimmers, but they cannot live on land, so it remains puzzling why ailing or dying whales sometimes beach themselves in a purposive manner (although the stranding of schools of pilot whales appears to be by accident). Cetaceans lack hind limbs, and their front limbs act mainly as stabilizers, with extremely short arm bones and long hand bones supporting *flippers*. Their heavy muscular tails terminate in fleshy, transverse *flukes* that are moved up and down for propulsion. Streamlining is improved by the male's lack of scrotum and recession of the female's nipples into a pair of abdominal grooves. Hair is lacking, except for a few bristles on the snout, and a thick layer of subcutaneous fat substitutes for fur as insulation. Sebaceous and sweat glands are missing. Cetaceans have no ear pinna, and the external auditory canal is quite small. Nevertheless, these mammals have excellent hearing and respond to tones as high as 130,000 Hz; some orient by echolocation. Vision out of water is better than was once supposed, to judge by the accuracy with which conditioned dolphins snatch fish from the trainer's hand in midleap. Olfaction is almost lacking.

Some cetaceans dive deeply (1100 m for sperm whales, which can submerge for an hour while hunting giant squid); during diving, *valves* along the bronchi prevent air from being forced out of the compressed lungs, the heart rate slows drastically (from 110 to 50 in bottle-nosed dolphins), and blood is shunted from the viscera to the muscles and brain while much venous blood is trapped in the lower body, preventing its high CO_2 level from stimulating the respiratory centers in the brain. When a cetacean surfaces, it exhales quickly and forcibly, the condensing moisture from the lungs forming a telltale "spout." All cetaceans breathe through one or two *blowholes* located atop the head that are closed during submersion; their larynx extends up through the back of the throat to open directly into the nasopharynx, so that the mouth can be used only for feeding and no water can enter the respiratory passages during ingestion. Their asymmetrically convoluted nasal passages may help in sound production as air is shunted about through confined spaces; some cetaceans, notably dolphins and humpback whales, are sophisticated vocalizers, but the significance of such communication remains unsolved.

> **Example 23:** At least when aggregated for breeding, male humpback whales produce sounds of many minutes' duration. These sounds actually are *songs*, not random compilations of tones, because once the "song of the year" has been worked out by modification of the previous year's song, all males in the breeding group sing the same song repeatedly (but never in unison), without further change.

The cetacean brain is large, and its cerebral cortex is as highly convoluted as that of humans. Learning tests with captive dolphins have demonstrated considerable *reasoning* capacity (e.g., one investigation showed that the subject readily grasped the principles that to receive a food reward she had to think up some novel behavior during each trial, and would be rewarded for only one new behavior per session). Bottle-nosed dolphins have been reliably reported to engage in altruistic behavior requiring *cooperative* interactions (e.g., a distressed individual may be supported at the surface by successive pairs of dolphins that buoy it up from below for the duration of a held breath, being immediately replaced by another pair when they must veer off to breathe). Commonly one or two females hover near one giving birth to her single young, ready to ward off sharks and help the baby surface to breathe. Other mammalian babies are usually born head first, but cetaceans are born *tail first*, reducing the chance of drowning. Newborns may be one-third their mother's length and can swim rapidly at once. Since cetaceans are incapable of facial expression, they also are unable to suck upon a nipple: when the baby grasps a nipple between its jaws, the mother muscularly squirts milk into its mouth, each feeding taking only a few seconds. For some time, mother and baby may be closely accompanied by an "aunt" that helps guard the baby.

The two suborders of cetaceans are Odontoceti and Mysticeti.

22.39. What are the characteristics of odontocetes?

Suborder Odontoceti includes all toothed cetaceans. Their teeth are alike and adapted for prey capture. The blowhole is single.

(*a*) Family *Physeteridae* includes *sperm whales*, with a narrow lower jaw bearing 20–25 pairs of 20-cm teeth that fit into sockets in the toothless upper jaw when the mouth is closed. Male sperm whales grow to 20 m, females being under 10 m. The massive head makes up one-third the body length and contains a huge *spermaceti* organ that stores a liquid that solidifies into a soft white wax of fine quality for which this whale has long been hunted. The intestine of occasional sperm whales contains a large mass of *ambergris* that is still highly valued as a perfume base; ambergris is thought to be a secretion that solidifies protectively around the beaks of ingested squid; when egested it floats and makes a lucrative find for beachcombers.

(*b*) Family *Ziphiidae* includes the 10-m *beaked whales*, which have only one large tooth (uncommonly 2) on each side of the lower jaw.

(*c*) Family *Delphinidae* includes most smaller odontocetes, which have numerous recurved teeth in both jaws. Largest are the 9-m killer whale, which hunts in packs and may even attack giant baleen whales, and the 5.5-m false killer and pilot whales. Littler odontocetes are known as dolphins, with the name porpoise being applied to small species having a triangular rather than recurved dorsal fin and no rostrum protruding below the forehead.

22.40.   What are mysticetes?

Suborder *Mysticeti* includes all whales with two blowholes, no functional teeth, and fringed plates of horny *baleen* suspended from the roof of the mouth and used in trapping small prey. The prey, usually planktonic crustaceans (krill), are enmeshed when the whale forces water from its mouth through the baleen fringe and then are licked off by the fleshy tongue. Nearly exterminated for their flesh and blubber, these giant creatures cannot submerge for long and so are helpless victims of whaling fleets with spotter airplanes. Mysticeti consists of the following families: (*a*) *Balaenopteridae*, the rorquals, rapid swimmers such as the 30-m blue whale, 24-m finback, and 16-m humpback whale, have a dorsal fin, short baleen plates, and deep longitudinal furrows on the throat, which allow it to be inflated, sucking in great mouthfuls of food. (*b*) *Balaenidae*, the right whales, to 21 m, lack throat grooves and dorsal fin, swim slowly, and have bow-shaped mouths with baleen plates up to 2.4 m long. (*c*) *Eschrichtidae*, the gray whale, mainly breeds in shallow coastal lagoons.

22.41.   What are the characteristics of order Carnivora?

Members of *order Carnivora* are mainly flesh eaters with teeth adapted for tearing flesh and crushing bone. All have incisors that are shorter than the sharp, fanglike *canine* in each half-jaw. The premolars and molars have pointed crowns. The articulation of mandible with skull forms a wide half-cylinder that provides power in biting but precludes lateral movements during chewing. They have at least four, but usually five, digits on each foot, and terrestrial species may be plantigrade or digitigrade. The clavicles have nearly been lost, bringing the shoulders close together. The uterus is bicornuate and the placenta zonary. Many have rich fur and have been hunted ruthlessly for their pelts.

The order is divisible into those with paws, adapted for walking on land (fissipeds: "split foot"), and those with flippers, adapted for an aquatic existence (pinnipeds: "feather foot"). Pinnipeds are separately recognizable in the fossil record since the Miocene, and some zoologists consider them a separate order. However their affinities to other carnivores, especially bears, are strong, and here they will be considered a suborder.

22.42.   What characterizes the fissipeds?

Suborder Fissipedia (Fig. 22.13*c*) includes carnivores that are usually terrestrial with unwebbed toes on their paws. Their deciduous teeth are well developed, and in the permanent set nearly all have three small incisors in each half-jaw and differently shaped premolars and molars. Their young are altricial. There are seven families:

(*a*) *Canidae*, the dog family, includes dogs, wolves, and foxes. They have long, bushy tails, are digitigrade, walking on four toes, and have the fourth upper premolar and first lower molar opposing one another as a pair of shearing *carnassial* teeth. Wolves and wild dogs live in social groups.

(*b*) *Ursidae*, the bears, contains the largest members of the suborder (to 750 kg). Chunky, short-tailed, and plantigrade, bears *amble* when walking (move both legs on one side together) and gallop for speed. Many are omnivorous, lacking carnassials and having round-cusped, flat-crowned molars for grinding plant material. Ursids are solitary except for females with cubs. In temperate regions bears den up and sleep through the winter, when the twin cubs are born. Mothers fiercely defend their young against roving males.

> **Example 24:**   The giant panda, which on the basis of DNA analysis seems to be an odd bear, feeds in the wild only on bamboo; a handlike use of the forepaws in holding plucked leaves in a bundle for consumption is made possible by elongation of one carpal as a fake opposable thumb.

(*c*) *Procyonidae* includes American raccoons, coatimundis, ringtails, and the Himalayan red panda. Most have banded tails and walk plantigrade, leaving tracks like little hands with five fingerlike digits.

> **Example 25:**   The black-masked *raccoon* is omnivorous but frequents streams and marshes where fish, frogs, and crayfish may be caught. Food and novel items are thoroughly manipulated by the sensitive hands, and food is often washed before it is eaten. *Coatis* live in bands of up to 20 females and young, while males live solitarily except when mating. They prey on such terrestrial invertebrates as tarantulas and land crabs and also dig out lizards and caecilians. *Ringtails* are nocturnal predators on rodents; crossing moonlit clearings, they arch their very long, bushy tails over the body, possibly confusing great horned owls that prey on these squirrel-size creatures. "Panda" simply means "bamboo eater," and that is what *red pandas* eat; they resemble red-haired, woolly raccoons.

(*d*) *Mustelidae*, the weasel family, includes weasels, stoats, otters, badgers, skunks, ferrets, and wolverines. Most are strict carnivores, but skunks also favor fruit. Mustelids have especially thick fur; many are trapped for their pelts, but mink are successfully raised on farms. They have in common particularly short legs with five toes on each foot, well-developed scent glands, and only one molar on each side of the upper jaw, but two below. Many have slender bodies well-adapted for pursuing rabbits and rodents into their burrows. The powerful, pugnacious *badger* simply digs marmots out of their burrows.

> **Example 26:** The amphibious *otters* are known for their playful dispositions; their digits are all fully webbed and their tails flattened. The hind feet of sea otters are particularly huge. Sea otters lack blubber and rely on the insulative qualities of their extremely dense fur, which must be meticulously groomed; if the fur is contaminated by spilled oil, the otters quickly die of hypothermia. California sea otters live in the surf zone and almost never come ashore. They dive after invertebrates such as sea urchins, crabs, abalone, and clams, often collecting a rock to be used as an anvil on which the otter beats its prey while floating on its back.

(*e*) *Viverridae*, the civet family, is confined to Afroasia except where introduced. Like mustelids they are somewhat long-bodied and short-legged and have musk glands. When diluted, secretions of the civet, *Viverra*, make excellent perfume. Civets are solitary, nocturnal, and omnivorous.

> **Example 27:** The best-known viverrids are *mongooses*, which range from 0.3–1.2 m and resemble rather scruffy stoats or weasels but are more omnivorous and diurnal, much enjoying a sunbath. Mongooses make good pets where venomous snakes abound, because they fearlessly tackle even king cobras by lightning lunges and retreats, biting at the back of the snake's head; they have great resistance to venom, and each bite increases their immunity. Recklessly imported into the Hawaiian and Caribbean islands to control rats, the diurnal mongooses imperil birds more than they do nocturnal rodents.

(*f*) *Hyaenidae* includes the Afroasian hyenas, which resemble dogs with heavy heads and exceptionally powerful jaws, thick necks, and a back that slopes from taller foreparts to shorter hindquarters. They live in packs and dig subterranean nests for their young. Striped and spotted hyenas scavenge lion kills, but the 80-kg spotted hyena is also an efficient predator in its own right. The brown hyena frequents beaches, feeding on stranded carcasses from crabs to whales.

(*g*) *Felidae*, the cat family, is distinguished for especially sharp, flesh-cutting cheek teeth and retractile claws (except in cheetahs). Cat claws are kept sharp by the desquamation of outer horny layers, thus felids typically have favorite scratching posts. The claws are protected from blunting by being folded back into horny sheaths at the tip of each digit. Cats are digitigrade with four toes contacting the ground. Felids include many large predators such as lions, tigers, leopards, snow leopards, cheetahs, jaguars, and cougars, some of which are seriously endangered for their attractive pelts, others being persecuted for attacking livestock.

22.43. What characterizes the pinnipeds?

Suborder Pinnipedia includes the marine sea lions, seals, and walruses. Most are piscivorous, but walruses eat only shellfish, while elephant seals also relish squid, and leopard seals devour fish, penguins, squid, and krill. All have very short tails and bearlike skulls, suggesting their descent from some Miocene ursid, but instead of having three incisors in each half-jaw as in fissipeds, pinnipeds have three above but only two below. In addition, the premolars and molars of pinnipeds are alike with conical crowns, and the deciduous teeth are vestigial and sometimes shed before birth. All four limbs are modified into flippers, but the nature and use of these flippers differ considerably among the three families. All have five digits per limb and are unusual in having digits I and V longer than the rest. Although pups are well-furred, only fur seals have heavy pelts as adults; insulation is derived mainly from a thick coat of blubber. Pinnipeds have reduced or absent ear pinnae and tiny external ear openings, and detection of underwater vibrations is much aided by their stout vibrissae. Some navigate and detect prey by echolocation. Pinnipeds typically come ashore to rest, and breeding and parturition also take place on land, often after a long migration to a population's traditional breeding site.

> **Example 28:** Typically, dense breeding aggregations form in which males battle for territories, then herd arriving females into harems. Within a day or two, females impregnated the previous year give birth to a single, precocial pup; this yearlong pregnancy is much longer than that of modern bears (to 250 days) and may be due to delayed implantation. (Since the female mates again when her pup is only a few days old, development of the embryo might possibly be delayed by around 100 days.)

Sharks, killer whales, and humans are the major predators on pinnipeds, but many are heavily parasitized, especially by lung flukes.

(a) Family *Otariidae* includes pinnipeds with ear pinnae: sea lions and fur seals. These so-called eared seals are agile on land because their hindquarters can be tucked under the trunk, allowing them to use their hind flippers in walking and galloping. The skin of the hind flippers extends well beyond the end of the digits, and claws, used in scratching dog-style, occur only on the three middle digits. The winglike front flippers are clawless and are used in "flying" underwater; in fact, otariids build up enough speed to "porpoise" in bounds that clear the water. The third upper incisor is fanglike but smaller than the canines. Adult male Steller's sea lions may exceed 3 m and weigh a ton. Females weigh only one-third this much. Both sea lions and fur seals have been trained to participate in "seal shows."

(b) Family *Odobenidae* includes only one species of walrus, circumpolar arctic pinnipeds that can weigh over 1600 kg when well loaded with blubber. Their upper canines form 1-m-long tusks, which seem more useful in social fighting and hauling out than in digging up the mollusks on which they feed. (The family name means to walk with one's teeth.) Despite their remote habitat, walruses are imperiled by ivory-seeking humans. The short, flat muzzle bears a "walrus moustache" of short, thick bristles. Ear pinnae are absent. The front flippers are broad and oarlike and about 25 percent the body length, but they end in strong claws useful for traction and clam digging. The hind flippers can be tucked under the body for support and thrust during locomotion ashore, but its portliness makes a walrus ungainly on land.

(c) *Phocidae* is the family of earless or "true" seals; these lack ear pinnae and cannot rotate their hind limbs under the trunk. This makes movement on land laborious: they pull with the five-clawed front flippers, which resemble spatulate paws, while advancing the torso in a wormlike humping manner with hind limbs dragging behind. Phocids swim mainly by lateral undulations of their huge webbed hind feet, which are held vertically and spread like fans, thus resembling the caudal fin of a fish. The largest phocids are elephant seal males, which may exceed 6.5 m and weigh over 3600 kg, with females reaching about 3.6 m and 900 kg. During the breeding season the male's nose is inflated by blood pressure and muscular action and in the northern species pendulously overhangs the mouth so that snorts through the nose are amplified into a roar by the resonating chamber formed by the open mouth. Conversely, males of the southern species roar through their mouths and use their shorter noses to amplify the sound. Other phocids include harbor, gray, and hooded seals, Antarctic leopard and Weddell's seals, and harp seals, whose snow-white, furry babies are slaughtered to make fur coats.

22.44.   What is an aardvark?

The African *aardvark*, or earth pig, is the sole living member of order Tubulidentata. It has erect, donkeylike ears, a stout, piglike body so sparsely furred that it looks naked, and a blunt piggish muzzle bearing only a small terminal mouth from which a slender, nearly 0.5-m tongue can be protruded to lick up termites from nests torn into by its powerful claws. Aardvarks are nocturnal and sleep by day in subterranean chambers at the end of burrows 3–4 m long. They live alone except for mothers with their single young, which remain dependent for 6 months.

22.45.   What are the characteristics of elephants?

Two species of modern elephants are the only surviving representatives of *order Proboscidea*, which until 10,000 BP included the even larger mammoths and mastodons. Elephants today reach 3.5 m in height and can weigh 6 tons. Fossils tell us that proboscideans played around for some time with a combination of pendulous nose ("trunk") and various pairs of teeth elongated into curved tusks, which could operate in conjunction for uprooting trees and similar mischief by which foliage out of reach can be put in reach. The most successful proboscideans had two upper incisors enlarged as tusks. The trunk-tusk combination has made the male Asian elephant a useful domestic animal for carrying hardwood logs out of rain forest terrain too difficult for wheeled transport. The trunk is also used to pick up small food items and to suck up water to be squirted into the mouth or over the back. Aside from their two upper incisors, modern elephants have only one or two large, ridged molars in each half-jaw at a time; as the more anterior tooth wears down, it is gradually pushed forward by the one behind and eventually falls out, this process continuing with new molars moving into position as if on conveyor belts, until seven sets have been used up, after which the elephant will starve if it has not already died at the end of a normal life span of 50 years (African) or 70 years (Asian).

Elephants are sometimes called "subungulates" because their three, four, or five toes bear small hooflike nails, but they are actually subdigitigrade, for they walk on a pad behind the toes (like a human on "tiptoe"), with the wrist or ankle joint elevated above the ground. The skeleton gains strength by having spongy bone rather than marrow filling the interior of the bones. The skin is sparsely haired, loose, and thick (warranting the name "pachyderm"). Mature males join the females only for mating, and a single precocial 90-kg baby is born after a gestation period of 510–720 days.

Example 29:   The larger African elephant is distinguished from the Asian by having tusks in both sexes rather than only in males, longer limbs with only three hind toenails instead of five, larger ears, a concave rather than convex back, and two "fingers," instead of one, projecting from the tip of the trunk. It is far less domesticable than the Asian species and is seriously endangered by poachers seeking ivory, as well as by confinement to preserves where any increase in their numbers may necessitate killing off animals in excess of what the preserve can sustain. Unfortunately, elephants have strong social bonds, living in matriarchal herds of females and juveniles with a strong sense of group protection, so it is sometimes considered more humane to kill an entire group than leave survivors, which seem to mourn the fallen.

22.46.   What are hyraxes?

Hyraxes are tailless Afroasian mammals of the order *Hyracoidea*. They resemble short-eared rabbits and live in colonies of up to 50, communicating by shrill screams and barks. Rock hyraxes live amongst boulders and eat low vegetation; tree hyraxes climb trees after foliage, using rubbery footpads since hyrax toenails are hooflike and maladapted for climbing. The forelimbs have four digits, the hind limbs three. Certain aspects of the skeleton, especially the feet, are remarkably elephantine, and these small mammals are considered to share a close common ancestor with proboscideans. The incisors (one above, two below, on each side) are oddly shaped, the upper ones being pointed like little tusks; they are separated from the cheek teeth by a diastema as in rodents. A litter of two to three is born after an amazingly long gestation period of 225 days (compared with 30–35 days for rabbits, which are of comparable size).

22.47.   What are the characteristics of sirenians?

Order Sirenia includes *manatees*, which frequent the Atlantic coastlines and coastal rivers of equatorial Africa and the Americas from Guyana to South Carolina, and *dugongs*, now restricted mainly to the north coast of Australia. The gigantic 7.5-m *Steller's sea cow* of the arctic Pacific may be extinct since 1768, after just 46 years of being exploited for meat and blubber. Sirenians resemble cetaceans in living mostly in social groups and in overall shape, having streamlined, fusiform bodies with a neckless look, horizontal tail flukes that are moved up and down, no ear pinnae or hind limbs, flipperlike arms, chambered stomachs, and absence of hair except for muzzle bristles. In both groups the single calf is precocial. Sirenians differ from cetaceans in being herbivores with a mobile upper lip used for drawing in vegetation and in having only six neck vertebrae, remarkably elephantine skulls with a shortened lower jaw and a pair of stout, peglike upper incisors, nostrils on the snout instead of blowholes atop the head, a pair of nipples that are not abdominal but located in the armpit, and a movable elbow joint that allows the female to clasp her youngster to her breast (reportedly a dugong habit that may possibly have caused woman-hungry, myopic sailors to originate the mermaid legend). The fossil record suggests that sirenians, proboscideans, and hyraxes originated in Africa from a common ancestor early in the Cenozoic, at a time when that continent was isolated from other land masses.

22.48   What are the characteristics of order Perissodactyla?

Order Perissodactyla includes odd-toed hoofed mammals that are unguligrade, walking on the keratinous hoof that covers the terminal phalanx of each functional toe; they are herbivorous and have simple stomachs.

(*a*) *Equidae*, the horse family, originated with little *Eohippus* (*Hyracotherium*) that lived in Eurasia as well as in North America, but further evolution of the family took place on the latter continent, where a fossil series displays progressive toe reduction from the Eocene *Eohippus*, with four front and three rear digits, to modern *Equus*, with a single much-elongated third digit (see Fig. 5.8). Before becoming extinct in North America only a few thousand years ago, horses had spread across an ice age land bridge to Eurasia and diversified into several *Equus* species: horse, ass, and zebras, all inclined to live in herds.

(*b*) *Tapiridae*, the tapirs, once inhabited Eurasia and both Americas but require a warm climate and have come to be confined to the tropics of Asia and America, where they browse in wet forests, mostly solitarily. Like archaic horses, their front feet have four digits, the hind feet three. Superficially they resemble large furry swine, with an elongate snout overhanging the mouth, a very short tail, and large erect ears, but pigs belong to an entirely different order.

(*c*) *Rhinocerotidae* was once widespread throughout Eurasia, Africa, and North America, and fossil forms show toe reductions from four to three, with the median toe becoming enlarged. Rhinoceroses today are massive, the Indian species standing 2 m tall and weighing 2.5 tons, but the gigantic Oligocene *Baluchitherium* stood over 5 m tall, like a ponderous giraffe. Modern rhinoceroses are limited to two African and three Asian species, all severely endangered because human activities are preempting their habitats and because their horns (which consist of hairlike strands cemented together) find a lucrative market in the orient, where they are pulverized and sold as an aphrodisiac (a claim that fails scientific testing). The Indian and Javan species have a single long horn on the nose, the Sumatran rhino has two small horns, and the two African species have one long nose horn and a second shorter one on the forehead.

22.49.   What are the characteristics of artiodactyls?

Order Artiodactyla includes even-toed hoofed mammals. These diverse ungulates have in common two or four digits per foot with the functional axis of the leg passing between digits III and IV, premolars that are usually simpler than molars, and a typically trilobate rear molar. They are native to all continents except Australia, ranging from the hare-sized mouse deer to the 4-ton hippopotamus, the largest of all living terrestrial mammals after the elephants. Artiodactyls of two suborders have chambered stomachs (Fig. 22.14) and when at leisure regurgitate food to be remasticated at length; these cud chewers profit by the activities of bacteria that digest cellulose in the stomach chambers.

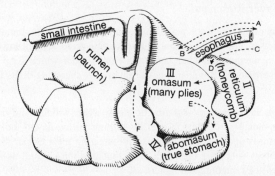

Fig. 22.14   Chambered stomach of a ruminant, the cow. Grasses and roughage (A) pass first into the rumen (I), while grains (C) pass straight into the reticulum (II). Roughage is regurgitated (B) from the rumen into the mouth for further mastication (cud chewing) and, when reswallowed (C), passes into the reticulum (D). Subsequently, the food being digested passes successively to the omasum (III) and abomasum (IV) and finally goes (F) into the small intestine. This leisurely passage allows symbiotic bacteria ample time to digest the cellulose to sugars. *(From Storer et al.)*

22.50.   What characterizes members of suborder Suiformes?

The pigs, peccaries, and hippopotamuses of suborder Suiformes have four well-developed toes, but in pigs and peccaries only the middle two contact the ground in walking. Their canines are enlarged into curved tusks of sometimes formidable length (up to 1 m in hippos). They have simple stomachs and do not chew a cud. Family *Suidae* includes the African warthog and the European wild boar (progenitor of domestic swine). *Tayasuidae* includes the American peccaries. Pigs and peccaries are omnivorous and will eat anything they can catch, the latter being immune to rattlesnake venom. Even the herbivorous warthog will scavenge off carcasses.

Family *Hippopotamidae* includes only two species, both African, one king-sized and one pygmy. Both are amphibious and eat water plants but also come ashore at night to graze. When frightened, hippos flee into the rivers in which they live; they can submerge for up to 30 minutes, though the usual time is 3–4 minutes. Pygmy hippos frequent forest streams and when alarmed run ashore into undergrowth. Both species are nearly hairless except for muzzle bristles, and the skin is kept moist and glossy with an oily exudate of cutaneous glands. A thick layer of blubber insulates and provides buoyancy, making the hippo remarkably graceful as it walks along the river bed.

22.51. What are camels?

Suborder Tylopoda includes the two-humped Asiatic (Bactrian) camel and one-humped Arabian camel (or dromedary), and the humpless American vicuna, guanaco, llama, and alpaca. Camels are cud chewers but differ from true ruminants in having a pair of upper incisors and a stomach with only three chambers, and in actually lacking real hooves. Instead they walk on the soft pads of the tips of two toes that have heavy toenails at the front. Camels evolved in North America and spread to Asia and the mountains of South America before becoming extinct in their homeland only a few thousand years ago. That so many large mammals (saber-toothed cats, mastodons, mammoths, imperial elephants, horses, camels, giant ground sloths) died out in North America *after* the rigors of this last ice age were safely past and in the space of only a few thousand years may well have been because of hunting pressure by prehistoric humans armed with throwing spears, who entered North America over the Bering straits about 12,000 BP.

22.52. What characterizes suborder Ruminantia?

Suborder Ruminantia is made up of cud chewers lacking upper incisors and usually the upper canines as well, while the lower canines are shaped like incisors, providing a row of four cutting teeth in each lower half-jaw. They typically have four-chambered stomachs, with the first chamber being the *rumen* from which food is regurgitated for further mastication. All but chevrotains are herbivores. Ruminants have true hooves sheathing digits III and IV. Most bear horns or antlers, at least in the male.

(*a*) *Tragulidae* includes the African and Asian *chevrotains*, or mouse deer, which stand only about 30 cm tall at the shoulder and differ from other members of the suborder by lacking horns, having only three stomach chambers, and sporting fanglike upper canines and premolars adapted for shearing meat (when these shy forest dwellers can get it). They have four visible toes with only the central two touching the ground, as in pigs, but curiously, they walk on the tips of their hooves, like quadrupedal ballerinas.

(*b*) *Cervidae*, the deer family, includes all ruminants with branching *antlers* that are shed and regrown even larger the following year. Except for reindeer (caribou), antlers occur only in males, which employ them in social fighting during the mating season. The antler consists of solid dermal bone outgrown from the frontal bone of the cranium and covered during growth by living epidermis (velvet) that dies and sloughs off when the antler has reached full size. Caribou and wapiti (American elk) live in migratory herds, while deer and moose are more solitary. Moose, to 2 m at the shoulder and over 600 kg, favor aquatic plants and like to stand submerged to head height.

(*c*) *Giraffidae* includes only the African giraffes and okapi, both with elongated necks used in reaching into the crowns of trees after foliage to be pulled off by the long, prehensile tongue. Both sexes have one to three pairs of skin-covered horns (bony outgrowths on top of the head, covered permanently by living skin).

(*d*) *Antilocapridae* is known by a single living species, the North American *pronghorn*. Both sexes have one pair of horns made up of a bony core encased by a two-pronged sheath of horny keratinized epidermis. The horny sheath is shed and regrown annually, but the bony core is permanent.

(*e*) *Bovidae* includes cattle, sheep, goats, buffalo, bison, and antelopes, which have one pair of unbranched horns that are never shed and usually occur in both sexes (but are larger in males). The horn consists of a bony outgrowth of the frontal bone, permanently covered by a sheath of very dense keratinized epidermis, and both core and sheath grow continuously from the base outward.

Chapter 23

Animal Nutrition

All creatures need *nutrients* obtained from the environment, for energy, growth, and reproduction. *Nutrition* includes all activities directly concerned with taking in and assimilating nutrients. Animal nutrition is *heterotrophic:* organic nutrients must be consumed, since animals cannot synthesize organic compounds from inorganic substrates. Most organic molecules are too large to be taken into cells without preliminary digestion, after which, unless the animal is tiny or rather inactive, it must possess means of internal transport to distribute digested food and oxygen to the tissues and carry away metabolic wastes, as well as special respiratory and excretory structures to expedite gas exchange and waste removal. Thus nutritional processes in animals commonly include *ingestion, digestion, internal transport, respiration,* and *excretion.* Animals have evolved a number of alternative adaptive pathways by which these processes are accomplished. In this chapter *common* features of these nutritional processes will be emphasized, with some additional reference to mammals.

The above processes are mere adjuncts to the central aspect of nutrition, namely, *metabolism*, the chemical reactions by which nutrients are utilized within cells, by which cells obtain energy, grow, and multiply. The *metabolic rate* (as measured by the rate at which the body uses oxygen) is the mean of the metabolic rates of different tissues as they carry on energy-yielding oxidative reactions. The activities of an organism result from the metabolic activities of its cells: generation of impulses by nerve cells, contraction of muscle cells, secretion by gland cells, and more. An organism grows by enlargement and multiplication of the cells that make up its body, and reproduction stems from the capacity of cells and cell components such as DNA to reproduce themselves.

23.1. What inorganic nutrients do animals need?

Inorganic nutrients include *water* and *minerals*.

(*a*) Heterotrophs cannot use water as a source of hydrogen atoms for photosynthesis, as autotrophs can, but it nevertheless makes up 50–90 percent of an animal's body, serving as a solvent and medium for metabolic processes, and as the principal constituent of bodily fluids such as blood. It is vital for internal transport, urine formation, and operation of hydrostatic skeletons (as in burrowing earthworms and digging clams).

Water participates in many chemical reactions, such as chemical digestion, or *hydrolysis*, in which one water molecule is used up (as —H and —OH) for each chemical bond broken; e.g., hydrolyzing a disaccharide molecule such as sucrose to two molecules of simple sugar uses up one molecule of water: $C_{12}H_{22}O_{11} + H_2O \rightarrow 2C_6H_{12}O_6$.

(*b*) *Minerals*, mostly in ionic form, play many vital roles (see Table 23.1). Some act as enzyme *cofactors* (prosthetic groups) that must attach *permanently* to enzyme molecules to activate them.

23.2. What organic nutrients do animals require?

Organic nutrients may or may not have to be digested, according to their size when ingested. Solids must be liquefied, and molecules too large to absorb as is must be broken down to smaller ones. *Protein, lipid, nucleic acid, polysaccharide,* and *disaccharide* molecules require hydrolysis; *amino acids, monosaccharides,* and *vitamins* do not, because they are already small enough to be handled by cellular membrane transport systems.

Vitamins usually have to be ingested by animals, although some can be synthesized by the tissues (e.g., vitamin D by human skin) or by symbiotic gut bacteria. Vitamins are small organic molecules that play many specific metabolic roles (Table 23.2). They often serve as *coenzymes* that attach *temporarily* to enzyme molecules during some particular phase of their activity. The clinical syndromes of vitamin deficiency are given as in humans, but the basic metabolic deficit would affect the health of any animal incapable of synthesizing the vitamin in question.

Table 23.1   Micronutrient Elements in Animal Metabolism

Element		Function
S	sulfur	Constituent of proteins, vitamins (thiamine, biotin)
P	phosphorus	Constituent of nucleotides, DNA, RNA, and phospholipids; as phosphate ion forms part of skeletal matrix
Ca	calcium	Mainly as calcium ion contributes to skeletal materials; enzyme cofactor; needed for blood clotting and nerve and muscle functioning
Mg	magnesium	Cofactor for activation enzymes, carboxylases, and dehydrogenases
K	potassium	Main intracellular cation; involved in action potentials; promotes protein synthesis
Na	sodium	Major extracellular cation; involved in action potentials; promotes intranuclear protein synthesis
Cl	chlorine	Major inorganic anion; helps maintain isotonicity of body fluids; component of HCl secreted by vertebrate gastric glands; catalyst in electron transfers
Fe	iron	Component of hemoglobin, cytochromes, myoglobin, catalase, hemerythrin
Mn	manganese	Cofactor of many enzymes (dipeptidases, carboxylases, phosphorylases, etc.)
Cu	copper	Constituent of hemocyanin; needed for synthesis of iron-containing compounds; enzyme component
Zn	zinc	Cofactor of carbonic anhydrase, catalase, phosphatases; mitotic accelerator, concentrated in spindle
Si	silicon	Component of feathers, siliceous sponges, radiolarian tests
Co	cobalt	Constituent of vitamin B_{12}; promotes heme synthesis; activates peptidases and other enzymes
Mo	molybdenum	Cofactor for flavoprotein enzymes
I	iodine	Constituent of thyroid hormones, spongin, coral skeletons, and the purple dye of some mollusks
Ni	nickel	Needed for insulin synthesis; enzyme activator; antianemic factor

23.3. How do animals take in nutrients?

Nutrients are taken in by absorption, phagocytosis, and oral ingestion.

(a) *Absorption* is the process by which water, ions, and molecules (even macromolecules) can be taken directly into cells; it involves both movement across cell membranes by diffusion, osmosis, and carrier-mediated transport, and engulfing of larger molecules by pinocytosis (see Questions 2.20–2.23). *Transcellular transport* is characteristic of metazoan cells that engage in absorption: taking in materials on their free surface, they then discharge them, modified or unmodified, through their opposite surface into the tissue fluids or bloodstream for distribution.

(b) Most animals are *holozoic*, ingesting food in *solid* as well as liquid form. Solid particles of subcellular dimensions can be ingested by phagocytosis, enclosed by a food vacuole that pinches off into the cytoplasm; this is important to animals that rely on intracellular digestion: protists and such metazoans as sponges, cnidarians, flatworms, and certain mollusks. Larger morsels are taken in by oral ingestion, through a mouth.

> **Example 1:**   Many animal adaptations assist oral ingestion: teleost jaws that shoot out and create a strong suction; a muscular sucking pharynx or proboscis (turbellarian flatworms, leeches); a sticky, protrusible tongue (anteaters, chameleons, anurans); an eversible proboscis (nemerteans); cilia that draw particles to the mouth (lophophorates, clams, sedentary polychaetes); tentacles with stinging capsules (cnidarians), adhesive cells (ctenophores), or suckers (cephalopods) that capture prey and bring it to the mouth; piercing, pinching, and manipulative mouthparts (arthropods), and so forth.

23.4. Where does digestion take place?

Digestion may be *intracellular* or *extracellular* (Fig. 23.1).

Table 23.2   The Vitamins and Their Characteristics

Name, Formula, and Solubility	Important Sources	Functions	Result of Deficiency or Absence (in humans, except as noted)
Lipid-soluble vitamins: **A** ($C_{20}H_{30}O$), anti-xerophthalmic	Plant form (carotene, $C_{40}H_{56}$) in green leaves, carrots, etc.; is changed in liver to animal form ($C_{20}H_{30}O$), present in fish-liver oil (shark); both forms in egg yolk, butter, milk	Maintains integrity of epithelial tissues, especially mucous membranes; needed as part of visual purple in retina of eye	Xerophthalmia (dry cornea, no tear secretion), phrynoderma (toad skin), night blindness, growth retardation, nutritional roup (hoarseness) in birds
D ($C_{28}H_{44}O$), antirachitic	Fish-liver oils, especially tuna, less in cod; beef fat; also exposure of skin to ultraviolet radiation	Regulates metabolism of calcium and phosphorus; promotes absorption of calcium in intestine; needed for normal growth and mineralization of bones	Rickets in young (bones soft, yielding, often deformed); osteomalacia (soft bones), especially in women of Asia
E, or tocopherol ($C_{29}H_{50}O_2$), antisterility	Green leaves, wheat-germ oil and other vegetable fats, meat, milk	Antioxidative; maintains integrity of membranes	Sterility in male fowls and rats, degeneration of testes with failure of spermatogenesis, embryonic growth disturbances, suckling paralysis and muscular dystrophy in young animals
K ($C_{31}H_{46}O_2$), antihemorrhagic	Green leaves, also certain bacteria, such as those of intestinal flora	Essential to production of prothrombin in liver; necessary for blood clotting	Blood fails to clot
Water-soluble vitamins: **B** complex Thiamine (B_1) ($C_{12}H_{17}ON_4S$), antineuritic	Yeast, germ of cereals (especially wheat, peanuts, other leguminous seeds), roots, egg yolk, liver, lean meat	Needed for carbohydrate metabolism; thiamine pyrophosphate an essential coenzyme in pyruvate metabolism (stimulates root growth in plants)	On diet high in polished rice, beriberi (nerve inflammation); loss of appetite, with loss of tone and reduced motility in digestive tract; cessation of growth; polyneuritis (nerve inflammation) in birds
Riboflavin (B_2) ($C_{17}H_{20}O_6N_4$)	Green leaves, milk, eggs, liver, yeast	Essential for growth; forms prosthetic group of FAD enzymes concerned with intermediate metabolism of food and electron-transport system	Cheilosis (inflammation and cracking at corners of mouth), digestive disturbances, "yellow liver" of dogs, curled-toe paralysis of chicks, cataract
Nicotinic acid, or niacin ($C_6H_5O_2N$), antipellagric	Green leaves, wheat germ, egg yolk, meat, liver, yeast	Forms active group of nicotinamide adenine dinucleotide, which functions in dehydrogenation reactions	Pellagra in humans and monkeys, swine pellagra in pigs, blacktongue in dogs, perosis in birds

(Table continues on page 376)

Table 23.2   The Vitamins and Their Characteristics (*continued*)

Name, Formula, and Solubility	Important Sources	Functions	Result of Deficiency or Absence (in humans, except as noted)
Folic acid ($C_{19}H_{19}O_6N_7$)	Green leaves, liver, soybeans, yeast, egg yolk	Essential for growth and formation of blood cells; coenzyme involved in transfer of single-carbon units in metabolism	Anemia, hemorrhage from kidneys, and sprue (defective intestinal absorption) in humans; nutritional cytopenia (reduction in cellular elements of blood) in monkeys; slow growth and anemia in chicks and rats
Pyridoxine (B_6) ($C_8H_{12}O_2N$)	Yeast, cereal grains, meat, eggs, milk, liver	Present in tissues as pyridoxal phosphate, which serves as coenzyme in transamination and decarboxylation of amino acids	Anemia in dogs and pigs; dermatitis in rats; paralysis (and death) in pigs, rats, and chicks; growth retardation
Pantothenic acid ($C_9H_{17}O_3N$)	Yeast, cane molasses, peanuts, egg yolks, milk, liver	Forms coenzyme A, which catalyzes transfer of various carboxylated groups and functions in carbohydrate and lipid metabolism	Dermatitis in chicks and rats, graying of fur in black rats, "goose-stepping" and nerve degeneration in pigs
Biotin (vitamin H) ($C_{10}H_{16}O_3N_2S$)	Yeast, cereal grains, cane molasses, egg yolk, liver, vegetables, fresh fruits	Essential for growth; functions in CO_2 fixation and fatty acid oxidation and synthesis	Dermatitis with thickening of skin in rats and chicks, perosis in birds
Cyanocobalamin (B_{12}) ($C_{63}H_{90}N_{14}O_{14}PCo$)	Liver, fish, meat, milk, egg yolk, oysters, bacteria and fermentations of *Streptomyces;* synthesized only by bacteria	Formation of blood cells, growth; coenzyme involved in transfer of methyl groups and in nucleic acid metabolism	Pernicious anemia, slow growth in young animals; wasting disease in ruminants
C, or ascorbic acid ($C_6H_8O_6$)	Citrus fruits, tomatoes, vegetables; also produced by animals (except primates and guinea pigs)	Maintains integrity of capillary walls; involved in formation of "intercellular cement"	Scurvy (bleeding in mucous membranes, under skin, and into joints) in humans and guinea pigs

Source: T.I. Storer, R.L. Usinger, R.C. Stebbins, and J.W. Nybakken, *General Zoology,* 6th ed., McGraw-Hill, New York, 1979.

23.5.   How does intracellular digestion take place?

After a food vacuole is formed, enzymes are transferred into it from the lysosomes and hydrolysates (digestive products) are absorbed through the vacuolar membrane into the cytoplasm. Indigestible residues are *egested* by fusion of the vacuole wall with the plasma membrane, which opens up and allows the vacuolar contents to be expelled.

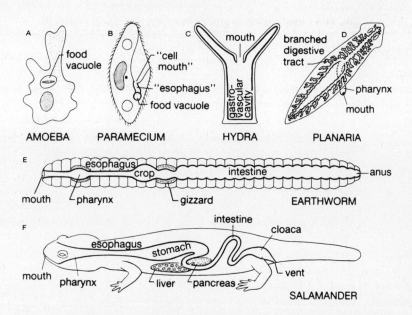

Fig. 23.1 Schematic representation of the digestive apparatus of various animals. An amoeba (A) ingests food through any part of the cell
surface and digests it intracellularly in a food vacuole; a paramecium
(B) also digests food in vacuoles, but can ingest only by way of a
definite mouth and gullet. A gastrovascular cavity with mouth but
no anus subserves both digestive and distributive functions in hydras
(C) and planarians (D); digestion is completed intracellularly, in
vacuoles within the cells lining the cavity. Digestion is extracellular
within a tubular digestive tract, complete from mouth to anus and
differentiated into various functional regions, in the earthworm (E)
and salamander (F); the latter also has a liver and pancreas that
assist digestion. (*From Storer et al.*)

23.6.   Where does extracellular digestion take place?

Extracellular digestion takes place within a *digestive tract* into which enzymes and other secretions are released. In
the *gastrovascular cavities* of radiates and flatworms, both ingestion and egestion take place by way of the mouth, and
no specialized multicellular digestive glands are present. Simple *tubular* digestive tracts extending from mouth to *anus*,
with only a single layer of columnar epithelium making up the gut wall for most of its length, are seen in the acoelomate
nemerteans and the pseudocoelomate nematodes.

The eucoelomate digestive tract is muscularized and regionalized, its parts including mouth, buccal cavity, pharynx,
esophagus, stomach, intestine, and anus. The buccal cavity may contain a tongue (vertebrates, radulate mollusks) and
teeth (vertebrates) and receive the ducts of salivary glands (vertebrates, certain mollusks, and arthropods). The esophagus
may develop specialized portions, such as the dilated crop of birds. The stomach may be divided into two or more
functional regions, such as the crop and gizzard of earthworms, the proventriculus and gizzard of birds, and the four
chambers of ruminant stomachs (see Fig. 22.14); the stomach lining may invaginate forming numerous *gastric glands*.
The intestine may be similar throughout its length or differentiated into sections such as the mammalian *duodenum*,
jejunum, and *ileum*, together making up the *small intestine*, and the *cecum, colon*, and *rectum* or *cloaca* forming the
posterior part. The duodenum secretes digestive enzymes, and food is absorbed through fingerlike *villi* lining the entire
small intestine. The blind-ending cecum is especially large in certain herbivorous birds and mammals, holding partly
digested plant matter while bacteria break down the cellulose. The colon (large intestine) itself is the site of final water
and mineral absorption, bacterial synthesis of certain vitamins, and fecal storage until defecation. The terminal chamber
of the digestive tract is either a rectum (therians and many invertebrates) or a cloaca, which also receives urinary and
genital ducts (nonmammalian vertebrates).

Table 23.3   Mammalian Digestive Enzymes and Their Functions

Source and Enzyme	Action
Salivary glands	
Ptyalin (Salivary amylase)	Breaks alternate bonds in polysaccharides, yielding disaccharides and oligosaccharides
Gastric glands	
Pepsin	Breaks internal bonds of proteins, especially those next to methionine, leucine, phenylalanine, tyrosine, and tryptophan, yielding peptide chains
Gastric lipase	Hydrolyzes fats to fatty acids and glycerol or monoglycerides
Rennin	In young ruminants only, coagulates milk
Pancreas	
Trypsin	Attacks internal bonds of proteins and peptides at carboxyl end of lysine and arginine only
Chymotrypsin	Breaks internal bonds of proteins and peptides at carboxyl end of phenylalanine, tyrosine, and tryptophan only
Carboxypeptidase	Breaks off terminal amino acids from the end of a peptide chain exposing a —COOH
Pancreatic lipase	Breaks down fats to fatty acids and glycerol or monoglycerides, aided by emulsifying action of bile from the liver
Pancreatic amylase	Breaks every other bond in polysaccharides and oligosaccharides, yielding disaccharides
Deoxyribonuclease	Hydrolyzes DNA, freeing deoxyribose nucleotides
Ribonuclease	Hydrolyzes RNA, freeing ribose nucleotides
Duodenal glands	
Aminopeptidase	Removes terminal amino acids from the end of a peptide chain exposing an amino group (—NH$_2$)
Tripeptidases	Remove one amino acid from a chain of three
Dipeptidases	Separate final two amino acids
Sucrase	Hydrolyzes sucrose to glucose and fructose
Maltase	Hydrolyzes maltose to two glucose molecules
Lactase	Hydrolyzes milk sugar to glucose and galactose
Nuclease	Splits nucleotides into nitrogenous bases and 5-carbon sugars

23.7.   What are some important accessory digestive glands?

Large accessory digestive glands open into the eucoelomate alimentary tract. In mammals, ducts from *parotid*, *submandibular*, and *sublingual* glands open into the mouth, delivering *saliva* that dissolves the food, glues the masticated food into a sticky *bolus* for swallowing, and contains a digestive enzyme (amylase). The *pancreas* secretes into the duodenum an alkaline juice containing a number of enzymes (see Table 23.3). The liver has many functions (see Question 18.34), but its main digestive contribution is *bile*, which enters the duodenum by way of the *common bile duct*.

23.8.   How does the gut musculature assist digestion?

The gut wall of eucoelomates contains layers of longitudinal and circular muscle fibers that work oppositely (see Figs. 2.4 and 14.2c); oblique fibers may also occur, as in the vertebrate stomach. Food is propelled along the tract by *peristalsis*, ringlike waves of contraction that sweep forward, pushing the food ahead. *Sphincters* (constricting muscles such as the *cardiac sphincter* at the anterior end of the mammalian stomach and the *pyloric sphincter* at its lower end) keep the food from being moved along the tract too quickly, while *segmenting* and *pendular* (back and forth) contractions of short portions of the intestine also hold back the food and mix it with digestive juices.

23.9.   How are digestive processes regulated in vertebrates?

Regulation of digestive activities involves both hormones and the nervous system. *Gastrointestinal hormones* are secreted by cells in the gut lining, mostly in response to the presence of food or distension of that part of the tract.

Gastrin, secreted by cells in the gastric mucosa especially when meat enters the stomach, circulates throughout the body, eventually returning to the stomach, where it causes the gastric glands to secrete *hydrochloric acid* and *pepsinogen*. Secreted by the duodenal lining, *enterogastrone* inhibits gastric motility in response to fat in the duodenum; *secretin* prompts the pancreas to secrete alkaline fluid; and *cholecystokinin* reduces gastric motility, makes the gall bladder contract, stimulates the flow of pancreatic enzymes, and perhaps curbs appetite by its action on the brain.

Motility is brought about by two intrinsic nerve plexuses (networks of nerve cells) that lie within the gut wall: the *myenteric plexus* between the two main muscle layers and the *submucous plexus* that innervates the muscle fibers in the submucosa and intestinal villi. These two nerve networks receive input from the autonomic nervous system, with sympathetic fibers inhibiting gut motility during emergencies. Parasympathetic nerves innervate the salivary glands, causing salivation in response to food in the mouth or even the odor or thought of food.

23.10.   What does physical digestion involve?

Physical, or *mechanical, digestion* includes all processes that alter the physical state of ingested material without changing its molecular structure. Such devices as teeth (in mammals, cutting incisors, ripping canines, grinding premolars and molars) and the *gastric mill* (chitinous stomach ridges) of certain arthropods carry out *trituration* (pulverization) of solid food. The pulverized food is then liquefied by digestive juices. In the vertebrate stomach, *hydrochloric acid* dissolves the calcium phosphate matrix of bone and cartilage; this acid also provides the pH (1.6–2.4) needed for pepsin to function. Later, the pH is made alkaline by the addition of pancreatic juice to the semiliquid food (*chyme*) in the duodenum. *Emulsification* of lipids is another important aspect of physical digestion; in vertebrates this is accomplished by *bile*, which coats fat droplets like soap, keeping them from coalescing.

23.11.   What does chemical digestion involve?

Chemical digestion consists of the *hydrolytic* breakdown of molecules, so called because one molecule of water is added, as—H and—OH, for each chemical bond broken (Fig. 23.2). For example, proteins are hydrolyzed into shorter chains of amino acids (peptides) and finally to free amino acids; if a protein molecule contains 1000 amino acids, then 999 water molecules will be broken down and incorporated into the amino acids set free.

23.12.   What types of hydrolytic enzymes bring about chemical digestion?

(*a*)   *Carbohydrases* hydrolyze carbohydrates. *Amylases* digest polysaccharides to disaccharides. *Disaccharidases* split these into monosaccharides. Thus, *sucrase* splits *sucrose* (cane or beet sugar) into glucose and fructose, *lactase* digests *lactose* (milk sugar) to glucose and galactose, and *maltase* breaks down *maltose* (grain sugar) into two molecules of glucose (see Fig. 3.5).

(*b*)   *Lipases* hydrolyze lipids (see Fig. 3.6) into monoglycerides, free fatty acids, and glycerol. In vertebrates, this takes place partly within the intestinal lumen and partly within the mucosal epithelium, which takes up microscopic fat droplets by pinocytosis; interestingly, the products of fat digestion are built up into fats again within those same epithelial cells, then secreted from their inward surface, to be taken up into the *lymph capillaries* (since they are too large to enter the blood capillaries against the opposing force of blood pressure).

(*c*)   *Ribonucleases* hydrolyze RNA into ribose nucleotides, while *deoxyribonucleases* break down DNA into deoxyribose nucleotides. *Nucleases* then separate nucleotides into their constituent sugars and nitrogenous bases (see Fig. 3.9).

(*d*)   *Proteases* hydrolyze proteins, eventually to amino acids, but this must be done in stages through the agency of a number of *proteolytic* enzymes. This is so because proteins contain some 20 kinds of amino acids, each of which has a different side chain (see Fig. 3.7).

> **Example 2:**   An enzyme's activity depends on the shape of its *active site*, into which only certain substrate molecules will fit; obviously, no one (proteolytic) enzyme will be able to form an *enzyme-substrate complex* with just any two amino acids to break the bond that joins them. Thus, in vertebrates, certain bonds internal to a protein are first attacked by the *endopeptidases pepsin, chymotrypsin,* and *trypsin*, reducing the protein molecule to a number of shorter chains (*peptides*). Pepsin mainly breaks bonds adjacent to the amino acids methionine, leucine, phenylalanine, tyrosine, and tryptophan, whereas chymotrypsin breaks bonds only at the carboxyl (—COOH) end of phenylalanine, tyrosine, and tryptophan, and trypsin can break bonds only at the carboxyl end of lysine and arginine. Next, *exopeptidases* lop off one amino acid at a time from the end of each peptide chain. The terminal amino acid at one end of the chain exposes its amino group (—NH$_2$), and so the enzyme that attacks the

(a) amylase action:

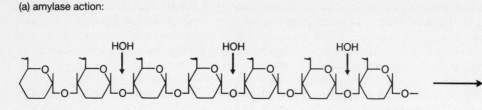

maltose and larger
fragments

some glucose

(b) lipase action:

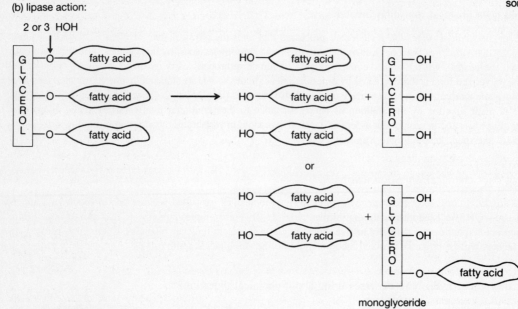

monoglyceride

(c) protein digestion

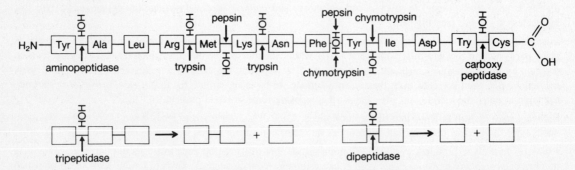

Fig. 23.2 Actions of major digestive (hydrolytic) enzymes. (*a*) Amylase splits polysaccharides at every other bond. (*b*) Lipase separates fatty acids from glycerol. (*c*) A number of proteases are needed to complete the digestion of proteins, because a given enzyme can break bonds only between certain amino acids, or at certain places in the peptide chains: trypsin, pepsin, and chymotrypsin each break specific bonds within the chain; aminopeptidase and carboxypeptidase cleave off single amino acids from opposite ends of the chain; and an assortment of di- and tripeptidases separate the last two or three amino acids left in a peptide fragment.

chain at this end is known as an *aminopeptidase*. The terminal amino acid at the opposite end exposes its carboxyl (—COOH) group, and the enzyme *carboxypeptidase* removes amino acids one by one from this end of the chain. Finally, units composed of two or three amino acids remain, and their separation is accomplished by a busy flock of *tripeptidases* and *dipeptidases* geared to deal with the specific configurations presented by various combinations of the remaining amino acid side chains. Table 23.3 summarizes mammalian digestive enzymes, their sources and functions.

23.13.   How do animals accomplish respiration?

As used here, *respiration* refers to the exchange of gases between the organism and its environment, also termed "external respiration," with "internal respiration" referring to the movement of gases between cells and the surrounding tissue fluids or bloodstream. Both phases involve *diffusion* of O_2 and CO_2 along their concentration gradients. Gaseous exchange between an animal and its environment takes place across *respiratory membranes* that must be moist, exposed to a flow of fresh air or water, and (excepting radiates and flatworms) supplied with blood capillaries or coelomic fluid through which gases are carried between the tissues and respiratory surfaces (Fig. 23.3). Respiratory systems include not only the respiratory organs themselves, which provide an extensive area of moist membrane, but also passageways for air or water leading to those organs, and *breathing muscles* that bring air or water through the respiratory organ.

23.14.   What are the respiratory structures of aquatic animals?

Aquatic animals exchange gases through the integument, integumentary extensions (parapodia, tentacles, radioles), external gills, and internal gills (see Fig. 20.2). When the flow of water past the gills is directed oppositely to the flow of blood in the gill capillaries, oxygen uptake can be highly efficient (to 80 percent in teleosts). In fishes, muscles of the jaws, gill arches, and pharynx draw water in through the mouth and out the pharyngeal clefts. The surface area of the gill filaments correlates with, and may delimit, a species' activity level; e.g., the gill area of the sluggish toadfish is only 151 mm^2/g body weight, whereas in the speedy mackerel it amounts to 1040 mm^2/g, a sevenfold increase.

23.15.   How do terrestrial animals respire?

Terrestrial animals keep their respiratory membranes moist by internalizing them as tracheal systems (see Fig. 16.9*b*), book lungs (see Figs. 15.5*b* and 15.6), or alveolar lungs as seen in mammals or as modified in birds (see Fig. 21.14). Expiratory and inspiratory muscles compress and expand the cavity in which the respiratory organs lie, expelling air and pushing or sucking it in (see Fig. 21.2 for reptiles).

The mammalian respiratory system (see Fig. 22.6*a*) consists of air passageways extending from the exterior to the alveolar sacs of the lungs, together with the muscles of respiration, chiefly the diaphragm (see Question 22.10). Air can enter or leave by way of either external nares or *mouth*, since the nasal and buccal passages meet at the nasopharynx into which the internal nares open. Nasal breathing is preferable, because air is filtered, humidified, and warmed before entering the lungs. Olfactory cells in the nasal lining detect odors. Air passes from the nasopharynx through the glottis into the cartilaginous larynx. The flaplike *epiglottis* closes off the glottis during swallowing, and *vocal folds* within the larynx produce sound by setting up vibrations in air being exhaled. Below the larynx, the air passageway continues as the trachea, which bifurcates into two primary bronchi. These subdivide progressively to form a bronchial tree. A *ciliated mucous membrane* lines this portion of the passageway, with the ciliary beat moving mucus and inhaled particles toward the throat. Cartilage rings keep these air passages from collapsing, but they leave off at the microscopic, muscularized bronchioles, which open into the *alveolar sacs*, clusters of alveoli where gases are exchanged.

> **Example 3:**   In an average human, the passageways from exterior to bronchioles constitute about 150 ml of air-filled *anatomic dead space* in which no gas exchange takes place; life depends on each breath being deep enough that not only the air passageway but the *alveoli* are adequately ventilated. A resting person is sustained by a *tidal volume* of some 500 ml per breath, which allows 350 ml to enter the alveoli. However, during exercise respiratory centers in the hindbrain increase the rate and depth of breathing, so that a larger quantity of air is exchanged, up to the person's *vital capacity* (the total amount of air that can be forcibly exhaled after the deepest possible inhalation), which is about 4500 ml; any respiratory tract in which air is inhaled and exhaled along the same route leaves *residual air* in the system; in humans this amounts to about 1500 ml, so that total lung capacity is about 6000 ml.

Alveoli consist of a squamous epithelial layer offering little impedance to diffusion of gases between the air and pulmonary capillaries, but oxygen uptake is also facilitated by (1) a *surfactant* that makes the alveolar membrane more permeable to O_2 and (2) a specific *cytochrome* in the epithelial cytoplasm that binds absorbed O_2 until it can be carried away in the bloodstream, thus maintaining a steep concentration gradient for the gas.

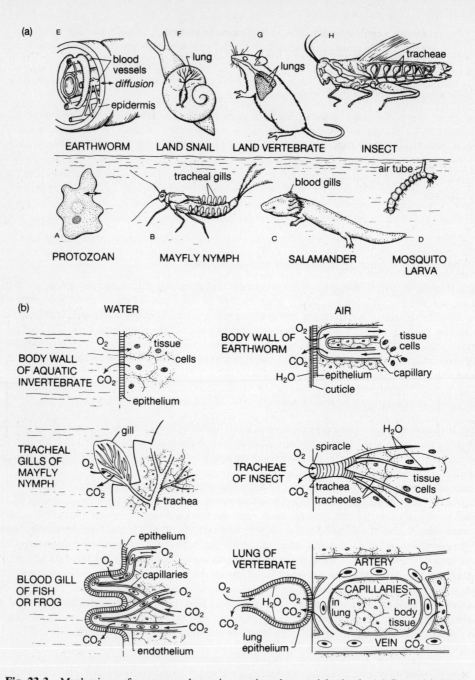

Fig. 23.3 Mechanisms of gaseous exchange in aquatic and terrestrial animals. (*a*) General features of respiratory mechanisms; A, gases diffuse through plasma membrane of a protozoan. B, mayfly nymph has thin-walled extensions from is air tubules (tracheae) known as "tracheal gills." C, aquatic salamander has feathery external gills filled with capillaries. D, mosquito lava breathes air by way of a "snorkel." E, gases diffuse through the moist integument of an earthworm and into the blood vessels for transport. F, blood vessels ramify in the wall of a saclike lung formed by the mantle of a pulmonate land snail. G, mammal has lungs filled with microscopic alveoli where gases diffuse between air and bloodstream. H, insect has internal system of air sacs and tubules (tracheae) that conduct air between spiracles at the body surface and deep-lying body tissues. (*b*) Microscopic detail (schematic) of various gas-exchange mechanisms. [(*a*) *from Storer et al.*]

23.16. Why do most animals require a means of internal transport?

Digested nutrients and oxygen must be transported throughout the body and, when possible, preferentially distributed to tissues most in need. Metabolic wastes must be carried away from the tissues, since metabolic processes are quickly blocked if such wastes are allowed to accumulate. Hormones from endocrine glands must be circulated to distant tissues that they affect. Transport systems involve specialized cells, body fluids, or both.

23.17. How are cells involved in internal transport?

(*a*) Cells that transport gases contain *respiratory pigments*; e.g., erythrocytes filled with hemoglobin occur in vertebrates and certain invertebrates such as annelids. Other annelids instead possess pink corpuscles loaded with hemerythrin. Cells containing respiratory pigments engage in O_2 transport. They occur in the bloodstream of some animals (vertebrates, annelids) and in the coelomic fluid of others (sipunculids). Mammalian erythrocytes transport not only O_2 but CO_2 in loose combination with hemoglobin (the former as *oxyhemoglobin*, the latter as *carbaminohemoglobin*); they also take care of CO_2 by converting it to bicarbonate ion (HCO_3^-): an enzyme (carbonic anhydrase) bonds CO_2 with H_2O, making carbonic acid (H_2CO_3), which dissociates to H^+ and HCO_3^- in the plasma. The properties of respiratory pigments are adapted to different animals' ways of life.

> **Example 4:** The lugworm, *Arenicola*, burrows in mud where very little oxygen is available; it survives under these conditions because its hemoglobin has a high affinity for O_2 and is 90 percent saturated at O_2 tensions of only 5–10 mmHg, whereas human hemoglobin is 90 percent saturated at 80 mmHg.

(*b*) Nutrient-transporting cells include annelidan eleocytes, detached chloragogen cells loaded with glycogen and fats, which migrate to sites of regeneration and healing. In sponges, amoebocytes store food and transport it through the mesohyl to sites of need, such as maturing ova. Ascidian vanadocytes concentrate the element vanadium and migrate into the tunic, where they disintegrate, freeing the vanadium needed for cellulose deposition.

(*c*) *Excretophores* ("waste bearers") are cells, often mobile, that collect and transport wastes. Coelomic amoebocytes perform this function in echinoderms, removing the concentrated wastes by suicidally creeping right out of the animal's body. Sipunculid *urns* (see Question 12.12), vase-shaped clusters of a few peritoneal cells, break free and wander about the coelom, collecting waste particles that they transport to the ciliated openings of the kidney tubules.

23.18. How are bodily fluids used for transport?

(*a*) The gastrovascular cavities of radiates and flatworms provide channels for the distribution of nutrients as well as their digestion; they may be highly branched so that no tissue lies far from a source of food. Pseudocoelomates and eucoelomates that lack blood-vascular systems rely on the fluid in the body cavity for internal distribution. Food absorbed through the thin walls of the alimentary tube diffuses throughout the body cavity, reaching both internal organs and the inner surface of the body wall.

(*b*) Circulatory systems are used for fluid transport and may be open or closed (Fig. 23.4*a*).

23.19. How does an open circulatory system operate?

The open circulatory system, as seen in arthropods and most mollusks, consists of a heart that collects blood from the tissue spaces (hemocoel) by sucking in blood through ostia in the heart wall; the heart then pumps the blood out into arteries that simply open up at their ends, allowing blood to flow out into the hemocoel (see Figs. 13.3, 15.5*b*, 16.2*c*). Venous channels or vessels sometimes collect oxygenated blood from the respiratory organs and carry it to the heart, but true capillaries are missing. In very active animals such as insects an open circulatory system transports nutrients and wastes, but it does not meet their needs for oxygen, which reaches the tissues by way of a separate tracheal system of air tubules (see Fig. 16.9*b–c*).

23.20. How does a closed circulatory system function?

Closed circulatory systems occur in nemerteans, annelids, cephalopods, echiurids, cephalochordates, and vertebrates. Blood, propelled by a pulsatile heart (or hearts) or by peristaltic vessels, passes through a series of arteries that eventually

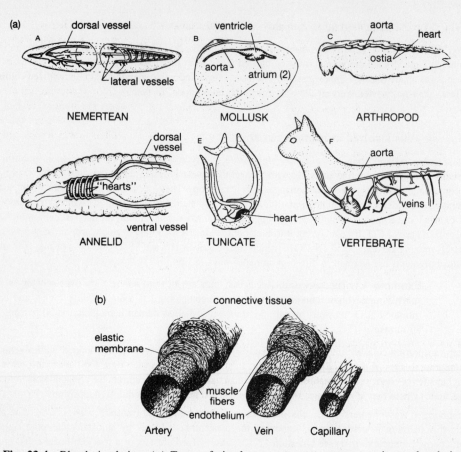

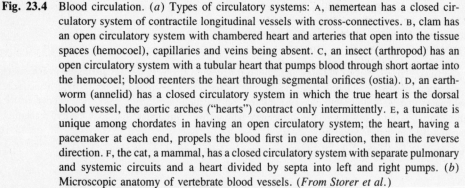

Fig. 23.4 Blood circulation. (*a*) Types of circulatory systems: A, nemertean has a closed circulatory system of contractile longitudinal vessels with cross-connectives. B, clam has
an open circulatory system with chambered heart and arteries that open into the tissue
spaces (hemocoel), capillaries and veins being absent. C, an insect (arthropod) has an
open circulatory system with a tubular heart that pumps blood through short aortae into
the hemocoel; blood reenters the heart through segmental orifices (ostia). D, an earthworm (annelid) has a closed circulatory system in which the true heart is the dorsal
blood vessel, the aortic arches ("hearts") contract only intermittently. E, a tunicate is
unique among chordates in having an open circulatory system; the heart, having a
pacemaker at each end, propels the blood first in one direction, then in the reverse
direction. F, the cat, a mammal, has a closed circulatory system with separate pulmonary
and systemic circuits and a heart divided by septa into left and right pumps. (*b*)
Microscopic anatomy of vertebrate blood vessels. (*From Storer et al.*)

lead into capillaries, where materials are exchanged between the bloodstream and surrounding tissue fluids. Veins then
collect the blood, and the circuit continues.

To review the organization and general structure of vertebrate and mammalian circulatory systems, see Questions
18.36–18.44 and 22.11. Blood and lymph vessels are lined by a continuous layer of squamous epithelium known as the
endothelium, encircled by connective tissue (elastic and collagenous) and smooth muscle (Fig. 23.4*b*). Endothelium alone
makes up the microscopic blood and lymph capillaries, the only sites at which materials can move between the tissue
fluids and the circulatory system. Blood capillaries are so minute that erythrocytes must squeeze through in single file;
they present such an enormous cross-sectional area that all the blood in the body could not possibly fill all of them at
once; e.g., in a resting dog only 5 percent of the blood is in the capillaries at any one time, with 20 percent in the arteries
and 75 percent in the veins.

The above-noted disparity between arteries and veins is partly due to (1) arteries having a smaller bore than the veins
that often parallel them and (2) the presence of extra veins, which form a *collateral* venous circulation. However, when

we consider that a well-vascularized tissue such as fat may contain some 700 km of capillaries per kg, it seems incredible that only 5 percent of the blood supply could be found in the capillaries. The answer lies in a sophisticated system of *vasomotor control* that operates mainly at the small arteries and arterioles.

> **Example 5:** Many capillary beds can be closed off by a ring of smooth muscle (*precapillary sphincter*) at their arteriolar end and the blood shunted directly from arteriole to venule by way of a cross-connecting *anastomosis*. A capillary bed cannot be closed off indefinitely, or the cells which it serves would perish, so the trick is for precapillary sphincters to open and close as needed. Local factors such as accumulation of CO_2 and wastes cause the precapillary sphincter to open briefly, then it closes down again.

Vasomotor nerves* innervate the smooth musculature of the vessels, especially the smaller arteries. *Vasoconstrictor nerves* that reduce blood flow by diminishing the vessel's diameter serve all parts of the body, but only in skeletal muscles are the arterioles *doubly innervated*, both with vasoconstrictor fibers and with sympathetic *vasodilator* nerves that make the vessels expand.

> **Example 6:** During strenuous muscular activity, blood distribution changes tremendously from the resting situation: in humans the quantity of blood to the skeletal muscles increases 10-fold (1000 percent), and to the skin (involved in cooling) about 400 percent. Conversely, blood flow to the abdominal organs decreases to about 40 percent of the usual amount. On the other hand, the blood supply to the brain remains remarkably steady.

23.21. How can the heart adjust to changing body needs?

The vertebrate heart (see Figs. 22.7 and 18.12) responds in two ways to the demands imposed by strenuous activity. During exercise, venous return to the heart is speeded up, and the incoming blood overfills the heart, stretching it. Stretched cardiac muscle responds with a *stronger* contraction, increasing the *stroke volume* (the amount of blood expelled from the heart per beat). The *rate* of the heartbeat can also be adjusted, for although the heart beats independently of nervous stimulation, it is still supplied with both *parasympathetic* and *sympathetic* fibers. The parasympathetic fibers of the vagus nerve act mainly on the pacemaker, slowing it down. Sympathetic nerves, on the other hand, penetrate throughout the entire myocardium (heart muscle) and cause the heart to beat *both* more powerfully and more rapidly. This effect is enhanced and prolonged by epinephrine (adrenalin) from the adrenal glands. At all times two *reflex arcs* fine-tune heart action against rate of venous return.

> **Example 7:** The two reflex arcs work as follows: pressure receptors in the wall of the right atrium respond when the atrium is stretched by venous blood entering more rapidly than it is being pumped out; nerve impulses from these receptors to the brain stimulate the *cardiac accelerator* center, and sympathetic impulses are relayed back to the heart, speeding it up. Then as blood is pumped more rapidly and forcibly into the aortic arch, pressure receptors in the wall of that vessel send impulses to the *cardiac decelerator* center in the hindbrain, which initiates parasympathetic vagal output, slowing the heart down, thus helping keep its action continuously responsive to need by way of these two opposing reflexes.

23.22. How can hemorrhage be controlled?

Closed circulatory systems present a problem: the blood they contain is under some degree of pressure, and if any vessels rupture, a *hemorrhage* may ensue. In both invertebrates and vertebrates the initial response to hemorrhage is for the damaged vessel itself to contract and close off. This usually halts blood loss in invertebrates and lower vertebrates, which have low blood pressures, but higher vertebrates require two further measures: production of *platelet plugs*, and blood *coagulation*. *Platelets* are enucleated bits of cytoplasm fragmented from cells in mammalian red bone marrow; their functional equivalents in nonmammals are nucleated *thrombocytes*. They first combat hemorrhage by aggregating at any break in a blood vessel, agglutinating, and plugging the leak with their sticky little bodies. In addition, they liberate chemicals (serotonin and epinephrine) that promote localized vasoconstriction. Last, they expose a surface phospholipid that aids coagulation.

> **Example 8:** Coagulation of human blood results from a complex sequence of events involving some 13 factors, mostly plasma proteins in the form of inactive proenzymes (e.g., the antihemophiliac factor). The sequence commences when damaged cells release a substance that activates a proenzyme

known in humans as the *Hageman factor*. This sets off an *enzyme cascade*, in which each enzyme serves to activate the next until finally the inactive enzyme *prothrombase* is activated to *thrombase*, which fragments the plasma protein *fibrinogen* into *fibrin* threads that form the clot. The adaptive value of such a cascade is *amplification:* a pyramid effect that need begin with only a few activated molecules of the Hageman enzyme, but snowballs rapidly to the simultaneous activation of the countless thrombase molecules needed to form a clot so rapidly that it will not be swept away by the flowing blood. Each enzyme in the cascade can activate many thousands of molecules of the next enzyme in the series.

23.23.   What is involved in excretion?

Excretion is removal of waste products of metabolism, crucial since accumulated end products affect the *direction* of reactions, and thus impede further metabolism. Furthermore, when nitrogenous molecules are degraded, quite toxic wastes are often formed. However, excretory structures in animals are concerned not just with waste removal, but also with *homeostasis*, the maintenance of a dynamic equilibrium with respect to the chemical composition of body fluids. Thus, they may function also in *osmoregulation*, maintenance of water balance between organism and environment, and in excretion of *any* solute present in *excess*, even when useful in lesser concentrations. (For example, glucose may appear in the urine of a nondiabetic, after a candy binge.) At the same time, the excretory systems even of invertebrates appear widely capable of *conserving useful materials* so that the urine contains little or nothing of these.

23.24.   How do animals get rid of metabolic wastes, osmoregulate, and maintain homeostasis?

In protists, many aquatic invertebrates, and teleosts, the main nitrogenous waste is *ammonia*, a gas that readily passes out through the body surface, so that for these *ammonotelic* (ammonia-excreting) species osmoregulation may be the greatest problem, especially in freshwater habitats. Protists rely on contractile vacuoles that collect excess water and forcibly pump it out. Even when protected by water-permeable integuments, freshwater metazoans must produce copious amounts of dilute urine from their kidneys. In the sea, some animals (most invertebrates) are *osmoconformers*, their tissues being as salty as seawater, but others are *osmoregulators* and expend much energy conserving water and excreting salts. Cartilaginous fishes tolerate such high tissue concentrations of urea that their bodies are saltier than the sea and they urinate copiously, but marine bony fishes must constantly excrete salts through their gills and void as little urine as possible. Marine birds and reptiles that drink seawater rid themselves of excess salts by way of *nasal salt glands* that secrete a solution of NaCl about twice as concentrated as seawater.

23.25.   What cells are involved in selective excretion?

Cells concerned with selective excretion and retention must respond differentially to various substances, sometimes quite differently from the way other body cells respond to those substances. These choosy cells are of two main types: (1) Excretophores (see Question 23.17) are cells, often amoeboid, that concentrate wastes given off by other tissues, sometimes converting these into nontoxic storage compounds. (2) Renal tubule epithelium engages in selective transcellular transport to secrete wastes and reabsorb useful materials. In vertebrates at least, renal tubule cells are responsive to certain hormones that modify their permeability and transport functions.

23.26.   What types of kidneys occur in invertebrates?

We have seen three basic types of kidneys in invertebrates: protonephridia, metanephridia, and Malpighian tubules (Fig. 23.5). *A metanephridium* drains fluid from the coelom by way of a ciliated funnel, modifying this fluid into urine by transferring useful materials back into the bloodstream. A *protonephridium* ends internally in a duct closed off by some type of flagellated cap cell, such as a flame cell. The cap cell joins the renal tubule at a fenestrated region, where the lower fluid pressure within the protonephridium (caused by the beating of the flagella, which drives fluid along the tubules) can draw more fluid across the thin membranes into the renal tubule. Urine is therefore formed only by passage of substances *across a membrane*, either at the fenestrated region or across the membranes of the renal tubule cells themselves, which should allow the properties of those membranes to regulate urine composition at the moment of its formation, rather than by subsequent reclamation of specific materials. This is also true of the blind-ended *Malpighian tubules* of terrestrial arthropods; these tubules lie bathed in blood in the hemocoel and produce urine by transcellular

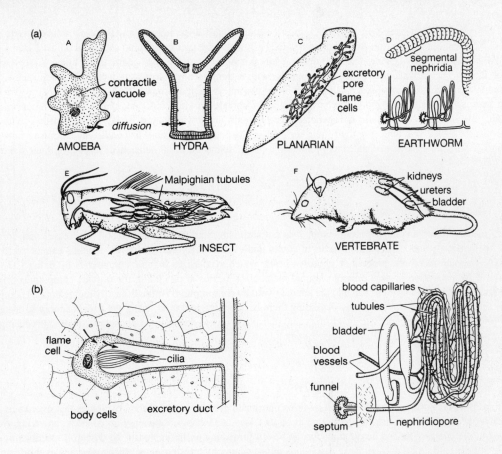

Fig. 23.5 Excretion: the removal of metabolic wastes. (*a*) Excretory mechanisms of various animals. A, in protists (e.g., amoeba) wastes diffuse through the cell surface or are expelled along with excess water by the contractile vacuole. B, diffusion through the moist body surface is adequate for radiates (e.g., hydra), which have no special excretory organs. C, flatworms (e.g., planarians) have protonephridia that end blindly in flagellated flame cells that propel the urine through the duct system to the excretory pores. D, the earthworm has segmental pairs of metanephridia that collect coelomic fluid through ciliated funnels, then modify this fluid to urine by selective reabsorption and secretion during its passage through the lengthy renal tubule. E, terrestrial arthropods (e.g., insects) have Malpighian tubules that end blindly in the hemocoel, take up wastes from the blood, and drain into the hindgut, which reabsorbs much water from the urine before it is voided from the anus. F, vertebrates have complex kidneys containing nephrons that take up a filtrate of blood plasma from glomeruli and modify this filtrate by tubular secretion and reabsorption. (*b*) Urine formation in a protonephridium (*left*) and metanephridium (*right*). (*From Storer et al.*)

transport of substances from the hemocoel into the interior of the tubules; osmotic movement of water into the urine is somewhat counteracted by the water-reabsorbing properties of the hindgut into which the Malpighian tubules open.

23.27. How does the mammalian kidney form urine?

Vertebrate kidneys contain microscopic functional units known as "nephrons," about 1 million per human kidney. A nephron ends internally as a closed, double-walled Bowman's capsule surrounding a ball of capillaries known as a *glomerulus* (see Fig. 22.8). Urine is formed initially as a filtrate of blood plasma driven out of the glomerular capillaries and into the Bowman's capsule by the force of blood pressure. In humans some 180 L of glomerular filtrate is formed per day, yet only 1–2 L of urine will be voided; thus the glomerular filtrate consists mostly of substances needing to be reclaimed. Reclamation is accomplished by the epithelial cells that make up the proximal and distal tubules, Henle's loop, and the collecting ducts. To review the structure and functioning of vertebrate and mammalian kidneys, see Questions 18.45 and 22.13. The metanephric kidney of mammals, with its long Henle's loops that cycle chloride in such a manner that mammalian urine can become much saltier than blood plasma, is the most sophisticated excretory organ in the animal kingdom and is responsible for much of the success of mammals on land, since mammals (like elasmobranchs and amphibians) are *ureotelic*: their main nitrogenous waste is toxic *urea*, which must be voided in aqueous solution. By contrast, many reptiles and birds improve their survival in dry habitats by being *uricotelic*, metabolizing nitrogenous wastes into less toxic, solid *urates* and thus producing a pasty, semisolid urine rather than relying on their rudimentary Henle's loops for water conservation.

23.28.   What characterizes animal metabolism?

Since metabolism is a cellular phenomenon, it is not surprising that the major metabolic processes seen in animal cells are fundamentally the same as those seen in plants and even prokaryotes. However, animals cannot perform the great energy-trapping metabolic process of photosynthesis, which synthesizes organic compounds by fixing CO_2 and H atoms excited by absorption of light energy. Animal metabolism begins at this point: the organic compounds originally made by autotrophic plants find their way into the animal world by consumption. Animals mostly juggle organic compounds into new combinations or obtain energy by degrading them back to CO_2 and H_2O, using the excess energy of the excited H atoms found in those compounds to bond inorganic phosphate (P_i) to adenosine diphosphate (ADP), producing high-energy adenosine triphosphate (ATP) (see Fig. 3.8a).

23.29.   Why do cells need ATP and enzymes?

ATP, a little nucleotide with a kite's tail of three phosphate groups, is one of the best things that ever came along, from the standpoint of life on earth. As profligate as Santa Claus, ATP hands out energy on all sides, powering synthetic reactions, muscle contraction and active transport and even furnishing an initial "push" to downhill reactions by which more ATP will be built. ATP does all these marvelous things simply by transferring to other molecules its third phosphate group, the one at the end of the kite's tail, together with much of the bonding energy by which this group was tied onto the end of the tail. Breaking the $\sim$(P) bond by which the third phosphate group is bonded to ADP to make ATP yields roughly *twice* the energy liberated by breaking an ordinary covalent bond such as the O—P bond that holds the second phosphate in place. When $\sim$P is transferred from ATP to some other molecule (a transfer that requires an enzyme, of course), the recipient is *activated*, that is, energized to a level that makes it *reactive* (more apt to enter into chemical reactions).

Whether a reaction takes energy (is *endergonic*) or liberates it (is *exergonic*), the reacting molecules in living systems must have available both *enzymes* and some source of *activation energy*. ATP is not the only, but certainly the most ubiquitous, source of activation energy. Enzymes reduce the amount of activation energy needed, by holding reactants in specific positions that favor the reaction (in other words, A and B are more likely to form the compound A-B if they are held together in the active site of an enzyme molecule than if they are bouncing around all over the place in solution). Within cells, some enzymes operate in solution, as during glycolysis, but the majority are fixed into membranes, forming mosaic assembly lines. Enzymes are quite specific in terms of the reactions they catalyze, because only substrates of certain shapes can conform to the shape of an enzyme's active site: a site comfortable for A and B might not accommodate A and C or F and G at all. However, enzymes are quite indifferent as to which way the reaction goes: an enzyme that facilitates the bonding of A and B to make A-B just as readily facilitates the breakdown of A-B to A and B; the reaction direction is determined by the relative quantities of A and B (the substrates) compared with A-B (the product). Enzymes speed up reactions tremendously.

> **Example 9:**   In a single minute one molecule of the enzyme *catalase* is able to break down 5 million molecules of poisonous hydrogen peroxide, while one molecule of *cholinesterase*, which operates at nerve synapses, can demolish 20 million molecules of the neurotransmitter acetylcholine.

Such enzymatic pyrotechnics, however, cannot come to pass without the energizing input of ATP: even a Roman candle first requires a match. Vital activities take a lot of matches: a single bacterium growing fast enough to divide in about a half hour has to use up some 2.5 million ATPs *per second* over this period of time for the happy event to take place on schedule. If you like to think big, consider the size of the human body, with its trillions of cells, and make a stab at how many ATP molecules *you* use and rebuild each day; although many human body cells are not growing as actively as bacteria can, we have additional needs, one being simply to maintain our body temperature.

23.30.   How do animal cells make ATP?

When ATP gives away its third phosphate group, it becomes a piteously deenergized ADP (adenosine diphosphate). In this case, it is easier to give than to receive. For the subdued ADP to regain its identity as peppy ATP, either it can be donated a ‿(P) from some high-energy compound with even greater energy content than ATP, or else it must be bonded with an inorganic phosphate group (P_i), which is thermodynamically a steeply uphill reaction (meaning it takes a lot of energy). This energy comes indirectly from the sun, for it resides within the excited electrons of H atoms in organic compounds. If you carried a bowling ball to the top of Mt. Whitney and started it rolling downhill, it would transfer a lot of energy on the way down—crack some rocks, smash some trees, break a head or two—before it came to rest peacefully at the bottom. So when these excited electrons are started "downhill," a fair amount of the energy *they* release can also be transferred, rather than being dissipated as heat. This is made possible because exergonic metabolic pathways do things one little step at a time, instead of all at once. If you burn a kilogram of sugar completely to CO_2 and H_2O in a calorimeter, a goodly portion of the solar energy trapped in the sugar molecules will be liberated as heat: 4000 Calories (kilocalories), in fact—as much as one day's diet for a physically very active man. At the cellular level, the heat released by breaking down sugar directly to CO_2 and H_2O in one moment of combustion would be lethal as well as useless. Instead, each energy-releasing step in a metabolic pathway liberates such a tiny amount of energy that only a tolerable quantity of heat is generated and most of the energy "packet" is transferred to bind P_i to ADP, restoring ATP to active status once more.

Two major metabolic pathways are used by plant and animal cells in breaking down organic compounds and transferring energy to make ATP: (1) *glycolysis* and the *Krebs citric acid cycle* produce most of the energy needed for muscle contraction; (2) the *pentose phosphate cycle* produces up to 90 percent of the energy needed by such vertebrate organs as brain and liver.

23.31.   How do these pathways transfer energy to make ATP?

Both of these pathways converge by transferring electrons (from H) to a series of electron-transfer agents (enzymes and cytochromes), located in the inner mitochondrial membranes (see Fig. 2.5) close to the tiny stalked spheres (F_1 particles, commonly nicknamed "lollipops,") that protrude into the inner mitochondrial compartment and contain the enzymes of the Krebs cycle. Electrons carried along this mitochondrial transport chain end up stripped of their excess energy: now quite unexcited, each is reunited with a proton (H nucleus) and enzymatically bonded with oxygen to make water (Fig. 23.6). For each molecule of glucose ($C_6H_{12}O_6$) that enters the glycolytic–Krebs cycle–electron-transport pathway, two molecules of ATP will be used to get things started, but about 38 will be made, for a net maximum gain of some 36 ATP molecules produced per glucose molecule broken down to 6 CO_2 and 6 H_2O. (The reason these figures are not as precise as those obtained by balancing chemical equations is that living systems do not always operate at peak efficiency, so that the yield can be less, just as your car does not always get its top gas mileage.) Of the 38 ATPs that can be produced, 4 are made during glycolysis, 2 by the Krebs cycle, and 32 by the electron chain.

Similarly, a glucose molecule broken down to 6 CO_2 and 6 H_2O by way of the pentose phosphate cycle–electron-transport pathway yields 32 ATPs, all formed by electron transport, which, subtracting 1 ATP used to provide activation energy, furnishes a net possible yield of 31 ATP.

The best way to understand these pathways is to look carefully at each step shown in Figs. 23.6 and 23.7. The latter shows only the pentose phosphate cycle, and not the electron-transport chain to which it is coupled, but Fig. 23.6 provides a blow-by-blow review of each reaction in the glycolytic pathway and Krebs cycle, together with the steps in the mitochondrial electron-transport chain as currently known. Take time to examine the structural formulas closely, and you will see something interesting: H atoms from the organic compounds being broken down are first transferred to a *hydrogen carrier* [flavin adenine dinucleotide (FAD), containing the vitamin riboflavin, or nicotinamide adenine dinucleotide (NAD) or NAD phosphate (NADP), containing the vitamin nicotinic acid (niacin)]; count up the H atoms transferred from either the Krebs or pentose phosphate cycle to the hydrogen carriers, and you will see that they number 24 per glucose molecule broken down, while a glucose molecule contains only 12. What is the source of the extra 12 H atoms? Notice that for each molecule of glucose broken down, 6 molecules of *water* also enter into these reaction cycles. Thus the additional 12 H atoms are transferred from intermediates in the cycles, after they have undergone *hydration* reactions, to wit, have taken on —H and —OH from the 6 molecules of H_2O used up. The fate of all 24 H atoms per glucose is reviewed in Questions 23.32–23.35.

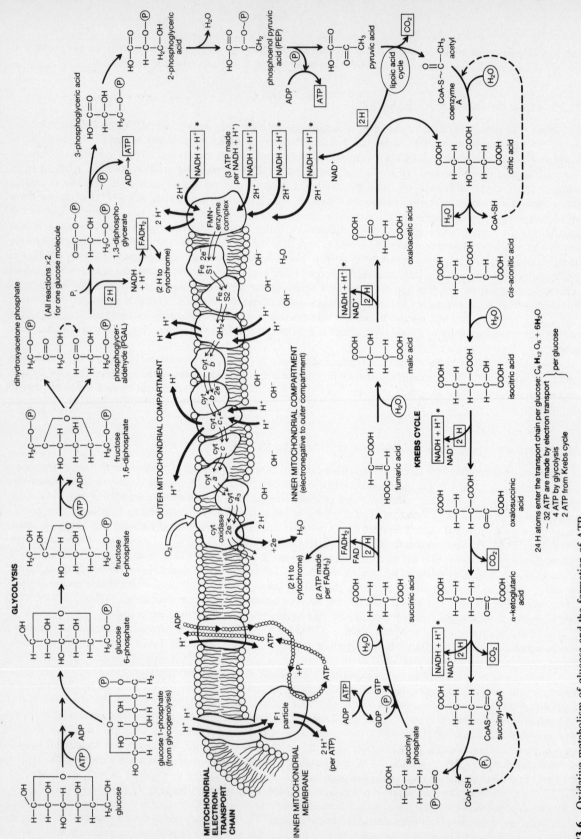

Fig 23.6 Oxidative metabolism of glucose and the formation of ATP.

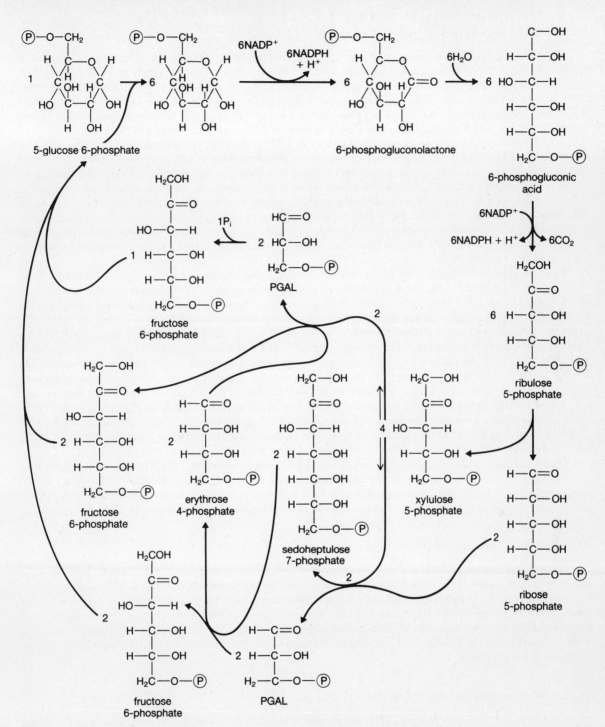

Fig. 23.7 The pentose phosphate cycle. This energy-yielding pathway is more important in many animal tissues than its alternative, the glycolytic–Krebs cycle pathway, although it can yield only 32 ATP per glucose molecule, all by the mitochondrial electron-transport chain that is common to both. Pentoses (five-carbon sugars) are intermediates in this cycle, which closely resembles the sugar-building Calvin cycle of photosynthesis, run in reverse.

23.32.   What happens during glycolysis?

Glycolysis breaks down one molecule of glucose to two 3-carbon molecules of *pyruvic acid* as the end product, in the process making 4 ATPs per glucose molecule by *substrate-level phosphorylation*, which means that high-energy phosphate compounds are generated in the reaction pathway and transfer $\sim$(P) directly to ADP from the substrate molecules (intermediates in the pathway). Glycolysis also transfers 4 H per glucose to NAD^+, making 2 $(NADH + H^+)$. NAD^+ is positively charged because it has lost an electron; when it takes up 2 H, it actually holds on to one complete H atom and one H nucleus, which collects an electron when transferred to FAD as shown, making $FADH_2$.

Glycolysis commences with an expenditure of either 1 or 2 ATPs. If *glucose* (blood sugar) is the starting compound, it must first be phosphorylated to glucose 6-phosphate (i.e., a phosphate group is added to carbon atom 6). But animal tissues, especially muscle, store up the polysaccharide *glycogen*, a branching polymer of many hundred glucose units. Unlike polysaccharide digestion in the alimentary tract, which is by hydrolysis, the intracellular breakdown of glycogen is by *phosphorolysis*, with a *phosphatase* enzyme inserting P_i into the bond being broken. Thus, *glycogenolysis* yields not glucose, but glucose 1-phosphate, which becomes rearranged into its isomer, glucose 6-phosphate, which in the next step is rearranged once more into another isomer, fructose 6-phosphate, which now takes on a $\sim$(P) from ATP and, thus, energized, readily splits into two 3-carbon molecules; phosphoglyceraldehyde (PGAL) and an isomer that is promptly reorganized into a second PGAL.

From this point on in Fig. 23.6, the fate of only one PGAL is traced through the balance of glycolysis and the Krebs cycle, so each event should be thought of as doubled: whatever happens to one PGAL happens to the other as well. Each of the 2 PGALs gives off 2 H to NAD^+, which, if enough oxygen is present for *aerobic glycolysis*, transfers them to FAD, making $FADH_2$, which will pass them into the mitochondrial electron-transport chain, as we will see below. The oxidation of PGAL by its loss of 2 H provides energy for it to bond with P_i, forming 1,3-diphosphoglycerate (1,3-DPG), a phosphate compound considerably richer in energy than ATP. In the next reaction, 1,3-DPG passes $\sim$(P) downhill to ADP, yielding 1 ATP for each of the two 1,3-DPGs. Two more steps along the way, 2-phosphoglycerate (2-PGA) undergoes a dehydration reaction, losing H_2O. This causes the molecule to undergo an internal redistribution of electrons (an *intramolecular oxidation-reduction*) that converts its O—P bond to an O~P bond and transforms it into the most energetic of all phosphate compounds: phosphoenolpyruvic acid, quite appropriately dubbed PEP. Once more, ATP is produced, this time by a downhill transfer of $\sim$(P) from each of the 2 molecules of PEP, which are thereby converted to pyruvic acid, the end product of the glycolytic pathway. The glycolytic reactions take place in solution in the cytoplasm; now, pyruvic acid must move into the mitochondria, where a great deal more energy can be recovered.

During strenuous exercise the rate of glycolysis may outstrip the oxygen supply, and ATP must be replenished by *anaerobic glycolysis*, because $(NADH + H^+)$ cannot pass on its H atoms to the electron-transfer chain, which requires oxygen; therefore, it simply gives them back to pyruvic acid, at the end of the glycolytic pathway. This converts pyruvic acid to *lactic acid*, which is toxic enough to cause muscle pain and fatigue if strenuous muscular activity is sustained.

23.33.   What happens before pyruvic acid can enter the Krebs cycle?

Acetyl is an important two-carbon moeity produced as an intermediate in many exergonic metabolic pathways. It is formed by combination of pyruvic acid with an intermediate containing the vitamin *thiamin*, followed by *decarboxylation*, i.e., elimination of CO_2 (which diffuses out of the mitochondrion as a waste). The two-carbon residue, acetyl, is bonded first with a form of the vitamin *lipoic acid*, and next with *coenzyme A* (CoA), with 2 H being transferred to make $(NADH + H^+)$. *Acetyl-CoA* is a very significant combination: not only glucose but all sorts of other organic molecules being broken down end up as acetyl-CoA, for a while. As acetyl-CoA, they can continue into the Krebs cycle or enter a pathway of lipid synthesis.

23.34.   What happens in the Krebs cycle?

The *Krebs citric acid cycle* (see Fig. 23.6) accepts one acetyl at a time, bonding it to oxaloacetic acid to produce citric acid, obtaining bonding energy from the S—C bond by which acetyl is linked to CoA. CoA (containing the vitamin *pantothenic acid*) is thus freed to pick up another acetyl. Besides moving acetyl into the cycle, CoA is also instrumental in producing the single ATP directly formed at a later point along this pathway: CoA bonds transiently with succinyl, forming the high-energy S ~ C bond that allows P_i to be incorporated by succinyl; the resulting $\sim$(P) is then transferred to GDP (guanosine diphosphate), making GTP, which in turn passes the $\sim$(P) on to ADP. More important energetically than this direct yield of ATP by substrate-level phosphorylation is the fact that the Krebs cycle transfers 8 H atoms to NAD^+ or FAD, which passes them on into the electron-transport chain. For each two "turns" of the cycle, as needed to complete the degradation of the original glucose molecule, 2 acetyls enter, 2 ATPs are generated, 4 CO_2 are given off as waste, and 16 H atoms are donated to NAD^+ or FAD. Thus the 24 H atoms mentioned above are passed to hydrogen carriers, 4 (2×2) by glycolysis, 4 (2×2) from the conversion of pyruvate to acetyl, and 16 (8×2) from the Krebs cycle.

23.35. What is the fate of these hydrogen atoms?

Eventually, the 24 H atoms per glucose will be combined with gaseous oxygen, making water, but first, both their protons and electrons will perform work that will result in the production of ATP from ADP and P_i.

Electron transport occurs within the inner mitochondrial membranes that separate the inner mitochondrial compartment from the outer compartment enclosed by the outer mitochondrial membrane. NAD^+ and FAD are *hydrogen* carriers, but the other components of the chain transport only *electrons*. In Fig. 23.6 for simplicity the fate of only 2 H atoms is traced. $(NADH + H^+)$ delivers its 2 H to an enzyme containing FMN (flavin mononucleotide, another derivative of the vitamin riboflavin), at the beginning of the pathway, so that a maximum of 3 ATPs can be formed per $(NADH + H^+)$. $FADH_2$ transfers in its 2 H at a later point along the pathway, so that no more than 2 ATP can be made per $FADH_2$. The FMN-enzyme dissociates each H atom into a *proton* (H^+) and an *electron* (e^-), and transfers only the electrons to FeS_1, the first *electron carrier* in the chain, while the two protons are liberated into the *outer* compartment. When 2 e^- are transferred to enzyme Q, this becomes electronegative and attracts 2 protons, becoming the uncharged QH_2; the attracted protons come from the inner compartment, in which water exists (as always) in both molecular (H_2O) and ionic $(H^+$ and $OH^-)$ form. Next, QH_2 transfers 2 e^- to *cytochrome b* and moves the 2 H^+ into the outer compartment. The energy of the excited electrons being transported is thus used to run a *"proton pump"* that continuously transfers protons from the inner to the outer mitochondrial compartment, leaving OH^- (hydroxyl) ions behind. Because the electrons travel in pairs, two protons at a time are transferred across. The 2 e^- are now passed on to a second cytochrome *b*, then to cytochrome c_1, at this point providing energy to move two more protons across the membrane from inner to outer compartment. The electrons are now transferred to cytochromes *c*, *a*, and a_3 and are discarded into the inner compartment, their excess energy pretty well drained. Here, they reassociate with protons and, once more in the form of H atoms, are bonded with oxygen by the enzyme *cytochrome oxidase*, making H_2O, which diffuses out of the mitochondrion as "metabolic water." One molecule of H_2O is made for each 2 H entering the electron-transport chain, or 12 per glucose broken down. Thus each pair of electrons transported furnishes enough energy to move up to 6 protons across the membrane against an osmotic and electrochemical gradient.

23.36. How does transporting protons relate to making ATP?

Anything being kept on the "wrong" side of such a gradient as mentioned above has potential energy that becomes available to do work when the substance in question is permitted to go in the opposite direction: thus protons are allowed to leak back from the outer to the inner compartment, following their concentration gradients and electrostatic attraction to OH^-, but *only* through the membrane of the F_1 lollipops, where their free energy can be tapped to bond P_i to ADP, at the rate of 1 bond per 2 protons. The maximum yield of 32 ATPs per glucose, on top of the 2 ATPs produced in the Krebs cycle together with the 4 made during glycolysis, is actually not fully realizable, because the free energy of some protons must be used to transfer ADP into the inner compartment, and ATP out, at the rate of 1 proton for each such exchange.

23.37. Are substances other than glucose broken down by the Krebs cycle?

Yes; not only does fat degradation yield acetyl, but many amino acids, stripped of their amino groups by *deaminase* enzymes, are converted into one of the organic acids in the Krebs cycle, thus entering the cycle at that point. For instance, the amino acid aspartic acid is deaminated to fumaric acid, and glutamic acid to α-ketoglutaric acid; as shown in Fig. 23.6, both these organic acids are intermediates in the Krebs cycle. However, these catabolic pathways cannot deal with the—NH_2 groups: in most aquatic animals these are converted to NH_3 (ammonia) for excretion, but in mammals amino groups from the deamination of amino acids are transferred into the *urea cycle*, where they are synthesized into the waste product *urea*: $O{=}C{-}(NH_2)_2$.

23.38. What happens in the pentose phosphate cycle?

The *pentose phosphate cycle*, interestingly, is essentially the same as the sugar-building *Calvin cycle* of photosynthesis, run backward. Remembering that enzymes can catalyze reactions in either direction, we find ourselves with a chicken-egg problem: which came first, the pentose phosphate cycle or the Calvin cycle? Probably the former, for in the very early days of life on earth, primordial cells could be *heterotrophic*, obtaining needed materials and energy by digesting one another or abiotically produced organic compounds that could lie around indefinitely in the absence of atmospheric oxygen. The lack of gaseous oxygen need not have bothered these early organisms, since even today anaerobic organisms use some element other than oxygen (e.g., sulfur) as a hydrogen acceptor at the end of the electron-transport chain. Once the enzymes that carry out the pentose phosphate cycle had evolved in primordial heterotrophs, the stage would have been set for the advent of photosynthesis, which involves essentially the same series of five-carbon sugars (pentoses). Furthermore, cytochromes, first evolved as electron carriers in energy-liberating oxidative reactions, would have been on hand to serve in the light-powered photophosphorylations that produce the ATP needed for the Calvin cycle. It would seem that the metabolic machinery for photosynthesis was present before the advent of chlorophyll, with its remarkable capacity to trap light energy in the form of excited electrons. With a constant flow of these peppy little electrons on hand to provide energy for making bonds instead of breaking them, the pentose phosphate cycle could run in the opposite direction, putting the living world on a profit economy from that time on.

The pentose phosphate cycle itself (see Fig. 23.7) takes in one molecule of $C_6H_{12}O_6$ at a time, and in six "turns" of the cycle gives off 6 CO_2 as waste and passes 12 H from glucose and 12 more H made available by the entry of 6 H_2O molecules into the cycle to the H carrier NADP (a phosphorylated form of NAD), which transfers them all to the electron-transport chain. This cycle still serves well to meet the energy requirements of many tissues, but it is a little less efficient than the glycolytic–Krebs cycle pathway, since no ATP is made by substrate-level phosphorylation, the entire maximum yield of 32 coming from electron transport.

23.39. How do animal cells use energy?

Energy is required for biosynthesis, active transport across cell membranes, and muscle contraction.

23.40. What are a few biosynthetic pathways?

Cells build many kinds of molecules, and the reaction pathways involved are too numerous and complex for more than a few to be considered here.

(*a*) *Glycogen synthesis* (Fig. 23.8*a*) involves adding one glucose molecule at a time; glucose is first phosphorylated by ATP, then bonded to the polymer via another nucleotide, UTP.

(*b*) *Transamination* (Fig. 23.8*b*) produces new amino acids by transferring—NH_2 groups from preexisting amino acids to appropriate organic acids. Animals cannot build amino acids from scratch as plants do; they can only carry out transaminations. Some 10 essential amino acids must always be included in the diet, for they cannot be produced by transamination.

(*c*) *Protein synthesis* (see Fig. 3.11) is an example of *template synthesis*, because the new molecule is assembled on a pattern provided by a preexisting molecule, in this case, messenger RNA, which in turn was formed on the template of a gene (see Fig. 3.9). Each amino acid to be incorporated in the protein is first bonded with its particular *transfer RNA*, which bears a three-base sequence (anticodon) that H-bonds with a reciprocal base triplet (codon) on the mRNA. Since this interaction takes place on the surface of a ribosome, it results in the amino acid being fitted into a ribosomal active site, in which it is freed from its tRNA and bonded onto the growing peptide chain.

23.41. How do muscles contract?

Muscle contraction uses large amounts of ATP to produce sliding movements of actin filaments along myosin filaments within cells called "muscle fibers" (Fig. 23.9).

The cytoplasm of each muscle fiber is fully packed with long, slender *myofibrils*. Each myofibril is made up of a linear series of tin-can–shaped units called *sarcomeres* that are attached end to end with cross partitions ("Z lines") between. From both sides of each Z line, thin *myofilaments* of the protein *actin* protrude their free ends toward the center of each sarcomere. The actin myofilaments are so arranged that in cross section each forms one angle of a hexagon. Centered within each hexagonal ring of six actin filaments is a stouter *myosin* myofilament, attached by *cross bridges* to the six encircling actin threads. Conversely, the hexagonal arrangement allows each actin to form cross bridges with three adjacent myosin filaments. The myosin filaments are only about 60 percent as long as the sarcomere, and are suspended midway between the partitions by their cross bridges with actin.

The actin-myosin cross bridges are formed by the club-shaped heads of the myosin molecules, which protrude at regular intervals from the surface of the myosin filament. The heads contain enzymes that break down ATP, providing

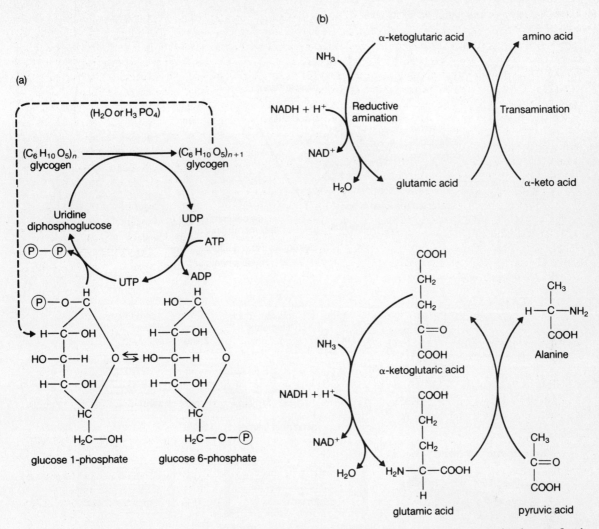

Fig. 23.8 Simple biosynthetic reactions. (*a*) Synthesis of glycogen from glucose. Each glucose molecule must first be made reactive by being phosphorylated, after which still another molecule of ATP must be broken down for the glucose to be bonded to the chain. (*b*) Amino acid synthesis. Plants can synthesize amino acids by using ammonia to aminate various organic acids; animals must consume certain essential amino acids, but synthesize others by transamination, the transfer of an amino group from one amino acid to another. The bottom reaction illustrates the formation of alanine.

energy for the cross bridges to form and then oscillate, i.e., to lengthen, pulling the attached actins from either end toward the midpoint of the myosin filament in the center of the sarcomere, then to detach from the actin, shorten, and reattach to new sites on the actin filaments, and so forth. For each cross bridge oscillation, one ATP is used up, and a second ATP is used to let the myosin head detach from its previous site and reattach farther along to form a new bridge. Thus by a ratcheting action, the actin filaments slide along the myosin filaments, pulling with them the partitions to which they are anchored, thereby shortening the entire sarcomere. Since this happens simultaneously to all the sarcomeres in all the myofibrils in the muscle fiber, the entire cell shortens. This shortening can continue only until the actin filaments, pulled together from opposite directions, butt into each other at the midline of the sarcomere. When the muscle relaxes, the cross bridges detach and the actin filaments passively slide back to their original positions.

The contractile process starts when the muscle fiber is stimulated by a release of neurotransmitter at the neuromuscular junction. This sets off an action potential similar to that of a nerve cell (see Fig. 24.3) that sweeps across the entire membrane of the muscle fiber. This membrane not only covers the outside of the fiber but penetrates inward at regular intervals by *transverse (T-) tubules* that reach far into the cytoplasm. As the propagated wave of excitation travels inward along the membrane of the T-tubules, it reaches the membrane of the *sarcoplasmic reticulum* (endoplasmic reticulum,

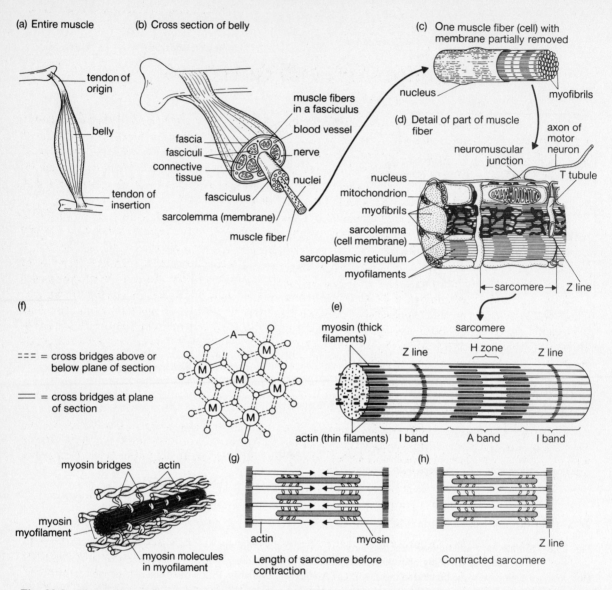

Fig. 23.9 The structure of vertebrate striated muscle. (*a*) Structure of an entire muscle, which is anchored to one part of the skeleton by a tendon of origin and attached to the bone that it moves by way of a tendon of insertion. (*b*) Internal arrangement of tissues within a muscle. (*c*) Part of a single cell (muscle fiber). (*d*) Portion of three myofibrils within a muscle fiber, showing relation to cell surface, nucleus, and motor nerve cell ending. (*e*) Enlarged portion of a myofibril showing one sarcomere and part of those adjacent. (*f*) Each myosin filament forms cross bridges with the six actin filaments surrounding it (*below*), and each actin connects with three myosin filaments (*above*). (*g*) Portion of one sarcomere showing position of myosin and actin filaments before contraction. (*h*) Portion of sarcomere showing shortening brought about by myosin bridges pulling the actin filaments toward the midline of the sarcomere.

see Question 2.24) and alters its permeability, so that stored calcium ions escape into the cytoplasm. These ions attach to a complex of two proteins, *tropomyosin* and *troponin*, that twine about the actin filament, preventing it from forming cross bridges with myosin. Once the tropomyosin-troponin complex is rendered ineffectual, the myosin-actin cross bridges form and oscillate, with the expenditure of ATP as stated above. When the muscle fiber is no longer stimulated to contract, calcium returns to the sarcoplasmic reticulum and the tropomyosin-troponin filaments again wind around the actin threads, inhibiting cross bridge formation in the relaxed muscle. Muscle contraction is one of the major products of cell metabolism in the animal body, and the more active the animal is, the greater will be its metabolic demands for ATP.

Chapter 24

Integration and Response

Integration is essential for the survival of both cells and organisms, not to mention societies. The organism must be integrated *internally* and with its *environment*. Integration relies initially on the capacity of cells to detect stimuli of both external and internal origin. Cells specialized to respond to stimuli are known as *receptors* and are usually *sensory neurons*. But even plants and sponges, which lack identifiable receptors and nerve cells, respond to stimuli, for *irritability* (sensitivity) is a universal property of life. This means that not only specialized receptor cells detect stimuli, but other cells do so as well.

Detection of stimuli is only the first step in integration: sensory input must be processed within a nervous system, and appropriate responses selected. The quality of response depends on an organism's *effectors*, which are chiefly muscles and glands. Responses may be *localized* or *systemic:* localized opening and closing of capillary beds occurs in response to wastes accumulating at that immediate site. By contrast, response of tissues to epinephrine (adrenalin) is systemic and gears up the body as a whole to deal with some emergency.

Responses may be *physiological* or *behavioral*, although the latter nearly always involve the former. Physiological responses maintain *homeostasis,* adjusting pH, solute and water concentrations, rate of heartbeat and breathing, and the like. Behavioral responses are what an organism *does* when stimulated, such as moving or ceasing to move, orienting, or vocalizing. Since most behaviors require energy, their performance depends on the effective correlation of physiological responses.

24.1. How do endocrine cells bring about bodily coordination?

Endocrine cells secrete into the bloodstream powerful regulatory chemicals known as *hormones*. In higher invertebrates these control such events as molting, metamorphosis, and color change (see Questions 16.4 and 16.5 and Fig. 16.10), but at this time we will concentrate on the mammalian body, where endocrine cells of many kinds exist, which may or may not be clustered into recognizable glands.

24.2. What are some endocrine cells and organs?

Scattered endocrine cells that do not form definite glandular masses occur at a number of sites. Those found in the lining of the digestive tract secrete *gastrointestinal hormones* such as gastrin and cholecystokinin; in the thyroid gland such cells secrete *calcitonin*, a hormone quite different from those produced by the thyroid follicles; little clusters of *interstitial cells* between the seminiferous tubules in the testes secrete male sex hormones; *neurosecretory neurons* in certain brain centers secrete neurohormones. In fact, we now know that a wide variety of cells and tissues never before suspected of secreting hormones in fact do so.

> **Example 1:** Cardiac muscle cells in the atria of the vertebrate heart secrete a peptide hormone called *atrial natriuretic factor* (ANF). ANF has several effects: it lowers renal blood pressure by relaxing the muscles of the arterioles, promotes sodium (and therefore water) excretion by increasing glomerular permeability, prevents the kidney from secreting the enzyme renin, and reduces the flow of the adrenal hormone aldosterone.

Many kinds of cells secrete hormones named *prostaglandins* (because they were first isolated from prostate glands).

> **Example 2:** *Platelets* (Question 23.22) produce a prostaglandin called *thromboxane* that not only causes platelets to stick together to form a platelet plug, but also induces nearby arteries to squeeze shut. On the other hand, the endothelium of undamaged blood vessels secretes *prostacyclin* that inhibits clots from forming and keeps blood vessels open.

If it is suspected that some bodily response is governed by an unidentified hormone, one useful trick is to conjoin the circulations of two animals to see if the common blood flow promotes changes in one or the other.

Example 3:   The *parabiotic union* of an unoperated rat and one with a liver regenerating after partial removal has shown that the liver of the intact animal enlarges abnormally; thus, some hormone still unknown stimulates liver regeneration. The nature and source of this hormone remain obscure.

Luckily, the source of a hormone is usually less elusive, for most are secreted by cells grouped into readily distinguishable glands ranging from the quite conspicuous thyroid to the microscopic islets of Langerhans in the pancreas.

24.3.   What are the major hormones of mammals?

Table 24.1 summarizes the major hormones that help maintain homeostasis; regulate growth, maturation, and reproduction; and affect behavior in mammals. Some (e.g., insulin, parathormone, aldosterone) are essential to life itself. Since hormones are critical to so many vital activities, their secretion must be precisely controlled.

24.4.   How is hormone secretion regulated?

Secretion of hormones can be controlled by direct response of independent effectors to chemical or mechanical stimuli, by direct receptor-effector mechanisms, by some other hormone, and by the nervous system.

24.5.   How do independent effectors respond to the internal chemical environment?

Some endocrine cells are both *chemoreceptors* and *independent effectors*; that is, they sense the chemical composition of the surrounding tissue fluid and secrete (or withhold) their hormones in direct response.

Example 4:   *Parathormone- and calcitonin-secreting* cells are sensitive to fluctuations in Ca^{2+} concentrations in the tissue fluids. The *parathyroid glands* respond to a drop in Ca^{2+} by secreting *parathormone*, which increases renal retention and intestinal absorption of Ca^{2+} and stimulates *osteoclasts* to liberate Ca^{2+} from bone matrix. Conversely, calcitonin-secreting cells respond to rising levels of Ca^{2+} by stimulating deposition of bone matrix, decreasing intestinal absorption of Ca^{2+}, and increasing its renal excretion. The two types of glandular cells are found very close together, which allows Ca^{2+} level to be monitored by both at the same time and place. Because of their opposing actions in calcium regulation, these two types of endocrine cells are known as the body's "calcistat."

Similarly, the microscopic islets of Langerhans embedded in the pancreas are sensitive to glucose concentration and produce two hormones, *insulin*, and *glucagon*, which affect glucose concentration in opposite directions, making up a "glucostat."

Example 5:   Any fall in blood sugar prompts glucagon secretion, which increases *glycogenolysis* in the liver, so that glucose pours out into the bloodstream. Any increase in circulating glucose stimulates the secretion of insulin, which promotes cellular absorption of glucose throughout the body; *diabetes mellitus*, fatal if unmanageable, results from inadequate absorption of glucose, with high blood sugar levels leading to glycosuria (sugar in the urine) and even to retinal detachment.

Diabetes can result either from insulin deficiency (alleviated by insulin injection) or from a deficiency of *insulin receptor sites* in the plasma membranes of body cells, a more refractory situation. This illustrates the important point that not all tissues are equally susceptible to regulation by a given hormone, even in health: thus the metabolism of cardiac and skeletal muscle, liver, and adipose tissue is much affected by thyroid hormone, while brain, spleen, and testis, lacking a thyroid hormone–binding protein in the mitochondria, are little affected.

24.6.   How may mechanical stimuli control hormone secretion?

Independent effectors that respond to *mechanical stimuli* (e.g., pressure and tension) include the cardiac muscle fibers that secrete ANF in response to the mechanical stimulation of being stretched. A small rise in atrial pressure triggers hypersecretion of ANF, which by lowering blood volume (by increasing urinary output) reduces blood pressure.

Table 24.1 Mammalian Hormones

Hormone and Main Source	Effects
Pineal gland	
Melatonin	Regulation of annual reproductive cycles by suppressing LHRH secretion
Hypothalamus	
Growth-hormone-releasing-hormone	Secretion of pituitary growth hormone
Somatostatin	Inhibition of GH secretion
Thyrotrophin-releasing-hormone	Secretion of pituitary thyrotrophin
Adrenocorticotrophin-RH	Secretion of pituitary ACTH
Follicle-stimulating-hormone-RH	Secretion of pituitary FSH
Luteinizing-hormone-RH	Secretion of pituitary LH
Prolactin-RH	Secretion of prolactin
Prolactin-inhibiting hormone	Inhibition of prolactin secretion
Hypothalamus via posterior pituitary	
Oxytocin	Contraction of uterine muscles (during labor and orgasm) and mammary duct myoepithelium (milk ejection)
Antidiuretic hormone (ADH)	Increase of water retention in kidney by making collecting ducts more porous to water
Anterior pituitary (adenohypophysis)	
Growth hormone (somatotrophin)	Somatic growth (especially skeletal), glycogenolysis, amino acid transport
Thyrotrophin	Secretion of thyroid hormones
Adrenocorticotrophin (ACTH)	Secretion of glucocorticoids by adrenal cortex
Follicle-stimulating hormone (FSH)	Maturation of sperm and ovarian follicles, secretion of estrogens
Luteinizing hormone (LH)	Secretion of sex hormones; ovulation; transformation of follicle to corpus luteum and secretion of progesterone
Prolactin	Milk production
Pars intermedia of pituitary	
Melanocyte-stimulating hormone (MSH)	Skin-darkening chromatophore responses; melanin synthesis
Thyroid	
Thyroxin, triiodothyronine	Stimulation of oxidative metabolism; promotion of somatic growth
Calcitonin	Decrease of blood calcium level by promoting urinary excretion and deposition in bone matrix, and inhibiting intestinal absorption of calcium
Parathyroids	
Parathormone	Increase of blood calcium level by decreasing urinary excretion and promoting intestinal uptake; stimulation of osteoclastic action on bone matrix
Adrenal cortex	
Glucocorticoids (cortisol, etc.)	Increase of gluconeogenesis (glucose formation from proteins and fats); decrease of inflammation
Mineralocorticoids (aldosterone, etc.)	Increase in renal retention of chloride, sodium ion, and water; promotion of excretion of potassium ion and hydrogen ion
Androgens	Male secondary sexual characteristics, especially in females; female sex drive
Estrogens	Female secondary sexual characteristics, especially in males

(Table continues on page 400)

Table 24.1   (*continued*)

Hormone and Main Source	Effects
Adrenal medulla Epinephrine (adrenalin) Norepinephrine (noradrenalin)	Reinforcement and prolongation of the action of sympathetic nervous system in "fight-or-flight" response to emergencies
Islets of Langerhans Glucagon (from alpha cells)	Raising of blood glucose by increasing glycogenolysis in liver
Insulin (from beta cells)	Lowering of blood glucose by increasing glucose uptake into cells
Somatostatin (from delta cells)	Same action as hypothalamic somatostatin?
Growth-hormone-releasing-H	Same action as hypothalamic GHRH?
Ovarian follicle Estrogens (estradiol, etc.)	Female secondary sexual characteristics, growth of mammary ducts, inhibition of skeletal growth, preovulatory thickening of uterine lining (endometrium)
Corpus luteum of ovary Progesterone	Postovulatory thickening and secretory activity of endometrium; inhibition of uterine musculature
Placenta Chorionic gonadotrophin	Continuation of pregnancy by promoting growth of corpus luteum of pregnancy
Progesterone	Continuation of pregnancy by keeping uterine muscles quiescent and lining thick
Relaxin	Relaxation of pubic symphysis before parturition
Interstitial cells of testes Testosterone	Male sex drive and sexual behaviors; male secondary sexual characteristics
Thymus Thymosin	Differentiation of T lymphocytes
Gastric epithelium Gastrin	Secretion of HCl and pepsinogen
Duodenal epithelium Enterogasterone	Inhibition of flow of gastric juice
Secretin	Secretion of alkaline component of pancreatic fluid; inhibition of gastric motility
Cholecystokinin	Secretion of pancreatic enzymes; contraction of gall bladder; inhibition of stomach motility; maybe suppression of appetite
Cardiac muscle cells Atrial natriuretic factor	Increase in renal excretion of sodium ion, thus increasing urinary water loss
Many tissues Prostaglandins	Numerous effects (see text)

24.7.   What is an example of hormone regulation by direct receptor-effector links?

Direct receptor-effector links control certain hormones without involving the nervous system. For example, in mammalian kidneys (see Fig. 22.8) part of each distal convoluted tubule lies right against the afferent arteriole entering the glomerulus. Together, cells in the wall of the distal tubule and cells in the arteriolar wall constitute a *juxtaglomerular complex* that regulates blood pressure and renal retention of NaCl. The distal tubule cells monitor urinary salt content, and when this is too low, they signal the arteriolar cells to secrete an enzyme, *renin*, into the bloodstream. The arteriolar cells also secrete renin in response to any drop in renal blood pressure, for this threatens life itself. Renin converts an inactive hormone in the blood plasma, *angiotensinogen,* to *angiotensin I,* which another enzyme converts to *angiotensin II.* This hormone has two main effects: generalized vasoconstriction of peripheral arteries, which immediately elevates renal blood pressure, and secretion of aldosterone by the adrenal cortex. Aldosterone increases chloride transport in Henle's loop and by promoting NaCl retention also increases water retention; the resulting rise in blood volume restores blood pressure to safe levels.

24.8. What hormones control the secretion of other hormones?

We have just seen that angiotensin II stimulates the secretion of aldosterone from the adrenal cortex. A more complex example is the *hypothalamic-hypophyseal control axis* that governs the functioning of several major endocrine glands and is in turn controlled by the hormones of these, by way of *negative feedback loops.* In brief, the hypothalamus of the brain secretes neurohormones, some of which regulate the flow of hormones from the nearby hypophysis, or pituitary gland (Fig. 24.1). Hormones from nerve cells in the anterior hypothalamus travel by way of the hypophyseal portal vein directly to a second capillary bed in the anterior pituitary (adenohypophysis); here, some act as *release factors* promoting the secretion of certain pituitary hormones, while others serve as *inhibiting factors* [e.g., *prolactin-inhibiting hormone* (PIH), which inhibits *prolactin* secretion in mammals except during lactation]. A dual hormonal mechanism controls the secretion of pituitary growth hormone (*somatotrophin*). Growth-hormone-releasing hormone (GHRH) invokes pituitary secretion of somatotrophin, which stimulates skeletal and bodily growth, promotes fat and glycogen breakdown, and facilitates amino acid transport. The flow of GHRH itself is reduced by a rise in blood sugar, probably directly sensed by the hypothalamic cells themselves, but in addition, the hypothalamus also secretes a growth-hormone-inhibiting hormone (GHIH), also called *somatostatin,* which actively inhibits the secretion of somatotrophin. Malfunction of these control mechanisms in childhood can lead to pituitary *gigantism* or *dwarfism.*

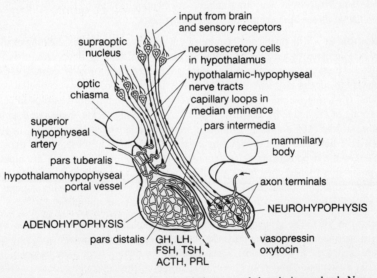

Fig. 24.1 Relations between the hypothalamus and the pituitary gland. Neurosecretory cells in the supraoptic nucleus of the hypothalamus secrete into the hypophyseal portal blood system neurohormones that control secretion of hormones from the anterior pituitary (adenohypophysis); other hypothalamic neurons send fibers down into the posterior pituitary (neurohypophysis) where they secrete neurohormones directly into the bloodstream for general distribution throughout the body. (*From Storer et al.*)

In similar fashion, releasing hormones from the hypothalamus stimulate the pituitary to secrete thyrotrophin, adrenocorticotrophin (ACTH), and two *gonadotrophins:* luteinizing hormone (LH) and follicle-stimulating hormone (FSH). In these cases, the flow of the hypothalamic and hypophyseal hormones is diminished by negative feedback, the inhibiting factors being hormones from the thyroid, adrenal cortex, and gonads, respectively.

Example 6:   Hypothalamic neurons secrete thyrotrophin-releasing hormone (TTRH), which causes the hypophysis to liberate thyrotrophin (TTH); thyrotrophin induces the thyroid glands to release thyroid hormones stored within bubblelike follicles. Thyroid hormones, circulating throughout the body, promote glycogenolysis and oxidative metabolism and suppress amino acid catabolism; when they reach the hypothalamus and pituitary, they also act to shut down the flow of TTRH and TTH. This negative feedback loop can go awry when the diet contains too little iodine: the thyroid then puts out too little hormone to shut down the hypothalamic control mechanism. The latter continues to pour out TTRH, and the pituitary TTH, to which the poor bedeviled thyroid responds by hypertrophy, swelling into a *simple goiter*. If the control mechanism itself somehow becomes unresponsive to circulating thyroid hormone when plenty of dietary iodine is available, the unhappy thyroid cannot shut off but goes on pouring out more hormone, so that the victim becomes thin, jumpy, hallucinative, and pop-eyed, a condition known as *exophthalmic goiter*.

24.9.   How does nervous stimulation control hormone secretion?

Nerve fibers can release excitatory or inhibitory neurotransmitters directly onto endocrine cells.

Example 7:   Cells in the *pars intermedia* of the pituitary gland secrete *melanocyte-stimulating hormone* (MSH) when stimulated by neurons that lie in the hypothalamus but send long fibers down through the pituitary stalk to reach the glandular cells. MSH promotes pigment synthesis in mammalian skin and triggers color changes in amphibians and fishes.

Example 8:   Epinephrine and norepinephrine are discharged from the *adrenal medulla* in response to sympathetic nerve input. This is scarcely surprising since the adrenal medulla is simply an enlarged *sympathetic ganglion* that secretes as hormones the same substances that other sympathetic neurons liberate at synapses as neurotransmitters, augmenting the sympathetic fight-or-flight response by disseminating those same chemicals to *all* body cells.

Example 9:   Milk flow from the mother's mammary glands during suckling results from two different reflexes initiated by mechanoreceptors in the nipples: sensory fibers from these receptors go to two different parts of the hypothalamus. Some go to the posterior hypothalamus, where they cause certain neurons to secrete *oxytocin*, which provokes contraction of the myoepithelial cells in the mammary ducts, triggering the *milk ejection reflex*. This reflex is especially vigorous in cetaceans, where mere grasping of the nipple by the baby, which cannot actually suckle, causes milk to spurt from the nipple right into the baby's mouth. (Although secreted by hypothalamic neurons, oxytocin and the related ADH are liberated into the bloodstream at capillary beds in the posterior pituitary, where the long axons of the hypothalamic cells terminate.) Other sensory fibers from the nipple go to the anterior hypothalamus, where they block secretion of the prolactin-inhibiting hormone, allowing an increased flow of prolactin, which accelerates milk production even while nursing is going on. When suckling ceases, PIH secretion recommences and oxytocin secretion diminishes.

Example 10:   The production of oxytocin in another situation involves *positive feedback* control: when uterine contractions commence at the onset of labor, sensory nerve fibers in the wall of the uterus signal the hypothalamus, which responds by putting forth more oxytocin, which stimulates the uterine musculature to contract even more strongly. This causes more intense reflex stimulation of the hypothalamus with still greater output of oxytocin, until finally birth takes place and the runaway control system comes to a screeching halt.

24.10.   How do environmental cues, sense organs, and the brain interact in regulating hormone secretion?

The coordination of reproductive physiology and behaviors shows beautifully how external stimuli, brain, and hormones all interact. First, the third eye of vertebrates is thought to monitor photoperiod changes throughout the year. In mammals, the *pineal organ* is no longer eyelike, but glandular, but it still receives information on photoperiod by way of a sympathetic relay from eye muscles that close down the pupil in bright light and expand it in the dark. The pineal responds by suppressing reproductive activity until the time of year when such activity would allow optimum survival of young. Until the length of daylight shortens or lengthens to the "right" point, the pineal secretes *melatonin* that prevents the hypothalamus from secreting luteinizing-hormone-releasing hormone. Then, at the proper season, melatonin secretion ceases, and LHRH begins to flow, stimulating the pituitary to release LH, which among other things prompts the gonads to secrete sex hormones.

In birds, photoperiod, the sight and sound of a singing mate, and the presence of nest materials all affect the hypothalamus by way of relays from sensory-analyzing parts of the brain, interacting to induce nest building, egg laying, incubation, and subsequent care-giving behaviors.

> **Example 11:** Female budgerigars can be made to lay eggs simply by being played recordings of the male's "soft warble" vocalization. This shows that signals from the acoustic part of the brain are relayed to the hypothalamus, which then secretes FSHRH (follicle-stimulating-hormone-releasing hormone), causing the pituitary to secrete FSH, which promotes maturation of ovarian follicles.

24.11.   How do hormones affect the brain?

The brain, which controls so many hormones, in turn is affected by hormones, especially sex hormones that stimulate reproductive behaviors. Male sexual behavior is *testosterone-dependent*, as demonstrated by the full restoration of mating activity in castrated males given testosterone injections. *Aggressive behaviors* are also stimulated by testosterone, as shown by injecting subordinate animals, such as hens at the bottom of a social hierarchy, which then enjoy a combative and meteoric rise to the peak of the pecking order. Sex hormones, either testosterone or estrogen, injected into newborn female mammals have the startling effect of permanently masculinizing their brains, suppressing fertility by blocking the estrus cycle, and making such females as aggressive as males of their kind.

24.12.   What is the chemical nature of hormones?

Hormones represent an astonishing variety of organic compounds: thyroxine is an amino acid, epinephrine is an amine, prostaglandins are fatty acid derivatives, oxytocin and ADH are oligopeptides containing only 9 amino acids, atrial natriuretic factor is a peptide with 28 amino acids, insulin and pituitary hormones are proteins, and adrenocortical and sex hormones are steroids. With all this diversity, hormones nevertheless appear to operate in only two or three basic ways. Also, some are effective only in concert with another.

24.13.   How do hormones exert their effects?

(*a*)   Most *water-soluble hormones* fit into receptor sites in the membranes of the cells they affect. Except for thyroid hormones, which are bound by a receptor protein in the mitochondrial membranes, nearly all water-soluble hormones occupy sites in the plasma membrane and thus seem never to enter the cell at all. Strange to say, most such receptor sites are coupled to the activation of a single enzyme, *adenyl cyclase*, which converts ATP to *cyclic AMP* (see Fig. 3.8), a nucleotide remarkably versatile in activating various metabolic pathways. Specifically, cAMP activates an enzyme known as a *protein kinase*. Protein kinases are a family of enzymes that activate other enzymes by providing them with phosphate from ATP. Thus the different actions of various hormones, and the different effects of the same hormone on different tissues, relate to *which* protein kinase is coupled to the hormone receptor site, and which metabolic pathway is thereby activated. The cardiac atrial natriuretic factor acts in a slightly different way: it occupies a receptor site coupled to the activation of the enzyme *guanylate cyclase*, which converts the nucleotide GTP (guanosine triphosphate) to *cyclic GMP*, which serves as the intracellular messenger for this hormone; at the same time, ANF actually *inhibits* activation of adenyl cyclase (thus blocking the effect of hormones dependent on cyclic AMP).

(*b*)   Insulin is a water-soluble hormone that is dealt with differently: at the plasma membrane it is covalently bonded with a protein, and when a number of insulin-protein complexes have accumulated, the cell gulps them all in by forming a vacuole around the whole aggregation. What happens next is not yet known.

(*c*)   *Fat-soluble* hormones readily penetrate cells by oozing between the phospholipids of the plasma membrane. Once in the cell, however, their effectiveness depends on their being taken up by a *cytoplasmic binding protein*. The combined hormone-protein complex then enters the nucleus and binds to a chromosomal receptor, activating a specific gene. Steroid hormones have been radioactively traced into the nuclei of cells in specific organs, which soon thereafter show new peaks of RNA production and produce new proteins. For example, administration of hydrocortisone causes a prompt rise in liver RNA synthesis, quickly followed by activity of a particular enzyme (tyrosine-glutamate transaminase). Thus hormones represent a crucial factor in the regulation of *genes*.

(*d*)   Some hormones operate *synergistically:* together they produce an effect that neither can bring about singly. For example, neither thyroxine nor epinephrine alone is effectual in mobilizing stored fat out of adipose cells; when both are present, as when epinephrine levels rise during exercise, fat is mobilized effectively.

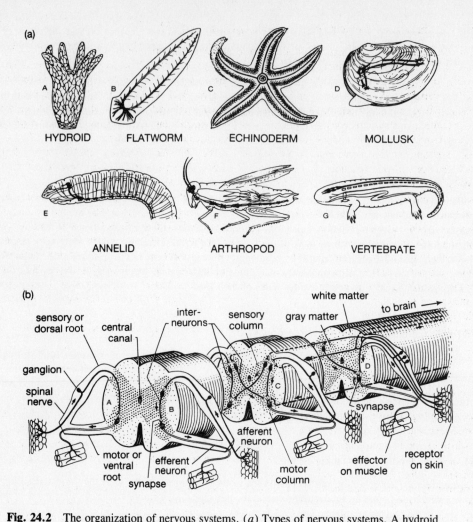

Fig. 24.2 The organization of nervous systems. (*a*) Types of nervous systems. A hydroid (A) lacks a central nervous system, having only a diffuse nerve plexus. A planarian flatworm (B) has a primitive brain and sensory concentration in the head, and two main lateral nerve cords giving off peripheral nerves. An echinoderm (C) has a circumoral nerve ring, from which come radial nerves, one to each arm. A bivalve mollusk (D) has three pairs of ganglia with connecting nerve cords. Annelids (E) and arthropods (F) have paired ganglia in the head, and single or paired ventral nerve cords with a ganglion in each body segment. A vertebrate (G) has a hollow central nervous system, with dorsal brain and spinal cord giving off segmental pairs of cranial and spinal nerves. (*b*) Portion of the vertebrate spinal cord showing afferent (sensory), association, and efferent (motor) neurons and their distribution. Association neurons, or interneurons, lie entirely within the brain or spinal cord. Simple spinal reflex arcs may include (A) only afferent and efferent fibers and no interneurons (such as a knee jerk response to a blow on the patellar ligament); (B) sensory neurons, interneurons, and motor neurons (such as a pain-withdrawal reflex); cross-connections to the opposite side of the cord (C), allowing a bilateral reflexive response (such as the crossed-extension reflex, in which one leg straightens as the other bends); and perhaps (D) connections to and from the brain (as in the control of skeletal muscles via the gamma loop). (*From Storer et al.*)

24.14.   How do nervous systems bring about integrated responses?

Nervous systems (Fig. 24.2a) range in complexity from the brainless nerve plexus of radiates to the elaborate system of vertebrates, which is divided anatomically into a *central nervous system* (CNS) and a *peripheral nervous system* (PNS), and functionally into *somatic sensory-motor* and *autonomic* (parasympathetic-sympathetic) divisions. To review the organization of the vertebrate nervous system and brain, see Questions 18.21–18.28 and Fig. 18.9. Bilateria possess central nervous systems having a brain or central ganglia and ganglionated nerve trunks, such as the vertebrate *spinal cord* (Fig. 24.2b). Peripheral nervous systems consist of nerves, peripheral ganglia, and sense organs. *Peripheral ganglia*, such as sensory and autonomic ganglia, are nodular masses or networks of nerve cells located outside the CNS. Nerve cell bodies and synaptic junctions between nerve cells occur only within the CNS and peripheral ganglia. *Nerves* do not contain neuron cell bodies or synaptic junctions, but are simply bundles of nerve fibers: *sensory*, or *afferent*, fibers carrying messages to the CNS and *motor*, or *efferent*, fibers carrying them from the CNS to the effectors.

The somatic sensory-motor system deals mainly with nonvisceral sensory reception and perception, and with the control of voluntary muscles that bring about bodily movement and locomotion. The autonomic system is concerned with physiological integration; its parasympathetic division receives sensory input from internal organs and promotes the normal functioning of glands and involuntary muscles such as those of the gut. The sympathetic division is an evolutionary newcomer, confined to vertebrates and probably making their lives more vivid than those of invertebrates: it evokes responses to emergency situations by accelerating heart rate and breathing, diverting blood from organs to skeletal muscles, stimulating glycogenolysis, and triggering a flow of adrenal hormones. Most vertebrate organs are doubly innervated by parasympathetic and sympathetic fibers that act in opposition; e.g., parasympathetic fibers slow down the heart while sympathetic impulses speed it up.

The operation of nervous systems must be understood in terms of how nerve cells (*neurons*) behave.

24.15.   How do nerve cells function?

A *neuron* (see Fig. 2.3c) is adapted to be especially excitable, to conduct messages in the form of nerve impulses, and to transmit excitation to other nerve cells or effectors across junctions known as "synapses." A neuron has a *cell body* from which extend two types of *nerve fibers* of variable length: *dendrites* which conduct excitation *toward* the cell body and an *axon* which conducts excitation *away from* the cell body to a *synaptic ending*. At the ends of axons a chemical neurotransmitter is ordinarily secreted, which affects the postsynaptic cell.

Neurons that lie entirely within the CNS are known as *interneurons*, or *association neurons*. They form multiple connections within the CNS and also interconnect sensory and motor neurons, as in an involuntary *spinal reflex arc* (see Fig. 24.2b). Sensory neurons relay sensory information from receptors to the CNS; their dendritic endings often *are* receptors. In vertebrates, the cell bodies of sensory neurons are located in separate *sensory ganglia* lying outside the brain or spinal cord, and only their axons enter the CNS. By contrast, motor neurons transmit excitation from the CNS to the effectors; in vertebrates, the dendrites and cell bodies of somatic motor neurons lie within the brain and spinal cord, and their axons travel outward in cranial and spinal nerves to terminate on skeletal muscle fibers. Motor neurons of the autonomic system have their cell bodies located in peripheral ganglia, from which axons extend to the organs being served.

24.16.   On what does nerve function depend?

The *irritability* of a nerve cell depends on an unequal distribution of ions across its plasma membrane. When this membrane is not being excited by a stimulus, its carrier-mediated transport mechanisms are at work, concentrating potassium ions (K^+) within the cytoplasm and ejecting sodium ions (Na^+), by way of a *sodium/potassium ion-exchange pump*. In maintaining this highly unequal distribution of Na^+ and K^+, the nerve cell also creates an *electrochemical gradient*, the *resting potential*, across its membrane, with the cell interior being some -60 millivolts (mV) electronegative to the outside of the membrane (Fig. 24.3). This internal electronegativity has two main causes: first, cytoplasmic proteins that cannot pass through the plasma membrane dissociate as negatively charged *anions*; second, K^+ is such a tiny particle that a certain amount tends to leak out through the pores of the plasma membrane, despite the opposing action of the Na^+/K^+ pump, until a balance is struck between the outward diffusion gradient of K^+ and the internal attractive force exerted on K^+ by the cytoplasmic anions. Na^+ is a substantially larger ion, and the plasma membrane is more successful at keeping it out against the combined forces of concentration gradient and electrostatic attraction.

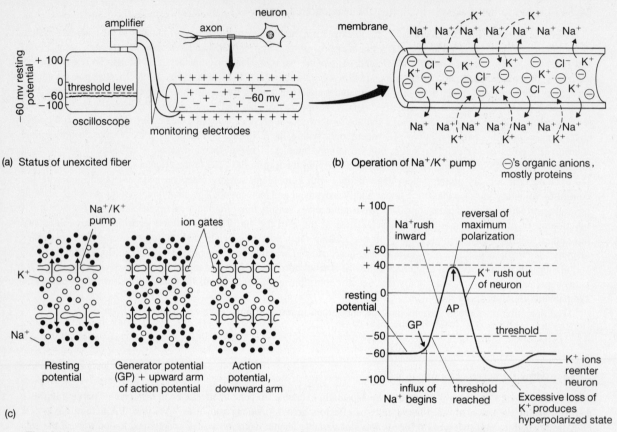

(a) Status of unexcited fiber

(b) Operation of Na^+/K^+ pump $\ominus$'s organic anions, mostly proteins

(c)

Resting potential

Generator potential (GP) + upward arm of action potential

Action potential, downward arm

Fig. 24.3 The nerve impulse. (*a*) An unexcited nerve fiber possesses a resting potential of about -60 to -70 mV internal electronegativity. (*b*) The "resting" fiber is actually expending a great deal of energy operating a sodium-potassium pump. (*c*) When the nerve cell membrane is stimulated to a threshold of -50 mV, the voltage-controlled Na^+ activation gates open fully and Na^+ rushes in, reversing the electrical potential across the membrane. At $+40$ mV internal electropositivity (the *action potential*), the Na^+ inactivation gate closes and a voltage-controlled K^+ gate opens, allowing K^+ to leave the cell, thereby restoring its original electronegativity. Each action potential opens the Na^+ activation gates in the membrane just ahead, so that a new action potential may be triggered in this region. Thus a zone of excitation, the nerve impulse, moves along the fiber at a steady rate.

24.17. How does a neuron respond to stimulation?

(*a*) When some point on the neuron's plasma membrane is stimulated, this uneasy imbalance is overturned, for the cell has to work hard to keep Na^+ out, since positive ions are strongly attracted to the cell's electronegative interior. The stimulus interferes with the cell's ability to do this job, and a few Na^+ ions manage to sneak in; this reduces the internal electronegativity a bit, an event visible on an oscilloscope as the *generator potential*. Like ripples on a pond, this effect dies away unless a certain threshold of intensity is reached. If the stimulus is strong enough (i.e., of *threshold intensity*), the Na^+ leakage that causes the generator potential persists until the cell's internal electronegativity drops locally from -60 to about -50 mV and a sudden dramatic reversal of membrane polarity, the action potential, ensues.

(*b*) The *action potential*, an all-or-nothing event of about 1 millisecond's duration, is triggered only when Na^+ influx depolarizes the membrane to about -50mV, which causes a "sodium activation gate" to open: Na^+, suddenly free to follow its concentration and electrochemical gradients, rushes inward, reversing the polarity of the membrane, so that the cell's interior at that point momentarily becomes some $+40$ mV *electropositive* to the exterior. Viewed by oscilloscope, this reversal of charge traces the upward arm of a spike, which halts abruptly when a voltage deviation of about 100 mV (from -60 to $+40$ mV) has taken place, and a "sodium inactivation gate" closes. The downward arm of the spike is then traced, reflecting a restoration of internal electronegativity, by way of an abrupt efflux of K^+. This rapid exit of K^+ is due not only to the steep concentration gradient for this ion, but also to the decrease in the attractive force of the cytoplasmic anions, which are momentarily balanced by the intruding Na^+.

24.18.   How do sodium and potassium ions get across the neuron's membrane?

The facility with which Na^+ and K^+ cross the neuron's plasma membrane during production of the action potential is due to *pores* through the membrane known as *voltage-regulated ion gates*. The K^+-ion gates open only at $+40mV$ and close again almost instantaneously, but not before enough K^+ has escaped for the resting potential to be restored. The Na^+-ion gates are more complicated, for they act like a double door: when the *Na^+ activation gate* opens at $-50mV$, Na^+ is allowed to rush inward; at $+40mV$, the *Na^+ inactivation gate* slams shut and no more Na^+ can enter. The generator potential results from a slight opening of Na^+ activation gates in response to a localized stimulus; if the inward leakage of Na^+ brings about the requisite reduction of internal electronegativity (partial depolarization), the Na^+ activation gates open fully, Na^+ enters freely, and the membrane polarity is reversed, as mentioned above. The K^+ gate then opens, and the loss of K^+ restores the resting potential. Although the original internal electronegativity has now been restored and, with it, the cell's ability to generate another action potential at that same spot, the original Na^+-K^+ distribution has *not* been restored; sooner or later this must be remedied by operation of the Na^+/K^+ pump, or the neuron would eventually become unresponsive as the gradients of the two cations leveled out.

24.19.   What is a nerve impulse?

A *nerve impulse* is a linear sequence of action potentials occurring at adjacent points along the length of a nerve fiber. An action potential does *not* travel anywhere, but a nerve impulse does. This forward propagation results from the fact that each action potential affects the Na^+ activation gates in the membrane just adjacent, so that the membrane becomes depolarized to the point at which yet another action potential is triggered, and so on. Thus a succession of action potentials, in the aggregate making up the nerve impulse, progresses along the entire length of the nerve fiber without slowing down or diminishing in intensity. Meanwhile, the resting potential is restored by loss of K^+ in the wake of each action potential, so that repeated stimulation produces an entire train of nerve impulses traveling along the fiber, spaced apart by the few milliseconds needed for recovery from the preceding action potential (the membrane's *refractory period*). Since each action potential involves the exchange of only about 1 ion in 10 million, quite a number of nerve impulses could be generated before the Na^+-K^+ distribution evened out; if this were to happen, the cell would no longer be able to conduct impulses, but long before that point can be reached, the Na^+/K^+ pump is reactivated, restoring the original distribution of the two types of cations.

24.20.   What is saltatory conduction?

Saltatory conduction occurs in myelinated vertebrate nerve fibers. In peripheral nerve fibers, the fatty *myelin sheaths* around such fibers are formed by the encircling membranes of *Schwann cells;* in the CNS, where myelinated fiber tracts are known as *white matter*, neuroglial cells known as *oligodendroglia* form the myelin sheaths. (Peripheral autonomic fibers and fibers in the *gray matter* of the CNS lack myelin sheaths and are said to be *amyelinated.*) The myelin sheath is interrupted at *nodes*, where one sheath cell gives way to the next. This allows conduction of nerve impulses to be significantly accelerated, and saves the cell a great deal of ATP, since excitation can jump (saltate) from node to node without affecting the intervening membrane. When an action potential is produced at one node, a minute electrical field spreads quickly through the medium *outside* the fiber and sets off an action potential at the next node; by contrast, in amyelinated fibers the impulse must travel uninterruptedly along the neuron's membrane. Heavily myelinated fibers conduct at a velocity of about 100 m/s, whereas amyelinated fibers have a conduction rate of only 0.3–1.6 m/s. Furthermore, an amyelinated fiber must use about 5000 times more ATP than a myelinated fiber does, since the former must keep operating its Na^+/K^+ pump and ion gates at all points along its membrane.

24.21.   What are self-excitatory neurons?

Self-excitatory neurons fire patterned bursts of nerve impulses at regular intervals even when experimentally cut off from other nerve connections. In invertebrates, such spontaneously firing cells often are *pacemakers* that control the activity of motor neurons when the animal is performing some repetitive behavior such as swimming or flying. Self-excitatory neurons must possess some endogenous oscillatory mechanism that operates their ion gates in a rhythmic manner. Other neurons may act to inhibit this spontaneous self-excitation except under particular circumstances.

24.22. How do neurons communicate with other cells?

Neurons affect other cells across junctions known as *synapses* (Fig. 24.4). A neuron's axon ends in one or more swollen *synaptic knobs* packed with mitochondria and vesicles full of neurotransmitter chemicals. Each knob lies almost in contact with the plasma membrane of a *post-synaptic cell*, which may be another neuron or an effector cell (muscle fiber, gland cell). Between the two plasma membranes lies a narrow *synaptic cleft* into which the axonal vesicles discharge neurotransmitter molecules each time a nerve impulse reaches this point. The postsynaptic membrane contains receptor sites that the neurotransmitter molecules occupy for only a split second before enzymes destroy them. Instant destruction of the neurotransmitter molecule allows the receptor site to be occupied afresh with each new spate of neurotransmitter secretion, and each occupancy of the site constitutes a discrete stimulus upon the postsynaptic membrane. These stimuli serve to open *chemically gated ion channels* in the postsynaptic membrane (in contrast to the voltage-regulated ion gates described above). *Excitatory* neurotransmitters cause the opening of Na^+ channels, and the resulting influx of Na^+ sets off action potentials that initiate the response of the postsynaptic cell, such as muscle contraction. *Inhibitory* neurotransmitters reduce the excitability of the postsynaptic membrane by *hyperpolarizing* it, that is, *increasing* its internal electronegativity. This is accomplished by opening *chloride-ion channels* with the result that Cl^- enters the cell, thereby increasing its cytoplasmic content of anions, or by opening K^+-*ion channels* through which K^+ escapes from the cell, with similar results. This hyperpolarization raises the threshold of excitation, so that a stronger stimulus will then be required to activate the postsynaptic cell.

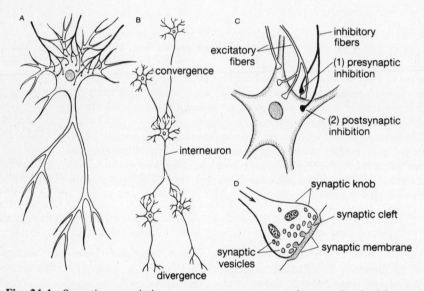

Fig. 24.4 Synaptic transmission. A, many axons may terminate on the dendrites and cell body of a given neuron. B, synaptic convergence occurs when two or more neurons synapse upon one postsynaptic cell, while synaptic divergence takes place when the branching axon of a presynaptic cell terminates on two or more postsynaptic cells. C, synapses may be excitatory or inhibitory; in the latter case, the inhibitory fibers may terminate on the synaptic knob of the excitatory fiber (1) or on the membrane of the postsynaptic cell (2). D, detail of a synapse showing liberation into the synaptic cleft of a chemical neurotransmitter from vesicles within the synaptic knob. (*From Storer et al.*)

The cell bodies and entire dendritic trees of many neurons in the CNS are literally encrusted with synaptic knobs, indicating how greatly *synaptic convergence* affects those neurons, for they are receiving messages from many other parts of the nervous system. Of the hundreds of axons terminating on one neuron, some may be excitatory and others inhibitory. *Synaptic divergence* is also common: the axon of a single neuron may branch and end on a number of postsynaptic cells. Even though neurons do not undergo mitosis once they have differentiated, new fibers and synaptic knobs continue to proliferate after birth, and to some extent throughout life.

24.23.   What is the chemical nature of neural secretions?

(*a*)   *Neurotransmitters* are usually synthesized in the cell body close to the nucleus and transported along cytoskeletal channels through the axoplasm. If they are used up faster than they can be synthesized, *synaptic fatigue* results: no further transmission can occur until the supply is replenished. Chemically, most neurotransmitters are either *amines* or *neuropeptides* with 2–40 amino acids. Amines include dopamine (deoxyphenylalanine amine), norepinephrine, epinephrine, and serotonin. The hallucinogenic effects of certain drugs seem to result from their competing with neurotransmitters for receptor sites. Figure 24.5 compares the formulas of norepinephrine with mescalin, and serotonin with psilocin; note that the drug and neurotransmitter molecules differ only in side groups; these differences may keep the enzymes that break down the neurotransmitters from disposing of the drugs, or cause the drugs to jam in the receptor sites.

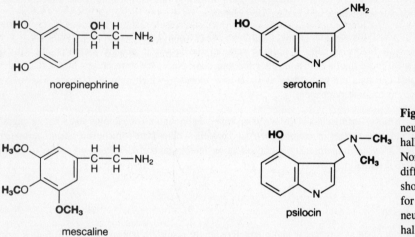

norepinephrine

serotonin

mescaline

psilocin

Fig. 24.5   Comparison of neurotransmitters and hallucinogenic drugs. Norepinephrine and mescaline differ only in the side groups shown in boldface; this is also true for serotonin (a brain neurotransmitter) and the hallucinogen psilocin.

Peptide neurotransmitters include *enkephalins* that bring about reduction of pain perception. Axons of enkephalin-secreting neurons are thought to end on the *axons* of sensory neurons involved in pain reception, and by way of *axon-axonal synapses* to inhibit the pain receptors from liberating their own special transmitter (substance P) onto neurons in the pain-perceiving centers. Enkephalins are made up of short amino acid sequences that also occur as portions of the larger molecules called *endorphins*, which also alleviate pain. The discovery of endorphins and enkephalins began with a search for receptor sites into which opiate molecules might fit; discovery of such sites prompted a successful hunt for endogenous morphinelike substances, or "end/orphins." A natural rise in endorphin secretion occurs in mammals just before parturition, and the pain-relieving properties of acupuncture are also thought to stem from stimulation of endorphin production.

(*b*)   *Neurohormones* that are disseminated from the brain to distant parts of the body are secreted by special *neurosecretory neurons*. Insect molting and metamorphosis are triggered by a brain neurohormone, and the vertebrate hormones oxytocin and ADH are also products of neurosecretory neurons, as are the releasing and inhibiting hormones that regulate the anterior pituitary.

24.24.   How do neurons store information?

Memory and behavioral modification through *learning* depend on permanent or self-perpetuating changes in neurons. Such changes are known to involve modified patterns of RNA synthesis in affected parts of the brain, which implies gene regulation and subsequent synthesis of new proteins.

(*a*)   Brain RNA, transferred from an animal that has undergone a particular learning experience into the brain of a subject lacking such experience, can affect the performance of the recipient.

   Example 12:   Electrode monitoring of the olfactory lobes of salmon brains shows that bursts of electrical activity are evoked only when water from the home river is perfused through the fish's nostrils; if RNA is extracted from the olfactory lobes of a salmon from river A and injected into the brain of a salmon from river B, soon the brain of the latter salmon will respond just as vigorously to water from river A.

(b) *Memory retention* in regenerating flatworms depends on preservation of neural RNA from the original worm. Trained planarians regenerating new heads from severed tail ends remember their conditioning *unless* the cut surface is treated with an enzyme that destroys RNA; apparently memory in these animals is stored throughout the central nervous system, not only in the brain, but cannot be transmitted to the new brain if RNA has been destroyed at the exposed surface from which neural regeneration will be initiated. These and many other experiments indicate that the formation of lasting memories involves changes in RNA synthesis by nerve cells. Since RNA is produced by genes, it would appear that learning involves changes in gene activity, and since mRNAs carry codes for building specific proteins, learning appears to be associated with the synthesis of new brain proteins.

(c) *Consolidation* must take place if a memory is to become permanent. Training can be obliterated if an animal is injected with chemicals that suppress either RNA synthesis or brain protein synthesis, but only if the injection takes place within a few minutes of the acquisition of a conditioned response.

> **Example 13:** Acetyloxycycloheximide suppresses 90 percent of cerebral protein synthesis; if it is given to mice just before or after a training session, the lesson is quickly forgotten, but if not administered until more than a half hour after learning took place, the drug proves ineffectual. Such findings suggest that data are first stored as *short-term memory*, which requires a process known as "consolidation" to become permanent.

The lasting change in the memory-storing cells is known as an *engram* although no one knows just what an engram might be. The experientially induced changes in neuronal RNA and protein synthesis are somehow rendered permanent during consolidation, whereas they are impermanent if consolidation fails to take place. *Attention* seems to be one essential prerequisite for the consolidation process.

24.25. What is sensory reception?

Reception is a process that takes place at a receptor and characteristically results in the energy of a stimulus being transformed into the energy of nerve impulses. A *stimulus* is a *change* in the environment of a receptor, *to which the receptor is competent to respond*. A stimulus that becomes a constant condition ceases to excite its receptors and thus is no longer a stimulus. A form of energy that lies outside the range of sensitivity of receptors cannot constitute a stimulus either, even though it may have significant effects upon the organism: for example, humans possess no receptors capable of detecting radar or radio and television waves, which are all around us, nor can we detect x-rays or radioactive emissions, even though heavy exposure may sicken or kill us.

24.26. What are sensory receptors and sense organs?

Sensory receptors are cells specialized to respond to some particular form of stimulus energy. Although many body cells are receptors in the sense of responding to immediate stimulation, we will restrict the definition here to cells that respond to stimuli by causing *sensory nerve impulses* to be generated. Receptors in invertebrates are simply *sensory neurons:* either their dendritic endings (see Fig. 2.3c) or their modified cell bodies. Among vertebrates, most receptors either are the dendritic endings of sensory neurons or are ciliated *neuroepithelial* cells (e.g., olfactory cells, rods, cones), but others are *epithelial* cells that excite the endings of sensory neurons across a synaptic gap. Epithelial receptors in taste buds bear long microvilli at their free ends; epithelial receptors in the inner ear are *hair cells* bearing nonmotile cilia, which differ from neuroepithelium in not having their opposite ends drawn out into elongated axons.

A *sense organ* exists whenever one or more receptors are associated with accessory tissues that abet the receptors' function; e.g., in the vertebrate eye, a complicated sense organ, only the rods and cones are receptors, all other structures being auxiliary (see Fig. 18.10a and b). Table 24.2 summarizes known senses of mammals. Note that what we often think of as a single "sense" may be multiple, if more than one specific type of receptor is present.

> **Example 14:** Vision is a dual sense: rods function only in dim light and provide *scotopic* vision; cones function only in bright light and furnish *photopic* vision; furthermore, mammals such as primates have three specific kinds of cone cells that, respectively, absorb light most strongly in the red, green, and blue parts of the spectrum, providing such mammals with *color vision*. Similarly, taste buds fall

Table 24.2   Senses of Mammals

Sense	Location of Receptors	Receptor Type
Scotopic vision (nocturnal)	Rods in retina of eye	Radioreceptor
Photopic vision (day vision)	Cones in retina of eye	Radioreceptor
Taste (4 acuities in humans)	Taste buds in tongue	Chemoreceptor
Olfaction (7 acuities in humans)	Neuroepithelium in nasal lining	Chemoreceptor
Hearing	Hair cells in organ of Corti of cochlea in inner ear	Mechanoreceptor
Gravity, linear acceleration and deceleration	Hair cells in utriculus and sacculus of inner ear	Mechanoreceptor
Angular acceleration and deceleration	Hair cells in ampullae of semicircular canals in ear	Mechanoreceptor
Touch:		
Hairless skin	Meissner's corpuscles in dermis of extremities and lips, and in tip of tongue (rapidly adapting); Merkel's discs in dermis of tips of digits (slowly adapting)	Mechanoreceptor
Hairy skin	Nerve fiber baskets around hair follicles	Galvanoreceptor?
Heat	Unknown, in skin, wall of gut, hypothalamus	Radioreceptor
Cold (reduction in kinetic energy)	Unknown, probably in skin, gut, and hypothalamus	Mechanoreceptor?
Muscle tension	Spindle organs in striated muscles	Mechanoreceptor
Joint position and tension	Ruffini's end organs in joint capsules; Golgi tendon organs in and around joints	Mechanoreceptor
Pressure	Pacinian corpuscles in dermis and under mucous and serous membranes	Mechanoreceptor
Distension of hollow organs	Stretch receptors in walls of gut, lungs, blood vessels (e.g., carotid sinus)	Mechanoreceptor
Blood pH, CO_2, and O_2	In carotid and aortic bodies	Chemoreceptor
Thirst	In lining of rear of pharynx	Osmoreceptor?
Blood concentration	Hypothalamus	Osmoreceptor
Magnetism	Unknown, perhaps in brain	Magnetoreceptor
Pain: sharp, localized	Dendritic endings of A-delta sensory neurons (myelinated)	Chemoreceptor?
Pain: burning, diffuse	Dendritic endings of C sensory neurons (nonmyelinated)	Chemoreceptor?
Electricity	In muzzle skin of platypus	Galvanoreceptor

into four distinct types, each primarily excited by one type of solution that elicits only one of the four primary taste sensations: sweet, sour, bitter, and salty. Olfaction probably includes seven distinct senses.

24.27.   What types of receptors are known in animals?

Receptors can be classified according to the type of stimulus to which they are sensitive: *chemoreceptors* respond to molecules and ions; *mechanoreceptors* detect mechanical stimuli (touch, pressure, tension, vibration); *radioreceptors* respond to energy in the electromagnetic spectrum; *magnetoreceptors* detect magnetic fields; and *galvanoreceptors* are sensitive to even very weak electrical fields.

Receptors can also be described in terms of their location, or the source of stimulation: *exteroceptors* detect stimuli from outside the body, while *interoceptors* detect stimuli from within the body. Two types of interoceptors are *visceroceptors* located in the walls of organs and *proprioceptors* located in skeletal muscles, tendons and joints.

24.28.   What are chemoreceptors?

Chemoreceptors bear receptor sites in their plasma membranes, into which molecules or ions may fit.

> **Example 15:**   Vertebrate olfactory receptors are ciliated neuroepithelial cells scattered among the columnar epithelial cells that line the nasal cavity; their inner ends are prolonged as axons reaching to the olfactory bulb of the brain, while their free ends bear immobile cilia having receptor sites into which molecules of certain shapes can be accommodated. Humans seem to have seven kinds of olfactory receptors, resulting in only seven primary odor sensations (floral, pepperminty, ethereal, camphoraceous, musky, putrid, and pungent). Supporting this idea is the fact that the odor of a synthetic organic compound, if any, can be predicted provided that its molecular shape conforms to any of seven configurations associated with these basic odor sensations.

Internal chemoreceptors include cells which monitor blood oxygen level and osmoreceptors which respond to dehydration.

24.29.   What are some kinds of mechanoreceptors?

Hearing in vertebrates depends on sensory hair cells in the inner ear. In terrestrial species, stimulation of these receptors depends on propagation of pressure waves from air to the eardrum and ossicles and thence into the perilymph of the cochlea, where they cause vibratory displacements of the basilar and tectorial membranes that pull on the cilia of the hair cells (see Fig. 22.4). Gravity reception in general involves the mechanical distortion of ciliated hair cells by some weighty material that is displaced by gravity, such as the calcareous otoliths in the inner ear of fishes (see Fig. 18.11a). The semicircular canals monitor rotational head movements in three coordinates, by the rotation of fluid within the canals, stimulating a cluster of hair cells in each swollen ampulla. Exteroceptive cells known as *rheoreceptors* detect water currents, allowing freshwater animals to move mainly upstream. *Thigmoreceptors* are concerned with touch; two major types in mammals are (1) corpuscles found just below the epidermis and most abundant in hairless skin, which detect the texture of objects being felt and (2) baskets of nerve fibers around the bases of hair follicles, which detect when the hairs are being touched. The latter may really be galvanoreceptors, since bent hairs generate electricity. *Proprioceptors* ("self sensors") are responsible for *kinesthesia*, the sense of position and movement of the body and its parts. Proprioceptors are essential for postural maintenance and muscle coordination. *Stretch receptors* in the walls of large vessels initiate reflexes that modify heart rate. Stretch receptors in the stomach wall provide sensations of fullness and inhibit further ingestion, while those in the colon and urinary bladder initiate defecation and urination.

24.30.   What are radioreceptors?

Radioreceptors respond to radiant energy: *photoreceptors* detect light, and *thermoreceptors* infrared radiation (heat). Sensitive thermoreceptors occur in the facial pits of pit vipers and lip pits of boas. Photoreception involves the absorption of light by *visual pigments* in the retina cells. These represent a combination of an enzyme ("opsin") with an aldehyde form of vitamin A called *retinal* that keeps the enzyme inactive until it is illuminated. When occupying the active site of the opsin molecule, retinal has a bent form; light absorption causes the bent molecule to straighten out and separate from the opsin. An enzyme cascade is believed to ensue, in which activated opsin serves to activate the next enzyme, and this one the next, until a pyramiding amplification enables the entire cell to respond to as little as one photon of light. For the cell to continue to function, the retinal-opsin complex must become reestablished; in the human rod pigment *rhodopsin*, such regeneration can occur only in dim light, but for the cone pigments (*iodopsins*), the regenerative mechanism is not inhibited by light.

24.31.   What photoreceptive organs have evolved in animals?

Eyes have evolved along only a few basic patterns (Figs. 18.10a and b and 24.6). *Cup eyes* (see Fig. 10.2b) are shaped like open cups, with light entering the open end and a pigment screen preventing light from penetrating through the cup wall. (Such pigment screens are important in many eyes, including those of vertebrates, in which the black-pigmented choroid coat underlying the retina performs this function and is continuous with the iris, which keeps light from entering the eye's posterior chamber except through the pupil.) *Vesicular eyes* as seen in vertebrates and most cephalopods are closed vesicles filled with fluid; in some nocturnal species the spherical lens is so large as to fill nearly the entire space within the eyeball. The photoreceptors of *compound eyes* (Fig. 24.7) are clustered at the inner end of cylindrical functional units known as *ommatidia*, and each ommatidium provides a single point in a mosaic field of vision.

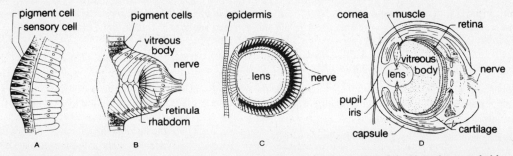

Fig. 24.6   Invertebrate eyes: A, Jellyfish; B, beetle larva; C, snail; D, cuttlefish. Note the remarkable resemblance of the cuttlefish eye to the vertebrate eye, an outcome of convergent evolution. (*From Storer et al.*)

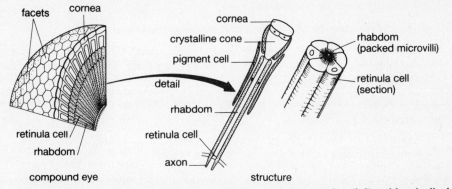

Fig. 24.7   The arthropod compound eye. General structure in longitudinal section (*left*) and longitudinal and cross sections of a single ommatidium (*right*). The focusing apparatus consists of cornea and crystalline cone, and the sensory rhabdom consists of the packed microvilli of a cluster of photosenstive retinula cells. During the daytime a pigment sleeve encloses each ommatidium so that only light rays aligned to its long axis enter, allowing each ommatidium to provide one separate dot in a mosaic image; at night the pigment sleeve retracts, increasing acuity in dim light at the probable cost of producing blurred and overlapping images, since the sensory cells can then be affected by light rays penetrating from various angles.

Whatever the overall form of the eye, only two main types of *photoreceptors* exist in the animal world. Invertebrate photoreceptors bear numerous long, pigment-filled microvilli that protrude at right angles to the longitudinal axis of the cells; vertebrate rods and cones are derived from ciliated neuroepithelium of the embryonic brain wall: a single mammoth cilium develops a stack of pigment-filled discoidal projections as outgrowths of its covering membrane.

24.32.   What are magnetoreceptors?

Magnetoreceptors detect magnetic fields as weak as that of the earth. They seem to depend on the presence of *magnetite* in certain sensitive cells.

> **Example 16:**   Magnetic sensitivity has long been recognized in marine snails that have magnetite in their radular teeth. A homing pigeon trained to fly under total overcast can still home successfully, but not if a small bar magnet or electromagnetic coil is attached to its head. Even blindfolded and disoriented human subjects told to face in a homeward direction can usually do so within a few degrees, but like birds are confused by bar magnets taped to their heads.

24.33.   What are galvanoreceptors?

Galvanoreceptors detect electric currents. They occur in electric fishes that orient by generating weak radar fields and in certain teleosts and elasmobranchs that can locate prey by the extremely tenuous electrical currents produced by living organisms. Galvanoreceptors of sharks occur in small cutaneous pits on the snout, known as *ampullae of Lorenzini*. Galvanoreceptors used in locating prey also occur in the skin-covered snout of the duck-billed platypus.

24.34. What general principles apply to sensory input?

(*a*) An organism cannot detect stimuli if they lie outside the range of sensitivity of its receptors; thus, we detect as light only a narrow band of wavelengths ranging from about 400–750 nm, and our eyes are blind to all other portions of the electromagnetic spectrum.

(*b*) Different species may have different sensory worlds.

> **Example 17:** Dogs, dolphins, and bats can hear sounds far above our upper limit of hearing. Although the visible spectra for people and honeybees overlap, bees are blind to red light that we see, but they see ultraviolet, which we cannot. Certain moths not only see light as we know it, but even see x-rays: when a room is lighted, they fail to detect x-rays, but they jump up and fly when the beam is suddenly turned on in a darkened room.

(*c*) Different *individuals* of the same species may have different sensory worlds.

> **Example 18:** A number of genetic *taste polymorphisms* exist among humans; e.g., some detect minute amounts of phenylthiocarbamide (PTC) as very bitter, while for others it is tasteless in any concentration.

(*d*) Different species may have differing *thresholds* of sensory acuity. Definition of such thresholds must be deduced from animals' responses, e.g., feeding.

> **Example 19:** Honeybees reject weakly sweetened water acceptable to butterflies; this prevents them from bringing back to the hive nectar too dilute to convert into honey.

(*e*) An individual organism may possess receptors with the same range of sensitivity but having different *thresholds* of acuity.

> **Example 20:** Noctuid moths have an ear on each side of the body that is marvelously economical, for it has only two sensory cells, one a low-threshold cell and the other a high-threshold cell. Both are most sensitive to sound frequencies emitted by hunting bats. The low-threshold cell in each ear begins to fire when the bat is far away and its cries faint, allowing the moth to steer a course directly away from the predator. The high-threshold cell responds only when a bat is close by, causing the moth to plunge toward the ground.

(*f*) An individual organism may possess several kinds of receptors that divide up a range of stimulus energy. Such *range fractionation* can greatly increase the sensory information obtained, as when the red-, green-, and blue-absorbing cones of the human eye make color vision possible.

(*g*) *Sensory adaptation* takes place when a stimulus becomes a constant condition.

> **Example 21:** When a microelectrode is inserted into a sensory nerve fiber and a constant stimulus (such as pressure) applied to its receptive end organ, fewer and fewer impulses pass along the nerve fiber as the stimulus persists. Human subjects report that sensation wanes when a stimulus is maintained.

24.35. What is meant by perception?

Perception refers to the *analysis and interpretation* of sensory information, mainly in the brain. Rods and cones in the retina constitute our visual *receptors;* the visual areas of the brain are where nerve impulses from the retina are first resolved into shape, movement, and color. Further perception may involve recognition and evaluation of what we see; e.g., redness may translate into a traffic light, and we apply the brakes. Heredity and learning can both affect perception. We tend to notice and remember stimuli that are meaningful in our culture or personal experience. When we try to analyze the behavior of nonhuman species, we cannot ask them what they feel or understand, we can go only by what they *do*. Here, we run up against some remarkable and puzzling phenomena that hold promise of giving us significant insights into how brains function.

24.36.   What are sign stimuli?

In the animal world, certain stimuli known as *sign stimuli,* or *releasers,* are so potent that they tend to trigger specific responses even when presented in the form of simplified models. The responses provoked are usually in the category of more or less *fixed action patterns* (FAP) that tend to be performed in a rather stereotyped, species-typical manner. Hypothetically, a given FAP is genetically organized in the brain, but cannot be performed until an appropriate signal (the releaser) neutralizes some inhibitory mechanism (the *innate releasing mechanism*). Simplified models with releaser properties can elicit aggressive, sexual, and feeding responses in a wide variety of animals.

> **Example 22:**   Birds, which have exceptionally keen eyesight and are perfectly capable of learning many things such as what is good to eat, act like complete idiots when presented with a relevant sign stimulus in the form of a simplified model. Redwing blackbird males flail away at stuffed black socks with red patches sewn on where a male's shoulder epaulets should be, although remaining unmoved by socks that are only black. English robins rip into small tufts of red feathers mounted on wires in their territories, while ignoring stuffed juveniles that lack the red breasts of adult males. Furthermore, certain stimuli (*supernormal stimuli*) have been identified that have even greater releaser potential than the real thing, as when a gull ludicrously ignores its own eggs to try to tuck under its body a realistically painted giant egg.

Many similar instances reinforce the conclusion that sign stimuli represent a strange contradiction between sensory reception and perception: it is as if they enjoy a key-in-lock relationship with respect to some neural stimulus analyzer that causes other relevant but contradictory stimuli to be totally discounted. A similar key-in-lock phenomenon seems to occur in some learning situations, where *what* can be learned and *when* learning can occur seem to be genetically restricted.

> **Example 23:**   Mockingbirds can learn new songs throughout their lives, but this is not true of most birds. White-crowned sparrows and American bluebirds raised in isolation never sing full songs as adults if as nestlings they do not get to hear the song of their own species; given this opportunity, the following spring they spend some time trying out different tonal sequences, and each that "fits" is set unchangeably into the vocal repertory until the species-specific song is complete. American bluebirds have been reported to sing a normal adult song after hearing it only *once* as juveniles, and even if played backward! If, however, young birds of these species are allowed to hear only the song of a different but related species during the *sensitive period* when such learning can take place, they prove totally incapable of learning that song, and in adult life give voice to only the simple subsong they develop when not exposed to any type of birdsong at all. Such constraints on learning imply limitations on the perception of those stimuli that cannot be learned, while the effective stimulus (the proper song) seems to have the qualities of a sign stimulus.

24.37.   Does genetic programming affect behavior?

Certain behaviors appear to be strongly influenced by *heredity,* which seems to indicate that response as well as perception can be influenced by genes. *Instincts* can be defined as behaviors that are performed in an effective, functional manner the first time they are elicited. Although many instincts are subject to modification with experience, as when birds become more dexterous in building their species-typical nests, others seem remarkably resistant to modification by relevant experience. This is admittedly an extreme situation, for most forms of behavior involve an interplay of both genetic predisposition and learning, but the very existence of behaviors that seem to owe nearly everything to genes and little or nothing to learning poses a great riddle as to how genes in nerve cells can do anything so remarkable as to pattern a behavioral response. (When you come to think of it, learning is equally remarkable, for it too seems to involve gene regulation, since it results in altered patterns of brain RNA synthesis.)

> **Example 24:**   A simple behavioral response in humans is which thumb one puts on top when clasping one's hands together. For nearly everybody, there is only one way that feels "right." This is a simple monogenic Mendelian trait with dominance: persons who clasp their hands with the right thumb on top carry the dominant allele and can be homozygous or heterozygous; those who put the left thumb on top are homozygous for the recessive gene. The two phenotypes are so clear-cut as to make it easy to work out a family pedigree for this trait: try it. Thumb-on-top has nothing to do with hand preference in writing, and there appears to be no difference in survival value between the two phenotypes.

24.38. What other behaviors owe much to genetic programming?

(*a*) *Perinatal behaviors* occur right around birth and are unlikely to be modified by experience when they are performed only once.

> **Example 25:** Newly hatched birds of parasitic species may be programmed to dispose of nest mates mercilessly: an infant honeyguide, naked and blind, simply kills any other hatchling with its fiercely hooked bill; a baby cuckoo, equally naked and blind, gropes about, takes onto its back any egg or chick, and pushes it up the side of the nest and out.

(*b*) *Reproductive behaviors* often have a substantial component of genetic programming.

> **Example 26:** When primed by injected sex hormones, young animals such as turkey chicks precociously perform courtship and mounting behaviors in a remarkably adult fashion; this indicates that the behavior patterns are already organized in the infant brain and require only the hormone to lower the threshold for their performance.

(*c*) *Avoidance behavior* is usually based on learning, but in some cases it may be genetically determined.

> **Example 27:** Motmots raised in captivity refuse to bite rods banded in red and yellow, a reaction that would protect them in the wild from trying to eat poisonous coral snakes.

(*d*) *Habitat preference* can be so strongly defined by heredity that not even captive raising in alternative habitats can reverse the innate preference.

> **Example 28:** Prairie deer mice still select open grassland when given a choice, even when raised for several generations in laboratory cages or in a woodland setting. In the wild, blue tits strongly prefer to perch in oak trees as opposed to pines, whereas coal tits much prefer pines to oaks, a difference that reduces competition between these species; birds raised in captivity without foliage retained their species' preferences with little weakening.

Innate habitat preference seems to be a *perceptual* phenomenon: in a species-typical manner the organism simply perceives one habitat as more desirable than another. The unanswered question is, How do the stimuli presented by the genetically preferred habitat operate on the brain to induce the observed responses?

24.39. Can responses be modified by individual experience?

We have just seen that heredity can powerfully affect certain behaviors and that some responses owe little or nothing to that kind of relevant personal experience on which learning is based. However, no animal can survive in an unpredictable environment without some behavioral plasticity, which we call *learning*. The individual's behavior may be modified by forms of restricted learning such as imprinting, and by the less restricted forms of learning known as classical conditioning, instrumental learning, allelomimetic learning, and cognitive learning.

24.40. What is meant by restricted learning?

Much learning is restricted: genes exert constraints on learning, so that some learning may take place only at a certain sensitive period in early life; also, some things may be learned readily, while others cannot be learned at all. Restricted learning usually relates to the performance of certain behaviors characteristic of the whole population or species that are particularly vital for survival or successful reproduction. Thus, adult white-crowned sparrows sing the adult song correctly only if they are allowed to hear it at least once while nestlings: they are unable to learn it at a later time, and they cannot learn the song of any other species, at any time at all. *Imprinting* is a form of restricted learning that can take place only during a limited sensitive period, often in early life.

> **Example 29:** Baby chicks and goslings imprint on whatever slowly moving object they get to see and follow when about 16 hours old; this has the normal effect of fixating their following response upon the mother, but experimentally they can be imprinted on a wide range of alternatives, even boots on slowly walking human feet. Once established, such imprinting cannot be undone.

Imprinting during early life may exert latent effects on adult life. Birds that acquire their songs in the manner of white-crowned sparrows can be said to become imprinted to their father's song, heard while they are nestlings. However, some birds can imprint on the song of foster parents.

Example 30: Zebra finches raised by Bengalese finches sing a Bengalese finch song when mature, even though this is quite different from the song of their own species; once imprinted on their foster father's song, they become quite unable to acquire the zebra finch song when grown. Similarly, imprinting on one's parents often shapes mate preference in adulthood, and animals, especially birds, raised by foster parents of another species often prefer when adult to woo females like their foster mother, ignoring females of their own kind.

24.41. What is classical conditioning?

Classical conditioning, or *associative learning,* takes place when a secondary or, *conditional, stimulus* comes to elicit a response because it has occurred along with the natural or, *unconditional, stimulus* often enough for an association to become established. Experimentally demonstrated as *Pavlovian* conditioning, such association may even involve *autonomic conditioning:* evocation of a physiological reflex by the conditional stimulus. Thus, dogs given food right after a bell is rung not only come to behave as if expecting food but also salivate copiously when the bell is rung even when food is withheld.

24.42. What is instrumental learning?

Instrumental learning involves the determination of behavior by its consequences: behavior that is *positively reinforced* (rewarded or successful) is repeated, whereas behavior that is *negatively reinforced* (punished) is abolished. Thus a toad soon learns which insects are tasty and which will nauseate or sting. In animal training, however, it has been found that these effects do not yield symmetrical results: the role of reward in strengthening a response is much greater than the role of punishment in abolishing it. Thus a powerful tool for trainers is *operant conditioning,* the coupling of desired response with reward. Instrumental learning, also called "trial and error," is the most universal form of behavioral modification. If a previously rewarded behavior ceases to be reinforced, eventually it will no longer be performed; this phenomenon, *extinction,* does not imply forgetting, because the response is quickly resumed if reinforcement is restored. A similar and very common process is *habituation,* the abolition of a response that is not reinforced at all: the situation comes to be ignored as meaningless (e.g., songbirds that mob living birds of prey soon habituate to a stuffed hawk or owl, because their mobbing produces no response).

24.43. What is allelomimetic learning?

Allelomimetic learning, or *aping,* results from the tendency of many vertebrates to behave as others of their kind do. Not all allelomimesis involves learning: birds fly in a coordinated flock, and fishes in a school swim without colliding, simply through keen awareness of what others in the group are doing. Many young mammals learn to feed by watching what their elders do; conversely, the habits of eating sweet potatoes and throwing grain into water to separate it from sand spread gradually through a rhesus monkey troop after being originated by an inventive youngster, showing that elders can learn from the young as well. When aluminum foil was adopted in England for capping milk bottles, blue tits attracted by the glint found they could peck through the cap and drink the cream; since this behavior spread like ripples on a pond from several different points, apparently other blue tits learned by watching those who first stumbled onto this trick.

Allelomimesis can *inhibit* learning as well as facilitate it: a rat not trained to a bar-pressing response to get a drink will learn the response more quickly when caged where it can watch another rat pressing its own lever and drinking than when it is kept in isolation; conversely, if caged next to a rat provided with neither lever nor water, the subject rat is considerably slower to press the lever in its own cage than if it had no neighbor at all to watch.

24.44. What is cognitive learning?

Cognitive learning, or *insight,* entails some grasp of relationships that allows a new behavior to be performed without going through random trial and error. What is involved in insight and how widespread it is among nonhuman species remain in dispute.

Example 31: Chimpanzees handle things; faced with the problem of obtaining fruit hung out of reach, they quickly stack up boxes and use them to climb up on. Is this cognition, or simply the result of their manipulative tendencies? Have they previously manipulated other objects and suddenly perceived the possibilities inherent in boxes? Dogs are superior to raccoons in solving *detour problems* to get to food, to wit, going *away* from the food to get around an intervening barrier, which seems an insightful approach to the problem. In humans, insight learning often involves applying to a new situation something learned in a different context, or visualizing the outcomes of various courses of action as a mental exercise in trial and error, so to speak.

We should keep in mind that comparative studies of animal intelligence are usually both misleading and unfair: different species perceive and learn in different ways, and a species' behavior is as much a part of its total package of adaptations as its physiology and form. If individual numbers and durability through geologic time are taken as criteria of biological success, cockroaches and roach behaviors might be judged far more successful than humans and cognitive behaviors.

24.45. What effector mechanisms may animals possess?

An animal usually responds to stimulation by moving, but other responses are possible, depending on its effector mechanisms.

(a) *Muscular contractions* (see Question 23.41) bring about the majority of behavioral responses, including orientation, movement, posturing, and locomotion.

(b) *Photogenesis* is production of light, often in patterned signals, as by fireflies.

(c) *Electrogenesis* is production of electrical discharges, either of stunning force (e.g., electric ray) or as a weak radar field (e.g., gymnotid teleosts).

(d) *Chromatophore responses* by pigment cells controlled by nerve fibers or hormones can bring about color changes in response to background (e.g., chameleons) or social situations (e.g., cuttlefish).

(e) *Secretion* of defense chemicals (e.g., skunks), or of communicative pheromones (e.g., sex attractants, territory markers), is a widespread phenomenon.

(f) *Sound production*, by setting air or water into vibration (e.g., by way of vocal folds in the larynx or by muscles in the avian syrinx), is useful in communication.

Chapter 25

Animal Ecology

No organism is an island: each, no matter how solitary, is a part of an *ecosystem*. Each embodies a complex of adaptations that enable it to endure, exploit, and effectively interact with other components of that ecosystem.

25.1. What is an ecosystem?

Eco means "household." *Ecology*, "science of the household," is the study of interrelationships among living things and between life and its physicochemical environment. "Ecosystem" ("placing the household together") is a term that implies the principle that all organisms sharing a common habitat relate to one another and to their habitat in complex ways that knit together a cyclic use of mass and a spiraling flow of energy. Thus an *ecosystem* may be defined as a *biotic community* and its *physicochemical environment*, in which mass and energy pass through the system in a series of stages, with mass cycling indefinitely, while energy flows thermodynamically downhill, passing from component to component in the system as it does so. A biotic community consists of *all living things*, in terms of both species and individuals, that live *sympatrically*, sharing a common habitat.

25.2. What are closed and open ecosystems?

(*a*) A *closed ecosystem* is unproductive in the sense that nearly everything produced within it remains there and is recycled, often for a very long time. No ecosystem can be completely closed, because energy must be replenished from some external source, usually the sun. Earth itself represents a closed planetary ecosystem, sustained by the sun. Smaller ecosystems exist within larger ones, so that the planetary ecosystem contains both marine and continental ecosystems stable enough to be considered "closed," but with the latter requiring precipitation as well as solar energy, and the former receiving runoff from the land, giving back as clouds vast quantities of water evaporated from its surface.

(*b*) An *open ecosystem* is like a whirlpool in a river: *both* mass and energy flow into it, are cycled, and flow out. Open ecosystems are productive in the sense that much of the mass produced is exported to some other ecosystem.

> **Example 1:** *Estuaries* and salt marshes are highly productive open ecosystems: on the landward side they receive a flow of fresh water laden with silt and solutes, and from the sea, young fish and larval invertebrates that concentrate in these "nurseries of the sea." These ecosystems export to the sea a flood of dissolved organic nutrients from decaying vegetation and a wealth of maturing marine fishes and shrimp, while still sustaining their own populations of indigenous estuarine fishes and other marsh species including waterfowl. In terms of fish production alone, estuarine productivity is 50–200 times greater than that of the neighboring open sea.

25.3. What are successional ecosystems?

Successional ecosystems are unstable and transitional stages in a process of *ecological succession* toward a stable, self-perpetuating climax state. They exist where disturbances such as lava flow, fire, farming, or flooding have destroyed preexisting communities. Various successional ecosystems (*seres*) prevail in sequence, each providing the seeds of its own destruction by creating conditions suitable for the next sere. *Primary succession* begins on bare rock, as after a volcanic eruption or landslide, where soil must first be built by pioneer species such as lichens, primitive plants that erode rocks with acid. *Secondary succession* occurs where soil already exists, as after a forest fire, or when farmland is abandoned, and requires fewer seres to regain the climax stage.

Lakes and ponds are successional ecosystems doomed by siltation and *eutrophication*. The latter is a runaway process in which rotting aquatic vegetation provides nutrients for the growth of more vegetation, which in turn decomposes, nourishing more plant growth, and so on. Eutrophication is much accelerated in lakes that receive sewage and agricultural runoff; so much plant growth occurs that decomposition of dead vegetation exhausts the dissolved oxygen, fish die, and the lake becomes barren. Without contamination from human processes, succession can be so slow that certain large, deep lakes have existed since the Mesozoic.

Subclimax seres may become relatively *permanent* features, in a perpetuated instability if succession is arrested short of the climax.

> **Example 2:**   California's renowned redwood and sequoia forests represent successional ecosystems maintained by fire: without it, their seeds fail to germinate and seedlings from the neighboring climax forest start growing up under the giants. Managed burning now replaces natural fires that preserved these ancient trees through countless generations.

25.4.   What are climax ecosystems?

Climax ecosystems manifest a state of internal stability and natural balance that resists perturbation, like a gyroscope. If a climax ecosystem is destroyed, as by fire, secondary succession ordinarily leads to the reestablishment of the original climax community, even though this may take centuries. An exception to this is *desertification:* replacement of the former climax ecosystem by deserts, a process much aggravated by human activities. Extensive forests modify climate by transpiring enormous quantities of water vapor, prevent erosion, allow water to sink in and contribute to the water table, and reduce *albedo,* or reflectance of the land surface. Deforestation and overgrazing reduce humidity and increase albedo so that the climate may change irreversibly. When the abused land becomes unproductive for human usage, the original plant cover may never grow back, and the indigenous fauna, dependent on the flora, must vanish too.

Spared such extreme disturbance, major climax ecosystems may endure even for many millions of years, during which their biotic communities have time to evolve many coadaptations through which each species comes to occupy a particular *ecological niche* within the ecosystem. A *niche* can be defined as the role a given species plays in its community: the sum total of its requirements, tolerances, and interactions with other species. Climax ecosystems are characteristically complex, including a wide variety of species adapted for many different niches.

25.5.   What are biomes and biogeographic realms?

Biomes are major terrestrial climax ecosystems characterized by a certain type of dominant vegetation, which is largely determined by *mean temperature* and *annual precipitation* (Fig. 25.1). These physicochemical factors determine where in the world certain biomes can occur, in terms of both latitude and altitude. Altitudinal life zones, as seen in mountainous regions, echo latitudinal zonation and produce equivalent biomes on a more restricted scale (Fig. 25.2).

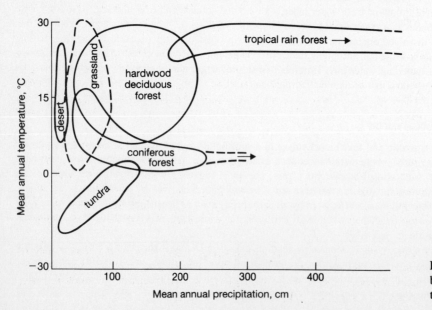

Fig. 25.1 Occurrence of major biomes as determined by mean temperature and precipitation.

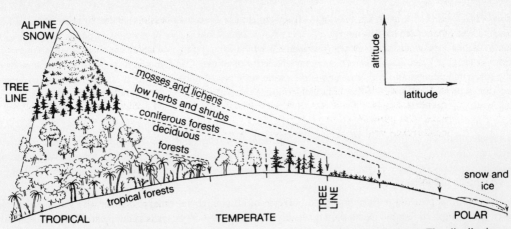

Fig. 25.2　Equivalent altitudinal and latitudinal zonation in the northern hemisphere. The distribution of plant associations, here determined mainly by temperature, influences the distribution of animals. (*From Storer et al.*)

In various geographic regions, the particular *species* forming the biotic community of a given biome will of course differ, especially when equivalent biomes occur in different *biogeographic realms.* These realms represent regions isolated from one another by major geographic barriers, mostly the sea, for the better part of the past 65 million years while living things were rediversifying following the mass extinction event which ended the Mesozoic Era. This long-standing isolation commenced with the mid-Mesozoic breakup of Pangaea, so that by the beginning of the Cenozoic Era the *Neotropical* (South American), *Ethiopian* (African), *Oriental* (south Asian), and *Australian* realms were already separated from one another by sea, and from the northern supercontinent (Laurasia) that later fragmented into the *Palearctic* (Eurasian) and *Nearctic* (North American) realms. Since then, the isthmus of Panama has risen out of the sea, connecting the Nearctic and Neotropical realms by a narrow neck that has allowed a certain amount of faunal exchange. Also, the Indian subcontinent has collided with Eurasia, bringing the Oriental and Palearctic realms into proximity, but the Himalayan mountains, thrust up in the force of that collision, have maintained a geologic and climatic barrier to faunal distribution.

25.6.　What are the major types of terrestrial biomes?

(*a*)　*Deserts* occur wherever precipitation is less than 25 cm/yr. Natural deserts often occupy a *rain shadow* in the lee of mountains tall enough to form a cloud barrier, so that rain falls mainly on the windward side. Plants and animals must be *xerically* adapted to withstand water scarcity, animals often by burrowing or being nocturnal, and losing little water as urine.

(*b*)　*Tundra* and high-altitude *alpine* ecosystems occur above or north of the timberline, where temperatures remain so low throughout the year that growing seasons are very short and trees cannot mature. Tundra vegetation grows in shallow soil atop *permafrost* that remains unmelted throughout the year. In summer many animals migrate northward into the tundra, or vertically to alpine meadows, taking advantage of rapid seasonal plant growth. Smaller nonmigratory mammals such as marmots may hibernate or, as pikas do, den up with plenty of winter stores. The Arctic tundra is vital to many migratory water and shore birds, which breed there in the unending summer daylight.

(*c*)　*Grasslands,* the most recently evolved biome, are well adapted to hold soil against erosion owing to the dense fibrous root systems of grasses. They mostly occur where annual precipitation is 25–75 cm and have invaded many formerly wooded regions. The North American plains and Russian steppes are treeless, while in eastern Africa tree-studded acacia-grass *savannas* are prominent. These tropical savannas experience up to 150 cm/yr of rainfall, but this is concentrated in such a short wet season that the remainder of the year is so dry that grass fires are common. Grassland animals are often adapted to run rapidly, especially such fleet-footed herbivores as bison and antelope, and many migrate in great nomadic treks.

(*d*)　*Chaparral,* evergreen scrub forest, is limited to coastal regions, especially in southern California and around the Mediterranean, where temperatures are subtropical and the rainfall (25–75 cm/yr) highly seasonal. The long-lived dwarf trees bear reduced leaves with heavy cuticles. Many are fire-resistant. Being diminutive, chaparral mainly shelters small xerically-adapted animals, including arthropods, reptiles, and rodents.

(e) *Coniferous forests,* also evergreen, survive where winters are too severe for grasslands and other types of forests to endure. The foliage, reduced to moisture-conserving, wax-covered scales or needles, is not shed, and so, conifers can make the best of the short growing season. Precipitation is highly variable, but even when abundant, during much of the year water may be frozen and unavailable for plant use. Coniferous forests occur in mountainous regions just below the treeless alpine biome, and across wide stretches of northern Eurasia and North America, where they are known as *taiga.* Animals of the taiga include grizzly and black bears, deer, moose, wapiti, wolves, porcupines, and lynx.

(f) *Deciduous hardwood forests* occur in temperate zones where precipitation of 100 cm or more is rather evenly distributed throughout the year. Plants become dormant in winter, leaf drop building a rich humus that nourishes thick undergrowth. Deer, tree squirrels, and raccoons still frequent these disappearing forests, which are also the breeding grounds of many migratory songbirds, but countless acres have been cut down in favor of more rapidly growing pine plantations.

(g) *Tropical rain forests,* the most populous and diverse of all continental ecosystems, are also the most endangered. Once girdling the earth in all equatorial regions blessed with 250–450 cm/yr of rainfall, this biome has already been greatly reduced by deforestation, and the remainder is now disappearing at the rate of 1 percent per year. Unfortunately, the soil sustaining this lush growth is quite poor: it loses fertility and bakes like clay when the forest is removed. Native plants survive through the agency of *symbiotic root fungi* that trap nutrients before they can be leached away, but attempts to convert these regions to agriculture have spelled a story of repeated failure. Rain forest fauna, including many insects, amphibians, reptiles, birds, sloths, and primates, are mostly arboreal and doomed to disappear as the jungle that sustains them is destroyed.

25.7.   What is the nature of aquatic ecosystems?

Aquatic organisms may be *planktonic, nektonic,* or *benthonic.* Drifting and weakly swimming animals make up the *zooplankton,* while diatoms and other autotrophic protists constitute the *phytoplankton.* Animals that live as drifters all their lives (e.g., radiolarians, salps, medusas, ctenophores, and krill) are known as *holoplankton. Meroplankton* includes larval invertebrates and newly hatched fishes that will not be planktonic as adults. *Nekton* consists of strong swimmers, e.g., fishes, squid, and marine mammals. *Benthonic* organisms are bottom dwellers; they may live on the bottom (as *epifauna*) or dug into the bottom (as *infauna*).

Aquatic animals are protected from the trials of life on land, but they have problems of their own. Freshwater animals *osmoregulate* endlessly in their hypotonic milieu; they must also avoid downstream displacement, which they counteract by *positive rheotaxis,* i.e., moving against the current. They also may face seasonal drought that can leave them dead if they cannot encyst or estivate, and if they inhabit lakes or ponds, they are subject to displacement as ecological succession advances on their habitats, which eventually are converted first to marshes, then to boggy meadows.

Marine animals have other problems. Most marine invertebrates remain in osmotic balance with seawater, but marine vertebrates have the problem of living in a medium saltier than vertebrate blood normally would be. Intertidal animals must endure exposure to air, wide fluctuations in temperature and salinity, and the destructive force of surf. Further out, they are limited by the vertical distribution of light, which reaches a depth of only about 200 m; just the upper 100 m of this *photic zone* is light enough for photosynthesis; in the *aphotic zone* below, creatures dwell in darkness except for the eerie glow of bioluminescence. Many small marine animals perform energy-costly *vertical migrations* that take them nearly to the surface at night and down into darkness by day.

In both the sea and deep lakes great differences in temperature can exist between the surface and deeper waters, which are separated by a stable zone of rapid thermal change, the *thermocline.* This thermal barrier is so effective that in summer lake water above the thermocline (the *epilimnion*) may be 20°C warmer than that below (the *hypolimnion*). In lakes, the thermocline also impedes oxygen circulation into the cold, stagnant water below. Except in the tropics, lakes undergo fall and spring *overturns* when the thermocline disappears as the surface waters are chilled or rewarmed, briefly allowing water circulation to carry nutrients up from the bottom and oxygen down to the depths. Nutrient *upwelling* also occurs in the sea, wherever deep-running cold currents are deflected upward by a land mass, while warmer surface waters may be blown aside by prevailing coastal winds. Since nutrient upwelling causes prolific growth of phytoplankon and attached algae, inshore waters are far richer in life than the open sea. In fact, most marine life occupies the shallow *neritic province* from the shore to the edge of the *continental shelf.* Lying almost entirely within the photic zone, and therefore supporting a rich crop of attached seaweeds, the neritic province also takes in the *continental seas* that inundate low-lying land whenever rising sea levels permit. Today enough water is still tied up in the polar ice caps to limit the continental seas, but time and again during earth's past, those shallow seas have extensively overlain the continents and harbored a great abundance of life.

Once the seafloor drops off the edge of a continent, descending a steep *continental slope* into an ocean *basin*, the *oceanic province* commences. This includes *pelagic* (open-water) and *benthic* habitats. The floor of ocean basins (at an average depth of 4000 m) constitutes the *bathyal benthos*, and the floors of rifts make up the *abyssal benthos* (to 6000 m) and the *hadal benthos* below that. The lighted waters of the ocean surface from shore to shore to a depth of 200 m make up the *epipelagic* region, which is most richly populated with zooplankton and nekton, for it is only near the surface that autotropic phytoplankton can survive. Below lie the aphotic *mesopelagic* (to 1000 m), *bathypelagic* (to 4000 m), and *abyssopelagic* (below 4000 m) regions.

Example 3: Robot deep submersibles have revealed abyssopelagic and benthonic fauna residing at even the greatest depths. Deep-sea animals living under tremendous pressures can still be delicately constructed, for they are no more burdened by those pressures than we are by our blanket of atmosphere; in fact they are so perfectly adapted to their dark, cold, pressurized environment that they cannot survive being hauled to the surface, even if they have no internal, gas-filled spaces that would explode upon depressurization. They are also metabolically adapted to remain active at temperatures on the edge of freezing, as epipelagic fishes do in wintry polar seas.

25.8.   How do populations behave in ecosystems?

A *population* consists of all sympatric individuals belonging to a given species. During ecological succession, populations rise and fall as old seres give way to new, but as the climax ecosystem starts to mature, the species that will comprise the biotic community of that ecosystem become established. The success of populations in ecosystems is influenced by toleration of habitat qualities, competitive exclusion, population growth characteristics, and exogenous and endogenous population-limiting factors.

25.9.   How does toleration of habitat qualities affect populations?

Toleration of various physicochemical factors that characterize a given habitat—pH, salinity, temperature, availability of water and oxygen—is the first prerequisite for survival in that habitat. The *limiting factor* for each species is that for which it has the *narrowest tolerance*.

Example 4: *Euryhaline* animals tolerant of wide fluctuations in salinity succeed well in estuaries and the intertidal zone, where *stenohaline* animals intolerant of salinity changes cannot survive. Similarly, *eurythermal* animals tolerate a wide range of temperature, whereas *stenothermal* animals are adapted to only some narrow temperature range, but not necessarily the same range: placed in tepid water a stenothermal fish from Arctic waters would die of overheating whereas a tropical fish would suffer hypothermia.

Successful reproduction occurs only within a zone of *optimal tolerance*, outside of which some individuals just get by in a zone of *physiological stress*, but rarely breed. For example, lake trout can survive in water to 18°C (65°F), but spawning requires temperatures below 10°C (50°F).

Only those species possessing the requisite environmental tolerances have even have a chance of becoming permanent members of the ecosystem, but factors relating to other species also influence success.

25.10.   How does competition affect populations in a developing ecosystem?

Gause's principle of competitive exclusion operates whenever two different species with seriously overlapping needs try for a toehold in the same ecosystem. Either (*a*) competition will eventually exclude one species or the other from that ecosystem, or (*b*) one species must alter its requirements. The latter is an evolutionary possibility, provided that a diversified gene pool allows selection to progress toward some innovative strategy, such as adopting a new diet. Such diversification provides a happy solution, for after it has taken place, the two populations will occupy *different* niches and the competitive exclusion principle will no longer apply. However, evolution cannot always come to the rescue, and interspecific competition may be to the death. Competitive exclusion operates mainly when ecosystems are still maturing, or when new species invade an established community, or when an established population in a long-lived climax community hits upon some splendid evolutionary innovation that upsets the former balance of power.

After competitive exclusion has taken its toll, or after competing populations have undergone natural selection to the point of niche alteration, interspecific competition within an ecosystem no longer is a major factor determining survival, because most species are *narrow-niched*. By having specialized needs (i.e., occupying a narrow niche) potentially competitive species manage to subdivide the habitat and minimize competition for the commodities of life. In Question 24.38 we saw that a number of animals display unlearned preferences for specific habitats. Taking this a step further, preferential use of space has been found to be a key factor in allowing related sympatric species to coexist.

> **Example 5:** A field study of five species of warblers in North American spruce forests showed that each species of these insectivorous birds has its own preferred feeding zone in the spruce trees, where at least 50 percent of its foraging time is spent, with each zone having some areas of nonoverlap with that of the other species. In Jamaica a number of species of *Anolis* lizards coexist compatibly without undue competition for insects and living space because some favor sun and others shade, some shelter in low shrubs, others live in trees, and of the arboreal species, some stay to main trunks and large limbs, while others seek terminal twigs.

By contrast, *broad-niched* species are tolerant of a wide range of environmental factors and can often exist well in a number of different ecosystems. The cougar, or puma (*Felis concolor*), now sadly depleted by hunters, once ranged throughout the Americas, inhabiting deserts, prairies, mountains, and coniferous, deciduous, and rain forests. Although puma are predatory, many broad-niched species are omnivorous and opportunistic. When a broad-niched species invades a new ecosystem or realm, it is very likely to succeed, as when the common starling was introduced from Europe into New York and spread across the entire nation within 60 years. Freed from constraints present in its native land, the invader may multiply uncontrollably, seriously competing with more desirable indigenous species and gaining opprobrium as a "pest." In North America, starlings are endangering native woodpeckers, for they usurp the woodpeckers' nest holes. On the other hand, when a narrow-niched species penetrates a new ecosystem, if it survives at all, it may do so by occupying a "vacant niche," that is, fulfilling a role in that ecosystem not assumed by any native species. In this event the invader simply enriches the fabric of the ecosystem.

> **Example 6:** Cattle egrets (*Ardeola ibis*) traditionally migrate between Africa and southern Europe, where they associate commensally with wild and domestic grazing animals, catching insects that the mammals scare up as they forage. When a flock of these handsome, white, long-legged birds found its way across the Atlantic to South America and subsequently adopted a migratory pathway between there and North America, they introduced themselves into two new biogeographic realms: Neotropical and Nearctic, in which no indigenous birds regularly exploit this particular feeding strategy. Accordingly, this narrow-niched species apparently has been successfully assimilated into the American landscape, competing with no native fauna and adding interest to the meadow scene as the egrets hunt for insects almost under the hooves of sheep and cattle.

Homo sapiens occupies a unique position in being so extremely broad-niched that humans can successfully live nearly anywhere, eat nearly anything, and transform natural ecosystems into pastures, tillage, and cities. A current danger to the survival of countless species of plants and animals, the human population, today numbering 5 billion and increasing at a world rate of *2 percent per year*, may well exceed 8 billion persons by 2015, and 16 billion by 2050. Obviously, this rate of increase cannot and will not be sustained much longer, but if it continues for even a few more decades, the planetary ecosystem may become so simplified as to contain little more than humans, their crop plants, and domesticated animals. What many do not realize is that in endangering other forms of life by this unprecedented expansion, human beings are also endangering themselves. Why this is true may be clarified below.

25.11. How do a population's growth characteristics affect its success?

Population growth curves in a developing ecosystem are characteristically sigmoid: after hanging on for some time by the skin of its teeth, the population begins to reproduce logarithmically, its numbers approximately doubling in each generation, after which the growth rate slows and a plateau, the *asymptote*, is reached (Fig. 25.3a). Thereafter, population size remains remarkably stable over long periods of time or oscillates about the asymptote in a fashion that may range from quite mild fluctuations to a rather drastic roller coaster effect. The natural balances that maintain the stability of climax ecosystems admit of population fluctuations that cancel each other out.

> **Example 7:** The boom-bust strategy is typical of migratory locusts that sporadically multiply into flying hordes that may ravage the countryside for up to 6 years, after which a recession of approximately equal length takes place during which only a few solitary locusts can be found. Fluctuation about the asymptote is usually less dramatic in character; lynx and snowshoe hare populations in the Canadian taiga oscillate together: periodic increases in the number of hares are quickly followed by a rise in

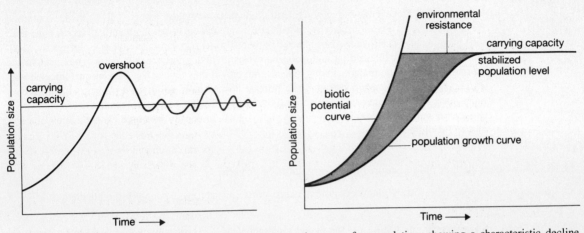

Fig. 25.3 Population dynamics. (*a*) Typical sigmoid growth curve of a population, showing a characteristic decline when the carrying capacity of the habitat is exceeded, followed by a stable fluctuation about the carrying capacity thereafter. (*b*) The difference between a population's maximum reproductive rate (biotic potential) and its actual growth curve represents a measure of environmental resistance. If environmental resistance mainly relates to food supply, the population should stabilize at the carrying capacity for the habitat; however, if environmental resistance also includes such intraspecific factors as a compelling psychological need for large territories, the population growth curve may stabilize well below the carrying capacity.

the lynx population, which reduces the hare population and then itself declines again. Thus, despite the ups and downs, a balance is maintained and neither predator nor prey species is imperiled.

The sigmoid growth curve shows that populations cannot long proliferate to the extent of their full reproductive potential, reflected by the exponential portion of the curve. As growth slows down, a widening difference is seen between the potential and actual growth curves (Fig. 25.3*b*). This difference represents the extent of *environmental resistance* to unlimited expansion. Environmental resistance includes both exogenous and endogenous population-limiting factors.

25.12.   What are exogenous population-limiting factors?

Exogenous population-limiting factors are those imposed on a population by other species or by its habitat. They include carrying capacity, predator pressure, parasites, and habitat modification.

25.13.   How does carrying capacity limit a population's growth?

Carrying capacity refers to the number of animals of one kind that can be fed adequately within the area occupied. Insufficient food can limit populations by starvation, failure to reproduce, and desertion of young (the fate of many lion cubs).

25.14.   How does predator pressure limit a population's growth?

Predator pressure controls the population growth of prey species, usually with a culling effect because predators select the easiest victims, and thus weed out senescent, immature, injured, and ailing individuals. In the absence of predators, prey species may multiply to exceed the carrying capacity and suffer starvation.

Prey species with low reproductive rates may increase the number of young born when predator pressure is intensified, as when moose preyed upon by wolves bear twin calves while single births are the rule where wolves are absent.

25.15.   How do parasites limit population growth?

Parasites ranging from viruses to lampreys may kill susceptible hosts and weaken others to the point of breeding failure. *Population density* much increases the *communicability* of certain parasites, as in human influenza epidemics.

Example 8:   Honeybees ordinarily benefit from their combined labors, but when the foulbrood bacterium strikes a hive, it usually spreads riotously through the honeycomb until all larvae have been killed and the hive destroyed. Also, infected individuals in a population can serve as vectors by which others are indirectly infected through parasitic cycles involving intermediate hosts. Thus, anopheline mosquitos disseminate malaria among humans, but only if they themselves first become infected with *Plasmodium* by sucking blood from some primate host, usually an infected person, which becomes much more likely where host populations live densely. Similarly, human helminthic infections tend to increase cumulatively in regions where high population density increases contamination of water and soil with human feces.

25.16.   How does habitat modification limit a population's growth?

Habitat modifications may destroy entire ecosystems when gross, or may merely modify the population levels of certain species when subtle.

Example 9:   In California, the Death Valley pupfish, which requires permanent water, is nearing extinction as agricultural water use progressively lowers the water table, further limiting the habitat of this species and barring it from traditional breeding ledges. Conversely, seasonal rains create vernal ponds and puddles within which many small invertebrates flourish briefly until the dry season brings death, encystment, cryptobiosis, and the blowing away of hard-walled eggs like dust.

Habitat modification by man endangers many species but has provided new opportunities for others, such as the Norway rat, roof rat, and house mouse, which favor urban environments.

25.17.   What are endogenous population-limiting factors?

These factors operate *within* the individual species and often serve to keep the population well below the habitat's carrying capacity, even when parasites and predators are few. This allows the population as a whole to be well-nourished, uncrowded, and comparatively unstressed. Such endogenous factors include intraspecific competition, territoriality, hierarchical systems, caste systems, evolved restraints on fecundity, and self-toxification.

25.18.   How does intraspecific competition limit a population's growth?

Competition among conspecifics cannot be alleviated by competitive exclusion, as can competition between different species. Cutthroat sibling rivalry may doom weaker juveniles to die of malnourishment, and those that survive are often driven off by their parents, to perish or to establish their own territories in marginal regions.

25.19.   How does territoriality limit a population's growth?

Territoriality forces individuals or groups to be spaced out in their habitat and tends to inhibit breeding by nonterritorial individuals or to compel them to emigrate. A *territory* is an area from which an animal or social group excludes conspecifics. Territorial defense may include active social fighting (usually noninjurious or at least not lethal), but often visual or acoustic threats suffice.

Usually territories are defended only against conspecifics, but some especially pugnacious species may also attempt to drive off members of other species, sometimes even when much larger, as when badgers badger bears. The psychological advantage belongs with the territory holder, and the trespasser usually vacates the premises. Territories used for breeding or display can be vital to reproductive success, since nonterritorial males may be unable to secure mates. Strange to say, even when vacant land appears available, a population does not necessarily expand indefinitely.

Example 10:   Seabirds that crowd their nesting territories into dense aggregations may require intense social interaction to achieve the physiological arousal needed to attain the breeding state; peripheral individuals, lacking the arousal value of neighbors all around, may simply not reproduce at all.

25.20.   How do hierarchical systems affect a population's growth?

Reproductive success may depend on one's rank in a social hierarchy, and hierarchical systems may limit breeding to only a few individuals in the population.

Example 11: In small wolf packs (which are extended families) only the dominant male and his mate usually produce cubs, which are fed by all pack members. Being consanguineous, the other pack members are not acting in true altruism, for by aiding the survival of each year's cubs (their cousins and younger siblings), they help pass along genes that they hold in common. Also, all benefit by the limitation of population growth achieved by restricting mating to the dominant pair.

25.21. How do caste systems affect a population growth?

Caste systems place certain limitations on reproduction in social insects by limiting fertility to one or a few individuals in the society, while the vast majority are sterile workers.

25.22. What evolutionary constraints limit family size?

Why do gulls ordinarily lay three eggs, instead of two or four? Sight of the clutch of eggs laid in the nest serves a stimulus that (by way of the hypothalamus and pituitary) turns off egg laying and triggers incubation. Brood size in many animals seems to be the outcome of opposing selective forces: maximal reproduction versus parental investment. Gulls that lay three eggs raise more surviving young than those that lay two eggs or four; four are too many to feed, and two do not make up for perinatal mortality.

25.23. How does population density limit the population's growth?

Population density can activate mechanisms that curtail reproductive success until the population thins out. The suicidal emigration of lemmings seems to be an example of this in nature, but the phenomenon can also be seen in captive animals.

Example 12: Captive rats provided with ample food and water nevertheless desert and kill youngsters when crowding reaches a critical point. Mice in crowded colonies seldom engage in infanticide: females simply cease to become pregnant, owing to an *olfactory pregnancy block* caused by a pheromonal chemical in male urine.

25.24. How can self-toxification limit population growth?

Self-toxification can cause populations to plateau or die off, even when food is plentiful; it results from excessive accumulation of waste materials generated by that population. This is a familiar event in bacterial cultures but is not likely to affect animals unless they are unable to escape from a confined habitat that has become overly polluted. Unfortunately, it is beginning to apply with increasing force to the 5 billion humans now alive, to the detriment of a good many living things besides ourselves, since the wastes of *Homo sapiens* come not only from bodily processes, but also from the industries, transportation, and power plants on which civilization relies. Sewage composed only of human wastes can be problem enough if untreated to kill pathogens and parasitic ova, but sewage also contains industrial and household chemicals that may be both toxic and nonbiodegradable, including heavy metals and polychlorinated organic compounds (e.g., DDT and PCB) that pass through food chains, progressively concentrating in consumers higher up the line. Eutrophication by excess phosphate is destroying major lakes, including the lower Great Lakes. *Acid rain* containing nitric and sulfuric acids from industrial and transport emissions is killing forests and fishes and dissolving marble statuary. Various components of polluted air, e.g., ozone, threaten health. Also, atmospheric carbon dioxide has edged up by 15 percent over the past century, mostly owing to increased combustion of wood and fossil fuels combined with destruction of forests that could have helped fix the excess CO_2. Just this small increase has already set in motion a *greenhouse effect*, and nobody can tell how serious this might become. Atmospheric CO_2 acts as a heat trap by reducing the amount of solar radiation reflected into space. A fairly mild greenhouse effect could melt the polar ice caps and alter topography interestingly; a more drastic effect might make equatorial regions, or possibly the entire planet, too hot for life to endure.

25.25. How do mass and energy move through ecosystems?

Mass is used cyclically in ecosystems, while *energy* flows spirally downward. Atoms keep their own identity and will go on entering into new molecular combinations for as long as the universe can furnish energy for such bonding. At the planetary ecosystem level, nothing but hydrogen gas can escape from the atmosphere, and only radioactive elements ever decay. Within less inclusive ecosystems any loss of atoms must be replenished, or the ecosystem will suffer. Energy enters ecosystems mainly as sunlight and is trapped in chemical bonds. As mass passes cyclically through an ecosystem, it is accompanied by the transfer of its bonding energy, but with each such transfer the amount of energy available to do work diminishes, thus the pathway of energy can be thought of as a thermodynamically downhill spiraling flow.

25.26.   How do animals gain the atoms they need from their ecosystem?

Animals obtain vital elements—chiefly nitrogen, carbon, hydrogen, and oxygen—in two ways: (1) by taking them directly from the nonliving environment (absorbing oxygen and drinking water, also gaining dissolved minerals); (2) by participating in *biogeochemical cycles* (Fig. 25.4) in which animal life depends on plants and microbes to convert certain elements into forms that animals can metabolize.

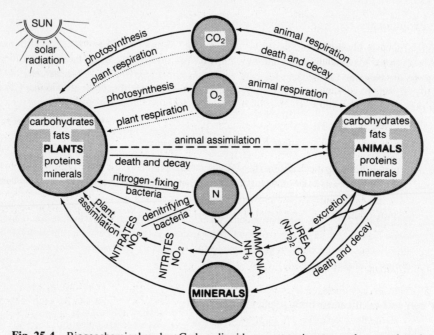

Fig. 25.4   Biogeochemical cycles. Carbon dioxide, oxygen, nitrogen, and many minerals pass cyclically between the living (plants, animals, microbes) and nonliving (atmosphere, soil, water) components of the system, being used and reused indefinitely. (*From Storer et al.*)

25.27.   How is nitrogen made available for animals?

Nitrogen enters the living world by a very narrow gateway. Nitrogen gas, N_2, makes up 78 percent of dry air: strange to say, animals breathe it in and out but cannot utilize it at all, although they must obtain nitrogen to synthesize such organic nitrogenous compounds as proteins and nucleic acids. In fact, *no* eukaryotes can directly tap this huge reservoir of atmospheric nitrogen. N_2 can be utilized as such only by certain bacteria and cyanophytes, referred to as *nitrogen-fixers*. An estimated 45 million metric tons of nitrogen per year is added to the soil by nitrogen-fixers, which are able to use N_2 in making nitrogenous organic molecules for their own uses; eventually, however, the fixed nitrogen becomes available to other organisms, mainly as ammonia (NH_3), which readily ionizes in water to the ammonium ion, NH_4^+. NH_4^+ is readily usable by plants when released directly into their tissues by endosymbiotic bacteria, but if liberated into the soil, ammonium ions bind electrostatically to clay particles and so become largely unavailable to plants. The *nitrogen cycle* must therefore continue with the *nitrification* of NH_3 by *nitrite* and *nitrate* soil bacteria. These *chemoautotrophs* obtain energy for their own metabolism by oxidizing NH_3 to nitrite ions (NO_2^-), and NO_2^- to nitrate ion (NO_3^-), which plants readily absorb and use in synthesizing nitrogenous compounds. Animals tap into the cycle at this point by eating plants directly as herbivores, or indirectly as carnivores. Animal digestive processes hydrolyze proteins to amino acids, and nucleic acids to nucleotides, and these smaller molecules are absorbed into the tissues and used in building components of the animal body. As nitrogenous compounds are catabolized, urea, urates, and ammonia are excreted as wastes, mostly to be recycled locally by nitrifiers before they can escape from the ecosystem. Decay of plant and animal tissues and wastes by *reducers* (bacteria and fungi that bring about putrefaction) involves the *ammonification* of nitrogenous compounds: their reduction to NH_3. This is once more oxidized by nitrite and nitrate bacteria, but excessive ammonia is converted back to nitrogen gas (N_2) and returned to the atmosphere by *dentrifying* bacteria.

25.28. How do animals obtain carbon atoms?

Carbon enters the living world almost exclusively as the gas carbon dioxide, CO_2, which animals cannot use directly and must in fact excrete. However, plants fix CO_2 to the tune of an awesome 75 billion metric tons of carbon per year, by way of the complex photosynthetic process, which can be simply stated in terms of substrates and end products as:

$$6CO_2 + 12H_2O \rightarrow C_6H_{12}O_6 + 6O_2 + 6H_2O$$

Animals obtain the products of photosynthesis by direct or indirect consumption of plants.

Chemoautotrophism allows certain organisms to fix CO_2 without need for light, obtaining the necessary energy from the oxidation of some inorganic substance, such as the nitrification of ammonia or the oxidation of sulfide. Deep-sea ecosystems totally dependent on chemoautotrophism exist in darkness on ocean bottoms wherever submarine geysers of boiling mineral water rich in H_2S and CO_2 erupt from rifts along seafloor spreading zones.

Example 13: The Galápagos rift community includes animals unknown elsewhere: strange "spaghetti worms" and giant pogonophorans 3 m long. Clams, bright red with hemoglobin, grow here 500 times faster than related cold-water species. The water is turbid with chemoautotrophic sulfur bacteria that fix CO_2 using the energy of the sulfide-to-sulfate reaction. These bacteria are therefore the *producers* of this ecosystem, fed upon by filter-feeding animals that constitute the *primary consumers*, and these, in turn, are devoured by carnivores (*secondary consumers*). The gutless pogonophorans, also red with hemoglobin, probably obtain nourishment entirely from endosymbiotic sulfur bacteria kept supplied with O_2 by their hosts.

25.29. How do hydrogen and oxygen cycle in ecosystems?

The *hydrogen* and *oxygen* cycles are linked to the carbon cycle: since CO_2 is the form in which carbon enters the living world, it follows that two atoms of oxygen are incorporated into organic compounds for each atom of carbon fixed, and for each CO_2 molecule fixed by either photosynthesis or chemosynthesis, two atoms of hydrogen (chiefly from H_2O) must also be incorporated. Subsequently, for each CO_2 liberated as a waste of aerobic cell respiration (mitochondrial oxidations), two H atoms will be recombined with oxygen, making water. Autotrophic plants use H_2O as their source of H atoms, and give off oxygen gas, O_2, as a waste. The accumulation of this waste product to a present level of about 21 percent of the atmosphere sustains the *aerobic* metabolism characteristic of nearly all plants and animals.

25.30. What is a food chain?

Mass moving cyclically through an ecosystem can be portrayed in simplified fashion as following a *food chain* (Fig. 25.5). A food chain can be thought of as beginning with autotrophs, or producers, chiefly photosynthesizing plants that convert inorganic substrates to *biomass* (the dry weight of living matter). In aquatic communities autotrophs include planktonic protists such as diatoms, dinoflagellates, euglenids, and volvocids. The producers are ingested in aquatic habitats by such *primary consumers* as copepods and other tiny crustaceans, and on land by fruit, seed, root, and foliage eaters. These in turn are devoured by *secondary consumers:* predators and scavengers. The cycle is closed by *decomposers:* bacteria and fungi that cause decay of dead tissues, liberating CO_2, water, nitrates, phosphates, and sulfates needed by autotrophs.

25.31. What do trophic pyramids show?

Mass can cycle endlessly within an ecosystem, as the food chain concept implies, but with each transfer through a food chain, only about 10 percent of the biomass consumed will be represented as biomass at the next step. Some 90 percent of the food consumed, with its mass and chemical energy, will be catabolized, with the energy then transferred to meet metabolic needs other than biosynthesis, to wit, movement, active transport, thermogenesis, and the like. Thus

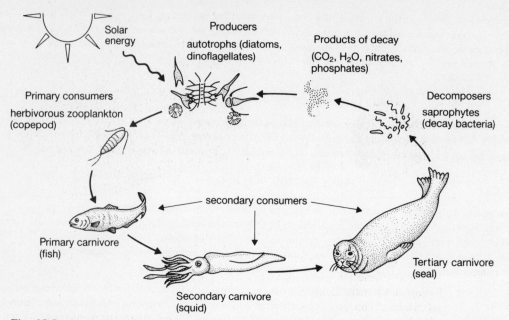

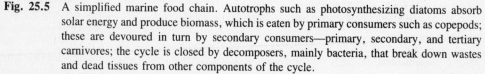

Fig. 25.5 A simplified marine food chain. Autotrophs such as photosynthesizing diatoms absorb
solar energy and produce biomass, which is eaten by primary consumers such as copepods;
these are devoured in turn by secondary consumers—primary, secondary, and tertiary
carnivores; the cycle is closed by decomposers, mainly bacteria, that break down wastes
and dead tissues from other components of the cycle.

a food chain can also be pictured as a *trophic pyramid* with a broad base and tiny apex, as both biomass and available
energy diminish at each step.

Comparing the two trophic pyramids shown in Fig. 25.6, notice that one has three levels and the other five. The
fewer trophic levels in a pyramid, the more efficient it is.

> **Example 14:** Roughly speaking, if 1 kg of human flesh is gained by eating 10 kg of tuna, which
> entailed consumption of 100 kg of anchovies, which fed upon 1000 kg of zooplankton, the autotrophic
> base of a five-step pyramid would amount to some 10,000 kg of phytoplankton required for the 1 kg
> of human flesh. By contrast, 1 kg of baleen whale can be built by consumption of 10 kg of krill,
> which fed upon only 100 kg of phytoplankton.

25.32.   How do sympatric organisms interact in biotic communities?

All living things sharing a common habitat make up the biotic community of an ecosystem. Many animal adaptations
have to do with relating to organisms of other species, and also to individuals of the same species. Questions 25.33–
25.40 deal with such interactions among sympatrics.

25.33.   How do different species interact within communities?

The web of relationships that exists among indigenous species found only in particular climax ecosystems indicates
that these ecosystems have endured for a long enough time that such relationships could evolve and stabilize. Many
species native to long-lived climax ecosystems appear to be "fully evolved" in having reached an adaptive plateau at
which selection operates in a stabilizing mode, maintaining the population on that plateau (see Fig. 5.10). Most species
are specialists in their community, each adapted to occupy a particular niche, which is partly defined by the species'
relationships with other sympatric species. These relationships fall into several major types—including neutralism, competition, cooperation, trophic relationships, mutualism, commensalism, and parasitism—although intergradations occur.
Ecosystem analyses, vital to minimizing traumatization caused by human enterprises, must include ferreting out even
those interspecific relationships that remain obscure to the casual eye.

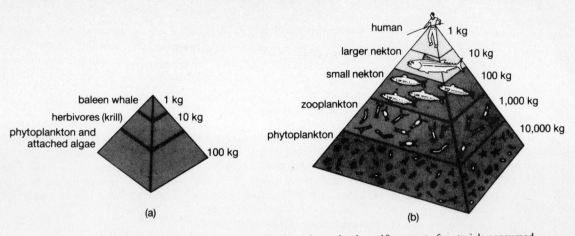

Fig. 25.6   Trophic pyramids. Biomass pyramids show that only about 10 percent of materials consumed is retained as biomass at the next higher level. The other 90 percent is mostly used up to meet the energetic demands of life. A three-step marine biomass pyramid (*left*) can be seen to be much more productive than a five-step pyramid (*right*). In the Antarctic, 100 kg of phytoplankton translates to 10 kg of krill, which sustains 1 kg of growth for a baleen whale. When a human body gains 1 kg by consuming tuna, this gain rests upon a foundation of some 10,000 kg of phytoplankton, since tuna devour anchovies, which feed upon zooplankton, which eat the phytoplankton.

25.34.   What is neutralism?

Neutralism refers to absence of known connections between two sympatric species; some species may really be neutral to one another, but this conclusion must be reached cautiously, for as ecosystem analysis becomes more sophisticated, threads of dependency are found to link even quite unlikely species.

> **Example 15:**   What possible link can explain the disappearance of three types of Amazonian rain forest frogs, when white-lipped peccaries vanish? Piggishly, peccaries love to wallow, creating mud puddles that happen to be these frogs' only breeding site!

Part of the aim of modern ecology is discovering these dependencies before, instead of after, the fact.

25.35.   How important is interspecific competition?

The principle of competitive exclusion operates while an ecosystem is maturing, eliminating one species or another if competition is too severe. Thus, in a climax ecosystem interspecific competition has already been reduced to tolerable levels. Animal species with similar requirements can alleviate competition when one is *nocturnal* and the other *diurnal* (e.g., owls and hawks). *Stratification* is common in forests, reducing competition, as some animals favor the canopy, others low branches and shrubbery, and still others the ground, including subterranean (fossorial) diggers. Sometimes competition exists but can be eased by behavioral modifications.

> **Example 16:**   A study in Panama of 11 species of sympatric freshwater fishes found that when food is abundant, all species use the same resources, but when food becomes scarce, each switches to its own highly specialized diet. Thus competition is limited to times when competing harms none.

25.36.   How do sympatric species cooperate?

Cooperation is a mutually advantageous relationship between species, which is not so intimate or obligate as to be symbiotic.

> **Example 17:**   Although quite unaware that they are doing oak trees a good turn, western gray squirrels bury acorns here and there for winter use and then forget where most of them are hidden. Certain insects, birds, and bats feed on nectar and inadvertently spread pollen from blossom to blossom,

thus not only aiding the plants but assuring a food supply for future animal generations. Small songbirds happening upon a resting predator, such as an owl, give sharp, repetitive *mobbing calls* so similar from species to species that little birds of several kinds may rise to the occasion and join forces to drive the predator away.

25.37. What is involved in trophic relationships?

Trophic relationships exist when one species feeds on another. For brevity, we shall consider only predator-prey relationships among animals, which have greatly influenced the evolution of both prey and predator species.

Predatory adaptations include a variety of behavioral strategies and the physical equipment to execute them. Prey may be located by odor, vision, or sound and then actively stalked or pursued, sometimes by predators acting in social concert (as lions and wolves). Or, the predator may wait in ambush or construct a snare such as a web. The prey may be subdued by a venomous sting or bite, or seized by jaws, talons, or claws.

Antipredator adaptations may be behavioral, structural, or both:

1. *Escape responses* involve moving actively away from the predator by running, leaping, flying, volplaning, or swimming.

2. *Autotomy* is the shedding of some body part, such as a tail or leg, which is actively snapped off and later regenerated (e.g., crabs autotomize their claws, certain lizards their tails, and ophiuroids their rays).

3. *Threat and counterattack* can be effective options. Threat behavior usually precedes counterattack, sometimes rendering the latter unnecessary (e.g., striped skunks stamp their feet and make jerky forward movements, spraying only as a last resort). Counterattack often consists of *secretion* of some noxious matter (such as a vinegaroon spraying acid, or a skunk musk). But all too often threat is only bluff, easily seen through by the experienced predator.

4. *Masking behavior* is especially characteristic of certain crabs, which cover themselves with extraneous materials; after each molt, these masking crabs must reaffix small pieces of sponge and seaweed onto their body spines until they are quite hidden from overhead view.

5. *Frightening displays* combine some possibly fearful aspect of the prey with a threatening movement or posture. Certain tropical caterpillars much resemble the head ends of tree snakes and bend their bodies in a striking motion against anything that touches them. The effectiveness, against birds, of eyespots on butterfly wings has been shown by the simple experiment of comparing predation by caged birds upon butterflies with eyespots versus the same butterflies with their eyespots rubbed off.

6. *Crypsis* is background matching, at least by camouflage but sometimes also by cryptic body form. Twiglike caterpillars and orthopteran insects that resemble twigs or leaves are masters of crypsis. Cryptic animals are better protected if they act the part, moving slowly, swaying like a leaf when touched, or remaining utterly immobile. Certain twiglike caterpillars are *cataleptic* by day, sticking out rigidly at an angle to the branch to which they cling by hind feet only. A similar immobility, usually less prolonged, is seen in various mammals and birds that freeze when alarmed, such as bitterns that stand erect with bill pointed skyward, invisible among the surrounding reeds.

7. *Warning coloration*, usually brilliant combinations of red, orange, or yellow with black, is flaunted by noxious insects and other animals that can punish would-be predators with a sting, bite, or dreadful taste. The striking color and pattern serves as a reminder to predators that have undergone aversive conditioning. The similarity of warning colors in unrelated species suggests that some stimulus generalization takes place in a conditioned predator: having a bad time with one gaudy morsel, the predator tends to avoid other potential meals merely on the basis of similar coloration.

8. *Batesian mimicry* is an evolved resemblance between the edible species, the *mimic*, and a second species, the *model*, which escapes predation because it is bad-tasting or has means of self-protection, such as venom. Examples are the edible viceroy and inedible monarch butterflies of California, and several species of red-black-and-yellow-banded harmless snakes that mimic venomous coral snakes. In the first case, protection of the mimic depends on the predator's being negatively conditioned by having a nauseating experience with the model. In the second case (see Chap. 1, Example 4), experiments have shown that snake-eating motmot birds avoid wooden rods painted like coral snakes even when they have never seen a coral snake, and so an evolved avoidance response rather than conditioning appears to protect both model and mimics alike.

9. *Mullerian mimicry* is an evolved resemblance among several sympatric species that are all noxious to predators; tropical butterflies present the best examples of Mullerian complexes including mimics belonging even to different families. Mixed Batesian-Mullerian complexes also exist: sorting out the two types of mimics is easily accomplished by the mean trick of offering them to hungry, inexperienced cagebirds.

25.38. What is mutualism?

Mutualism is a symbiotic relationship that benefits both species, referred to as *host* and *symbiont*. (*Symbiosis* is any interspecific relationship between two species, which is *obligate* for at least one of the participants and usually involves some intimacy of coexistence.)

> **Example 18:** *Cleaning symbiosis* is a mutualistic association in which shrimp or small fishes groom larger fishes, removing parasites and damaged tissue, even entering the mouths of their temporary hosts to clean around the teeth and gills. Wood-eating insects depend on the cellulose-digesting capabilities of symbiotic zooflagellates that can themselves survive nowhere except in the intestinal tract of their hosts.

25.39. What is commensalism?

Commensalism is a symbiotic relationship essential for the symbiont but neither significantly harmful nor helpful to the host. Short of pathogenicity, the impact of symbiont on host is most objectively defined in terms of energy saved or lost.

> **Example 19:** Some judgment is required here, for the pea crab that scrapes its food off the gills of a bivalve host is in fact robbing the host of a certain amount of nourishment, while a similar pea crab residing in the cloaca of a fish or sea cucumber feeds only on unabsorbed residues and apparently costs its host nothing; yet pea crabs are all judged commensal, not parasitic. The commensal shark sucker (remora) that rides a shark's belly, holding tight by its modified dorsal fin, breaks free to share scraps of the host's kill, which might cost the shark a few calories, but other species of remoras that hitchhike on whales and sea turtles do not share those hosts' food supply at all.

25.40. What is parasitism?

Parasitism is a symbiotic relationship that is obligate for the *parasite*, which takes its nourishment from the host to an extent that is usually measurably harmful to the host.

(*a*) *Ectoparasites* remain on the host's exterior but penetrate the skin and commonly suck blood. They may remain on the host only long enough to take a meal (e.g., leeches, mosquitos, ticks, vampire bats) or live on the host's body most of the time (e.g., fleas). The ectoparasite's saliva may have anticoagulant properties and can serve as a medium for transfer of endoparasitic pathogens such as *Plasmodium*.

(*b*) *Endoparasites* (e.g., tapeworms, digenetic flukes, acanthocephalans, and many nematodes) live within the host's body during at least some stage of life. Their life cycles may be complex, involving the exploitation of both intermediate and definitive hosts, and they must be highly fecund to balance the heavy mortality experienced during passage from host to host.

(*c*) *Kleptoparasites* rob other species, often of food. Jaegers and frigate birds are inefficient predators and get most of their fish by harrying other birds until they drop or disgorge their catch. Certain ant species have no worker caste of their own, but remedy this lack by raiding nests of other species, carrying off larvae that will grow up as "slaves," performing the duties of workers for the parasitic species.

(*d*) *Nest parasites* such as widow birds, honeyguides, and certain cuckoos build no nests and provide no parental care, simply wishing their offspring on another species. Cuckoos and honeyguides, not satisfied with just costing the foster parents the energy required to forage for an extra mouth, also reduce the reproductive success of the host species. The female cuckoo tosses out one of the host's eggs before laying her own in the nest, and the newly hatched cuckoo chick completes the demolition by pushing all other eggs or chicks out of the nest. The infant honeyguide simply slaughters its foster siblings.

25.41. How do animals interact with conspecifics?

Interaction with conspecifics includes competition, cooperation, and communication.

25.42. What are the effects of intraspecific competition?

Competition is more acute within a species than among different species because conspecifics share the same niche. Many creatures produce more offspring than the habitat can support. Predator pressure may eliminate the surplus, but if not, conspecific interactions may complete the job, or the population may outstrip its food supply. Conspecifics compete for mates, food, and territory—behavior that may result in emigration and agonistic behavior.

25.43. How may emigration alleviate intraspecific competition?

Emigration is mostly the fate of the young, which are often exiled from a region populated by territorial adults; emigrants may find new places to settle, thereby expanding the species' range. Benthonic marine invertebrates usually produce planktonic larvae, which emigrate passively as they are swept away by currents, few surviving to find a place to settle. Sporadic *mass emigrations* occur in certain species (e.g., lemmings), seemingly when crowding becomes stresssful. Emigrants are always at risk, but their exodus leaves the established population in a less competitive situation.

25.44. How does agonistic behavior function in competitive situations?

Agonistic behavior helps determine possession of territories, position in dominance hierarchies, and access to mates and food. This behavior includes not only intraspecific *aggression* (threat and social fighting), but also *appeasement*, *submission*, and *avoidance*. Such behaviors between conspecifics are ordinarily quite different from attacks on prey or self-defense against predators, and are controlled by different parts of the brain. Agonistic encounters rely on *communication*. A *threat display* alone is often enough, especially when given by a territory owner or hierarchical superior. Among socially living birds and mammals the existence of hierarchies is often betrayed simply by the manner in which dominant individuals confidently *displace* subordinates, which move away without ado from food or a resting place. Since in a social group individuals recognize one another, such displacement can take place with no overt threat at all. On the other hand, if threat proves ineffectual, it may be followed by social fighting, which as a rule has the odd quality of being *ritualized* into a rather formalized confrontation in which the combatants' major weapons are not put to use. Therefore, serious injury is usually not inflicted, and fighting becomes a contest of endurance.

> **Example 20:** Venomous snakes arguing over potential mates do not use their fangs: they wrestle, coiling their bodies together and thrusting until one is upset (see Fig. 21.10c). Male tortoises could inflict damage with claws and horny beak, but instead use a projection of the plastron (underplate) to overturn the rival. Oryx antelopes gore lions with their lethal horns, but fighting males use these horns only to fence and are quite unable to gore an opponent even when a vulnerable flank lies bare.

Acts of *submission* and *appeasement* terminate or avert combat. A territorial battle ends when the loser quits the scene. In hierarchical combat, attack breaks off when the one bested adopts a submissive posture, typically crouching down and looking small and pitiful. This posture serves as an inhibitory sign stimulus that prevents the victor from continuing its attack. Similarly, in hierarchical social groups subordinates spontaneously direct gestures of appeasement toward dominant individuals.

> **Example 21:** "Presenting" the buttocks to a dominant individual is an appeasement gesture common to baboons, which human observers may mistake for sexual invitation; females often present to mothers with new infants, as an overture to being allowed to touch the baby. Appeasement postures also ameliorate the tensions of courtship, when personal distance must be violated in the cause of intimacy; e.g., in the "facing away" display, courting gulls abruptly turn their heads away, hiding their menacing beaks and (in some species) black face masks as well.

25.45. What is the overall effect of agonistic behaviors?

Agonistic behaviors reduce intraspecific competition at the expense of the losers, at least in terms of reproductive success. In territorial species, males unable to hold a territory fail to attract mates; they may even be forced to emigrate. In hierarchical social groups, dominant males not only have the best opportunity to sire offspring, but they may also doggedly break up subordinates' romances. Sometimes subordinates appear stressed.

> **Example 22:** Captive mice reared in colonies show enlargement of their adrenal glands in inverse correlation with social rank, i.e., mice at the top have the smallest adrenals, those at the bottom the

largest. Any increase in stress, as in overcrowding or food deprivation, may trigger death of the lowliest subordinates by exhaustion of their overworked adrenals.

25.46. What are some ways in which conspecifics cooperate?

Cooperation among conspecifics includes division of labor, pair-bonding with shared parental duties, communal parenting, group defense, and social hunting.

(*a*) *Division of labor* is clearly illustrated by polymorphic cnidarian colonies (see Figs. 9.3*a* and 9.5) and social insects. The latter often have castes, e.g., termite workers, soldiers, young reproductives, and the adult king and queen, which carry on caste-specific behaviors. Soldiers defend the termitary; workers build, gather food, and care for eggs, young, and the royal pair; the king inseminates the queen as needed, the queen lays eggs, and both may give off pheromones that inhibit the sexual maturation of juveniles; the winged reproductives leave the termitary, fly a short way and alight, snap off their wings, court, and scurry off in pairs to found a new settlement in which they will provide parental care to their first offspring until these mature into workers that take over such menial tasks. Tactile and chemical communication are crucial for the integration of insect societies.

(*b*) *Pair-bonding*, most common among birds, is preceded by courtship rituals that confirm species identity, sex, and readiness to mate. During pair-bonding personal distance is overcome and individual recognition established. The bonded pair then shares parental duties, at least for that year, but in some species for life.

(*c*) *Communal parenting* has several aspects. (1) *Colonial nesting* is seen in weaverbirds: each pair builds its own nest and cares for its own young, within a dense nesting colony that provides group protection and mutual arousal. (2) *Communal nesting* by ostriches begins when all females in a harem lay eggs into a single nest; the completed clutch is then incubated by only the male and the dominant female. A communal nest is shared by two mated couples of Mexican jays, together with their unmated helpers. (3) *Helpers* are unmated individuals that assist the actual parents. As seen in Florida scrub jays, helpers are usually young adults that assist with the care of their parents' newest brood; dominant helpers have the best chance of inheriting the role of a deceased parent. "Aunts" that assist mothers are common among primates and cetaceans.

(*d*) *Group defense* involves various strategies: individuals may trade off the role of sentry (e.g., prairie dogs); adults may form a tight ring about juveniles, horns outward (e.g., musk oxen); individuals may sleep in bodily contact (e.g., the sleeping rings of bobwhite quail); adult males may threaten and even counterattack predators while females and young flee (e.g., baboons); young penguins huddle together in *creches*, protected from cold, wind, and the predatory attacks of gull-like skuas.

(*e*) *Social hunting* is seen in wolves and hunting dogs, and stalking of prey is often divided between two lionesses.

25.47. How is communication important among conspecifics?

Communication is the key to effective social interactions. Species-specific signals, crucial as they are to reproductive success, are often programmed genetically and not subject to much individual variation. A drake that thinks up some brand-new courtship routine may catch the eye of a lot of ducks, but he is most unlikely to catch their hearts. This is because *signal and receiver coevolve*: many animal signals operate as sign stimuli (see Question 24.36) that lock into some stimulus analyzer in the brain of the perceiver. Changing either "key" or "lock" makes communication ineffectual. Thus, the genetic basis of communicative behavior—which may be based on chemical, tactile, electric, acoustic, or visual signals—is highly conservative (that is, not readily changeable).

Although communication is primarily intraspecific, some signals are equally effective with other species: the intentions of a growling dog with hackles erect are as meaningful to other mammals as they are to dogs themselves.

25.48. What are some instances of chemical communication?

Chemical communication involves secretion of pheromones, organic compounds with sign stimulus properties. Pheromones serve as sex attractants and territorial markers. Ant pheromones include a *trail-marking substance* laid down by foragers, an *alarm pheromone* that recruits workers to a site of disturbance, and a *death pheromone* that a dying ant gives off in prompting others to carry it out of the anthill to a "graveyard." Insect sex attractants provide good means of controlling certain pest species, since males in quest of love can be lured into traps; in fact, they cannot avoid such traps, for their directional movement (positive chemotaxis) is as automatic as a reflex.

25.49. What does tactile communication involve?

Tactile communication stems from bodily contact. Animals of *noncontact species* require some personal distance between adult individuals, and in such species the pleasures of touch are reserved for interactions between mates, or parents and offspring. Other animals belong to *contact species,* which enjoy bodily contact with conspecifics throughout life, even in nonsexual contexts. Most primates spend hours each day in gentle mutual grooming that not only removes parasites but is vital to social integration.

> **Example 23:** Rhesus monkeys raised in isolation with only cloth or wire artificial mothers for company demonstrated the importance of *contact comfort:* the young monkeys snuggled up to the cloth mother for hours each day, even when their feeding bottles were mounted on the wire mother, and gained confidence from the cloth mother's presence when frightening objects were introduced. Monkeys deprived of such cuddlesome contact came to display extreme anxiety and later failed to relate socially to other monkeys.

> **Example 24:** Wild chimpanzees communicate by touch in ways that suggest deep biological roots for some of our own "body language": finger-to-finger hand touching in appeasement, resembling a rudimentary handshake; lip-to-lip contact reminiscent of kissing; touching lips to the back of another's hand, as in hand kissing, a subservient gesture used in certain human cultures; and gentle back-patting (mostly of juveniles by adults, even males). Have you ever felt an ungovernable impulse to pat a crying person on the back?

25.50. Does electrical communication exist?

Electrical communication probably occurs in certain fish that generate weak electrical fields used in orientation. Each species' signal has unique characteristics, and when the signals can be modified, mates modulate their output and thus avoid jamming each others' signals.

25.51. What are the benefits of acoustic communication?

Acoustic communication, limited to animals having means of sound production—orthopteran insects (crickets, locusts), anuran amphibians, birds, and certain reptiles and mammals—is the most versatile communicative mode, for signals can be varied with infinite variety in terms of amplitude, frequency, rate, and sequencing. Specificity of mating calls promotes *ethological isolation,* which means that intermating between similar species is prevented by behavioral differences, including calls.

> **Example 25:** Western and eastern meadowlarks look just alike, but their distinctive songs prevent interbreeding where their ranges overlap in Mexico and the southwest.

25.52. What is the nature of visual signals?

Visual signals may involve color and bodily appearance, movement, and posturing. They are of particular interest to students of animal communication, for they lend themselves to being tested in the form of simplified models. Such models can be designed to present different qualities separately in order to define the parameters of the sign stimulus or releaser.

> **Example 26:** Gull chicks induce their parents to regurgitate food by pecking at a red spot on the parent's beak. The importance of this red spot was investigated by presenting chicks with a variety of models of seagull heads: three-dimensional or flat; with red spot present, absent, or displaced; with bill normally shaped, thick, or thin; with bill variously colored; with head shape normal, distorted, or absent; with model moved or held still; and so on. Properties unimportant in inducing the chick to peck at the model included head shape and three-dimensionality; significant properties included movement, redness, a properly positioned red spot, and bill shape, which surprisingly was even more effective if thinner than normal (a *supernormal stimulus* effect).

The posturing of courting birds provides intriguing insights into the evolution of communicative signals. A courting drake engages in remarkable contortions: while swimming, he may oscillate like a rocking horse, stand straight up in the water with wings spread, rise up with his bill down-pointed, dunk beak and head and then poke his dripping bill into his shoulder feathers, and so forth. Comparing the courtship of various duck species, one sees many actions that are quite similar but with some differences in form, frequency, and rate, suggesting variations upon behavior patterns first evolved in a common ancestor. These postures are strikingly stereotyped and overblown, yet they echo behaviors performed under noncommunicatory, maintenance situations, such as bathing, grooming, and preparing to fly. It seems as if, by some neural mechanism not yet understood, behaviors can be *emancipated* from their everyday contexts and become grouped together, so that actions ordinarily performed in quite different contexts end up welded together into a communicative routine. Once this emancipation has taken place, acts now performed in communicative situations undergo selection toward becoming more effective signals: *exaggerated, theatrical, stereotyped, and repetitive*.

Behavioral emancipation seems to be favored by *motivational conflict*, that is, quandary states. One outcome of a quandary state is performance of an *incomplete* action: a bird on its nest, faced by an approaching person, shows *flight intention* movements as it vacillates between the drive to escape and the urgency to remain on its eggs; a number of bird displays, both in courtship and in agonistic contexts, appear to be modified and frozen versions of the takeoff crouch and leap. A second product of motivational conflict is performance of a behavior that is somehow *intermediate* between two mutually exclusive actions. "Facing away" exemplifies this, as a display that combines conflicting tendencies to approach and to get away from the potential mate, as the birds stand with bodies facing, but heads swiveled in opposite directions. A third outcome of motivational conflict is seen when an animal faced with alternatives A and B does C instead. This is known as *displacement behavior*. The behavior performed seems quite irrelevant to the dilemma, but may serve to reduce nervous tension. Displacement behaviors often involve grooming, but in its quandary state the animal may groom itself so mercilessly that the difference between displacement and genuine grooming is easy to see. Wiping beak on perch is a common displacement behavior in songbirds; a courtship display derived from displacement beak wiping is seen in spice finches, performed by courting males as part of a "low twist" display in which the beak is moved in an incomplete wipe that does not actually touch the perch.

25.53. What is the future for earth's ecosystem?

We are living today in a planetary ecosystem that has withstood many tribulations and evolved over billions of years; it is a dynamic world in which continents drift, collide, and separate again and mountains and sea levels rise and fall. Life, decimated by recurrent major catastrophes such as cometary impacts, has nevertheless recovered resiliently, weaving a biosphere of ever-increasing diversity and complexity—until now. Within the last 100 years, unprecedented changes have come to pass: human premature death rates have declined dramatically, while worldwide birthrates have not declined commensurately. In the mid-1800s the earth supported a billion people, already a population much more robust than that of most other mammals larger than rabbits. By 1950 the human population had increased to 3 billion, and by 1976 to 4 billion. Only 11 years later, in 1987, the world population reached 5 billion, and addition of another billion people should take place in less than 10 years, since an ever larger base is reproducing. This means that 8 billion people may well be alive by 2010, provided that agricultural yields can be nearly doubled over current levels, and provided that we have not by then poisoned ourselves and the rest of the world, on purpose by chemical or nuclear warfare, or inadvertently by way of our accumulating waste products. So far as we can tell, never before in earth's history has the multiplication of one form of life come to threaten the survival of so many others.

Ecology has vital lessons for those willing to learn:

(*a*) No species can multiply indefinitely; a plateau *will* be reached, but whether or not human population densities at that plateau will be compatible with survival above subsistence levels remains to be seen. A worldwide slum accommodating 16 billion or more half-starved people looms as a distinct option.

(*b*) Some species maintain quite stable population levels once a plateau has been reached, while others ride the roller-coaster of boom and bust; it is not hard to figure out which population profile is more compatible with culture and civilization, or which we are likely to be stuck with if things keep going as they now are.

(*c*) A species can be poisoned by its own wastes, which now include industrial emissions toxic to others than humans alone. Since atmospheric oxygen is mainly replenished by marine phytoplankton, an ocean killed by pollution might well spell a planet stripped of aerobic life. Furthermore, life as we know it can exist only in the temperature range of liquid water; a runaway greenhouse effect caused by excess carbon dioxide production could prove a good contender for converting Earth into Venus.

(*d*) Desertification is claiming ecosystems used and degraded by human activities; ecosystems simplified into agricultural and urban monocultures are both successional and at risk of degradation to derelict status. Much energy is needed to keep a monoculture going (*monoculture* meaning that only one species is being cultured, such as cattle on grazing land, wheat on a farm, or people in a city). Roughly speaking, mechanized agriculture invests one calorie of energy (fuel and the like) per calorie of food energy harvested. Furthermore, unless eroded soil is replaced, lost fertility restored, and salts removed that accumulate from "hard" irrigation water, more and more land will decline to manmade desert, supporting neither human enterprises nor native fauna and flora. Natural ecosystems have remained viable and rich for thousands and millions of years; now millions of hectares of once fertile land are being degraded and lost each year.

(*e*) Through evolved means of endogenous population regulation, many nonhuman species keep their numbers well below the carrying capacity of the habitat, and thus lead lives of plenty instead of want. Species lacking endogenous controls may starve and crash, or be limited by predator pressure or disease. Today, predatory attacks on humans seldom occur, but increased population densities promote epidemic disease. There is no reason to suppose that *Homo sapiens* will escape these exogenous controls, if a stabilized population cannot be achieved by endogenous means.

(*f*) Species that have reached a plateau of optimum adaptation can persist little changed for millions of years; given an unpolluted planet and a world human population stabilized at numbers low enough to allow natural ecosystems to recover and spread once more, the animal life that enriches our lives today can continue to enrich the lives of future generations into the distant future. Who knows? Some day giraffes and elephants may be prized for parks in the Alpha Centauri system! In the meantime, committed people are floating a thousand arks to save wildlife from extinction. With luck, the flood will subside before the arks all sink.

Beyond the dimensions of science lie the realms of philosophy: science can discover facts, formulate principles, work out the technology of solutions, but the honesty to examine facts and the commitment to implement solutions lie in those parts of the mind that deal with ethics and know compassion. A fascinating legacy of plant and animal life has come up to us from the past: what we do today will determine the future of earth's ecosystem, and our place within it.

Index

Page numbers in italics indicate locations of illustrations. Generic names are italicized; those of more inclusive taxa are given in capital letters. Most items are indexed in their singular form, even when plural in the text.

Aardvark, *356, 360,* 369
Abalone, *170*
Abdomen:
  in arthropods, 196, 197, 198, 200,
    201, 202, *203,* 205, 208, 210,
    215, 216, *218,* 219, 220, 222,
    223, 225, *227,* 228
  in mammals, 348
Abducens nerve, 268
Abiotic, 7
Abiotic formation of organic molecules,
  38
Abomasum, *371*
Aboral pole, *10, 237*
Absorption, 3, 62, 134, 173, 182, 197,
  214, 272, 374, 376, 377, 398,
  399
Abyssal benthos, 423
Abyssopelagic, 423
Acanthobdella, 188
ACANTHOBDELLIDA, 186, 188
ACANTHOCEPHALA, 106, 143
Acanthocephalan, 140, *141,* 143, 433
ACANTHODIA, 257, *258,* 280, *286,*
  294
Acanthodian, 280, 286–287, 294,
  298
ACANTHOPTERYGII, 303
Acanthor, 143
ACARINA, *201,* 207, *208*
Acarine, 208–209
Accessory nerve, 268
Accommodation of eye, 270, 332
Acetabulum, *311, 323*
Acetaldehyde, *28*
Acetate, methyl, *28*
Acetyl, 392, 393
Acetylcholine, 388
Acetyloxycycloheximide, 410
Acetone, *28*
Aciculum, *180,* 181
Acid, 26
  acetic, 26, *28,* 204
  amino (*see* Amino acid)
  ascorbic, 376
  caprylic, 204
  carbonic, 349, 383
  citric, 27
  fatty, *30,* 31, 38, 376, 378, 379,
    380, 403
  hyaluronic, 27
  hydrochloric, 26, 374, 379, 400
  nitric, 427

Acid (*continued*)
  nucleic, 31–32
    (*See also* DNA; Nucleic acid; RNA)
  organic, 26, 27, *28,* 351
  sulfuric, 427
  tricarboxylic, 27
    (*See also* Krebs cycle)
Acid rain, 427
Acineta, 80, 81
Acipenser, 301
Acoel, 125, 126
ACOELA, 126
Acoeloid ancestor, 125
ACOELOMATA, 106
Acoelomates, 98, *99,* 125–139
  evolutionary status, 125
Acontium, 121
Acorn worm, *247,* 250
Acoustic communication, 436
Acoustic nerve (*see* Otic nerve)
ACRASIOMYCOTA, 65
ACROTHORACICA, 215, 219
ACTH (*see* Adrenocorticotrophin)
Action potential, 374, 395, *406, 407*
Actin, 73, 394, 395, *396*
ACTINIARIA, *121*
Actinophrys, 75
ACTINOPODA, 68
ACTINOPODEA, 68
Actinopodean, 75
ACTINOPTERYGII, *258,* 298, 300,
  301
Actinopyga, 243
Actinosphaerium, 10, 73, 75
Actinotroch, 157
Actinula, 112, *113,* 114
Action potential, *406,* 407
Activation energy, 388
Activation of egg, 93, 94, *97,* 102
Active site (of enzyme), 379, 388
Active transport, 18, 151, 161, 394,
  429
Actomyosin (*see* Actin; Myosin)
Acupuncture, 409
Adaptability, 2, 3
Adaptation, 3, 54, 55
  radiative, *265, 337–338*
  sensory, 414
Adaptive plateau, 54, 430, 438
Adaptive radiation, 54, 281
Additive genes (*see* Gene, duplicate)
Adductor, 157, *171, 216,* 217, *218,*
  300

Adenine (A), *32, 34*
Adenohypophysis, 399, *401*
  (*See also* Pituitary)
Adenosine diphosphate (*see* ADP)
Adenosine monophosphate (*see* AMP)
Adenosine triphosphate (*see* ATP)
Adenyl cyclase, 403
ADH (antidiuretic hormone), 309, *351,*
  399, 402, 403, 409
Adhesive structures, 114, 140, 154,
  159, 169, 205, 216, *307,* 374
Adipose tissue, 403
  (*See also* Fat cell)
ADP (adenosine diphosphate), 32, 388,
  389, *390*
Adrenal gland, *266, 347,* 385, 397
  cortex, *30,* 399, 401, 403
  medulla, 400, 402
Adrenalin (*see* Epinephrine)
Adrenocorticotrophin (ACTH), 399,
  401
Adrenocorticotrophin-releasing hormone
  (ACTH-RH), 399
Advanced (in phylogeny), 64
Aedes, 234
Aerobe, 39
Aerobic metabolism, 429
Aesthete, 167
Aestivation (*see* Estivation)
Afferent nerve fiber, 267, *404,* 405
Aftershaft, *330*
AGAMIDAE, 327
Age:
  of amphibians, 41
  of earth, 37
  of eukaryotic life, 39
  of fishes, 41
  of prokaryotic life, 38
  of reptiles, 43
  radioisotope determination of, 37–38
Agelenid, 207
Aggregation, 368, 426
Aggression, *326,* 403, 415, 426,
  434
Aglaura, 112
AGNATHA, 257, *258,* 280
Agnathan, 272, 280
Agonistic behavior, 434–435
Agriculture, 437
Air capillary, 335
Air sac:
  bird, 314, *334,* 335, 352
  grasshopper, *229,* 230

Air tubule, 196
    (*See also* Trachea; Tracheole)
AISTOPODA, 303
Aistopod, 305
Albatross, 330, 339
Alanine, 395
Albedo, 420
Albinism, 3, 59, 359
Albumen, *170, 278,* 313, 335
Alcohol, 27, *28*
ALCYONACEA, 119
ALCYONARIA, *108,* 119, *120*
Aldehyde, 27, *28*
Alderfly, 104
Aldosterone, *30,* 31, 397, 398, 399,
    401
Algae:
    blue-green, 39
    coralline, 122
    green, 39, 62
    symbiotic, 176
        (*See also* Zoochlorella;
            Zooxanthella)
    (*See also* Kelp)
Alimentary canal, *99*
    (*See also* Digestive system)
Alkali (*see* Base)
Allantois, *313,* 314, 354
Allele, 45, 50, 51, 58, 59, 60
    beneficial, 60, 61
    codominant, 45, 58
    detrimental, 60
    dominant, 45, 51, 52, 53
    neutral, 59
    recessive, 45, 51, 52
Allelic frequency:
    determination of, 59
    effects of selection upon, 60
    effects of genetic drift upon, 61
Allelomimesis, 417
Alligator, *266, 314, 321,* 325,
    348
Alligator gar, 302
Alligator lizard, 327
Allopatric, 54
ALLOTHERIA, *259*
Alopias, 228
Alpaca, 360, 372
Alpha-androstenol, 341
Alpha cell, 400
Alpha particle, 22
Alpine, *421*
Alternation of generations, 74, *111,*
    112, 254, *255*
Altitudinal zonation, *421*
Altricial young, 335, 341, 365, 367
Altruistic behavior, 366, 427
Alula, *330,* 331
Alveolar sacs, 381
Alveolus, *308, 347,* 381
Ambergris, 366
Amble, 367

Ambulacrum, 238, 241, 243, 245,
    246
Ambulacral groove, *237,* 241, 242,
    244, 247
Ambystoma, 256, *306*
AMETABOLA (*see* APTERYGOTA)
Amia, 302, 305
Amictic egg, 143
Amine, *409*
Amino acid, 19, *31,* 33, *35,* 36, 38,
    58, 352, 373, 378, 379, *380,* 394,
    395, 399, 402, 403, 409
Amino group, *31,* 184, 378, *380,* 395
Aminopeptidase, 378, *380*
Ammocoetes, *256–257,* 282, *283,* 284
Ammodiscus, 73
Ammonia, 3, 4, 26, 63, 129, 184, 210,
    254, 351, 386, 393, *395, 428,* 429
Ammonification, 429
Ammonite, *9, 42, 43,* 164, 173, 177
Ammonium ion, 27, 428
AMMONOIDEA, 177
Ammonotelic, 38
Amnion, 260, 280, *313,* 314
Amniote, *259,* 260, 264, 273, 277,
    278, 309, 312–372
Amniote egg, 280, 312, *313,* 354
Amniotic cavity, *313*
Amoeba, 72, *73,* 74, *377, 387*
    social, 65
Amoebic dysentery, 74
Amoebid, 74
AMOEBIDA, 72
Amoebocyte, 86, *87,* 88, 89, 90, 117,
    119, 139, 142, 150, 154, 165,
    239
Amoeboid movement, 72–73
AMP, cyclic, *32,* 33, 65, 403
AMPHIBIA, 57, *252,* 257, *258*
Amphibian, 7, 42, 218, 260, 270, 280,
    294, 298, 302–311, 332
    adaptive limitations, 303
    brain, *266*
    circulatory system, *274, 275,* 310
    development, 94, *95,* 101, 102, 103,
        305, 306, *307,* 310, 311
    diet, 305, 315
    evolutionary history, *42,* 57, 303–
        306, 316
    hearing, 309
    integumentary glands, 260
    life cycle, 303, 306–*307,* 312
    limbless, 305, 310
    osmoregulation, 309
    reproduction, 310, 311
    reptilelike, *304,* 305–306, 312
    respiration, 273, 306, *307, 308*
    skeleton, 303–305, *311,* 317, 345
    skin, *275,* 303, 307, 308–309
    tongue, 310
    (*See also* Caecilian; Frog;
        Salamander)

Amphiblastula, *87, 89*
Amphid, 146
Amphinomid, 181
Amphioxus, *95,* 96, 98, *252, 255,*
    256–257
Amphipod, 197, *211,* 221
AMPHIPODA, 215, 221
AMPHISBAENIA, 320, 327
Amphisbaenid, 315, 325, *326,* 327
Amphistylic, 288 (*see* Euamphistyly)
Amphitrite, 184
Ampulla:
    in ear, *271, 272, 344,* 411, 412
    of Lorenzini, 293, 414
    of tubefoot, 238, *239, 244,* 247
Amylase, 378, 379, *380*
Anabolism, 3
    (*See also* Biosynthesis)
Anadromous, 284, 296
Anaerobe, 39, 63, 131, 137
Anaerobic glycolysis, 392
Anal gland/sac, *159,* 160
Anal pore, 79, 124
Anal valve, 147
Analges, 208
Anamniote, 260, 277, 278, 280
ANAMORPHA, 224
Anaphase, *48,* 49, 50
Anaplasmosis, 209
Anapsid, *318,* 319
ANAPSIDA, *259*
ANASPIDA, *281*
Anastomosis, 385
Anatomy, 7
Anchovy, 430
Ancylostoma, 148
Andalusian fowl, *46,* 58
Androctonus, 204
Androgen, 279, 399
Anemia, 148, 376
    pernicious, 136, 376
    sickle-cell, 58, 76
Anemone, *111*
    beadlet, 12
    sea, 109, 112, *121–122*
    tube, 109, 120
Anemone fish, 122
Angel shark, *288,* 290
Angiotensinogen, 401
Angiotensin, 401
Anglaspis, 281
Anglerfish, 303
ANGUIDAE, 327
Anguilla, 296, *301*
Animal, 4
    architecture of, 9–20
    compared with plant life, 4
    size, *9*
Animal kingdom, *40*
    classification of eumetazoans, 106
    geologic history of, 41–44, *42*
ANIMALIA, *40*

Anion, 26, 405, 408
(*See also* Ion)
Anisogamete, 63, 72, 74, 75
Ankle (*see* Tarsal; Tarsus)
Ankylosaur, 323
ANKYLOSAURIA, 320
Ankylosaurus, 317, 319
Annelid, 41, *42,* 76, 166, 178–190,
214, 383
body organization, 179–181, 182–
185
circulatory system, 182–*183,* 383,
384
classification, 186–190
compared with onychophoran, 191–
192
development, *99, 151*
digestive tract, 182
excretion, *183,* 184
feeding adaptations, 181–182
gas exchange, 182, 184
locomotion, 181, 188
nervous system, *183,* 184–185, 188,
404
reproduction, 179, 184, 185, 186,
188, *189*
ANNELIDA, 106, 151, 186
ANNULATA, 320
Anodonta, 171
Anole, 326
Anolis, 424
ANOMALODESMATA, 177
ANOMURA, 215, 222
Anomuran, *211,* 215, 221, 222
Anopheles, 76, *234*
ANOPLA, 138
ANOPLURA, *233,* 235
ANOSTRACA, 215
Anostracan, 216
ANSERIFORMES, *338,* 339
Ant, 194, 195, 230, *234,* 235, 433, 435
Ant lion, *234,* 235
Antedon, 237, 242
Anteater, 374
banded, *358*
scaly, 364
spiny, 357
true, 360, 364
Antechinus, 357
Antelope, 372, 421
Antenna, *190,* 191, 197, 198, 210,
212, 214, *216,* 217, *218,* 219,
220, 221, 222, *223,* 225, 235
Antennal gland, 195, 210, 217, 221
Antenniform leg, 204
Antennule, *197, 212*
Anterior, *10,* 11, 96
ANTHOZOA, *108,* 113, 119, *120,*
121, 123
Anthozoan, 112, 119–123
Anthracosaur, *304,* 305–306, 309, 312
ANTHRACOSAURIA, 303

Anthropoid, 362–364
ANTHROPOIDEA, 360, 362
Antiarch, *286,* 287
ANTIARCHI, 287
Antibody, 5, 273
Anticoagulant, 182, 184, 397, 433
Anticodon, *35,* 36, 394
Antidiuretic hormone (ADH), 309,
351, 399, 402, 403, 409
Antihemophiliac factor, 385
ANTILOCAPRIDAE, 372
ANTIPATHARIA, 120
Antipredator adaptations, 432
Antler, 260, 372
ANURA, *259, 306,* 311
Anuran, 306, 307, 308, 309, 311, 374,
436
Anus, 94, 97, 107, 137, *138, 151,*
152, *153, 156, 159, 190, 249,*
250, 377
arthropod, *201,* 202, 210, *212, 218,*
229
chordate, 253, *255*
echinoderm, *237,* 242, 243, *244,*
245, 248
vertebrate, *260,* 292, 295, 310, 325,
347, 353
Anvil, 342
Aorta, *166,* 167, *168, 171,* 172, 226,
229, 230, 310, *334, 348*
dorsal, *255,* 256, *274, 275, 283,*
292, 295
ventral, 256, 257, *274, 275, 283,*
292, 295
Aortic arches, 104–*105,* 182, *183,*
274–275
Aortic body, 349
Ape, *42,* 360, 362, 363
APHASMIDA, 146
Aphid, 94, 194, 195, *233,* 235
Aphotic zone, 422, 423
Aphrodite, 187
APHRODITIDAE, 186
Apical complex, 75
Apical sense organ, 123, 151, 178
APICOMPLEXA, 68, 75, *77*
Aping, 417
Aplysia, 176
Apocrine, 18
APODA, *259, 306,* 310
Apodid, 243
APODIFORMES, *338,* 339
Apopyle, *87*
Appeasement behavior, 434, 436
Appendage, 10, 178
arthropod, *193,* 197, 198, 210, 211,
212, 216, 218
biramous, 197, 198, 211
uniramous, 178, 197, 222
vertebrate, 263, 264–*265,* 298, *299*
Appendicular skeleton (*see* Skeleton,
appendicular)

Appetite, 379, 400
Apposition image, 213
APTERYGOTA, *233,* 235
Aquaculture, 162
Aquatic animals, convergence in, *56*
Aquatic ecosystems, 422–423
Aqueduct of brain, *266*
Aqueous humor, *269*
Arachnid, *42,* 198, 200–209
ARACHNIDA, *193,* 197, 200
Arachnoid mater, 265
Aragonite, 164
ARANEAE, *201,* 206
ARANEOMORPHAE, 207
Arapaima, 296
Arboreal mammals, 360, 362, 364, 422
ARCELLINIDA, *73*
ARCHAEBACTERIA, *40*
Archaeopteryx, 328–329, 337
ARCHAEORNITHES, 337
Archelon, 319
Archenteron, *95,* 96, *97,* 98, 152, 253
Archeocyte, 88
ARCHEOGASTROPODA, *175,* 176
Archeozoic Era, *42*
Archiannelid, *187,* 188
ARCHIANNELIDA, 186, 188
Archipallium, 265, 317, *336,* 342, *343*
Archipelagoes, speciation in, 54, 226
Architeuthis, 173
Archoophoran, 126–127
Archosaur, 315, *317,* 322–325, 328,
340
ARCHOSAURIA, *259,* 320, 322,
323
Arcualia, 305, 306
Ardeola, 424
Arenicola, 185, 383
ARENICOLIDAE, 187
Argasine tick, 208, 209
Arginine phosphate, 236
Argon, 37
Argonaut, 177
Argonauta, 174
Aristotle's lantern, 245
Ark, 176
Arm:
cephalopod, *174,* 177
echinoderm, *237, 239,* 240, 241,
242, *244,* 246, 247
pterobranch, *249,* 250
tetrapod, *265*
Armadillo, 70, 260, 360, 364
Armor, 260, 261, 262, 280, 281, 282,
284, 287, 319
Arrectores pilorum, 341
Arrectores plumarum, 333
Arrow worm, 246, *249*
Arteriole, *350,* 385, 397, 400
Artery, 165, *212,* 214, 273, *308,* 383,
384, 397
(*See also* Aorta; Carotid artery)

ARTHRODIRA, *287*
Arthrodire, *286*, 287
Arthropod, 179, 193–235
  adaptive radiation, *193*, 197
  characteristics, 195–197
  chelicerate, 198–209
  circulatory system, 195, 196, 207,
    212, 214, 230, 383, *384*
  classification, 197
  compared with annelid, 195–197
  compared with onychophoran, 191–
    192
  development, 195
  digestive system, 196–197, 214,
    230, 377
  excretory organs, 195, 196, 199,
    200, *201*, 202, 204, 205, 206,
    207, 209, 210, 217, 223, 224,
    229
  exoskeleton, 195
    (*See also* Cuticle)
  mandibulate, 210–235
  mouthparts (*see* Mouthparts)
  nervous system, 195, 202–*203*, 204,
    214, *229*, 230, *404*
  phylogenetic status, 151
  relationships with humans, 193–195
  respiratory organs, 196, 198, 199,
    200, *201*, *202*, 204, 205, 206,
    207, *212*, 216, 220, 221
  reproduction, 199, 200, 201, 203,
    204, 205, 206, 207, 208, 214,
    217, 218, 220, 221, 224, 225,
    226, 230–231
  sense organs, 196, 203, 204, 205,
    206, 207, 214, *227*, 228
    (*See also* Ear; Eye)
ARTHROPODA, 106, 151, 193, 197,
  210
Articular process, intervertebral, 303
Articulare, *271*, 342, 345
ARTICULATA, 152, 157, 241
Articulate brachiopod, *156*, 157, 158
Artiodactyl, *356*, 371–372
ARTIODACTYLA, *346*, 371
Ascaphus, 311
Ascaris, 144, *145*, 147
ASCIDIACEA, 254
Ascidian, 253, 254, *255*, 383
Ascon(oid), *85*, 86, 88, 89
Ascorbic acid, 376
ASCOTHORACICA, 215
Ascothoracican, 219
Asexual reproduction, 48, 84, 85, 132,
  179, 247, 253
  (*See also* Budding; Fission;
    Fissiparity; Strobilation)
ASPIDOBRANCHIA (*see*
  ARCHAEOGASTROPODA)
Asplanchna, 104
Ass, 370
Assassin bug, *233*

Association neuron, *404*, 405
Assocative learning, 417
Assortative mating, 53, 54
Assortment, law of independent, 51–*52*
Astacia, 69
Astacid, 222
ASTACIDEA, 215, 222
Aster, 48
Asterias, *237*
Asterina, *237*
Asteroid, *237*, 246, 247–248
ASTEROIDEA, *237*, 241, 246, 247–
  248
ASTEROZOA, 241, 246
Astragalus, *311*
Astrangia, *108*
Asymptote, 424
ATHERINOMORPHA, 303
Atlas, *263*, *311*, 317, 329
Atmosphere, primeval, 38
Atoke, 185
Atoll, 122, *123*
Atom, 21, 427
  reactivity of, 22
Atomic mass, 21
Atomic nucleus, 21, *22*
Atomic number, 21
ATP (adenosine triphosphate), 19, *32*,
  71, 73, 388, 389, *390*, 394, *395*,
  396, 407
Atrial natriuretic factor (ANF), 397,
  398, 400, 403
Atriopore, 254, *255*, 257
Atrioventricular node, *348*
Atrium, 106, *171*, *175*, 254, *255*, 257
  genital, *170*, 185, 224
  heart, *105*, 165, *260*, *274*, *275*, 282,
    292, 310, 333, *348*
    (*See also* Auricle)
  tracheal, 224
Auditory meatus, 315
  (*See also* Ear canal, external)
Auditory nerve (*see* Otic nerve)
Auger, 169, 176
Auk, 339
"Aunt", 435
Aurelia, *108*, 112, 117, 118
Auricle:
  of ear, 342
    (*See also* Pinna)
  of heart, 165, *166*, 167, *168*, 169, 176
    (*See also* Atrium)
Auricularia, 238, 245
Australian realm, 421
Australopithecus, 44, 363
Autogamy, 79
Autonomic conditioning, 417
Autonomic nervous system, 230, *266*,
  267, 268, 379, 405, 407
Autosome, 46
Autotomy, 153, 187, 205, 213, 240,
  247, 327, 432

Autotroph, 63, 66–69, 429, *430*
Autotrophic nutrition, 4, 40, 62, 63,
  388
Aversive conditioning, 432
AVES, 57, *252*, *259*, 260, 323, 328,
  337
Avicularium, 155, *156*
Avocet, 339
Avoidance, 2, 416, 432, 434
Axial cell, 92
Axial gradient, 11
Axial rod, 75
Axial skeleton, 262, 264
Axis, body, *10*, 11, 94, 96, 99, 107
Axis (bone), *263*, 317
Axolotl, 106, 256
Axon, *14*, 302, *396*, *401*, 402, 405,
  408, 409, 410, 412, *413*
Axoplasm, 409
Axopod, 75
Axostyle, *69*, 71
Aye-aye, 362

B lymphocyte, 277
Babesia, 63
Baboon, 341, 363, 434, 435
Backbone, 7, 264, 267, 305
  (*See also* Vertebral column)
Bacteria, 16, 40, 90, 216, 230, 333,
  371, 375, 377, 388, *428*–429,
  430
Baculum, 352
Badger, *361*, 367, 368, 426
BALAENIDAE, 367
BALAENOPTERIDAE, 367
Balanus, *211*, 218
Baleen, 210, 221, 260, 360, 367
Baluchitherium, 370
Bandicoot, 350, 357, 358
Barb, 290, *330*, 331
Barbel, 285, 300
Barbicel, 330
Barbule, *330*, 331
Barnacle, 197, *211*, 214, 215, 218–
  220
  parasitic, 215, *219*, 220
Barr body, 47
Barrier reef, 122, *123*, 163
Basal body, 17, 71, 79
  (*See also* Kinetosome)
Base, 26, 27
  nitrogenous, 26, 32, *34*
  number of pairs, 45
  pairing in DNA, *34*
Base substitution mutation, 58
Base triplet, *35*, 36, 394
Basement membrane, 127
Basibranchial, *286*
Basihyal, *286*, 310
Basilar membrane, *344*, 345, 412
Basket star, 241, 246, 247
Basking shark, 290, 291

Basophil, 277
Bass, 303
Bat, *265*, 354, *356*, 360, 361, 362, 414
Batesian mimicry, 432
Bath sponge, 86, *87*, 90
Bathyal benthos, 423
Bathypelagic, 423
Batoid, *288–290*
BATOIDEI, *258*, 289, 290
Batrachotoxin, 309
Beach hopper, *211*, 221
Beachworm, 188
Beak:
  bird, 333, 337, *338*
  cephalopod, 173
Beaked whale, 366
Bear, 341, 350, 360, 367, 368, 422, 426
Beardworm, *159*, 160
Beaver, *346*, 360, *361*, 365
Bedbug, 194
Bee, 195, 228, 230, *234*, 235
  (*See also* Honeybee)
Bee-eater, 339
Beef tapeworm, *135*, 136–137
Beer eel, 144
Beetle, 6, 143, 194, *229*, 232, *234*, 235, 413
Behavior, 7, 397
  aggressive, *326*, 403, 415, 426, 434
  agonistic, 434–435
  allelomimetic, 417
  altruistic, 366, 427
  anti-predator, 414
  avoidance, 2, 416, 432, 434
  breeding, 53, 169, 175, 196, 200, 203, 204, 205, 206, 207, 209, 224, 226, 230, *283*, *307*, 337, 355, 400, 403
  care-giving, 196, 205, 230, 298, *306*, 307–308, 316–317, 335, 341, 435
  communicative, 435–437
  defensive, *301*
  emancipation, 437
  evolution, 436–437
  genetic basis, 336, 415–416
  habitat-selecting, 416
  hormones and, 398, 400, 403
  ingestive, 267
  innate, 340
  instinctive, 196, 230
  learned (*see* Learning)
  perinatal, 416
  reproductive (*see* Behavior, breeding)
  ritualized, 434
  testosterone-dependent, 103, 403
Belemnite, 177
Benthic habitat, 423

Benthonic, 64, 422, 423
Benthos, 423
Beriberi, 375
Bernoulli's law, 331
Beroë, *123*, 124
Beta cell, 400
Beta particle, 22
Bicarbonate ion, 23, 27, 349, 383
Biceps brachii, 15, *263*
Bichir, 300, *301*
Bilateral symmetry, *10*, 125
BILATERIA, 106, 109
Bile, 378, 379
Bile duct, 292, 333, 379
Bile pigments, 272
Bile salts, 272
Binary fission, 63, *66*, 75, 79
Binomial nomenclature, 57
Biocontrol, 5, 6, 163
Bioelectricity, 293
Biogeochemical cycle, *428–429*
Biogeographical realm, 421
Biological clock, 337
Bioluminescence, 66, 67, 124
Biomass, 38, 429
  pyramid, *431*
Biome, 420, 421–422
Biosynthesis, 394, *395*
Biotic, 7
Biotic community, 57, 419, 421, 423, 430
Biotic potential, *425*
Biotin, 374, 376
Bipedality, 317, 320, 323, 324, 329
Bipinnaria larva, 245, 248
Bird, *42*, 43, 44, *56*, 57, *252*, 260, 272, 317, 328–339
  adaptive radiation, 337, *338*
  air sacs, *334*, 335
  behavior, 336, 415, 416, 417, 426, 433, 434, 435, 436–437
  brain, *266*, *334*, *336*
  circulatory system, 333, 348
  classification, 337–339
  development, *95*, 96, *313*, 354
  diet, 337
  digestive system, 333, *334*, 377
  ear, 333
  evolutionary history, 280, 305, 317, 320, 323, 324, 328–329, 337
  excretion, 335
  eye, 270, 332
  feathers, 328, 329, *330–331*
  flight, 329, 331–332
  fossil, 328, 329, 337
  inheritance of plumage, *46*
  internal anatomy, *334*
  migration, 336–337
  perching mechanism, 331, *332*
  reproduction, *334*, 335
  respiratory system, *334*, 335
  senses, 333

Bird (*continued*)
  sex-determination, 47
  sex reversal, 335
  skeleton, *265*
  skin, 333
  song development, 415, 416, 417
  teeth, 329, 331, 337
  thermoregulation, 333
  urogenital system, *334*, 335
  vocalization, 335
  wing span, 329
  wing structure, *56*, *330*, 331–*332*
Birth, 366
  (*See also* Parturition)
Bison, 372, 421
Bitter Springs formation, 39
Bittern, 331, 432
Bivalent (*see* Tetrad)
Bivalve mollusk, 91, 162, 163, 169, 170–173, 247
  body organization, 170–173
  internal structure, *171*
  pearl formation, 162
  subclasses, 176–177
BIVALVIA, *164*, 170, 175, 176
  (*See also* PELECYPODA)
Black coral, 120–121
Black widow, 194, 195, 206
Blackbird, 339
Bladder:
  gall, *256*, 257, *260*, 295
  swim (*see* Swim bladder)
  urinary, 131, 133, 172, *180*, *183*, 191, 210, *260*, *266*, 277, *278*, *295*, 309, 314, 316, 327, *347*, *352*, *353*, *387*, 412
Bladderworm, 134, 136
  (*See also* Cysticercus)
Blastocoel, *95*, 96, 97
Blastocyst, 354
Blastoid, 241
BLASTOIDEA, 241
Blastomere, 94, 96, 102
  (*See also* Macromere; Micromere)
Blastopore, 96, *97*, 98, 101, 106, 125, 152, 161, 236, 253
  dorsal lip, 103
Blastula, *89*, *95*, 96, *97*, *111*, 115, *118*, 124
Blastulation, 93, *95*, 96
Blood, vertebrate, 273
  coagulation, 374, 375, 385–386, 397
  composition, 165, 276–277, 284, 285, 349, 411, 412
  distributional changes in exercise, 385
  gas transport by, 383
  pH, 27
  plasma, 276
  volumetric distribution (in dog), 384
Blood circulation (*see* Heart; Circulation; Circulatory system)

Blood corpuscle, 182, 272, 276–277
Blood fluke, 133, 164
Blood pressure, 285, 397, 398, 400, 401
Blood-vascular system, 273
  (*See also* Circulatory system)
Blood vessel, 137, *138, 159, 180,* 182, *183,* 244, 254, *255*
  vertebrate, *261,* 411
  (*See also* Aorta; Artery; Capillary; Circulatory System; Vein)
Bloodworm, 181
Blowhole, 360, 366, 367, 370
Blubber, 367, 368, 369, 370, 372
  (*See also* Fat)
Bluebird, 415
Blue whale, *9,* 221, 354, 367
Bluegreen algae, 16, 39, 80, 90, 428
  (*See also* Cyanophyte)
Boa, 261, 328, 412
Bobcat, 361
Body cavity, 236, 383
  benefits of, 140
  formation of, 93, 98–*99,* 236
  (*See also* Coelom; Pseudocoel)
Body organization, 9–20
Body size, *9,* 11
Body wall, 86, *138,* 140, 142, 143, 146, 157, 181, 191, 248
  (*See also* Muscle, body wall)
BOIDAE, 328
Bolas spider, 207
Boll weevil, 195
Bolus, 378
Bombyx, 194
Bond, chemical, 21
  covalent, *22*
  double, 24
  electrostatic, 22
  hydrogen, 25–26
  hydrophobic, 28
Bone, 13, 261, 262, *263,* 280, 281, 287, 291, 294, *295*
  cartilage, 261
  deposition and erosion, 398, 399
  dermal (*see* Dermal bone)
  endochondral (*see* Bone, cartilage, replacement)
  membrane, 261
  replacement, 261
  structure, 262–*263*
Bone cell, *263*
  (*See also* Osteoblast; Osteocyte)
Bonefish, 303
Bonellia, 104, 160
Bony fish, *42, 172,* 270, 294–303
  (*See also* Fish; Osteichthyan)
Booby, 339
Book gill, 199, 200
Book lung, 196, 200, *201, 202,* 204, 207
Boomslang, 328

Boophilus, 208
Botfly, 194
Bovid, 360
BOVIDAE, 372
Bowfin, 302
Bowman's capsule, 277, *278,* 316, 387
Brachiator, 363
Brachiolaria, 248
Brachiole, 240, 241
Brachionus, 104
Brachiopod, *42,* 98, 152, *156,* 157–158
BRACHIOPODA, 106, 151, *156,* 157
BRACHYURA, 215, 222
Brachyuran, *211*
Bract, 112
Brain, *404, 405,* 409
  amphibian, *266*
  annelid, 179, 181, *183,* 184, *189*
  arthropod, 195, *201, 203, 212,* 230
  bird, *266, 334, 336*
  cephalopod, 173, *174*
  chordate, 252
  fish, *266,* 282, *283, 292,* 409–410
  flatworm, 410
  hormonal effects on, 403
  mammalian, *266,* 342, *343,* 358
  onychophoran, *190,* 191
  reptile, *266*
  tunicate, 254, *255*
  vertebrate, *260,* 265–267, *343,* 389, 413
  (*See also* Ganglion; Nervous system)
Branchia, *212, 239,* 248
  (*See also* Gill)
Branchial arch, 296
  (*See also* Gill arch)
Branchial artery, *292*
Branchial chamber, 213, 221, 222
Branchial heart, 173, *174*
Branchial sac, 255
  (*See also* Pharyngeal basket)
Branchial skeleton, 284, 285
Branchiopod, 197, 214, 216–217
BRANCHIOPODA, *211,* 215, 216
Branchostegite, *212*
BRANCHIURA, 215, 218
Branchiuran, 218
Breastbone (*see* Sternebra; Sternum)
Breathing, 381
  alligator, *314*
  bird, *334, 335*
  costal, 264
  fish, 284, 285, 289, 290, 294, 296, *297,* 299, 301, 302
  frog, 308
  mammal, 348
  turtle, *314,* 315, 320–321
  (*See also* Respiration; Respiratory system)

Brine shrimp, 215, 216
Bristle, 179, 181, *249, 330,* 331, 339, 341, 369, 370, 371
  sensory, 203, 228
Bristleworm, 179
Brittle star, 91, 219, *237,* 241, 246
  (*See also* Ophiuroid)
Broad fish tapeworm, 136
Broad ligament, 354
Broad-niched species, 424
Bronchial tree, 314, 347, 381
Bronchiole, *347,* 381
Bronchus, *308,* 314, 315, *334,* 335, 366, 381
Brontosaurus, 9, 319, 323
Brood chamber, 140, 204, 205, 214, 217, *219,* 220, 242, *306*
Brood pouch, 215, 220, 357
Brood size, limitation of, 427
Brooding of young, *153,* 155, 168, 169, *172,* 185, 204, 205, 214, 216, 217, 218, 224, 239, 244, 246, 248, 298, 307
Brown fat, 349
Brown recluse spider, 195, 206
Brucke's muscle, 270
BRYOZOA, 151
  (*See also* ECTOPROCTA)
Bryozoan, 98, 155, *156*
Bubonic plague, 365
Buccal cavity, 138, 164, 196, 230, 245, 250, 292, *297*
Buccal funnel, *283,* 284
Buccal pumping, 308
Buccopharyngeal cavity, 308
Bud, 85, 94, *110, 113,* 255
  limb, 97
  medusa, *111*
  tail, *307*
Budding, 63, 79, *123,* 155, 157, 179, 254
Budgerigar, 403
Buffalo, 5, 372
Bufo, 311
BUFONIDAE, 311
Bug, 47, *233,* 235
  kissing, 70
  reduviid, 195, *233*
Bugula, 156
Bulbourethral gland, 352
Bull shark, 291
Bullfrog, *308,* 311
Bumblebee, *234*
Bunting, 339
Buoyancy regulation, 115, 164, 173, 177, 273, 296
Bursa, 239, 248
Bush baby, 362
Bushmaster, 328
Butterfly, 104, 228, 229, *232, 234,* 235, 414, 432
  mimicry in, 61, 432

Butterfly (*continued*)
  sex determination, 47
Byssus, 173, 176
By-the-wind sailor, 115, *116*, 176

Cacops, 304
Caddis fly, *234*, 235
Caecilian, 302, *306*, 307, 308, 310
Caecum (*see* Cecum)
CAENOLESTIDAE, 358
Caiman, 325
Calcaneum, 311
CALCAREA, 89
  (*See also* Calcispongiae)
Calciferol (*see* Vitamin D)
Calciferous gland, 182
Calcisponge, 89
CALCISPONGIAE, *85*, 89
Calcistat, 398
Calcite, 164
Calcitonin, 399
Calcium (Ca), 1, 37, 41, 213, 214,
    340, 374, 375
  blood, 263
  carbonate, 27, 74, 86, 89, 107, 122,
      155, 157, 162, 164, 195, 217,
      225, 238, 262, 313
  ion, 27, 182, 396, 398
  phosphate, 27, 157, 261, 262, 379
Calico coloration, 47
Calorie, 389
Calvin cycle, *391*, 393, 394
Calyx, 152, *153*, 242, *350*
Camarasaurus, 317
Cambarus, 211, 222
Cambrian explosion, 41
Cambrian (Period), 41, *42*, 74, 82,
    162, 166, 177, 178, 197, 217,
    238, 240, 241, 242, *258*
Camel, 5, 360, 372
Cameo, 162
Camouflage, 262, 341, 432
  (*See also* Crypsis)
Campanularia, 113, 115
Canaliculi (of bone), *263*
Cancer, 211, 222
CANIDAE, 57, 367
Canine (tooth), 57, 322, *346*, 369,
    371, 372, 379
Caninus, 342
Canis, 57
Cannibalism, 311
Cap cell, 129, 386
Capillary:
  air, 335
  blood, *15*, 173, *183*, 273, *297*, *347*,
      350, 351, 376, 383, *384*, *387*,
      401
    regulation of blood flow through,
        385
  lymph, 273, 384
Capitulum, 208

Caprella, 211, 221
CAPRIMULGIFORMES, *338*, 339
Captacula, *169*
Captorhinomorph, 316, 317
CAPTORHINOMORPHA, 320
Capybara, 360, 365
Carapace, 198, 199, 204, *212*, 213,
    215, *216*, 217, 218, 219, 220,
    221, *301*, 319, 321, 322
Carbaminohemoglobin, 349, 383
Carbohydrase, 379
Carbohydrate, 27, *29*, 378, 379
Carbon (C), 21, *22*, 23, 24, 429
Carbon cycle, *428*, 429
Carbon dioxide, 4
  and greenhouse effect, 427, 437
  atmospheric concentration, 427
  internal transport, 383
  photosynthetic fixation, 4, 427, *428*,
      429
  production by decarboxylation, *390*,
      391, 392
Carbonic anhydrase, 349, 374, 383
Carboniferous (Period), 41, *42*, 200,
    258, 287, 302, 304, 305, 312,
    316, 322
Carboxyamino linkage, *31*
Carboxyl group, 26, *31*
Carboxylase, 374
Carboxypeptidase, 378, *380*
CARCHARINIFORMES, 290
Carcharinus, 288
Carcharodon, 288
Cardiac accelerator center, 385
Cardiac decelerator center, 385
Cardiac muscle, 13, *14*, 254, 348, 397,
    398, 400
Cardinal vein, *105*, *275*, *292*
Caribou, 372
CARIDEA, 215, 222
Caridid, 222
Caridoid facies, 220
Carina (*see* Keel)
CARNIVORA, 57, *346*, 360, 367
Carnivore, *42*, 353, 354, *356*, 367–
    369, *430*
Carotene, 66, 375
Carotid artery, *105*, *266*, *334*, *347*,
    349
Carotid body, 349, 411
Carotid sinus, 411
Carp, 303
263, 264, *265*, 298, 303, 331, 367
Carpoid, 240
Carpometacarpus, *330*
Carpus, *311*
Carrier-mediated transport, *17*, 18,
    374, 405
Carrying capacity, *425*, 426, 438
Cartilage, 13, *14*, 261, 262, 263, *281*,
    287, 291, 300, 306, 342, *347*, 381
Cartilage bone, 261

Cartilages, laryngeal, *308*, 315, 345
Cartilaginous fish (*see* Chondrichthyan;
    Elasmobranch; Fish, cartilaginous;
    Shark)
Carybdea, 117
Cassia, 162
Cassiopeia, 119
Cassowary, 337
Caste, social, 104, 235, 427, 435
CASUARIIFORMES, 337
Cat, 41, 47, 57, *263*, *266*, *347*, 352,
    354, 360, 368
Catabolism, 3 (*see* Glycolysis; Krebs
    cycle; Pentose phosphate cycle)
Catadromous, 296
Catalase, 20, 374, 388
Catalepsy, 432
Catalyst, 23
  (*See also* Enzyme)
CATARRHINI, 360, 363
Caterpillar, 228, *232*, 432
Catfish, 303
Cation, 26, 407
  (*See also* Ion)
Cattle, 80, 372
  twinning in, 277
  (*See also* Cow)
Cattle egret, 424
Caudal fin (*see* Fin, caudal)
Caudal organ, 141
Caudal ramus, 217
CAUDATA, *259*, *306*, 310
  (*See also* Salamander)
CAUDOFOVEATA, 167
Caudofoveate, 166, 167
CAVIIDAE, 365
Caviar, 300
Cayeuxina, 73
CEBOIDEA, 360, 363
Cebid, 363
Cecum, 155, *170*, 173, *174*, 182, 189,
    200, *201*, 204, 214, *229*, 230,
    239, *244*, 247, 248, *249*, 250,
    257, 333, *334*, 377
Celestial navigation, 337
Cell, 7, 12
  cycle, 33, *35*
  daughter, 48, 49
  division, 47, *48*
    (*See also* Fission; Mitosis)
  eukaryotic, 16
    origin of, 39
  germ, *49*, 93, 100
  meiotic division of, 48–50
  metabolism, 372, 388–396
  organization, 16–20
  prokaryotic, 16, 38, 39
  sex, 48, *49*
    (*See also* Egg; Gamete; Sperm)
  somatic, 100
  specialization, 84
  structure, *17*

Cell (*continued*)
  suicide, 20
  surface of, 16–17
  surface markers, 82, 83
  surface receptors for hormones, 403
Cell wall (in plants), 4, 67
Cellulose, 4, 28, *29,* 71, 253, 254,
    371, 377, 383, 433
Cement gland, *125, 141*
Cement, intercellular, 27, 83, 376
Cement(um), *346*
Cenozoic (Era), *42,* 44, *258,* 355, 370,
    421
Centipede, 143, 178, 191, 192, *193,*
    194, 195, 197, 222, *223–224*
Central nervous system (CNS), 405
  advantages of hollow, 252–253
  organization of vertebrate, 265–267
  (*See also* Brain; Nervous system)
Centriole, *17,* 19, *48,* 71
Centrolethical, 96, 199
Centromere, 20, 45, 47, *48,* 50
Centrosome (*see* Centriole)
Centrum, 288, 300, 301, 303, 305
  arch, 306
  directly ossifying (husk), 306
  inter-, 303, 305, 306
  lepospondylous, 305
  pleuro-, 303, 305, 306
Centruroides, 204
Cephalad, 11
Cephalaspid, 282
CEPHALASPIDOMORPHI, 257, *258,*
    280, 282, *283*
Cephalic grooves, 139
Cephalic shield, *249,* 251
Cephalization, 11, 126
CEPHALOCARIDA, 215, 216
Cephalocaridan, 214, 216
CEPHALOCHORDATA, *252,* 256
Cephalochordate, 253, 256–257, 280,
    383
  (*See also* Amphioxus)
Cephalon, *197,* 198
Cephalopod, 91, 163, 164, 165, 173–
    175, 177, 268, 374, 383, 412,
    413
CEPHALOPODA, *164,* 173, 175, 177
Cephalothorax, 197, 198, *201,* 203,
    204, 205, 208, 210, 213, 217,
    221, 222
Ceratobranchial, 285, *286*
Ceratohyal, *286*
Ceratium, 66
CERATOPSIA, 320
Ceratopsian, 323
Cercaria, *131,* 132, 133
CERCOPITHECOIDEA, 360, 363
Cercopod, 216
Cercus, *227,* 228
Cerebellum, *14,* 265, *266,* 267, *269,*
    282, 342, *343*

Cerebral cortex, 265, *266,* 315, 336,
    342, *343,* 366
Cerebral basal nuclei, 265, *266,* 267,
    342
  (*See also* Corpus striatum)
Cerebral ganglion (*see* Ganglion,
    cerebral)
Cerebral hemisphere, 265, *266,* 269,
    282, 315, *336, 343,* 358
Cerebral organ, 139
Cerebrospinal fluid, 100, 253, 265,
    336
Cerebrum, 265, *347*
  evolution of, *343*
CERIANTHARIA, 120
CERIANTIPATHARIA, *108,* 120
Cerianthid, 109, 120
Cerianthus, 108
Ceruminous gland, 341
CERVIDAE, 372
Cervix, *353,* 354
Cestid, 124
CESTODA, 133
Cestode, 133–137
Cestum, 123, 124
CETACEA, 360, 365
Cetacean, 12, 352, *356,* 365–367, 370,
    402, 435
Chaetognath, 246–250, 253
CHAETOGNATHA, 106, 246, *249*
Chaetopterus, 182, 187
Chagas's disease, 70, 71, 195
Chalk, 5, 74
Chameleo, 326
Chameleon, *326,* 327, 374, 418
  American, 326
CHAMELEONIDAE, 327
Chamid, 176
Chaparral, 421
Character displacement, 55
CHARADRIIFORMES, 339
Cheetah, 368
Cheilosis, 375
Chela, 215
  (*See also* Cheliped)
Chelate, 198, 204, 205
Chelicera, 197, 198, 199, 200, *201,*
    204, 205, 206, 208, 209
CHELICERATA, 197, 198, *199, 200,*
    201
Chelicerate, *193,* 198–209, 210
Cheliped, 222
  (*See also* Chela; Pincher)
CHELONIA, 320
Chelonia, 321
CHELONIIDAE, 321
CHELYDRIDAE, 321
Chemical bond (*see* Bond, chemical)
Chemical communication, 435
  (*See also* Pheromone)
Chemical cycle (*see* Biogeochemical
    cycle)

Chemical element, 21, 22
  occurrence in living matter, 21, 374
Chemical reaction (*see* Reaction,
    chemical)
Chemoautotroph(ism), 428, 429
Chemoreceptor, 126, 185, 398, 411,
    412
Chemotaxis, 64, 435
Chevrotain, 360
Chickadee, 339
Chicken, 339
  behavior, 103, 403, 416
  brain, *266*
  comb shape, 52–*53*
  egg, 313
  embryo, 100, *313*
  heredity in, *46,* 52–*53*
  incubation time, 314
  sex reversal in, 335
Chiggers, 207
Chilaria, 198
CHILOGNATHA, 226
Chilomonas, 67
Chilomycterus, 301
CHILOPODA, *193,* 197, 223
Chimaera, 287, *288,* 289, 293
Chimaera, 288
Chimpanzee, 362, 363, 417, 436
Chinchilla, 365
Chipmunk, 365
Chironex, 117
CHIROPTERA, 360, 362
Chitin(ous), 27, 74, 75, 114, 154, 155,
    157, 160, 161, 164, 166, 167,
    171, 173, 179, 181, 185, 186,
    188, 190, 191, 192, 262
  in arthropods, 193, 195, 196, 199,
    209, 213, 214, 219, 225, *229,*
    230, 231, 313, 316, 379
Chiton, *164,* 166, 167–*168*
CHLAMYDOSELACHIFORMES, 290
Chloragog(en) cell, *180,* 182, 184,
    188, 383
Chloride ion, 23, 25, 296, *351,* 388,
    399, 401, *406,* 408
Chlorine (Cl), *22,* 23, 374
Chlorocruorin, 182
Chlorohydra, 115
CHLOROPHYTA, 62, 63, 67
Chlorophyte, 39
Chlorophyll, 23, 66, 67, 68
Chloroplast, 4, 39, 40, 63, 67, 68, 69
Choana, 299, 300
Choanocyte, *69,* 82, 86, *87,* 88, 89, 90
Choanoflagellate, 69, 70, 82, 90
Cholecystokinin, 379, 397, 400
Cholinesterase, 388
Chondrichthyan, 280, 281, 285, 287–
    293, 296, 298, 313
  characteristics, 289–293
  evolutionary history, 287–*289*
  subclasses, 287

Chondrichthyan (*continued*)
  (*See also* Batoid; Chimaera;
    Elasmobranch; Fish,
    cartilaginous; Shark)
CHONDRICHTHYES, *252, 257, 258,*
  280, 287, *288*
Chondrification, 262
Chondroblast, 262
Chondroclast, 262
CHONDROPHORA, 115
Chondrophore, *116*
Chondrostean, 300, 302
CHONDROSTEI, *258,* 300, *301*
CHORDATA, 57, 99, 106, 250, *252*
Chordate, *99,* 179, 238, 248, 253–372
  ancestry, 251
  characteristics of phylum, 252–253
  subphyla, *252*
CHORDEUMIDA, 226
Chorioallantoic placenta, 354
Chorion, 95, 201, 231, *313,* 314, 354,
  357
Chorionic gonadotrophin, 354, 400
Choroid, *269,* 412
Choroid plexus, 265, *266*
Chromatid, 47, *48,* 50
  (*See also* Chromosome, daughter)
Chromatin (*see* Chromosome)
Chromatophore, 20, 188
  cephalopod, 173
  crustacean, 213
  responses of, 184, 399, 418
  vertebrate, *261*
Chromatophorotropic hormone, 213
Chromoplast, 63, 66, 67
Chromosome, *17,* 20, 94, 403
  daughter, 47, 49
    (*See also* Chromatid)
  dinoflagellate, 67
  diploid state (2n), *35, 45,* 47, 48, 50
  haploid state (n), 48
  homologous, 45, 48, *49,* 50
  inversion in, 59
  mapping, 50
  meiotic behavior, *49,* 50, 51
  mitotic behavior, *48*
  monoploid state, 47
  number, 45
  polynemic, 101
  polyploid state, 79
    (*See also* Polyploidy)
  proteins in, 20, 45
  puffs, 101, 231
  rearrangements, 59
  structure of, 45, 50
  translocation of, 59
  transmission of, 47
  sex, 46, 47
  virus, 38
Chrysaora, 118
Chrysemys, 321
Chrysomonad, 66, 67

CHRYSOPHYTA, 67
Chuckwalla, 326
Chyme, 379
Chymotrypsin, 378, 379, *380*
Cicada, *233,* 235
Cichlid, 303
CICONIIFORMES, 337, *338,* 339
Ciliary body, *269*
Ciliary muscle, 268, *269,* 270, 332
CILIATA, 68, 80
Ciliate, 62, 77–81
CILIOPHORA, 68
Ciliophoran, 77, 80
Cilium, *17,* 18, 137, 138, 150, *151,*
  155, 157, 161, 249, 250, 374
  absence of, 147, 191, 196, 202
  action of, 71–72
  in annelids, 178, 181
  in chordates, 254, 273, 277, *278,*
    353, 381
  in echinoderms, 239, 242, 245, 248
  in flatworms, 126, 129
  in mollusks, 163, 165, 169, 170,
    171, 172, 173, 176
  in protists, *70,* 71, 77, *78,* 79, 80, 81
  in pseudocoelomates, 140, *141,* 142
  in radiates, *123,* 124
  sensory, 71, 269, 271, 272, 344,
    410, 412, 413
    (*See also* Hair cell)
  structure of, *70*
Circadian rhythm, 337
Circulation:
  double, 274, *275,* 310
  fetal, 349
  pulmonary, 274
  single, 274, *275,* 310
  systemic, 274
Circulatory system, 13, 383–386
  amphibian, 274, *275,* 310
  amphioxus, 256
  annelid, 182, *183,* 383, *384*
  arthropod, 196, 199, 200, *201,* 202,
    207, *212,* 214, 230, 383, *384*
  bird, *265, 274,* 333
  brachiopod, 157
  closed, 138, 159, 173, 182, 273,
    383–385
  echinoderm, 239, 244
  echiurid, 160
  enteropneust, 250
  fish, *105, 274, 275*
  human, *105*
  mammal, *274, 275, 348–349*
  mollusk, 165, *166,* 169, *171,* 173,
    174, 383, *384*
  nemertean, 138–139, 383, *384*
  onychophoran, 191
  open, 138, 157, 165, *166,* 196, *212,*
    214, 250, 254, 383, *384*
  phoronid, 157
  tunicate, 254, *384*

Circulatory system (*continued*)
  pogonophoran, 161
  vertebrate, 273–277, 383, *384*
Cirratulid, 187
Cirratulus, 187
CIRRIPEDIA, *211,* 215, 218
Cirrus, 79, 81, 132, *141, 178, 180,*
  181, 186, *187,* 219, 242, *255*
Citric acid cycle, 27, 389
  (*See also* Krebs cycle)
Civet, 57, 360, 368
CLADOCERA, 215
Cladoceran, 105, 216–217
Cladogenetic speciation, 54, *55*
Cladoselache, 287, 288, 289
CLADOSELACHII, *258*
Clam, 67, *164, 171, 172,* 176, 207,
  374, *384*
Clamworm, *180, 187*
Clasper, *216,* 287, *288,* 292, 293, 296
Class, 57
Classical conditioning, 417
Classification, 7
  hierarchical system of, 57–58
  DNA analysis and, 337, 363, 367
Clavicle, *263,* 264, 294, *299,* 329,
  331, *332,* 367
Claw, 161, 191, 200, 208, 262, 314,
  362, 368
  poison, 195, *223, 227*
  (*See also* Talon)
Cleaning symbiosis, 433
Cleavage, *89,* 93, *94,* 95
  bilateral, 152
  determinate, 124, 143, 152, 165,
    195
  discoidal, *95,* 96, 297, 354
  holoblastic, *95,* 96, *111,* 155, 192,
    195, 199, 204, 214, 231, 306,
    307, 354
  indeterminate, 135, 152, 253
    (*See also* Regulation,
      indeterminate)
  meroblastic, *95,* 96, 204
  mosaic, 124, 153
    (*See also* Cleavage, determinate)
  radial, *94,* 152, 154, 155, 157, 158,
    236
  spiral, *94,* 125, 137, 143, 152, 153,
    158, 160, 165, 236
  superficial, 96, 192, 214, 231
Cleithrum, *299*
Climatius, 286
Climax ecosystem, 420, 423, 424, 430
Clingfish, 303
CLITELLATA, 186
Clitellum, 185, 186, 188, *189*
Clitoris, 279, *353*
Cloaca, 144, 147, 239, 243, *260,* 277,
  278, 292, 293, 310, 316, 333,
  334, 335, 353, 355, 377
Cloacal bud, 316

Clock, biological, 336, 337
  (*See also* Rhythm)
Clone, 102, 121
Clonorchis, 132
Clotting (*see* Blood, coagulation of)
CLUPEOMORPHA, 303
Clypeus, *227*
Cnida, 109
CNIDARIA, 106, 107, 108, 113
Cnidarian, 108–123, 176, 374
  body organization, *110, 111, 413*
  cell types, 109–111
  characteristics, 108
  development, 108–109
  life cycles, *111*, 112, *113*
  polymorphism in, *111*–112
  taxonomic groups, 113–123
Cnidoblast, 67, *108*, 109
Cnidocil, *108*, 109
Cnidocyte, 109, 114
CNIDOSPORA, 68, 77
Cnidosporan, 77
Coatimundi, 367
Cobalamin, 136
Cobalt (Co), 374
Cobra, 204, 327, 328, 368
Coccidiosis, 75, 76
Cochlea, *271*, 333, *344*, 345, 411,
  412
Cochlear duct, *271*
  (*See also* Scala media)
Cockle, 176
Cockroach, 71, 143, 228, 235, 418
Cocoon, 185, 188, *189*, 206, 226, 230,
  234, 235, 300
Cod, 298, 303
Codominance, *46*, 58
Codon, *35*, 36, 394
Codosiga, 69, 70
Coelacanth, 298, *299*, 313
COELENTERATA (*see* CNIDARIA)
Coelenterate, *42*, 107, 108, 109
  (*See also* Cnidarian)
Coelepid, 281
Coeliac artery, *347*
Coeloblastula, 96
Coelom, *98, 99*, 140, 150, 151, 152,
  154, 155, *156*, 157, *159*
  annelid, 179, *180*, 181, 182, *183*,
    184, 185, 188, *189*
  arthropod, 202, 210
  chaetognath, 249
  chordate, 253, *255*
  echinoderm, 236, 238, 239, 243,
    244, 248
  hemichordate, *249*, 250, 251
  molluscan, 165, *166*
  vertebrate, *260*, 277, 278, 282, *283*,
    285, *295*, *334*, *347*
Coelomic funnel, 151, 157, 277, *278*,
  285
  (*See also* Nephrostome)

Coelomocyte, 160, 182, 184, 238, 239,
  243, 244, 246
Coelomoduct, 277
Coelurosaur, 324, 328, 329, 331
Coenenchyme, 199, *120*
Coenosarc, *111*
Coenzyme, 373, 375, 376
Coenzyme A, 376, *390*, 392
Cofactor, 373, 374
Cognitive learning, 417–418
Cohesion model, 82–83
Coiling, 169
Cold, sensing, 411
COLEOIDEA, 177
COLEOPTERA, *234*, 235
COLIIFORMES, 339
Collagen, 3, 83–84, 86, 88, 140, 142,
  261, 262, 313
Collar cell, *69*, 70, 82, 85, 86, *87*, 90
  (*See also* Choanocyte)
Collecting duct (tubule), 316, *350*,
  351, 399
COLLEMBOLA, *233*, 235
Collemolan, 231
Collencyte, 88
Colloblast, 124
Colloidal change of state, 72
Collum, *223*, 225
Colon, *266*, 377
Colonial nesting, 435
Colonialism, 70, 82, 83, 86, *111*, 112,
  116, 133, *153*, 155, *156*, 251, 253
Colony (*see* Colonialism)
Color blindness, 47, 58, 340
Color vision, 58, 228, 270, 410, 414
Coloration:
  changes in, 173, 188, 309, 326, 327,
    397, 402, 418
  communicative value of, 262, 329
  cryptic, 3, 309, 330, 432
    (*See also* Camouflage)
  inheritance of, *46*, 47, *51*, 58–59
  warning, 176, 262, 309, 432
  (*See also* Pigmentation)
Colpoda, 77
COLUBRIDAE, 328
Colugo, *356*, 360
COLUMBIFORMES, *338*, 339
Columella, *271*, 309, 315, 333, 342,
  345
Comatulid, *237*, 242
Comb (of chickens), 52–53
Comb jelly, 107, *123*, 124
  (*See also* Ctenophore)
Comb plate, *123*, 124
Commensal, 64, *80*, 90, 160, 218, 424
Commensalism, 64, 433
Commissure, 127, *128*
Communal nesting, 435
Communal parenting, 435
Communication, 196, 230, 262, 341,
  434, 435–437

Communication (*continued*)
  acoustic, 329, 370, 418
  chemical, 435
    (*See also* Pheromone)
  evolutionary conservatism of, 435
  electrical, 436
  tactile, 436
  visual, 329, 341, 436–437
Community, biotic, 57, 419, 421, 430
Competition:
  interspecific, 55, 423–424, 431
  intraspecific, 426, 434
Competitive exclusion principle, 423,
  431
Compound, chemical, 21
  covalent, 21
  inorganic, 23
  ionic, 21
  organic, 23–24, 27–36
    abiogenetic origin of, 38
Compound eye, 196, *197*, 198, 199,
  210, 213, *216*, 217, 218, 220,
  221, 224, *227*, 228, *413*
Concentration gradient, 18, 381,
  406
Conch, 162, 176
Conchiolin, 164, 166
CONCHOSTRACA, 215
Conchostracan, 105, *216*
Conditioned reflex, 417
Conditioning:
  autonomic, 417
  aversive, 432
  classical, 417
  operant, 175, 417
  Pavlovian, 417
Condor, 329, 330, 337, 339
Cone, 71, 163, 176, 267, *269*, 270,
  410, 411, 412, 413, 414
Coniferous forest, *420, 421*, 422
Conjugation, 63, 72, 78, 79
Connective tissue, 3, 13, *14*, 83, 260,
  261, 262
Consciousness, 267
Consolidation (in memory), 410
Conspecific interactions, 433–437
Consumer, 194, 429, *430*
Contact comfort, 436
Contact inhibition, 82
Contact species, 436
Continental drift, 42, 43
Continental seas, 43, 422
Continental shelf, 422
Continental slope, 423
Continuous variation, curve of, 58, *60*
Contour feather, *330*, 331
Contractile vacuole, 19, 63, *73, 78,*
  79, *80*
Contraction of muscle, 374, 389, 394-
  396, 418
Conus, 163, 165
Conus arteriosus, 292

Convergence:
    evolutionary, 55, *56*, 57, 98, 173,
        300, 324, 337, 364
    synaptic, *408*
Convoluted tubule, *350*, 351, 400
Convolutions, 342, 366
Cooperation:
    interspecific, 431, 432
    intraspecific, 435
Coordination, muscular, 267, 342, 412
Coot, *338*, 339
Copepod, 5, 148, 194, 195, 197, 210,
    217, 429, *430*
COPEPODA, *211*, 215, 217
Copepodid, 217
Copper (Cu), 165, 374
Copperhead, 328
Copulation, 130, 132, 134, 141, 142,
    144, 169, 185, *189*, 206, 214,
    231, 287, 292, 293, 321
Copulatory organ, *125*, 137, 277, 310,
    311
    (*See also* Clasper; Hemipenis; Penis;
        Stylet)
Coracidium, 136
CORACIIFORMES, *338*, 339
Coracoid, 264, 294, 331, *332*, 355
Coral, 41, 67, 112, 167, 374
    black, 120–121
    gorgonian, *120*
    horny, 120
    hydrozoan, *113*, 115
    soft, 120
    stinging, 115
    stony, 116, 121, 122, *123*, 219
Coral reef, 90, 122–*123*, 176
Coral snake, 2, 328, 416, 432
Corallimorph, 121, 122
CORALLIMORPHARIA, 122
Corallium, 119
Cormorant, 339
Cornea, 177, 267, *269*, 270, 332,
    413
Cornified layer of skin, *261*
    (*See also* Stratum corneum)
Corona, *141*, 142
CORONATAE, 118
Coronatid, 118
Cornuspiroides, 73
Corpora cavernosa, 316, 352
Corpus allatum, *231*
Corpus callosum, 342, *343*, 358
Corpus cardiacum, *231*
Corpus luteum, *353*, 354, 399, 400
Corpus spongiosum, 352
Corpus striatum, 265, *266*, 267, *336*,
    342, *343*
Corpuscle:
    blood, 139, 157, 272
        (*See also* Erythrocyte; Leukocyte;
            Red blood corpuscles)
    sensory, *261*

Cortex:
    adrenal, 399, 401
    cerebral, 253, 265, *266*
    egg, 101
    renal, 350, 351
Corti, organ of (*see* Organ of Corti)
Cortisol, 399
Corynactis, 122
Cosmine, 287
Cosmoid scale, 262, 287
Cotylosaur, 312, 316, 317, 318, 322
COTYLOSAURIA, *259*, 320
Cougar, 368, 424
Counterattack, 432, 435
Countercurrent exchange, 292–293,
    297, 350
Countershading, 330
Courtship, 169, 175, 196, 203, 204,
    205, 224, 226, 230, 310, 321,
    331, 337, 416, 434, 435, 436–437
Covalent bonding, 22, 23
Cow, *353*, 371
Cowper's gland, *347*
Cowrie, 162, 176
Coxa, 198, 199, 200, 220
Coxal gland, 191, 199, 202, 204, 205,
    207, 209, 210
Coxal sac, 225
Coyote, 57
Crab, 133, 193, 194, 195, 197, 214,
    215, *219*, 220, 221, 432
    coconut, 222
    development of, 105
    fiddler, *211*, 222
    ghost, 222
    hermit, 122, 194, 196, 197, *211*,
        215, 222
    horseshoe, 197, 198–199
    king, 222
    land, 222
    lithode, 215, 222
    mole, 215, 222
    pea, 160, 433
    porcelain, 215, 222
    red, 222
    sand, 222
    true, *211*, 214, 215, 222
Crampton's muscle, 270, 332
Crane, 339
Cranial capacity, 363–364
Cranial nerve, *266*, 267, 268, *343*
Cranioamphistyly, 345
Cranium, 262, *263*, 264, *286*, 287,
    294, 298, 302, 305, 309, *311*, 345
Craspedacusta, *113*, 114
Crayfish, 133, *193*, 195, 210, *211*,
    212, 214, 221, 222
    appendages, *212*
    body organization, *212*, 220
Creatine, 351
Creatine phosphate, 236
Creche, 435

Cretaceous (Period), *42*, 43, 177, 258,
    317, 319, *324*, 337
Cribrostomum, 73
CRICETIDAE, 365
Cricket, 143, *233*, 235, 436
Crinoid, *42*, 219, *237*, 240, 241, 242,
    245
CRINOIDEA, *237*, 241, 242
CRINOZOA, 240
Crinozoan, 241
Crithidia, 62, 69, 70
Crocodile, 6, *265*, *317*, 325
CROCODILIA, *259*, 320, *321*, 325
Crocodilian, 314, 315, 316, 323, 325,
    348
    sex determination in, 316
Crocodylus, 325
Cro-Magnon man, 364
Crop, *170*, 182, *183*, *189*, 199, *229*,
    230, 333, *334*, 339, *377*
Cross-bridge, 71, 394, 395, *396*
Cross-current exchange, 335
Crossing-over, *49*, 50
Crossopterygian, 296, 298, *299*
CROSSOPTERYGII, *258*, 298
Crotalus, *326*
Crow, 337, 339
Crowding, effects of, 427, 434
Crown, 184, 186, 188
Crown (of tooth), 181
Crown-of-thorns starfish, 123
CRUSTACEA, *193*, 210, 215
    taxonomy, 215
Crustacean, *42*, 193, 196, 197, 210–
    222
    appendages, 210, 211, *212*
    body organization, 210–214
    characteristics, 210
    chromatophores of, 213
    circulatory system, 214
    development, 105, *213*, 214–215
    excretory organs, 210
    exoskeleton, 213
    food-getting, 214
    head organization, 210
    molting in, 213
    nervous system, 214
    reproduction, 214
    subgroups of, 215–222
Crypsis, 432
Cryptobiosis, 150, 154, 426
CRYPTODONTA, 176
CRYPTODIRA, 320, 321
CRYPTOPHYTA, 67
Cryptomonad, 66, 67
Crystalline cone, *413*
Crystalline clays, 38
Crystalline style, 165, *172*
Ctenidium, 163
Ctenoid scale, 262, 294, *295*
CTENOPHORA, 106, 107, 124
Ctenophore, *123*, 124, 125, 374, 422

Ctenoplana, 124
Cubomedusa, 110, 112, 116, *117*
CUBOZOA, 113, 116, *117*
Cuckoo, 339, 416, 433
CUCULIFORMES, 339
Cud, 371, 372
Culex, 76, *234*
Cupula, 272
Cuticle:
    annelid, *180,* 181, *183,* 184
    arthropod, 195, 196, 213
        (*See also* Exoskeleton)
    chaetognath, 248
    ectoproct, 155, *156*
    entoproct, 152, *153*
    molluscan, 166, 167
    onychophoran, 191
    pentastomid, 161
    pogonophoran, 161
    priapulid, 154
    pseudocoelomate, 140, 142, 146,
        150
    tardigrade, 154
Cuttlebone, 162, 164, 173, 177
Cuttlefish, 162, 164, 173, 177, *413,*
    418
    (*See also Sepia*)
Cuvierian tubules, 243
Cyanea, 117, 118
Cyanocobalamin, 376
Cyanophyte, 39, 40
    (*See also* Bluegreen algae)
Cyclic AMP, *32,* 33, 65, 403
Cyclic GMP, 403
Cycloid scale, 262, 294, *295,* 302
Cyclopia, 103
Cyclops, 148, *211,* 217
CYCLOSTOMATA, *252,* 282
Cyclostome, *269,* 281, 282, *283*
Cydippid, 124
Cynognathus, 322
Cyprinid, 303
Cyprinodont, 303
Cypris, 214
Cypris, 214, *218, 219,* 220
Cyrtocyte, 140
Cyst, 66, 67, 74, 75, 77, *131,* 133,
    135, 136, 184
Cysticerciasis, 135
Cysticercus, 134, *135,* 136
Cystoid, *42,* 241
Cytochrome, 349, 374, 381, *390,* 393
Cytochrome oxidase, *390,* 393
Cytodifferentiation, 93
Cytology, 7
Cytopharynx, 79
Cytoplasm, 16, 19
    components of, *17,* 19–20
    unequal division in oogenesis, *49*
Cytoplasmic binding protein, 403
Cytopyge, 63, *78,* 79
Cytosine (C), 32, *34*

Cytoskeleton, 19, 75
Cytostome, 63, 69, 72, *78,* 79
Cyzicus, 216

Dactylozooid, 112, *113,* 115, *116*
Daddy longlegs, 205, 206
Damselfish, 122, 303
Damselfly, *232,* 233, *235*
Daphnia, 211, 216, 217
Dart sac, *170*
Darwin, C., 57, 70–71
Darwin's finches, 54
Dasypeltis, 326
Dasyurid, 357, 358
DASYURIDAE, 358
Datum, 1–2
DDT, 6, 427
Deamination, 184
Death, 84–85, 175, *428,* 435
Decapod, *211,* 214, 221–222
DECAPODA, 215, 221
Decarboxylation, 376, *390, 391,* 392
Decay, *428,* 429, *430*
Deciduous hardwood forest, *420, 421,*
    422
Decomposer, 194, 429, *430*
Deep-sea animals, 90, 160, 167, 179,
    199, 242, *289,* 296, 423, 429
Deer, *346,* 360, 372
Deer mouse, 416
Defecation, 412
Defensive behavior, 190, 204, 414,
    418
    group, 435
Defensive cells, 276, 277
Deficiency diseases, 375–376
Definitive host, *135,* 143, 146, 433
Deforestation, 420, 422
Dehydration:
    tolerance of, 142, 143, 150, 154,
        184, 204, 217, 426
        (*See also* Egg, thick-walled;
        Encystment)
    protection from, 196, 262
Dehydrogenase, 374
Delamination, *95,* 96
DELPHINIDAE, 366
Delta cell, 400
Demodex, 208
Demosponge, 90
DEMOSPONGIAE, *85,* 89, 90
Dendrite, *14*
DENDROBATIDAE, 311
DENDROBRANCHIATA, 215, 222
Dendrocystites, 240
Denitrifying bacteria, *428,* 429
Density, population, 427
Dental formula, 346
Dentalium, 169
Dentary, 345
Denticle, 281, 287, 291, *292, 295*
Dentine, 262, *292,* 298, *299, 346*

Dentition, 280, 316
    heterodont, 322
    mammalian, 340, *346*
    shark, 288, *289*
Deoxyadenylic acid, *34*
Deoxyribonuclease, 378
Deoxyribonucleic acid (*see* DNA)
Deoxyribose, 32, *34*
Depth perception, 362
Dermal armor, 261, 262
    (*See also* Armor)
Dermal bone, 262, 285, 287, 294, 295,
    300, 309, 372
Dermal branchia, *239,* 248
Dermal denticle, 262
Dermal ostium, 86
DERMAPTERA, *233,* 235
Dermatitis, 376
Dermatome, 97, *98*
Dermis, 138, 260, *261, 292, 294, 295,*
    330
DERMOCHELYIDAE, 322
Dermocranium, *318,* 319
DERMOPTERA, 360
Desertification, 420, 437
Deserts, 420, 421
Desmosome, 16
Determinate cleavage
    (*see* Regulation, determinate)
Determinate growth, 4
Determination, 93
Detorsion, *175,* 176
Detour problems, 175, 417
Deuterostome, 94, 96, 97, 98, 99, 102,
    150, 179, *249,* 253
    compared with protostomes, 129,
        152, 236
    phylogenetic linkage with
        protostomes, 152
DEUTEROSTOMIA, 106
Development:
    sponge, *89*
    eumetazoan, 93–106
        (*See also* Embryonic development)
Devonian (Period), 41, *42,* 167, 240,
    246, *258,* 281, 286, 287, 289,
    293, 298, 299, 300, 302, 303,
    304, 305
Diabetes mellitus, 398
Diadectes, 305
Diadectid, 305, 312
Diapause, 232
Diaphragm, *347,* 348, 381
Diaphragmaticus, *314,* 315
Diaphysis, 262, *263*
Diapsid, *318,* 320, 322, 325
DIAPSIDA, 322
Diarthrognathus, 340
Diastema, 346, 364, 370
Diatom, 5, 40, 194, 422, 429, *430*
DIBRANCHIA (*see* COLEOIDEA)
Dicyemid, *91*

DIDELPHIDAE, 358
Didelphis, 357
Didinium, 79, *80*
Diencephalon, 103, 265, *266*, 267, 269
Differentiation (*see* Cytodifferentiation)
Difflugia, *73*, 74
Diffusion, 18, 374, 381, 382, *387*, 405
   facilitated, 18
DIGENEA, 130
   life cycle, 132
Digestion, 3, 373
   chemical, 373, 378, 379–*380*
   external, 191, 201, 204, 205, 206
   extracellular, 62, 107, 108, 127,
      131, 147, 155, 167, 173, 182,
      248, *377*, 378
   intracellular, 63, 88, 110, 125, 127,
      147, 155, 158, 165, 172, 200,
      248, 374, 376–377
      (*See also* Vacuole, food;
         Phagocytosis)
   physical (mechanical), 379
   regulation of, 378–379
Digestive enzymes, 376, 377, 378, 379
Digestive system, 13
   amphibian, 310, *377*
   amphioxus, *255*, 256, 257
   annelid, 182, *183*, *377*
   arthropod, 196–197, 199, 214, 230
   bird, 333, *334*
   brachiopod, *156*, 158
   bryozoan, 155, *156*
   chaetognath, *249*
   ctenophore, *123*, 124
   earthworm, *183*, *377*
   echinoderm, 239, *244*, 245, 247
   echiurid, *159*, 160
   enteropneust, 250
   entoproct, *153*
   flatworm, 127–*128*, 131, *377*
   gastrotrich, 140, *141*
   hydra, *377*
   kinorhynch, *141*, 142
   mammal, *347*, *371*, 378
   mollusk, *164*, *165*, 167, *172*, 173
   nematode, *145*, 147
   nemertean, 137, *138*
   onychophoran, *190*, 191
   phoronid, *156*, 157
   priapulid, 154
   radiate (*see* Gastrovascular cavity)
   rotifer, *141*, 142
   sipunculid, 158, *159*
   tardigrade, 155
   tunicate, 254, *255*
   vertebrate, 268, 284, 285, *292*
Digitigrade, 329, 337, 354, 367, 368
Digit, 264, *265*, 303, 308, *311*, 327,
      329, *330*, 331, *332*, 367, 368,
      370, 372
   reduction in horse family, *56*
Dihybrid cross, *52*

Dileptus, 79, 80
Dimetrodon, *322*
Dimorphism, sexual, 203
Dinichthys (see Dunkleosteus)
Dinoflagellate, 40, 62, 66–67, 75, 163,
      194, 429, *430*
DINOFLAGELLIDA, 66
Dinopsis, 207
DINORNITHIFORMES, 337
Dinosaur, *42*, 43, 315, *317*, 320, 323,
      329
   ostrichlike, 324
Dioctophyma, 149
Dioecious, 88, 121, 165, 167, 169,
      172, 175, 185, 188, 203, 214,
      219, 244, 251
Dipeptidase, 374, 378, *380*
Dipeptide, 143
Diphycercal tail, 287
Diphyllobothrium, 136
Diploblastic, 97, 107
Diploid(y) (2n), *35*, 45, 47, 48, 50
Diplomonad, 69, 71
DIPLOPODA, *193*, 197, 225
Diplopore, 241
DIPLOPORITA, 241
DIPLORHINA, 282, 285
Diplosegment, 225, 226
Diplostracan, 216
DIPNEUSTI, 298, 299
DIPNOI, *258*
   (*See also* DIPNEUSTI)
Diprotodon, 357
Dipsosaurus, 315
DIPTERA, 101, *234*, 235
Dipteran, 228, 230
Dipylidium, 136
Directional selection, *60*
Disaccharidase, 379
Disaccharide, 28, *29*, 373, 378, 379
Discontinuous variation, 58
Disk, *237*, 240, 246, 247
Displacement behavior, 437
Displays, communicative, 434, 436–
      437
Disruptive selection, *60*–61
Distal, 11
Diurnal habit, 431
Divergence:
   evolutionary, 55, 56
   synaptic, *408*
Diverticulum, *172*, 201, 204, 257
Diving animals, 339, 366
Division of labor, 230, 341, 435
DNA, 3, 20, 32, *34*, 35, 45, 101, 374
   analysis in taxonomy, 337, 363, 367
   production of RNA by, 33, *34*
   replication errors in, 53, 58
   synthesis (self-replication) of, 33,
      34, 47, *49*, 50, 94, 373
DNA ligase, 33, *34*
DNA polymerase, 33, *34*

Dobsonfly, 235
Docodont, 355
Dodecaceria, 179
Dog, 5, 57, 333, *346*, 352, 354, 360,
      367, 414, 417, 435
Dogfish, *288*, 290
Doliolaria larva, 245
Dolphin, *56*, 360, 365, 366, 414
Dominance:
   genetic, 45–46, *51*, *52*, *53*, 415
   social, 341, 426, 427, 434
      (*See also* Social hierarchy)
Donkey, 5
Dopamine, 267, 409
Dorsal, *10*
Dorsal lip of blastopore, 103
Dorsal root, 282, *404*
Dorsal root ganglion, *404*
Dove, *338*
Down, 5, *330*, 331
Down's syndrome, 47
Dracunculus, 148, 195
Dragonfly, 228, 229, 232, 235
Drake, 435, 436–437
Drepanaspis, *281*
Drift, genetic, 61
Drill, 176
Drosophila, 45, 47
   speciation in Hawaii, 226
Duck, 335, *338*, 339, 435, 436–437
Duck-billed platypus (*see* Platypus)
Ductus arteriosus, 275, 349
Ductus deferens, *128*, 131, *141*, *170*,
      189, 278, 292, *334*, *347*, *352*
      (*See also* Sperm duct, Vas deferens)
Dugong, 360, 370
Dunkleosteus, *286*, 287
Duodenum, 377, 378, 379, 400
Duogland adhesive complex, 127, 140
Duplicate genes, 58
Dura mater, 265
Dwarfism, 103, 401

Eagle, 339
Eagle ray, 290
Ear:
   amphibian, *271*, 309
   bird, 333
   fish, *271*, 284, 285, 412
   insect, 414
   mammalian, *271*, 342, *344*, 345
   reptilian, *271*
   vertebrate, *256*, 270–272, 412
Ear wax, 260, 341
Eardrum, 228, *271*, 309, 327, 333,
      344
Earthworm, 179, *180*, 184, 185, 188,
      189, *377*, 382
Earwig, *233*, 235
Ecdysiotropin, 231
Ecdysis, 213
   (*See also* Molting)

Ecdysone, 31, 101, 103, *231*
Echidna, 355, *356,* 357
Echinarachnius, 237
Echinococcus, 136
Echinoderm, *42,* 99, 236–248, 253, *404*
  characteristics, 238–240, 383
  phylogenetic status, 236–238, 250, 251
  classification, 240–248
  development, 239
ECHINODERMATA, 106, 237, 240
Echinoid, *237,* 245–246
  (*See also* Sea urchin)
ECHINOIDEA, *237,* 241, 245
Echinopluteus, 246
ECHINOZOA, 241, 242
ECHIURA, 106, 151, *159*
Echiurid, 159–160, 178, 383
Echolocation, 360, 362, 366, 368
Ecological equivalence, 57, 337
Ecological niche, 420, 423, 424
Ecological succession, 419–420, 422, 423
Ecology, 7, 419–438
Ecosystem, 419, 424, 438
  aquatic, 422–423
  climax, 420, 423, 424
  closed, 419
  future of, 437–438
  open, 419
  successional, 419–420
Ectoderm, 96, *97, 98, 102,* 107, 196, 253, *313*
Ectoparasite, 130, 146, 433
  (*See also* Amphipod;
    Ascothoracican; Branchiuran;
    Bug; Copepod; Flea;
    HEMIPTERA; HOMOPTERA;
    Isopod; Lamprey; Leech; Louse;
    Mite; Mosquito; Screwworm;
    Tick; Tsetse fly; Vampire bat;
    Wasp)
Ectoplasm, 72, *73, 78, 80*
Ectoproct, *99,* 155, *156*
  (*See also* Bryozoan)
ECTOPROCTA, 106, 151, 155, *156*
Ectothermal, 312, 315
  (*See also* Poikilothermal)
EDENTATA, 360, 364
Edentate, 345, *356,* 364
Edest(o)id, 287, 288, *289*
Ediacara Hills, 41
EDRIOASTEROIDEA, 241, 242–243
Edwardsia, 108
Eel, 296, *301,* 303
Effector, 107, *183,* 405, 418
  independent, 398
  (*See also* Gland; Muscle; Muscle contraction)
Efferent ductules, *352*

Efferent nerve fiber, 267, *404,* 405
Egestion, 3, 63
Egg, 50, *51, 52,* 68, 82, *87,* 88, *91,* 110, 175, *180,* 185, 200, *232, 234*
  activation of, 93, 94, 102
  amictic, 143
  amphibian, *306, 307,* 310, 311
  amniote, 280, 312, *313,* 354, 355
  axis of, *97*
  bird, 313, 335
  centrolecithal, 96, 199, 214
  cleidoic, 231, 313
  cytoplasm, regionalization of, 94, 101–*102*
  diploid, 217
  ectolecithal, 127
  endolecithal, 126
  fertilization of, 94, *97*
  fish, 279, 284, 285, 292, 293, 296, 297, 298, 300, 313
  maturation of, *49*
  mictic, 143
  telolecithal, 96
  thick-walled, 130, 142, 143, 150, 155
  thin-walled, 130, 142, 150, 155
Egg capsule, 169, 175
Egg case, 177, 285, 292
Egg-eating snake, *326*
Egg membranes, 313
Egg sac, 216
Egg-laying:
  stimulation of, 403
  limitation of, 427
Eggshell, 312–313
Egret, cattle, 424
Eimeria, 76
Eimeriid, 75, 76
Ejaculation, 205, 207, 335
Ejaculatory duct, 192, 231
ELAPIDAE, 328
Elasmobranch, 289–293, 413
ELASMOBRANCHII, 287, 289
Elasmosaurus, 319, 322
Elastic connective tissue, *14*
Elbow, 317, 365, 370
Electric catfish, 296
Electric eel, 296
Electric fish, 288, 290, 296, 303, 413, 436
Electric organ, 290
Electric ray, *288,* 290, 418
Electrical communication, 436
Electrical receptor, 261, 355, 411, 413
Electrogenesis, 290, 303, 412, 418
  (*See also* Electric fish)
Electron, 21, *22,* 374
  excitation by light energy, 23
  transport of, 375, 389, *390,* 393
Electroreceptor, 293
  (*See also* Galvanoreceptor)

Electrostatic bonding, 22
Elements, chemical, 21, 22
  use as micronutrients, 374
Eleocyte, 182, 383
Elephant, *5, 9,* 11, *42,* 350, 354, *356,* 360, 369–370, 372
Elephant seal, 368, 369
Elephant shrew, 359–360
Elephantiasis, 145, 148
Elk, 372
ELOPOMORPHA, 303
Elytra, 186, 235
Emancipation, behavioral, 437
Embryo:
  amniote, *315*
  anamniote, *95, 307*
  metabolism in, 93
Embryogenesis (*see* Embryonic development)
Embryology (*see* Embryonic development)
Embryonic development:
  amphibian, 94, *95,* 96, *97, 307*
  amphioxus, *95*
  bird, *95, 313*
  eumetazoan, 93–106
  mammal, *95*
  processes in, 93
  regulation of, 100–104
  stages of, 93–100
  sponge, *89*
Embryonic cells, 83
Embryonic plate, 96
Emigration, 426, 427, 434
Emu, 337
Emulsification, 272, 379
EMYDIDAE, 321
Enamel, *292, 346,* 364
Encephalitis, 195, 209
Enchytraeus, 188
Encystment, 64, 65, 67, 75, 77, 132, 134, 217, 426
  (*See also* Cyst)
Endamoeba, 74
Endergonic reaction, 388, 394, *395*
Endocrine gland, 279, 397, 399–400
Endocrine system, 13, 276, 397–403
Endocuticle, 213
Endocytosis, 18
Endoderm, 96, *97, 98, 102,* 107, 150, 178, 196, *313*
Endolymph, *271, 272, 344*
Endometrium, 354, 400
Endoparasite, 433
  (*See also* Acanthocephalan;
    Copepod; Fluke; Linguatulid;
    Nematode; Nematomorph;
    Rhizocephalan; Tapeworm)
Endopeptidase, 379
Endoplasm, 72, *73, 78*
Endoplasmic reticulum, *17,* 19, 396
Endopodite, 211, *212,* 216, 220, 221

ENDOPTERYGOTA, *234,* 235
Endorphin, 409
Endoskeleton, 238, 253, 261, 262, 300
　(*See also* Skeleton)
Endostyle, 254, *255,* 256
Endosymbiotic, 67
Endothelium, *382, 384,* 397
Endothermal, 315
Endothermy, 315, 349
Energy:
　activation, 388
　chemical bonding, 39, 388, 429
　free, 38, 39
　light (photic, solar), 39, 388, 389
　　role in excitation of electrons, 23
　passage through ecosystems, 419,
　　427–428, 429–430
　radiant, 412
　transfer of, 32, 388, 389, 390
　use by cells, 394–396
Engram, 410
Eniophyes, 208
Eniwetok atoll, 122
Enkephalin, 409
ENOPLA, 138
Ensis, 171
Entamoeba, 9, 74
Enterobius, 148
Enterocoel(ous), 98, *99,* 106, 150,
　152, 236, 253
Enterocoelomate, 98, *99*
Enterogasterone, 379, 400
ENTEROPNEUSTA, 250, 252
Entodiniomorph, 81
　(*See also Epidinium*)
Entomology, 7
Entoproct, 150, 152–*153,* 178
ENTOPROCTA, 106, 151, *153*
Entropy, 38
Environment, 7, 419
Environmental resistance, *425*
Enzyme, 3, 23, 373, 388
　active site, 379, 388
　deaminase, 393
　digestive, 110, 378, 379
　hydrolytic, 19, 379
　proteolytic, 127
　reaction speed of, 388
　transaminase, 403
Enzyme cascade, 386, 412
Enzyme-substrate complex, 379
Eocene (Epoch), *42,* 56, 370
EOCRINOIDEA, 240, 241
Eohippus, 55, 56, 370
Eosinophil, 277
EOSUCHIA, *259,* 320
Eosuchian, 322
EPHEMEROPTERA, *233,* 235
Ephyra, 117, *118*
Epiactis, 108
Epibenthonic, 176
Epiboly, *95,* 96, *97,* 124, 125

Epibranchial, 285, *286*
Epicuticle, 195, 196, 200, 213, 223,
　228
Epidermal seal, 309, 314
Epidermis, 97, 98, *99,* 107, *110,* 138,
　180, 181, *183,* 188, 213, 248,
　256, 382
　vertebrate, 260, *261,* 262, 280, 291,
　　292, 294, *295,* 309–310, *330,*
　　333, 339, 372
Epididymis, 185, *352*
Epidinium, 80, 81
Epifauna, 422
Epiglottis, 345, 381
Epilimnion, 422
Epimere, 97
EPIMORPHA, 224
Epimysium, 15
Epinephrine, 385, 400, 402, 403, 409
Epipelagic, 423
Epiphyseal cartilage, 262, *263*
Epiphysis, 262, *263*
Epipodite, 220
Epipubic bones, 358
Epithalamus, *266,* 267
Epithelium, 13, *14, 15,* 107, 181, 377,
　382, 400, 410
　stratified squamous, *14,* 248, 260
Epitheliomuscular cell, 109, *110*
　(*See also* Myoepithelium)
Epitoke, 185
Epitoky, 185, 186
Equational division, 50
EQUIDAE, 370
Equilibrium, 267, 268
Equus, 55, 56, 370
Erector muscle, 333, 341, 350
ERETHIZONTIDAE, 365
Erosion, 420, 421
ERRANTIA, 186, *187*
Erythrocyte, 13, 76, 276, 349, 383
　human, 349
　hummingbird, 349
　invertebrate, 157
　ostrich, 349
　turtle, 349
Escape response, 432
ESCHRICHTIDAE, 367
Esophagus:
　invertebrate, 138, *153, 156,* 165,
　　168, 172, 174, 182, *183, 190,*
　　196, 201, 206, *212,* 230, *239,*
　　244, *245, 249,* 254, *377*
　vertebrate, *260, 283,* 292, 296, *297,*
　　334, 371, 377
Ester, 27, *28*
Esthetasc, 214
Estivation, 184, 299–300, 309
Estradiol, 400
Estrogen, 277, 279, *353,* 354, 399,
　400, 403
Estrone, 29, *30*

Estrus cycle, 352, 354, 403
Estuaries, 419, 423
Ethanol, *24*
Ether, dimethyl, *24*
Ethiopian realm, 421
Ethological isolation, 436
Ethology, 7
Euamphistyly, 286
Euautostyly, 286
EUCARIDA, 215, 220, 221
Eucaridan, 221–222
EUCOELOMATA, 106
Eucoelomate, *99*
Eucypris, 211, 217
Euglena, 9, 62, *66,* 68, 69
Euglen(o)id, 40, 68
EUGLENOPHYTA, 62, 63
EUKARYOTA, *40*
Eulamellibranchiate, 176
EUMALACOSTRACA, 215, 220
EUMETAZOA, 82, 93, 106
Eumetazoan:
　diversification of, 106
　embryonic development, 89, 93–106
　(*See also individual phyla by name*)
EUMYCETOZOEA, 65
Euoplocephalus, 319, 323
EUPHAUSIACEA, 215, 221
Euphausid, 194, 197, 210, 221
Euplotes, 80, 81
Euryapsid, *318,* 322
EURYAPSIDA, *259,* 320, 322
Euryhaline, 423
Eurythermal, 423
Eurypterid, 198, *199,* 200, 281
EURYPTERIDA, 197
Eustachian tube, *271,* 309, *344*
Eusthenopteron, 298
Eutely, 140, 146
EUTHERIA, *259,* 357
Eutherian, 340, 349, 354, 357
　orders, 359–372
　placenta of, 354
Eutrophication, 419–420, 427
Euviviparous, 192, 204, 248, 292, 296,
　310, 327
Evisceration, 240, 243
Evolution, 7, 53–61, 104
　characteristics of, 53
　convergent, 55–*56*
　definition of, 53
　divergent, 54
　macro-, 57, 312
　mechanisms of, 58–61
　parallel, 173
　rate of, 53, 54
　theory of, 1
Excretion, 3, 373, 386
　(*See also* Excretophore; Excretory
　　system)
Excretophore, 13, 200, 254, 383, 386
　(*See also* Coelomocyte)

Excretory duct, *260*
  (*See also* Ureter; Urinary duct)
Excretory system, 13
  acanthocephalan, 143
  annelid, 184
  arthropod, 196, 202, 205, 207
  brachiopod, 158
  cephalochordate, 257
  echiurid, 160
  entoproct, 153
  flatworm, *125,* 126, 127, 128, 129,
    131, 134
  gastrotrich, 140
  kinorhynch, 142
  mammalian, *278, 350–351*
  molluscan, 165
  nematode, *145,* 147
  nemertean, 139
  phoronid, 157
  priapulid, 154
  pseudocoelomate, 140
  rotifer, 142
  sipunculid, 158
  tardigrade, 155
  vertebrate, 257, 277, *278*
  (*See also* Kidney; Malpighian tubule;
    Mesonephros; Metanephridium;
    Metanephros; Nephridium;
    Opisthonephros;
    Protonephridium)
Exergonic reaction, 388, 389–394
Exocytosis, 18
  (*See also* Gland; Secretion)
Exopeptidase, 379
Exophthalmic goiter, 402
Exopodite, 211, *212,* 216
Exopterygota, *233,* 235
Exoskeleton, 191, 193, 197, 213, 287,
  316
Exotic animals, 195, 424
Experimentation, controlled, 1–2
Extensor muscle, 196, *212*
External auditory canal (meatus), 315,
  333, 341, *344*
Exteroceptor, 411
Extinction:
  behavioral, 417
  mass, 42, 43–44, 57, 322, 340, 421
Extracolumella, 315
Extraembryonic coelom, 313
Extraembryonic membranes, 95, *313–*
  314
Eye, 179, 412, *413*
  annelid, 184, 186, *187,* 249
  arthropod, *190,* 200, *201, 203,* 204,
    205, 207, 210, *212,* 213, *216,*
    217, 218, 220, 221, *223,* 224,
    225, *227,* 228, *413*
  compound (*see* Compound eye)
  cup, 184, 199, 412
  dark-adaptation of, 213

Eye (*continued*)
  direct, 203, 204, 207
  flatworm, 127, 129
  human, *269, 270*
  indirect, 203, 205, 207
  inverse pigment cup, *128,* 129, 139,
    142, 184, 210
  median, 254, *255,* 267, *269,* 280,
    285, 325, *326*
  medusan, 112, 114, 116, *413*
  molluscan, 169, *170,* 173, *174,* 176,
    177, 412, *413*
  nauplius, 210, *216,* 217, 218
  pinhole-camera, 177
  vertebrate, *256,* 257, *266,* 267, 268–
    270, 282, 284, 290, 291, *292,*
    332, 410, 412
    retinal structure of, *269*
  vesicular, 177, 184, 268, 412, *413*
Eyelash, 331, 341
Eyelid, 290
Eyespot, 146, *153,* 243, 248, 432
Eyestalk, 213

Facial expression, muscles of, *342*
Facial nerve, 268
Facial pit, 261, 412
FAD (flavin adenine dinucleotide), 375,
  389, *390,* 392, 393
Fairy shrimp, *211,* 215, 216
Falcon, 329, *338,* 339
FALCONIFORMES, 337, *338,* 339
Family, 41, 57
Fang, 206, 322, 327, 328
Fascia, *396*
Fasciculus, 15, *396*
Fasciola, 131, 132
Fasciolopsis, 133
Fat, 4, *14,* 182, 214, *261,* 272, 349,
  354, 365, 378, 379, 403
  (*See also* Adipose tissue; Blubber;
    Lipid)
Fate mapping, 101, *102*
Fatty acid, *30,* 31, 38, 376, 378, 379,
  380, 403
Fauna (regional animal life), 422
Feather, 56, 260, 262, 315, 328
  evolution, 329
  functions, 329–331
  primary, *330*
  secondary, *330*
  structure, *330,* 333, 374
  tertiary, 330, 331
  types of, *330–331*
Feather erector muscle, 333
Feather follicle, 329, 330, 333
Feather mite, *208*
Feather star, *237,* 242
Feather tract, 329, 333
Feces, 133, 135, 377
Feedback, 401, 402

Feet (*see* Foot)
FELIDAE, 41, 57
Felis, 424
Female:
  genetic determination of, 47
  regulation of gene dosage in, 47
  reproductive system of, *141, 145,*
    352–353
  sexual selection by, 60
Femur, 227, *263,* 264, 298, 303, *311,*
  323, 332, 334
Fenestra, 309
  antorbital, 323
  mandibular, 323
Fer-de-lance, 328
Ferret, 367
Fertilization, 48, 51, 94, 97,
  165
Fetus, 354
Fiber:
  collagenous (*see* Collagen)
  connective tissue, 13, *14*
  muscle (*see* Muscle fiber)
  nerve, *14*
    amyelinated, 407
    excitatory, *408*
    inhibitory, *408,* 409
    myelinated, *14,* 407
  spindle, 19, 20
Fibril:
  collagen, 83, 86, 88
  contractile, *110*
  motile, 17
  myo-, 394, *396*
  retractile, *80*
Fibrin, 386
Fibrinogen, 386
Fibroblast, 13
Fibula, *263,* 264, 298, 303, *332*
Fig Tree formation, 37, 38
Fight-or-flight response, 268, 400,
  402
Filament (in muscle), 394, 395, *396*
Filaria worm, *145*
Filariasis, 148
Filariform larva, 148
Filibranchiate, 176
Filoplume, *330,* 331
Filopod, 74
Filter-feeding
  (*see* Amphioxus; Basking shark;
    Bivalve; Brachiopod; Ectoproct;
    Entoproct; Euphausid;
    Leptostracan; MYSTICETI;
    Phoronid; SEDENTARIA;
    Sponge; THORACICA;
    Tunicate; Whale shark)
Filtration, 165, 173, 210, 273, 276,
  277, 296, 351, 387
Fin, 280, *281,* 282, 291
  anal, *281,* 290, *295,* 296

Fin (*continued*)
  caudal, 246, 248, *249*, 256, 282,
    286, 287, *288*, 290, 294, *295*,
    302, 303, *304*, 306, 310, 322
  dorsal, *255, 256, 281*, 282, *288*,
    289, 290, *292, 295*, 300, 322,
    366, 367
  evolution in fishes, 286, 287
  fleshy, 298
  lateral, 177, 246, *249, 281*
  lobed, 298, 299
  median, *283*
  pectoral, 256, 287, *288, 289*, 290,
    292, 294, 300, 301, 303
  pelvic, 256, 287, *288, 289*, 290,
    292, 294, *295*, 300, 303
  tail, *288*, 291
  (*See also* Fin, caudal)
Fin ray, *255*, 294, *295*, 298, 300, 303,
  304
Finback whale, 367
Finch, 54, *338*, 339, 417
Fingers, modification in wings, 56
Fireworm, 181
Firefly, 194, 235, 418
Fish louse, 215
Fish, 7, *10*, 41, *42, 56*, 253, 280–303,
  422, *430*, 431
  annual catch, 302
  armored, 57, 257
  (*See also* Ostracoderm;
    Placoderm)
  as parasitic hosts, 130, 132, 136,
    149, *172*, 188, 215, 217, 221,
    283
  bony, *42*, 264, 272, 280, 285, 291,
    292, 294–303, 386
  brain, *266, 343*
  cartilaginous, 257, 264, 272, 280,
    287–293, 385
  circulatory system, *274, 275*
  digestive system, *292*
  ear, *271*, 284, 285, 412
  egg, 292, 293
  electrogenic, *288*, 290, 296, 303,
    413, 436
  excretory system, *278*
  eye, *269*, 270, 280, 282, 284, 285,
    290, 291
  fleshy-finned, 294, 298
  heart, *274*, 282, *283*
  integumentary glands, 260
  jawless, 257
  (*See also* Hagfish; Lamprey;
    Ostracoderm)
  lateral line system, 270, 284, 285,
    291, 294, *295*, 309
  lobe-finned, *42*, 57, 257, 280, 298
  nervous system, *266*, 282
  osmoregulation, 285, 291, 296
  ray-finned, 257, 298, 294, 300–303

Fish (*continued*)
  respiration, 284, 285, 289, 290, 294,
    296, *297*, 299, 301, 302
  reproduction, 292, *283*, 284–285,
    292, 296, 297–298
  skeleton, *286, 288, 289*
  tongue, 310
  urogenital system, *278*
  warm-blooded, 293
Fission, 121, 130, 157
  binary, 63, *66*, 75, 79
  multiple, 63, 74, 75, 76, *77*, 79, 83
Fissiparity, 247
Fissiped, *356*, 367–368
FISSIPEDIA, 360, 367
Fixed action pattern, 415
FLAGELLATA (*see*
  MASTIGOPHORA)
Flagellate, 62, 66–71
Flagellated pit, 257
Flagellum, *17, 18*
  structure and action of, *70*, 71
  occurrence, 68, *69, 70*, 71, 86, *87*,
    89, 90, 109, *110*, 129, 140
Flame bulb, 140, 142
Flame cell, 127, *128*, 129, 137, 139,
  140, 143, 386, *387*
Flatfish, 303
Flatworm, 125–137, 374, *404*
  biological advances in, 125–126
  evolutionary status, 125
  parasitic, 82, 91, 130–137
  memory in, 129, 410
Flavin adenine dinucleotide (*see* FAD)
Flavin mononucleotide (*see* FMN)
Flavoprotein, 379
Flea, 194, 195, *234*, 235, 433
Flexor muscle, 196, *212*
Flight (*see* Bat; Bird; Insect; Pterosaur)
Flipper, *265*, 321, 322, 360, 365, 368,
  369, 370
Flounder, 303
Fluke, 130–133, 164, 433
Fluke, tail, 12, 360, 365, 370
Fly, two-winged, 228, *234*, 235
  (*See also* DIPTERA)
Flycatcher, 339
Flying dragon, 327
Flying fish, 303
Flying fox (*see* Fruit bat)
Flying phalanger, 358
Flying squirrel, *361*
Foettingeria, 64
Folic acid, 376
Follicle:
  feather, 329, 330, 333
  hair, 208, *261*, 411, 412
  ovarian, 352, *353*, 354, 399, 400,
    403
  thyroid, 397, 402
Follicle mite, *208*

Follicle-stimulating hormone (*see* FSH)
Follicular phase, 353
Following response, 416
Food chain, 5, 193, 427, 428, *429*
Food supply, 425, 434
Food vacuole (*see* Vacuole, food)
Foot:
  bird, 329, *332*, 337, *338*
  horse, evolution of, 55, *56*
  mammalian, *263*, 355
  molluscan, 163, *166*, 167, *168, 169,
    170, 171*, 172, 173
Foramen (Foramina), 74
Foramen magnum, 345
Foramen ovale, 349
Foraminiferan, 54, 63, 64, 72, *73*, 74
FORAMINIFERIDA, *73*
Forebrain, 257, 265
  (*See also* Cerebrum; Diencephalon;
    Telencephalon)
Foregut, 196, 230
Forelimb:
  crossopterygian and labyrinthodont,
    299
  radiative adaptation, *265*
Forests, *420, 421*, 422
Forewing, 228
Formula:
  empirical, 24
  structural, 24
  type, 28
Fossils, 7, 37, 38
  oldest known, 38
  radioactive dating of, 37
  (*See also listings by phylum*)
Four-eyed fish, 303
Fovea, *269*
Fox, 12, 57, 341, 350, *361*, 367
Fragmentation, 139, 179
  (*See also* Autotomy; Pedal
    laceration)
Frame-shift mutation, 58
Free-living habit, 64
Freezing, 432
Freshwater habitats, 422
Frigate bird, 339, 433
Frightening displays, 432
Frilled shark, 290
Frog, 72, 146, 302, 306, 311, 431
  African clawed, 102, 311
  arrow-poison, 307, 309, 311
  brain, *266*
  breathing cycle, 308
  Darwin's, 307
  development, 94, 95, *97, 307*
  gastric, 307
  kokoi, 309
  skin, *14, 261*
  tailed, 311
  tongueless, 311
  tree, 307, 311

Frons, 227
Frontal bone, 318, 372
Frontal gland, 125, 127, 138
Frontal lobe, 343
Frontal organ, 199
Frontal plane, 10
Frontalis muscle, 342
Fructose, 352, 378, 379
Fruit bat, 362
Fruit fly (see Drosophila)
FSH (follicle-stimulating hormone),
    279, 353, 399, 401, 403
FSH-releasing hormone, 399, 403
Fungus, 40, 422, 429
Funnel, 151, 157, 173, 180, 183, 184,
    185, 189
Fur, 5, 315, 340, 355, 367, 368
    (See also Hair)
Fur seal, 368, 369
Furcula, 329

Galactose, 378, 379
Galapagos finches, 54, 57
Galapagos islands, 54, 321
Galapagos rift community, 429
Galapagos tortoise, 321
Galathea, 222
Galatheoid, 222
Gall, 208
Gall bladder, 256, 257, 260, 295, 333,
    334, 347, 379, 400
Gall mite, 208
GALLIFORMES, 339
Gallop, 325, 365, 367, 369
Galvanoreceptor, 411, 412, 413–414
Gamete, 48, 49, 51, 52, 53, 72
    (See also Egg; Sperm)
Gametocyte, 77
    (See also Gamont)
Gametogenesis, 49, 77, 279
Gametogonium, 126, 139
Gamma rays, 22
Gamont, 74, 75, 76, 77
Ganglion, 141, 153, 169, 170, 203,
    249, 254, 255, 268
    autonomic, 405
    cerebral, 129, 140, 165, 169, 171,
        175, 184, 249
    cerebropleural, 172
    parasympathetic, 266, 268
    pedal, 165, 169, 171, 172, 175
    peripheral, 267, 405
    pleural, 165, 169, 175
    segmental, 155, 161, 179, 184, 189,
        191, 195, 203, 204, 214, 223,
        224, 229, 230
    sensory, 404, 405
    subesophageal, 155, 229, 230
    supraesophageal, 229, 230
    suprapharyngeal, 155
    sympathetic, 266, 268, 402
    visceral, 165, 169, 171, 172, 175

Ganglion cell, retinal, 269, 270
Ganoid scale, 262, 294, 295, 301, 302
Ganoin, 262, 295
Gap-junction organelle, 18
Gar, 262, 301, 302
Garden centipede, 197
Gas gland, 115
Gas transport, 276, 349
    (See also Erythrocyte; Hemoglobin;
        Respiratory pigment; Trachea;
        Tracheole)
Gastric gland, 374, 377, 378, 379
Gastric juice, 261, 400
Gastric mill, 379
Gastric shield, 172
Gastrin, 379, 397, 400
Gastrocoel, 95
Gastrodermis, 97, 99, 107, 110
Gastrointestinal hormones, 378–379,
    397
Gastropod, 165, 167, 217, 222
    characteristics, 169
    internal structure, 170
    subclasses, 175–176
GASTROPODA, 164, 169, 170, 175
Gastrotrich, 140, 141, 142, 147, 150
GASTROTRICHA, 106, 140
Gastrovascular cavity, 107, 109, 110,
    111, 113, 114, 120, 127, 128,
    377, 383
Gastrozooid, 111, 112, 113, 114, 115,
    116
Gastrula, 89, 95, 96, 97, 108, 118,
    124, 125, 307
Gastrulation, 93, 95, 96, 97, 102, 103,
    108, 124
Gate, ion, 17, 406, 407
Gause's principle, 423
Gavial (see Gharial)
GAVIIFORMES, 337
Gecko, 326
GEKKONIDAE, 326
Gel, 72
Gelation, 73
Gemmule, 88
Gena, 227
Gene, 20, 32, 45, 50, 51, 93, 394
    crossing-over of, 50
    dosage, regulation of, 47
    duplicate, 58
    flow, impediments to, 54
    linkage of, 47, 51
    locus, 45
    mapping of loci, 50
    mutation of, 3, 53, 58
    pool, 52, 53, 54, 55, 59
    regulation of activity, 100, 104, 403,
        409, 410, 415
    sex-linked, 47
    structural, 100
    transmission of, 47–51
    (see also Allele; DNA)

Generator potential, 406
Genital atrium, 170
Genital canal, 242
Genital duct, 260, 277, 293
    (See also Gonoduct)
Genital pore, 134, 170, 171, 201, 244
    (See also Gonopore)
Genital tract, 277, 279, 280
Genitalia, 277, 279
Genetic code, 58, 84
Genetic drift, 61
Genetic mosaic, 47
Genetic recombination, 51–53
Genetics:
    animal, 45–53
    population, 59
    zoological, 7
Genome, 57
    quantitative augmentation of, 58
Genotype, 45
    heterozygous, 45, 51, 52, 53, 59,
        415
    homozygous, 45, 51, 52, 53, 59,
        415
Genus (Genera), 57
Geographic isolation, 54, 55
Geologic record, 37–44
Geologic time table, 42
GEOMYIDAE, 365
GEOPHILOMORPHA, 224
Geotaxis, 185, 208
Gephyrostegid, 304, 305, 316, 317
Gerbil, 365
Germ cells, 49
    origin of, 93, 100
Germ layer, 93
    (See also Ectoderm; Endoderm;
        Mesoderm)
Germinal zone, 178, 179
Germinative layer, 261
Germovitellarium, 143
Gessnerium, 67
Gestagen, 279
Gestation period, 354, 357, 368, 369,
    370
Gharial, 325
GH (see Growth hormone,
    Somatotrophin)
GHIH (see Growth-hormone-inhibiting
    hormone)
GHRH (see Growth-hormone-releasing
    hormone)
Giant kidney worm, 149
Giardia, 69, 71
Giardiasis, 71
Gibbon, 363
Gigantism, pituitary, 11, 401
Gila monster, 327
Gill, 182, 187, 249, 381, 382
    amphibian, 306, 307, 308, 310, 311,
        382
    area of surface, 381

Gill (*continued*)
  arthropod, 197, 198, 213, 214, 220,
    221, *382*
  book, 199, 200
  countercurrent exchange in, 296, *297*
  echinoderm, *244*, 245
  elasmobranch, *297*
  external, 197, 273, 306, *307, 308,*
    310, 311, *382*
  internal, 273, 306, *307, 308,* 310,
    311
  molluscan, 163, *166*, 167, *168,* 169,
    171, 172, 173, *174, 175,* 176,
    177
  osteichthyan, *172,* 296, *297*
  tracheal, 230, 235, *382*
  vertebrate, *274, 275,* 280, 292, 294,
    295, 296, *297,* 306, *307*
Gill arch, 105, 273, 285, *286, 297,*
  345
Gill bar, *171, 255,* 256
Gill filament, 172, 176, *283,* 291, 296,
  297
Gill lamella, 257, 296, *297*
Gill opening, 250, *281, 283,* 284,
  285
Gill pouch, *256,* 273, 282, 284, 285,
  291, 309
Gill raker, *297*
Gill slit, 104–105, 238, *249,* 250, 252,
  255, 260, 273, 287, *288,* 289,
  290, 291, *292, 293, 297*
Giraffe, 360, 372
GIRAFFIDAE, 372
Girdle:
  chiton, 167, *168*
  ciliary, 150, 155, 161, 178
  clitellate, 185
  limb, 264, 291, *299,* 303, 310, 317,
    331
    (*See also* Pectoral girdle; Pelvic
      girdle)
Gizzard, 182, *183,* 199, 230, 333, *344,*
  377
Glaciations, Precambrian, 41
Gland, 13
  adhesive, 127, 140, 216
  albumen, *170*
  alveolar, *14*
  anal, 204
  antennal, 195, 210, 217, 221
  atrial, 130
  axial, 239
  bulbourethral (*see* Gland, Cowper's)
  calciferous, 182
  cement, *125, 141,* 200, *218,*
    219
  ceruminous, 260, 341
  clitellar, 185
  Cowper's, *347*
  coxal, 191, 195, 199, 202, 204, 205,
    206, 207, 209, 210

Gland (*continued*)
  digestive, 165, *166,* 167, *168, 170,*
    171, 172, 197, 199, 214, 247,
    378
    (*See also* Cecum; Hepatopancreas;
      Pancreas; Liver)
  duodenal, *15,* 378
  endocrine, 398, 399–400
  eyelid, 260
  frontal, *125,* 127, *138*
  gas, 115, 296
  gastric, 374, 377, 378, 379
  green, 210, *212*
  integumentary, 260
  intestinal, *15*
    (*See also* Gland, duodenal)
  lacrimal, 260
  mammary, 18, 260, 262, 340
  mandibular, *223*
  maxillary, 195, 210, 217, 221
  Mehlis's, *131,* 132
  mucous, *170,* 256, 260, *261,* 285,
    308
  musk, 368
    (*See also* Gland, scent)
  oil, 260, 372
    (*See also* Gland, sebaceous)
  pedal, 142, *170*
  photogenic, 217
  poison, *201,* 202, 205, 206, 260,
    261, 262, 290
  preen, *334*
    (*See also* Gland, uropygeal)
  prostatic, 147, *347*
  prothoracic, *231*
  rectal, 230, 291, *292*
  repugnatory, 225
  salivary, *14,* 76, *77,* 142, *164,* 165,
    170, 174, 190, 223, 224, 225,
    229, 230, 260, *266,* 377, 378,
    379
  salt, 291, 335, 339, 386
  scent, 260, 262, 368
  sebaceous, 18, 260, *261,* 341,
    365
  seminal, 126
  shell, *334*
  silk, *201,* 202, 205, 206, 226,
    235
  sinus, 213
  slime, *190, 283*
  subpharyngeal, 253, *256*
  sweat, 18, 260, *261,* 262, 341,
    365
  tear, 260
  tube-secreting, *249*
  tubular, *14,* 260
  unicellular, *14, 15,* 90, 110, *128*
  uropygeal, 333
  venom, 189, 309, 311, 327, 355
    (*See also* Gland, poison)
  vitelline, 132, 134

Gland (*continued*)
  yolk, 126, *128,* 130, *131, 134*
Glans penis, 279, 316, 352
Glider, *356,* 358
Globigerina, 74
Glochidium, *172*
Glomerulus, *249,* 250, 276, 277, *278,*
  296, *350,* 351, 387, 397, 400
Glomus, *278*
Glossina, 70
Glossopharyngeal nerve, 268
Glottis, 273, *308, 334,* 345, 381
Glucagon, 398, 400
Glucocorticoid, 399
Gluconeogenesis, 399
Glucose, 27, 28, *29,* 143, 378, 379,
  380, 394, *395*
  oxidative metabolism of, 389–
    393
  regulation of blood level, 398, 399,
    400
Glucostat, 398
Glutathione, 109
Glutinant, *108,* 109
Glycera, 178, 181
GLYCERIDAE, 186
Glycerol, 378, 379, *380*
Glycogen, 27, 28, *29,* 182, 214, 272,
  383, 392, 394, *395*
Glycogenolysis, 392, 398, 399, 400,
  402, 405
Glycolysis, 388, 389, *390,* 392
  anaerobic, 392
Glycoprotein, *17,* 19
Glycosuria, 398
GMP (guanosine monophosphate),
  cyclic, 403
Gnathobase, 198
GNATHOBDELLIDA, 186, 190
Gnathochilarium, *223,* 225
Gnathopod, 221
GNATHOSTOMATA, 257, *258,* 280
Gnathostome, 260, 272
Gnathostomulid, 125, 137, 140, 141,
  142
GNATHOSTOMULIDA, 106, 125,
  137
Goat, 372
Goblet cell, *15*
Goby, 303
Goiter, 402
Golden jackal, 57
Golgi apparatus, *17,* 19
Golgi tendon organ, 411
Gonad, 13, *98,* 110, *111, 113,* 155,
  159, 185, 192
  arthropod, 200, 203, 214
  echinoderm, *239, 244,* 246, 247,
    248, 250, *255*
  meiosis in, 49, 50
  molluscan, *166, 168, 169, 171,* 172,
    174

Gonad (*continued*)
　vertebrate, *260,* 277, 282, 285, 293,
　　295, 311, 401, 402
　(*See also* Ovary; Testis)
Gonadotrophin, *353,* 401
　chorionic, 400
　(*See also* FSH; LH)
Gonadotrophin-release hormone, 279
Gonangium (*see* Gonozooid)
Gondwanaland, 43, 299
Gonionemus, 108, 112, *113*
Gonocoel, 150
Gonoduct, 167, 185, 244
　(*See also* Genital duct; Oviduct;
　　Sperm duct)
Gonophore, 112, *113,* 115, *116*
Gonopod, 224, 226
Gonopore, *125, 139,* 153, *168,* 185,
　　188, 192, 200, 204, 206, 207,
　　209, 217, 226, 231, *244,* 246,
　　248, *249,* 250
Gonotheca, *111, 113,* 114
Gonozooid, *111,* 112, *113, 116*
Goose, 335, 339, 416
Gopher, 365
GORDIOIDEA, 144
GORGONACEA, 199
Gorgonia, 108, 120
Gorgonin, 120
Gorilla, 363
Graafian follicle, 353
　(*See also* Follicle, ovarian)
Gradient:
　axial, 11
　concentration, 18, 406
　electrochemical, 405, *406*
Grand Canyon, 37
Grasshopper, *193, 227, 229, 232,* 235
Grassland, 363, *420, 421*
Gravity, 11
　sensing, 107, 172, 214, 254, 270,
　　271, 309
　　(*See also* Sacculus; Statocyst;
　　　Utriculus)
Gray crescent, *97,* 101, 102
Gray matter, 253, 265, *343,* 407
Gray whale, 367, *404*
Grebe, 339
Green turtle, 321
Greenhouse effect, 427, 437
Gregarine, 63, 75, 76
Gribble, 195, 221
Group defense, 435
Grouse, 339
Growth, 4
　cellular, 33, *35*
　organismal, 398, 399, 400, 401
　population, 424–425
Growth hormone (GH), 103, 399, *401*
Growth-hormone-inhibiting hormone
　　(GHIH), 401
　(*See also* Somatostatin)

Growth-hormone-releasing hormone
　　(GHRH), 399, 400
GRUIFORMES, 339
Grunion, 303
GTP (guanosine triphosphate), 33, *390,*
　403
Guanaco, 360, 372
Guanine (G), 32, *33,* 200, 207
Guanylate cyclase, 403
Gubernaculum, 352
Guinea pig, *356,* 365, 376
　heredity in, *51, 52,* 58
Guinea worm, 148–149, 195
Guitarfish, *288,* 290
Gull, 339, 415, 427, 434, 436
Gullet, *78, 121*
Guppy, 296
Gut, *125, 223,* 411
　(*See also* Alimentary canal;
　　Archenteron; Digestive system;
　　Intestine)
GYMNOPHIONA, 306, 310
Gymnotid, 303, 418

Habitat, 419
　innate preference for, 416
　modification of, 420, 422, 426
　tolerance of factors in, *420, 421,*
　　423, 424
Habituation, 417
Hadal benthos, 423
Hadrosaur, 323
HADROSAURIA, 320
Haematococcus, 67
Hageman factor, 386
Hagfish, 257, 264, 272, 280, 281, 282,
　　283, 284–285
Hair, 9, 260, *261,* 262, 341, 412
　sensory, 214
　　(*See also* Hair cell; Sensillum)
Hair cell, 71, 107, 110, 254, 272, 333,
　　344, 345
Hair erector muscle, *261,* 341, 350
Hair follicle, *261,* 341, 411, 412
Hair-fan, 249
Half-life, 21, 37–38
Halibut, 303
Haliclona, 85
Haliclystus, 108, 118
Hallucinogenic drug, *409*
Haltere, 228, 235
Hammer, 342
Hammerhead, *288,* 290, 291
Hamster, 365
Hand, *265,* 330, 331
Haploid(y) (n), 48, 50
HAPLOTAXIDA, 186, 188
Hardy-Weinberg equilibrium principle,
　　59
Hare, 12, 341, 360, 364, 365, 424–425
Harem, 368, 435
Harp seal, 369

Harvestman, *201,* 205
Haversian system, 262, *263*
Hawaiian fauna, speciation in, 54, 227
Hawk, 339, 431
Hawksbill, 321
HCl (*see* Hydrochloric acid)
Head, *164, 170*
　annelid, 179, *180,* 184, 186, 191
　arthropod, 196, 197, 200, 210, 216,
　　217, 220, *223,* 225, 227
　vertebrate, 286, 305, 317
Head capsule, 223
Head-foot complex, 163, 169
Head shield, 282, 287
Hearing, 268, 309, 333, 345, 411,
　　412, 414
　(*See also* Ear; Lateral line system)
Heart, 236, 383
　accessory, 284, 285
　amphibian, *274,* 310
　annelid, 182, *183*
　arthropod, 196, 199, *201,* 202, 207,
　　212, 214, *216,* 220, 221, *223,*
　　224, 225, *229,* 230, 383, *384*
　bird, *274, 275,* 333, *334*
　booster, 230
　branchial, 173, *174*
　chordate, 253
　cyclostome, 282, *283*
　echinoderm, 244
　enteropneust, 250
　fish, *274, 275,* 295
　lymph, 273
　mammalian, *266, 274, 275,* 333,
　　348, 349, *384*
　molluscan, 165, 167, *168, 169, 170,*
　　172, 173, *174, 175,* 384
　onychophoran, 191
　reptilian, *274,* 316
　tunicate, 254, *255,* 384
　vertebrate, *105,* 256, 257, *260,* 268,
　　273–274, *292,* 397
Heart muscle fibers, *14*
　(*See also* Cardiac muscle)
Heart urchin, 238, 245
Heartbeat, *348*
　rate of, 333, 349, 350, 366
　control of, 385
　propagation of, 348
Heat:
　generation of, 293, 349, 389
　receptors of, 184, 261, 411, 412
　regulation of body, 262, 349–350
　retention of body, 293, 331, 350
　tolerance of, 315
Hectocotylized tentacle, 175
Hedgehog, 359, 360
Heleosphaera, 10, 73, 75
Heliaster, 247
HELICOPLACOIDEA, 241, 243
Helicoplacus, 24, 243
Helicotrema, 344

HELIOZOA, 72
Helix, 169, *170,* 176
Helix, double, 32, *34*
Helmet, 162, 176
Helminthology, 7
HELODERMATIDAE, 327
Helper, 435
Hemal arch, 287
Hemal system, 238, 239, 246
Heme, 349, 374
Hemerythrin, 154, 157, 158, 160, 182,
    374, 383
HEMICHORDATA, 106, *249,* 250,
    252
Hemichordate, 238, 250–251
Hemicyclaspis, 281
Hemidiphyodont, 347, 358
HEMIMETABOLA (*see*
    EXOPTERYGOTA)
Hemipenis, *316,* 326
HEMIPTERA, *233,* 235
Hemocoel, 150, 157, 161, 165, 170,
    191, 196, 202, 210, 214, 230,
    383, *384,* 386
Hemocyanin, 165, 173, 182, 199, 202,
    204, 374
Hemocyte, 244
Hemoglobin, 3, 13, 374, 383
    adult and fetal mammal, 349
    invertebrate, 157, 160, 182, 184,
        239, 244, 429
    sickle, 58, 76
    vertebrate, 257
Hemorrhage, control of, 276, 385–
    386
Hemorrhagic fever, 209
Hemosporidian, 75, 76–77
Hemotumescent, 316
Henle's loop, *350, 351,* 388, 401
Hepatic cirrhosis, 276
Hepatic portal, *275, 292, 347*
Hepatic vein, *275, 347*
Hepatopancreas, *212*
Herbivorous, 164, 224, 225, 226, *227,*
    235, 305, 316, 320, 321, 322,
    323, 360, 363, 365, 369, 370,
    371, 372
Heredity, 7, 45–53
    effects on behavior, 415
Hermaphroditism, 88, 130, 137, 141,
    165, 167, 172, 185, 188, 216,
    218, 239, 248, 249, 254, 277,
    285, 297
    (*See also* Monoecious)
Hermione, 187
Hermit crab, 122, 194, 196, 197, *211,*
    215, 220, 222
Heron, 331, 333, *338,* 339
Herpetology, 7
Herring, 303
Hesperornis, 56, 337
Heteroadhesion, 83, 93

Heterocercal tail, 282, 286, 288, 289,
    291, 300
Heterochromatin, 45
HETERODONTA, 176
HETERODONTIFORMES, 290
Heterodonty, 322, 346
HETEROMYIDAE, 365
Heteronomous metamerism, 179, 196
Heteropod, *170,* 176
Heterosome, 46
    (*See also* X chromosome; Y
        chromosome)
HETEROSTRACI, *281*
Heterotrich, 81
Heterotroph:
    facultative, 63
    obligate, 63
Heterotrophic nutrition, 4, 39, 62, 373
Heterozygous, 45
HEXACTINELLIDA, 90
    (*See also* Calcispongiae)
HEXANCHIFORMES, 290
Hexapod, 223
    (*See also* Insect)
Hexose, 27, 28, *29*
    (*See also* Glucose)
Hibernation, 333, 349–350, 359, 365,
    421
Hierarchy, 426–427, 434
Hindbrain, 257, 265, 271, 302, 385
    (*See also* Medulla oblongata)
Hindgut, 196, 200, 202, 224, 389
Hindwing, 228
Hinge, 157, *171,* 217
Hinge teeth, 157, 217
Hippocampus, 301
Hippoid, 222
HIPPOPOTAMIDAE, 371
Hippopotamus, 6, 360, 371–372
Hirudin, 182
HIRUDINEA, 186, 188
Hirudo, 189, 190
Histamine, 277
Histology, 7
Histomonas, 70
Histone, 20
Holapsis, 198
Holoblastic cleavage (*See* Cleavage,
    holoblastic)
Holocephalan, *288,* 293
HOLOCEPHALI, *258,* 287, *288,* 293
Holocrine, 18
HOLOMETABOLA (*see*
    ENDOPTERYGOTA)
Holoplankton, 422
Holostean, 302, 305
HOLOSTEI, *258,* 300, *301,* 302
Holothuria, 243
Holothurin, 243
Holothuroid, *237,* 243–245
HOLOTHUROIDEA, *237,* 241, 243
Holozoic nutrition, 62, 374

HOMALOZOA, 240
Homarus, 210, 222
Homeostasis, 3, 386, 397, 398
Homeotherm, facultative, 333
Homeothermal, 315, 328, 333, 340,
    354
Homeothermy, 349
Homing, 167
Hominid, 44, 360, 362, 363–364
HOMINIDAE, 363
HOMINOIDEA, 360, 363
Homo, 44, 363, 364, 424, 427,
    438
Homoadhesion, 83, 93
Homocercal tail, 288, 294, 302
HOMOIOSTELA, 240
Homologous, 45, *49,* 50, 56, 212, 264,
    265, 298
Homonomous metamerism, 179, 196
HOMOPTERA, *233,* 235
HOMOSTELA, 240
Homozygous, 45
Honey, 5, 414
Honeybee, 77, 196, 228, 357, 414,
    426
Honeycreeper, 54
Honeyguide, 339, 416, 433
Hood, *174, 249*
Hoof, 55, 260, 262, 355, 370, 371,
    372
Hoofed mammal, 341, 354, 360, 370–
    372
Hookworm, 148
Hoopoe, 339
HOPLOCARIDA, 215, 220
Hormone, 3, 397–403
    behavioral effects, 403, 416
    chemical nature, 403
    in developmental regulation, 103
    invertebrate, 213, *231*
    mammalian, 398, 399–402
    modes of action, 403
    regulation of secretion of, 398, 401–
        403
    sex, *30,* 31, 45, *353,* 354, 400, 403
    steroid, *30,* 31, 403
    synergistic action, 403
Horn, 260, 371, 372
Hornbill, 339
Horned lizard, 326
Horned shark, 290
Horse, 5, *42,* 55, *56, 346,* 354, 360,
    370, 372
Horsefly, 229, *234*
Horsehair worm, 143–144
Horseshoe crab, 197, 198-*199*
Host, 64, 143, 146, 433
Housefly, 235
Human, *42*
    benefits from animals, 5–6
    blood, 349, 383
    brain, *343*

Human (*continued*)
  chromosomes, 45, 48
    number of mRNAs made by, 45
  ecological changes caused by, 44,
      123, 419–420, 422, 424, 426,
      427, 437–438
  evolution, 44
  eye, *269*, 270, 414
  genetic traits, 45, 46, 47, 58, 59, 415
  gestation period, 354
  kidney, *350*
  magnetic sense, 413
  menstrual cycle, *355*
  milk, 340
  musculature, facial, *342*
  nervous system, *266*
  nutritional deficiencies, 375–376
  population growth, 44, 424, 437
  reproductive system, *352, 353*
  respiratory system, *347*
  size compared with animals, *9*
  skeleton, 264, *265*, 305
  skin, 58–59, *261*
  teeth, *346*
  (*See also* Homo)
Humerus, 56, *263*, 264, *265*, 298, *299*,
    303, *311, 330, 332, 334*
Hummingbird, 12, 329, 333, *338*, 339,
    349
Humpback whale, 366, 367
Hyalella, 211, 221
Hyaline cap, 72, *73*
Hyaline cartilage, *14*
  (*See also* Cartilage)
Hyalosponge, 90
HYALOSPONGIAE, *85,* 89, 90
Hyaluronidase, 327
Hybodont, *289*
Hybodus, 289
Hybrid:
  genetic, 51, *52*
  interspecific, 53, 54, 55
Hydatid cyst, *136*
Hydra, *108, 110,* 112, *113,* 115, *377,*
    387
Hydranth, *111,* 114
Hydration shell, *25*
Hydrochloric acid, 26, 374, 379, 400
Hydrocoral, *113,* 115
HYDROCORALLINA, 115
Hydrocortisone, 403
Hydrogen (H), 21, *22,* 389, *390, 391,*
    392, 393, 394
Hydrogen acceptor, 394
Hydrogen (H—) bond, 25–26, 33, *34*
Hydrogen carrier, 389, 393
  (*See also* FAD; NAD; NADP)
Hydrogen cycle, *428,* 429
Hydrogen ion, 26, 351, 399
  (*See also* Proton)
Hydrogen peroxide, 20, 388
Hydrogen sulfide, 429

Hydroid, *111, 113,* 114–115, 167, *404*
HYDROIDA, 114, 115
Hydrolagus, 288
Hydrolysis, 373, 378, 379
Hydrolytic enzymes, 378, 379, *380*
Hydromedusa, 112, *113,* 114
HYDROPHIIDAE, 328
Hydrophilic, 31
Hydrophilic pore, *17*
Hydrophobic, 28, 31
Hydrospire, 241
Hydrostatic skeleton, 140, 147, 170,
    179, 181, 252, 373
Hydrotheca, *111,* 114
Hydroxyapatite, 262
Hydroxyl group, 25
Hydroxyl ion, 26
Hydroxyproline, 84
HYDROZOA, *108*
Hydrozoan, *113–116*
Hyena, 360, 368
HYENIDAE, 368
Hyla, 311
HYLIDAE, 311
HYMENOPTERA, *234,* 235
Hymenostomatid, 81
Hyoid, *263,* 339
Hyoid arch, 286, 300, 302, 309, 345
Hyomandibula, *286,* 300, 302, 309,
    345
Hyostyly, 286
Hypermastigidan, 69, 71
Hyperosmotic, 17, 291
Hyperpolarization, *406,* 407
Hypertrophy, 402
Hypertonic, 17, 147
  (*See also* Hyperosmotic)
Hypobranchial element, *286*
Hypobranchial muscles, 310
Hypodermic impregnation, 143, 185
Hypoglossal nerve, 268
Hypolimnion, 422
Hypomere, 97, *98*
Hypopharynx, 224, *227,* 228, *229*
Hypophyseal canal (pouch), *256, 257*
Hypophyseal portal system (vein), 276,
    401
Hypophysis, *266,* 276, *343,* 401, 402
  (*See also* Pituitary)
Hyposmotic, 18
Hypostome, *110,* 115
Hypothalamic-hypophyseal control axis,
    401
Hypothalamus, *266,* 267, 276, *343,*
    349, 399, 400, *401,* 402, 403,
    411, 427
Hypothesis, 1
Hypotonic, 18, 422
Hypotrich, 81
HYRACOIDEA, 360, 370
Hyracotherium, 55, 370
Hyrax, 356, 360, 370

Ibis, 339
Ichthyology, 7
Ichthyophis, 306
ICHTHYOPTERYGIA (*see*
    ICHTHYOSAURIA)
Ichthyornis, 337
Ichthyosaur, *42, 56*
ICHTHYOSAURIA, *259,* 320, 322
Ichthyostega, 299, 303, *304*
Ichthyostegan, 303–305
ICHTHYOSTEGALIA, 303
Iguana, 316, 326
  desert, 315
IGUANIDAE, 326
Ileum, 377
Iliac vessels, *347*
Ilium, *263,* 264, 305, *311, 323,* 311,
    345
Imago, 228
Immune response, 277
Implantation, 354
  delayed, 368
Imprinting, 336, 416, 417
In vitro, 7
In vivo, 7
INARTICULATA, 157
Inarticulate brachiopod, 157, 158
Incisor, 322, *346,* 358, 360, 364, 367,
    368, 369, 370, 372, 379
Inclusion, 20
Incurrent canal, *85, 87*
Incus, *271, 342, 344,* 345
Independent assortment, law of, 51–*52*
Indeterminate, 102
Indeterminate growth, 4
Inducer (of operon), 100
Induction, 103
Infauna, 422
Infraciliature, 68, 71, 72, 77, *78, 79,*
    80
Infrared, 412
Infundibulum, *266*
Infusorian, 92
Infusoriform larva, *91,* 92
Ingestion, 3, 373, 374
Inguinal canal, 352
Inheritance:
  monogenic, *51, 52,* 58, 93, 415
  polygenic (quantitative), 58–59, 60
  (*See also* Heredity)
Inhibition:
  of learning, 417
  synaptic, *408,* 409
Ink gland/sac, 173, *174,* 177
Innate behavior, 415–416
Innate releasing mechanism, 415
Inner cell mass, *95*
Inner ear, *256,* 270–271, 282, 303,
    344–345, 410, 411, 412
Innominate bone, 345
Inorganic compound, 23
Inorganic nutrient, 373

Insect, 6, 7, *42*, 76, 178, *193*, 194,
    196, 197, 222, 223, 225, 226–
    235, 302, 313, 316
  anatomy, *227–228*, *229–231*, *382*,
    384
  behavior, 230
  beneficial, 194
  carriers of parasites, 70, 76, *77*, 143,
    145, 148
  factors in success of, 226
  flight, *227*, 228–229
  hemimetabolous, 222
  holometabolous, *234*, 235
  life cycles, *232*
  metamorphosis, *231*, *232*, 235
  molting, *231*
  mouthparts, 226, *227*, 228, 235
  orders of, *233*, *234*, 235
  resistance to insecticides, 76, 226
  social, 196, 230, 427, 435
  wing structure, *227*, 228, 235
INSECTA, *193*, 197, 226
INSECTIVORA, *346*, 359, 360, 361
Insectivore, *42*, 354, *356*, 359–360
Insertion of muscle, 15, *396*
Insight, 417
Inspiration, muscles of, *314–315*
Instinct, 415
Instrumental learning, 417
Insulin, 5, 374, 398, 400, 403
Integration, mechanisms of, 397–415
Integument, functions in vertebrates, 262
  (*See also* Skin)
Integumentary system, 13
Intelligence, *418*
Interambulacrum, 238, 241, 245
Interbrachial web, 177
Intercentrum, 303, 305, 306
Interclavicle, *299*, 303, 331
Intercostal muscles, 348
Intermediate host, *135*, *136*, 143, 146,
    433
Intermediate metabolite, 27, *28*
Internal transport, 3, 373, 383–386
Interneuron, *404*, 405, *408*
Interoceptor, 411
Interphase, *35*
Interspecific relationships, 430–433
Interstitial cell, *110*, 352, 397, 400
Interstitial fluid, 273
Intertemporal, 305
Intertidal, 422, 423
Intestinal gland, *15*
Intestine, *98*, *128*, 138, *141*, *145*, *153*,
    156, *159*, 249, 250, *255*
  annelid, *180*, 182, *183*, 184, *189*,
    377
  arthropod, 200, *201*, *212*, 230
  echinoderm, 239, *244*, 245, 248
  structure of wall, *15*
  vertebrate, *256*, *260*, *283*, 284, *292*,
    295, 300, 333, *334*, 377

Intraspecific relationships, 433–436
Introvert, 158, *159*
Invagination, *89*, *95*, 96, *97*
Inversion, 59, *89*
Invertebrate, 7
Involuntary muscle (*see* Muscle,
    smooth)
Involution, *95*, 96, 97
Iodine (I), 37, 253, 254, 307, 374,
    402
Iodopsin, 270, 412
Ion, 26, 27, 405
  ammonium, 26, 428
  bicarbonate, 23, 27, 349, 383
  calcium, 27, 374, 396, 398, 399
  carbonate, 27
  chloride, 23, 27, 285, 296, *351*,
    374, 399, 401
  hydrogen, 26, 351, 399
    (*See also* Proton)
  hydroxyl, 26, *390*
  magnesium, 71
  negatively charged, 26
  nitrate, 27, *428*
  nitrite, *428*
  nutrient roles of, 374
  positively charged, 26
  phosphate, 27, 374
  potassium, 27, 238, 273, 285, 296,
    351, 374, 399, 405
  sodium, 23, 27, 273, 285, 296, 351,
    374, 397, 399, 405
Ion channel, 408
Ion gate, *18*, *406*, 407
Ion pump, 405, *406*, 407
Ion regulation, 210, 291, 296
Ionic bonding, 22
Ionic compound, 23, 26
Ionization, 22, 26
Iridia, *73*
Iridium, 43
Iris, 177, 268, *269*, 412, *413*
Iron (Fe), 154, 182, 254, 272, 349,
    374
Irritability, 3, 405
Ischial callosities, 363
Ischium, *263*, 264, 303, *323*, 345
Islands, speciation on, 54, 226
Islets of Langerhans, 398, 400
Isogamete, 63, 75
Isolation, 54, 55
Isomer, 24, 28
Isopod, 195, *211*, 220, 221
ISOPODA, 215, 221
ISOPTERA, *233*, 235
Isosmotic, 18
Isotonic, 18
Isotope, 21, 22
  radioactive, 37–38
Ivory, 5, 369, 370
IXODIDES, 208
Ixodine tick, 208

Jackal, 57
Jackrabbit, 60, 350, 365
Jacobson organ (*see* Vomeronasal
    organ)
Jaeger, 433
Jaguar, 368
Janthina, 176
Jaw:
  invertebrate, 137, *141*, 142, 247
    annelid, 181, 182, 186, *187*, 190
    arthropod (*see* Mandible)
    molluscan, *170*, 173, *174*, *178*
  vertebrate, 268, 280, 281, 285–*286*,
    287, 291, 294, 298, 300, *301*,
    307, *318*, *326*, 327, 331, 339, 345
Jaw musculature, 316, *318*
Jaw suspension, 286, 309, 345
Jay, 339, 435
Jaymotius, *281*
Jejunum, 377
Jellyfish, 41, 97, 112, 301, 322, *413*
Jingle shell, 176
Joint, skeletal, *263*, 411
Jugal, *318*
Jugular vein, *334*
Juliform millipede, *223*
Jurassic (Period), *42*, 43, *258*, 317,
    319, *324*, 325, 328, 340
Juvenile hormone, *231*
Juxtaglomerular complex, 400

Kala azar, 70
Kangaroo, *356*, 357, 359
Kangaroo rat, *361*
Karyotype, 45, 53
Katydid, 194, 235
Keel, 331, *332*, 339, 362
Kelp, 66, 67, 245
Keratin(ous), 20, 86, 260, 292, 309,
    313, 314, 319, 329, 331, 333,
    341, 360, 370, 372
Ketone, 27, *28*
Keyhole limpet, 176
Kidney, 4, *170*, *171*, *174*, 386, 387
  amniote, 277, *278*, 316, *334*, 335
  anamniote, 227, *278*, 282, *295*
  mammalian, *266*, *347*, *350–351*,
    387–388, 397, 399, 400
  morphogenesis of, 103
  mesonephric, 277, *278*, 316
  metanephridial (*see* Metanephridium)
  metanephric, 277, *278*, 316, *334*,
    335, *347*, *350–351*
  opisthonephric, 277, *278*, 293
  pronephric, *256*, 257, 277, *278*
  protonephridial (*see*
    Protonephridium)
  vertebrate, *275*, *276*, 277, *278*, *292*,
    309, *387*
  embryonic development, 99, *98*,
    277, *278*
  (*See also* Nephridium)

Killer whale, 366
Kilocalorie, 389
Kinase, 403
Kinesthesia, 412
Kinetodesma, 71, 72, 79
Kinetofragminophoran, 80
Kinetoplast, 70
Kinetoplastid, 69, 70
Kinetosome, 17, 19, 70, 71, 72
King crab, 222
King snake, 328
King termite, 104, *233,* 435
Kingdom, *40,* 57
Kingfisher, *338,* 339
Kinorhynch, 140, *141,* 142, 144, 150, 154
KINORHYNCHA, 106, 142
Kiwi, 335, 337
Kleptoparasite, 433
Knee, 317, 362
Koala, *356,* 357, 358
Komodo dragon, 316, 327
Krebs cycle, 389, *390,* 392–393
Krill, 215, 221, 367, 422, 430, *431*
Kupffer cell, 272

La Brea tar pits, 364
Labia majora (Labium major), 279, *353*
Labial cartilage, *286*
Labial palp, 170, 177
Labium, 218, 224, *227,* 228, *229*
Labor, 399, 402
Labrum, 199, 217, 218, 224, 225, *227,* 228, *229*
Labyrinth:
    crustacean, 210
    vertebrate, *271,* 309
Labyrinthodont, 280, 298, *299,* 302, 303, *304,* 305, 306, 312, 313, 317
LABYRINTHODONTIA, *259,* 303
LACERTILIA, 320, 326
Lacewing, *234,* 235
Lachrymaria, 80
Lactase, 378, 379
Lacteal, *15*
Lactic acid, 392
Lactose, 28, 340, 379
Ladybird beetle, 6, 194, 235
Lagena, *271,* 309, 315, 333, 344
Lagomorph, 354, *356,* 364–365
LAGOMORPHA, 360, 364
Lake, 419, 422
Lake Turkana fossil beds, 54
Lamellibranch, 172, 176
Lamellisabella, 159
LAMNIFORMES, 290
Lamp shell, 157
    (*See also* Brachiopod)
Lamprey, *252, 256,* 257, 264, 272, 280, 282–284
Lancelet, *252, 256*

Land animals, 41
    adaptations of, 200–201
Land-adapted egg, 201, 312–313
Lantern, *244*
    (*See also* Aristotle's lantern)
Lantern fish, 303
Large intestine, *347*
    (*See also* Colon)
Lark, 339
Larva, 95, 148, *208, 231, 232, 234,* 235, 303, 306–*307, 308,* 422, 434
    (*See also* Acanthor; Actinotroch; Actinula; Ammocoetes; Amphiblastula; Auricularia; Bipinnaria; Cercaria; Cydippid; Cypris; Cysticercus; Doliolaria; Echinopluteus; Leptocephalus; Microfilaria; Miracidium; Nauplius; Oncosphere; Parenchymula; Pilidium; Planula; Plerocercoid; Procercoid; Protapsis; Protonymphon; Protozoea; Tadpole; Tornaria; Trilobite; Trochophore; Veliger; Vitellaria; Zoea)
LARVACEA, 254
Larvacean, 106, 254, 256
Larynx, 273, *308,* 345, *347,* 366, 381, 418
Lateral aspect, *11*
Lateral canal, 238, *239,* 247
Lateral line system, 270, 284, 285, 291, 294, *295,* 309
Lateral rectus muscle, 268
Latimeria, 296, 298, *299,* 313
Latitudinal zonation, *421*
Latrodectus, 206
Laurasia, 43, 421
Leafhopper, 235
Leakey, L., M., and R., 363
Learning, 7, 416
    allelomimetic, 417
    associative, 417
    chemical changes in, 129, 409
    cognitive, 417–418
    genetically restricted, 415–416
    instrumental, 417
Leather, 5, 260, 291
Leatherback, 322
Lecithotrophic larva, 239
Leech, 70, 178, 179, 182, 184, 185, 188, *189*–190, 374, 433
Leg:
    arthropod, *212,* 228
        antenniform, 204
        insect, *227*
        ovigerous, 200
        raptorial, 220
        walking, 198, 199, 200, 201, 204, 205, *208,* 211, 212, 215, 220, 221, 222, 226
    onychophoran, *190,* 191

Leg (*continued*)
    tetrapod, *265,* 303
Leishmania, 70
Lemming, 365, 427, 434
Lemur, 360, 362
    flying, 360
Lens, 103, 110, 177, 184, 267, 268, *269,* 270, 284, 332, 412, *413*
Leopard, 368
Leopard seal, 368, 369
Lepas, 218
Lepidodiscus, 240, 243
LEPIDOPTERA, *234,* 235
LEPIDOSAURIA, *259,* 320, 322, 325, *326*
Lepidosiren, 299, 300
Lepidurus, 211, 216
Lepisosteus, 301
Lepospondyl, 305, 306
LEPOSPONDYLI, *259,* 303
Leptocephalus, 303
LEPTOMEDUSAE, 114
LEPTOSTRACA, 215
Leptostracan, 220
Leptosynapta, 237
Leucon(oid), *85,* 88, 89
Leucosolenia, 86, 87
Leukocyte, 13, 276, 277
Levator labii superioris, *342*
Levels of organization, 12
Leydig cell, 352
LH (luteinizing hormone), 279, 352, *353,* 399, *401,* 402
LH-releasing hormone (LHRH), 399, 402
Life:
    characteristics of, 2–3
    chemical basis of, 21–36
    history of, 37–44
    origin of, 38–39
Ligament, *141,* 143
    broad, 354
    hinge, *171*
    round, *353,* 354
    skeletal, 262, *263*
    suspensory, 270
Light, 414
    photosynthetic use of energy (*see* Photosynthesis)
    production of, 262
        (*See also* Bioluminescence; Photophore)
    sensitivity to, 412, *413,* 414
        (*See also* Eye)
Ligia, 211, 221
Limax, 176
Limb, 264, *265*
    bud, 97, 213
    radiative adaptation of, *265*
    regeneration, 213
    rotation in reptiles, 317
    tribasal articulation of, 289
        (*See also* Appendage; Arm; Leg)

Limestone, 5, 63, 74, 242
Limiting factor, 423
Limnoria, 221
LIMNOMEDUSAE, 114
Limpet, *151, 170,* 176
Limulus, 42, 198, *199*
Linea alba, 264
Linguatulid, *159,* 161
Lingula, 157
Linkage, 47, 51
Linnaeus, C., 57
Linnaean hierarchy, 57
Lion, 341, 368, 425, 432, 435
  marsupial, 357
Lipase, 378, 379
Lipid, *17,* 27, 28, *30–*31, 213, 273,
  373, 379, 380
Lipoic acid, *390,* 392
LISSAMPHIBIA, 306
Lithium, 103
LITHOBIOMORPHA, 224
Lithobius, 223
Liver, *168, 169, 170,* 173, *174, 175,*
  201, 212, 218, 255
  vertebrate, *256,* 257, *260, 266,* 272,
    275, 276, *283, 292, 295, 314,*
    315, *334, 347, 377,* 378, 389,
    398, 400, 403
Liver fluke, *131,* 132–133
Lizard, 205, *252,* 269, 270, 316, 317,
  320, 322, 325, *326–327,* 423
Llama, 360, 372
Lobefin, *42,* 296, 298
Lobes of mammalian brain, *343*
Lobopod, 72, 74
Lobster, 105, 194, 197, 210, 214, 221
  large-clawed, 215, 222
  spiny, *211,* 215, 222
  squat, 222
Locomotion:
  amoeboid, 72–73
  annelid, 181
  arthropod, 196
  ciliary, 71
  flagellar, 71
  muscular (*see* Muscle contraction)
  onychophoran, 191
Locus coeruleus, 267
Locust, 195, 424
Loligo, 174
Loon, 339
Lophocyte, 88
Lophophorate, 98, *99,* 152, 155–158,
  374
Lophophore, 151, 155, *156,* 157, 251
Lorica, *141,* 150
Loris, 360, 362
Louse, 194, 232, *233,* 235
  fish, 215
  whale, 221
Loxosceles, 206
Lugworm, 185, 187, 383

LUMBRICIDAE, 188
LUMBRICULIDA, 186, 188
Luminescence (*see* Bioluminescence;
  Fluorescence; Photophore)
Lung:
  amphibian, *275, 308,* 310
  bird, *275, 334,* 335
  book, 200, *201, 202,* 207, 381
  fish, 296, 298, 299, 301, 302
    (*See also* Lungfish)
  mammalian, *266, 275, 347,* 381,
    382, 411
  molluscan, 163, *170, 175,* 291,
    382
  reptilian, *314,* 327
  vertebrate, *260,* 269, 273, 280
Lungfish, 272, 273, 280, 296, 299–*300*
Lung fluke, 133
Lungworm, 146
Lure, 173, 207, 262
Luteal phase, 353
Luteinizing hormone (*See* LH)
Lycaenops, 322
Lycosid, 207
Lymnea, 176
Lymph, 273
Lymph heart, 273
Lymph node, 273
Lymph vessel, *15,* 148, 384
Lymphatic system, 273
Lymphocyte, 273, 277, 400
Lynx, 422, 424–425
Lyriform organ, 207
Lysosome, *17,* 19, 376

Mackerel, *301,* 303, 381
Mackerel shark, 290, 291
Macroconjugant, 79
Macrocyst, 65
Macroevolution, 57, 312
Macrogamont, 76
Macromere, *89,* 96, *97*
Macronucleus, 72, 77, *78,* 79, *80*
MACROPODIDAE, 359
Macrostomid, 125, 127
Macrostomoid ancestor, *125*
Macruran, *211,* 215, 222
Madreporite, *237,* 238, *239,* 242, 243,
  244, 248
Magelona, 167
Magnesium (Mg), 71, 374
Magnetic sense, 167, 337, 411, 413
Magnetite, 167
Magnetoreceptor, 411, 413
Magpie, 339
Malacology, 7
MALACOSTRACA, *211,* 215, 220
Malacostracan, *212,* 220–222
Malaria, 6, *9,* 75, *77,* 195, 426
Male:
  haploid, 143
  dwarf, 160, 177

Male (*continued*)
  reproductive system of, *141,* 143,
    352
  supplemental, 219
Male-determining factor, 47
Malleus, *271,* 342, *344,* 345
MALLOPHAGA, *233,* 235
Malpighian tubule, *153,* 155, 191, 196,
  197, 200, *201,* 202, 204, 205,
  207, 209, *223,* 224, 225, *229,*
  230, 386, *387*
Maltase, 378, 379
Maltose, 378, 379, *380*
Mamba, 328
Mammal, 7, *56,* 340–372
  biological contributions of, 340–354
  brain, *266,* 315, 342, *343,* 358
  characteristics, 340–354
  chorioallantoic placenta, 354
  circulatory system, *274, 275, 348–*
    349
  classification, 360
  dentition, *346, 347,* 355
  development, *95,* 354
  digestive system, *347,* 360, 370,
    371, 372
  diversification, 355–372
  ear, 309, 342, *344–*345
  evolutionary history, *42,* 57, 280,
    305, 317, 322
  excretory system, *278, 350–351*
  eye, *269,* 270
  fetal circulation, 349
  gene dosage regulation, 47
  hormones, 398, 399–402
  lactation, 307, 340
  marine, 12, 340, 366–367, 368–369,
    370, 422
  nervous system, *266,* 342, *343*
  oviparous, 340, 355, *356*
  reproductive system, *278, 279, 352–*
    354
  respiratory system, *347,* 348, 381,
    382
  senses, 411
  sex-determination in, 47
  skeleton, *263,* 264, 345, *346,* 369
  skin, *261,* 341
  social relationships, 341
  thermoregulation, 349–350
  urogenital system, *278*
MAMMALIA, 57, *252, 259,* 260
Mammal-like reptile, 320, 322, 340
Mammalogy, 7
Mammary gland, 18, 260, 262, 340,
  357, 399, 402
Mammary line, 340, 355
Mammoth, 369, 372
Manatee, 6, 360, 370
Mandible, 191, 197, 198, 210, 214,
  216, 217, *218,* 220, 222, *223,*
  224, 225, 227, 228, 229, *332,* 367

Mandibular arch, 285, *286*, 291
Mandibular cartilage, 285
 (*See also* Meckel's cartilage)
MANDIBULATA, 197, 210, 215
Mandibulate arthropods, 210–235
Manganese (Mn), 374
Mange, 207
Mange mite, *208*
Manta, 288, 290
Manta ray, *288*, 290
Mantis, *233*, 235
Mantis shrimp, *211*, 215, 220
Mantle, *156*, 157, 162, 163, *166*, 167,
 169, 170, 171, 173, *174,* 175,
 176, 218, 219, 253, 255
Mantle cavity, 165, *166*, 167, *169,*
 171, 172, 173, 177
Manubrium, *111,* 112, *113,* 114
Marble, 5
Marine habitats, 422, 423
Marlin, 303
Marmoset, 341, 363
Marmot, 365, 368, 421
Marrow bone, *263,* 385
Marsh, 419
Marsupial, *42,* 340, 349, 354, *356,*
 357–359
MARSUPIALIA, 357
Marsupium, 357
Masking behavior, 432
Mass:
 atomic, 21
 bio-, 38, 429
 pathway through ecosystems, 419,
  427–430
Mastax, *141,* 142
Mastigamoeba, 69, 70, 72
MASTIGOPHORA, 66, 68
Mastigoproctus, 204
Mastodon, 369, 372
Mate preference, 417
Matrix, 13, *14,* 86, 97, 262, 374, 398,
 399
Mauthernian system, 302
Maxilla:
 arthropod, 210, 214, *216,* 217, 218,
  220, 222, 224, 225, *227,* 228
 vertebrate, 299, *301,* 302, 325, 328,
  345
Maxillary gland, 195, 210, 217, 221
Maxilliped, 214, 218, 220, 221, *223*
Mayfly, 230, *233,* 235, *382*
Meadowlark, 339, 436
Mechanoreceptor, 411, 412
Meckel's cartilage, *286, 318,* 345
Medial, 11
Median plane, *10*
Medulla:
 adrenal, 400, 402
 kidney, *350, 351*
Medulla (oblongata), *266,* 267, 268,
 343, 347

Medusa, *111,* 112, *113,* 115, 422
 attached, *113, 116*
  (*See also* Gonophore)
Medusoid, 111, 113, 114
MEGACHIROPTERA, 360, 362
Megalops, 105, 215
Megapodes, 339
Mehlis's gland, *131,* 132
Meiosis, 47, 48–50, 51, 53, 59, *65,*
 66, 67, 75, 76, 79, 185, 217
Meissner's corpuscle, 261, 411
Melanin, 59, 262, 399
Melanocyte-stimulating hormone, 399,
 402
Melatonin, 399, 402
Membrane:
 basement, 127
 basilar, *344,* 345, 412
 differentially permeable, 16
 epithelial, 13, *14*
 extraembryonic, *313*–314
 mitochondrial, 19, *390*
 nictitating, 332
 oval, 342
 plasma, *17,* 18, 403, 405, *406*
 round, 344
 shell, 313
 synovial, 263
 tectorial, *344,* 345, 412
Membrane bone, 261
Membrane polarity, 405, *406,* 407
Membranelle, 71, 79, *80,* 81
Memory, 129, 409, 410
Mendel, G., 51, 57
Mendelian laws of heredity, 51–52
Meninx (meninges), 148, 265
Menstrual cycle, 352, *353*
Merganser, *338*
Merkel's disc, 261, 411
Meroblastic cleavage, *95,* 96, 204
Merocrine, 18
Meroplankton, 422
MEROSTOMATA, 197, 198
Merostomate, 198–*199*
Merozoite, 75, 76, *77*
Merychippus, 56
Mescalin, *409*
Mesencephalon, *256,* 265, *266,* 267
Mesenchyme, *14,* 96, 97, 98, *99, 102*
Mesentery, *98, 99,* 150, *156, 180*
 radial, 119, *120*
Mesocoel, 152, 157
Mesoderm, 96, *97,* 107, 140, 150, *313*
 contributions of, 96, 126, 238, 253
 differentiation of, *97*
 formation of, 93, 96–*97,* 152, 155,
  158, 160, 161
 origin of, 98, 102, 152, 195, 236
 parietal, 98, *99,* 152, 313
 somatic, 313
 splanchnic, 313
 visceral, 98, *99,* 152, 313

MESOGASTROPODA, 176
Mesoglea, 107, *111,* 112, *113,* 114,
 118, 119, 124
Mesohyl, 86, *87,* 88, 89
Mesomere, 97, *98,* 103
Mesonephric duct, *278,* 316
Mesonephros, 277, *278,* 316
Mesopelagic, 423
Mesopterygium, 298
MESOSAURIA, 320
Mesosome, 161
Mesothorax, 228
Mesovarium, 354
MESOZOA, 82, 91
Mesozoan, *91*–92
Mesozoic (Era), *42,* 43, 173, *258,* 287,
 302, 305, 315, 316, *319,* 322,
 340, 420, 421
Metabolic rate, 373
Metabolic water, 393
Metabolism, 3, 253, 373, 399, 402
 aerobic, 429
 animal, 374, 388
 cell, 388–396
Metacarpal, *263,* 264, *265,* 298, 303,
 311, 331, *332*
Metacercaria, 132, 133
Metacoel, 152, 157, 161
Metacrinus, 237, 242
Metagenesis (*see* Alternation of
 generations)
Metamere, 178, 179, 188
Metamerism, 158, 160, *166,* 167, 178,
 179, 195, 196, 253
Metamorphosis, 238, 254, *255,* 256,
 282, 397
 amphibian, 280, *307,* 310, 311
 hormonal control of, 103, *231,* 409
 types in insects, *232*
Metanephric duct, 277, 316
Metanephridium, 151, 157, 158, 160,
 191, 202, 210, 257, 277
 annelid, *180, 183,* 184, 185, 195,
  387
 molluscan, 165, *166,* 167, *169,* 172,
  173
Metanephros, 277, *278,* 316, *350*
Metaphase, *48,* 49, 50
METAPHYTA, *40,* 67
Metapterygium, 298
METASTIGMATA, 208
Metatarsal, *263,* 264, 298, 303, *311,*
 329
METATHERIA, *259,* 357
Metathorax, 228
Metatroch, 178
Metatrochophore, *178*
Metautostyly, 345
METAZOA, *40,* 82
Metencephalon, 265, 268
Methanol, *28*
MICROCHIROPTERA, 360

Microciona, 85
Microconjugant, 79
Microelectrode, 414
Microfilament, 18, *19*, 73, 79
Microfilaria, *145*, 148
Microgamont, 76
Micromere, *89*, 96, *97*, 103
Micrometer, 9, 11–12
Micronucleus, 72, 77, *78*, 79, 80
Micronutrients, 374
Micropyle, 88
MICROSPORA, 68
Microsporidian, 77
Microtubule, *17*, 48, *70*, 71, 73, 75
Microvilli, 12, *15*, 16, 17, *18*, 70, 82, 86, 109, 127, 129, 137, 161, 181, 272, 326, 410, *413*
Mictic egg, 143
Midbrain, 257, 265, *266*, 267, 268, *269*
Middle ear, *271*, 309
Midge, 229, 231
Midgut, 196, 201, 206, 214, *216*, 225, 230, 257
Migration:
  benefits of, 336
  bird, 336–337, 424
  fish, 296
  mechanisms, 336–337
  vertical, 422
Miliammnia, 73
Milk, 57, 307, 340, 355, 378, 399
  pigeon's, 339
  uterine, 310
Milk ejection, 340, 366, 399, 402
Milk sugar (*see* Lactose)
MILLEPORINA, *113*, 115
Millipede, 143, 178, *193*, 194, 197, 222, *223*, 225–226
Mimicry, 61, 432
Mineral nutrients, 373, 374
Mineralocorticoid, 399
Mink, 368
Miocene (Epoch), *42*, 56, 368
Miohippus, 56
Miracidium, *131*, 132
Mite, 194, 195, 197, 202, 207
Mitochondrion, *17*, 19, 70, 349, *390*, 398, 403, 408, 429
  ATP production in, 389, *390*, 392–394
  binding of thyroid hormone by, 403
Mitosis, *35*, 47, *48*, 50, 93, 94
  in protozoans, 63
Mitotic accelerator, 374
Mixed nerve, 267, 268
Moa, 337
Mobbing, 417, 432
Mobility, 3, 4
Mockingbird, 339, 415
Model, 61, 432
Mola, 130

Molar, *346*, 358, 364, 367, 368, 369, 371, 379
Mole, 57, *265*, *346*, 358, 359, 360, 361
  marsupial, 57, 358
Molecule, 21
  polar, 24–25
MOLLUSCA, 106, 151, 163
Mollusk, 7, *42*, 162–177, 178, 198, 374
  ancestral body plan, *166*
  body organization, *164*, *404*
  characteristics, 163–165
  classification, 166–177
  development, 98, *99*, *151*, *165*
  usefulness to humans, 162–163
  venomous, 163, 173
Molly, 303
Molting, 142, 154, 155, 161, 191, 195, 204, 205, 213, 214, 220, 221, 228, 231, 232, 309, 320, 326, 329, 351, 409
Molting hormone, 213, *231*
  (*See also* Ecdysone)
Molt-inhibiting hormone, 213
Molybdenum (Mo), 374
Monarch butterfly, 432
MONERA, *40*
Mongoose, 368
Monitor, 316, 327
Monkey, 341, 360, 362–363, 436
Monoculture, 437
Monocyte, 277
Monoecious, 88, 121, 124, 169, 185, 219
  (*See also* Hermaphroditism)
MONOGENEA, 130
Monogenic trait, *51*, *52*, 58, 93
Monoglyceride, 378, 379, *380*
Monohybrid cross, *51*
Monophyletic, 68, 69
Monophyodont, 358, 359
MONOPLACOPHORA, *164*, 166, 167, *168*
Monoplacophoran, 167
Monoploid(y), 47
MONORHINA, 282
Monosaccharide, 27, 28, *29*, 38, 373, 429
MONOTREMATA, 355
Monotreme, 340, 349, 355, *356*, 357
Monstrilloid, 217
Moon shell, 176
Moose, 372, 422, 425
Mormyrid, 303
Morphogenesis, 72, 83, 93, 103
Morphogenetic movements, 93, 96
Morphological zoology, 7
Moschops, 340

Mosquito, 6, 76, *77*, *145*, 148, 194, 195, 230, *234*, 235, *382*, 426, 433
Mosquitofish, 303
Mother-of-pearl, 162
Moth, 47, 230, 235, 414
  noctuid, 207, 414
  silkworm, 194, *231*
  sphinx, 229
  yucca, 53
Motility, endogenous, 3
Motivational conflict, 437
Motmot, 2, 339, 416, 432
Motor cortex, 342, *343*
Motor nerve, 267, 268
Motor neuron, *14*, 302, *396*, *404*, 405, 407
Motor root, *404*
Motor unit, 15
Mouse, 45, 83, 270, 341, *356*, 357, 360, 365
  marsupial, 357
Mouse deer, 371, 372
Mousebird, 339
Mouth, 94, *138*, *141*, *144*, *151*, *153*, *156*, *159*, 249, 377
  annelid, *178*, 179, 182, *183*, *187*, *189*
  arthropod, 198, 200, *201*, *212*, *218*
  chordate, 254, *255*, *260*, 282, *283*, 286, 287, *288*, 291, *292*, 295, 297, 300, *301*, 369
  echinoderm, *237*, 242, 243, *244*, 245
  flatworm, *125*, *128*, *131*
  hemichordate, *249*, 250
  molluscan, *164*, *165*, *168*, *169*, *170*, *171*, 173
  radiate, *110*, *111*, *113*, *116*, *120*, *121*, *123*
Mouthparts, 196, 208, 210, 226, 227, 228, *307*, 374
Mucopolysaccharide, 181, 191, 195
Mucosa, *15*
Mucus, 19
Mullerian duct, 354
Mullerian mimicry, 432
Mullet, 303
Multicellularity:
  advantages, 84–85
  possible origin of, 82–84
Multicilia, 70, 72
Multituberculate, 355
Murex, 176
  (*See also* Rock shell)
MURIDAE, 365
Murre, 339
Muscle:
  abductor, 196
  adductor, 157, *171*, *216*, 217, *218*, 300
  appendicular, 317
  biceps, *263*

Muscle (*continued*)
　body wall layers, 126, *128, 138,*
　　140, 143, *145,* 161, *180, 183,*
　　191, 248, 256, *283, 292, 295*
　breathing, 381
　cardiac (*see* Cardiac muscle)
　ciliary, 268, *269,* 270, 332
　dorsal (epaxial), 256
　erector, *261,* 333
　extensor, 196, *212*
　eyeball, 268
　facial, *342*
　flexor, 196, *212*
　flight, *227,* 228–229, 331–*332*
　gut, *14,* 378, 379
　heart (*see* Cardiac muscle)
　hypobranchial, 310
　inspiratory, 315
　integumentary, 261
　involuntary, 405
　protractor, *164, 180,* 196
　radial, 173
　red, 293
　retractor, 143, 151, 155, 158, *159,*
　　164, 166, 167, *169, 180,* 196
　segmental, *168,* 179, 253, *255, 256,*
　　310
　skeletal, *14,* 385, 405, 411
　smooth, 13, *14,* 15, 124, 268, 354
　spiracular, 230
　striated, *14,* 114, 181, 196, 411
　structure, 394–*396*
　tail, *295*
　triceps, *263*
　voluntary, 405
Muscle contraction, 374, 389, 394–
　396, 418
Muscle fiber, 13, 394, 395, *396*
　multiply innervated, 196
Muscle tissue, 13, *14*
Muscular system, *13*
Muscularis mucosae, *15*
Musk ox, 435
Muskrat, 360, 365
Mussel, 172, 176
MUSTELIDAE, 57
Mutability, 3
Mutagenic agents, 3
Mutation, genetic, 3, 53, 58
　base substitution, 58
　frame-shift, 58
Mutation pressure, 59
Mutualism, 64, 71, 80, 122, 176, 433
Mya, 171
Myelencephalon, 265, 267
Myelin sheath, *14,* 265, 407
Myelinated fiber, 407
Myenteric plexus, 379
MYGALOMORPHAE, 207
Mynah, 339
Myocardium, 385
Myocyte, 13, 88

Myoepithelium, 140, 147, 155, 399,
　402
Myofibril, 394, *396*
Myofilament, 394
Myogenic contraction, 229, 254, 273
Myoglobin, 374
Myomere, 253, *255,* 256
Myometrium, 354
Myoneme, 79
Myosin, 73, 394, 395, *396*
Myoseptum, 253, 256, 264
Myriapod, *223*
MYRMECOBIIDAE, 358
Mysid, 210, 221
MYSIDACEA, 215, 221
Mysis, 105, *213,* 214
Mysis, 105, *211,* 221
MYSTACOCARIDA, 215, 218
MYSTICETI, 360, 367
Mytilus, 171
MYXINI, 257, *258,* 280, 282,
　283
MYXOMYCOTA, 65
Myxosporidian, 77
MYXOZOA, 68

Nacre(ous), 162, 164
NAD, 375, 389, *390,* 392, 393
NADP, 389, *391,* 394
Naegleria, 74
Naiad, 232, 235
Nail, 260, 362, 372
Nanometer, 73
Nanomia, 115
Nanoplankton, 194
Naris, 345, 381
　(*See also* Nostril)
Narrow-niched species, 424
Nasal canal, 285
Nasal capsule, *286*
Nasal cavity, 322, 325, 345,
　412
Nasal sac, 285, 291
Nasal salt gland, 335, 339, 386
Nasopharynx, 345, 381
Natantian, 221
Natrix, 321
Natural balance, 424
Natural selection, 60, 423
Nauplius, 105, 197, 210, *213,* 214,
　215, 217, *218, 219,* 221,
　222
Nautiloid, 164, 173, 177
NAUTILOIDEA, 177
Nautilus, 173, *174,* 177
Nautilus:
　chambered, *174,* 177
　paper, 177
　　(*See also* Argonaut)
Navigation, 293
　magnetic, 337, 413
　sidereal, 336–337

Navigation (*continued*)
　sun-compass, 337
Neanderthal man, 364
Neanthes, 180, 187
Nearctic realm, 421, 424
Nebalia, 211, 220
Necator, 148
Nectophrynoides, 311
NECTRIDIA, 303
Nectridian, 305
Needlefish, 303
Nekton, 422, 423, *431*
Nektophore, 112, 115
NEMATHELMINTHES (*see*
　NEMATODA)
Nematocyst, 67, *108,* 109, 112, 119,
　122, 124, 127, 176
NEMATODA, 106, 144
Nematode, 98, 140, 144–149, 155,
　377, 433
Nematogen, *91*
Nematomorph, 140, 143–144
NEMATOMORPHA, 106, 143
NEMERTEA, 137, *138*
　(*See also* RHYNCHOCOELA)
Nemertean, 91, 98, 137–139, 374,
　383, *384*
NEMERTINEA (*see* NEMERTEA,
　RHYNCHOCOELA)
Neoceratodus, 299, *300*
NEOGASTROPODA, 176
NEOGNATHAE, 337
Neoophoran, 127
Neopallium, 265, 315, *336,* 342,
　343
Neopilina, 164, 167, *168*
NEORHABDOCOELA, 127
NEORNITHES, 337
Neoteny, 106, 114, 254
Neotropical realm, 421, 424
Nephridiopore, 129, 151, *159,* 165,
　166, 168, 172, 173, *180, 183,*
　184, *190,* 210, *387*
Nephridium, *125, 153, 156, 159, 166,*
　167, *168, 169,* 176, 177, 179,
　180, 189, 255
　(*See also* Metanephridium;
　　Protonephridium)
Nephrocyte, 254
Nephrogenic plate, 103, 277,
　316
Nephron, 316, 387–388
　cortical, *350*
　evolution of, 277
　medullary, *350, 351*
Nephrostome, 151, 158, *166,* 184,
　257, 277, *278*
Nereid, 186
NEREIDAE, 186
Nereis, 179, *187*
Nerillidium, 187
Neritic province, 422

Nerve, 129, 267, 268, 405
  cranial, *266*, 267, 268, *343*
  radial, 239, 242, 243, 246
  spinal, 267, 282, *404*
Nerve cell, *14, 110,* 405–410
  (*See also* Neuron)
Nerve cord, *128,* 161, *190*
  annelid, *180, 183,* 184, *189*
  arthropod, 191, 204, *212,* 214, *223,*
    224, *229,* 230
  chordate, 236, *249, 255, 260, 283,*
    292, 334
  molluscan, 165, *166, 168*
  onychophoran, *190*
  vertebrate (*see* Spinal cord)
Nerve fiber, 13, *14,* 110, *183,* 405,
  406
  afferent, 267, *404,* 405
  amyelinated, 407
  efferent, 267, *404,* 405
  excitatory, *408*
  inhibitory, *408,* 409
  myelinated, 407
  sensory, 414
Nerve impulse, 406–407, 410
  rate of conduction, 407
Nerve plexus, 107, 114, 124, 157,
  165, 184, 379
Nerve polarization, 405, *406*
Nerve ring, 159, 165, *166, 168,* 184,
  203, 242, 243, 246
Nerve tissue (*see* Nervous tissue)
Nerve trunk, 129, 195, 405
  (*See also* Nerve cord)
Nervous system, 4, 13, 398, 401, 402,
  404, 405–418
  acanthocephalan, 143
  amphioxus, *255,* 257
  annelid, *183,* 184–185, 195, *404*
  arthropod, 195, 202–*203,* 204, 214,
    229, 230, *404*
  autonomic, 230, *266,* 267, 268, 379,
    405, 407
  bird, *266, 336*
  central, 405, 407, 408, 410
    (*See also* Brain; Nerve cord;
      Spinal cord)
  centralization of, 126
  chaetognath, 249
  chordate, 252–253
  echinoderm, 242, *404*
  ectoproct, 155
  entoproct, 153
  fish, *256,* 257, *266*
  flatworm, 126, 127, *128*
  human, *266, 343*
  mammalian, 342, *343*
  molluscan, 165, 167, *168, 169,* 173,
    175, 404
  nematode, 145
  nematomorph, 143
  nemertean, 139

Nervous system (*continued*)
  onychophoran, 191–192
  parasympathetic, 230, *266,* 267–268,
    379, 385, 405
  peripheral, 265, 267, 268, 405
  priapulid, 154
  radiate, 107, 124, *404*
  reptile, *266*
  somatic sensory-motor, 267, 268,
    405
  sympathetic, *266*–268, 379, 385,
    402, 405
  tardigrade, 155
  vertebrate, 265–268, *404*
Nervous tissue, 13, *14*
  (*See also* Neuron)
Nest, 298, 337, 402, 403, 415, 435
Nest parasite, 339, 433
Nesting, 435
  colonial, 435
  communal, 435
Neural arch, 284, 285, 287, 288, 306
Neural fold, *97,* 307
Neural plate, *97,* 99
Neural tube, *98,* 99, 250, 252, 253,
  257, 264
Neurilemma, *14*
Neurocranium, 302, *318*
Neuroepithelial cell, 268, 410, 411,
  412, 413
Neuroglia, 407
Neurohormone, 13, 184, 188, 213,
  276, 397, *401,* 409
Neurohypophysis, *401*
Neuromast, 270
Neuromuscular junction, 395, *396*
Neuron, 252, *409*
  functioning, 13, 405–410
  structure, *14*
  types:
    association, *404,* 405
    bipolar, 110, *269*
    inhibitory, 196
    inter-, *404,* 405, *408*
    giant, 302
    motor, 302, *396, 404,* 405, 407
    multipolar, 110
    neurosecretory, 184, 231, 397,
      401, 402, 409
    pacemaker, 407
    phasic, 196
    self-excitatory, 407
    sensory, 203, 268, *404,* 405, 409,
      410, 211
    tonic, 196
Neuropeptide, 409
Neuropodium, 181
NEUROPTERA, *234,* 235
Neurotransmitter, 107, 110, 267, 402,
  408, *409*
Neurulation, *97,* 99
Neurulizing factor, 103

Neutralism, 431
Neutron, 21
Neutrophil, 277
Newt, 310
Niacin, 375, 389
Niche, 420, 423, 424, 430
Nickel (Ni), 374
Nicotinamide adenine dinucleotide (*see*
  NAD)
Nicotinic acid, 375, 389
Nictitating membrane, 332
Night blindness, 375
Nighthawk, *338,* 339
Nipple, 340, 355, 357, 370, 402
Nitrate bacteria, 428, 429
Nitrate ion, 27, *428*
Nitrification, 428
Nitrite bacteria, 428, 429
Nitrite ion, *428*
Nitrogen (N), 21, *22,* 164
  -fixers, 39, *428*
Nitrogen cycle, *428*–429
Nitrogenous base, 26, 32, *34,* 378, 379
Nitrogenous wastes (*see* Ammonia;
  Urea; Uric acid)
Noctiluca, 66
Noctuid moth, 207, 414
Nocturnal, 431
Node, 407
  of Ranvier, *14*
Nomenclature, binomial, 57
Nonionella, 73
Noradrenalin, 267
  (*See also* Norepinephrine)
Norepinephrine, 400, 402, *409*
"North Pole", Australian, 39
Nose, 369
  (*See also* Nostril, Proboscis)
Nostril, *256,* 257, 272, 282, *283, 284,*
  285, 288, 292, 295, 325, 331,
  334, 337, 339, 341, 363
  (*See also* Naris)
NOTORYCTIDAE, 358
Notochord, 57, *97, 98, 102,* 250, 252
  253, 254, *255, 256,* 257, *260,*
  264, *281,* 282, *283,* 287, 288,
  289, 292, 300, 301, 305
Notopodium, 181
NOTOSTRACA, 215
Notostracan, 216
Notum, 228
Nubeculina, 73
Nuchal organ, 184
Nuclear envelope, *17,* 20, 48
Nuclease, 378, 379
Nucleic acid, 3, 27, 38, 93, 373, 378,
  379
  (*See also* DNA, RNA)
Nucleolus, *17,* 20
Nucleoprotein, 20
Nucleotide, *32,* 33, *34,* 36, 38, 58,
  374, 378, 379

Nucleotide sequences, concordance of, 39, 53
Nucleus:
  atomic, 21
  cell, *17, 20, 73,* 403
    transfer of, 102
  cerebral basal, 265, 267
  raphe, 267
Nucula, 171
NUDA, *123,* 124
Nudibranch, *170,* 176
Numbat, *356,* 358
Nummulites, 74
Nut clam, 176
Nuthatch, 339
Nutria, 365
Nutrients, 373–376
Nutrition, 3, 4, 40, 62, 63, 373–396
Nymph, 104, 161, *208, 232, 233,* 235, *382*
Nymphon, 200

Obelia, 111, 112, 115
Obliquus muscle, *314,* 315, 321
Occipital bone, 345
Occipital condyle, 317, 329, 345
Occipital lobe, *343*
Oceanic productivity, 422
Oceanic province, 423
Ocellus, 107, 110, 112, 114, 129, 132, 142, 170, 210, 224, 225, *227,* 228, *229,* 248, 249
Octocoral, 119
Octopod, 173, 177
Octopus, 91, 162, 163, 173, *174,* 177
Octopus, 175
Oculomotor nerve, 268
ODOBENIDAE, 369
ODONATA, *233,* 235
ODONTOCETI, 360, 366
ODONTOGNATHAE, 337
Odontophore, *164*
Odor-detection, 268, 381, 411, 412
Odor sensations, primary, 412
Oecium, 156
  (*See also* Ovicell)
Oikomonas, 9
Oils, 4
Okapi, 374
Okazaki fragments, 33, *34*
Olfaction, 265, 272, 285, 411
Olfactory bulb, *266,* 268, 272, 282
Olfactory cell, 381, 410
Olfactory lobe, 409
Olfactory nerve/tract, *266,* 268, 272, *343*
Olfactory pregnancy block, 427
Olfactory receptor, 411, 412
Oligocene (Epoch), *42,* 56, 370

OLIGOCHAETA, 186, 188
Oligochaete, 178, 179, 181, 182, 184, 185, 188
Oligodendroglia, 407
Oligopeptide, 403
Oligosaccharide, 378
Olive, 176
Omasum, *371*
Ommatidium, 198, 199, 210, 213, 224, 228, 412
Omnivorous, 206, 424
Onchid, 176
Onchomiracidium, 132
Oncosphere, 134, *135,* 136, 137
Ontogenetic extension, 105–106, 215
Ontogeny, 100, 104
ONYCHOPHORA, 106, 151, *190*
Onychophoran, 178, *190*–192, 210, 225
  compared with annelids, 191
  compared with arthropods, 192
  phylogenetic status, 190, 222
Oocyst, 75, 76, *77*
Oocyte, *49*
Oogenesis, *49,* 111
Oogonium, *49,* 88, 100
Ookinete, 76, *77*
Ootid, *49*
Opalina, 70, 72
OPALINATA, 68
OPALINIDA, 68
Operator gene, 100, 101
Operculum, 164, 188, 291, 293, 294, *297,* 300, 306, *307, 308*
  genital, 198
  middle ear, 309
Operon, 100
OPHIDIA, 320, 327
OPHIOCISTIOIDEA, 241, 246
Ophiopluteus, 248
Ophiura, 237
Ophiuroid, *237,* 239, 240, 241, 245, 246, 247–248, 432
OPHIUROIDEA, *237,* 241, 246, 247–248
Ophthalmosaurus, 319, 322
OPILIONES, *201,* 205
Opilionid, 197, *203,* 205–206
Opisthaptor, 130, 132
Opisthobranch, *175,* 176
OPISTHOBRANCHIA, *170, 175,* 176
Opisthonephros, 277, *278,* 284, 285, 293
Opisthorchis, 132
Opisthosoma, 152, *159*
Opossum, *356, 357,* 358
Opossum shrimp, *211,* 215, 221
Opsin, 412
Optic chiasma, *266, 401*
Optic cup, 103, 267
Optic lobe, *229, 266*

Optic nerve, 203, 268, *269,* 270
Optimal tolerance, zone of, 423
Oral, 10
Oral-aboral axis, 10, 107, 124, 238, 243
Oral arm, 118
Oral disk, 119, *121*
Oral groove, *78,* 79, 81
Orangutan, 363
Orbicularis oris, *342*
Orbit, *286,* 311, *318, 332,* 362
Orbital, 22, 23
  bonding, *22,* 23
Orchestia, 211, 221
Order, 57
Ordovician (Period), 41, 197, 241, 243, 245, 246, *258,* 280
ORECTOLOBIFORMES, 290
Organ, 7, 12, 13, *15,* 405, 411
Organ of Corti, *271,* 309, *344,* 345
Organ system, 12, 13
Organelle, 19
Organic compounds, 23–24, 27–36
  abiotic formation of, 38, 39
  in ecosystems, *428,* 429
  polychlorinated, 427
Organic nutrients, 373–376
Organism, 7, 9, 12
Orgasm, 399
Oriental realm, 421
Orientation:
  by echolocation, 360, 362, 366, 368
  by light, 216, 337
  by magnetism, 413
  by radar, 413
Origin of muscle, 15, *396*
Oriole, 339
ORNITHISCHIA, *259,* 320, *323*
Ornithischian, *317,* 323
Ornithology, 7
Ornithomimus, 319
Ornithorhynchus, 355
Orthonectid, 91
ORTHOPTERA, *233,* 235, 432
Oryx, 434
Oscillating universe model, 38
Oscilloscope, *406*
Osculum, *85, 86, 87,* 88
Osmoconformer, 386
Osmoreceptor, 411, 412
Osmoregulation, 63, 129, 147, 210, 291, 296, 298, 309, 386, 422
Osmoregulator, 386
Osmosis, *17,* 18, 374
Osmotic equilibrium, 18
Osmotic pressure, 26, 109, 147
Osphradium, 165, *166,* 169, 170
Osprey, *338*
Ossicle:
  dermal, 327

Ossicle (*continued*)
  echinoderm, 238, *239*, 241, 242, 243, *244*, 245, 246, 247
  middle ear, 309, 333, 342, *344*, 412
  Weberian, 303
Ossification, 262
OSTARIOPHYSI, 296, 303
Osteichthyan, 280, 294–303
  body organization, *295*
  classification, 298–303
  distinguished from other bony fish, 294
  distinguished from elasmobranchs, 294–298
  respiration, 296, *297*
OSTEICHTHYES, *252*, 257, *258*, 280, *294*, 302
Osteoblast, 13, 262, 263
Osteoclast, 263, 398, 399
Osteocyte, 263
OSTEOGLOSSOMORPHA, 303
Osteomalacia, 375
Osteon (*see* Haversian system)
OSTEOSTRACI, *281*, 284
Ostium, 86, 191, 196, 199, *201*, 202, 207, *212*, 214, 220, *223*, 224, 225, 226, 383, *384*
Ostracion, 301
Ostracod, 214, *216*, 217
OSTRACODA, *211*, 215, 217
Ostracoderm, *42*, 257, 261, 264, 280, *281*, 282, 284, 286, 287, 294
OSTRACODERMI, *258*, 280
Ostrea, 171
Ostrich, 9, 331, 337, 349, 435
OTARIIDAE, 369
Otic capsule, *271*, *286*, 309
Otic nerve, 268, 302, *344*, 345
Otic vesicle, 271
Otolith, 254, *271*, 412
Otter, 367, 368
Outer ear canal, *271*
  (*See also* External auditory canal)
Oval membrane, 342
Ovarian follicle, 231, 279, 352, *353*, 354
Ovariole, 230
Ovary, 50
  invertebrate, *110*, 111, 112, *113*, *125*, *128*, *131*, *134*, *138*, *141*, *145*, *153*, *156*, *174*, *183*, 184, 188, *189*, *190*, 192, *201*, 204, 207, 209, *212*, *218*, 225, 226, *229*, 230
  vertebrate, *255*, *278*, 279, 293, 296, 311, *334*, 335, 352, *353*, 354
Overturn, 422
Ovicell, 155
Oviduct:
  invertebrate, *125*, *128*, *145*, 169, *170*, 185, *189*, *190*, 192, 207, *212*, *218*, 226, *229*, 230, *249*

Oviduct (*continued*)
  vertebrate, *278*, 292, 293, 310, 312, 313, *334*, 335, *353*, 354
Ovigerous leg, 200
Oviparous, 204, 233, 292
Ovipositor, 206, *227*, *229*, 231, 235
Ovisac, 185, 216, 217, 218
Ovotestis, *170*
Ovoviviparous, 192, 204, 292, 296, 317
Ovulation, 279, 352, *353*, 399, 403
Ovum, 50, 313
  (*See also* Egg)
Owl, 329, *338*, 339, 431
Ox, 5
Oxidation-reduction, intramolecular, 392
Oxidative phosphorylation, *390*, 393
Oxygen (O), 4, 21, *22*
  cyclic pathway in ecosystems, *428*, 429
  gaseous, 39, 41, 422, 437
  internal transport, *154*, 182, 184, 207, 244, 349, 383, 412
  as H-acceptor in metabolism, 3, 373, 389, *390*, 393, 394
Oxyhemoglobin, 383
Oxytocin, 340, 399, *401*, 402, 403, 409
Oyster, 162, 176
Oystercatcher, 339
Ozone, 427

Pacemaker:
  heart, 105, 254, 273, 274, 333, 348, 384, 385
  neural, 114, 407
Pachyderm, 369
Pacinian corpuscle, 261, 411
Paddlefish, 300, 301
Paedomorphosis (*see* Neoteny)
Paguroid, 222
Pagurus, 211, 222
Pain:
  alleviation, 409
  receptors, 261, 409, 411
Pair-bond(ing), 341, 435
Palate, *318*, 322, 325, 345, *347*, 364
Palatoquadrate cartilage, 285, *286*, *318*, 345
Palearctic realm, 421
Paleocene (Epoch), *42*
Paleogenesis, 104
PALEOHETERODONTA, 176
Paleoniscoid, 300, 302
Paleontology, 7
Paleopallium, 265, *336*, 342, *343*
PALEOTAXODONTA, 176
Paleozoic (Era), 41, *42*, 109, 166, 197, 241, 242, *258*, 287, 288, 289, 298, 302, 305, 313, 315, 316, 322
PALINURA, 215, 222

Palinurid, 222
Pallial groove, 168
Pallial line, *168*
Pallium, 265
Palolo worm, 5, 185
Palp, 170, *171*, 177, 179, 181, 186, *187*, 200, *201*, 214, *216*, 217, 227
Palpar organ, 207
Pancreas, 173, *174*, *260*, *266*, 292, 333, *334*, *377*, 378, 379, 398
Pancreatic juice, 400
Pancreozymin (*see* Cholecystokinin)
Panda, 367
Pangaea, 42, 43, 289
Pangolin, *356*, 360, 364
Pantothenic acid, 376, 392
Panulirus, 211, 222
Papilla, *159*, 184, *190*, 242, *261*, *330*, *350*
Papula, 239, 248
Parabiotic union, 398
Parabronchus, 335
PARACANTHOPTERYGII, 303
Paragonimus, 133
Paralithodes, 222
Parallel evolution, 173
Paralysis, tick, 209
Paramecium, 9, 64, 72, *78*, 79, 80, 81, 84, *377*
Parapodium, 176, *178*, 179, *180*, 181, 182, 186, *187*, 188, 191, 196, 381
Parapsid, *318*, 322
Parasite, 433
  effect on host population, 425
  selection against pathogenicity, 64
  (*See also* Endoparasite; Ectoparasite)
Parasitic animals, 64, 70, 74, 75, 76, 77, 80, 91, 92, 104, 161, *172*, 188, *190*, 207–209, 215, 217, 218, *219*, 221, 235, *283*, 284, 339
Parasitism, 64, 161
  types of, 433
Parasympathetic, 230, *266*, 267–268, 379, 385, 405
Parathormone, 398, 399
Parathyroid, 398, 399
PARAZOA, 82, 90
Parchment worm, 182, 187
Parenchyma, *99*, 127, 150
Parenchymula, *89*
Parental behavior, 205, 230, *283*, 298, *306*, 307–308, 316–317, 335, 341, 435
Parenting, communal, 337
Parietal bone, *318*
Parietal eye, *269*
Parietal lobe, *343*
Parkinsonism, 267
Parotid gland, 378
Parotoid gland, 311
Parrot, 339
Parrotfish, 123, 303

Pars intercerebralis, 231
Pars intermedia, 399, *401*, 402
Parthenogenesis, 94, 130, 142, 150, 155, 214, 217, 225, 226, 235, 327
Parturition, 354, 400, 409
PASSERIFORMES, *338*, 339
Passerine, 331, 339
Passive transport, 18
(*See also* Diffusion; Osmosis)
Patagium, *330*
Patella, *263, 332*
Patella, 151
Pathology, 7
Pauropod, 197, 222, *223*, 225
PAUROPODA, 197, 225
Pauropus, 223
Pavlovian conditioning, 417
Peacock, 60, 330
Peanut worm, 158, *159*
Pearl formation, 162
Peccary, 360, 371, 431
Peck(ing) order, 403
(*See also* Social Hierarchy)
Pecten, 171
Pecten, 332
Pectin, 75
Pectine, 204
Pectoral fin (*see* Fin, pectoral)
Pectoral girdle, 264, *289*, 291, 294, *299*, 303, 331, 355, 359
Pectoralis muscle, 331, *332*, 362
Pedal disk, *121*
Pedal ganglion, 165, *169, 171*, 172, *175*
Pedal groove, 167
Pedal laceration, 121
Pedal retractor muscle, *166*, 167
Pedal shield, 167
Pedalium, 116, *117*
Pedicellaria, 238, *239, 244*, 245, 248
Pedicellina, 153
Pedicle, 157
Pedipalp, 197, *201*, 203, 204, *205*, 206, 207, 208
(*See also* Palp)
Peduncle, *156*, 157
Pelagia, 118
Pelagic, 423
PELECANIFORMES, 339
PELECYPODA, 170, *171*, 176
Pelican, 339
Pellagra, 375
Pellicle, 16, 63, *78, 79*
PELOBATIDAE, 311
Pelomyxa, 74
Pelvic fin (*see* Fin, pelvic)
Pelvic girdle, 264, *289*, 294, 303, 305, 317, *323*, 345
Pelvis, *332, 334*, 345
renal, *350*
Pelycosaur, 315, *222*
PELYCOSAURIA, *259, 320*

Pen (of squid), 173, *174*, 177
Pen shell, 176
Penaeid shrimp, 222
Penaeus, 211, 213, 222
Penella, 217
Penetrant, *108*, 109
Penguin, *56*, 339, 435
Penial spicule, 142
Penis, 126, *128*, 130, 132, 143, *170*, 185, *189*, 206, 209, 214, *218*, 226, 231, *278*, 316, 321, 335, *347, 352*, 355
PENNATULACEA, 119
Pennatulacean, 120
Pennsylvania (Period), 242
(*See also* Carboniferous Period)
Pentacrinoid, 242
Pentastomid, 161
PENTASTOMIDA, 151, *159*, 161
Pentose, 27, *29*, 30, *391*
Pentose phosphate cycle, 389, *391*, 393–394
Pentremites, 240, 241
PEP (phosphoenol pyruvic acid), *390*, 392
Pepsin, 3, 261, 378, 379, 380
Pepsinogen, 379, 400
Peptidase, 374
Peptide, 403
PERACARIDA, 215, 220
Peracaridan, 211, 220–221
PERAMELIDAE, 358
Peranema, 69
Perception, 414–415, 416
Perch, 303
surf-, 296
yellow, *295*
Perching mechanism, 331, *332*
PERCIFORMES, 303
Pericardial cavity, 165, 167, 172, 196, 199, *212*, 224
Pericardial coelom, 165, 172
Pericardial sinus, 191, 196, 214
Pericardium, *168, 171, 201, 212*, 222, 260
Periderm, *330*
Perihemal channel, 238, 239
Perilymph, 271, 309, *344*, 412
Periosteum, *263*
Periostracum, 157, 164
Peripatus, 190, 191
Peripheral nervous system, 265, 267–268, 405
Periphylla, 108
Periproct, 246
Perisarc, *111*, 114
Perissodactyl, *356*
PERISSODACTYLA, 360, 370
Peristalsis, 181, 182, 378
Peristome, 81, *244*
Peristomial membrane, 245
Peristomium, 179, *187*

Peritoneum, 97, 150, *180*, 181, 182, 185, 239, 242, 248, *260*
Peritrich, 81
Peritrophic membrane, 225, 230
Permafrost, 421
Permian (Period), 41, *42*, 241, *258*, 300, 302, 304, 305, 312, 316, 322
Peroxisome, 20
Personal distance, 435, 436
Pest control, 6
Pesticides, 6
Petalloid, 245
Petrel, 339
Petrifaction, 37
pH, 27, 379
PHAEOPHYTA, 66
Phagocyte, 85, 88, 191, 262, 277
Phagocytosis, *17*, 62, 374
Phalanger, 358
great flying, 358
PHALANGERIDAE, 358
PHALANGIDA, 205
Phalanx (phalanges), *263*, 264, *265*, 298, 303, *311*, 331, 370
Pharyngeal basket, 254
(*See also* Branchial sac)
Pharyngeal cleft (slit), 250, 252, 254, 273, 274
(*See also* Gill slit)
Pharyngeal pouch, 257
(*See also* Gill pouch)
Pharyngeal pumping, 280, 282
PHARYNGOBDELLIDA, 186, 190
Pharyngobranchial element, *286*
Pharynx:
invertebrate, 119, *123*, 124, *125*, 127, *128*, 140, *141, 145*, 154, *156, 170, 178*, 181, 182, *183*, 186, *187, 189*, 190, 200, 201, *229*, 245, *249*, 250, 254, *255*, 256, 257, 374, *377*
vertebrate, 257, *260*, 282, *283, 292*, *295, 297*, 302, 303, 309, *377*
PHASCOLARCTIDAE, 358
Phascolomis, 357
PHASCOLOMYIDAE, 358
Phasmid, 146
PHASMIDA, 146
Pheasant, 339
Phenotype, 45, *51, 52, 53*, 60, 415
Phenotypic ratio, *51, 52, 53*
Phenotypic stability, plateau of, 54
and stabilizing selection, 60
Phenylalanine, 378, 379, *380*
Phenylthiocarbamide, 414
Pheromone, 65, 103, 104, 147, 160, 207, 217, 218, 226, 310, 341, 418, 427, 435
Philodina, 142
PHOCIDAE, 369
Pholadid, 163, 176
Pholas, 171

Pholcid, 206
PHOLIDOTA, 360, 364
Phoronid, 98, 152, 157
PHORONIDA, *156,* 157
Phoronis, 156
Phosphatase, 392
Phosphate, *30, 32, 34,* 374, 388, 403
    calcium, 27, 157
Phospholipase, 327
Phospholipid, 16, *17, 30,* 374, 385,
    403
Phosphorolysis, *392*
Phosphorus (P), 374, 375
Phosphorylase, 374
Phosphorylation, 143
    energy transfer and, 388, 395, 403
    electron transport and, *390,* 393
    photo-, 23, 39
    proton pump and, *390, 393*
    substrate-level, 392, 394
Photic zone, 422, 423
Photogenesis, 217, 418
    (*See also* Bioluminescence)
Photoperiod, 232, 267, 402
Photophore, 217, 221, 262
Photophosphorylation, 23, 39
Photopic vision, 270, 410, 411
Photoreceptor, 126, 129, 167, 184,
    203, 207, 253, 257, 267, *269,*
    270, 332, 412, *413*
    (*See also* Cone; Eye; Ocellus;
        Retina; Rod)
Photosynthesis, 4, 5, 23, 39, 67, 119,
    373, 388, *428,* 429
Phototaxis (phototactic), 64, 114, 116,
    139, 184, 225
Phrynoderma, 375
Phyletic speciation, 54, *55*
PHYLLOCARIDA, *211,* 215, 220
Phyllopodium, 216
Phylogenetic recapitulation, 104–*105*
Phylogenetic trees, 57
Phylogeny, 104, 105, 150–152
    arthropod, 190, 210
    chordate, 251
    echinoderm, 236, 238
    recapitulation in ontogeny, 104–105
    vertebrate, 254, 256
Phylum, 41, 57, 106
Physalia, 108, 112, *116*
PHYSETERIDAE, 366
Physicochemical factors in habitat, *420,
    421,* 423
Physiological stress, zone of, 423
Physiology, 7
Phytoflagellate, *66–69*
PHYTOMASTIGOPHOREA, 66, 68
Phytoparasitic, 146
Phytoplankton, 194, 216, 221, 422,
    423, *430, 431,* 437
Phytosaur, 323, 324–325
PHYTOSAURIA, 320

Pia mater, 264
PICIFORMES, 339
Pig, 5, 371
Pigeon, 337, 339, 413
Pigment, 20, 90, 213, 402, 412
    respiratory, 13, 165, 182
    visual, 412
Pigment cell, *14,* 269, *413,* 418
    (*See also* Chromatophore)
Pigment granules, 213
Pigment sleeve, 213, 413
Pigmentation, heredity of human skin,
    58–59
    (*See also* Coloration)
Pika, 60, 360, 364, 365, 421
Pilidium, 130, 137, 151
Pill bug, 221
Pilot whale, 365, 366
Pinacocyte, 88, 89
Pincher, 204, 205, 222
    (*See also* Chela; Cheliped)
Pineal body, *266,* 284
Pineal eye, *269,* 280, 282
Pineal gland, *269,* 402
Pinna, 342, *344,* 355, 362, 365, 368,
    369, 370
Pinniped, *356,* 368–369
PINNIPEDIA, 360, 368
Pinnule, *120,* 161, 181, 242
Pinocytic vesicle, *17,* 18, 72, 130
Pinocytosis, *17,* 18, 143, 374
Pinworm, 148
Pipa, 311
Pipefish, 303
PIPIDAE, 311
Piranha, 303
Piroplasmid, 75
PISCES, 260
Pit viper, 261, 328, 412
Pituitary, 11, *266,* 276, 284, 307, 309,
    340, 352, 399, *401,* 402, 403,
    409, 427
Placenta, 275, 292, 327, 340, 354,
    357, 367
Placoderm, *42,* 262, 280, *286,* 287,
    288, 294
PLACODERMI, 257, *258,* 280, 287
Placoid scale, 262, 291, *292,* 294
PLACOZOA, 82, 90
Plague, bubonic, 195
Planaria(n), 127, *128, 377*
    learning, 129
    regeneration, 129
Planigale, 358
Plankton, 64, 422
Plankotrophic, 239
Plants:
    compared with animals, 4
    flowering, 194
    in biogeochemical cycles, *428,* 429
PLANTAE, *40,* 67
Plantigrade, 355, 359, 367

Planula, 108, *111,* 112, *113,* 114, *118,*
    119
Planuloid, 109
Plasma, blood, 272
Plasma membrane, *17,* 18, 403, 405,
    406, 412
Plasmagel, *73*
Plasmalemma, *72*
Plasmasol, *73*
Plasmodium, 63, *65,* 66
Plasmodium, 63, 75, *76,* 426
Plastron, 319, 321
Plate:
    bony, 280, 282, 287, 294, 300, 315,
        320, 327, 364
    skeletal, 238
    horny, 314, 319, 364
Platelet, 276, 385, 397
Platycten(o)id, 124, 125
PLATYHELMINTHES, 106, 125
    (*See also* Flatworm)
Platypus, *353,* 355, *356,* 411, 414
PLATYRRHINA, 360, 363
Pleistocene (Epoch), *42,* 56
PLEOCYEMATA, 215, 222
Pleopod, 220, 221, 222
Plerocercoid, 136
Plesiosaur, *42,* 322
PLESIOSAURIA, 320
Plethodontid, 307, 310, 312
Pleura, *347*
Pleurobrachia, 123, 124
Pleurocentrum, 303, 305, 306
PLEURODIRA, 320, 321
Pleuron, 228
Plexus:
    choroid, 265, *266*
    nerve, 107, 114, 124, 157, 165,
        184, 250
Pliocene (Epoch), *42*
Plover, 330, 336, 339
Plumage, 330, 331, 333
    inheritance of, *46*
    (*See also* Feather)
Plumularia, 108
Pluteus, 238
Pneumatophore, 112, 115, *116*
PODICIPEDIFORMES, 339
Podium, 238, *239,* 241, 242, 246, 247
Poecilid, 303
POGONOPHORA, 106, 151, *159,*
    160–161
Pogonopohoran, 150, 152, 160–161,
    429
Poikilothermal, 315
Poison claw, 195, *223*
Poison gland, *201,* 202, 205, 206
Poison spur, 355
Poisonous and toxic animals, 67, 109,
    117, 118, 181, 209, 245, 290, 293
    (*See also* Venomous animals)
Polar body, *49,* 50, *95, 97, 102,* 111

Polar capsule, 77
Polar filament, 68
Polarity, 11
  membrane, 405, *406*, 407
  molecular, 24–25
Polian vesicle, 238, *239*
Pollen, 431
Pollicipes, 211, 218
Pollination, 5, 194
Pollution, environmental, 5, 427
POLYCHAETA, 186
Polychaete, 178, 181, 182, 184, 185,
    186–188, 195, 217, 374
Polyclad, 127, 130
POLYCLADIDA, 127
Polydactyly, 93
Polygenic inheritance, 58–59
  operation of selection upon, *60*
Polygordius, 151
Polymer, 27
Polymorphism, 111, 435
POLYNOIDAE, 186
Polyp, 111, 112, *113*, 122, *123*, 200
  anthozoan and hydrozoan compared,
    119
Polypeptide, *35*, 327
Polyphyletic, 68, 69
POLYPLACOPHORA, *164*, 167, *168*
Polyploid(y), 58, 79, 101, *111*, 112,
    113, 114
Polypterus, 296, 300, *301*
Polysaccharide, 27, *28*, 373, 378, 379,
    380
PONGIDAE, 363
Pons, *343*
Poorwill, 333, *338*, 339
Population, 82, 423
  density, 427
  genetics, 59
  growth, 424–*425*
  human, 44, 424, 437, 438
  limiting factors, 423–427
    endogenous, 425, 426–427, 438
    exogenous, 425–426
Population dynamics, 7
Porcellio, 211, 221
Porcupine, 341, 360, *361*, 422
Porcupine fish, *301*
Pore, *17, 85*, 86, *87*
PORIFERA, 82
  (*See also* Sponge)
Pork, 136, 148
Pork tapeworm, 135, 136
Porocephalus, 159
Porocyte, 86, *87*, 88
Porphyrin, 182
Porpoise, 366
Portal system, 276
  hepatic, 276
  hypophyseal, 276
  renal, 276
Portuguese man-of-war, 109, 112, *116*

Postabdomen, 198, 201, 204
Postcardinal vein (*see* Cardinal vein)
Postcava, 105, *347, 348*
Postfrontal, *318*
Postorbital, *318*
Posterior, *10*, 11, 96
Postsynaptic cell, 408
  hyperpolarization of, 408
Potassium (K):
  ion, 27, 238, 285, 296, 351, 374,
    399, 405, *406*, 407
  radioactive, 37
Potassium-ion gate, *406*, 407
Potential:
  action, *406*, 407
  generator, *406*
  membrane, 405, *406*
  resting, 405, *406*, 407
Poterion, 85
Pouch:
  abdominal, 298, 340, 357
  cheek, 363, 365
  throat, 311, 339
Poultry (*see* Chicken)
Powder feather, 331
Prairie dog, 365, 435
Prawn, 221
Praying mantis (*see* Mantis)
Preabdomen, 198, 201, 204
Precambrian, 178
Precapillary sphincter, 385
Precava, 105, *275, 347, 348*
Precipitation, 419, *420*, 421, 422
Precocial, 335, 341, 365, 368, 370
Predation, protection against, 432
Predator pressure, 60, 425
Predator-prey relationships, 425,
    432
Predatory adaptations, 432
Pregnancy, 352, 354, 400
  blockage of, 427
Premolar, *346*, 358, 367, 368, 371,
    372, 379
Preoral ciliary organ, 250
Prepuce, 352
Pressure:
  blood, 285, 397, 398, 400, 401
  in deep-sea habitat, 423
  receptor for, 411
Prey, protective adaptations of, 432
Priapulid, 150, 154
PRIAPULIDA, 106, *153*, 154
Priapulus, 153
Primate, 270, 341, 354, *356*, 362–364,
    376, 410, 422, 435
PRIMATES, 360, 362
Primitive (in phylogeny), 64
Primitive gut, 96
  (*See also* Archenteron)
PRISTIOPHORIFORMES, 290
Pristis, 288, 290
PROBOSCIDEA, 360, 369

Proboscis, 137, *138, 141*, 143, 144,
    145, 159, 160, 181, 182, 186,
    190, 200, 235, *249*, 250, 374
PROCELLARIIFORMES, 339
Procercoid, 136
Procoracoid, 355
Proctodeum, 196
PROCYONIDAE, 367
Producer, 429, *430*
  (*See also* Autotroph)
Proenzyme, 385
Progesterone, 30, 31, 279, *353*, 354,
    399, 400
Proglottid, 133, *134, 135*
Prokaryote (prokaryotic), 37, 38, 39,
    40, 84
Prolactin, 307, 340, 399, *401*
Prolactin-inhibiting hormone (PIH),
    399, 401, 402
Prolactin-releasing hormone (PRH),
    399
Proloculum, 74
Pronephros, 277, *278*, 284, 285
Pronghorn, 360, 372
Prophase, *48*, 49, 50
Proprioceptor, 207, 271, 412
Propterygium, 298
Prorodon, 80
Prosencephalon (*see* Forebrain)
PROSIMII, 360, 362
Prosobranch, *175*
PROSOBRANCHIA, *170, 175*, 176
Prosoma, 201
Prosopyle, 86, *87*, 90
Prostacyclin, 397
Prostaglandin, 397, 400, 403
Prostate gland, 147, *278, 347, 352*,
    397
Prostatic vesicle, 130
Prosthetic group, 373
Prostomium, *178*, 179, 181, *183*, 184,
    187
PROTACANTHOPTERYGII, 303
Protandrous, 119, 121, 153, 165, 172,
    248, 297
Protapsis, 198
Protease, 327, 379, *380*
Protective resemblance (*see*
  Camouflage; Crypsis; Mimicry)
Protein, 3, 27, 31, 53
  anionic, 405, *406*
  digestion, 378, 379–380
  synthesis, 19, 33, *35*–36, 45, 58,
    93, 374, 394, 409, 410
  unwinding, 33, *34*
Protein kinase, 403
Proteolytic enzyme, 379
Proteroblastus, 240, 241
Proterospongia, 69, 82
Proterozoic Era, *42*
Prothoracic gland, *231*
Prothoracicotropic hormone, *231*

Prothorax, 228
Prothrombase, 386
Prothrombin, 375
  (*See also* Prothrombase)
Protist, 57, 62–81, 82, 84, 374
PROTISTA, *40, 67*
Protoavis, 328, 329, 337
Protobranch, 176
Protocerebrum, *223*
Protocoel, 152, *249*
Protogyny, 297
Protomesosome, *159*
Protomollusk, *166*
Proton, 21, *22,* 389, *390, 393*
Proton pump, mitochondrial, *390, 393*
Protonephridium, 126, 129, 137, 139,
      142, 143, 151, 153, 154, 184,
      188, 257, 386, *387*
  types of cap cells, 140
Protonymphon, 200
Protoplasm (*see* Cell; Cytoplasm)
Protopodite, 211, *212*
PROTOROSAURIA, 320
Protopterus, 299, *300*
Protosome, 161
Protostomes, 94, 96, 98, 99, 102, 150,
      151, 152, 178
  compared with deuterostomes, 129,
      152, 236
PROTOSTOMIA, 106
Protostyle, *166*
PROTOTHERIA, *259,* 355
Prototroch, *178*
PROTOZOA, *9,* 68
Protozoan, 40, *42,* 230
  general characteristics, 62–64, *377,*
      382
  phylogeny of, 68, 69, 82
  types of, 65–81
Protozoea, 105, *213,* 214
Protozoology, 7
Proventriculus, 230, 333, *334,* 377
Proximal, 11
PSELAPHOGNATHA, 226
Pseudepipodite, 216
Pseudocoel, 98, *99,* 106, 140, 142,
      145, 150
PSEUDOCOELOMATA, 106
Pseudocoelomate, 98, *99,* 140–149,
      248
Pseudoplasmodium, *65*
Pseudopod(ium), 12, *17,* 88, 94
  in protozoans, 62, 66, 67, 68, 72,
      73, 74, 75
Pseudoscorpion, 197, *201,* 202, 205
PSEUDOSCORPIONIDA, *201,* 205
Pseudotrachea, 221
Psilocin, *409*
PSITTACIFORMES, 339
Psychology, animal, 7
Pteranodon, 319, *324*
Pterichthyodes, 286

PTERIOMORPHA, 176
Pterobranch, *249,* 250–251
PTEROBRANCHIA, 250
Pterodactyl, *56,* 57
Pteropod, *170,* 176
Pterosaur, 12, 322, 323, *324*
PTEROSAURIA, *259,* 320, *324*
Pterygiophore, *289, 295,* 298, 300
Pterygoideus muscle, 316, 317
PTERYGOTA, 235
Ptyalin, 378
  (*See also* Amylase)
Ptychocyst, 109, 120
Puberty, 103
Pubic symphysis, 305, 320, 323, 324,
      329, 345, 400
Pubis, *263,* 264, 303, *323,* 345
Puffs, chromosomal, 101
Puffer, 303
Puffin, 339
Pulmocutaneous artery, 275, 310
Pulmonary arch (*see* Aortic arch VI)
Pulmonary artery, 105, *274,* 275, *308,*
      347, 348
Pulmonary circuit, 384
Pulmonary surfactant, 381
Pulmonary trunk, *274,* 316, 349
Pulmonary vein, *201, 308, 347, 348*
PULMONATA, *175,* 176
Pulp cavity, *292, 346*
Puma, 424
Pump, ion exchange, 405, *406,* 407
Punctuated equilibria, 54, 57
Punnett square, *51, 52, 53*
Pupa, 231, *232, 234,* 235
Pupation, 230
Pupil, *269,* 402, 412, *413*
Purine, 32, *34,* 38, 200, 207
Purkinje cell, *14*
Pycnogonid, 198, 199–*200*
PYCNOGONIDA, 197
Pygidium, 178, 179, *197,* 198, *223*
Pygostyle, 331, *332*
Pyloric cecum, *295*
Pyloric sphincter, 292, 296
Pyramid, trophic, 429–430, *431*
Pyridoxine, 376
Pyrimidine, 32, *34,* 38
Pyrrophyte, 63, 66
Pyruvic acid, *390,* 392, 395
Python, 261, 328

Q fever, 195, 209
Quadrate, 325, 342, 345
Quadratojugal, *318*
Quail, 339, 435
Quantitative inheritance, 58–59
Quantum level, 22–23
Quaternary (Period), *42,* 258
Queen termite, 104, 228, *233,* 435
Quetzal, 339
Quill, 260, 262, 329, *330,* 341

Quinones, 164, 225
Quokka, 359

Rabbit, 341, 352, 353, 354, 360, 364,
      365, 370
Rabies, 362
Raccoon, 360, 367, 417, 422
Racer, 328
Rachis, 329, 330, 331
Radar, 413, 418
Radial canal, *85,* 86, *87,* 90, 112, *113,*
      117, 238, *239,* 242, *244,* 247
Radial cleavage (*see* Cleavage, radial)
Radial symmetry, *10,* 86, 107
Radiant energy, 412
RADIATA, 106, 107–124
Radiate, 97, 126
  advances over sponges, 107
  characteristics common to, 107, 377
Radiative adaptation, 265, 337, *338*
Radioactive dating methods, 37
Radioactive decay, 21, 37–38
Radioactive isotope, 21–22, 37–38
RADIOLARIA, *73*
Radiolarian, 64, 72, 75, 374, 422
Radiole, 181, 182, 186, 188, 381
Radioreceptor, 412
Radioulna, *311*
Radius, *56, 263,* 264, 265, 298, 299,
      303, *330, 332*
Radula, *164–165, 166,* 167, *168, 169,*
      170, 173, *174,* 176, 413
Radula sac, 164, 165
Rail, 339
Rain forest, *420,* 422
RAJIFORMES, 290
  (*See also* BATOIDEI)
Rana, 102, *306*
Range fractionation, 414
RANIDAE, 311
Ranvier, node of, *14*
Raphe nucleus, 267
Ratfish, *288,* 293
Ratite, 331, 335, 337
Rat, *252,* 354, 365, 398, 417, 426,
      427
Ratio, phenotypic, *51, 52, 53,* 59
Rattle, 260
Rattlesnake, 261, *326,* 328, 371
Ray, 271, 287, *288*
  (*See also* Batoid)
Reabsorption, tubular, 351
Reaction, chemical:
  amination, *395*
  biosynthetic, 394, *395*
  deamination, 393
  decarboxylation, *390, 391,* 392
  endergonic, 388
  exergonic, 388, 389–394
  glycolytic, *390,* 392
  hydration, 389, *390, 391*
  hydrolytic, 378, 379, *380*

Reaction, chemical (*continued*)
  oxidative, *390, 399*
  phosphorylative (*see*
      Phosphorylation)
  transamination, 394, *395*
Realm, 421
Rear-fanged snake, 328
Reasoning, 366
Recapitulation in development, 89,
    104–105, 178
Reception, sensory, 410–414
Receptor, *183,* 405, 410, 414
  -effector links, 398, 400–401
  epithelial, 268, 272
  neuroepithelial, 268, 272
  types, 411–414
Receptor field, 270
Receptor site (in membrane), 272, 398,
    403, 408, 409, 412
Recessive, 3, 45, 46, 47, *51, 52*
Recombination, genetic, 51, 52
Rectal gland, 230, 291, *292*
Rectrices, 331
Rectum, *141,* 147, *153, 168, 171,* 172,
    173, *189, 190,* 196, *201, 229,*
    230, 245, *278,* 333, *334, 347,*
    352, 353, 377
Red blood cell, *9, 77*
  (*See also* Erythrocyte)
Red water fever, 75
Redia, *131,* 132
Reducer, 429
Reductional division, 50
Redwing blackbird, 415
Redwood forest, 420
Reef, coral, 90, 122–*123,* 163
Reflex:
  arc, 340, *404,* 405
  conditioned, 417
  crossed-extension, *404*
  dorsal light, 228
  heartbeat-regulating, 385
  knee jerk, *404*
  milk-ejection, 340, 399, 402
  pain-withdrawal, *404*
  righting, 161, 243
  rooting, 340
  shadow-withdrawal, 184
  suckling, 340
Refractory period, 407
Regadrella, 85, 90
Regeneration, 84, 129, 153, 160, 182,
    184, 213, 240, 247, 327, 383,
    398
Regulation:
  cytoplasmic factors in, 101–102
  determinate, 102, 236
    (*See also* Cleavage, determinate)
  developmental, 100–104
  indeterminate, 102, 236
  intercellular factors in, 103
  interspecific influences upon, 104

Regulation (*continued*)
  intraspecific influences upon, 103–104
  mosaic, 102
  of gene action, 100, 403, 409
Regulator gene, 100
Regulatory factors, 101
Reindeer, 372
Reinforcement, 417
Relapsing fever, 209
Relaxin, 400
Release factor (releasing hormone), 401
Releaser, 415
Remiges, 330, 331, 337
Remora, 303, 433
Renal artery, 276
Renal epithelium, 151
  (*See also* Renal tubule)
Renal portal system/vein, *275,* 282,
    316
Renal tubule, 129, 199, 202, 210, 276,
    296, 316, 386, *387,* 388
Renette cell, 147
Renilla, 120
Renin, 397, 401
Rennin, 378
Renopericardial duct (opening), *171,*
    173
Repression, gene, 100
Repressor protein, 100, 101
Reproduction, 3, 48, 63, 67, 79, 84,
    85, 88–89, 130, 132, 359, 398,
    400, 402, 423
  (*See also* Budding; Fission;
      Reproductive system)
Reproductive cycle, 352, *353,* 399, 402
Reproductive success, 60, 426, 427,
    434, 435
Reproductive system, 13
  annelid, 185
  arthropod, 200, 203, 206, 207, 209,
      214, 224, 225, 230
  flatworm, 126, *128, 131*
  mammalian, 352–354
  molluscan, 165, 168, 172, 175
  nemertean, 139
  onychophoran, 192
  pseudocoelomate, *141,* 142, 143,
      145, 147
  vertebrate, 277–280
Reptantian, 222
Reptile, 7, 41, *42,* 161, 272, 280,
    312–328
  advances over amphibians, 312–318
  brain, 265, *266,* 315
  classification, 320
  development, 95, 96
  dietary innovations, 315–316
  ear, 315
  excretory system, 316
  evolutionary history, 57, 294, 305,
      319
  heart, 316

Reptile (*continued*)
  mammal-like, 317, 320, 322, 340, *343*
  reproduction, 316–317
  skeleton, *317–318, 323*
  skin, 314
  thermoregulation, 315
REPTILIA, 57, *252, 259,* 260
Repugnatory gland, 225
Requiem shark, *288,* 290
Reservoir, 67, 68
Residual air, 381
Resistance to insecticides, 6, 76
Respiration, 3, 373, 381, *428*
  cellular, *390,* 429 (*see* Electron
      transport)
  external, 381
  gill, 381, *382*
  integumentary, 381, *382*
  internal, *381*
  percutaneous, 273, 308–309
  pulmonary (*see* Lung; Book lung)
  tracheal, 381, *382*
Respiratory membrane, 381
Respiratory pigment, 154, 165, 182, 383
  (*See also* Chlorocruorin;
      Hemerythrin; Hemocyanin;
      Hemoglobin)
Respiratory pore, *170*
  (*See also* Spiracle)
Respiratory system, 13, 381
  amphibian, 306, *307, 308, 382*
  arthropod, 191, 196, 198–199, 202,
      207, *229,* 230, *382*
  bird, *334,* 335
  echinoderm, *239, 244,* 245
  fish, 296, *297*
    (*See also* Fish, respiration)
  human, *347*
  mammalian, *347–348, 382*
  molluscan, *166,* 167, *168,* 169, *170,*
      171, 172, 173, *174, 382*
  onychophoran, 191
  reptile, *314*
  vertebrate, 273
Respiratory tree, 239, 243, *244*
Response, 397, 398, 400–401, 415–
    418
Resting potential, 405, *406,* 407
Restlessness, premigratory, 336
Rete mirabile, 293
Reticular activating system, 267
Reticular formation, *266,* 267
Reticulopod, 72, 74
Reticulum:
  endoplasmic, *17,* 19
  of ruminants, *371*
  sarcoplasmic, *396*
Retina, 103, *269,* 412, *413,* 414
Retinal, 412
Retinula, *413*
Retractor (*see* Muscle, retractor)
Retrograde evolution, 91

Rhabdias, 146
Rhabdiform organelle, 134
Rhabdite, 125, 127, *128,* 134, 138
Rhabditogen cell, 125
Rhabdocoel, 127, 133
Rhabdom, *413*
Rhabdopleura, 249
Rhagonoid (*see* Leuconoid)
Rhamphorhynchus, 324
Rhampotheca, 260, 319, 333, 337, 339
Rhea, 337
RHEIFORMES, 337
Rhenanid, 287
RHENANIDA, 287
Rheoreceptor, 126, 412
Rheotaxis, 422
Rhincodon, 288
Rhinobatos, 288, 290
Rhinoceros, 341, *356,* 360, 370
RHINOCEROTIDAE, 370
RHIPIDISTIA (*see* Rhipidistian)
Rhipidistian, 298–*299,* 302, 303, 305, 316
Rhizocephalan, *219,* 220
RHIZOMASTIGIDA, 72
Rhizomastigidan, 69, 70
RHIZOPOD(E)A, 68
Rhizopodean, 74
RHIZOSTOMAE, 118
Rhizostomid, 118–119
Rhodopsin, 270, 412
Rhombencephalon (*see* Hindbrain)
RHOMBIFERA, 240, 241
Rhombopore, 241
Rhopalium, 110, 116, *117, 118*
Rhopalura, 91
RHYNCHOBDELLIDA, 186, 190
RHYNCHOCEPHALIA, *259,* 320, 325, *326*
Rhynchocoel, 137, *138*
RHYNCHOCOELA, 106, 125, 137, *138,* 151
Rhythm, endogenous:
  circadian, 337
  circennial, 336
Rib, *263,* 264, *295,* 303, *317,* 327, 331, *332, 334*
  sacral, 305
Ribbonworm, 137–139
Riboflavin, 375, 389
Ribonuclease, 378, 379
Ribose, 27, *29, 32, 34*
Ribonucleic acid (*see* RNA)
Ribose nucleotide, *32,* 33
Ribosome, *17, 19,* 33, *35,* 394
Rickets, 375
Rift zone ecosystem, 160, 429
Right whale, 367
Righting response, 161, 243
Ring canal, *111,* 238, *239,* 242, *244*
Ring nerve, 239
Ringtail, 367

Risorius muscle, *342*
RNA, 3, 32, 33, *34,* 35, 374, 377, 378
  and learning, 129, 409, 410, 415
  heterogeneous nuclear, 33
  messenger, 33, *35,* 36, 45, 93, 101, 103, 394
  ribosomal, 33
  synthesis of, 33, *34,* 403, 409, 410
  transfer, 33, *35,* 394
RNA polymerase, 33, *34*
Roach (*see* Cockroach)
Roadrunner, 339
Robin, *252*
  English, 415
Rock shell, *170*
Rock slater, *211*
Rocks, age of, 37
Rod, 71, 267, *269,* 270, 410, 411, 412, *413,* 414
Rodent, *356*
  characteristics, 353, 365
  compared with lagomorph, 364–365
  dentition, *346*
RODENTIA, *346,* 360, 365
Roller, 339
Rorqual, 367
Romeriid, 316, 317, 320
Root (of tooth), *346*
Rooting reflex, 340
Rostrum, 222, *286,* 290, 300, 366
Rotalia, 73, 74
Rotation of limbs, 317
Rotifer, 104, 140, *141,* 142–143, 150, 154, 155
ROTIFERA, 106, 142
Round ligament, *353,* 354
Round membrane, 344
Roundworm, 144–149
Royal purple, 163, 374
Ruffini's end organ, 411
Rumen, 80, *371,* 372
Ruminant, 346, *356,* 371, 372, 377, 378
RUMINANTIA, 360, 372

Sabellaria, 187
Sabellariid, 188
Sabellid, 188
Saber-toothed cat (tiger), 357, 372
Saccoglossus, 249
Sacculina, 219
Sacculus, 271, 344
Sacral rib, 305, 317
Sacral plexus, 323
Sacral vertebra, *263,* 264, 305, *311, 317,* 345
Sacroiliac articulation, 305, 331, 345
Sacrum, 264, 311, 317, 331, 345
Sagitta, 249
Sagittal plane, *10*

Salamander, *252,* 256, 264, *265,* 269, 302, 305, *306, 308,* 309, *343, 377, 382*
  characteristics, 310
  neoteny in, 106, 310
  types, 307, 310
Salamandra, 310
SALIENTIA, 306
  (*See also* ANURA)
Salinity, 423
Saliva, 173, 209, 230, 359, 378
Salivary gland (*see* Gland, salivary)
Salivation reflex, 417
Salmon, *56,* 85, *252,* 296, 303, 409–410
Salp, 254, *255,* 422
Salt, 26, 27
  regulation of, 210, 291, 296, 335, 339, *351,* 400, 401
  (*See also* Ion)
Salt gland, 291, *292,* 335, 339
Salt marsh, 419
Saltatory conduction, 407
Salticid, 207
Sand dollar, *237,* 241, 245
Sand flea, 221
Sand tiger shark, 291
Sandpiper, *338,* 339
Saprobiotic, 62, 67
Saprophytic, 62, 430
Saprozoic, 62
SARCODINA, 68
Sarcodine, 70, 72, *73*
Sarcolemma, *396*
SARCOMASTIGOPHORA, 67, 68
Sarcomere, 394, 395, *396*
Sarcoplasmic reticulum, *396*
SARCOPTERYGII, *258,* 298
Sarcoptes, 207, 208
SAURIA, 320, 326
SAURISCHIA, *259,* 320, *323*
Saurischian, *317,* 323
Sauropod, 323
SAUROPODOMORPHA, 320
SAUROPTERYGIA, 320
Savanna, 421
Saw shark, 290, 291
Sawfish, 281, 290
Scabies, 207
Scala media, *344*
  (*See also* Cochlear duct)
Scala tympani, *344*
  (*See also* Tympanic canal)
Scala vestibuli, *344*
  (*See also* Vestibular canal)
Scale:
  dermal, 308, 310
  epidermal, 260, 314, 326, 327, 329, 360
  fish, 262, 287, 291, *292,* 294, *295,* 300, 301, 302, 303, 304
  molluscan, 166, 167

Scale (*continued*)
  reptilian, 260
Scale insect, 6, 195
Scale worm, 160, 186
Scallop, 170, 176
Scaphopod, 162, 163, 168–169
SCAPHOPODA, *164*, 168, *169*
Scapula, *263*, 264, 294, *299*, *332*, 355
Scapulocoracoid, *289*
Scarab, 194
Scavenger, 194, 339, 429
Sceloporus, *321*
Schistic speciation, 54, *55*
Schistosoma, 132, 133, 164
Schistosomiasis, 133
Schizammnia, *73*
Schizocoel(ous), 98, *99*, 106, 152, 236
Schizocoelomate, 98, *99*
Schizogony, 75, 76, *77*
Schizont, 74, *77*
Schwann cell, *14*, 407
Science:
  basic commitment of, 1
  limitations of, 39, 438
Scientific methods, 1
Scientific names, 57
SCINCIDAE, 327
SCIURIDAE, 365
Sclera, *269*, 270, 332
SCLERACTINIA, 122
Sclerite, 228
Scleroblast, *87*, 88
Scleroid bones, 270
Scleroseptum, 122, *123*
Sclerosponge, 90
SCLEROSPONGIAE, 90
Sclerotic bones, 332
Sclerotic cartilage, *281*
Sclerotic coat (*see* Sclera)
Sclerotic ring, 332
Sclerotome, 97, *98*
Scolex, 133, *134*, 179
Scolopendra, 224
Scolopendromorph, *223*
SCOLOPENDROMORPHA, 224
Scomber, *301*
Scorpion, 6, *42*, 193, 194, 195, 197,
    198, 200, *201*, 202, *203*
  pseudo-, 197, *201*, 202, 204
  water, 197, 198, *199*, 261, 281,
      282
  whip, 197, *201*, 204
SCOPIONIDA, *201*, 204
Scorpionfish, 303
Scotopic vision, 270, 410, 411
Screwworm, 6–7, 194
Scrotum, 279, *352*, 355, 365
Scurvy, 376
Scute, 327
Scutigera, 224
Scutigerella, *223*, 225
SCUTIGEROMORPHA, *223*

Scypha, *85*, 86, *87*, 89
Scyphistoma, 112, 117, *118*
Scyphomedusa, 112, 113
  compared with hydromedusa, 117
SCYPHOZOA, *108*, 113, 117
Scyphozoan, 110, *118*–119
Sea anemone, *10*, 64, 67, 112, *121*–
    122
Sea butterfly, 176
Sea cow, Steller's, 370
Sea cucumber, *237*, 241, 243–245
Sea fan, *120*
Sea gooseberry, 12 (*see Pleurobrachia*)
Sea gull (*see* Gull)
Sea hare, *170*
Sea horse, 298, *301*, 303
Sea krait, 328
Sea lily, *237*, 242
Sea lion, 368, 369
Sea mouse, 186
Sea nettle, 109, 118
Sea otter, 245, 368
Sea pansy, 120
Sea pen, 120
Sea porcupine, 238, 241, 245
Sea slug (*see* Nudibranch; Onchid; Sea
    hare)
Sea snail, violet, 176
Sea snake, 327, 328
Sea spider, 197, 199–200
Sea squirt, 253, 254
Sea star, *42*
  (*See also* Starfish)
Sea turtle, *56*, 321, 433
Sea urchin, 94, 219, *237*, 239, 241,
    244, 245–246
Sea wasp, 109, 117
Sea whip, 120
Seafloor spreading zone, 42
Seal, *56*, 360, 368, 369, *430*
Sebaceous gland, 18, 341
Secondary sexual characteristics, 399,
    400
Secretin, 400
Secretion, 18, 418, 432
  tubular, 351
Secretory phase, *353*
SEDENTARIA, 186, *187*
Segmental ganglion (*see* Ganglion,
    segmental)
Segmental muscle (*see* Muscle,
    segmental)
Segmentation (*see* Metamerism)
Segmented worm, 41, 178–190
  (*See also* Annelid)
Segregation, law of, 51, 52
Selachian, *288*, 289–290
Selection:
  directional, *60*
  disruptive, *60*–61
  natural, 60, 423
  sexual, 60

Selection (*continued*)
  stabilizing, *60*, 430
Selection pressure, 60
Selective factors, 60
Self-toxification, 427
SEMAEOSTOMAE, 118
Semaeostomid, 118
Semen, 201, 203, 205, 207, 209, 226,
    292, 311
Semicircular canals, *271*, 272, 282,
    284, 285, 287, *344*, 411, 412
Semiplume, 331
Seminal receptacle, 132, *134*, 141,
    170, *183*, 185, *189*, 192, *201*,
    203, 209, 219, 220, 226, 229
Seminal vesicle, 130, 147, *183*, 185,
    189, 192, 207, *223*, 231, *249*, 352
Seminiferous tubules, *352*, 397
Sensation, 414
Sense, 411
Sense organ, 405, 410, 412–413
  annelid, 179, 184
  arthropod, 199, 203, 207, 214, 222
  cnidarian, 107, 110, 116, *117*, *118*
  ctenophore, *123*, 124
  echinoderm, 239, 243, 248
  flatworm, 126, *128*, 129
  molluscan, 165, 169
  nematode, 146
  radiate, 107, 112, 116, *118*, *123*
  vertebrate, 261, 268, 284, 291, 293,
      411
  (*See also* Ear; Eye; Ocellus;
      Statocyst)
Sensilla, 228
Sensitive period, 415, 416
Sensory adaptation, 414
Sensory bristle, 228
Sensory cell, 13, 107, *110*
  (*See also* Receptor)
Sensory ganglion, *404*, 405
Sensory input, principles of, 414
Sensory nerve, 267, 268
Sensory neuron, 203, 268, *404*, 405,
    409, 410, 411
Sensory perception, 405, 414–415
Sensory pit, 284, 293
Sensory reception, 405, 410–414
Sensory root, *404*
Sensory worlds (of different species), 414
Sepia, 164, *174*
  (*See also* Cuttlefish)
Septibranch, 177
Septum, 249, 250
  annelid, 179, 181, *183*, 184
  atrial, 105, 106, 310
  cephalopod, 173, *174*, 177
  heart, *274*, 275
  interbranchial, 291, *297*
  oblique, 348
  radial, *121*
  sclero-, 122, *123*

Septum (*continued*)
  septibranch, 177
  transverse, *292, 314,* 315
  ventricular, 105, 106, 316, 333,
    348
Sequoia forest, 420
Sere, 419, 423
SERIATA, 127
Serosa, *15,* 150
Serotonin, 267, 385, *409*
Serpent star, 246
SERPENTES, 320, 327
Serpula, 187
Serpulid, 188
Sessile, 63
Seta, 146, 152, 157, 159, 160, 179,
  180, 181, 186, *187,* 188, 191,
  203, 209, 211, 216, 219, 220,
  221
Sewage, 132, 133, 245, 419, 427
Sex attractant, 147, 205, 207, 217,
  418, 435
Sex cell, 48, *49*
  (*See also* Gamete)
Sex chromosomes, 46, 47
Sex determination, 46, 47, 160,
  316
Sex drive, 399, 400
Sex hormones, *30,* 31, 103, 279, 397,
  399, 400, 402, 403
  effects on behavior, 403, 416
  effects on gene dominance, 46
Sex organ (*see* Ovary; Reproductive
  system; Testis)
Sex reversal, 277, 296, 311, 335
Sex-influenced traits, 46
Sex-limited traits, 46
Sex-linked genes, 47
Sexual dimorphism, 203
Sexual indifference, 277
Sexual reproduction, 48
  (*See also* Reproduction;
    Reproductive system)
Sexual selection, 60
Seymouria, 304
Seymouriamorph, 305
Shaft, 329, *330*
Shagreen, 291
Shark, *42, 56,* 252, 287, *288,* 294,
  298, 375, 433
  body organization, 291–*292*
  brain, *266*
  characteristics, 290–293
  circulatory system, *105,* 274
  classification, 96, 290
  ear, *271*
  eye, *269,* 270
  reproduction, 292
  senses, 261, 414
  skeleton, *286*
  skin, 291, *292*
  types, 290–291

Shark (*continued*)
  warm-blooded, 293
Sheep, 11, *131*–132, 372
Sheep liver fluke, *131,* 132–133
Shell, *218*
  brachiopod, *156,* 157
  egg-, *278,* 280, 312–313, 335
  mollusk, 54, *164, 165, 166,* 167,
    168, 169, 170, 171, 173, *174,*
    176, 177
  turtle, 319–320
  (*See also* Test)
Shell gland, *334*
Shell membranes, 313
Shipworm, 163, 176
Shorebird, *339*
Shrew, 12, 354, 355, 359, 360,
  361
  elephant, 359–360
  pouched, *356,* 358
  tree, 362
Shrike, 339
Shrimp, 193, 194, 197
  bean, *211, 215,* 217
  brine, 105, 215, 216
  caridid, 222
  clam, 215, *216*
  commensal, 90
  coral, 215
  decapod, 105, *211, 213,* 215,
    221
  fairy, *211, 214,* 215, 216
  ghost, 222
  mantis, *211,* 215, 220
  moustache, 215
  mud, 222
  opossum, *211,* 215, 221
  penaeid, 222
  skeleton, *211,* 221
  stenopodid, 222
  tadpole, *211,* 215, 216
  thallasinoid, 222
Siamang, 363
Sickle cell anemia, 58, 76
Sieve plate, 248
SIGALIONIDAE, 186
Sight (*see* Eye; Photoreception;
  Vision)
Sign stimulus, 415, 435
Siliceous (*see* Silica)
Silica, 67, 74, 75, 86, *87,* 88, 90
Silicon (Si), 374
Silk, 5, 194, 202, 206, 207, 224, 226,
  230
Silk gland, *201,* 205, 206, 226,
  235
Silkworm, 194
  American, *231*
Silurian Period, 41, *42,* 177, 200, *258,*
  281, 286, 287
Silverfish, 228, *232, 233,* 235
Sinoatrial node, 274, *348*

Sinus, blood, 169, 199, *212,* 214, 239,
  246, 253, 285, 354
  (*See also* Hemocoel)
Sinus gland, 213
Sinus venosus, 105, 106, *274, 275,*
  282, *292,* 310, 333, *348*
Siphon, 163, 170, *171,* 173, *174,* 176,
  177, *244,* 245, 254, 255
SIPHONAPTERA, *234,* 235
Siphonoglyph, 119, *120, 121*
SIPHONOPHORA, 115
Siphonophore, 113, 115–*116*
Siphuncle, *174*
SIPUNCULA, 106, 158, *159*
Sipunculid, 158, 178, 383
Sipunculus, 159
SIRENIA, 370
Sirenian, 12, 345, *356,* 370
Size:
  advantages of greater, 84
  comparisons of, *9*
  limitations on, 11–12
Skate, *288,* 290
Skeletal muscle (*see* Muscle)
Skeletal system, 13
Skeleton, 27, 374
  amphibian, 303–305, *311*
  appendicular, 262, 263, 264–*265,*
    298, *299*
  arthropod, 191, 193, 195, 196, 197,
    213, 287, 316
  axial, 262, 264
  bird, 331, *332*
  brachiopod, 157
  cnidarian, *111,* 119, *120,* 122, *123*
  echinoderm, 238
  endo-, 238, 253, 261, 262, 300
  exo- (*see* Exoskeleton)
  fish, *286*
  hydrostatic, 140, 373
  mammalian, *263,* 345
  reptilian, *317–318*
  rhipidistian, 298–299
  tetrapod, *263, 299,* 303–305
  vertebrate, 262–265
Skin, 13, *56*
  amphibian, *261, 275,* 308–309
  bird, 333
  mammalian, *261,* 341, 369, 411
  nemertean, 138
  shark, 291, *292*
  vertebrate, 260–262
  (*See also* Epidermis; Integument;
    Tegument)
Skink, 327
Skogsbergia, 216
Skull, 173, *174,* 294, *295,* 311, *318,*
  346
  (*See also* Skeleton)
Skunk, 341, *361,* 367, 432
Sleep, 267
Sleeping ring, 435

Sleeping sickness, 70, 195
Sliding filament theory, 73, 395
Slime gland, *190*
Slime mold, 62, 63, *65*, 74
Slime net, 182
Slime tube, *189*
Slipper shell, *170*
Slit sense organ, 203, 207
Sloth, 345, 360, 364, 372, 422
Slug, 162, 163, 169, 176
Small intestine, *266, 347, 371*
   (*See also* Duodenum; Ileum;
     Jejunum)
Smell (*see* Olfaction)
Smooth muscle, 13, *14*, 15, 124, 268,
   354
Snail, 162, 163, *164, 170, 175, 382,*
   413
   body organization, *170*
   characteristics, 169
   classification, 175–176
   development, 169
   parasitic host, *131*, 132
   speciation, 54
Snake, *42*, 272, 314, 315, 316, 317,
   320, 322, 325, *326*, 327, 328, 434
Snapping turtle, 321
Snipe, *338*
Snowshoe hare, 60, 424–425
Social behavior, 434–437
   (*See also* Behavior, breeding;
     Courtship)
Social fighting, 220, *326*, 341, 369,
   372
Social hierarchy, 341, 403, 426–427
Social hunting, 432, 435
Social insects, 196, 230, 427, 435
Society, criteria defining, 341
Sodium (Na), 23
   chloride, 23, 285
   hydroxide, 26
   ion, 23, *25*, 273, 296, 351, 374,
     397, 399, 400, 405, *406*, 407
Sodium/potassium ion-exchange pump,
   405, *406*, 407
Sodium-ion activation gate, *406*, 407
Sodium-ion inactivation gate, *406*, 407
Sodium-ion channel, 408
Sol, 72
Solar panel, 228
Solaster, 237
Solation, 73
Soldier, 235, 435
Sole, 303
Solemyid, 176
Solenocyte, 140, 142, 154, 184, 188, 257
Solenogaster, 167
SOLENOGASTRES, 167
SOLIFUGAE, 205
Solpugid, 197, 205
SOLPUGIDA, *201*, 205
Solvent, water as, *25*

SOMASTEROIDEA, 241
Somatic cells, 100
Somatic motor-sensory system, 267,
   268, 405
Somatocoel, 238
Somatopleure, 313
Somatostatin, 399, 400, 401
Somatotrophin, 401
Somatotrophin-inhibiting hormone (*see*
   GHIH)
Somatotrophin-releasing hormone (*see*
   GHRH)
Somite, *180*, 220
   (*See also* Metamere)
Song:
   bird, 402–403, 415, 416, 417
   whale, 366
Songbirds, 339, 422, 432, 437
   Australian, 337
Sound production, 418, 436
   amphibian, 302, 311, 436
   bird, 418, 436
   fish, 296
   insect, 436
   mammal, 366, 369, 370, 381, 418
Sound reception (*see* Ear; Lateral line
   system)
Sow bug, 194, *211*, 221
Sparrow, 1, 339, 415, 416
Speciation, 53, 59
   cladogenetic, 54, *55*
   in freshwater snails, 54
   in Hawaiian *Drosophila*, 226
   phyletic, 54, *55, 56*
   rate of, 54
   schistic, 54, *55*
Species, 54, 57
   criteria of, 53
   origin of, 53, 54–55, 57
   sibling, 54
Spectacle, 326, 327
Spectrograph, sound, 1
Sperm, *9*, 19, *49*, 50, *51, 52*, 68, 88,
   91, 94, *97*, 175, 185, 399
   unflagellated, 147, 214, 352
Sperm duct, 185, 188, 231, 277, *278*
   (*See also* Ductus deferens)
Sperm funnel, *189*
Sperm receptacle, 207
   (*See also* Seminal receptacle)
Sperm tubule, 231
Sperm whale, 173, 366
Spermaceti, 366
Spermatheca, 230, 231
Spermatid, *49*
Spermatocyte, *49*
Spermatogenesis, *49*, 375
Spermatogonium, 88, 100, 110
Spermatophore, 161, 175, 185, 192,
   201, 203, 204, 205, 209, 214,
   217, 221, 224, 225, 226, 231,
   249, 310

Spermatozoa (*see* Sperm)
SPHENISCIFORMES, 339
Sphenodon, 325, *326*
Sphincter, 119
   cardiac, 378
   precapillary, 385
   pyloric, 292, 378
Sphyrna, 288
Spicules, 84, 86, *87*, 89, 119, 120,
   167
Spider, *193*, 197, 200, *201*, 202, 206–
   207
Spider mite, *208*
Spinal accessory nerve (*see* Accessory
   nerve)
Spinal column, 264
   (*See also* Vertebral column)
Spinal cord, *256*, 264, 265, 282, 287,
   302, 345, *347, 404*, 405
Spinal nerve, 267, 282, *404*
Spinal reflex arc, *404*, 405
Spindle, 48, 50, 374
Spindle fibers, 19, 20, 33
Spindle organ, 411
Spine, 104, 142, 143, 154, 218, *237*,
   238, *239*, 243, *244*, 245, 246,
   247, 249, 262, 286, 290, *292*,
   293, 301, 314, 357, 359
Spinneret, *201*, 206, 224
Spiny anteater, 357
Spiny lizard, *321*
Spiny lobster, *211*
Spiny-headed worm, *141*, 143
Spionid, 187
Spiracle, 191, 200, 202, 205, 206,
   223, 224, 225, 226, *227, 229*,
   230, *271, 288*, 289, 290, 291,
   292, 293, *382*
Spiral cleavage (*see* Cleavage, spiral)
Spiral fold, 284
Spiral valve, *292*, 296, 300, 310
Spirocyst, 109, 121
Spiroloculina, 73
Spirostomum, 80
Spirotrich, 81
Spirula, 164, 173
Spittle bug, *233*
Splanchnocranium, *318*
Splanchnopleure, 313
Spleen, *260, 266, 292, 295, 347*
Sponge, *42*, 68, 82, 84, *85*–90, 107–
   108
   bath, 86, *87*, 90
   boring, 90
   calcareous, 89
   characteristics, *85*–86
   embryonic development, *89*
   glass, 90, 374
   hard, 90
   microscopic organization, 86–88
   nutrition, 88
   reproduction in, 88–89

Sponge (*continued*)
  symbionts in, 90
Spongia, 85
Spongin, 86, *87*, 88, 90, 374
Spongioblast, 88
Spongocoel, *85*, 86, *87*, 90
Spoonworm, 104, *159*, 160
Sporangium, *65*, 66
Spore, *65*, 75, 76, 77
Sporocyst, *131*, 132
Sporogony, 76, 77
SPOROZOA, 68, 75
Sporozoan, 75–77
Sporozoite, 75, 76, 77
Spotted fever, 195, 209
Springtail, *233*, 235
Sprue, 376
SQUALIFORMES, 290
Squalus, *286*, 288
SQUAMATA, *259*, 320, 321, *326*
Squamosal, *318*, 320, 322
Squatina, *288*, 290
SQUATINIFORMES, 290
Squid, *9*, 162, *164*, 173, *174*, 175,
  177, 366, 422, *430*
Squilla, *211*, 220
Squirrel, *356*, 360, *361*, 365, 422, 431
Squirrelfish, 303
Stabilizing selection, *60*
Stalk, *218*, 240, 241, *249*, 251
  (*See also* Pedicle; Peduncle)
Stapes, *271*, 309, 342, *344*
Starch, 4, 28
  (*See also* Polysaccharide)
Starfish, 121, 163, 219, *237*, *239*, 241,
  244, 245, 246
Stargazer, 300
Starling, 329, 339, 424
Statoacoustic nerve (*see* Otic nerve)
Statoblast, 155
Statocyst, 107, 110, 112, 114, *123*,
  124, *125*, 126, 127, 128, 172,
  174, 185, 214, 221, 243, 254
Statolith, 221
STAUROMEDUSAE, 118
Steady state (*see* Homeostasis)
Stegosaur, 323
STEGOSAURIA, 320
Stegosaurus, 315, *319*, 323
STELLEROIDEA, 241, 246
Stenohaline, 423
STENOPODIDEA, 215, 222
Stenothermal, 423
Stentor, 79, *80*, 81
Stephalia, 115
Stereoblastula, 96
Stereogastrula, 108, 109, 125
Stereoisomer(ism), *24*
Sternebra, 264
Sternum, 213, *223*, *227*, 228, 264,
  317, 324, 327, 331, *332*, *334*,
  337, 339, *347*, 362

Steroid, *30*, 31, 231, 403
Stickleback, 298, 303
Stigma, 67, 68
Stilt, 339
Stimulus, 410, 414
  chemical, 411
  conditional, 417
  electromagnetic, 411
  mechanical, 411
  sign, 415, 435
  supernormal, 415, 436
  threshold, 406
  unconditional, 417
Stimulus generalization, 432
Sting, 195, 204, 290
Stinger, 235
Stinging capsule, 108
  (*See also* Nematocyst)
Stinging coral, 115
Stingray, *288*, 290
Stoat, 367
Stolon, 120, 152, *153*, 251, *255*
STOLONIFERA, 119
Stoloniferan, 120
Stomach, 107, 138, *141*, 153, 156,
  182, *190*, 196, *249*, *255*
  arthropod, 196, *201*, 206, *212*, 214,
    218
  echinoderm, 239, *244*, 247, 248
  molluscan, 165, *168*, *169*, *170*, *171*,
    172, *174*
  ruminant, *356*
  vertebrate, *260*, *266*, 292, *295*, *347*,
    360, 370, *371*, 372, *377*, 400,
    412
Stomatopod, 220
Stomodeum, *111*, 196, 239
Stomphia, 121
Stone canal, 238, *239*, 242, *244*
Stork, 337, 339
Stratification, 431
Stratum, sedimentary, 37, 38, 41, 312
Stratum germinativum, 260, *261*
Stratum corneum, 260, 262, 314, 326
Stretch receptor, 411, 412
Striated muscle, 13, *14*
  (*See also* Muscle)
Stridulation, 226
STRIGIFORMES, *338*, 339
Strobila, *118*, 133, 134
Strobilation, 112, 117, 130, 133, 134,
  135, 179
Stromalolite, 38, 39
Strongylocentrotus, *237*
Strontium, 75
STRUTHIONIFORMES, 337
Sturgeon, 262, 300, *301*
Stylaroides, *187*
Stylatula, *108*
Style, crystalline, 165, *172*
Style sac, 165, *166*, 169, *172*
Stylet, 137, *138*, 146, *153*, 190

Stylonychia, 81
STYLOPHORA, 240
Subclavian artery, 105, 275, 333, 348
Subesophageal ganglion, *229*, 230
Submucosa, *15*
Sublingual gland, 378
Submandibular gland, 378
Submission, 434
Submucosa, *15*
Submucous plexus, 379
Subpharyngeal gland, 253, *256*
Subphylum, 57
Subradula sac, 167
Subsong, 415
Substance P, 409
Substrate, 379, 388, 429
Subungulate, *356*
Succession, 419–420, 422, 423
Sucker, 130, *131*, 173, *174*, 177, 188,
  189, 218, *239*, 247, 307, 374
Sucking cone, 218
Suckling, 240, 357
Sucrase, 378, 379
Sucrose, 28, *29*
Suctorian, *80*, 81
Sugar, 25, 27, 28, *29*, 373
Sugar glider, 358
SUIDAE, 371
SUIFORMES, 360, 371
Sulfur (S), 374
  as H-acceptor for anaerobes, 394
Sulfur bacteria, 40, 429
Sun spider, *201*, 205
Sun star, 247
  (*See also* Solaster)
Sun-compass reaction, 337
Sunfish, 130, *301*, 303
Sunlight (*see* Light)
Suoid, *356*
Superficial, 11
Superior oblique muscle, 268
Supernormal stimulus, 415, 436
Superposition image, 213
Supportive tissues, 13, *14*
Suprabranchial chamber, *171*
Supracleithrum, *299*
Supracoracoideus muscle, 331, *332*
Supraoptic nucleus, *401*
Suprarenal gland (*see* Adrenal)
Supratemporal, *318*, 322
Surface-to-volume ratio, 11–12
Surfactant, 381
Surfperch, 296, 303
Surgeonfish, 303
Survival of the fittest, 60
Suture, 238, 262
Swallow, 339
Sweat gland, 18, 260, *261*, 262, 341,
  365
Swift, 330, 339
Swim bladder, 273, 291, *295*, 296,
  300, 301, 302, 303

Swimmer's itch, 133
Swimming:
  in batoids, 290
  in cetaceans, 365
  in crocodilians, 325
  in "ghost" paramecia, 71
  in medusae, 112
  in mollusks, 173, *174*, 176
  in nemerteans, 138
  in osteichthyans, 294, 302
  in polychaetes, 179
  in seals, 369
  in sharks, 290
  in sirenians, 370
Swine, 354
Swordfish, 303
Swordtail, 303
Sycon (see *Scypha*)
Sycon(oid), *85*, 86, 88, 89, 90
Symbiont, 64, *80*, 90, 104, 433
Symbiosis, 64
  cleaning, 433
  commensalistic, 64
  mutualistic, 64, 67, 122, 176
  parasitic (see Parasite; Parasitic
    animals; Parasitism)
Symbiotic relationships, 53, 64, 115,
  176, 433
Symmetry, *10*, 94, 106
  bilateral, *10*, 125
  biradial, 124
  hexamerous, 119, 121
  pentamerous, 238
  octamerous, 119
  radial, *10*, 86, 106, 107
Sympathetic nervous system, *266–268*,
  379, 385, 402, 405
Sympathetic ganglion, *266*, 268, 402
Sympatry (sympatric), 54, *55*, 419,
  430, 431
SYMPHYLA, 197, 224
Symphylan, 222, *223*, 224–225
Symphysis, pubic, 305, 323, 329
Synapse, 107, 388, *404*, 405, *408*
  axon-axonal, 409
  bidirectional, 107
  of homologous chromosomes (*see*
    Synapsis)
Synapsid, *318*, 322
SYNAPSIDA, *259*, 320, 322
Synapsis, *49*, 50, 59
Synaptic cleft, *408*
Synaptic convergence, *408*
Synaptic divergence, *408*
Synaptic fatigue, 409
Synaptic knob, *408*
Synaptic transmission, *408*, 409
Synaptic vesicle, *408*
Syncytial model, 83
Syncytium, 90
Synergism, 403
Syngamy, 63

Synovial membrane, 263
Synthesis:
  DNA, 33, *34*, *35*
  glycogen, 394, *395*
  protein, *35*, 36, 394
  RNA, 33, *34*
  template-, 394
Syrinx, *334*, 335, 418
Systematics, 7, 57
Systemic arch, 275, 333, *348*, 349
Systemic circuit, 384
Systemic (aortic) trunk, *274*, 275, 316

T cell, 109
T lymphocyte, 277
Tachyglossus, 357
Tactile communication, 435, 436
Tactile receptor, 261
Tadpole, 307, *308*, 310, 311, 315
"Tadpole" larva, 253, 254, *255*
Taenia, 135, 136
Taeniarhynchus, *135*, 136
Tagelus, *171*
Tagmatization, 196, 210
Taiga, 421
Tail:
  amphibian, 280, 310
  autotomy of, 432
  bird, 331
  cetacean, 365
  fish, *281*, 282, 290
    (*See also* Fin, caudal)
  post-anal, 248, 249, 253, *255*, *256*
  prehensile, *301*, 327, 363, 364
Tail bud, *307*
Tail fan, 60, 220, 222, 331
Talon, 329, 337, 339
Tanager, 339
Tanning, 146, 164, 195, 213, 260
  sun, 262
Tapeworm, 133–137, 179, 433
Tapir, 360, 370
TAPIRIDAE, 370
Tarantula, 207
TARDIGRADA, 106, *153*, 154
Tardigrade, 150, 154–155, 161
Tarpon, 303
Tarsal, *263*, 264, 298, 303, 311
Tarsier, 360, 362
Tarsometatarsus, *332*
Tarsus, 207, *227*
Tasmanian devil, 358
Tasmanian wolf, 57, *356*, 357, 358
Taste, 272, 410, 411
  genetic polymorphisms in humans,
    414
Taste bud, 268, 272, 300, 410, 411
TAYASUIDAE, 371
Taxis (*see* Chemotaxis; Phototaxis;
  Rheotaxis)
Taxodont, 176
Taxonomy (*see* Classification)

Tectorial membrane, *344*, 345, 412
Tectum, optic, *266*
Teeth, 186, *187*, 200, 249, 379
  acanthodian, 287
  bird, 329, 331, 337
  bony fish, 295
  carnassial, 367
  cladodont, 287
  deciduous, 347, 367, 368
  echinoderm, *244*, 245
  elasmobranch, 291
    (*See also* Shark)
  hagfish, 282, 285
  hinge, 157, 176, 177
  holocephalan, 293
  human, *346*
  labial, *307*
  labyrinthine, 298, *299*, 303
  lamprey, 282, *283*, 284
  mammalian, *346*, 347, 355, 379
  maxillary, 300, 322
  palatine, 298
  placoderm, 287
  premaxillary, 300, 322
  radular, *164*, 165, 167, 176, 413
  reptilian, 320, 323, 327
  shark, 287, 288, *289*
  vertebrate, 262
Tegument, 130, 134, 143
TEIIDAE, 327
Telencephalon, 265
Teleost, *301*, 302, 374, 386, 413
TELEOSTEI, *258*, 300, *301*, 302, 303
Telolecithal egg, 96
Telophase, *48*, 49
Telotroch, 178
Telson, 198, 199, 210, *212*, 216, 220,
  222, *223*, 224
Temnospondyl, *304*, 305
TEMNOSPONDYLI, 303
Temperature:
  body, 315, 333, 349
    control of, 12, 60, 315, 322, 333,
      349
  environmental, *420*, *421*, 423
Template synthesis, 394
Temporal bone, 345
Temporal fossa, 317–*318*, 320, 322
Temporal lobe, *343*
Temporalis muscle, 316, 317, *318*
Temporal notch, *318*, 319
Tendon, *14*, 15, 331, *332*, 411
Tentacle, 374, 381
  annelid, 179, *180*, 181, 182, 186,
    187, 188
  entoproct, 152, *153*
  holothuroid, *244*
    (*See also* Podium)
  lophophorate, 151, 155, *156*, 157,
    158
  molluscan, 163, *164*, 167, 169, *170*,
    173, *174*, 175, 176, 177

Tentacle (*continued*)
  pogonophoran, *159,* 161
  pterobranch, *249,* 250
  radiate, *108, 110, 111,* 119, *120, 123,* 124
  sipunculid, *159*
  vertebrate, *256, 283*
TENTACULATA, *123,* 124
Terebella, 187
Teredo, 163, *171*
Tergal plate, 225
Tergite, genital, *223*
Tergum, 213, *223, 227,* 228
Termite, 64, 71, 104, 194, 195, 228, 230, *233,* 235, 435
Tern, 336, 339
Terrapin, 321
Territoriality, 204, 321, 426
Territory, 341, 368, 415, 418, 426, 434
Tertiary (Period), *42*
Test:
  echinoderm, 238, 240, 245
  protozoan, 64, 72, *73,* 74, 75
  tunicate, *255*
Testacid, 74
Testis, 50
  invertebrate, *110,* 112, *125, 128, 131, 134, 156, 183,* 185, 188, *189,* 192, 207, 209, *218,* 219, 220, *223,* 225, 226, 231, *249*
  vertebrate, *278,* 279, *292,* 311, *334, 335, 347, 352,* 375, 397, 400
Testosterone, *30,* 31, 103, 277, 279, 400, 403
TESTUDINIDAE, 321
TESTUDINATA, *259,* 319, 320, 321
Tetrad, 50
Tetranychus, 208
Tetrapod, 264, 272, 299, 309, 310, 312, 342, 345
TETRAPODA, 260
Texas cattle fever, 208, 209
Thalamus, *266, 267,* 268, *336, 343*
THALIACEA, 254
Theca, 238, 240, 241
Thecodont, 323, 325
THECODONTIA, *259,* 320, 323
Theory, scientific, 1
Therapsid, 43, 315, *322,* 340, 355, 356
THERAPSIDA, *259,* 320
THERIA, *259,* 354, 357
Therian, 356, 357–372
Theriodont, 322, 340
Thermocline, 422
Thermodynamics, second law of, 38
Thermoreception, 328
Thermoreceptor, 184, 412
Thermoregulation, 12, 60, 315, 322, 333, 349

Theropod, 323, 324
THEROPODA, 320
Thiamine, 374, 375, 392
Thigmoreceptor, 126, 412
Thirst, 411
Thomasid, 207
THORACICA, 215
Thorax:
  arthropod, 196, *197,* 198, 210, 213, 220, 223, 225, *227,* 228
  mammalian, *347,* 348
Thrasher, 339
Threat behavior, 426, 432, 434
Thresher shark, *288,* 290, 291
Threshold:
  behavioral, 416
  neural, 406
  sensory, 414
Thrip, 194
Thrombase, 386
Thrombin (*see* Thrombase)
Thrombocyte, 276, 385
Thromboxane, 397
Thrush, 339
Thumb, 362, 367, 415
THYLACINIDAE, 358
Thylacoleo, 357
Thymine (T), 32, *34*
Thymosin, 400
Thymus, 400
Thyone, 237
Thyroid cartilage, 345
Thyroid gland, 253, 254, 256, *266,* 307, 397, 398, 399, 401, 402
Thyroid hormone, 103, 106, 307, 374, 398, 399, 402, 403
Thyroid hormone-binding protein, 398
Thyrotrophin, 399, 402
Thyrotrophin-releasing hormone, 399, 402
Thyroxin, 253, 399, 403
THYSANURA, *233,* 235
Tibia, *227, 263,* 264, 298, 303
Tibiofibula, 311
Tibiotarsus, *332*
Tick, 194, 197, *201,* 207, *208–209,* 433
  as disease vector, 75, 195
Tidal air volume, 381
Tiedemann body, *239*
Tiger, 6, 359, 368
Timberline, 421
TINAMIFORMES, 337
Tinamou, 337
Tinbergen, N., 2
Tissue, 7, 12, 13, *14,* 93, 107
Tit, 339, 416, 417
Toad, 72, 302, 306, *311*
  spadefoot, 309, 311
  Surinam, 307–308, 311
  tongueless, 311
Toadfish, 303, 381

Tocopherol, 375
Toe, *141,* 142
  (*See also* Digit)
Tolerance, environmental, 423
Tongue, 45, 58, *260,* 268, *283,* 284, 285, *295,* 303, 310, *326,* 333, *334,* 339, 358, 364, 369, 372, 374, 377
Tongue worm, *159,* 161
Tool, 363, 368
Tooth:
  median, 327
  structure, *346*
  (*See also* Teeth)
Tooth plate, 293, 299
Tooth replacement:
  diphyodont, 347
  elephant, 369
  fish, 287, 288, 291
  hemidiphyodont, 347, 358
  monophyodont, 358, 359
Top (shell), 176
Tornaria, 250
Torpedo, 288, 290
Torsion, *165,* 169, *175,* 176
Tortoise, 314, 321
Tortoiseshell, 321
Toucan, 339
Touch, 411, 412
Towhee, 339
Toxin, 67
  (*See also* Venom)
Toxocyst, 79
Toxoplasma, 76
Toxoplasmosis, 75, 76
Trachea:
  arthropod, 191, 196, 197, 200, *201,* 202, 205, 206, 207, 209, *223,* 224, 225, 226, *229,* 230, *381,* 382
  onychophoran, 191
  vertebrate, 273, 314, 315, *334,* 335, 381
Tracheal gills, 230, 235, *382*
Tracheole, 202, 228, *229,* 230, *382*
Trachydon, 319, 323
TRACHYLINA, 114
TRAGULIDAE, 372
Tragus, 362
Transamination, 376, 384, *385*
Translocation, 59
Transmission, genetic, 47
Transport:
  active, 18
  internal, 3, 373, 383–386
  membrane, *17*
  passive, 18
  transcellular, 386
Transverse plane, *10*
Transverse process, 264
Transverse tubule, *396*
Transversus muscle, *314,* 321

Tree line, *421*
Tree shrew, 360, 362
TREMATODA, 130
Trematode, 130–133
Triadobatrachus, 311
Triarthrus, 197
Triassic (Period), *42*, 43, 198, *258*, 305, 311, 322, 323, 324
Triceratops, 319, 323
Trichina worm (*see Trichinella*)
Trichinella, 148
Trichinosis, 148
Trichocyst, 66, 77, *78,* 79
Trichogramma, 104
Trichomonad, 69, 71
Trichomonas, 69, 71
Trichonympha, 64, *69,* 71
Trichoplax, 90
TRICHOPTERA, *234,* 235
Triclad, 127
Triconodont, 355
Tridacna, 176
Trigeminal nerve, 268
Triggerfish, 303
Triglyceride, *30,* 31, *380*
Trihalose, 230
Triiodothyronine, 399
Trillina, 73
TRILOBITOMORPHA, 197
Trilobite, *42,* 43
Trilobite larva (of *Limulus*), 199
Tripedalia, 116
Tripeptidase, 378, *380*
Triploblastic, 126
TRITUBERCULATA, *259*
Trituration, 196, 214, 379
Trochanter, *227*
Trochlear nerve, 268
Trochophore, 137, 153, 155, 158
  annelid, *151, 178,* 184, 186
  echiurid, 160
  entoproct, 150
  molluscan, *151,* 165, 168, 169, 172
  sipunculid, 158
TROGONIFORMES, 339
Trophic pyramid, 429–430, *431*
Trophic relationships, 432
Trophoblast, *95*
Trophozoite, 75, 76
Tropical rain forest, *420, 421,* 422
Tropomyosin, 396
Troponin, 396
Trout, 284, 303, 423
Truncus arteriosus, 106, *274,* 282, 310, 316
Trunk, 158, 159, 178, 179, 210, 223, 224, *249,* 250, 251, 369
  elephant, 360, 369, 370
Trunkfish, *301*
Trypanosoma, 69, 70
Trypanosome, 195
TRYPANOSOMATIDAE, *70*

Trypsin, 378, 379, *380*
Tsetse fly, 70, 195, *234*
TSH (thyroid-stimulating hormone, *see* Thyrotrophin)
Tuatara, 269, 320, 325, *326*
Tube foot, *237,* 238, *239,* 241, 243, *244,* 245
  (*See also* Podium)
Tube-mouthed fish, 303
Tubeworm (*see* SEDENTARIA)
Tubifex, 188
TUBIFICIDA, 186, 188
Tubularia, 108, 112, *113,* 115
TUBULIDENTATA, 360, 369
Tubulin, 19, 71
Tularemia, 209
Tun, 176
Tuna, 293, 303, 375, 430
Tundra, *420,* 421
Tunic, 253, 254, *255,* 256, 383
Tunicin, 253
TUNICATA (*see* UROCHORDATA)
Tunicate, 57, *252,* 253–256, *384*
  development, *102,* 106
  larva, 253, 254, *255*
Turban (shell), *170,* 176
Turbatrix, 147
TURBELLARIA, 126
Turbellarian, 126–130, 143, 151
  archoophoran, 126–127
  biological contributions (*see* Flatworms)
  body organization, 127, *128*
  classification, 126–127
  eggs, 126–127
  excretion, 129
  food-getting, 127–128
  neoophoran, 127
  nervous system, 129
  osmoregulation, 129
  reproduction, 130
  sense organs, 129
Turbinate bone, *272*
Turkey, 339, 416
Turret, 176
Turritellella, 73
Turtle, *42,* 260, 320, 325
  characteristics, 270, 314, 316, 318, 319–321, 349
  major families, 321–322
Tusk, 360, 369, 370, 371
Tusk shell, *164,* 168, *169*
Twinning, monozygotic, 102, 236
TYLOPODA, 360, 372
Tylosaurus, 319
Tympanal organ, 228
Tympanic canal, 344
  (*See also* Scala tympani)
Tympanic membrane, 228, *271,* 309, 311, 315
Tympanum, *227,* 344
Typhlosole, *172, 180, 183*

Typhus, Asian scrub, 207
Tyrannosaurus, 319, 324

Uca, 211, 222
Ulna, *56, 263,* 264, *265,* 298, 299, 303, *330, 332*
Ultraviolet, 262, 414
Ultrafiltration (*see* Filtration; Glomerulus)
Umbilical circulation, 354
Umbilical cord, 354
Uncinate process, 331, *332*
Undulating membrane, *69,* 70, 71
Ungulate (*see* Hoofed mammal)
Unguligrade, 355, 370
UNIRAMIA, 197, 210, 222
Uniramous mandibulate, 193, 222–235
Univalve (*see* Gastropod; Snail)
Upwelling, 422
Uracil (U), 32, *33*
Uranium, 37
Urate, 388
  (*See also* Uric acid)
Urea, 3, 4, 184, 291, 296, 298, 309, 314, 386, 388, 393
Urechis, 159, 160
Ureotelic, 388
  (*See also* Urea)
Ureter, 165, 277, *278,* 292, 295, 316, *334, 350, 352,* 387
Urethra, 277, *278,* 352, *353,* 355
  cavernous, 352
Uric acid, 230, 254, 326, 335
Uricotelic, 335, 388
Urinary duct, 277
  (*See also* Ureter)
Urinary bladder (*see* Bladder, urinary)
Urination, 412
Urine, 4, 184, 210
  production of, 200, 291, 296, 350–351, 387–388
Urn, 158, 383
UROCHORDATA, *252,* 253, 254
Urocyon, 57
URODELA, 306
  (*See also* CAUDATA)
Urogenital duct, *278*
Urogenital opening, *295*
Urogenital ridges, 97
Urogenital sinus, 282, *353,* 354, 357
Urogenital system, *278,* 352
  (*See also* Excretory system; Reproductive system)
Urolophus, 288, 290
Uropatagium, 362
Uropod, 220, 221, 222
UROPYGI, *201,* 204
Urostyle, *311*
URSIDAE, 57
Uterine mucosa, *353*

Uterus, 134, *145*, *191*, 192, *223*, *278*, 293, 352, *353*, 354, 367, 399, 400, 402
  types in mammals, *353*
UTP (uridine triphosphate), *395*
Utriculus, *271*, *344*, 411

Vacuole, *17*, 19, 403
  contractile, 19, 63, *73*, *78*, *80*, 386, *387*
  food, 63, *73*, *78*, 86, 125, 127, 374, *377*
Vagina, *128*, *134*, *170*, 185, 207, *229*, 230, 231, *249*, *353*, 354, 355, 357
Vaginal sinus, 357
Vagus nerve, 268, 385
Valence, 22
Valve:
  anal, 147
  bronchial, 366
  heart, 5, *274*, *348*
  intestinal ring, 290
  oral, *297*
  shell, *156*, 157, 164, 167, *168*, 170, 173, 176, *216*, 217
  spiral, *292*, 296, 300, 310
  stomodeal, 230
  vascular, 273
Vampire bat, 362, 433
Vampire squid, 177
Vampirolepis, 137
Vampyrella, 74
Van der Waals forces, 28
Vanadium, 254, 383
Vanadis, 184
Vanadocyte, 254, 383
Vane, *330*
VARANIDAE, 327
Variation, genetic:
  continuous, 58
  discontinuous, 58
Vas deferens, *134*, 185
  (*See also* Ductus deferens; Sperm duct)
Vasa efferentia (*see* Efferent ductules)
Vascular tissues, 13
Vasoconstriction, 401
Vasoconstrictor nerves, 385
Vasodilator nerves, 385
Vasomotor control, 385
Vasopressin, *401*
  (*See also* Antidiuretic hormone, ADH)
Vein:
  blood, 173, 273, *275*, 276, *283*, *308*, *384*
    (*See also* Cardinal vein; Hepatic portal; Precava, Postcava)
  wing, 228
Velarium, 116, *117*
Velella, 115, *116*, 176
Veliger, *165*, 168, 169, 172

Velum, *111*, *113*, 114, *165*, 167, *168*, 169, *255*, *256*, *283*
Velvet, 372
Vena cava (*see* Precava; Postcava)
Venom, 5, 163, 204, 206, 218, 223, 245, 309, 327, 355
Venom sac, 204
Venomous animals, 163, 173, 181, 195, 204, 206, 224, 309, 311, 327–328, 355, 359
  (*See also* Poisonous and toxic animals)
Ventral aspect, *10*
Ventral root, 282, *404*
Ventricle:
  brain, 99, 100, 253, 265, *336*, *343*
  heart:
    molluscan, 165, *166*, 167, *168*, *171*, 172, *175*
    vertebrate, *105*, 106, *260*, *274*, *275*, 282, *292*, 310, 333, *348*
Venule, 385
Venus, *171*
Venus's flower basket, 90
Venus's girdle, *123*, 124
Verdin, 339
Vermiform larva, *91*
Verongia, 90
Vertebra, 97
  amphibian, 303, 305, 306, 311
  aspondylous, 300
  bird, 331, *332*, 334
  caudal, *263*, 264, *295*, *317*, 331, *332*
  cervical, *263*, 264, *317*, 331, *332*, 345
  echinoderm, 247
  fish, 252, *260*, *263*, 264, 288, *292*, *295*, 300, 303
  heterocoelous, 331
  lumbar, *263*, 264, *317*
  mammalian, *263*, 345, *347*
  reptilian, *311*, 325
  sacral, *263*, 264, 305, *311*, *317*, 345
  spondylous, 301, 305
  thoracic, *263*, 264, *317*
Vertebral column, 257, 264, 284, 287, 291, 294, 302, *311*, 322
VERTEBRATA, 57, *252*, 257
Vertebrate, 7, 179, 198, 253, 256–372
  blood, 276–277
    (*See also* Blood)
  body organization, *260*–269
  chemical senses, 272
  circulatory system, 273–277, 288, 285, 397
  classification, 257–260
  cranial nerves, 266, *267*, 268
  digestive tract, 284, 285
  distinguishing characteristics, 260–279
  diversification, *258–259*

Vertebrate (*continued*)
  ear, *256*, 270–272, 284, 285
  embryo, *95*, *96–97*, *98*, 100–104, *105*, 106
  evolutionary tree, *258–259*
  excretory system, 277, *278*, 282, 309
  eye, *256*, 257, 268–270, 410
  liver, 272
  nervous system, 265–268, *404*
  origin, 254, 256
  reproductive system, 277–280
  skeleton, 262–265
  skin, 260–261
Vertex, *227*
Vesicle:
  eye, 412
  pinocytic, *18*
  otic, 271
  synaptic, *408*
Vestibular canal, 344
  (*See also* Scala vestibuli)
Vestibule, 249
  (*See also* Ear; Inner ear; Labyrinth)
Vestibulocochlear nerve (*see* Otic nerve)
Vestigial structures, 143, 328
Veterinary medicine, 7
Vibracula, 155
Vibrissae, 341, 368
Viceroy, 432
Vicuna, 360, 372
Villus:
  chorionic, 354
  intestinal, *15*, 148, 377
Vinegar eel, 144, 147
Vinegaroon, 204, 432
Violet sea snail, 176
Viper, 328, 412
VIPERIDAE, 328
Vireo, 339
Visceral arch (*see* Gill arch)
Visceral hump, 163
Visceral mesoderm, 98, *99*
Visceral muscle, 13, *14*
Visceral receptor, 268
Visceroceptor, 411
Vision, 412, *413*
  color, 58, 228, 270, 410, 414
  mosaic, 228, 412, *413*
  photopic, 270, 410, 411
  scotopic, 270, 410, 411
Visual pigment, 58, 412
Visual purple, 375
Visual signals, 436
  evolution of, 437
Vital capacity, 381
Vital dye, 100, *102*
Vitamins, 262, 272, 292, 352, 373, 374, 375–376, 377
Vitellaria, 242, 245
Vitelline blood vessels, 314

Vitelline membrane, 313
Vitreous body, *413*
Vitreous humor, *269*
Viverra, 368
VIVERRIDAE, 57, 368
Viviparity, 201, 298
Viviparous, 147, 192, 204, *233,* 303,
   307, 310, 311, 313, 328, 357
Vocal cord, *308,* 311
Vocal fold, 381, 418
Vocal pouch, 307
Voicebox, 273
  (*See also* Larynx)
Volchovia, 240, 246
Vole, 346, 365
Volplaning, 327
Voluntary muscle (*see* Muscle)
Volvent, *108,* 109
VOLVOCALES, 67
Volvox, 66, 67–68
Vomeronasal organ, 272, 327
Vorticella, 79, 81
Vulpes, 57
Vulture, 333, 337, 339
Vulva, 226
Vulvulina, 73

Walking stick, 235
Walking worm, 178, 190
Wallaby, 359
Walrus, 360, 369
Wapiti, 372, 422
Warbler, wood, 339, 424
Warm-bloodedness, 293, 315
Warning coloration, 432
Warthog, 371
Wasp, 2, 104, 194, 230, 235
Wastes:
  anthropogenic, 5, 427, 437
  metabolic, 4, 383, 386
Water, 27
  biotic uses, 373, 389, *390, 391*
  cyclic movement, 419
  metabolic, 393
  percutaneous absorption, 309
  properties, *25, 26*
  renal reabsorption, 309, 399, 401
  solvent action, *25*
Water balance, 199
  (*See also* Osmoregulation)
Water bear, *153,* 154–155, 190
Water boatman, *233*
Water flea, 215, 216
Water mite, 207

Water moccasin, 328
Water ring, 238, 239, 242
  (*See also* Ring canal)
Water scorpion, 197, 198, *199,* 261,
   281, 282
Water shrew, 359
Water snake, *321,* 328
Water-vascular system, 238, *239,* 242
Wax, 195, 200, 228, 366
Weasel, 56, 360, *361,* 367
Web:
  silk, 206, 207, 224, 226
  skin, 262, 311, 315, 339, 355, 360,
   362, 368, 369
Weberian ossicles, 303
Weevil, *234*
Whale, *9, 42,* 173, 210, 217, 218,
   221, *265, 354, 356,* 360, 365–
   367, 430, *431,* 433
Whale louse, 221
Whale shark, *9, 288,* 290, 291
Whalebone, 260, 341
  (*See also* Baleen)
Whelk, *170,* 176
Whip scorpion, 197, *201,* 204
Whiptail, 327
White ant (*see* Termite)
White corpuscle, 13
  (*See also* Leukocyte)
White-crowned sparrow, 415, 416
White matter, 253, 265, *404,* 407
White Sands National Monument, 3
White shark, *288,* 290, 291, 293
Widow bird, 433
Wildlife, human impact on, 5, 359,
   372, 420, 422, 437, 438
Wing:
  bat, *56, 265,* 362
  bird, *56, 265, 330,* 331–*332,* 337
  insect, 226, *227,* 228–229, 235
  pterosaur, *56, 324*
Winter sleep, 350, 367
Wishbone, 329, *332*
Wolf, 57, 341, 367, 422, 425, 427,
   432, 435
  marsupial, 57, *356,* 357, 358
Wolverine, 367
Wombat, 357, 358
Woodchuck, hibernation of, 349–350
Woodcock, 330
Woodpecker, 333, *338,* 339
Worker caste, 104, 235, 427, 433, 435
Wormshell, 176
Wrasse, 277, 296, 303

Wren, *338,* 339
Wrist (*see* Carpal; Carpus)
Wuchereria, 145, 148

X chromosome, 47
X-linked factor, 47
X organ, 213
X-rays, 414
Xanthophyll, 66
Xenacanth, 287, 288, *289*
Xenopus, 102, 311
Xerophthalmia, 375
Xiphosurid, 198, *199*
XIPHOSURIDA, 197

Y chromosome, 47, 58
Y-linked factor, 47
Y organ, 213
Yellow fever, 195
Yoldia, 171
Yolk, 93, *95,* 132, 175, 185, 192, 195,
   204, 231, 239, 292, 297, 313
Yolk gland, 126, *128,* 130, *131, 134*
Yolk sac, 95, 96, *313,* 354
  origin of germ cells from, 100
Yucca moth, 53

Z line, 394, *396*
Zaglossus, 357
Zebra, 370
Zebra finch, 417
Zinc (Zn), 374
ZIPHIIDAE, 366
ZOANTHARIA, *108,* 120
Zoea, 105, *213,* 214, 215, 220, 222
Zona pellucida, *95*
Zonation (of life), 420, *421*
Zoochlorella, 64, 80, 115
Zooecium, 155, *156*
Zooflagellate, 69–71, 72, 433
Zooid, 70, 111, *153,* 155, 251, 253,
   254
Zoology, 1, 7
ZOOMASTIGOPHOREA, 68
Zooparasitic, 146
Zooplankton, 5, 221, 422, 423, *430,*
   431
Zooxanthella, 67, 115, 119, 122, 123
Zygapophysis, 303
Zygomatic arch, *318*
Zygomaticus, *342*
Zygote, 48, 49, 75, *77,* 85, *89,* 90, 94,
   111
  totipotency of, 93